Micro- and Nanotechnologies-Based Product Development

Micro- and Nanotechnologies-Based Product Development

Edited by
Neelesh Kumar Mehra and Arvind Gulbake

CRC Press
Taylor & Francis Group
Boca Raton London New York

CRC Press is an imprint of the
Taylor & Francis Group, an **informa** business

First edition published 2022
by CRC Press
6000 Broken Sound Parkway NW, Suite 300, Boca Raton, FL 33487-2742

and by CRC Press
2 Park Square, Milton Park, Abingdon, Oxon, OX14 4RN

© 2022 Taylor & Francis Group, LLC

CRC Press is an imprint of Taylor & Francis Group, LLC

Reasonable efforts have been made to publish reliable data and information, but the author and publisher cannot assume responsibility for the validity of all materials or the consequences of their use. The authors and publishers have attempted to trace the copyright holders of all material reproduced in this publication and apologize to copyright holders if permission to publish in this form has not been obtained. If any copyright material has not been acknowledged please write and let us know so we may rectify in any future reprint.

Except as permitted under U.S. Copyright Law, no part of this book may be reprinted, reproduced, transmitted, or utilized in any form by any electronic, mechanical, or other means, now known or hereafter invented, including photocopying, microfilming, and recording, or in any information storage or retrieval system, without written permission from the publishers.

For permission to photocopy or use material electronically from this work, access www.copyright.com or contact the Copyright Clearance Center, Inc. (CCC), 222 Rosewood Drive, Danvers, MA 01923, 978-750-8400. For works that are not available on CCC please contact mpkbookspermissions@tandf.co.uk

Trademark notice: Product or corporate names may be trademarks or registered trademarks, and are used only for identification and explanation without intent to infringe.

Library of Congress Cataloging-in-Publication Data
Names: Mehra, Neelesh Kumar, editor. | Gulbake, Arvind, editor.
Title: Micro and nanotechnologies based product development / edited by Neelesh Kumar Mehra and Arvind Gulbake.
Description: First edition. | Boca Raton : CRC Press, 2022. | Includes bibliographical references and index.
Identifiers: LCCN 2021011182 | ISBN 9780367488451 (hardback) | ISBN 9781032050720 (paperback) | ISBN 9781003043164 (ebook)
Subjects: MESH: Nanostructures | Drug Delivery Systems | Drug Development | Nanotechnology | Microtechnology
Classification: LCC RM301.25 | NLM QT 36.5 | DDC 615.1/9—dc23
LC record available at https://lccn.loc.gov/2021011182

ISBN: 9780367488451 (hbk)
ISBN: 9781032050720 (pbk)
ISBN: 9781003043164 (ebk)

DOI: 10.1201/9781003043164

Typeset in Times
by codeMantra

Contents

Preface vii
Editors ix
List of Contributors xi

SECTION A Introduction

Chapter 1 Micro- and Nanotechnology Approaches: Concepts and Applications 3

Bhavana Valamla, Pradip Thakor, Neelesh Kumar Mehra, Arvind Gulbake, and Satish Shilpi

Chapter 2 Formulation by Design (FbD): An Emerging Approach to Design Vesicular Nanocarriers 15

Shiv Kumar Prajapati, Payal Kesharwani, Nishi Mody, Ankit Jain, and Swapnil Sharma

Chapter 3 Thermally Responsive Externally Activated Theranostics (TREAT) for On-Demand Multifunctional Drug Delivery Systems 33

Renuka Khatik, Aakanchha Jain, and Faheem Hyder Pottoo

Chapter 4 C_{60}-Fullerenes as an Emerging Cargo Carrier for the Delivery of Anti-Neoplastic Agents: Promises and Challenges 49

Nagarani Thotakura and Kaisar Raza

SECTION B Bioactive Delivery Systems

Chapter 5 Pharmaceutical and Biomedical Applications of Multifunctional Quantum Dots 65

Devesh Kapoor, Swapnil Sharma, Kanika Verma, Akansha Bisth, Yashu Chourasiya, Mayank Sharma, and Rahul Maheshwari

Chapter 6 PLGA-Based Micro- and Nano-particles: From Lab to Market 83

Akash Chaurasiya, Parameswar Patra, Pranathi Thathireddy, and Amruta Gorajiya

Chapter 7 Targeted Lipid-Based Nanoparticles for Nucleic Acid Delivery in Cancer Therapy 95

Prashant Upadhaya, Swapnil Talkar, Vinod Ghodake, Namrata Kadwadkar, Anjali Pandya, Sreeranjini Pulakkat, and Vandana B. Patravale

Chapter 8 Formulation Strategies for Improved Ophthalmic Delivery of Hydrophilic Drugs 115

Arehalli Manjappa, Popat Kumbhar, John Disouza, Suraj Pattekari, and Vandana B. Patravale

Chapter 9 Metal Nanoparticles as a Surrogate Carrier in Drug Delivery and Diagnostics 139

Sushama Talegaonkar, Debopriya Dutta, Namita Chaudhary, Surya Goel, and Ruchi Singh

Chapter 10 Resealed Erythrocytes: A Biological Carrier for Drug Delivery 163

Satish Shilpi, Kapil Khatri, Umesh Dhakad, Neelesh Kumar Mehra, and Arvind Gulbake

Chapter 11 Nanostructured Hydrogel-Based Biosensor Platform 177

Anand Singh Patel and Keerti Jain

Chapter 12 Multifunctional Carbon Nanotubes in Drug Delivery 187

Anamika Sahu Gulbake, Ankit Gaur, Aviral Jain, Satish Shilpi, Neelesh Kumar Mehra, and Arvind Gulbake

SECTION C Product Development, Toxicity and Scale-Up

Chapter 13 Liposomal-Based Pharmaceutical Formulations – Current Landscape, Limitations and Technologies for Industrial Scale-Up 209

Radha Rani, Neha Raina, Azmi Khan, Manupriya Choudhary, and Madhu Gupta

Chapter 14 Impact and Role of Stability Studies in Parenteral Product Development 225

Mohammed Asadullah Jahangir, Syed Sarim Imam, and Jayamanti Pandit

Chapter 15 Reverse Engineering in Pharmaceutical Product Development 235

Rishi Paliwal, Aanjaneya Mamgain, Rameshroo Kenwat, and Shivani Rai Paliwal

Chapter 16 Role of Polymers in Formulation Design and Drug Delivery 243

Satish Shilpi, Umesh Dhakad, Rajkumari Lodhi, Sonal Dixit, Kapil Khatri, Neelesh Kumar Mehra, and Arvind Gulbake

Chapter 17 Drug Delivery Systems for Targeting Blood Brain Barrier: Examples of nanomedicines for the treatment of neurodegenerative diseases 257

Iara Baldim, Adriana M. Ribeiro, João Dias-Ferreira, Wanderley P. Oliveira, Francisco M. Gama, and Eliana B. Souto

Chapter 18 Validation, Scale-Up and Technology Transfer in Product Development 269

Ankit Gaur, Anamika Sahu Gulbake, Neelesh Kumar Mehra, and Arvind Gulbake

Chapter 19 Nanotoxicology: Safety, Toxicity and Regulatory Considerations 285

Raja Susmitha, Mounika Gayathri Tirumala, Mohd Aslam Saifi, and Chandraiah Godugu

Index 305

Preface

In the current scenario, micro- and nanotechnology-based product development is a boon in the healthcare, pharmaceutical, research and innovation area. Product development is a rapidly growing and emerging area consisting of multifunctional micro- and nano-size engineered nanomaterials (ENM) useful for the delivery of biotherapeutics (large and small molecules).

As of now, very few micro- and nanotechnology-based products are available on the market. The development of complex sterile and non-sterile products is crucial and needs proper systemic approaches starting from pre-formulation to validation and scale-up including toxicology data.

The aim of this book is to provide exhaustive information to the formulators and researchers in product development with emphasis on academic and industry perspectives along with regulatory perspectives. This book exhaustively covers the current research input beginning from the product strategy, scale-up by applying the recent technologies including quality by design, process optimisation, validation, reverse engineering, process analytical technology, stability and regulatory toxicology including the pharmacokinetics and pharmacodynamics of product development. It is edited through contribution of vivid chapters from renowned formulators, researchers and academicians across the world with their specialised research area of interest.

Our primary target readers are postgraduates, and postdoctoral research fellows with formulators, researchers and academicians to extend their knowledge through this book and also those who are working in drug delivery area. The book contains (i) the fundamental concepts, theory and advancement of micro- and nanotechnologies; (ii) biomaterials and their applications; (iii) recent trends/innovations and developments; (iv) sterile product development, technology transfer and validation; and (v) toxicological perspectives.

Chapter 1 is about the use of micro- and nanotechnologies in product development of large and small molecules. **Chapter 2** exhaustively comprises the formulation-by-design approaches, which are required in the development of products. **Chapter 3** reviews the thermally responsive externally activated theranostics as multifunctional drug delivery systems. **Chapter 4** summarises the drug delivery aspects, promises and challenges of fullerenes for the delivery of anti-neoplastic agents. **Chapter 5** discusses the pharmaceutical and biomedical applications of multifunctional quantum dots to the healthcare sector. **Chapter 6** discusses and summarises the PLGA-based micro- and nanoparticles and challenges, technical aspects and commercial platform updated from research laboratory to market. **Chapter 7** reviews the targeted stable nucleic acid-based lipid nanoparticles in cancer therapy. **Chapter 8** summarises the formulation strategies for the delivery of hydrophilic drugs for ophthalmic diseases. **Chapter 9** reviews and elaborates on metal nanoparticles in drug delivery and diagnostics. **Chapter 10** elaborates on the resealed erythrocytes and their promising applications in drug delivery. **Chapter 11** reviews the nanostructured hydrogel-based biosensor platform for promising biomedical applications. **Chapter 12** discusses carbon nanotubes, physicochemical properties and drug delivery aspects for biomedical and pharmaceutical applications. **Chapter 13** exhaustively covers the stability studies as per the regulatory guideline requirements for parenteral product development. **Chapter 14** discusses the current landscape, limitations and technologies for industrial scale-up for liposome-based pharmaceutical formulations. **Chapter 15** covers reverse engineering and its principle and role in generic product development. **Chapter 16** covers the various polymers used in the design of formulation for drug delivery development. **Chapter 17** discusses the quantitative pharmacology and different statistical methods for drug safety in the ocular route for sterile ocular product development. **Chapter 18** covers validation, scale-up and technology transfer in the development of pharmaceutical products. **Chapter 19** covers the various safety, toxicity and regulatory considerations.

Editors
Neelesh Kumar Mehra, PhD
Arvind Gulbake, PhD

Editors

Dr. Neelesh Kumar Mehra is working as an Assistant Professor of Pharmaceutics & Biopharmaceutics at the Department of Pharmaceutics, National Institute of Pharmaceutical Education & Research (NIPER), Hyderabad, India. He earned his PhD from Dr. Harisingh Gour University, Sagar, and postdoc from Irma Lerma Rangel College of Pharmacy, Texas A&M Health Science Center, Kingsville, TX, USA. He served as Manager in Product Development, Sentiss Research Centre, Sentiss Pharma Pvt Ltd, Gurgaon, for development, scale-up and technology transfer of complex ophthalmic, inhalation and optic pharmaceutical products. He received 'TEAM AWARD' for successful commercialisation of an ophthalmic suspension product. He has authored more than 60 peer-reviewed publications in highly reputed international journals and more than 10 book chapter contributions. He has filed a patent on Pharmaceutical composition and manufacturing process of novel nanoformulations to improved therapeutic efficacy for topical delivery. He guided PhD and MS students for their dissertations/research projects. He has received numerous outstanding awards including Young Scientist Award and Team Award for his research output. He is a peer reviewer of various international journals and publications. He recently published one edited book, *'Dendrimers in Nanomedicine: Concept, Theory and Regulatory Perspectives'*, in CRC Press. Currently, he is editing books on biopharmaceutical and nanotechnology-based products with Elsevier Pvt Ltd. He has rich research and teaching experience in the formulation and development of complex, innovative ophthalmic and injectable biopharmaceutical products including micro- and nanotechnologies for regulated market.

Dr. Arvind Gulbake is working as an Assistant Professor at the Faculty of Pharmacy, School of Pharmaceutical & Population Health Informatics, at DIT University, Dehradun. Earlier, he worked as a Coordinator, Research & Development, Centre for Interdisciplinary Research, D.Y. Patil Education Society (Institution Deemed to be University), Kolhapur, Maharashtra, India. He earned his PhD with Prof. Sanjay K. Jain from Dr. Harisingh Gour University, Sagar. He has authored more than 40 peer-reviewed publications in highly reputed international journals, four book chapters and a patent contribution. He guided PhD and MS students for their dissertations/research projects. He successfully completed an extramural project funded by SERB, New Delhi, Government of India. He has received outstanding awards including Young Scientist Award and BRG Travel Award for his research. He is a peer reviewer of various international journals and publications. He is an assistant editor for the International Journal of Applied Pharmaceutics and an editorial board member of the Journal of Liposome Research. He is a member of various academic committees at the university level, i.e. Board of Studies (Masters and PhD), IBSC, IAEC, IQAC and Institutional Research Committee. He has more than 12 years of research and teaching experience in the formulation and development of nanopharmaceuticals.

List of Contributors

Iara Baldim
CEB – Centre of Biological Engineering
University of Minho
Braga, Portugal
Faculty of Pharmaceutical Sciences of Ribeirão Preto
University of São Paulo
Ribeirão Preto, Brazil

Akansha Bisth
Department of Pharmacy
Banasthali Vidyapith
Banasthali Niwai, Rajasthan, India

Namita Chaudhary
Department of Pharmaceutics
Delhi Pharmaceutical Sciences and
    Research University (DPSRU)
New Delhi, India

Akash Chaurasiya
Department of Pharmacy
Birla Institute of Technology and Science
Pilani, Hyderabad, India

Manupriya Choudhary
Department of Pharmaceutics
Delhi Pharmaceutical Science and
    Research University (DPSRU)
New Delhi, India

Yashu Chourasiya
Department of Pharmaceutics
Smriti College of Pharmaceutical Education
Indore, India

Umesh Dhakad
Department of Pharmaceutics
Ravishankar College of Pharmacy
Bhopal, Madhya Pradesh, India

João Dias-Ferreira
Faculty of Pharmacy
University of Coimbra (FFUC)
Coimbra, Portugal

John Disouza
Department of Pharmaceutics
Tatyasaheb Kore College of Pharmacy
Warananagar, Panhala, Kolhapur, Maharashtra, India

Sonal Dixit
Department of Pharmaceutics
Ravishankar College of Pharmacy
Bhopal, Madhya Pradesh, India

Debopriya Dutta
Department of Pharmaceutics
Delhi Pharmaceutical Sciences and
    Research University (DPSRU)
New Delhi, India

Francisco M. Gama
CEB – Centre of Biological Engineering
University of Minho
Braga, Portugal

Ankit Gaur
Sentiss Research Centre
Sentiss Pharma Pvt Ltd.
Gurgaon, Haryana, India

Vinod Ghodake
Department of Pharmaceutical Sciences and Technology
Institute of Chemical Technology
Matunga, Mumbai, India

Chandraiah Godugu
Department of Regulatory Toxicology
National Institute of Pharmaceutical Education and
    Research (NIPER)
Balanagar, Hyderabad, Telangana, India

Surya Goel
Department of Pharmacy
ABESIT College of Pharmacy
Ghaziabad, Uttar Pradesh, India

Amruta Gorajiya
R&D-Injectables
Amneal Pharmaceuticals
Ahmedabad, Gujarat, India

Anamika Sahu Gulbake
Faculty of Pharmacy, School of Pharmaceutical &
    Population Health Informatics
DIT University
Dehradun, Uttarakhand, India

Arvind Gulbake
Faculty of Pharmaceutics, School of Pharmaceutical & Population Health Informatics
DIT University
Dehradun, Uttarakhand, India
Centre for Interdisciplinary Research
D.Y. Patil Education Society
Kolhapur, Maharashtra, India

Madhu Gupta
Department of Pharmaceutics
Delhi Pharmaceutical Science and Research University (DPSRU)
New Delhi, India

Syed Sarim Imam
Department of Pharmaceutics, College of Pharmacy
King Saud University
Riyadh, Saudi Arabia

Mohammed Asadullah Jahangir
Department of Pharmaceutics
Nibha Institute of Pharmaceutical Sciences
Rajgir, Bihar, India

Aakanchha Jain
Department of Pharmaceutics
Bhagyoday Tirth Pharmacy College
Sagar, Madhya Pradesh, India
and
National Institute of Pharmaceutical Education & Research (NIPER)
Ahmadabad, India

Ankit Jain
Department of Materials Engineering
Indian Institute of Science
Bangalore, Karnataka, India

Aviral Jain
Solisto Pharma
Sagar, Madhya Pradesh, India

Keerti Jain
Department of Pharmaceutics
National Institute of Pharmaceutical Education and Research (NIPER)
Raebareli, Lucknow, Uttar Pradesh, India

Namrata Kadwadkar
Department of Pharmaceutical Sciences and Technology
Institute of Chemical Technology
Matunga, Mumbai, India

Devesh Kapoor
Department of Pharmaceutics
Dr. Dayaram Patel Pharmacy College
Bardoli, Gujarat, India

Rameshroo Kenwat
Nanomedicine and Bioengineering Research Laboratory, Department of Pharmacy
Indira Gandhi National Tribal University
Amarkantak, Madhya Pradesh, India

Payal Kesharwani
Department of Pharmacy
Ram-Eesh Institute of Pharmacy
Greater Noida, Uttar Pradesh, India
Department of Pharmacy
Banasthali Vidyapith
Panch Batti, Ashok Nagar, Rajasthan, India

Azmi Khan
Department of Pharmaceutics
Delhi Pharmaceutical Science and Research University (DPSRU)
New Delhi, India

Renuka Khatik
Postdoctoral Research Associate
Texas A&M Rangel College of Pharmacy
Kingsville, TX, USA

Kapil Khatri
Department of Pharmaceutics
Ravishankar College of Pharmacy
Bhopal, Madhya Pradesh, India

Popat Kumbhar
Department of Pharmaceutics
Tatyasaheb Kore College of Pharmacy
Warananagar, Panhala, Kolhapur, Maharashtra, India

Rajkumari Lodhi
Department of Pharmaceutics
Ravishankar College of Pharmacy
Bhopal, Madhya Pradesh, India

Rahul Maheshwari
School of Pharmacy and Technology Management
SVKM's NMIMS
Jadcherla, Hyderabad, India

Aanjaneya Mamgain
Nanomedicine and Bioengineering Research Laboratory, Department of Pharmacy
Indira Gandhi National Tribal University
Amarkantak, Madhya Pradesh, India

List of Contributors

Arehalli Manjappa
Department of Pharmaceutics
Tatyasaheb Kore College of Pharmacy
Warananagar, Panhala, Kolhapur, Maharashtra, India

Neelesh Kumar Mehra
Pharmaceutical Nanotechnology Research Laboratory, Department of Pharmaceutics
National Institute of Pharmaceutical Education & Research (NIPER)
Hyderabad, Telangana, India

Nishi Mody
Department of Pharmaceutical Sciences
Dr. Harisingh Gour Central University
Sagar, Madhya Pradesh, India

Wanderley P. Oliveira
Faculty of Pharmaceutical Sciences of Ribeirão Preto
University of São Paulo
Ribeirão Preto, Brazil

Rishi Paliwal
Nanomedicine and Bioengineering Research Laboratory, Department of Pharmacy
Indira Gandhi National Tribal University
Amarkantak, Madhya Pradesh, India

Shivani Rai Paliwal
SLT Institute of Pharmaceutical Sciences
Guru Ghasidas Vishwavidyalaya (A Central University)
Bilaspur, Chhattisgarh, India

Jayamanti Pandit
Department of Pharmaceutics
Women Scientist (WOS-C), IPR in Patent Facilitating Centre (PFC) TIFAC
New Delhi, India

Anjali Pandya
Department of Pharmaceutical Sciences and Technology
Institute of Chemical Technology
Matunga, Mumbai, India

Anand Singh Patel
Department of Pharmaceutics
National Institute of Pharmaceutical Education and Research (NIPER)
Raebareli, Lucknow, Uttar Pradesh, India

Parameswar Patra
Department of Pharmacy
Birla Institute of Technology and Science
Pilani, Hyderabad, India

Vandana B. Patravale
Department of Pharmaceutical Sciences and Technology
Institute of Chemical Technology
Matunga, Mumbai, India

Suraj Pattekari
Tatyasaheb Kore College of Pharmacy
Warananagar, Panhala, Kolhapur, Maharashtra, India
Department of Pharmaceutics
Annasaheb Dange College of B. Pharmacy
Ashta, Walwa, Sangli, Maharashtra, India

Anil Pethe
School of Pharmacy and Technology Management
SVKM'S NMIMS
Jadcherla, Hyderabad, India

Faheem Hyder Pottoo
Department of Pharmacology
College of Clinical Pharmacy
Imam Abdulrahman Bin Faisal University
Dammam, Saudi Arabia

Shiv Kumar Prajapati
Department of Pharmacy
Ram-Eesh Institute of Pharmacy
Greater Noida, Uttar Pradesh, India

Sreeranjini Pulakkat
Department of Pharmaceutical Sciences and Technology
Institute of Chemical Technology
Matunga, Mumbai, India

Neha Raina
Department of Pharmaceutics
Delhi Pharmaceutical Science and Research University (DPSRU)
New Delhi, India

Radha Rani
Department of Pharmaceutics
Delhi Pharmaceutical Science and Research University (DPSRU)
New Delhi, India

Kaisar Raza
Department of Pharmacy, School of Chemical Sciences and Pharmacy
Central University of Rajasthan
Bandarsindri, Ajmer, Rajasthan, India

Adriana M. Ribeiro
Faculty of Pharmacy
University of Coimbra (FFUC)
Coimbra, Portugal

Mohd Aslam Saifi
Department of Regulatory Toxicology
National Institute of Pharmaceutical Education and Research (NIPER)
Balanagar, Hyderabad, Telangana, India

Swapnil Sharma
Department of Pharmacy
Banasthali Vidyapith
Banasthali, Panch Batti, Ashok Nagar, Rajasthan, India

Mayank Sharma
School of Pharmacy and Technology Management
SVKM'S NMIMS
Shirpur, Dhule, Maharashtra, India

Satish Shilpi
Department of Pharmaceutics
Ravishankar College of Pharmacy
Bhopal, Madhya Pradesh, India

Ruchi Singh
Department of Pharmacy
Rajkumar Goel Institute of Technology
Ghaziabad, Uttar Pradesh, India

Eliana B. Souto
Faculty of Pharmacy
University of Coimbra (FFUC)
Coimbra, Portugal
CEB – Centre of Biological Engineering
University of Minho
Braga, Portugal

Raja Susmitha
Department of Regulatory Toxicology
National Institute of Pharmaceutical Education and Research (NIPER)
Balanagar, Hyderabad, Telangana, India

Sushama Talegaonkar
Department of Pharmaceutics
Delhi Pharmaceutical Sciences and Research University (DPSRU)
New Delhi, India

Swapnil Talkar
Department of Pharmaceutical Sciences and Technology
Institute of Chemical Technology
Matunga, Mumbai, India

Pradip Thakor
Pharmaceutical Nanotechnology Research Laboratory, Department of Pharmaceutics
National Institute of Pharmaceutical Education & Research (NIPER)
Hyderabad, Telangana, India

Pranathi Thathireddy
Department of Pharmacy
Birla Institute of Technology and Science
Pilani, Hyderabad, India

Nagarani Thotakura
Department of Pharmacy, School of Chemical Sciences and Pharmacy
Central University of Rajasthan
Bandarsindri, Ajmer, Rajasthan, India

Mounika Gayathri Tirumala
Department of Regulatory Toxicology
National Institute of Pharmaceutical Education and Research (NIPER)
Balanagar, Hyderabad, Telangana, India

Prashant Upadhaya
Department of Pharmaceutical Sciences and Technology
Institute of Chemical Technology
Nathalal Parekh Marg
Matunga, Mumbai, India

Bhavana Valamla
Pharmaceutical Nanotechnology Research Laboratory, Department of Pharmaceutics
National Institute of Pharmaceutical Education & Research (NIPER)
Hyderabad, Telangana, India

Kanika Verma
Department of Pharmacy
Banasthali Vidyapith
Banasthali Niwai, Rajasthan, India

Section A

Introduction

1 Micro- and Nanotechnology Approaches
Concepts and Applications

Bhavana Valamla, Pradip Thakor, and Neelesh Kumar Mehra
National Institute of Pharmaceutical Education & Research (NIPER), Hyderabad

Arvind Gulbake
DIT University
D.Y. Patil Education Society

Satish Shilpi
Ravishankar College of Pharmacy

CONTENTS

1.1 Introduction 3
    1.1.1 Structure and Classification 4
    1.1.2 Synthesis 6
1.2 Characterisation Techniques 6
1.3 Properties of Particles 7
1.4 Pharmacokinetics, Toxicity and Biodistribution 7
1.5 Applications 7
    1.5.1 Bioavailability Improvements 7
    1.5.2 Theranostic Agent 8
    1.5.3 Multidrug Resistance 9
1.6 Stability of Active Pharmaceutical Ingredients (API) 9
1.7 Marketed Approval of NP Drug Delivery and Regulatory Status 9
    1.7.1 Nanomedicines: Current Status and Future Perspectives in the Aspects of Drug Delivery and Pharmacokinetics 9
1.8 Future of Nanomedicines and Drug Delivery Systems 9
1.9 Conclusion and Future Perspectives 10
Acknowledgement 11
References 11

1.1 INTRODUCTION

In the modern era, specialised drug delivery systems involving micro- and nanotechnology are prominent in pharmaceutical research, clinical practice and food and other industries. The difference in dimensions between microparticles (MPs) and nanoparticles (NPs) can be described in micrometres and nanometres. The difference in size measurements explains the real differences, from formulation to *in vivo* studies. To decide the size of delivery system, various parameters such as intended target, margination, interaction with biomembranes, biodistribution, toxicity and shape are to be considered. At present, there are no general rules to compare release pattern, efficacy, tissue retention, toxicity and distribution of micro- and nanoscale drug delivery systems.

Nanotechnology deals with materials and interactions at the nanoscale, size less than 100 nm. The small size of NPs, large surface area-to-volume ratio and changes in their physicochemical properties compared to those of their macromolecules offer many advantages that determine solubility, stability and mechanical, physical, chemical, electrical and optical properties. Improved drug delivery, targeted treatment and non-invasive diagnostics with reduced systemic and adverse effects can be attributed to the unique properties of NPs. However, the advantage of MPs over NPs is that they do not travel into the interstitium and transported through the lymph resulting in local action (Yang and Forrest 2016).

Particulate delivery systems are easy to inject intravenously or into tissue, facilitating systemic and local drug delivery. By passive or active targeting, particles can be delivered to local and systemic sites. High local drug level for an extended period with low systemic toxicity can be obtained by directly injecting or depositing the desired particles at the site of action with the aid of micro- and nanotechnology. To produce the effect of drug throughout the body, particles can be injected as depot for slow release of the drug. They are also easily adapted for topical use in an appropriate vehicle or inhalation as dry powder.

Micro- and nanotechnology have been predominantly advanced in the design of structures involving therapeutics, diagnostics and imaging. Special emphasis on studies addressing the major technical challenges including cargo release and removal from the site of action to be studied. Nonetheless, the same differences in the physical and chemical properties of NPs could lead to serious and unpredictable side effects for the human body and global ecosystem, including accumulation, re-circulation, and inflammatory, mutagenic and oncogenic potential.

Micro- or nano-sized delivery system is not the ultimate drug delivery system for optimal and efficient delivery of drugs because both have advantages and disadvantages. Often case-by-case evaluation is needed to determine the appropriate choice system. This chapter focuses on concepts and applications of micro- and nanotechnology in drug delivery systems with more emphasis placed on properties, synthesis, classification, characterisation, pharmacokinetics, toxicity, biodistribution and applications of MPs and NPs.

1.1.1 Structure and Classification

MPs' sizes range from 1 to 1,000 μm and usually exist in matrix or reservoir structures. MPs possess homogeneous or heterogeneous structures depending on the formulation requirement. Spheroid shape is the mostly preferred since it makes the further processing easier. Microspheres are matrix systems in which the drug is dispersed or dissolved or suspended homogeneously. The core is surrounded by membrane shell in heterogeneous reservoir structures (Lengyel et al. 2019), and various other MP structures are illustrated in Figure 1.1.

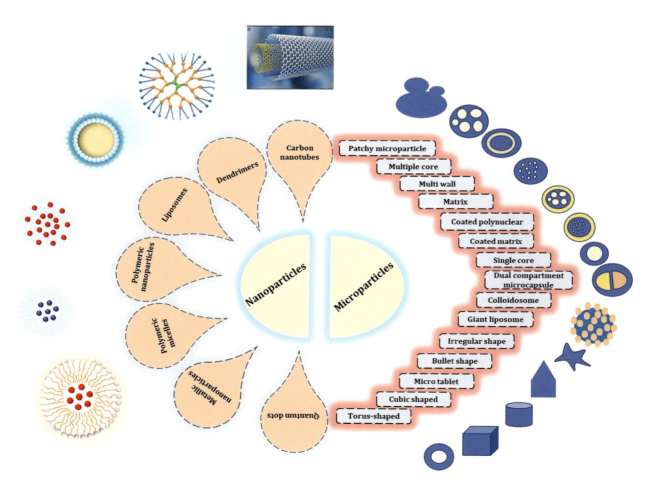

FIGURE 1.1 Illustration of the various nanoparticles and microparticles.

NPs' sizes range from 1 to 100 nm, and they may be described as the particles composed of three layers: the surface layer, which could be easily altered using a variety of metal ions, small molecules, polymers and surfactants; the shell layer; and the core, which is essentially the central portion of the NP and usually refers to the NP itself (Tiwari, Tiwari, and Kim 2012). Based on their shape, size and properties, NPs are broadly divided into various categories as shown in Table 1.1.

Siegel classified nanomaterials into zero-, one-, two- and three-dimensional based on their shape (Murr 2021). They can exist in single, aggregate or agglomerate form, or in fused form with spherical, irregular and tubular shapes. Some of the well-known common types of nanomaterials used in drug delivery are included in Figure 1.1. Zero-dimensional nanostructures have zero dimensions outside the nanoscale (spheres and clusters). One-dimensional nanostructures have one dimension

TABLE 1.1
Most Commonly Used Nanostructures with Size, Characteristics, Advantages and Disadvantages

S. No.	Nanostructure	Description	Inference	References
1	Dendrimers	Dendrimers are hyperbranched mono-dispersed, 3D polymer systems with three main parts: core, branch and surface	The globular, compact structure enables dendrimers both to encapsulate drug molecules in the interior and to attach or bound to the surface groups. The distinct applications include solubilisation; gene therapy; long circulatory, controlled and targeted delivery of bioactive compounds; immunoassay; and MRI contrast agents	Singh et al. (2016), Kavand et al. (2020), Chopdey et al. (2015)
2	Polymeric micelles	Polymeric micelles are synthetic amphiphilic blocks that are spherical vesicular bodies containing an aqueous solution with high biostability, drug entrapment and payload	Polymeric micelles are long circulatory NPs used for active and passive targeting of drug with diagnostic value	Khan et al. (2020), Nishiyama and Kataoka (2003)
3	Lipid NPs	Lipid NPs are phospholipid vesicles consisting of lipophilic molecules matrix and lipid solid core stabilised with emulsifiers or surfactants stabilised the external core of these NPs	Lipid NPs are versatile, long circulatory, biocompatible nanostructures with good entrapment efficiency that aid in passive and active delivery of gene, protein, peptide and various other	Rawat, Jain, and Singh (2011), (Khan, Saeed, and Khan (2019)
4	Carbon nanotubes	Carbon-based nanomaterials that belong to third allotropic crystalline form of carbon, arranged in hexagonal forming graphitic sheets either in single layer or in multiple layers and further rolled up into the tubular structure	The most widely used nanovehicle in therapeutic targeting, carrier for peptide and gene delivery with remarkable strength and unique electrical properties.	Mehra, Mishra, and Jain (2013), Klumpp et al. (2006)
5	Polymeric NPs	Mostly polymeric NPs are biodegradable, biocompatible with nanosphere or nanocapsular shapes	Polymeric NPs are used as drug delivery vehicles due to their improved bioavailability, encapsulation, controlled release and minimum toxic properties and targeted drug delivery	Kumari, Yadav, and Yadav (2010), Martis, Badve, and Degwekar (2012), Khan et al. (2020)
6	Metallic nanoparticles (NPs)	Noble metals such as gold and silver have many colours in the visible region due to localised surface plasmon resonance	Unique plasmon absorbance features of these noble metals have been used for a wide range of applications including biosensors, chemical sensors, drug and gene delivery, thermal ablation, sensitive diagnostic assays and radiotherapy enhancement	Unser et al. (2015), Khan, Saeed, and Khan (2019)
7	Quantum dots	Semiconducting nanocrystals with inorganic core (CdSe) and an aqueous organic coated shell (e.g. ZnS) mainly synthesised with II–VI and III–V column elements. Size ranges from 10 and 100 Å	Quantum dots (QDs) possess optical and fluorescence properties. QDs have been adopted as therapeutic vehicles for transport of DNA, proteins, drugs and cells, and as diagnostic tools for time-graded fluorescence imaging of tissues, labelling of breast cancer marker, DNA hybridisation, receptor-mediated endocytosis and immunoassay	Zhao and Zhu (2016), Pons et al. (2010), Khan et al. (2020)
8	Ceramic NPs	Inorganic nonmetallic solids, synthesised using alumina, silica and titania	Ceramic NPs can be found in polycrystalline, amorphous, dense, hollow or porous forms. Photodegradation of dyes, photocatalysis, drug and biomedical delivery imaging applications	Thomas et al. (2015), Sigmund et al. (2006), Cherian, Rana, and Jain (2000)

outside the nano-range (nanotubes, nanorods and nanowires). Two-dimensional nanostructures have two dimensions outside the nano-range (nanofilms, nanolayers and nanocoatings) and can be used in electronics, sensors and optoelectronics. Two-dimensional-type NPs can be used as single or multilayer structures that are deposited on a substrate and fused with the surrounding matrix material. Three-dimensional nanostructures have three dimensions outside the nano-range and include multi-nanolayers, dispersions of NPs, nanotubes, bundles of nanowires, quantum dots, dendrimers and fullerenes with different orientations (Chavan n.d.).

1.1.2 Synthesis

Bottom-up and top-down approaches are the conventional methods used for the fabrication of NPs. The top-down approach is the destructive method: larger molecules are decomposed into smaller units and further converted to suitable NPs. In the bottom-up approach, NPs are with the aid of simple techniques such as sedimentation, reduction techniques, sol–gel, green synthesis, spinning and biochemical synthesis (Khan, Saeed, and Khan 2019). Numerous biomedical applications and fabrication strategies have been developed for the synthesis of NPs. They can be mainly divided into self-assembly, microemulsion methods and layer-by-layer (LbL) assembly, as well as techniques based on instruments such as particle replication in non-wetting templates (PRINT), electrospraying and microfluidics (Fang et al. 2020).

Self-assembly, as a spontaneous bottom-up approach, is the most widely used and practical strategy for preparing organic NPs with building blocks, including amphiphilic block copolymers, host–guest complexes, polyelectrolytes or small organic molecules (Luo, Yan, and Wang 2015; Li, Luo, and Zhao 2018). Different types of self-assembly strategies are nanoprecipitation-based self-assembly, coordination-driven self-assembly, template-based self-assembly and thin-film hydration-based self-assembly (Fang et al. 2020). LbL assembly is a custom-optimised flexible method for preparing multifunctional NPs with coatings of polymers, biomolecules, colloids, etc. (Richardson et al. 2016). Electrostatic interaction, biological recognition, hydrogen bonding and host–guest interactions are involved in LbL assembly. Stimuli-responsiveness and targeting functionality can be introduced to multifunctional NPs produced by LbL assembly (Zhao et al. 2019).

Microfluidic technology handles single-phase or multiple-phase fluids in microchannels with small volumes. It has a wide range of chemical or biological applications in drug delivery, micromaterial and nanomaterial synthesis and analyte detection. High production efficiency, precise control of reactions, good reproducibility, uniform heat and mass transfers and high level of integration are the merits of microfluidics (Ahn et al. 2018). Furthermore, the microfluidic systems can be categorised as single-phase, multi-phase and integrated flow microfluidic systems (Wang and Song 2017). Electrospraying is a flexible one-step and cost-effective method. It includes single-needle, coaxial, multi-axial, single and multiplexed electrospray. The material solution is ejected from a syringe in the form of NPs under the application of a high voltage in electrospraying (Bock, Dargaville, and Woodruff 2012; Mehta et al. 2019). This technique produces NPs with controlled morphology, excellent reproducibility and effective encapsulation to deliver drugs, proteins, vitamins and growth factors (Sridhar et al. 2015).

DeSimone's group proposed the PRINT method to fabricate well-defined MPs and NPs (20 nm to 100 μm; Xu et al. 2013). The precise control of size, composition and geometry with a non-wetting perfluoropolyether-based elastomer are the advantages of PRINT (Fu et al. 2018). NPs prepared by PRINT can be used for loading of chemotherapeutics, proteins, siRNAs and magnetic resonance contrast agents (Hasan et al. 2012; Parrott et al. 2010; Perry et al. 2011; Wang et al. 2012). The surface properties of PRINT particles can be easily functionalised with ligand to improve the targeting of therapeutics (Reuter et al. 2015). Many robust strategies have been explored for organic NP preparation. Self-assembly can be regarded as the most representative bottom-up approach for NP synthesis in bioimaging, drug delivery and treatment of diseases. Nanoprecipitation is the most widely used one, due to its simplicity and low cost. Industrial production, clinical transformation and functionalisation of NPs and preparation of multifunctional NPs are the issues to be considered in the future.

1.2 CHARACTERISATION TECHNIQUES

Several analytical techniques have been practised for the characterisation of physicochemical properties of MPs and NPs. The most commonly used techniques for structural characterisation are transmission electron microscopy (TEM), scanning electron microscopy (SEM), environmental scanning electron microscopy (ESEM) and light scattering. SEM, TEM and polarised optical microscopy are used for the morphological characterisation of the particles. SEM is based on the electron scanning principle and TEM is based on the electron transmittance principle to study bulk properties from very low to high magnification. To some extent, techniques such as Coulter counter, sedimentation, optical light microscopy (micron-sized particles), atomic force microscopy (AFM) and capillary hydrodynamic fractionation are used (Peltonen and Hirvonen 2008).

SEM, TEM, AFM and XRD (X-Ray diffraction) can also be used for the particle size determination. The DLS/zeta potential size analyser can give a better idea regarding the size at low level. BET is the best technique to determine the surface area of NPs. The optical properties of metallic and semiconductor NPs can be studied with aid of ultraviolet–visible (UV–Vis), null ellipsometer and photoluminescence (PL) for photo-related applications.

1.3 PROPERTIES OF PARTICLES

MPs and NPs possess a wide variety of features affecting physiological, anatomical with different applications. The broad array of physicochemical properties such as surface area, mechanical, optical and chemical properties makes them unique and suitable applicants for diverse biomedical applications. One of the most important physicochemical properties is the surface/volume ratio that influences the dissolution rate of the particles.

The optoelectrical properties of metallic NPs are widely used in bioimaging applications due to localised surface plasmon resonance (LSPR) (Eustis and El-Sayed 2006; Unser et al. 2015). The properties and colours of particles vary with the shape and size. Twenty-nanometre gold, platinum, silver and palladium NPs have characteristic wine-red, yellowish grey, black and dark black colours as they possess a broad absorption band in the visible region of the electromagnetic spectrum (Dreaden et al. 2012). The distinct mechanical properties, elastic modulus, hardness, stress and strain, adhesion and friction of NPs can be investigated for novel applications in the fields of tribology, surface engineering, nanofabrication and nanomanufacturing (Guo, Xie, and Luo 2014). Superparamagnetic NPs with high magnetisation value, narrow particle size distribution, smaller size than 100 nm and surface functionalisation can be used for *in vivo* applications such as MRI, tissue repair, immunoassay, cell separation and drug delivery (Laurent et al. 2010). All these properties affect clinical performance in terms of efficacy, drug release, toxicity, biodistribution and choice of delivery system.

1.4 PHARMACOKINETICS, TOXICITY AND BIODISTRIBUTION

A number of toxicity and biodistribution studies are being conducted to evaluate the safety and efficacy of NPs and MPs. The deposition, accumulation and exposure of NPs and MPs may be different at various sites *in vivo* due to variation in surface area. The biodegradability of the particles results in lower toxicity if their degradation products are eliminated, or non-toxic products or fully degraded and do not harm the body. However, NPs can reach the tissues and target organs when compared to MPs. NPs possess limited clinical advantages over MPs in test animals. It has been shown that only 5% of the administered amount reaches the target site and the remaining amount reaches off-target sites. However, few studies show fivefold increases in the targeted delivery of drugs with NPs when compared with larger particles which can have both toxic and therapeutic effects (Fang et al. 2020).

Investigation of the size-dependent effect on toxicity, internalisation and biodistribution is crucial for the success of NPs. Most of the studies are performed with the same concentration of particles present in mass per unit volume instead of the same number of particles per unit volume for the different sizes, as the uptake of different-sized particles is dose-dependent. However, the size-dependent effect may be attributed to the different internalisation mechanisms.

1.5 APPLICATIONS

Particulate system applications range from bioavailability improvement, solubility increment, stability improvement, to theranostic agent with the aid of MP and NP drug delivery systems as shown in Table 1.2.

1.5.1 Bioavailability Improvements

In eradicating the limitations described above, nanotechnology can serve as an effective tool. Aqueous solubility and permeability across the biological membrane can be improved by reducing the size of phytomedicines into nano-phytomedicines and by changing the surface properties of phytomedicines (Aziz and Setapar 2020). It has been documented that various novel drug delivery systems such as liposomes, niosomes, nanospheres and phytosomes have the potential to deliver herbal drugs. The integration of herbal medicines into the delivery system also helps to increase solubility, improve stability, protect against toxicity, improve pharmacological activity, improve the distribution of tissue macrophages, maintain delivery and protect against physical and chemical degradation (Kesarwani and Gupta 2013; Rana, Kumar, and Rana 2020).

The beneficial bioavailability effects of drug nanonisation are mainly based on the reasoning that nanonisation improves the surface area of poorly soluble drugs. Consequently, one can predict the following:

i. An increase in the area of adhesion between the NPs and the mucin layer that coats the intestinal villus epithelium makes it easier for the nanonised drug to cross the mucin layer and the epithelial cells as a result of an increase in oral absorption of the nanonised drug.
ii. NPs typically have a higher surface curvature compared to large particles, which creates more dissolution pressure with a related increase in saturation solubility. In exchange, the increased solubility of saturation favours an increase in the gradient of concentration between the intestinal epithelial cells and underneath the mesenteric circulation. An increase in the rate of dissolution of the drug overcomes the limiting rate of absorption of the pharmaceutical product. Furthermore, the diffusion gap at the drug NP surface is decreased and the concentration gradient is increased. The increase in the area and concentration gradient leads, in comparison with the micronised product, to a more pronounced increase in the dissolution rate. The solubility and dissolution rate of saturation are important parameters that affect the bioavailability of drugs administered orally. Medicinal nanonisation can reduce erratic

TABLE 1.2
Applications of Particulate Drug Delivery Systems with Special Emphasis on Microparticle and Nanoparticle Systems

Drug Name	Drug Delivery	Problem in Drugs	Application	References
Peptides	Liposomes	Oral bioavailability	Improve bioavailability	Gradauer et al. (2013)
Probucol	Liposomes	Less absorption	Increase oral bioavailability by enhancing oral absorption	Ma et al. (2015)
Apatinib	Cyclic RGD peptide-modified liposomes	Oral bioavailability	Improve oral bioavailability	Song et al. (2017)
Insulin	Biomimetic thiamine- and niacin-decorated liposomes	Oral bioavailability	Improve oral bioavailability	He et al. (2018)
Deferoxamine mesylate	Micelles	Less oral bioavailability	Improve oral bioavailability	Salimi, Makhmal Zadeh, and Kazemi (2019)
Silybin	Micelles	Low solubility	Improve solubility	Zhu et al. (2016)
Magnolol	Mixed micelles and nanosuspension of Soluplus and Poloxamer-407	Low Solubility	Improve solubility and bioavailability	Li et al. (2020)
Erlotinib	Solid lipid NPs	Oral bioavailability	Improve oral bioavailability	Bakhtiary et al. (2017)
Doxorubicin	Iron oxide	Multidrug resistance	Overcome drug efflux from multidrug resistance C6 glioma cells.	Kievit et al. (2011)
Doxorubicin	Silica NPs	Multidrug resistance	Decrease MDR	Gao et al. (2011)
Doxorubicin	Polyrotaxane NPs	Multidrug resistance	Provide multidrug resistance	Wang et al. (2015)
Doxorubicin	Fucoidan-capped gold NPs	Drug delivery with imaging	Act as a theranostic agent	Manivasagan, Bharathiraja, Bui, Jang et al. (2016)
Paclitaxel	Chitosan-stabilised NPs	Drug delivery with imaging	Act as a theranostic agent	Manivasagan, Bharathiraja, Bui, Lim et al. (2016)
Indocyanine green (IG)	Polymeric NPs	Thermal stability, photostability and aqueous stability	Improve thermal stability, photostability and aqueous stability	Obinu et al. (2020)
Hypercin	Solid lipid NPs	Photostability	Increase photostability	Lima et al. (2013)
Octyl-methoxycinnamate	Lipid nanocarrier	Photostability	Protect drugs from ultraviolet-mediated photodegradation	Prado et al. (2020)

pharmaceutical absorption, thereby enhancing the adhesion of drug NPs to the mucosal surface. It is claimed that smaller particles of medicines are taken by macrophages more easily and achieve a higher rate of deposition and hence a better index of treatments (Date, Hanes, and Ensign 2016; Deshpande et al. 2015; Ensign, Cone, and Hanes 2012; Jia 2005; Patel et al. 2016; Da Silva et al. 2020).

1.5.2 Theranostic Agent

A NP with both therapeutic and diagnostic agents on a single platform is the most fundamental concept of 'theranostic' NPs (Ma et al. 2019; Xie, Lee, and Chen 2010; Xu, Bayazitoglu, and Meade 2019). Additional meanings of theranostics in general include efforts in the clinic using diagnostic testing for individual patients to develop personalised therapies and can include image-guided surgery and post-surgery evaluation. This can be achieved in two ways: (i) diagnosis accompanied by treatment to evaluate responses in order to treat and assess who has an effect and (ii) diagnosis accompanied by treatment to first classify the type of disorder in order to provide specific patients with a customised treatment (Silva et al. 2019; Zavaleta, Ho, and Chung 2018).

In view of the recent emphasis on accuracy and personalised medicines as well as advancement in nanochemical formulations and NPs, a range of physicochemical properties, including dimensions, surface load and deformability, should be carefully considered in developing smart and multifunctional NPs for cancer theranostics (Silva et al. 2019). A variety of organic and inorganic NPs, including lipids, synthetic and natural polymers, proteins, metals and other carbon structures, have been produced. Organic materials are especially involved in the preparation and synthesis of NPs, and organic NPs give additional benefits, in addition to size-dependent properties. In biological fluids, organic NPs are typically versatile and easily adjusted and extremely stable. Furthermore, some organic NPs display unique features for cancer or any other disease such as biodegradability and affinity to particular molecular biomarkers (Anselmo and Mitragotri 2015; Bae, Chung, and Park 2011; Chapman et al. 2013). In the imaging and therapy of tumours, the integration of nanotechniques and lipid systems plays a significant role.

1.5.3 MULTIDRUG RESISTANCE

The overexpression of pharmaceutical efflux carriers is one of the most prevalent phenomena in the clinic. These transporters of drug fluxes are also expressed in normal tissue to hold homoeostasis of different processes. Therefore, these transporters need to be blocked in tumour tissue and normal tissue exposure must be reduced. NPs are commonly used to do this and to improve therapeutic molecular position in tumour tissues. The usage of the technique to overcome drug efflux transporters includes the integration of therapeutic molecules into the NP system before internalisation (Callaghan, Luk, and Bebawy 2014; Choi and Yu 2014; Patel et al. 2013). However, most NP systems are designed to accumulate and release their cargo into the tumour microenvironment, thus optimising cell penetration and exposure (Patel et al. 2013). Also, powerful tools have resulted in identification of many non-toxic potent efflux transporter inhibitors as well as identification of RNAi, using which expression of drug efflux transporters can be inhibited. Therefore, several groups have focused on co-delivering therapeutic molecules with either drug transporter inhibitors or gene silencer inhibitors (Patil et al. 2010).

Certain of these nanomaterials were used by drug carriers to resolve drug efflux, thereby increasing the impact of medication retention in cancer cells. These systems involve the nanoformulation of siRNAs and medicinal products such as doxorubicin (DOX), particularly with liposomes, nanodiamonds and mesoporous silicon NPs, to impede the progression of cancer, inhibiting the drug detoxification, by suppressing cell defence mechanisms and activating apoptosis and DNA repair.

Other medications such as vincristine, verapamil, cyclosporine, paclitaxel, oxaliplatin, cisplatin and curcumin have also been supplied with polymeric (NPs) and surfactant nanoemulsions. Combining several drug and MDR inhibitors, this opens up a wide range of plans for targeting the extensive mechanisms of MDR, in particular through the EPR effect, and at the same time, this strategy poses some serious questions (Devalapally et al. 2007; Sánchez-López et al. 2019).

Most studies have shown that multiple agents such as siRNAs, chemotherapeutic drugs and antibodies and other MDR inhibitors assist in the same NPs, but these formulations are by now far from getting excellent performance (Ahmad et al. 2016; Conde, de la Fuente, and Baptista 2013).

1.6 STABILITY OF ACTIVE PHARMACEUTICAL INGREDIENTS (API)

Nanoparticulate-mediated drug delivery has been used for the protection of drug(s) against photolytic degradation, moisture degradation and acidic degradation. Specifically, NPs made from natural and synthetic polymers (biodegradable and non-biodegradable) have received more attention because adaptation of these systems can prevent endogenous enzymes from degrading the drug.

1.7 MARKETED APPROVAL OF NP DRUG DELIVERY AND REGULATORY STATUS

Several USFDA-approved drug products have been available in the market and numerous initiatives in pre-clinical and clinical studies are under pipeline, indicating that nanotechnology-based pharmaceutical products will create a boon and soon become available to the market. FDA-approved nanomedicinal formulations can be classified into the forms of polymeric nanomedicines, micelles, liposomes, antibody medicinal conjugates, protein NPs, inorganic NPs, nanocrystals and so on. Polymer nanomedicines are the most basic types of nanomedicines, which contain soft materials to improve solubility, biocompatibility, bioavailability and control the discharge of active nanomedicine pharmaceutical gradients into the body. Some of the FDA-approved nanomedicines classified by the type of carrier/material used in the preparation of the formulation are shown in Table 1.3 (Patra et al. 2018).

1.7.1 NANOMEDICINES: CURRENT STATUS AND FUTURE PERSPECTIVES IN THE ASPECTS OF DRUG DELIVERY AND PHARMACOKINETICS

In recent years, nanotechnology has grown dynamically, and investments in research and development are growing in all countries, regardless of whether they have evolved. However, those researchers dealing with functional nano-drug applications are facing high levels of uncertainty, such as the framing of the simple defiance of these products; the characteristic protection and toxicity of these nanomaterials; and the lack of effective control. The list of approved nanomedicines is very comprehensive, but its therapeutic potential ends up being hindered by the unavailability of regulatory guidelines to create and classify these nanomaterials.

Many researchers have often restricted their efforts to evaluate the toxic potential of nanomedicines in the early stages of studies because there are no standardised protocols for nontoxic materials at physicochemical, physiological or biological level. A closer collaboration between regulatory agencies is warranted in order to simplify and/or shorten the process for approval of nano-based products, drug delivery systems, etc. (Soares et al. 2018; Patra et al. 2018).

1.8 FUTURE OF NANOMEDICINES AND DRUG DELIVERY SYSTEMS

Nanomedicine technology is actually one of the most interesting research fields in the healthcare sector including drug delivery. A lot of studies have led to 1,500 patents and several dozen clinical trials completed in the last two decades.

TABLE 1.3
A Clinical Update of Marketed Formulations with Special Emphasis on Formulations and Applications

Name	Drug	Formulation	Approved Application	Approved Year
VYXEOS	Cytarabine/daunorubicin (5:10)	Liposomal formulation	Acute myeloid leukaemia	2016
ONPATTRO	RNAi	Lipid NPs	Transthyretin amyloidosis	2018
Ryanodex®	Dantrolene sodium	Nanocrystals	Malignant hypothermia	2014
Feraheme	Iron polyglucose sorbitol Carboxymethyl ether colloid	Iron NPs	Iron-deficient anaemia	2009
Cimzia®	Certolizumab	PEGylated antibody fragment	Crohn's disease; rheumatoid arthritis; psoriatic arthritis and ankylosing spondylitis	2008
Abraxane	Paclitaxel	Albumin NPs	Advanced non-small-cell lung cancer	2005
DepoDur®	Morphine sulphate	Liposomes	Loss of pain	2004
Emend®	Aprepitant	Nanocrystals	Antiemetics	2003
Estrasorb™	Estradiol	Micelles	Menopause hormone therapy	2003
Avinza®	Morphine sulphate	Nanocrystals	Mental stimulant	2002
Pegasys®	Interferon-alpha (IFN-α2a)	PEGylated IFN-α2a protein	Hepatitis B and C	2002
Eligard®	Leuprolide acetate	Polymer (PLGA – poly (D,L-lactide-co-glycolide))	Prostate cancer	2002
AmBisome®	Amphotericin B	Liposomes	Fungal and/or protozoal infection	1997
DaunoXome	Daunorubicin	PEGylated liposome	HIV-associated Kaposi's sarcoma	1996
DOXIL	Doxorubicin	Liposomes	Ovarian cancer	1995

The best example of illnesses in which their diagnosis and treatment have both gained from non-medical technology is cancer, as illustrated in the different sections above. The application of nanomedicines and the nanodrug delivery method is definitely a trend that will remain the future field of research and development, using different types of smart and multifunctional NPs to provide precise concentration of the delivered/targeted drugs cells affected, such as cancer or tumour cells, without disrupting normal cell physiology.

While it recognises the potential prospect of nanomedicines and the nanosupply system overwhelmingly, its true effect on the health system is still very minimal, also in cancer treatment/diagnosis. This is a particular area of science with just two decades of actual scientific study, and many essential characteristics are still unknown. One of the main research areas in the future is the fundamental markers of tissues, which include key organic markers that permit absolute targeting without altering the normal cellular process.

The simplistic view about the production of nanodevices, which work with a complete external control mechanism in tissue diagnosis and repair mechanism, was enthusiastic (Patra et al. 2018). It is not yet a reality and is therefore a futuristic investigation that humanity might maybe achieve in the very near future. However, the possible danger to humans and the environment as well as their advantages requires long-term studies. Proper impact analysis on people and the environment should be investigated for potential acute or chronic effects of emerging nanomaterials on toxicity. With nanomedicines gaining prominence, they would also be affordable to study in another field.

Due to the impacts of nanomaterials on the organism that range from cytotoxicity to hypersensitivity, the biocompatibility of nanomaterials is extremely significant. Consequently, it is important to implement cost-effective, stronger and safer nanobiomaterials that efficiently load and monitor drug releases of a variety of difficult drug carriers that do not yet have an adequate supply.

In the future, artificial intelligence (AI)-based technologies with micron-scale and nanoscale range would play a pivotal role in drug delivery. The new approaches of research that lead to identification and treatment of cancer cells are ligand- or antibody-conjugated nanoformulation and biphonic and multifunctional NPs. Nanomachines have also largely been investigated and developed, but some primitive molecular machines have been tested. In diagnostics for early detection and prevention from the deadly diseases the nanorobots that penetrate into the complex biological barriers of the human body can be used to detect the infections cells. Therefore, nanodrug systems will soon be a pioneer in nanomedicine in the prevention and diagnostic areas (Mukherjee et al. 2014).

1.9 CONCLUSION AND FUTURE PERSPECTIVES

Micro- and nanoparticulate systems for drug delivery have become prominent in research and clinical practice. In spite of the fact that the two size classes are essential for a continuum, an assortment of physiological, anatomical and physicochemical elements give them unique properties, and in this manner wide applications. Thus, in this chapter we summarised an overview of particulate systems (both micro- and nano-range), their types, synthesis, characterisations, properties, fundamental concepts and biomedical and pharmaceutical applications. Nanotechnology has proven to be beneficial in the treatment of cancer and many other diseases with advances in diagnostics and cures. In this way, quick advancement in particulate systems is anticipated in the coming decades.

In the future, particulate system-based drug delivery systems can be improved in various other fields such as gene therapy, antitumour therapy, radiotherapy, antibiotics and vaccines, and delivery of proteins for the sustainable and precise cure of diseases. The mechanism and fate of MPs and NPs need to be studied using animal models to enable scientists to develop targeting, transporting, releasing, drug loading, interactions with the barriers, low toxicity and safe conditions. The understanding of drugs when targeted to sensitive organelles, such as the nucleus and mitochondria, will help to improve the use of particulate systems. The development of multifunctional systems that are capable of detecting cancer cells and infections, visualising the location by imaging agents and delivering drugs at the same time with minimum side effects is expected. These particles may be improved and developed to cure and treat diseases. NPs in conjunction with a computer programming system can be used to automatically regulate the homoeostasis in human beings for better future and quality of life. These NPs can also be improved as powerful protectors in the body towards foreign particles in the future. Legitimate impact analysis of the possible acute or chronic toxicity effects of new nanomaterials on humans and the environment must be conducted. Finally, the regulation of nanomedicines, as elaborated in the section, will continue to evolve alongside the advances in nanomedicine applications.

ACKNOWLEDGEMENT

The authors would like to thank the National Institute of Pharmaceutical Education & Research (NIPER), Hyderabad, for providing opportunity to write this chapter (Research communication No. NIPER-H/2020/BC-09). Dr. Gulbake would like to thank SERB, New Delhi, for extending facilities to write this chapter (EEQ/2016/000789).

REFERENCES

Ahmad, Javed et al. 2016. "Engineered Nanoparticles Against MDR in Cancer: The State of the Art and Its Prospective." *Current Pharmaceutical Design* 22(28): 4360–73.

Ahn, Jungho et al. 2018. "Microfluidics in Nanoparticle Drug Delivery; From Synthesis to Pre-Clinical Screening." *Advanced Drug Delivery Reviews* 128: 29–53.

Anselmo, Aaron C., and Samir Mitragotri. 2015. "A Review of Clinical Translation of Inorganic Nanoparticles." *AAPS Journal* 17(5): 1041–54.

Aziz, Zarith Asyikin Abdul, and Siti Hamidah Mohd Setapar. 2020. "Nanotechnology: An Effective Approach for Enhancing Therapeutics and Bioavailability of Phytomedicines." In: Thangadurai, D., Sangeetha, J., and Prasad, R. (eds) *Functional Bionanomaterials. Nanotechnology in the Life Sciences*, Cham: Springer, 47–71.

Bae, Ki Hyun, Hyun Jung Chung, and Tae Gwan Park. 2011. "Nanomaterials for Cancer Therapy and Imaging." *Molecules and Cells* 31(4): 295–302.

Bakhtiary, Zahra et al. 2017. "Microparticles Containing Erlotinib-Loaded Solid Lipid Nanoparticles for Treatment of Non-Small Cell Lung Cancer." *Drug Development and Industrial Pharmacy* 43(8): 1244–53.

Bock, Nathalie, Tim R. Dargaville, and Maria Ann Woodruff. 2012. "Electrospraying of Polymers with Therapeutic Molecules: State of the Art." *Progress in Polymer Science* 37(11): 1510–51.

Callaghan, Richard, Frederick Luk, and Mary Bebawy. 2014. "Inhibition of the Multidrug Resistance P-Glycoprotein: Time for a Change of Strategy?" *Drug Metabolism and Disposition* 42(4): 623–31.

Chapman, Sandra et al. 2013. "Nanoparticles for Cancer Imaging: The Good, the Bad, and the Promise." *Nano Today* 8(5): 454–60.

Chavan, Bhagwat Raghunathrao (2015), Synthesis And Characterization Of Nonlinear Optical Nanoparticles And Their Applications. http://hdl.handle.net/10603/74773

Cherian, Anitha K., A. C. Rana, and Sanjay K. Jain. 2000. "Self-Assembled Carbohydrate-Stabilized Ceramic Nanoparticles for the Parenteral Delivery of Insulin." *Drug Development and Industrial Pharmacy* 26(4): 459–63.

Choi, Young, and Ai-Ming Yu. 2014. "ABC Transporters in Multidrug Resistance and Pharmacokinetics, and Strategies for Drug Development." *Current Pharmaceutical Design* 20(5): 793–807.

Chopdey, Prashant K. et al. 2015. "Glycyrrhizin Conjugated Dendrimer and Multi-Walled Carbon Nanotubes for Liver Specific Delivery of Doxorubicin." *Journal of Nanoscience and Nanotechnology* 15(2): 1088–1100.

Conde, João, Jesús M. de la Fuente, and Pedro V. Baptista. 2013. "Nanomaterials for Reversion of Multidrug Resistance in Cancer: A New Hope for an Old Idea?" *Frontiers in Pharmacology* 4(October): 1–5.

Da Silva, Flávia Lidiane Oliveira, Maria Betânia De Freitas Marques, Kelly Cristina Kato, and Guilherme Carneiro. 2020. "Nanonization Techniques to Overcome Poor Water-Solubility with Drugs." *Expert Opinion on Drug Discovery* 15(7): 853–64.

Date, Abhijit A., Justin Hanes, and Laura M. Ensign. 2016. "Nanoparticles for Oral Delivery: Design, Evaluation and State-of-the-Art." *Journal of Controlled Release* 240: 504–26.

Deshpande, Rohan D. et al. 2015. "The Effect of Nanonization on Poorly Water Soluble Glibenclamide Using a Liquid Anti-Solvent Precipitation Technique: Aqueous Solubility, *In Vitro* and *In Vivo* Study." *RSC Advances* 5(99): 81728–38.

Devalapally, Harikrishna, Zhenfeng Duan, Michael V. Seiden, and Mansoor M. Amiji. 2007. "Paclitaxel and Ceramide Co-Administration in Biodegradable Polymeric Nanoparticulate Delivery System to Overcome Drug Resistance in Ovarian Cancer." *International Journal of Cancer* 121(8): 1830–38.

Dreaden, Erik C. et al. 2012. "The Golden Age: Gold Nanoparticles for Biomedicine." *Chemical Society Reviews* 41(7): 2740–79.

Ensign, Laura M., Richard Cone, and Justin Hanes. 2012. "Oral Drug Delivery with Polymeric Nanoparticles: The Gastrointestinal Mucus Barriers." *Advanced Drug Delivery Reviews* 64(6): 557–70.

Eustis, Susie, and Mostafa A. El-Sayed. 2006. "Why Gold Nanoparticles Are More Precious than Pretty Gold: Noble Metal Surface Plasmon Resonance and Its Enhancement of the Radiative and Nonradiative Properties of Nanocrystals of Different Shapes." *Chemical Society Reviews* 35(3): 209–17.

Fang, Fang, Min Li, Jinfeng Zhang, and Chun Sing Lee. 2020. "Different Strategies for Organic Nanoparticle Preparation in Biomedicine." *ACS Materials Letters* 2(5): 531–49.

Fu, Xinxin et al. 2018. "Top-down Fabrication of Shape-Controlled, Monodisperse Nanoparticles for Biomedical Applications." *Advanced Drug Delivery Reviews* 132: 169–87.

Gao, Yu et al. 2011. "Controlled Intracellular Release of Doxorubicin in Multidrug-Resistant Cancer Cells by Tuning the Shell-Pore Sizes of Mesoporous Silica Nanoparticles." *ACS Nano* 5(12): 9788–98.

Gradauer, K. et al. 2013. "Liposomes Coated with Thiolated Chitosan Enhance Oral Peptide Delivery to Rats." *Journal of Controlled Release* 172(3): 872–78.

Guo, Dan, Guoxin Xie, and Jianbin Luo. 2014. "Mechanical Properties of Nanoparticles: Basics and Applications." *Journal of Physics D: Applied Physics* 47(1): 013001.

Hasan, Warefta et al. 2012. "Delivery of Multiple SiRNAs Using Lipid-Coated PLGA Nanoparticles for Treatment of Prostate Cancer." *Nano Letters* 12(1): 287–92.

He, Haisheng et al. 2018. "Biomimetic Thiamine- and Niacin-Decorated Liposomes for Enhanced Oral Delivery of Insulin." *Acta Pharmaceutica Sinica B* 8(1): 97–105.

Jia, Lee. 2005. "Nanoparticle Formulation Increases Oral Bioavailability of Poorly Soluble Drugs: Approaches, Experimental Evidences and Theory." *Current Nanoscience* 1(3): 237–43.

Kavand, Alireza et al. 2020. "Synthesis and Functionalization of Hyperbranched Polymers for Targeted Drug Delivery." *Journal of Controlled Release* 321: 285–311.

Kesarwani, Kritika, and Rajiv Gupta. 2013. "Bioavailability Enhancers of Herbal Origin: An Overview." *Asian Pacific Journal of Tropical Biomedicine* 3(4): 253–66.

Khan, Azhar U., Masudulla Khan, Moo Hwan Cho, and Mohammad Mansoob Khan. 2020. "Selected Nanotechnologies and Nanostructures for Drug Delivery, Nanomedicine and Cure." *Bioprocess and Biosystems Engineering* 43(8): 1339–57.

Khan, Ibrahim, Khalid Saeed, and Idrees Khan. 2019. "Nanoparticles: Properties, Applications and Toxicities." *Arabian Journal of Chemistry* 12(7): 908–31.

Kievit, Forrest M. et al. 2011. "Doxorubicin Loaded Iron Oxide Nanoparticles Overcome Multidrug Resistance in Cancer In Vitro." *Journal of Controlled Release: Official Journal of the Controlled Release Society* 152(1): 76.

Klumpp, Cédric, Kostas Kostarelos, Maurizio Prato, and Alberto Bianco. 2006. "Functionalized Carbon Nanotubes as Emerging Nanovectors for the Delivery of Therapeutics." *Biochimica et Biophysica Acta – Biomembranes* 1758(3): 404–12.

Kumari, Avnesh, Sudesh Kumar Yadav, and Subhash C. Yadav. 2010. "Biodegradable Polymeric Nanoparticles Based Drug Delivery Systems." *Colloids and Surfaces B: Biointerfaces* 75(1): 1–18.

Laurent, Sophie et al. 2010. "Erratum: Magnetic Iron Oxide Nanoparticles: Synthesis, Stabilization, Vectorization, Physicochemical Characterizations, and Biological Applications (Chemical Reviews (2008) 108 (2064))." *Chemical Reviews* 110(4): 2574.

Lengyel, Miléna et al. 2019. "Microparticles, Microspheres, and Microcapsules for Advanced Drug Delivery." *Scientia Pharmaceutica* 87(3): 20.

Li, Guoyuan et al. 2020. "Enhanced Oral Bioavailability of Magnolol via Mixed Micelles and Nanosuspensions Based on Soluplus® – Poloxamer 188." *Drug Delivery* 27(1): 1010–17.

Li, Menghuan, Zhong Luo, and Yanli Zhao. 2018. "Self-Assembled Hybrid Nanostructures: Versatile Multifunctional Nanoplatforms for Cancer Diagnosis and Therapy." *Chemistry of Materials* 30(1): 25–53.

Lima, Adriel M. et al. 2013. "Hypericin Encapsulated in Solid Lipid Nanoparticles: Phototoxicity and Photodynamic Efficiency." *Journal of Photochemistry and Photobiology B: Biology* 125: 146–54.

Luo, Dan, Cong Yan, and Tie Wang. 2015. "Interparticle Forces Underlying Nanoparticle Self-Assemblies." *Small* 11(45): 5984–6008.

Ma, Qian et al. 2015. "Oral Absorption Enhancement of Probucol by PEGylated G5 PAMAM Dendrimer Modified Nanoliposomes." *Molecular Pharmaceutics* 12(3): 665–74.

Ma, Zhuoran et al. 2019. "A Theranostic Agent for Cancer Therapy and Imaging in the Second Near-Infrared Window." *Nano Research* 12(2): 273–79.

Manivasagan, Panchanathan, Subramaniyan Bharathiraja, Nhat Quang Bui, Bian Jang et al. 2016. "Doxorubicin-Loaded Fucoidan Capped Gold Nanoparticles for Drug Delivery and Photoacoustic Imaging." *International Journal of Biological Macromolecules* 91: 578–88.

Manivasagan, Panchanathan, Subramaniyan Bharathiraja, Nhat Quang Bui, In Gweon Lim et al. 2016. "Paclitaxel-Loaded Chitosan Oligosaccharide-Stabilized Gold Nanoparticles as Novel Agents for Drug Delivery and Photoacoustic Imaging of Cancer Cells." *International Journal of Pharmaceutics* 511(1): 367–79.

Martis, Elvis, Rewa Badve, and Mukta Degwekar. 2012. "Nanotechnology Based Devices and Applications in Medicine: An Overview." *Chronicles of Young Scientists* 3(1): 68.

Mehra, Neelesh Kumar, Vijay Mishra, and Narendra K. Jain. 2013. "Receptor-Based Targeting of Therapeutics." *Therapeutic Delivery* 4(3): 369–94.

Mehta, Prina et al. 2019. "Broad Scale and Structure Fabrication of Healthcare Materials for Drug and Emerging Therapies via Electrohydrodynamic Techniques." *Advanced Therapeutics* 2(4): 1800024.

Mukherjee, Biswajit et al. 2014. "Current Status and Future Scope for Nanomaterials in Drug Delivery." In *Application of Nanotechnology in Drug Delivery*. doi: 10.5772/58450.

Murr, Lawrence E. 2021. "Handbook of Materials Structures, Properties, Processing and Performance." In *Handbook of Materials Structures, Properties, Processing and Performance*, 1–29. doi: 10.1007/978-3-319-01905-5.

Nishiyama, Nobuhiro, and Kazunori Kataoka. 2003. "Polymeric Micelle Drug Carrier Systems: PEG-PAsp(Dox) and Second Generation of Micellar Drugs." *Advances in Experimental Medicine and Biology* 519: 155–77.

Obinu, Antonella et al. 2020. "Indocyanine Green Loaded Polymeric Nanoparticles: Physicochemical Characterization and Interaction Studies with Caco-2 Cell Line by Light and Transmission Electron Microscopy." *Nanomaterials* 10(1): 133.

Parrott, Matthew C. et al. 2010. "Tunable Bifunctional Silyl Ether Cross-Linkers for the Design of Acid-Sensitive Biomaterials." *Journal of the American Chemical Society* 132(50): 17928–32.

Patel, Harshil M., Bhumi B. Patel, Chainesh N. Shah, and Dhiren P. Shah. 2016. "Nanosuspension Technologies for Delivery of Poorly Soluble Drugs – A Review." *Research Journal of Pharmacy and Technology* 9(5): 625–32.

Patel, Niravkumar R., Bhushan S. Pattni, Abraham H. Abouzeid, and Vladimir P. Torchilin. 2013. "Nanopreparations to Overcome Multidrug Resistance in Cancer." *Advanced Drug Delivery Reviews* 65(13–14): 1748–62.

Patil, Yogesh B. et al. 2010. "The Use of Nanoparticle-Mediated Targeted Gene Silencing and Drug Delivery to Overcome Tumor Drug Resistance." *Biomaterials* 31(2): 358–65.

Patra, Jayanta Kumar et al. 2018. "Nano Based Drug Delivery Systems: Recent Developments and Future Prospects 10 Technology 1007 Nanotechnology 03 Chemical Sciences 0306 Physical Chemistry (Incl. Structural) 03 Chemical Sciences 0303 Macromolecular and Materials Chemistry 11 Medical and He." *Journal of Nanobiotechnology* 16(1): 1–33.

Peltonen, Leena, and Jouni Hirvonen. 2008. "Physicochemical Characterization of Nano- and Microparticles." *Current Nanoscience* 4(1): 101–7.

Perry, Jillian L., Kevin P. Herlihy, Mary E. Napier, and Joseph M. Desimone. 2011. "PRINT: A Novel Platform toward Shape and Size Specific Nanoparticle Theranostics." *Accounts of Chemical Research* 44(10): 990–98.

Pons, Thomas et al. 2010. "Cadmium-Free CuInS2/ZnS Quantum Dots for Sentinel Lymph Node Imaging with Reduced Toxicity." *ACS Nano* 4(5): 2531–38.

Prado, Alice Haddad Do et al. 2020. "Synthesis and Characterization of Nanostructured Lipid Nanocarriers for Enhanced Sun Protection Factor of Octyl P-Methoxycinnamate." *AAPS PharmSciTech* 21(4): 1–9.

Rana, Mahendra, Aadesh Kumar, and Amita J. Rana. 2020. "Drug Delivery through Targeted Approach with Special References to Phytosomes." In *Role of Novel Drug Delivery Vehicles in Nanobiomedicine [Working Title]*, 1–14. doi: 10.5772/intechopen.8664.

Rawat, Manoj K., Achint Jain, and Sanjay Singh. 2011. "Studies on Binary Lipid Matrix Based Solid Lipid Nanoparticles of Repaglinide: *In Vitro* and *In Vivo* Evaluation." *Journal of Pharmaceutical Sciences* 100(6): 2366–78.

Reuter, Kevin G. et al. 2015. "Targeted PRINT Hydrogels: The Role of Nanoparticle Size and Ligand Density on Cell Association, Biodistribution, and Tumor Accumulation." *Nano Letters* 15(10): 6371–78.

Richardson, Joseph J. et al. 2016. "Innovation in Layer-by-Layer Assembly." *Chemical Reviews* 116(23): 14828–67.

Salimi, Anayatollah, Behzad Sharif Makhmal Zadeh, and Moloud Kazemi. 2019. "Preparation and Optimization of Polymeric Micelles as an Oral Drug Delivery System for Deferoxamine Mesylate: *In Vitro* and *Ex Vivo* Studies." *Research in Pharmaceutical Sciences* 14(4): 293–307.

Sánchez-López, Elena et al. 2019. "Current Applications of Nanoemulsions in Cancer Therapeutics." *Nanomaterials* 9(6): 821.

Sigmund, Wolfgang et al. 2006. "Processing and Structure Relationships in Electrospinning of Ceramic Fiber Systems." *Journal of the American Ceramic Society* 89(2): 395–407.

Silva, Catarina Oliveira et al. 2019. "Current Trends in Cancer Nanotheranostics: Metallic, Polymeric, and Lipid-Based Systems." *Pharmaceutics* 11(1): 22.

Singh, Jaspreet, Keerti Jain, Neelesh Kumar Mehra, and N. K. Jain. 2016. "Dendrimers in Anticancer Drug Delivery: Mechanism of Interaction of Drug and Dendrimers." *Artificial Cells, Nanomedicine, and Biotechnology* 44(7): 1626–34.

Soares, Sara, João Sousa, Alberto Pais, and Carla Vitorino. 2018. "Nanomedicine: Principles, Properties, and Regulatory Issues." *Frontiers in Chemistry* 6(August): 360.

Song, Zhiwang et al. 2017. "Cyclic RGD Peptide-Modified Liposomal Drug Delivery System for Targeted Oral Apatinib Administration: Enhanced Cellular Uptake and Improved Therapeutic Effects." *International Journal of Nanomedicine* 12: 1941–58.

Sridhar, Radhakrishnan et al. 2015. "Electrosprayed Nanoparticles and Electrospun Nanofibers Based on Natural Materials: Applications in Tissue Regeneration, Drug Delivery and Pharmaceuticals." *Chemical Society Reviews* 44(3): 790–814.

Thomas, Shindu, Bentham Science Publisher Harshita, Pawan Mishra, and Sushama Talegaonkar. 2015. "Ceramic Nanoparticles: Fabrication Methods and Applications in Drug Delivery." *Current Pharmaceutical Design* 21(42): 6165–88.

Tiwari, Jitendra N., Rajanish N. Tiwari, and Kwang S. Kim. 2012. "Zero-Dimensional, One-Dimensional, Two-Dimensional and Three-Dimensional Nanostructured Materials for Advanced Electrochemical Energy Devices." *Progress in Materials Science* 57(4): 724–803.

Unser, Sarah, Ian Bruzas, Jie He, and Laura Sagle. 2015. "Localized Surface Plasmon Resonance Biosensing: Current Challenges and Approaches." *Sensors* 15(7): 15684–716.

Wang, He et al. 2015. "Folate-Mediated Mitochondrial Targeting with Doxorubicinpolyrotaxane Nanoparticles Overcomes Multidrug Resistance." *Oncotarget* 6(5): 2827–42.

Wang, Junmei, and Yujun Song. 2017. "Microfluidic Synthesis of Nanohybrids." *Small* 13(18): 1604084.

Wang, Yapei, James D. Byrne, Mary E. Napier, and Joseph M. DeSimone. 2012. "Engineering Nanomedicines Using Stimuli-Responsive Biomaterials." *Advanced Drug Delivery Reviews* 64(11): 1021–30.

Xie, Jin, Seulki Lee, and Xiaoyuan Chen. 2010. "Nanoparticle-Based Theranostic Agents." *Advanced Drug Delivery Reviews* 62(11): 1064–79.

Xu, Jing et al. 2013. "Future of the Particle Replication in Nonwetting Templates (PRINT) Technology." *Angewandte Chemie – International Edition* 52(26): 6580–89.

Xu, Xiao, Yildiz Bayazitoglu, and Andrew Meade. 2019. "Evaluation of Theranostic Perspective of Gold-Silica Nanoshell for Cancer Nano-Medicine: A Numerical Parametric Study." *Lasers in Medical Science* 34(2): 377–88.

Yang, Qiuhong, and Laird Forrest. 2016. "Drug Delivery to the Lymphatic System." In: Wang, B., Hu, L., and Siahaan, T. J. (eds) *Drug Delivery*, Hoboken, NJ: John Wiley & Sons, Inc, 503–48.

Zavaleta, Cristina, Dean Ho, and Eun Ji Chung. 2018. "Theranostic Nanoparticles for Tracking and Monitoring Disease State." *SLAS Technology* 23(3): 281–93.

Zhao, Mei-Xia, and Bing-Jie Zhu. 2016. "The Research and Applications of Quantum Dots as Nano-Carriers for Targeted Drug Delivery and Cancer Therapy." *Nanoscale Research Letters* 11(1): 1–9.

Zhao, Shuang et al. 2019. "The Future of Layer-by-Layer Assembly: A Tribute to ACS Nano Associate Editor Helmuth Mohwald." *ACS Nano* 13(6): 6151–69.

Zhu, Yuan et al. 2016. "*In Vitro* Release and Bioavailability of Silybin from Micelle-Templated Porous Calcium Phosphate Microparticles." *AAPS PharmSciTech* 17(5): 1232–39.

2 Formulation by Design (FbD)
An Emerging Approach to Design Vesicular Nanocarriers

Shiv Kumar Prajapati and Payal Kesharwani
Ram-Eesh Institute of Vocational and Technical Education

Nishi Mody
Dr. Harisingh Gour Central University

Ankit Jain
Indian Institute of Science

Swapnil Sharma
Banasthali Vidyapith

CONTENTS

Abbreviations 16
2.1 Introduction 16
2.2 FbD Terminology 16
2.3 The Methodology of FbD Optimisation 18
    2.3.1 Step I – Defining Formulation Objectives 18
    2.3.2 Step II – Selection of Significant Factors and Response Variables 18
    2.3.3 Step III – Formulation Development as per ED 19
    2.3.4 Step IV – Modelisation and Search for Optimum Formulation 19
    2.3.5 Step V – Validation Studies and Scale-Up 19
2.4 Design of Experiments 19
    2.4.1 Experimental Designs 20
    2.4.2 Selection of ED 20
        2.4.2.1 Comparative Objective 20
        2.4.2.2 Screening Objective 20
        2.4.2.3 Response Surface Method Objective 20
    2.4.3 Various EDs for the Optimisation of Nanovesicles 20
        2.4.3.1 Factorial Design 20
        2.4.3.2 Central Composite Design (CCD) 20
        2.4.3.3 Box–Behnken Design (BBD) 21
        2.4.3.4 Simple Mixture Design (SMD) 21
        2.4.3.5 Optimal Design 21
        2.4.3.6 Fractional Factorial Design (FFD) 21
        2.4.3.7 Taguchi Design (TgD) 21
        2.4.3.8 Plackett–Burman Design (PBD) 21
2.5 Application of QbD 21
    2.5.1 Liposomes 21
    2.5.2 Niosomes 22
    2.5.3 Transferosomes 25
    2.5.4 Pharmacosomes and Ufasomes 25
    2.5.5 Aquasomes and Polymeric Micelles 26

2.6 Conclusion and Prospects 27
Acknowledgement 27
Conflict of Interest 27
References..28

ABBREVIATIONS

CAA	Critical analytical attributes
BBD	Box–Behnken design
CCD	Central composite design
CFA	Critical formulation attribute
CMP	Critical method parameters
CN	Cinnarizine
CPP	Critical process parameter
CQA	Critical quality attribute
D-OD	D-optimal design
DoE	Design of experiments
DDS	Drug delivery system
ED	Experimental design
EE	Entrapment efficiency
FbD	Formulation by design
FCCD	Face-centred cubic design
FD	Factorial design
FDA	Food and Drug Administration
FFD	Fractional factorial design
ICH	International Conference on Harmonisation
NLP	Nanoliposome
PAT	Process analytical technology
PBD	Plackett–Burman design
PDI	Polydispersity index
QA	Quality attributes
QbD	Quality by design
QTTP	Quality target product profile
RSM	Response surface methodology
SMD	Simple mixture design
SSA	Specific surface area
TMC	Trimethyl chitosan
TgD	Taguchi design

2.1 INTRODUCTION

The concept of QbD was first developed by quality innovator Dr. Joseph M. Juran. He gave a theory that describes the quality of a product with reference to the problems and crises faced during its development. Woodcock said a superior drug product does not have contamination and consistently brings the therapeutic profit to the consumer (Woodcock 2004). It predominantly revolves around the principles of sound and risk-based development practice for expanding the understanding of product and process. As QbD shows its applications in various stages of product development, 'FbD', a branch of QbD, is used for the development of formulation practice at different stages of product development. The USFDA encourages risk-based tactics and implementation of QbD ideologies in drug product development, manufacturing and regulation.

Quality must be built into the product (Lawrence et al. 2014). QbD is a newer regulatory framework developed by ICH and USFDA for systematic drug product development. Fundamentally, it is about sound principles and risk-based development practice to supplement understanding of the product and process (Beg, Katare, and Singh 2017). These days, numerous QbD methodologies such as DoE, FbD and PAT are being meticulously used in diverse areas of science, for instance pharmaceutical formulation development and analytical chemistry. (Hasnain et al. 2019). In product development, QbD shows diverse applications. The FbD approach is categorised as the branch of QbD, in particular for formulation development (Beg, Katare, and Singh 2017). Figure 1a illustrates the important features carried out in FbD. FbD demands a scrupulous development of pharmaceutical products with the right choice of EDs (Ankita et al. 2013). The CFAs and CPPs applied in the design should meticulously define design and space utilising computer-aided optimisation followed by summarising the control strategies with effective improvement (Singh et al. 2011, Bai et al. 2020). Quality, output, price, cycle time and value are interconnected terms. The action of quality control must attempt to identify problems related to quality, early enough to allow the activities deprived of settlement in budget or schedule. The emphasis must be on precaution rather than redress of quality problems. The main purpose is to build the quality in the formulation so as to avoid the upcoming failure. Throughout the progression of development of vesicular formulation, numerous facets such as drug substances, excipients, processes and quality control tests, containers and distinctive atmospheric circumstances are risky for quality of product. It is a vital assignment to recognise the critical process variables and formulation characteristics for the assurance of quality (Beg, Katare, and Singh 2017).

2.2 FbD TERMINOLOGY

FbD is a statistical approach to unify the experiment, and it is a fashion to obtain precise and efficient information. During utilisation of the FbD approach, specific terminology both for technical and for anything else is used to assist improved clarity of rules of FbD. To simplify the clarity of principles of FbD for the vesicular system, vital definitions are given together in Table 2.1. It is necessary to be cognizant of FbD terminology and prior multifaceted information on several possible products and process variables (Singh et al. 2017). These variables are occupied by 'knowledge space' affecting the entire quality of the product. As a subcategory of knowledge space, 'design space' must be established

Formulation by Design

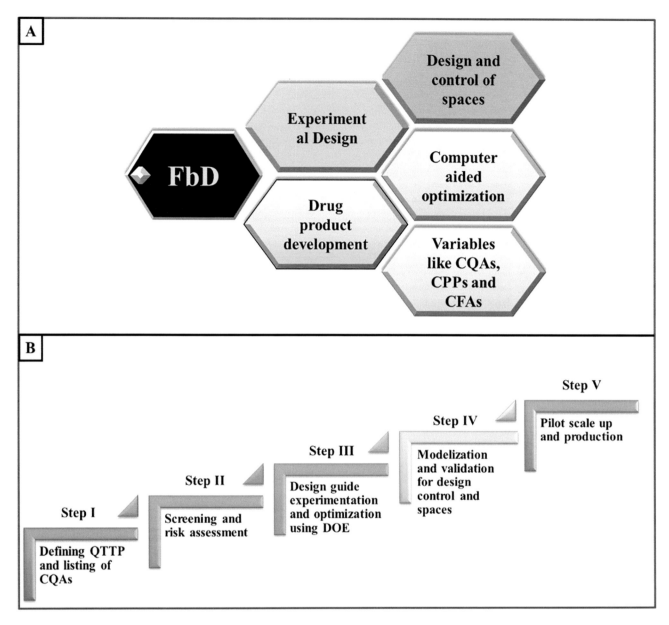

FIGURE 2.1 (a) Elements of FbD. (b) The five-step strategy of optimising drug delivery systems.

TABLE 2.1
Essential Terminologies Used during the FbD Protocols

Term	Definition
Antagonism	Undesired change due to interactions among the factors
Blocks	A fair set of experimental conditions, which defines the occurrence of each level of primary factor which is same as the number of each level in nuisance factor.
Categorical variables	Defined as the ones that cannot quantify the qualitative variables
Coding (normalisation)	The technique that allows the transformation of natural parameters into non-dimensional coded variables.
Confounding	Defined as the absence of orthogonality
Constraints	Diminution faced by each factor level
Contour plot	The geometric portrayal of a response acquired by plotting one independent variable in contrast to another, though holding the extent of response and other variables as constant

(*Continued*)

TABLE 2.1 (*Continued*)
Essential Terminologies Used during the FbD Protocols

Term	Definition
Control space	The field of design space designated for comprehensive controlled strategy
Critical formulation attributes	Formulation parameters disturbing critical quality attributes
Critical process parameters	Self-regulating process parameters almost certainly to affect the quality traits of a product or intermediates
Critical quality attributes	Parameters ranging within proper limits, which guarantee the anticipated product quality
Design matrix	The layout of experimental runs in matrix form as per experimental design
Design space	Multidimensional combination and interaction of input variables and process parameters, confirmed to provide quality assurance
Effect	The degree of variation in response to changing the factor level(s)
Empirical model	A mathematical term model relating factor–response by means of polynomial equations
Experimental domain	Factor space constituents that are optimised experimentally
Factor space	Dimensional space defined by coded variables
Factors	Independent variables, which have a tendency to affect the product/process features or output of the process
Independent variables	Input variables, which are directly under the control of product development scientist
Interaction	Lack of additivity of factor effects
Knowledge space	Scientific elements to be considered and explored on the basis of former knowledge as product attributes and process parameters
Level	Values assigned to a factor
Main effect	The effect of a factor averaged over all the levels of other factors
Nuisance factor	Factors which is used in blocking the effect of responsive variable but is of no interest to research topic
Optimisation	Execution of organised tactics to accomplish the optimal design of the product and/or characterisation of the process using FbD and software
Optimise	To make effective and functional use
Orthogonality	A state where the assessed effects are due to the main factor of interest, but independent of interactions
Quantitative variables	Variables that are identified by numeric values
Resolution	The extent to which the degree of confounding is defined
Response surface	Pictorial representation of mathematical ideas
Response surface plot	3D graphical portrayal of response plotted between two independent variables and one response variable
Response variables	Variables that refer to a subject of change within the experiment whose change is dependent on other variable
Runs or trials	The specified experimental conditions that are used to carry out the experiment
Synergism	The interaction between the two variables leading to a positive result

that approves execution of quality of product or process comprising few important variables. 'Control space' is the experimental domain kept for meticulous studies in course of studies within the sophisticated range of input variables denoted as 'control strategy'. 'Design space' applies a systematic approach to archival data to convert the 'knowledge space' to 'control space'. Runs or trials are the experiments conducted according to the selected ED. Factors (independent variables) influence the feature of product/process or output of the process, generally designated as response variables (Jain, Hurkat, and Jain 2019).

2.3 THE METHODOLOGY OF FbD OPTIMISATION

The methodology of FbD optimisation helps to identify the major factors affecting the process. The FbD optimisation method is organised in a five-step hierarchy that provides contemplative and comprehensive data on various FbD facets (Singh et al. 2017) (Figure 2.1b).

2.3.1 STEP I – DEFINING FORMULATION OBJECTIVES

It is the initial phase where an attempt is made to establish objective(s) for the drug delivery. In this step, many CQAs or response variables, which practically characterise the objective(s), are assigned for this purpose. Likewise, all the independent product variables or process variables are also planned (Kumari et al. 2019).

2.3.2 STEP II – SELECTION OF SIGNIFICANT FACTORS AND RESPONSE VARIABLES

In this step, the response variables that may significantly affect the quality of the product (e.g. particle size for nanoparticles, emulsification time for self-emulsifying systems) are selected. In the screening process, the selection of 'prominent few' significant factors amidst the 'possible many' input variables is accompanied by means of EDs. Afterwards, factors influencing studies are to be accompanied regularly to compute the effect of factors and the presence of interactions, if any. Experimental studies are also commenced to describe the broad range of factor levels (Singh et al. 2017, Jain, Kumari et al. 2018).

2.3.3 Step III – Formulation Development as per ED

In this step, an appropriate ED is accomplished to plan the responses based on study objective(s), responses being explored, number and the type of factor and its levels, viz. high, medium or low. As per ED, a design matrix is an arrangement of experimental runs in the form of a matrix and successively created to direct the drug delivery scientist. The nanovesicular formulations consistently prepared with the design matrix, and the response variables can be estimated precisely (Kumari et al. 2019, Bishnoi et al. 2020).

2.3.4 Step IV – Modelisation and Search for Optimum Formulation

On the basis of statistical significance of generated data, an appropriate model, either graphical or numerical model, is recommended. Then, RSM is involved to relate the response variable to levels of input variables. Furthermore, the optimum concentration of formulation within experiment is searched by employing optimisation techniques (Singh et al. 2017).

2.3.5 Step V – Validation Studies and Scale-Up

The final phase of FbD involves the validation of response predictability of the proposed design model. Various studies to evaluate the drug delivery performance are assessed and compared with those predicted by using RSM (Sangshetti et al. 2017).

2.4 DESIGN OF EXPERIMENTS

DoE techniques have broad applications in optimisation of various nanovesicular formulations (Rafidah et al. 2014, Saraf, Jain, Tiwari, Verma, Panda et al. 2020). To lessen the number of factors, various designs can be applied to develop the nanovesicles. Thus, the technique reduces the number of experiments without losing valuable information (Beg et al. 2019). Statistical models which are involved in the DOE help provide a promising assessment in optimization of formulations. Statistical DoE offers logical conformity that leads to an effectual and descriptive investigation. DoE techniques bring about desired objectives of nanovesicle formulations as a consequence of fewer experimental runs. Besides, the DoE screening process assists in distinguishing significant and insignificant variables; additionally, it requires minimum efforts, time and material and has low cost (Mishra et al. 2018, Politis et al. 2017, Jain et al. 2020). DoE has its widespread province of application out there even in the pharma sector; therefore, FbD has in recent times been announced for the development of improved nanovesicle formulations. Additionally, DoE plays a significant role in optimisation of the composition of nanovesicles. It has already been applied to various nanovesicular formulations of liposomes, niosomes, transferosomes, cubosomes, etc., to improve their drug delivery applications (Jain et al. 2019, Jain and Jain 2016b, 2018a, b). Optimisation and validation of nanovesicle formulation using different EDs provides a deep understanding of each attributes involved in an experiment that helps to achieve a better quality with high performance output of the formulation (Jain and Jain 2016a). For applying the FbD approach of nanovesicles, various software packages are used illustrated in Figure 2.2a.

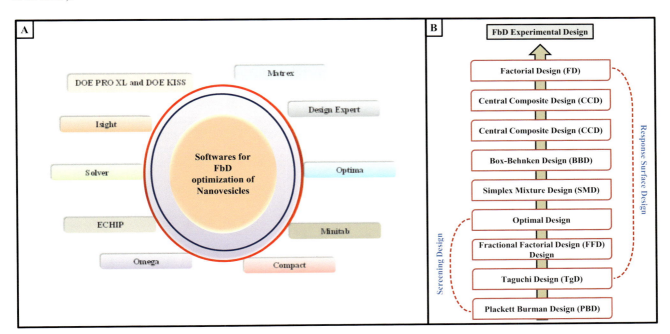

FIGURE 2.2 (a) Software packages for optimisation of nanovesicles. (b) Experimental design for FbD-based optimisation.

2.4.1 Experimental Designs

EDs are employed for 'screening' of influential variables as well as subsequent response surface analysis. Further, EDs FD, CCD and FFD have been extremely employed for systematic optimisation of DDSs. Figure 2.2b represents major EDs that are used for the optimisation of different vesicular system with cardinal remark. The factorial, composite and mixture designs are the most explored designs for optimisation of DDSs. Studied formulation variables exert a noteworthy impact on the response parameters. The desirability method can be utilised for numerical optimisation by fixing limits of the dependent and independent variables (Garg and Singhvi 2015, Debnath, Aishwarya, and Babu 2018).

2.4.2 Selection of ED

There are various parameters to be considered during the selection of EDs which influences the objective and goal of the experiment. Minimum-run design permits verification of curvature in a two-level screening design and benefits reserved properties to repeat the process and to avoid discrepant consequences. FDs, PBDs or TgDs are generally capable to do appropriate screening and support only linear responses. However, in order to detect non-linear responses or for a more accurate response surface, complex designs are suitable. Henceforth, response surface designs are used to evaluate the interactions leading to non-linear response or even quadratic effects or to envisage the shape of the response surface. It is important to screen out insignificant factors and to identify the important one using interaction effects (Singh et al. 2017, Trivedi et al. 2015). FDs using colour coding or Min-Run Res V designs are beneficial to understand key effects and to attain sample information about two-factor interactions (Jain, Hurkat, and Jain 2019). Therefore, based on the experimental objective, the EDs are selected as follows.

2.4.2.1 Comparative Objective

Here, the objective is to screen out the important factor among the existing ones and draw a conclusion to a relative problem which can be solved by taking on comparative designs.

2.4.2.2 Screening Objective

The goal of this design is to screen the more essential factors among the less important. In this, it is possible to select FFD or PBD.

2.4.2.3 Response Surface Method Objective

Response surface design is used when it is necessary to examine the interaction between the factors, when the conditions involve the expansion of new ideas regarding the shape of response surface. These designs are used to improve the problems of process and to make a product more robust. Consequently, the product will not be affected by uncontrollable variables. BBD and CCD are widely used in this category. Aside from all these, the selection of designs depends on the number of factors because every design has limitations that will not be accepted. For instance, in BBD minimum digit entered as numeric factor is 3 and maximum is 21 (Trivedi et al. 2015).

2.4.3 Various EDs for the Optimisation of Nanovesicles

2.4.3.1 Factorial Design

The factorial design technique for optimisation is the thoroughly used technique for statistical analysis and is a well-organised technique for nanovesicle optimisation. In this, all levels (x) are joined together with a given factor (k). This technique is effective to access the interactions and main effect (Figure 2.3a). The key deficit of FD is that it requires more experimental runs (Swarbrick 2018, Kesharwani et al. 2020).

2.4.3.2 Central Composite Design (CCD)

CCD is extensively applied for process optimisation. CCD model contains four–four full-factorial design points, axial points and one point at least at the centre of an experimental region that provides properties such as orthogonality and rotatability for fitting quadratic polynomials equation in CCD. The design is also called the Box–Wilson design (Figure 2.3b). This combines the advantages of FD and time-saving and cost-effective alternative to three-level full-factorial design with comparable performance (Aziz, Abdelbary, and Elassasy 2018b).

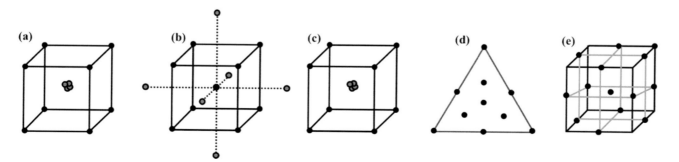

FIGURE 2.3 Examples of response surface designs. (a) FD, (b) CCD, (c) BBD, (d) mixture design and (e) optimal design.

2.4.3.3 Box–Behnken Design (BBD)

BBD is a self-determining, rotatable quadratic design. The treatment combinations are present at the process and the centre of experimental domain (Figure 2.3c). This design avoids conducting experiments under extreme conditions as it does not contain experiment of combinations of factors that are located at their highest or lowest levels. Therefore, a three-level (+, 0, −) and three-factor BBD was selected as a suitable approach to develop and optimise the formulation variables in order to explore the quadratic response surfaces and to bring into being polynomial models to enable optimisation with minimum experiments. It incapacitates the implicit pitfall of CCD where each factor has to be studied at five levels (Box and Behnken 1960, Pathak, Sharma, and Sharma 2016).

2.4.3.4 Simple Mixture Design (SMD)

SMD is the statistical design that involves mixture methodology, and it is an effective method to study products made from components at various levels. SMD can be lattice or centroid design (Figure 2.3d). Both designs are identical for first- and second-order models. SMD is recommended to test combinations with two or more compounds and interpolate effective mixtures from non-constant ratio combinations (Jain 2018).

2.4.3.5 Optimal Design

A significantly lower number of trial points are obtained by an optimisation approach that systematically converges to the optimal design using a significantly lower number of trial points (Figure 2.3e). The variance of the parameter can be minimised using optimal design applied by D-OD. The D-optimality concept can be applied for the design selection at the time when the conventional symmetrical designs cannot be applied, for instance when the experimental region is not consistent in shape, when the number of experiments selected by a conventional design is too large or when the applied model deviates from the normal first- or second-order ones (Jain, Hurkat, and Jain 2019).

2.4.3.6 Fractional Factorial Design (FFD)

FFDs are EDs consisting of a subdivision of the full-factorial design. The two-level design is expressed as 2^{k-p}, where k signifies the factors screened and p is the size of fraction representing the full-factorial design. A pre-determined (1/xr) fraction is completely explained by full FD. FFD uses identified properties of the design to selectively reduce the size of an experiment and can be used for a large number of factors or factor levels, but effects are puzzled with interaction terms. The trade-off of FFD is its inability to distinguish the main effect with high-order interaction between independent variables. FFD is suitable to investigate the impact of process variables on dependent variables using a single set of experiments (Gunst and Mason 2009, Montgomery 2017).

2.4.3.7 Taguchi Design (TgD)

It is also called 'offline quality control' EDs. The Taguchi method is a statistical approach to optimise the process parameter. TgD identifies proper control factors to acquire the optimal results of the process. To conduct a set of experiments, orthogonal arrays are used. Taguchi provided a method that studied all parameters with the minimum number of experiments using orthogonal array. Thus, to measure the performance features that are differing from the target value, Taguchi endorsed the usage of the missing function. The value of lost function is thus changed into signal-to-noise (S/N) ratio. The S/N ratio can be analysed by nominal-the-best, larger-the-better and smaller-the-good (Athreya and Venkatesh 2012).

2.4.3.8 Plackett–Burman Design (PBD)

The PBD is a two-level FFD and is especially useful in screening studies to analyse the main effects of variables. The variables screened can be optimised by statistical and mathematical optimisation tactics, e.g. RSM. It is a two-level and small-size factorial design to recognise critical physicochemical parameters from N number of variables in N + 1 experiments without recourse to the interaction effects between and among the variables (Ekpenyong et al. 2017). Since the sample size is traditionally small, the interaction effects are completely shrouded in the main effects. PBD therefore simply screens the design space to detect large main effects (Cavazzuti 2012, Plackett and Burman 1946).

2.5 APPLICATION OF QbD

2.5.1 Liposomes

Since the last era, liposomes have diverted scientists' attention towards the developing dormant in nanocarriers of drugs (Woodle 1995, Mehta, Narayan, and Nayak 2019, Saraf, Jain, Tiwari, Verma, and Jain 2020). They can be expounded as the amphiphilic in nature viewed as microscopic concentric spheres. They are fabricated as bilayer, formed by phospholipids and cholesterol to lodge inside the water- and lipid-soluble molecules (Torchilin 2005, Jain, Tiwari et al. 2018, Jain and Jain 2016c). The formulation of liposomes for hydrophilic drugs faces difficulties in low EE. This limits the development as a commercial product (Rathore and Winkle 2009). QbD can be an optimistic tool for recognising various parameters involved in manufacturing, designing and optimisation of liposomal composition through high-throughput screening methodologies (Patil and Pethe 2013). A theoretical aspect and risk-focused approach designate productive practical methodology that fulfils timely pharmaceutical developments (Gieszinger et al. 2017). In a study conducted by Pallagi et al., QbD in addition to risk assessment was constructed during the premature development phase of a new liposomal formulation for targeting the brain through nasal administration. The research was focused on four stages of QbD execution, i.e. elucidate the target profile, adopt various critical factors, implement risk assessment and a factorial design-based liposome preparation and evaluate the result. To effectuate the goal, QTTP elements were defined and liposomes were manufactured by the lipid film hydration method. The proposal of factorial design was fulfilled using CQAs, CMAs/CPPs and RA. The theoretical prophecy based on critical factors manifested a clear relation between product design (temperature, composition and functional parameters) and product characteristics. The parameters influencing the desired liposomal product in the existing study are shown by the Ishikawa diagram (Figure 2.4). This execution helps to assess the formulation process whether QbD approach utility has improved it and made it effective. The software used for the study was JMP 13 software (SAS Institute, Cary, USA). Following variables were contributed in the screening design: phospholipid concentration (X1; lower limit: 85% w/w, upper limit: 95% w/w), evaporation temperature during film formation (X2; lower limit: 55°, upper limit: 65°C), no of second filtration with the 0.22-μm membrane filter (X3; lower limit: 1, upper limit: 3). Size and size distribution were estimated for the response. An increase in phospholipid concentration decreased the vesicular size and increased Specific surface area (SSA). On the other side, increased filtration numbers also decreased the vesicle size but led to a moderate increase in SSA. The result demonstrated the effectiveness of QbD approaches which boost the formulation functioning and rationalisation of liposomes (Pallagi et al. 2019).

Pandey et al. developed and optimised chitosan-coated NLPs containing hydrophilic drug using QbD. A modern technique named modified ethanol injection technique was preferred to prepare the NLPs. The PBD was selected with 12 runs using Minitab version 16 as it conducts a smaller number of runs with more variables. The formulation factors chosen were cholesterol, lipid and drug concentration. The process variables focused were stirring speed, sonication time, organic/aqueous phase ratio and temperature. The outcome of the formulation variable was observed in particle size and %EE before and after coating of NLPs with chitosan. The effect of process variable was spotted on zeta potential and %EE before and after coating of NLPs with chitosan. The critical information obtained by risk assessment with respect to formulation and process variable was controlled. The high risk factor was distinguished by risk analysis which includes subjects such as particle size, %EE and % coating efficiency to set upper and lower limits. The robustness of predicted design space was investigated by distinct studies. The results obtained for particle size, %EE and % coating efficiency were 111.3 nm, 33.4% and 35.2%, respectively. The half-normal plot showed that some factors were above the line, which indicates that they were significant and had an impact on response. From this plot, it was observed that chitosan concentration, aqueous phase and drug concentration were significantly affecting the Quality attributes (QA) of the product (Pandey et al. 2014).

2.5.2 Niosomes

In the last three decades, niosomes have expanded appreciation as a rationalised approach for drug delivery (Gilani et al. 2019). These are self-assembled non-ionic vesicles composed of non-ionic surfactants in an aqueous environment. For drugs with poor solubility, it has been verified as a good carrier. Niosomes are similar to liposomes

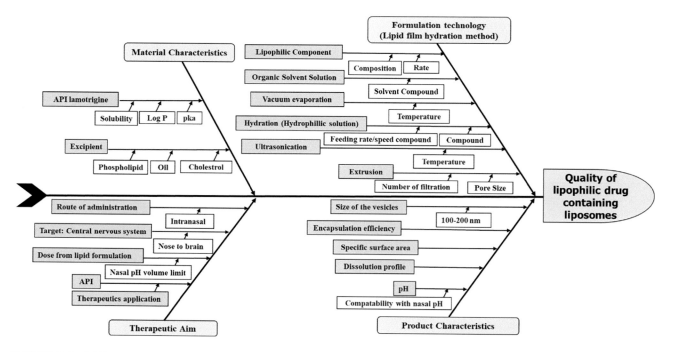

FIGURE 2.4 Ishikawa diagram showing the relationship between material and process variables to specify the quality of the aimed liposome product (Pallagi et al. 2019).

but noisome shows better stability, biocompatible and non-toxic when compared with liposomes. In a research work, Qumbar et al. prepared lacidipine-loaded niosomes by thin-film formation. For the developed formulation, statistical optimisation was carried by a four-factor, three-level BBD. The variables marked as independent variables were concentration of Span 60, cholesterol concentration, hydration time and sonication time. The factors were evaluated by vesicular size, EE and flux. The optimised formulation was found to have 676.98 ± 10.92 nm, $82.77\% \pm 4.34\%$ and 38.43 ± 2.43 mg/cm^2/h as vesicle size, EE and flux, respectively. A linear regression plot was constructed on the basis of comparison between the experimental value and the predicted value (Figure 2.5) (Qumbar et al. 2017).

Srikanth et al. obtained niosomes from capecitabine proniosomes to study the influence of four independent variables. To observe the influence of various independent variables, BBD was adopted as it provides a deep perception of plausible interactions among various levels. A second-order polynomial equation and contour plots were constructed from the information obtained from the design (Figure 2.6). The design consists of four factors and three levels for optimisation, and 29 runs were conducted. It also assists in discerning the best formulation with minimal time, effort and development cost. Important independent variables contributing to its effect were Span 60, cholesterol concentration, hydration volume and sonication time. Vesicle size and % EE were analysed as dependent variables. A full-model second-order polynomial equation was developed by setting up multiple regression subjected to transformed values of independent and dependent variables. Further, the reduced-model polynomial equation was established by calculating the F value. The equation prognosticates %EE and vesicle size of niosomes derived from proniosomes. The reduced-model polynomial equation was set by omitting insignificant terms from full-model equation. The impact of independent variables was characterised using 3D plots and contour plots. The F value was obtained to be 5,059, which implies that it is significant (Srikanth 2018).

Verma et al. optimised cationic surface-engineered mucoadhesive vesicles using DoE. A three-factor, three-level BBD was applied using Design-Expert® software (10.0; Stat-Ease, Inc., Minneapolis, USA). The amount of Span 60 (X1), the amount of cholesterol (X2) and TMC concentration (X3) were focused as independent variables. Each independent variable was marked as +1, 0 and −1, which represent high, medium and low level, respectively.

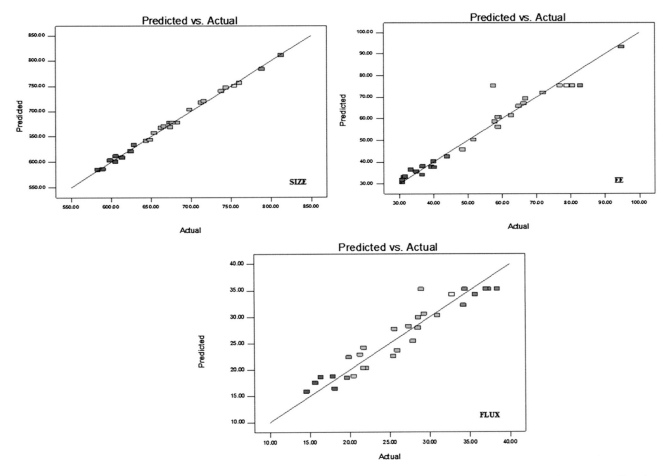

FIGURE 2.5 Linear correlation plots of actual and predicted values and the corresponding residual plots for all responses (Qumbar et al. 2017).

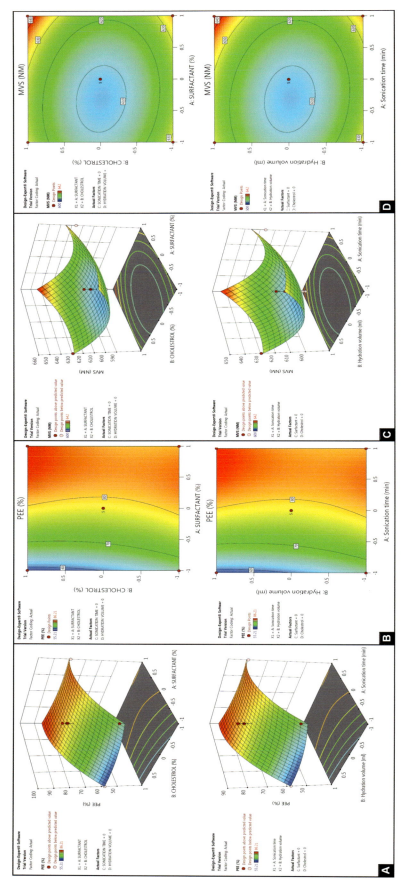

FIGURE 2.6 (a–b) Three-dimensional and (c–d) counter response surface plot: (a) influence of independent variable on (a) Y1 response and (B) Y2 response (Srikanth 2018).

Formulation by Design

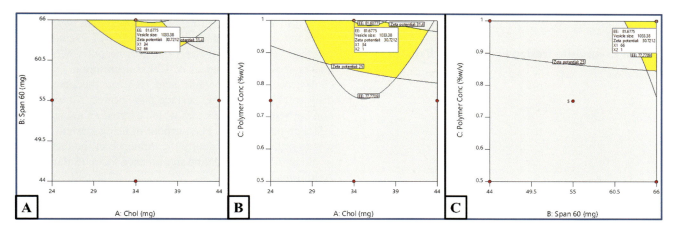

FIGURE 2.7 Design space depicted between (a) the amount of Span 60 and cholesterol amount, (b) polymer concentration and cholesterol amount and (c) polymer concentration and the amount of Span 60 (Verma et al. 2019).

The responses selected were encapsulation efficiency (Y1), vesicle size (Y2) and zeta potential (Y3). The design matrix generated a total of 17 runs. The model parameters such as multiple correlation coefficient (R^2), adjusted multiple correlation coefficient, correlation of variance and predicted residual sum of squares were analysed. 2D contour plots and 3D surface plots were designed using the software. Additionally, the required optimised region was confirmed by overlay plot in which the plot between the amount of Span 60 and cholesterol amount, polymer concentration and cholesterol amount, and polymer concentration and the amount of Span 60 exhibited the required region in the yellow shade and flag was pointed showing optimised formulation (Figure 2.7). The overlay plots gives the best possible result of feasible factors suited best in design space.. The overlay plot also represented a design space for directing the credible formulation development (Verma et al. 2019).

2.5.3 Transferosomes

Ultra-flexible vesicles known as transferosomes are composed of a lipid bilayer structure. They are ultra-deformable vesicles having an aqueous core. They show the promising property to penetrate across the skin, by squeezing through the intracellular lipid of the stratum corneum. They are fabricated by water, surfactant and phospholipid (Walve et al. 2011). The optimisation of various variables was implemented by Jangdey et al. using BBD. Rotary evaporation sonication technique having modern implementation was utilised to develop formulation. The RSM applied followed a three-factor, three-level design in 17 runs. The focus of the design incorporated in the study was to presume the adequate condition required for the formation of transferosomes of desired characteristics. The RSM assisted in describing the connotation of independent and dependent variables with their responses. The phospholipid/surfactant concentration ratio, sonication time and rotating speed were the independent variables. The dependent variables (response) chosen were vesicle size, drug loading and EE.

The range of experiments for each variable was chosen on the basis of preliminary experiments and feasibility in availability. For the investigation of maximum variables with less experimental runs was embedded by the combined effort of RSM and BBD (Jangdey et al. 2017).

2.5.4 Pharmacosomes and Ufasomes

Pharmacosomes are a colloidal dispersion of drugs covalently bound to lipids. They are ultrafine vesicles, identified as micellar or hexagonal aggregates. They form a complex with phospholipid bearing active hydrogen, which binds to phospholipid, thus forming amphiphilic substance (Prajapati et al. 2019, Jain et al. 2013). Udapurkar et al. formulated a complex of standardised Citrus limon extract and phospholipid. The study was directed towards the improvement of the bioavailability of its constituent. For the optimisation of various parameters, CCD was implemented. The formulation and process variables were phospholipid/drug ratio, reaction temperature and reaction time. The EE as a critical quality attribute was used to evaluate the influence of various variables. A design promoted a total of 20 runs with three levels (Table 2.2). One-way analysis of variance (ANOVA) was utilised for statistical analysis (Udapurkar et al. 2018).

Ittadwar et al. reported the CCD to formulate and optimise a novel salicin phospholipid complex (phytosome). The solvent evaporation method was utilised to formulate the salicin phytosomes. The basis of utilising CCD was to reduce the number of trials and attain the highest amount of information. Three levels were set for each variable (low, medium and high). The independent variables chosen were salicin–Phospholipon 90H ratio (X1; 1:05, 1:1.075 and 1:3), reaction temperature (X2; 40°C, 50°C and 60°C) and reaction time (X3; 1, 2 and 3 h). The dependent variables selected were the complexation rate (Y1) and partition coefficient (Y2). Three factors were selected for evaluation and nine trails were performed according to the design (Ittadwar, Bhojne, and Puranik 2018).

TABLE 2.2
Central Composite Design Formulation Batches

Formulation	X1	X2	X3	EE
1	−1	−1	−1	64.72 ± 1.5
2	1	−1	−1	82.21 ± 0.3
3	−1	1	−1	71.51 ± 1.6
4	1	1	−1	90.36 ± 0.9
5	−1	−1	1	89.52 ± 1.3
6	1	−1	1	86.41 ± 1.1
7	−1	1	1	83.23 ± 1.2
8	1	1	1	94.32 ± 1.4
9	−1.68	0	0	58.16 ± 1.0
10	1.68	0	0	96.24 ± 0.9
11	0	−1.68	0	79.42 ± 1.7
12	0	1.68	0	90.93 ± 1.2
13	0	0	−1.68	76.78 ± 0.8
14	0	0	1.68	94.45 ± 104
15–20	0	0	0	90.51 ± 0.9

In a study, ufasomes were developed by Salama and Aburahma for the intranasal delivery of CN. The ufasomes were prepared from oleic acid by encapsulating the CN using the thin-film hydration method. A $3^1\,4^1$ full-factorial design defines the result of varying drug concentration and cholesterol percentage on size, PDI and %EE of ufasomes. The design consists of the minimum number of experimental runs with two independent variables. The first factor (X1: CN concentration in oleic acid) was designed with four levels, while the other factor (X2: cholesterol percentage related to oleic acid) was designed with three levels. The effect was seen on dependent variables such as %EE (Y1), PS (Y2) and PDI (Y3). Using DoE, data were fitted into a linear model and a quadratic model by Design-Expert software version 7 (Stat-Ease, Inc., Minneapolis, MN). The results showed the impact of CN concentration (X1) and cholesterol relative to oleic acid (X2) on the percentage drug entrapped. The 3D surface plot showed communication between the two factors which exerted an antagonistic effect on particle size. The purpose of using Design-Expert was to source independently the influence of the factor followed by ANOVA to define significance factor (Salama and Aburahma 2016).

2.5.5 Aquasomes and Polymeric Micelles

Kaur et al. prepared recombinant human interferon-α-2β-loaded aquasomes. The optimisation was done by CCD and RSM. PEGylated phospholipid used in the formulation of aquasomes provided prolonged interferon-α-2b release and greater therapeutic index against ovarian cancer cells. Core-to-coat ratio, sonication power and time were used as independent variables that influenced the particle size. The range of experimentation was selected on the basis of previous experience. A total of 20 runs were designed that yield 20 experiments (Kaur et al. 2015). Daman et al. focused to prepare Gem C18-loaded poly(ethylene glycol)–poly(lactide) polymeric micelles as well as self-assembled nanoparticles for pancreatic cancer. The D-OD was preferred for optimisation. The factors chosen were the type of organic solvent, organic solvent-to-water ratio, initial drug/polymer ratio, total solid content and sonication time. The dependent variables were particle size, PDI, zeta potential and EE. The response was assessed by ANOVA, and the response surface plot was determined (Daman et al. 2014) (Table 2.3).

TABLE 2.3
Examples of FbD Methodology Used for Vesicular Drug Delivery Systems

Delivery System	Bioactive	Design Employed	References
Liposomes	siRNA	BBD	Zorzi et al. (2019)
Niosomes	Curcumin and quercetin	BBD	Ghadi et al. (2019)
Niosomes	Tamoxifen and doxorubicin	BBD	Kulkarni et al. (2019)
Niosomes	Ethionamide and D-cycloserine	BBD	Kulkarni and Rawtani (2019)
Liposomes	Insulin	BBD	Dawoud et al. 92019)
Niosomes	Natamycin	BBD	Verma et al. (2019)
Niosomes	Olanzapine	FFD	Khallaf, Aboud, and Sayed (2020)
Niosomes	Insulin	BBD	Hakamivala et al. (2019)
Bilosomes	Tizanidine hydrochloride	FFD	Khalil et al. (2019)
Flexisomes	Sertaconazole	FD	Mandlik, Siras, and Birajdar (2019)
Transethosomes	Progesterone	FFD	Salem et al. (2019)
Niosomes	Capecitabine	BBD	Srikanth (2018)
Transethosomes	Fisetin	BBD	Moolakkadath et al. (2018)
Liposomes	Pingyangmycin	BBD	Zhang et al. (2018)
Niosomes	Diacerein	CCD	Aziz, Abdelbary, and Elassasy (2018b)
Liposomes	Ginkgolides	CCD	Lv et al. (2018)

(*Continued*)

TABLE 2.3 (*Continued*)
Examples of FbD Methodology Used for Vesicular Drug Delivery Systems

Delivery System	Bioactive	Design Employed	References
Nanoliposomes	Pilocarpine hydrochloride	CCD	Zhao et al. (2018)
Proniosomes	Itraconazole	BBD	Ahmed et al. (2018)
Liposomes	Lornoxicam	CCD	Joseph (2018)
Aspasomes	Methotrexate	FFD	Ghosh et al. (2018)
Transferosomes	Felodipine	FFD	Kassem, Aboul-Einien, and El Taweel (2018)
Liposomes	Amisulpride	FFD	Shukr and Farid (2018)
Elastosomes	Diacerein	FFD	Aziz, Abdelbary, and Elassasy (2018a)
Pharmacosomes	Citrus limon extract	CCD	Udapurkar et al. (2018)
Pharmacosomes	Salicin phospholipid complex (phytosome)	CCD	Ittadwar, Bhojne, and Puranik (2018)
Nanolipid vesicles	Ursolic acid	Factorial design	Khan et al. (2018)
Transferosomes	Apigenin	BBD	Jangdey et al. (2017)
Transferosomes	Luteolin	CCD	Setyawati, Surini, and Mardliyati (2017)
Niosomes	Lacidipine	BBD	Qumbar et al. (2017)
Nanoethosomes	Triamcinolone acetonide	BBD	Akhtar et al. (2017)
Liposomes	Tilianin	CCD	Zeng et al. (2016)
Transferosomes	Sildenafil	CCD	Badr-Eldin and Ahmed (2016)
Ethosomes	Methoxsalen	CCD	Garg et al. (2016)
Transferosomes	Timolol	Taguchi	González-Rodríguez et al. (2016)
Ethosomes	Glimepiride	Draper–Lin small composite design	Ahmed et al. (2016)
Liposomes	Methotrexate	PBD	Vanaja, Shobha Rani, and Narendra (2016)
Ufasomes	Cinnarizine	Full-factorial design	Salama and Aburahma (2016)
Ethosomes	Ondansetron	BBD	Giram, More, and Parwe (2015)
Niosomes	Methotrexate	BBD	Abdelbary and AbouGhaly (2015)
Transferosomes	Sildenafil	PBD	Ahmed (2015)
Niosomes	Methotrexate	BBD	Abdelbary and AbouGhaly (2015)
Proniosomes	Risperidone	BBD	Imam et al. (2015)
Niosomes	Benzoyl peroxide	BBD	Goyal et al. (2015)
Niosomes	Dynorphin	Simplex-centroid design	Bragagni et al. (2014)
Liposomes	Paclitaxel	BBD	Rane and Prabhakar (2013)
Liposomes	5-Fluorouracil	Factorial design	Ankita et al. (2013)
Nanoethosomes	Valsartan	BBD	Ahad et al. (2013)
Niosomes	Piroxicam	CCD	Ahmed et al. (2013)
Pharmacosomes	Geniposide	CCD	Yue et al. (2012)
Niosomes	Sumatriptan succinate	TgD	González-Rodríguez et al. (2012)
Nanotransferosomes	Valsartan	BBD	Ahad et al. (2012)
Transferosomes	Antifungal agent	PBD	Patel and Parikh (2012)
Pharmacosomes	Geniposide	CCD	Yue et al. (2012)

2.6 CONCLUSION AND PROSPECTS

Various pharmaceutical processes such as drug development, analytical method development, formulation of dosage form and biopharmaceuticals have been successfully developed with the implementation of FbD technique. The regulatory requirements across the countries have made FbD mandatory for marketing. The design concept also replaces tedious approaches with more advanced qualitative design. The FbD helps to identify critical parameters, risk assessment and method optimisation of various factors. The risk assessment provides a better knowledge about the factors influencing design space which helps in understanding the relation between CMPs/CPPs with CAAs. Moreover, cost-effective formulation of a quality product fulfils the basic need of mass population. Thus, FbD helps to rationalise the vesicular preparation and make it effective. In the near future, it is going to be a stepping stone to integrate various fields such as pharmaceutics, biomedicine, biotechnology and nanomedicine to bring cost-effective products in the market.

ACKNOWLEDGEMENT

Dr. Ankit Jain acknowledges the financial assistance from the Department of Biotechnology (Govt. of India), Delhi, in the form of DBT-RA (Postdoc).

CONFLICT OF INTEREST

The authors report no conflict of interest.

REFERENCES

Abdelbary, A. A., and M. H. H. AbouGhaly. 2015. "Design and optimization of topical methotrexate loaded niosomes for enhanced management of psoriasis: application of Box–Behnken design, *in-vitro* evaluation and *in-vivo* skin deposition study." *International Journal of Pharmaceutics* no. 485 (1–2):235–243.

Ahad, A., M. Aqil, K. Kohli, Y. Sultana, and M. Mujeeb. 2013. "Enhanced transdermal delivery of an anti-hypertensive agent via nanoethosomes: statistical optimization, characterization and pharmacokinetic assessment." *Int J Pharm* no. 443 (1–2):26–38. doi: 10.1016/j.ijpharm.2013.01.011.

Ahad, A., M. Aqil, K. Kohli, Y. Sultana, M. Mujeeb, and A. Ali. 2012. "Formulation and optimization of nanotransfersomes using experimental design technique for accentuated transdermal delivery of valsartan." *J Nanomed Nanotechnol Biol Med* no. 8 (2):237–249.

Ahmed, A., M. Ghourab, S. Shedid, and M. Qushawy. 2013. "Optimization of piroxicam niosomes using central composite design." *Int J Pharm Pharm Sci* no. 5 (3):229–236.

Ahmed, M., A. SamyAfaf, S. M. Ramadan Amal, El-Enin Abu, and I. M. Yasmin. 2018. "Mortagi. Formulation and optimization of itraconazole proniosomes using Box Behnken design." *Int J Appl Pharm* no. 10 (2):41–51.

Ahmed, T. A. 2015. "Preparation of transfersomes encapsulating sildenafil aimed for transdermal drug delivery: Plackett–Burman design and characterization." *J Liposome Res* no. 25 (1):1–10.

Ahmed, T. A., K. M. El-Say, B. M. Aljaeid, U. A. Fahmy, and F. I. Abd-Allah. 2016. "Transdermal glimepiride delivery system based on optimized ethosomal nano-vesicles: preparation, characterization, *in vitro*, *ex vivo* and clinical evaluation." *Int J Pharm* no. 500 (1–2):245–54. doi: 10.1016/j.ijpharm.2016.01.017.

Akhtar, N., A. Verma, and K. Pathak. 2017. "Feasibility of binary composition in development of nanoethosomal glycolic vesicles of triamcinolone acetonide using Box-Behnken design: *in vitro* and *ex vivo* characterization." *Artif Cells Nanomed Biotechnol* no. 45 (6):1123–1131.

Ankita, S., V. Bharakhada, K. S. Rajesh, and L. L. Jha. 2013. "Experimental design and optimization studies of thermoreversible hydrogel containing liposomes for the controlled delivery of 5-fluorouracil." *Pharmagene* no. 1 (3):24–32.

Athreya, S., and Y. D. Venkatesh. 2012. "Application of Taguchi method for optimization of process parameters in improving the surface roughness of lathe facing operation." *Int Refereed J Eng Sci* no. 1 (3):13–19.

Aziz, D. E., A. A. Abdelbary, and A. I. Elassasy. 2018a. "Fabrication of novel elastosomes for boosting the transdermal delivery of diacerein: statistical optimization, *ex-vivo* permeation, *in-vivo* skin deposition and pharmacokinetic assessment compared to oral formulation." *Drug Deliv* no. 25 (1):815–826. doi: 10.1080/10717544.2018.1451572.

Aziz, D. E., A. A. Abdelbary, and A. I. Elassasy. 2018b. "Implementing central composite design for developing transdermal diacerein-loaded niosomes: *ex vivo* permeation and in vivo deposition." *Curr Drug Deliv* no. 15 (9):1330–1342. doi: 10.2174/1567201815666180619105419.

Badr-Eldin, S. M., and O. A. Ahmed. 2016. "Optimized nano-transfersomal films for enhanced sildenafil citrate transdermal delivery: *ex vivo* and *in vivo* evaluation." *Drug Des Devel Ther* no. 10:1323–1333. doi: 10.2147/DDDT.S103122.

Bai, G., Z. Chen, K. Raines, H. Chen, K. Dave, H.-P. Lin, and B. S. Jolnik. 2020. "Assessment of applications of design of experiments in pharmaceutical development for oral solid dosage forms." *J Pharm Innov* no. 15 (4):547–555.

Beg, S., O. P. Katare, and B. Singh. 2017. "Formulation by design approach for development of ultrafine self-nanoemulsifying systems of rosuvastatin calcium containing long-chain lipophiles for hyperlipidemia management." *Colloids Surf B Biointerfaces* no. 159:869–879 doi: 10.1016/j.colsurfb.2017.08.050.

Beg, S., S. Swain, M. Rahman, Md. S. Hasnain, and S. S. Imam. 2019. "Application of design of experiments (DoE) in pharmaceutical product and process optimization." In *Pharmaceutical Quality by Design*, edited by S. Beg and Md. S. Hasnain, 43–64. Elsevier: Amsterdam.

Bishnoi, M., A. Jain, Y. Singla, and B. Shrivastava. 2020. "Sublingual delivery of chondroitin sulfate conjugated tapentadol loaded nanovesicles for the treatment of osteoarthritis." *J Liposome Res*:1–15. doi: 10.1080/08982104.2020.1730400.

Box, G. E. P., and D. W. Behnken. 1960. "Some new three level designs for the study of quantitative variables." *Technometrics* no. 2 (4):455–475.

Bragagni, M., N. Mennini, S. Furlanetto, S. Orlandini, C. Ghelardini, and P. Mura. 2014. "Development and characterization of functionalized niosomes for brain targeting of dynorphin-B." *Eur J Pharm Biopharm* no. 87 (1):73–9. doi: 10.1016/j.ejpb.2014.01.006.

Cavazzuti, M. 2012. *Optimization Methods: From Theory to Design Scientific and Technological Aspects in Mechanics*. Springer Science & Business Media: Dordrecht.

Daman, Z., S. N. Ostad, M. Amini, and K. Gilani. 2014. "Preparation, optimization and *invitro* characterization of stearoyl-gemcitabine polymeric micelles: a comparison with its self-assembled nanoparticles." *Int J Pharm* no. 468 (1–2):142–151.

Dawoud, M. H. S., G. E. Yassin, D. M. Ghorab, and N. M. Morsi. 2019. "Insulin mucoadhesive liposomal gel for wound healing: a formulation with sustained release and extended stability using quality by design approach." *AAPS PharmSciTech* no. 20 (4):158.

Debnath, S., M. N. L. Aishwarya, and M. N. Babu. 2018. "Formulation by design: an approach to designing better drug delivery systems." *Pharma Times* no. 50:9–14.

Ekpenyong, M. G., S. P. Antai, A. D. Asitok, and B. O. Ekpo. 2017. "Plackett-Burman design and response surface optimization of medium trace nutrients for glycolipopeptide biosurfactant production." *Iran Biomed J* no. 21 (4):249–60. doi: 10.18869/acadpub.ibj.21.4.249.

Garg, B. J., N. K. Garg, S. Beg, B. Singh, and O. P. Katare. 2016. "Nanosized ethosomes-based hydrogel formulations of methoxsalen for enhanced topical delivery against vitiligo: formulation optimization, *in vitro* evaluation and preclinical assessment." *J Drug Target* no. 24 (3):233–246. doi: 10.3109/1061186X.2015.1070855.

Garg, R. K., and I. Singhvi. 2015. "Optimization techniques: an overview for formulation development." *Asian J Pharm Res* no. 5 (3):217–221.

Ghadi, Z. S., R. Dinarvand, N. Asemi, F. T. Amiri, and P. Ebrahimnejad. 2019. "Preparation, characterization and *in vivo* evaluation of novel hyaluronan containing niosomes tailored by Box-Behnken design to co-encapsulate curcumin and quercetin." *Eur J Pharm Sci* no. 130:234–246.

Ghosh, S., B. Mukherjee, S. Chaudhuri, T. Roy, A. Mukherjee, and S. Sengupta. 2018. "Methotrexate aspasomes against rheumatoid arthritis: optimized hydrogel loaded liposomal formulation with in vivo evaluation in Wistar rats." *AAPS PharmSciTech* no. 19 (3):1320–1336. doi: 10.1208/s12249-017-0939-2.

Gieszinger, P., I. Csoka, E. Pallagi, G. Katona, O. Jojart-Laczkovich, P. Szabo-Revesz, and R. Ambrus. 2017. "Preliminary study of nanonized lamotrigine containing products for nasal powder formulation." *Drug Des Devel Ther* no. 11:2453–2466. doi: 10.2147/DDDT.S138559.

Gilani, S. J., Md. Rizwanullah, S. S. Imam, J. Pandit, Md. Aqil, M. Alam, and S. Beg. 2019. "QbD considerations for topical and transdermal product development." In *Pharmaceutical Quality by Design*, edited by S. Beg and Md. S. Hasnain, 131–150. Elsevier: Amsterdam.

Giram, P. S., S. R. More, and S. P. Parwe. 2015. "Formulation development with Box–Behnken design study of ondansetron HCl ethosome for chemotherapy induced nausea vomiting." *World J Pharm Pharm Sci* no. 3:288–298.

González-Rodríguez, M. L., C. M. Arroyo, M. J. Cózar-Bernal, P. L. González-R, J. M. León, M. Calle, D. Canca, and A. M. Rabasco. 2016. "Deformability properties of timolol-loaded transfersomes based on the extrusion mechanism. Statistical optimization of the process." *Drug Dev Ind Pharm* no. 42 (10):1683–1694.

González-Rodríguez, M. L., I. Mouram, M. J. Cózar-bernal, S. Villasmil, and A. M. Rabasco. 2012. "Applying the taguchi method to optimize sumatriptan succinate niosomes as drug carriers for skin delivery." *J Pharm Sci* no. 101 (10):3845–3863.

Goyal, G., T. Garg, B. Malik, G. Chauhan, G. Rath, and A. K. Goyal. 2015. "Development and characterization of niosomal gel for topical delivery of benzoyl peroxide." *Drug Deliv* no. 22 (8):1027–1042. doi: 10.3109/10717544.2013.855277.

Gunst, R. F., and R. L. Mason. 2009. "Fractional factorial design." *Wiley Interdiscip Rev Comput Stat* no. 1 (2):234–244.

Hakamivala, A., S. Moghassemi, and K. Omidfar. 2019. "Modeling and optimization of the niosome nanovesicles using response surface methodology for delivery of insulin." *Biomed Phys Eng Express* no. 5 (4):045041.

Hasnain, Md. S., S. A. Ahmed, S. Beg, M. T. Ansari, and A. K. Nayak. 2019. "'Quality by design' approach for development of multiparticulate drug delivery systems." In *Pharmaceutical Quality by Design*, edited by S. Beg and Md. S. Hasnain, 351–365. Elsevier: Amsterdam.

Imam, S. S., M. Aqil, M. Akhtar, Y. Sultana, and A. Ali. 2015. "Formulation by design-based proniosome for accentuated transdermal delivery of risperidone: *in vitro* characterization and *in vivo* pharmacokinetic study." *Drug Deliv* no. 22 (8):1059–1070. doi: 10.3109/10717544.2013.870260.

Ittadwar, P. A., S. V. Bhojne, and P. K. Puranik. 2018. "Novel salicin phytosomal complex: development and optimization using central composite design." *World J Pharm Res* no. 7 (9):735–751.

Jain, A, and S. K. Jain. 2016a. "Ligand-mediated drug-targeted liposomes." In *Liposomal Delivery Systems: Advances and Challenges*, edited by A. Samad, S. Beg, and I. Nazish, 145. Future Medicine: London.

Jain, A., and S. K. Jain. 2016b. "*In vitro* release kinetics model fitting of liposomes: an insight." *Chem Phys Lipids* no. 201:28–40. doi: 10.1016/j.chemphyslip.2016.10.005.

Jain, A., and S. K. Jain. 2016c. "Liposomes in cancer therapy." In *Nanocarrier Systems for Drug Delivery*, edited by C. Jimenez, 1–42. Nova Science Publishers: New York.

Jain, A., and S. K. Jain. 2018a. "Advances in tumor targeted liposomes." *Curr Mol Med*. doi: 10.2174/1566524018666180416101522.

Jain, A., and S. K. Jain. 2018b. "Stimuli-responsive smart liposomes in cancer targeting." *Curr Drug Targets* no. 19 (3):259–270. doi: 10.2174/1389450117666160208144143.

Jain, A., A. Gulbake, S. Shilpi, A. Jain, P. Hurkat, and S. K. Jain. 2013. "A new horizon in modifications of chitosan: syntheses and applications." *Crit Rev Ther Drug Carrier Syst* no. 30 (2):91–181.

Jain, A., P. Hurkat, and S. K. Jain. 2019. "Development of liposomes using formulation by design: basics to recent advances." *Chem Phys Lipids* no. 224:104764. doi: 10.1016/j.chemphyslip.2019.03.017.

Jain, A., R. Kumari, A. Tiwari, A. Verma, A. Tripathi, A. Shrivastava, and S. K. Jain. 2018. "Nanocarrier based advances in drug delivery to tumor: an overview." *Curr Drug Targets* no. 19 (13):1498–1518. doi: 10.2174/1389450119666180131105822.

Jain, A, S. K. Prajapati, A. Kumari, N. Mody, and M. Bajpai. 2020. "Engineered nanosponges as versatile biodegradable carriers: an insight." *J Drug Deliv Sci Technol*. doi: 10.1016/j.jddst.2020.101643.

Jain, A., A. Tiwari, A. Verma, and S. K. Jain. 2018. "Ultrasound-based triggered drug delivery to tumors." *Drug Deliv Transl Res* no. 8 (1):150–164. doi: 10.1007/s13346-017-0448-6.

Jain, A., A. Tiwari, A. Verma, S. Saraf, and S. Jain. 2019. "Combination cancer therapy using multifunctional liposomes." *Crit Rev Ther Drug Carrier Syst*. doi: 10.1615/CritRevTherDrugCarrierSyst.2019026358.

Jain, N. K. 2018. *Pharmaceutical product development*. CBS Publishers & Distributors: New Delhi.

Jangdey, M. S., A. Gupta, S. Saraf, and S. Saraf. 2017. "Development and optimization of apigenin-loaded transfersomal system for skin cancer delivery: *in vitro* evaluation." *Artif Cells Nanomed Biotechnol* no. 45 (7):1452–1462. doi: 10.1080/21691401.2016.1247850.

Joseph, J. 2018. "Experimental optimization of Lornoxicam liposomes for sustained topical delivery." *Eur J Pharm Sci* no. 112:38–51 doi: 10.1016/j.ejps.2017.10.032.

Kassem, M. A., M. H. Aboul-Einien, and M. M. El Taweel. 2018. "Dry gel containing optimized felodipine-loaded transferosomes: a promising transdermal delivery system to enhance drug bioavailability." *AAPS PharmSciTech* no. 19 (5):2155–2173. doi: 10.1208/s12249-018-1020-5.

Kaur, K., P. Kush, R. S. Pandey, J. Madan, U. K. Jain, and O. P. Katare. 2015. "Stealth lipid coated aquasomes bearing recombinant human interferon-α-2b offered prolonged release and enhanced cytotoxicity in ovarian cancer cells." *Biomed Pharmacother* no. 69:267–276.

Kesharwani, P., A. Jain, A. K. Srivastava, and M. K. Keshari. 2020. "Systematic development and characterization of curcumin loaded nanogel for topical application." *Drug Dev Ind Pharm*:1–36. doi: 10.1080/03639045.2020.1793998.

Khalil, R. M., A. Abdelbary, S. K. El-Arini, M. Basha, and H. A. El-Hashemy. 2019. "Evaluation of bilosomes as nanocarriers for transdermal delivery of tizanidine hydrochloride: *in vitro* and *ex vivo* optimization." *J Liposome Res* no. 29 (2):171–182.

Khallaf, R. A., H. M. Aboud, and O. M. Sayed. 2020. "Surface modified niosomes of olanzapine for brain targeting via nasal route; preparation, optimization, and *in vivo* evaluation." *J Liposome Res* no. 30 (2):163–173. doi: 10.1080/08982104.2019.1610435.

Khan, K., M. Aqil, S. S. Imam, A. Ahad, T. Moolakkadath, Y. Sultana, and M. Mujeeb. 2018. "Ursolic acid loaded intra nasal nano lipid vesicles for brain tumour: formulation, optimization, *in-vivo* brain/plasma distribution study and histopathological assessment." *Biomed Pharmacother* no. 106:1578–1585 doi: 10.1016/j.biopha.2018.07.127.

Kulkarni, P., and D. Rawtani. 2019. "Application of Box-Behnken design in the preparation, optimization, and *in vitro* evaluation of self-assembly-based tamoxifen-and doxorubicin-loaded and dual drug–loaded niosomes for combinatorial breast cancer treatment." *J Pharm Sci* no. 108 (8):2643–2653.

Kulkarni, P., D. Rawtani, and T. Barot. 2019. "Formulation and optimization of long acting dual niosomes using Box-Behnken experimental design method for combinative delivery of ethionamide and D-cycloserine in tuberculosis treatment." *Colloids Surf A Physicochem Eng Asp* no. 565:131–142.

Kumari, N., B. Singh, G. Saini, A. Chaudhary, K. Verma, and M. Vyas. 2019. "Quality by design: a systematic approach for the analytical method validation." *J Drug Deliv Ther* no. 9 (3-s):1006–1012.

Lawrence, X. Y., G. Amidon, M. A. Khan, S. W. Hoag, J. Polli, G. K. Raju, and J. Woodcock. 2014. "Understanding pharmaceutical quality by design." *AAPS J* no. 16 (4):771–783.

Lv, Z., Y. Yang, J. Wang, J. Chen, J. Li, and L. Di. 2018. "Optimization of the preparation conditions of borneol-modified ginkgolide liposomes by response surface methodology and study of their blood brain barrier permeability." *Molecules* no. 23 (2):303.

Mandlik, S. K., S. S. Siras, and K. R. Birajdar. 2019. "Optimization and characterization of sertaconazole nitrate flexisomes embedded in hydrogel for improved antifungal activity." *J Liposome Res* no. 29 (1):10–20. doi: 10.1080/08982104.2017.1402926.

Mehta, C. H., R. Narayan, and U. Y. Nayak. 2019. "Computational modeling for formulation design." *Drug Discov Today* no. 24 (3):781–788. doi: 10.1016/j.drudis.2018.11.018.

Mishra, V., S. Thakur, A. Patil, and A. Shukla. 2018. "Quality by design (QbD) approaches in current pharmaceutical set-up." *Expert Opin Drug Deliv* no. 15 (8):737–758. doi: 10.1080/17425247.2018.1504768.

Montgomery, D. C. 2017. *Design and Analysis of experiments.* John Wiley & Sons: New York.

Moolakkadath, T., Md. Aqil, A. Ahad, S. S. Imam, B. Iqbal, Y. Sultana, Md. Mujeeb, and Z. Iqbal. 2018. "Development of transethosomes formulation for dermal fisetin delivery: Box–Behnken design, optimization, *in vitro* skin penetration, vesicles–skin interaction and dermatokinetic studies." *Artif Cells Nanomed Biotechnol* no. 46 (Suppl 2):755–765.

Pallagi, E., O. Jojart-Laczkovich, Z. Nemeth, P. Szabo-Revesz, and I. Csoka. 2019. "Application of the QbD-based approach in the early development of liposomes for nasal administration." *Int J Pharm* no. 562:11–22 doi: 10.1016/j.ijpharm.2019.03.021.

Pandey, A. P., K. P. Karande, R. O. Sonawane, and P. K. Deshmukh. 2014. "Applying quality by design (QbD) concept for fabrication of chitosan coated nanoliposomes." *J Liposome Res* no. 24 (1):37–52. doi: 10.3109/08982104.2013.826243.

Patel, R. B., and R. H. Parikh. 2012. "Preparation and formulation of transferosomes containing an antifungal agent for transdermal delivery: application of Plackett-Burman design to identify significant factors influencing vesicle size." *J Pharm Bioallied Sci* no. 4 (Suppl 1):S60–S61. doi: 10.4103/0975-7406.94140.

Pathak, K., V. Sharma, and M. Sharma. 2016. "Optimization, *in vitro* cytotoxicity and penetration capability of deformable nanovesicles of paclitaxel for dermal chemotherapy in Kaposi sarcoma." *Artif Cells Nanomed Biotechnol* no. 44 (7):1671–83. doi: 10.3109/21691401.2015.1080169.

Patil, A. S., and A. M. Pethe. 2013. "Quality by design (QbD): a new concept for development of quality pharmaceuticals." *Int J Pharm Qual Assur* no. 4 (2):13–19.

Plackett, R. L., and J. P. Burman. 1946. "The design of optimum multifactorial experiments." *Biometrika* no. 33 (4):305–325.

Politis, S. N., P. Colombo, G. Colombo, and M. Rekkas D. 2017. "Design of experiments (DoE) in pharmaceutical development." *Drug Dev Ind Pharm* no. 43 (6):889–901. doi: 10.1080/03639045.2017.1291672.

Prajapati, S. K., A. Jain, A. Jain, and S. Jain. 2019. "Biodegradable polymers and constructs: a novel approach in drug delivery." *Eur Polym J* no. 120:109191. doi: 10.1016/j.eurpolymj.2019.08.018.

Qumbar, M., Ameeduzzafar, S. S. Imam, J. Ali, J. Ahmad, and A. Ali. 2017. "Formulation and optimization of lacidipine loaded niosomal gel for transdermal delivery: *in-vitro* characterization and *in-vivo* activity." *Biomed Pharmacother* no. 93:255–266. doi: 10.1016/j.biopha.2017.06.043. https://www.sciencedirect.com/science/article/abs/pii/S0753332217312143?via%3Dihub

Rafidah, A., A. Nurulhuda, A. Azrina, Y. Suhaila, I. S. Anwar, and R. A. Syafiq. 2014. Comparison design of experiment (doe): Taguchi method and full factorial design in surface roughness. Paper read at *Applied Mechanics and Materials*.

Rane, S., and B. Prabhakar. 2013. "Optimization of paclitaxel containing pH-sensitive liposomes by 3 factor, 3 level Box-Behnken Design." *Indian J Pharm Sci* no. 75 (4):420–426. doi: 10.4103/0250-474X.119820.

Rathore, A. S., and H. Winkle. 2009. "Quality by design for biopharmaceuticals." *Nat Biotechnol* no. 27 (1):26–34. doi: 10.1038/nbt0109-26.

Salama, A. H., and M. H. Aburahma. 2016. "Ufasomes nanovesicles-based lyophilized platforms for intranasal delivery of cinnarizine: preparation, optimization, *ex-vivo* histopathological safety assessment and mucosal confocal imaging." *Pharm Dev Technol* no. 21 (6):706–15. doi: 10.3109/10837450.2015.1048553.

Salem, H. F., R. M. Kharshoum, H. A. Abou-Taleb, H. A. AbouTaleb, and K. M. AbouElhassan. 2019. "Progesterone-loaded nanosized transethosomes for vaginal permeation enhancement: formulation, statistical optimization, and clinical evaluation in anovulatory polycystic ovary syndrome." *J Liposome Res* no. 29 (2):183–194. doi: 10.1080/08982104.2018.1524483.

Sangshetti, J. N., M. Deshpande, Z. Zaheer, D. B. Shinde, and R. Arote. 2017. "Quality by design approach: regulatory need." *Arab J Chem* no. 10:S3412–S3425.

Saraf, S., A. Jain, A. Tiwari, A. Verma, and S. K. Jain. 2020. "Engineered liposomes bearing camptothecin analogue for tumor targeting: *in vitro* and *ex-vivo* studies." *J Liposome Res*:1–32. doi: 10.1080/08982104.2020.1801725.

Saraf, S., A. Jain, A. Tiwari, A. Verma, P. K. Panda, and S. K. Jain. 2020. "Advances in liposomal drug delivery to cancer: an overview." *J Drug Deliv Sci Technol* no. 56:101549.

Setyawati, D. R., S. Surini, and E. Mardliyati. 2017. "Optimization of luteolin-loaded transfersome using response surface methodology." *Int J Appl Pharm* no. 9:107–111.

Shukr, M. H., and O. A. A. Farid. 2018. "Amisulpride–CD-loaded liposomes: optimization and *in vivo* evaluation." *AAPS PharmSciTech* no. 19 (6):2658–2671.

Singh, B., R. Kapil, M. Nandi, and N. Ahuja. 2011. "Developing oral drug delivery systems using formulation by design: vital precepts, retrospect and prospects." *Expert Opin Drug Deliv* no. 8 (10):1341–1360.

Singh, B., S. Saini, S. Lohan, and S. Beg. 2017. "Systematic development of nanocarriers employing quality by design paradigms." In *Nanotechnology-Based Approaches for Targeting and Delivery of Drugs and Genes*, edited by K. P. K. Mishra and M. I. M. Amin, 110–148. Elsevier: San Diego, CA.

Srikanth, Y. Anand Kumar, and M. Setty. 2018. "Design and optimization of capecitabine proniosomes." *International Journal of Pharma Research and Health Sciences* no. 6 (4):2717–2722. doi: 10.21276/ijprhs.2018.04.13. https://ijpsr.com/bft-article/design-and-optimization-of-capecitabine-niosomes-derived-from-proniosomes/?view=fulltext

Swarbrick, B. 2018. "Quality by design in practice." In *Multivariate Analysis in the Pharmaceutical Industry*, edited by Ana Ferreira, Jose C. Menezes, Mike Tobyn, 125–171. Elsevier: USA. https://www.sciencedirect.com/book/9780128110652/multivariate-analysis-in-the-pharmaceutical-industry#book-description

Torchilin, V. P. 2005. "Recent advances with liposomes as pharmaceutical carriers." *Nat Rev Drug Discov* no. 4 (2):145.

Trivedi, D., V. V. S. R. Karri, A. K. M. Spandana, and G. Kuppusam. 2015. "Design of experiments: optimization and applications in pharmaceutical nanotechnology." *Chem Sci Rev Lett* no. 4 (13):109–120.

Udapurkar, P. P., O. G. Bhusnure, and S. R. Kamble. 2018. "Development and characterization of Citrus limon-phospholipid complex as an effective phytoconstituent delivery system." *Int J Life Sci Pharma Res* no. 8 (1):29–41.

Vanaja, K., R. H. Shobha Rani, and C. Narendra. 2016. "Application of statistical designs in the formulation of lipid carriers of methotrexate." *J Drug Res Dev* no. 2 (2). doi: 10.16966/2470-1009.113.

Verma, A., G. Sharma, A. Jain, A. Tiwari, S. Saraf, P. K. Panda, O. P. Katare, and S. K. Jain. 2019. "Systematic optimization of cationic surface engineered mucoadhesive vesicles employing design of experiment (DoE): a preclinical investigation." *Int J Biol Macromol* no. 133:1142–1155. doi: 10.1016/j.ijbiomac.2019.04.118.

Walve, J. R., S. R. Bakliwal, B. R. Rane, and S. P. Pawar. 2011. "Transfersomes: a surrogated carrier for transdermal drug delivery system." *Int J Appl Biol Pharm Technol* no. 2 (1):204–213.

Woodcock, J. 2004. "The concept of pharmaceutical quality." *Am Pharm Rev* no. 7 (6):10–15.

Woodle, M. C. 1995. "Sterically stabilized liposome therapeutics." *Adv Drug Deliv Rev* no. 16 (2–3):249–265.

Yue, P. F., Q. Zheng, B. Wu, M. Yang, M. S. Wang, H. Y. Zhang, P. Y. Hu, and Z. F. Wu. 2012. "Process optimization by response surface design and characterization study on geniposide pharmacosomes." *Pharm Dev Technol* no. 17 (1):94–102. doi: 10.3109/10837450.2010.516439.

Zeng, C., W. Jiang, M. Tan, X. Yang, C. He, W. Huang, and J. Xing. 2016. "Optimization of the process variables of tilianin-loaded composite phospholipid liposomes based on response surface-central composite design and pharmacokinetic study." *Eur J Pharm Sci* no. 85:123–31 doi: 10.1016/j.ejps.2016.02.007.

Zhang, L., F. Chen, J. Zheng, H. Wang, X. Qin, and W. Pan. 2018. "Chitosan-based liposomal thermogels for the controlled delivery of pingyangmycin: design, optimization and *in vitro* and *in vivo* studies." *Drug Deliv* no. 25 (1):690–702. doi: 10.1080/10717544.2018.1444684.

Zhao, F., J. Lu, X. Jin, Z. Wang, Y. Sun, D. Gao, X. Li, and R. Liu. 2018. "Comparison of response surface methodology and artificial neural network to optimize novel ophthalmic flexible nano-liposomes: characterization, evaluation, *in vivo* pharmacokinetics and molecular dynamics simulation." *Colloids Surf B Biointerfaces* no. 172:288–297. doi: 10.1016/j.colsurfb.2018.08.046.

Zorzi, G. K., R. S. Schuh, V. J. Maschio, N. T. Brazil, M. B. Rott, and H. F. Teixeira. 2019. "Box Behnken design of siRNA-loaded liposomes for the treatment of a murine model of ocular keratitis caused by Acanthamoeba." *Colloids Surf B Biointerfaces* no. 173:725–732. doi: 10.1016/j.colsurfb.2018.10.044.

3 Thermally Responsive Externally Activated Theranostics (TREAT) for On-Demand Multifunctional Drug Delivery Systems

Renuka Khatik
Texas A&M Rangel College of Pharmacy, Kingsville Texas

Aakanchha Jain
Bhagyoday Tirth Pharmacy College, Sagar; and
National Institute of Pharmaceutical Education and Research, Ahmedabad

Faheem Hyder Pottoo
Imam Abdulrahman Bin Faisal University

CONTENTS

3.1 Introduction 33
    3.1.1 Concept behind TREAT 34
3.2 Thermally Responsive Polymers and Nanocarriers 35
3.3 Diagnostic Approach 37
3.4 Optical Imaging 37
3.5 Magnetic Resonance Imaging (MRI) 37
3.6 Ultrasound (US) 38
3.7 Computed Tomography (CT) 40
3.8 Single-Photon Emission Computed Tomography and Positron Emission Tomography ... 42
3.9 Conclusions and Future Directions 42
Conflict of Interest 42
References 43

3.1 INTRODUCTION

Multifunctional smart drug delivery systems (MDDSs) have the potential to provide improved drug targeting efficacy for clinical diagnostic tools in a single nanosystem. Thermally responsive externally activated theranostics (TREAT) is multifunctional entities that act as tools for therapy and diagnosis. In the early 2000s, 'theranostics' or TREAT was created by Funkhouser and was defined as a tool of integrated treatment and medical imaging into one 'packaging' material to overcome adverse changes in biodistribution and therapeutic efficacy (Ding and Wu 2012; Wang, Chuang, and Ho 2012). 'Theranostics' (also known as 'theragnostics') is the emerging field that merges clinical diagnosis in the pharmaceutical industry. A special magazine called 'Theranostics' (www.thno.org/) opened a forum to exchange clinical and scientific information to help the therapeutic and diagnostic nanomedicine society and the professions involved in the integration of molecular imaging and molecular therapy (Chen 2011). TREAT is a combined technology with low adverse effects and cost-effectiveness, which will improve the potency of various disease such as cancer, type 1 diabetes, neurological disorders, inflammatory diseases and cardiovascular diseases (Chen 2011; Patel and Janjic 2015; Sriramoju et al. 2015; Wang, Lin, and Ai 2014). There are many advantages of using TREAT, such as the presence of large surface area associated with good percentage drug loading and large particle uptake, easy to functionalise through ligand or complexation, and allows simultaneously diagnose and provide therapeutic drug delivery.

Cancer was one of the leading causes of death in the twentieth century, and currently still exceeds 10 million cases each year (Ahmed et al. 2012). This fact related to chemotherapy failure and new translation failure developing a platform for the clinic (Maeda 2015). Early detection and enhanced treatment options with less adverse effects are the advantages of TREAT in cancer treatment (Moghimi, Hunter, and Murray 2005). The purpose of using TREAT and its agents is to prevent the failure of chemotherapy. Formulating theranostic agents requires an in-depth understanding of the diagnosis and treatment mechanisms, that is an understanding of molecular-level changes, disease biomarkers, diagnostic scheme, treatment principles and adverse effects. As a potential nanoplatform for detecting of clinical complications, various challenges are encountered in the development of TREAT as mentioned in Figure 3.1. The different factors that should be considered when developing theranostic agents are as follows:

- The chemicals should be biocompatible and biodegradable
- Preparation conditions should be well mentioned
- Toxicity study of materials used and their metabolites must be analysed prior to formulation
- The pharmacokinetic and pharmacodynamic parameters should be carefully evaluated.

Using engineered TREAT materials, therapy and diagnosis can be performed simultaneously, and diagnosis tools can be used to view treatment progress, such as fluorescent imaging, magnetic resonance imaging (MRI) and computed tomography (CT) (Janib, Moses, and MacKay 2010). Diagnosis is the initial step to confirm the disease. Early detection of abnormalities in the body is the finest way to prevent their development and treat them from the later stages. Due to their attractive optical/magnetic properties, different types of nano-delivery systems can be utilised for multimodal imaging/image-guided cancer therapy. The novel and MDDSs could gradually progress the potency of TREAT, which are totally depending upon the different factors such as tumour microenvironment (TME) (Caldorera-Moore, Liechty, and Peppas 2011). In this chapter, we have discussed TREAT as, on-demand multifunctional drug delivery systems in detail with their future prospects. As a potential nanoplatform for detecting clinical difficulties and treatment methods, different challenges were encountered during the development of TREAT (Figure 3.1).

3.1.1 Concept behind TREAT

Research has been engaged in stimuli-responsive polymers and their potential as viable site-explicit deliverables. Applied to fields, for example nanotechnology including biosensors and textiles, bioengineering and tissue building (Stuart et al. 2010), these polymers/drug delivery systems have contributed essentially to pharmaceutical research. These are also named as 'smart stimuli-responsive' carrier systems. These systems show unique properties of changing themselves in a flash of a second as per environmental changes (externally or internally applied) (González and Frey 2017). Stimuli-responsive or triggered responsive polymers/novel drug delivery systems may be said as polymers that show response with a small change in physical/chemical parameters of ingredients used in the formation (Hoffman 2013). Triggered-release drug delivery systems based on responsive polymers are classified into two broad categories: externally (exogenous) triggered-release systems and internally (endogenous) triggered-release systems. Endogenous environmental conditions include enzymes and other biomolecules, pH, redox potential, oxygen levels and disease-specific cytokines. External stimuli are those that affect the body from outside and are not present internally in the cells. The most imperative ones are magnetic-responsive systems, ultrasound (US)-responsive systems and electrical systems (Wang and Kohane 2017; Zhang et al. 2016).

Temperature-dependent systems are being effectively used since a long time for the conveyance of bioactives to the human body. Temperature has been utilised as a powerful external trigger for the structure of responsive nanocarriers because of its physiological importance. In a perfect world, a temperature-responsive polymeric nanocarrier should be effective in the naturally pertinent range of temperature between 10°C and 40°C (mostly between 37°C and 42°C; the latter temperature is helpful in reducing any toxic effects due to protein denaturation above 40°C).

Recently, the combination of hyperthermia treatment and chemotherapy showed promising results for cancer treatment. Hyperthermia has dual action of killing the tumour cell firstly, by high temperature (42.5°C–43.5°C) and secondly, by dilating the vessels causing change in membrane permeability in tumour cells, both factors cause increase in the delivery of drugs to tumour (Ding et al. 2016). Physiologically increased levels of prostaglandin E2 in the brain are reported in the literature to mediate the body temperature by altering the firing rate of neurons that control thermoregulation (Chaterji, Kwon, and Park 2007).

Temperature changes can be effectively observed and controlled both to and fro. A vital significant property that thermosensitive polymers present is 'critical solution

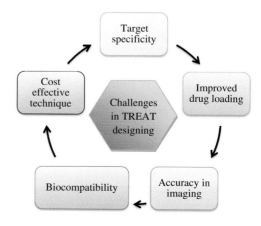

FIGURE 3.1 Schematic highlighting the challenges in TREAT designing.

temperature' (CST). CST is where the polymer experiences phase transition beneath or over a particular temperature. Contingent upon whether this point is at a low or high temperature, it is called a lower critical solution temperature (LCST) and higher or upper critical solution temperature (HCST or UCST), respectively (Roy, Brooks, and Sumerlin 2013; Calejo, Sande, and Nyström 2013). Naya et al. (2020), Zhang et al. (2017) and Victor et al. (2016) in different studies reported that the majority of thermosensitive nanocarriers are formulated from polymers having LCST because they form week hydrogen bonds with the solvent, i.e. water, and form clathrate-like structures through the free polar groups available. This hydrogen bond formation solubilises the polymer by exothermic enthalpy change ($\Delta H < 0$). The formation of clathrate structures leads to an unfavourable entropy of mixing (negative ΔS), and the ΔS has been reported to play an important role which increases its importance along with rising temperature (Naya et al. 2020).

Surpassing LCST, ΔH predominates, and it makes Gibbs free energy (ΔG) positive, resulting in phase separation:

$$\Delta G = \Delta H - T\Delta S$$

When $T > LCST$, the polymer converts itself into an insoluble globular form (which is compact hydrophobic) from the soluble linear/coil conformation (which is swollen hydrophilic) (Kamaly et al. 2016; Guisasola et al. 2015). The increase in swellability amid temperature is also termed as 'zipper effect' (Kikuchi and Okano 2002). This property has been utilised in a wide variety of biomedical controlled-release drug delivery applications (Dimitrov et al. 2007). Thermo-responsive polymers are reported to contain methyl, ethyl and propyl groups in preference. These render polymers the hydrophobic property (Richardson et al. 2005).

3.2 THERMALLY RESPONSIVE POLYMERS AND NANOCARRIERS

Additionally, another procedure for accomplishing temperature-related affectability is to encapsulate thermo-responsive materials inside nanocarriers. Many of them are being prepared, viz. liposomes (Ho et al. 2018; Chen et al. 2013; Liu et al. 2019), polymeric micelles (Li et al. 2017; Araki et al. 2018), nanocomposites (Hervault et al. 2016), nanocapsules (Lee et al. 2008), nanogels (Luckanagul et al. 2018; Almeida et al. 2017; Ruan et al. 2019) and vesicles (Park and Champion 2014; Zheng et al. 2019). The thermo-responsive nanovesicles are engendered from materials that show a change in their physical/chemical properties amid temperature variations (van Elk et al. 2014). The temperature-sensitive materials include poly(N-isopropylacrylamide) (PNIPAM) (van Elk et al. 2014; Mi et al. 2010), poly(N-vinylisobutyramide) (Kono et al. 2011), poly(2-oxazoline) (POx) (Osawa et al. 2017), poly[2-(2-methoxyethoxy)ethyl methacrylate] [PMEO2MA] (Yang et al. 2010), poly(N-isopropyl acrylamide), hydroxybutyl-grafted chitosan, poly(N-isopropylacrylamide-co-acrylamide), polyaniline (Wang et al. 2016), poly(N-isopropylacrylamide-co-N,N-dimethylacrylamide-b-lactide) and poly(N-isopropylacrylamide-co-N,N-dimethylacrylamide-b-ε-caprolactone) diblock copolymers (Li et al. 2011). A quantity of polymer has an opposite nature to LCST called UCST, below which they crumple, e.g. poly(N-vinylcaprolactam) (Vihola et al. 2008; Fernández-Barbero et al. 2009).

A huge extent of delivery systems targets inner physiological conditions for the bioactive delivery. This internal stimulus is usually selected over an externally applied one due to its expanded consistency and reproducibility *in vivo* and practicality during enormous manufacturing (Fernández-Barbero et al. 2009). The thermo-responsive liposomes, polymer micelles and nanoparticles (NPs) have been formulated based on the alteration in the thermal/sharp/sudden behaviour of any one of the formulation components in response to heat due to LCST (Mitra et al. 2015).

Kim et al. (2017) reported the preparation of sustained-release thermo-responsive PEGylated human α-elastin NPs encapsulating insulin and bovine serum albumin (BSA). The findings suggested that the formulation underwent sol-to-particle transition in water at a CST of 25°C–40°C. Also 72-h sustained release of insulin and BSA was observed from the same.

Jung et al. (2017) developed externally thermo-responsive intra-articularly injectable hydrogel for piroxicam. Briefly, they mixed hyaluronic acid (HA) with Pluronic F-127 (HP) in an aqueous medium. HA/HP micelles, 160 nm in size, were formed at a temperature below the LCST, and the thermo-responsiveness was verified as the size was found to increase at elevated temperature. The *in vivo* studies were done on beagle dogs, and the gel was found to be effective for arthritis.

The first and the most important block polymer for thermosensitive delivery is PNIPAM-based thermo-responsive system. However, due to its reported neurotoxicity and bioincompatibility, the medical application of its nanocarrier systems is yet a dilemma (Shakya, Kumar, and Nandakumar 2011). In the past decades, many researches have been done to extend the usage of thermo-responsive polymeric nanocarrier systems other than PNIPAM. For instance, poly(N-vinylcaprolactam)- and poly(oligo ethylene glycol)-based systems were investigated, and it has been reported that the temperature range at which the nanocarrier responds can be tuned by modulating the ratio between hydrophilic and hydrophobic moieties of these polymers. Shao et al. (2014) reported block and random copolymers consisting of oligo(ethylene glycol) and cholic acid pendant groups via ring-opening metathesis polymerisation of their norbornene derivatives. They reported that both the prepared copolymers are thermosensitive with the same cloud points. The micelles (encapsulating paclitaxel) were formed as a result of self-assembling of both the copolymers. However, the block polymer micelles showed higher per cent release in 24 h than random copolymers. PTX-loaded micelles exhibited apparent antitumour efficacy towards the ovarian cancer cells with a particularly low half-maximal inhibitory concentration (IC_{50}) value. Folic acid (FA)-conjugated Pluronic F-127–poly(D,L-lactic acid) (F127-PLA) copolymer micelles encapsulating doxorubicin (Dox) were formulated by Guo et al. (2014). The micelles that were fabricated with 100°C polymerisation of PLA segment reported 39.2°C as the LCST, which is very close to normal human body temperature. Also greater drug release was observed at 40°C. The *ex vivo* cell line study on fibroblast NIH/3T3 and HeLa cell lines has shown excellent cell viability and better uptake under hyperthermia than that at physiological temperature. Another research of FA-targeted Dox-encapsulated thermo-responsive micelles is reported (Panja et al. 2016). The authors of that research prepared micelles from block copolymers, pentaerythritol polycaprolactone-β-poly(N-isopropylacrylamide) (PE-PCL-β-PNIPAM) and pentaerythritol polycaprolactone-β-poly(N-vinylcaprolactam) (PE-PCL-β-PNVCL). High drug release rate was seen above LCST. The *ex vivo* cell line study in C6 glioma cells and the *in vivo* study suggested better tumour uptake (Panja et al. 2016).

Talelli and co-workers (2011) reported the fabrication and evaluation of biodegradable and thermo-responsive polymeric micelles with covalently bound Dox. They investigated the fabrication of thermosensitive polymeric micelles (50 nm in size) made of poly(ethylene glycol)-β-poly[N-(2-hydroxypropyl) methacrylamide-lactate] (mPEG-β-p(HPMAmLacn)) as a delivery system for therapeutic (40% drug release in 5 days with β-glucuronidase) and imaging agents. The MTT assay showed that Dox-loaded micelles when seeded onto 14C cells have the same cellular toxicity as free drug.

Liu et al. reported temperature-responsive polymersomes for the controlled delivery of Dox. The temperature-sensitive poly(N-vinylcaprolactam)–poly(dimethylsiloxane)–poly(N-vinylcaprolactam) (PVCL–PDMS–PVCL) copolymers with different poly(vinylcaprolactam) (PVCL) chain length was synthesised by reversible addition fragmentation chain transfer (RAFT) polymerisation. They prepared self-assembled vesicles at room temperature and found that the vesicle size and the Dox loading depended on the chain length of the PVCL in the temperature range of 37°C–42°C. They reported sustained release and time-dependent cytotoxicity of Dox polymersomes in human alveolar adenocarcinoma cells (Liu et al. 2015).

Thermosensitive amphiphilic block copolymer P-(N,N isopropylacrylamide-*co*-N-hydroxymethylacrylamide)-β-caprolactone with LCST about 38°C was prepared (Wang et al. 2014). The Dox-loaded micelles adequately restrained multiplication and incited apoptosis of QBC939 cells *in vitro* and altogether hindered tumour development in nude mice. In the same year, Luo and co-workers published an article reporting thermo-responsive poly(N-isopropylacrylamide)-β-hydroxyl-terminated polybutadiene (PNIPAM-β-HTPB) block copolymer micelles of the drug camptothecin (CPT). They have formulated two diverse polymeric structures, AB4 four-armed star multiblock and linear triblock copolymers with HTPB as a core block. They discovered that the multiblock copolymers were all the more immediately amassed into round centre shell nanoscale micelles than the straight triblock copolymers. CPT embodied in micelles and assessment for drug release indicated that copolymer micelles have thermal activated drug emancipation ability (Luo et al. 2014).

John et al. reported FA-tethered PNIPAM–phospholipid hybrid nanocarriers. Temperature-responsive lipopolymers were created by conjugating PNIPAM with an alternate level of polymerisation onto three distinct phospholipids by the mix of RAFT and azide-alkyne click responses. The prepared NPs showed temperature-responsive controlled release (John et al. 2015).

Fu and co-workers synthesised a series of poly(3-caprolactone)–poly(ethylene glycol)–poly(3-caprolactone) triblock copolymers (PCECs). Relatively high molecular weight PEG6000 was used for the ring-opening polymerisation of caprolactone, and it optimised the ratio of hydrophilic/hydrophobic chain to obtain a series of thermo-responsive injectable solutions of indomethacin in PCL–PEG6000–PCL copolymers with suitable gel–sol alteration character. The same was found to be effective for tumour targeting with Dox when tested in S180 xenograft tumour in mice (Fu, Lv, and Qiu 2015). Another flourishing combination of an mPEG-*g*-chitosan gel was reported by Tsao et al. (2015). The sol–gel transition of mPEG-*g*-chitosan is inferable from the fragile changes to be decided by hydrophilicity/hydrophobicity. The sol–gel conversion of mPEG-*g*-chitosan is affected by variation in concentration of salt/solute, temperature and pH.

Zeng et al. reported thermo-receptive centre shell-organised Pluronic-based nanocapsules exemplifying drug through the cross-linking response between paranitrophenyl-enacted Pluronic F-127 and hyaluronic acid or poly(ε-lysine). The ensuing shell cross-linked nanocapsules displayed bigger volume change over a temperature scope of 4°C–37°C because of the temperature-subordinate parchedness of cross-linked Pluronic chains. Controlled thermo-responsive release was observed from the nanocapsules in aqueous solution (Zeng et al. 2014).

FA-conjugated thermosensitive polymeric nanospheres were formulated by using hydroxypropyl cellulose (LCST at 41°C in water) (Metaxa et al. 2016). Thermo-responsive cyclodextrins (CDs) have been used to formulate thermally activated nanovesicles, micelles, NPs, nanocapsules, hydrogels, microgels and liposomes for various drug delivery purposes. Many PNIPAMs with CDs are utilised as a thermo-responsive carrier (Schmaljohann 2006; Yanagioka et al. 2003). These can be linear or branched by uniting with different polymers, or organised by polymerising with cross-linkers. The difficulty with thermally reactive hydrogel dependent on PNIPAM, for example, absence of explicit binding regions for drugs and non-biodegradable properties, can be overseen by the copolymerisation of vinyl-functionalised β-CDs with PNIPAM hydrogel (Constantin et al. 2014).

Zhang et al. (2008) arranged thermo-responsive semi-interpenetrating polymeric system hydrogels composed of PNIPAM and β-CD-grafted polyethylenimine by radical polymerisation. Then, a few analysts used different polymers to get various improvements (Sánchez-Moreno et al. 2018).

Miao et al. prepared and characterised self-healing, thermally active hydrogels, between alginate-*graft*-β-CDs and Pluronic F-108 (poly(ethylene glycol) block–poly(propylene glycol) block–poly(ethylene glycol)). Because of their modifiable mechanical properties, the double cross-linked, multi-stimuli-responsive hydrogels were proposed for different biomedical applications such as "drug delivery and cell transplantation" (Miao et al. 2015).

In a recent study, the siRNA-SS-PNIPAM conjugates could form siRNAsome by self-assembly at a higher temperature (>32°C) than LCST for phase transition. In different research, the nanocarriers with PNIPAM on the surface formed micellar networks (i.e. aggregates) at a temperature greater than LCST, but deaggregated at a lower temperature. Thus, the thermo-responsive nanocarriers also lead its footprint in plasmid-DNA (p-DNA) condensation, folding proteins, and encapsulating hydrophobic anticancer drugs (e.g. Doxorubicin). These micelles showed greater cellular internalisation at 42.5°C than at 37°C. They also showed lower survival than free Dox as tested on MCF-7 cells. In spite of the fact that much advances in creating temperature-responsive nanocarriers has been developed, only few of them really do exist and thus, needs further turn of events. The thermo-responsiveness of certain materials and nanocarriers was neither in the range of human body (e.g. 37°C–42°C) nor could be basically moved to another ideal temperature. It further has to be pointed out that some thermal-responsive nanocarriers were developed with non-biodegradable polymers (e.g. PNIPAM), which may be difficult for clinical translation. Thus, development of biodegradable and thermosensitive materials would be a future direction and the growth of nanocarrier in tumours is still fundamentally significant for accomplishing pinpoint thermo-activated drug release and treatment (Mi 2020).

3.3 DIAGNOSTIC APPROACH

The diagnosis in cancer treatment should consider many factors, including the patient's medical history, symptoms and physical examination. However, it can be confirmed by performing laboratory examinations, biopsies and biomedical imaging analysis. Biomedical imaging is not only in the early stages of disease detection, but also a promising revolution for the prognosis of patients and drug delivery systems. In addition, it helps to monitor real-time drug release in response to therapeutic, physiological and biological changes, tracking stem cells, organ biodistribution *in vivo* and gene delivery (Fass 2008; Bollineni, Collette, and Liu 2014).

The techniques currently available for biomedical imaging include optical imaging, MRI, CT, US and single-photon emission computed tomography (SPECT) (Frangioni 2008; Mora-Huertas, Fessi, and Elaissari 2010). Early diagnosis is crucial for positive results, so these bioimaging techniques make an important contribution to successful treatment. The main advantages and disadvantages of these modalities will be introduced, and their applications in TREAT are discussed further.

3.4 OPTICAL IMAGING

Optical imaging (OI) has been a powerful clinical tool for non-invasive research in molecular detection and imaging based on photons emitted by bioluminescence or fluorescent probes. It has several merits, including safety, ease of use, inexpensiveness, and great spatial resolution from visible to near-infrared light (NIR, 600–900 nm) spectra, and demerits such as poor penetration due to light scattering in tissue and more noisy sound during diagnosis and autofluorescence of the tissues (Janib, Moses, and MacKay 2010). OI has the capacity to be combined with other imaging modalities such as MRI, positron emission tomography (PET) and SPECT. For *in vivo* studies, bioluminescence or fluorescent probes (such as gold NPs and quantum dots) and fluorescent biomarkers (such as Cy5.5, Cy7 and luciferase) have been widely used for detection and are used in image-guided surgery (IGS) and tumour diagnosis in cancer treatment (Martelli et al. 2016; de Boer et al. 2015).

The combination of IO technology with cancer-targeting contracting agents (CAs) can prevent cancer recurrence due to the presence of tumour boundaries. The specificity of CAs for cancer cells can be combined with imaging dyes for biomolecules (e.g. antibodies, cell-penetrating peptides, cRGD) or cargo molecules that distinctively bind to cancer tissues (Ye and Chen 2011; Achilefu et al. 2002; Khatik et al. 2020). Bioluminescence technology for quantitative measurement of tumour burden, therapeutic response, immune cell transport and gene transfer testing has also been developed. Luminous properties depend on adenosine triphosphate and oxygen dependence enzymatic property within living tissues (Sweeney et al. 1999).

In TREAT, anticancer treatment can also be achieved through fluorescence imaging therapy with metal NPs. According to reports, gold and silver NPs with high metal-enhanced fluorescence can be used for tissue imaging (Li et al. 2015), gold NPs can be assembled with nanocarriers to develop their stability and targeting capabilities, and they may also be combined with therapeutic agents for image-guided therapy (Yang et al. 2017; Zhong et al. 2016). A similar technique is possible with quantum dots (Alibolandi et al. 2016) and fluorophores (Choi et al. 2012).

3.5 MAGNETIC RESONANCE IMAGING (MRI)

Traditional MRI allows detailed non-invasive visualisation/examination of the structure of the posterior cranial fossa and represents a basic step in the diagnostic examination of many cerebellum diseases and solid tumours. Due to its wide application in medical research, it has excellent soft tissue contrast characteristics and can provide three-dimensional (3D) anatomical images with high spatial resolution. MRI can basically display a detailed image of anatomical structure, can detect lesions and can also provide information about body functions (Wei et al. 2018). MRI produces two types of imaging, T_1-weighted images and T_2-weighted images, entirely depending on the characteristics of the MRI CA uptake, water protons and tissue structure. T_2-weighted MR images show a darker signal due to the higher value of transverse relaxivity (r_2) of the CA, while the T_1-weighted MR images show a brighter signal, because the longitudinal relaxation rate (r_1) of the CA is usually between 3 and 5 s^{-1}/mM (Galanaud et al. 2003; Xu et al. 2012).

Some agents having paramagnetic property in transition metal ions such as gadolinium (Gd^{3+}) and manganese (Mn^{2+}) that have been shown to effectively change T_1 and T_2. Gd^{3+} having seven unpaired electrons in the inner orbital shell donates with a high degree of paramagnetism potential, resulting in an increase in the T_1 relaxation rate of neighbouring liquid content. Superparamagnetic agents, such as iron oxide NPs or Fe/Mn composite metal cores coated in a polymer matrix, have great magnetic property than paramagnetic agents (Shokrollahi 2013).

In the past few decades, several research studies published on MRI has been evaluated as promising materials in the fields of cancer diagnosis and therapy. In 2018, Yan et al. reported the potential of superparamagnetic iron oxide nanoparticles (SPIONs) coated with protoporphyrin IX (PpIX) to be used as a CA in preclinical MRI. Intravenously injection of SPIONs coated with PpIX at the dose of 5 mg Fe/kg of body weight T_2-weighted MR images were obtained after 24 h. T_2-weighted MR images showed very little signal at the target tumour site, thus indicating that SPIONs have entered the tumour (Yan et al. 2018).

Unfortunately, the SPIONs are not conspicuous in the ruptured blood vessels. Therefore, gadolinium-based positive MRI CAs have been substituted for T_1 positive in clinical settings (Anderson, Lee, and Frank 2006; Aime et al. 2007). Gadolinium-based CAs are associated with severe nephrogenic fibrotic disease (NFD) (Viswanathan et al. 2010), which makes them non-preferred drugs. Incorporating MRI CAs with specific macromolecules (antibody or peptide) may ensure the disease diagnosis and treatment (Chung 2016). Combining gadolinium-based positive MRI CAs with specific antibodies and peptides may confirm the diagnosis and treatment of cancer diseases.

Khatik et al. described gadolinium arsenate nanoneedles (Gd-AsNDs) coated with RGD conjugated CH (RGD-CH-Gd-AsNDs) as a new type of multifunctional drug delivery system that effectively target tumours through ligand receptor-mediated endocytosis (Figure 3.2) and could improve treatment efficiency, and has the potential of arsenic as a drug targeting efficiency and T_1-weighted monitoring of positive CAs to enhance *in vivo* and *in vitro* MRI (Khatik et al. 2020).

The combinations of T_1/T_2 double contrast-enhanced MRI reagents have received widespread attention because they can improve the accuracy during diagnosis by providing two pieces of complementary diagnostic information (Lee et al. 2015; Choi et al. 2010). To date, several efforts have been made in preparation to combine T_1/T_2 dual CAs for MRI investigation. In the motivation to develop sophisticated T_1/T_2 dual CAs for MRI, Khatik et al. in 2018 combined Mn^{2+} species with iron oxide NPs which enabled better treatment of cancer through better diagnostics. Here, it has been determined that when the developed Magnus nanobullets are combined with GSH-sensitive T_1-weighted MRI performance established on the redox reaction after the exchange of GSH ions with Mn^{2+} ions. The great T_1/T_2 dual contrasting image and impressive biocompatibility make the developed Magnus nanobullets "smart" and the best strategy for selecting bioimaging evaluations as shown in Figure 3.3

3.6 ULTRASOUND (US)

Various cancer imaging techniques involve variable CAs. Although some of them have been clinically accepted, but the problem of toxicity must be carefully observed. The imaging method using US as a diagnostic method has been widely used owing to the inexpensive, non-invasive real-time imaging probe, the ability to penetrate tissues deeply, rapid examination, high spatial and temporal resolution, painless procedure and non-exposure to ionising radiation (Li, Su et al. 2016; Dwivedi et al. 2020). In addition, the fact that it uses a portable device has promoted its application and induced the instantaneous permeability of cell membranes, thereby increasing the cellular uptake of drugs (Martin and Dayton 2013; Pangu et al. 2010). Because US has the capability to interact with different types of NPs, it has the potential to design drug-loaded NPs that rupture in response to US and release a therapeutic payload drug. Thus, US can be used to release drugs from NPs, which is called US-mediated drug delivery (UMDD). US can also be used to regulate intracellular pressure and heat stroke. In fact, every mechanism of US can be applied to trigger controlled drug release from the carrier/NPs, which involves the design of NPs that are sensitive to more than one trigger (Tharkar et al. 2019). For example, copolymer-based NPs, such as high-intensity focused ultrasound (HIFU)-responsive self-assembled copolymer micelles containing polyethylene glycol (PEG) and polypropylene glycol (PPG), can rapidly decompose under the action of US. The release the drug target tissue quickly and accurately (Li, Xie et al. 2016).

The main disadvantage of cancer treatment using nanoparticulate drug delivery systems is that NPs absorb very less (about 0.7%) of the target site. If US is used, the uptake of NPs by the tumour can be increased, which is endorsed to the increased uptake of NP accumulation within the tumour. In addition, microbubbles (gas-like

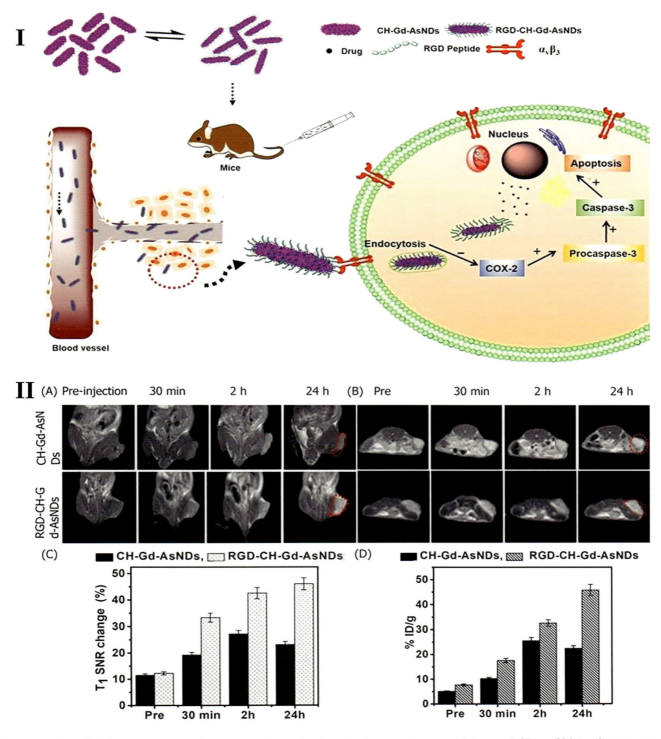

FIGURE 3.2 (I) Schematic diagram of nanoneedles becoming targeted through interactions between cRGD–αvβ3 integrin receptors on the cell matrix and activation of the caspase pathway. (II) (a) T_1-weighted coronal MR images of tumour models acquired at 3.0 T with 2 mg Gd/kg of mouse body weight under conditions before and after intravenous administration of CH-Gd-AsNDs (above) and RGD-CH-Gd-AsNDs (below), (b) T_1-weighted axial MR images of xenograft tumour models before and after intravenous administration of CH-Gd-AsNDs and RGD-CH-Gd-AsNDs, (c) SNR changes ration in tumour of (b) T_1-weighted images and (d) the ICP-AES analysis used for the measured amount of As ions in tumour uptake treated with CH-Gd-AsNDs and RGD-CH-Gd-AsNDs (*$p<0.05$, n=3/group). (Reproduced with copyright permission from Renuka et al. 2019.)

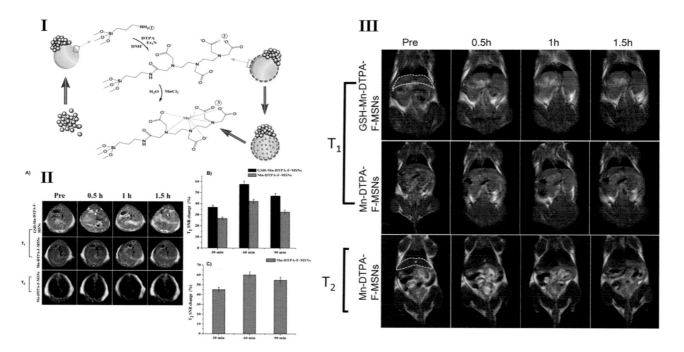

FIGURE 3.3 (I) Schematic diagram of plane functionalisation steps, (II) (a) T_1/T_2-weighted MR images of Balb/C mice before and after intravenous administration of GSH-Mn-DTPA-F-MSNs and Mn-DTPA-F-MSNs (at a dose of 2.5 mg Fe/kg of mouse body weight at 3.0 T (above – T_1 imaging; below, – T_2 imaging). SNR changes ration in the liver of (b) T_1-weighted images and (c) T_2-weighted images (n=3), (III) T_1/T_2-weighted coronal MR images of Balb/C mice at 3 T before and after intravenous administration of GSH-Mn-DTPA-F-MSNs (above) and Mn-DTPA-F-MSNs (below) (at a dose of 2.5 Fe/kg of mouse body weight) at 3.0 T. (Reproduced with copyright permission from Khatik et al. 2019.)

bubbles stabilised by surfactants or different types of polymer coatings) can be used in conjunction with US to improve the transport of NPs to the tumour. When irradiated with US, microbubbles applied in the blood capillary will vibrate and reason mechanical forces on the walls of the blood vessels, which can enhance the transport of NPs into the extracellular matrix of the tumour through the capillary wall (Wilhelm et al. 2016).

As shown in Figure 3.3, there are two interactions or effects (thermal and physical) between US and NPs, which means that both interactions affect different molecular pathways. The treatment time in UMDD should be evaluated by the time required for US produces the desired effect without adversely affecting other parts of the body. The duration of the US treatment depends on many factors, such as the condition and type of tissue being treated, US intensity/frequency and microbubbles (if needed) (Juffermans et al. 2006). The interaction of US with NPs and cells facilitates the cancer drug delivery system and collectively improves the therapeutic effect. Moreover, UMDD using NPs has the ability to overcome the limitations related to modern cancer treatments.

3.7 COMPUTED TOMOGRAPHY (CT)

Compared with MRI, CT is a fast and inexpensive operation. It has similar characteristics to MRI such as good spatial resolution and deep penetration and has excellent signal transmission capability by using CAs (Ma et al. 2017). This technique involves measuring the absorption rate of X-rays passing through the body tissue to obtain a cross-sectional tomography photo. CT has many advantages, but as will be realised later, there are still many disadvantages.

The main advantages are non-toxicity, less scanning time and identification of whole geometry. Disadvantages include the many difficulties faced during scanning of multifunctional material in one system, radiation exposure, reduced measurement capability of the form due to evaluation errors, and the inability to track the results of unknown samples.

The draft of the German standard VDI/VDE 2630 Part 1.2 (Villarraga-Gómez 2016) explained a series of processes, that is the way to achieve the results, including four different measurement functions. Figure 3.4 illustrates a general process sequence. Differences in membrane density will affect CT absorption, where denser tissues will cause higher absorption (Lusic and Grinstaff 2013). Low-molecular-weight CAs (gold, iodine, different lanthanides and bismuth sulphide) used in CT imaging have been clinically approved to be used in TREAT. Moreover, these CAs have a rapid clearance rate, so the image is not clear during the measurement process. The encapsulation of these CAs in polymeric NPs can control this problem because they have good blood circulation features (Baetke, Lammers, and Kiessling 2015; Naha et al. 2014). This strategy can be used in TREAT diagnostic methods to calculate treatment outcomes of different MDDSs (Figure 3.5).

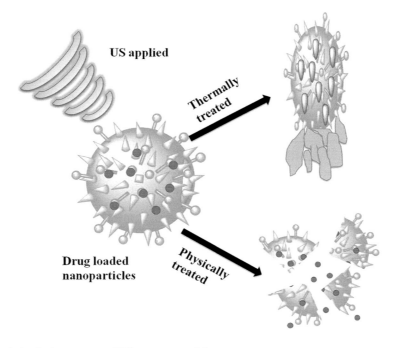

FIGURE 3.4 Thermal and physical treatment of US on nanoparticles.

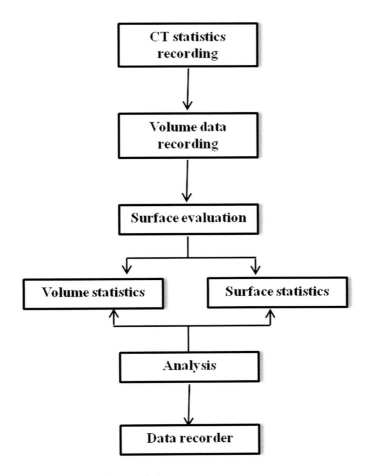

FIGURE 3.5 Process sequence for evaluation of CT statistics.

3.8 SINGLE-PHOTON EMISSION COMPUTED TOMOGRAPHY AND POSITRON EMISSION TOMOGRAPHY

The advanced radiopharmaceuticals with high-spatial-resolution tracking function of the latest PET and SPECT systems provide an opportunity to modify potential MDDS candidates. In the radiolabelled nanodrug delivery system, the biodistribution image and pharmacokinetic and pharmacodynamic data of the whole body can be monitored, so that the biochemistry of the compound and the accumulation of the target in the body can be realised (Kunjachan et al. 2012; Lammers et al. 2010). Both SPECT and PET are imaging technologies that work by the gamma rays generated by the attenuation of radioactive materials into our bodies. Therefore, these methods require the acquisition of radionuclides to produce signals for visible images (Velikyan 2012).

SPECT and PET technologies not only have important uses in early diagnosis, but can also observe the progress of treatment. In addition they have other advantages such as methods are more sensitive, non-invasive, low ambient noise and 3D image formation. Short half-lives of reagent, which shortens the investigation time and reduces radiation exposure, are the main disadvantage of these imaging technologies. PET is more sensitive than SPECT and is related with other technologies (such as CT) to acquire biological distinguish (Satterlee and Huang 2016).

General radioisotopes for PET are ^{89}Zr, ^{64}Cu, ^{18}F, ^{68}Ga, ^{15}O, ^{124}I and ^{13}N, while for SPECT, they are ^{123}I, ^{99m}Tc, ^{111}In, ^{166}Ho, ^{188}Re, ^{90}Y and ^{177}Lu (Polyák and Ross 2018; Key and Leary 2014).

NPs provide an excellent platform for the attachment of a radioactive agent, which can be used for the early detection of diseases and permits the image-guided therapy method by the EPR effect on the tumour target site (Abadjian, Edwards, and Anderson 2017; Chakravarty, Hong, and Cai 2014; Kunjachan et al. 2015). Different types of nanodrug delivery systems are good platforms for radioisotope attachment and can be used for early diagnosis of various diseases. The literature has reported several types of polymers (chitosan, gelatin and albumin) that can be modified by copolymerisation to complex the protein (Jain, Dandekar, and Patravale 2009; Khatik et al. 2015; Gwyther and Field 1966; Huang et al. 2010).

The combination of different imaging technologies (such as hybrid PET/MRI/SPECT) permits the development of protocols that can take advantage of the best of both. In order to obtain the best results of these combinations, it is very necessary to use dual probes. The combination of iron oxide NPs and ^{68}Ga radioisotope was synthesised using the microwave method and coupled with RGD peptide for targeted tumour therapy (Pellico et al. 2016). In vivo studies in mouse models found that through the accumulation of ^{68}Ga-C-IONP-RGD probes, PET and T_1-weighted MRI experiments established by the binding of integrin-binding receptors to tumours as shown in Figure 3.6.

Through the application of PET and SPECT imaging, the expected good efficiency can be achieved, and carefully planned imaging and treatment procedures are provided in an image-guided treatment method, so that nanotherapeutics can provide patients with personalised nanomedicine care (Polyák et al. 2014). The development of these TREAT or therapeutic techniques for diagnostic and treatment applications can enhance clinical therapeutic effects, and in the near future, they can become powerful multifunctional oncology probes.

3.9 CONCLUSIONS AND FUTURE DIRECTIONS

The field of theranostics is developing at a burgeoning speed given the unending endeavours in various fields of microbiology, biotechnology, nanotechnology and novel drug delivery system. Theranostic nanocarriers have the ability to modify the treatment of various conditions, for example allergies, irritations, neurological disorders and cardiovascular infections, by supporting early discovery, directed treatment and checking of restorative impact. In any case, theranostic nanocarriers despite everything have far to go for them being finalised for clinical interpretations. Definite examinations tending to their long-term stability and safety both *in vitro* and *in vivo*, pharmacokinetics, and dissemination and advancement of versatile creation approaches are basic to understand the clinical utility of theranostics. A mere combination of imaging agents, therapeutics and targeting ligands within a nanoplatform cannot suffice. Interaction and compatibility of the components with each other, and the effect of fabrication methodology on the stability of each component and the system need extensive evaluation. Moreover, the combination of therapeutic agents and diagnostic agents may prove many problems of incompatibilities.

We have talked about many researches in this chapter that have exhibited the current status of different exogenous improvements in thermo responsive micelles, nanocarriers and copolymer-based frameworks, to achieve a better therapeutic effect. Additionally, testing and keeping the stimuli-sensitivity in large-scale produced nanocarriers would be a potential challenge. Despite the fact that broad investigations on stimuli-sensitive nanocarriers has been done, just a couple of details have entered clinical interpretation requiring future confirmations. We hope that theranostics are ideally going to take an influential position in clinical applications by improving both the diagnosis and treatment for the fatal diseases.

CONFLICT OF INTEREST

The authors declare no competing financial interest.

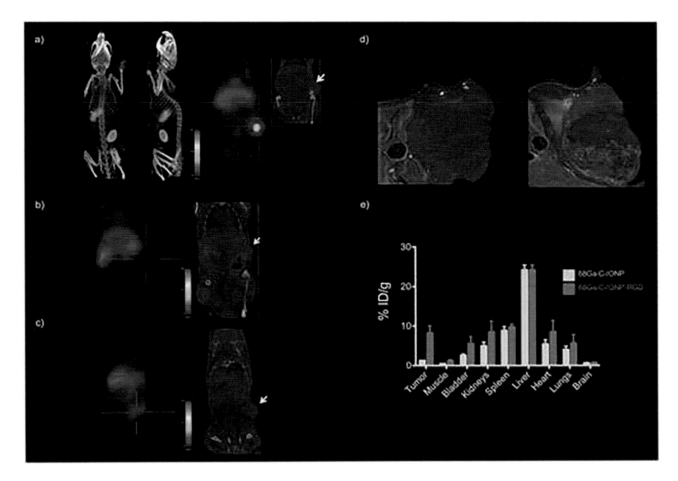

FIGURE 3.6 (a) PET/CT imaging of tumour-bearing mice 1 h after 68Ga-C-IONP-RGD injection, showing strong activity in the tumour; (b) control experiment; (c) blocking experiment; (d) 68Ga-injection axial T1-weighted spin-echo MRI of the tumour area of the mouse before C-IONP-RGD injection (left) and 24 h after injection (right). (e) The biodistribution after administration in rats was measured with a gamma counter (N=5).

REFERENCES

Abadjian, M. Z., W. B. Edwards, and C. J. Anderson. 2017. "Imaging the tumor microenvironment." *Adv Exp Med Biol* 1036:229–57.

Achilefu, S., H. N. Jimenez, R. B. Dorshow, J. E. Bugaj, E. G. Webb, R. R. Wilhelm, R. Rajagopalan, J. Johler, and J. L. Erion. 2002. "Synthesis, *in vitro* receptor binding, and *in vivo* evaluation of fluorescein and carbocyanine peptide-based optical contrast agents." *J Med Chem* 45 (10):2003–15.

Ahmed, H. U., R. G. Hindley, L. Dickinson, A. Freeman, A. P. Kirkham, M. Sahu, R. Scott, C. Allen, J. Van der Meulen, and M. Emberton. 2012. "Focal therapy for localised unifocal and multifocal prostate cancer: a prospective development study." *Lancet Oncol* 13 (6):622–32.

Aime, S., D. Delli Castelli, D. Lawson, and E. Terreno. 2007. "Gd-loaded liposomes as T1, susceptibility, and CEST agents, all in one." *J Am Chem Soc* 129 (9):2430–1.

Alibolandi, M., K. Abnous, F. Sadeghi, H. Hosseinkhani, M. Ramezani, and F. Hadizadeh. 2016. "Folate receptor-targeted multimodal polymersomes for delivery of quantum dots and doxorubicin to breast adenocarcinoma: *in vitro* and *in vivo* evaluation." *Int J Pharm* 500 (1–2):162–78.

Almeida, E. A. M. S., I. C. Bellettini, F. P. Garcia, M. T. Farinácio, C. V. Nakamura, A. F. Rubira, A. F. Martins, and E. C. Muniz. 2017. "Curcumin-loaded dual pH- and thermo-responsive magnetic microcarriers based on pectin maleate for drug delivery." *Carbohydr Polym* 171:259–66.

Anderson, S. A., K. K. Lee, and J. A. Frank. 2006. "Gadolinium-fullerenol as a paramagnetic contrast agent for cellular imaging." *Invest Radiol* 41 (3):332–8.

Araki, T., Y. Fuchi, S. Murayama, R. Shiraishi, T. Oyama, M. Aso, I. Aoki, S. Kobayashi, K. I. Yamada, and S. Karasawa. 2018. "Fluorescence tumor-imaging using a thermo-responsive molecule with an emissive aminoquinoline derivative." *Nanomaterials* 8 (10):782.

Baetke, S. C., T. Lammers, and F. Kiessling. 2015. "Applications of nanoparticles for diagnosis and therapy of cancer." *Br J Radiol* 88 (1054):20150207.

Bollineni, V. R., S. Collette, and Y. Liu. 2014. "Functional and molecular imaging in cancer drug development." *Chin Clin Oncol* 3 (2):6.

Caldorera-Moore, M. E., W. B. Liechty, and N. A. Peppas. 2011. "Responsive theranostic systems: integration of diagnostic imaging agents and responsive controlled release drug delivery carriers." *Acc Chem Res* 44 (10):1061–70.

Calejo, M. T., S. A. Sande, and B. Nyström. 2013. "Thermo-responsive polymers as gene and drug delivery vectors: architecture and mechanism of action." *Expert Opin Drug Deliv* 10 (12):1669–86.

Chakravarty, R., H. Hong, and W. Cai. 2014. "Positron emission tomography image-guided drug delivery: current status and future perspectives." *Mol Pharm* 11 (11):3777–97.

Chaterji, S., I. K. Kwon, and K. Park. 2007. "Smart polymeric gels: redefining the limits of biomedical devices." *Prog Polym Sci* 32 (8–9):1083–122.

Chen, K.-J., H.-F. Liang, H.-L. Chen, Y. Wang, P.-Y. Cheng, H.-L. Liu, Y. Xia, and H.-W. Sung. 2013. "A thermoresponsive bubble-generating liposomal system for triggering localized extracellular drug delivery." *ACS Nano* 7 (1): 438–46.

Chen, X. S. 2011a. "Introducing *Theranostics* journal – from the editor-in-chief." *Theranostics* 1:1–2.

Choi, J.-S., J.-H. Lee, T.-H. Shin, H.-T. Song, E. Y. Kim, and J. Cheon. 2010. "Self-confirming "AND" logic nanoparticles for fault-free MRI." *J Am Chem Soc* 132 (32):11015–7.

Choi, K. Y., E. J. Jeon, H. Y. Yoon, B. S. Lee, J. H. Na, K. H. Min, S. Y. Kim, et al. 2012. "Theranostic nanoparticles based on PEGylated hyaluronic acid for the diagnosis, therapy and monitoring of colon cancer." *Biomaterials* 33 (26):6186–93.

Chung, E. J. 2016. "Targeting and therapeutic peptides in nanomedicine for atherosclerosis." *Exp Biol Med (Maywood)* 241 (9):891–8.

Constantin, M., S. Bucatariu, P. Ascenzi, Bogdan C. Simionescu, and Gheorghe Fundueanu. 2014. "Poly(NIPAAm-co-β-cyclodextrin) microgels with drug hosting and temperature-dependent delivery properties." *Reactive and Functional Polymers* 84: 1–9.

de Boer, E., N. J. Harlaar, A. Taruttis, W. B. Nagengast, E. L. Rosenthal, V. Ntziachristos, and G. M. van Dam. 2015. "Optical innovations in surgery." *Br J Surg* 102 (2):e56–72.

Dimitrov, I., B. Trzebicka, A. H. E. Müller, A. Dworak, and C. B. Tsvetanov. 2007. "Thermosensitive water-soluble copolymers with doubly responsive reversibly interacting entities." *Prog Polym Sci* 32 (11):1275–343.

Ding, C., L. Tong, J. Feng, and J. Fu. 2016. "Recent advances in stimuli-responsive release function drug delivery systems for tumor treatment." *Molecules* 21 (12):1715.

Ding, H., and F. Wu. 2012. "Image guided biodistribution and pharmacokinetic studies of theranostics." *Theranostics* 2 (11):1040–53.

Dwivedi, P., S. Kiran, S. Han, M. Dwivedi, R. Khatik, R. Fan, F. A. Mangrio et al. 2020. "Magnetic targeting and ultrasound activation of liposome–microbubble conjugate for enhanced delivery of anticancer therapies." *ACS Appl Mater Interfaces* 12 (21):23737–51.

Fass, L. 2008. "Imaging and cancer: a review." *Mol Oncol* 2 (2):115–52.

Fernández-Barbero, A., I. J. Suárez, B. Sierra-Martín, A. Fernández-Nieves, F. Javier de las Nieves, M. Marquez, J. Rubio-Retama, and E. López-Cabarcos. 2009. "Gels and microgels for nanotechnological applications." *Advances in Colloid and Interface Science* 147–148:88–108.

Frangioni, J. V. 2008. "New technologies for human cancer imaging." *J Clin Oncol* 26 (24):4012–21.

Fu, J., X. Lv, and L. Qiu. 2015. "Thermo-responsive triblock copolymer micelles containing PEG6000 for either water-soluble or water-insoluble drug sustained release and treatment." *RSC Adv* 5 (47):37451–61.

Galanaud, D., F. Nicoli, Y. Le Fur, M. Guye, J. P. Ranjeva, S. Confort-Gouny, P. Viout, E. Soulier, and P. J. Cozzone. 2003. "Multimodal magnetic resonance imaging of the central nervous system." *Biochimie* 85 (9):905–14.

González, E., and M. W. Frey. 2017. "Synthesis, characterization and electrospinning of poly(vinyl caprolactam-co-hydroxymethyl acrylamide) to create stimuli-responsive nanofibers." *Polymer* 108:154–62.

Guisasola, E., A. Baeza, M. Talelli, D. Arcos, M. Moros, J. M. de la Fuente, and M. Vallet-Regí. 2015. "Magnetic-responsive release controlled by hot spot effect." *Langmuir* 31 (46):12777–82.

Guo, X., D. Li, G. Yang, C. Shi, Z. Tang, J. Wang, and S. Zhou. 2014. "Thermo-triggered drug release from actively targeting polymer micelles." *ACS Appl Mater Interfaces* 6 (11):8549–59.

Gwyther, M. M., and E. O. Field. 1966. "Aggregated Tc 99m-labelled albumin for lung scintiscanning." *Int J Appl Radiat Isot* 17 (8):485–6.

Hervault, A., A. E. Dunn, M. Lim, C. Boyer, D. Mott, S. Maenosono, and N. T. Thanh. 2016. "Doxorubicin loaded dual pH- and thermo-responsive magnetic nanocarrier for combined magnetic hyperthermia and targeted controlled drug delivery applications." *Nanoscale* 8 (24):12152–61.

Ho, Y.-J., C.-C. Wu, Z.-H. Hsieh, C.-H. Fan, and C.-K. Yeh. 2018. "Thermal-sensitive acoustic droplets for dual-mode ultrasound imaging and drug delivery." *J Control Release* 291:26–36.

Hoffman, A. S. 2013. "Stimuli-responsive polymers: biomedical applications and challenges for clinical translation." *Adv Drug Deliv Rev* 65 (1):10–6.

Huang, J., J. Xie, K. Chen, L. Bu, S. Lee, Z. Cheng, X. Li, and X. Chen. 2010. "HSA coated MnO nanoparticles with prominent MRI contrast for tumor imaging." *Chem Commun (Camb)* 46 (36):6684–6.

Jain, R., P. Dandekar, and V. Patravale. 2009. "Diagnostic nanocarriers for sentinel lymph node imaging." *J Control Release* 138 (2):90–102.

Janib, S. M., A. S. Moses, and J. A. MacKay. 2010. "Imaging and drug delivery using theranostic nanoparticles." *Adv Drug Deliv Rev* 62 (11):1052–63.

John, J. V., Y.-I. Jeong, R. P. Johnson, C.-W. Chung, H. Park, D. H. Kang, J. K. Cho, Y. Kim, and I. Kim. 2015. "Folic acid-tethered poly(N-isopropylacrylamide)–phospholipid hybrid nanocarriers for targeted drug delivery." *J Mater Chem B* 3 (42):8268–78.

Juffermans, L. J. M., P. A. Dijkmans, R. J. P. Musters, C. A. Visser, and O. Kamp. 2006. "Transient permeabilization of cell membranes by ultrasound-exposed microbubbles is related to formation of hydrogen peroxide." *Am J Physiol Heart and Circ Physiol* 291 (4):H1595–H601.

Jung, Y. S., W. Park, H. Park, D. K. Lee, and K. Na. 2017. "Thermosensitive injectable hydrogel based on the physical mixing of hyaluronic acid and pluronic F-127 for sustained NSAID delivery." *Carbohydr Polym* 156:403–8.

Kamaly, N., B. Yameen, J. Wu, and O. C. Farokhzad. 2016. "Degradable controlled-release polymers and polymeric nanoparticles: mechanisms of controlling drug release." *Chem Rev* 116 (4):2602–63.

Key, J., and J. F. Leary. 2014. "Nanoparticles for multimodal *in vivo* imaging in nanomedicine." *Int J Nanomed* 9:711–26.

Khatik, R., P. Dwivedi, V. R. Junnuthula, K. Sharma, K. Chuttani, A. K. Mishra, and A. K. Dwivedi. 2015. "Potential *in vitro* and *in vivo* colon specific anticancer activity in a

HCT-116 xenograft nude mice model: targeted delivery using enteric coated folate modified nanoparticles." *RSC Adv* 5 (21):16507–20.

Khatik, R., Z. Wang, D. Zhi, S. Kiran, and P. Dwivedi. 2020. "Integrin alphavbeta3 receptor overexpressing on tumor-targeted positive MRI-guided chemotherapy." 12 (1):163–76.

Kikuchi, A., and T. Okano. 2002. "Intelligent thermoresponsive polymeric stationary phases for aqueous chromatography of biological compounds." *Prog Polym Sci* 27 (6): 1165–93.

Kim, J. D., Y. J. Jung, C. H. Woo, Y. C. Choi, J. S. Choi, and Y. W. Cho. 2017. "Thermo-responsive human α-elastin self-assembled nanoparticles for protein delivery." *Colloids Surf B Biointerfaces* 149:122–9.

Kono, K., E. Murakami, Y. Hiranaka, E. Yuba, C. Kojima, A. Harada, and K. Sakurai. 2011. "Thermosensitive molecular assemblies from poly(amidoamine) dendron-based lipids." *Angew Chem Int Ed Engl* 50 (28):6332–6.

Kunjachan, S., J. Ehling, G. Storm, F. Kiessling, and T. Lammers. 2015. "Noninvasive imaging of nanomedicines and nanotheranostics: principles, progress, and prospects." *Chem Rev* 115 (19):10907–37.

Kunjachan, S., F. Gremse, B. Theek, P. Koczera, R. Pola, M. Pechar, T. Etrych et al. 2012. "Non-invasive optical imaging of nanomedicine biodistribution." *ACS Nano* 7:252–62.

Lammers, T., F. Kiessling, W. E. Hennink, and G. Storm. 2010. "Nanotheranostics and image-guided drug delivery: current concepts and future directions." *Mol Pharm* 7 (6): 1899–912.

Lee, N., D. Yoo, D. Ling, M. H. Cho, T. Hyeon, and J. Cheon. 2015. "Iron oxide based nanoparticles for multimodal imaging and magnetoresponsive therapy." *Chem Rev* 115 (19):10637–89.

Lee, S. H., S. H. Choi, S. H. Kim, and T. G. Park. 2008. "Thermally sensitive cationic polymer nanocapsules for specific cytosolic delivery and efficient gene silencing of siRNA: swelling induced physical disruption of endosome by cold shock." *J Control Release* 125 (1):25–32.

Li, F., C. Xie, Z. Cheng, and H. Xia. 2016. "Ultrasound responsive block copolymer micelle of poly(ethylene glycol)–poly(propylene glycol) obtained through click reaction." *Ultrason Sonochem* 30:9–17.

Li, H., H. Hu, Y. Zhao, X. Chen, W. Li, W. Qiang, and D. Xu. 2015. "Multifunctional aptamer-silver conjugates as theragnostic agents for specific cancer cell therapy and fluorescence-enhanced cell imaging." *Anal Chem* 87 (7):3736–45.

Li, W., J. Li, J. Gao, B. Li, Y. Xia, B. Meng, Y. Yu et al. 2011. "The fine-tuning of thermosensitive and degradable polymer micelles for enhancing intracellular uptake and drug release in tumors." *Biomaterials* 32 (15):3832–44.

Li, W.-P., C.-H. Su, Y.-C. Chang, Y.-J. Lin, and C.-S. Yeh. 2016. "Ultrasound-induced reactive oxygen species mediated therapy and imaging using a Fenton reaction activable polymersome." *ACS Nano* 10 (2):2017–27.

Li, W. S., X. J. Wang, S. Zhang, J. B. Hu, Y. L. Du, X. Q. Kang, X. L. Xu, X. Y. Ying, J. You, and Y. Z. Du. 2017. "Mild microwave activated, chemo-thermal combinational tumor therapy based on a targeted, thermal-sensitive and magnetic micelle." *Biomaterials* 131:36–46.

Liu, F., V. Kozlovskaya, S. Medipelli, B. Xue, F. Ahmad, M. Saeed, D. Cropek, and E. Kharlampieva. 2015. "Temperature-sensitive polymersomes for controlled delivery of anticancer drugs." *Chem Mater* 27 (23):7945–56.

Liu, K. C., A. Arivajiagane, S. J. Wu, S. C. Tzou, C. Y. Chen, and Y. M. Wang. 2019. "Development of a novel thermal-sensitive multifunctional liposome with antibody conjugation to target EGFR-expressing tumors." *Nanomedicine* 15 (1):285–94.

Luckanagul, J. A., C. Pitakchatwong, P. Ratnatilaka Na Bhuket, C. Muangnoi, P. Rojsitthisak, S. Chirachanchai, Q. Wang, and P. Rojsitthisak. 2018. "Chitosan-based polymer hybrids for thermo-responsive nanogel delivery of curcumin." *Carbohydr Polym* 181:1119–27.

Luo, Y.-L., X.-L. Yang, F. Xu, Y.-S. Chen, and B. Zhang. 2014. "Thermosensitive PNIPAM-b-HTPB block copolymer micelles: Molecular architectures and camptothecin drug release." *Colloids Surf B Biointerfaces* 114:150–7.

Lusic, H., and M. W. Grinstaff. 2013. "X-ray-computed tomography contrast agents." *Chem Rev* 113 (3):1641–66.

Ma, Y. Y., K. T. Jin, S. B. Wang, H. J. Wang, X. M. Tong, D. S. Huang, and X. Z. Mou. 2017. "Molecular imaging of cancer with nanoparticle-based theranostic probes." *Contrast Media Mol Imaging* 2017:1026270.

Maeda, H. 2015. "Toward a full understanding of the EPR effect in primary and metastatic tumors as well as issues related to its heterogeneity." *Adv Drug Deliv Rev* 91:3–6.

Martelli, C., A. Lo Dico, C. Diceglie, G. Lucignani, and L. Ottobrini. 2016. "Optical imaging probes in oncology." *Oncotarget* 7 (30):48753-87. doi: 10.18632/oncotarget.9066.

Martin, K. H., and P. A. Dayton. 2013. "Current status and prospects for microbubbles in ultrasound theranostics." *Wiley Interdiscip Rev Nanomed Nanobiotechnol* 5 (4):329–45.

Mi, P. 2020. "Stimuli-responsive nanocarriers for drug delivery, tumor imaging, therapy and theranostics." *Theranostics* 10 (10):4557–88.

Mi, P., X.-J. Ju, R. Xie, H.-G. Wu, J. Ma, and L.-Y. Chu. 2010. "A novel stimuli-responsive hydrogel for K+-induced controlled-release." *Polymer* 51 (7):1648–53.

Miao, T., S. L. Fenn, P. N. Charron, and R. A. Oldinski. 2015. "Self-healing and thermoresponsive dual-cross-linked alginate hydrogels based on supramolecular inclusion complexes." *Biomacromolecules* 16 (12):3740–50.

Mitra, A. K., V. Agrahari, A. Mandal, K. Cholkar, C. Natarajan, S. Shah, M. Joseph et al. 2015. "Novel delivery approaches for cancer therapeutics." *J Control Release* 219:248–68.

Moghimi, S. M., A. C. Hunter, and J. C. Murray. 2005. "Nanomedicine: current status and future prospects." *FASEB J* 19 (3):311–30.

Mora-Huertas, C. E., H. Fessi, and A. Elaissari. 2010. "Polymer-based nanocapsules for drug delivery." *Int J Pharm* 385 (1–2):113–42.

Naha, P. C., A. A. Zaki, E. Hecht, M. Chorny, P. Chhour, E. Blankemeyer, D. M. Yates et al. 2014. "Dextran coated bismuth-iron oxide nanohybrid contrast agents for computed tomography and magnetic resonance imaging." *J Mater Chem B* 2 (46):8239–48.

Naya, M., K. Kokado, K. B. Landenberger, S. Kanaoka, S. Aoshima, and K. Sada. 2020. "Supramolecularly designed thermoresponsive polymers in different polymer backbones." *Macromol Chem Phys* 221 (5):1900455.

Osawa, S., T. Ishii, H. Takemoto, K. Osada, and K. Kataoka. 2017. "A facile amino-functionalization of poly(2-oxazoline)s' distal end through sequential azido end-capping and Staudinger reactions." *Eur Polym J* 88:553–61.

Pangu, G. D., K. P. Davis, F. S. Bates, and D. A. Hammer. 2010. "Ultrasonically induced release from nanosized polymer vesicles." *Macromol Biosci* 10 (5):546–54.

Panja, S., G. Dey, R. Bharti, K. Kumari, T. K. Maiti, M. Mandal, and S. Chattopadhyay. 2016. "Tailor-made temperature-sensitive micelle for targeted and on-demand release of anti-cancer drugs." *ACS Appl Mater Interfaces* 8 (19):12063–74.

Park, W., and J. Champion. 2014. "Thermally triggered self-assembly of folded proteins into vesicles." *J Am Chem Soc* 136:17906–9.

Patel, S. K., and J. M. Janjic. 2015. "Macrophage targeted theranostics as personalized nanomedicine strategies for inflammatory diseases." *Theranostics* 5 (2):150–72.

Pellico, J., J. Ruiz-Cabello, M. Saiz-Alía, G. del Rosario, S. Caja, M. Montoya, L. F. de Manuel et al. 2016. "Fast synthesis and bioconjugation of 68Ga core-doped extremely small iron oxide nanoparticles for PET/MR imaging." *Contrast Media Mol Imaging* 11 (3):203–10.

Polyák, A., I. Hajdu, M. Bodnár, G. Dabasi, R. P. Jóba, J. Borbély, and L. Balogh. 2014. "Folate receptor targeted self-assembled chitosan-based nanoparticles for SPECT/CT imaging: demonstrating a preclinical proof of concept." *Int J Pharm* 474 (1–2):91–4.

Polyák, A., and T. L. Ross. 2018. "Nanoparticles for SPECT and PET imaging: towards personalized medicine and theranostics." *Curr Med Chem* 25 (34):4328–53.

Richardson, J. C., R. W. Bowtell, K. Mäder, and C. D. Melia. 2005. "Pharmaceutical applications of magnetic resonance imaging (MRI)." *Adv Drug Deliv Rev* 57 (8):1191–209.

Roy, D., W. L. Brooks, and B. S. Sumerlin. 2013. "New directions in thermoresponsive polymers." *Chem Soc Rev* 42 (17):7214–43.

Ruan, C., C. Liu, H. Hu, X.-L. Guo, B.-P. Jiang, H. Liang, and X.-C. Shen. 2019. "NIR-II light-modulated thermosensitive hydrogel for light-triggered cisplatin release and repeatable chemo-photothermal therapy." *Chem Sci* 10 (17):4699–706.

Sánchez-Moreno, P., J. de Vicente, S. Nardecchia, J. A. Marchal, and H. Boulaiz. 2018. "Thermo-sensitive nanomaterials: recent advance in synthesis and biomedical applications." *Nanomaterials (Basel)* 8 (11):935.

Satterlee, A. B., and L. Huang. 2016. "Current and future theranostic applications of the lipid-calcium-phosphate nanoparticle platform." *Theranostics* 6 (7):918–29.

Schmaljohann, D. 2006. "Thermo- and pH-responsive polymers in drug delivery." *Adv Drug Deliv Rev* 58 (15):1655–70.

Shakya, A. K., A. Kumar, and K. S. Nandakumar. 2011. "Adjuvant properties of a biocompatible thermo-responsive polymer of N-isopropylacrylamide in autoimmunity and arthritis." *J R Soc Interface* 8 (65):1748–59.

Shao, Y., Y.-G. Jia, C. Shi, J. Luo, and X. X. Zhu. 2014. "Block and random copolymers bearing cholic acid and oligo(ethylene glycol) pendant groups: aggregation, thermosensitivity, and drug loading." *Biomacromolecules* 15 (5):1837–44.

Shokrollahi, H. 2013. "Contrast agents for MRI." *Mater Sci Eng C Mater Biol Appl* 33 (8):4485–97.

Sriramoju, B., R. Kanwar, R. N. Veedu, and J. R. Kanwar. 2015. "Aptamer-targeted oligonucleotide theranostics: a smarter approach for brain delivery and the treatment of neurological diseases." *Curr Top Med Chem* 15 (12):1115–24.

Stuart, M. A. Cohen, W. T. S. Huck, J. Genzer, M. Müller, C. Ober, M. Stamm, G. B. Sukhorukov et al. 2010. "Emerging applications of stimuli-responsive polymer materials." *Nat Mater* 9 (2):101–13.

Sweeney, T. J., V. Mailander, A. A. Tucker, A. B. Olomu, W. Zhang, Ya Cao, R. S. Negrin, and C. H. Contag. 1999. "Visualizing the kinetics of tumor-cell clearance in living animals." *Proc Natl Acad Sci USA* 96 (21):12044–9.

Talelli, M., K. Morita, C. J. Rijcken, R. W. Aben, T. Lammers, H. W. Scheeren, C. F. van Nostrum, G. Storm, and W. E. Hennink. 2011. "Synthesis and characterization of biodegradable and thermosensitive polymeric micelles with covalently bound doxorubicin-glucuronide prodrug via click chemistry." *Bioconjug Chem* 22 (12):2519–30.

Tharkar, P., R. Varanasi, W. S. F. Wong, C. T. Jin, and W. Chrzanowski. 2019. "Nano-enhanced drug delivery and therapeutic ultrasound for cancer treatment and beyond." *Frontiers Bioeng Biotech* 7:324.

van Elk, M., R. Deckers, C. Oerlemans, Y. Shi, G. Storm, T. Vermonden, and W. E. Hennink. 2014. "Triggered release of doxorubicin from temperature-sensitive poly(N-(2-hydroxypropyl))-methacrylamide mono/dilactate) grafted liposomes." *Biomacromolecules* 15 (3):1002–9.

Velikyan, I. 2012. "Molecular imaging and radiotherapy: theranostics for personalized patient management." *Theranostics* 2 (5):424–6.

Vihola, H., A. Laukkanen, H. Tenhu, and J. Hirvonen. 2008. "Drug release characteristics of physically cross-linked thermosensitive poly(N-vinylcaprolactam) hydrogel particles." *J Pharm Sci* 97 (11):4783–93.

Villarraga-Gómez, H. 2016. X-ray computed tomography for dimensional measurements. In Conference: Digital Imaging 2016, An ASNT Topical conference at Mashantucket, CT (USA), pp. 44–57, The American Society for Nondestructive Testing, Inc. ISBN: 978-1-57117-384-3.

Viswanathan, S., Z. Kovacs, K. N. Green, S. J. Ratnakar, and A. D. Sherry. 2010. "Alternatives to gadolinium-based metal chelates for magnetic resonance imaging." *Chem Rev* 110 (5):2960–3018.

Wang, D., B. Lin, and H. Ai. 2014. "Theranostic nanoparticles for cancer and cardiovascular applications." *Pharm Res* 31 (6):1390–406.

Wang, F., G. Xia, X. Lang, X. Wang, Z. Bao, Z. Shah, X. Cheng, et al. 2016. "Influence of the graft density of hydrophobic groups on thermo-responsive nanoparticles for anti-cancer drugs delivery." *Colloids Surf B Biointerfaces* 148:147–56.

Wang, L. S., M. C. Chuang, and J. A. Ho. 2012. "Nanotheranostics – a review of recent publications." *Int J Nanomed* 7:4679–95.

Wang, X., S. Li, Z. Wan, Z. Quan, and Q. Tan. 2014. "Investigation of thermo-sensitive amphiphilic micelles as drug carriers for chemotherapy in cholangiocarcinoma *in vitro* and *in vivo*." *Int J Pharm* 463 (1):81–8.

Wang, Y., and D. S. Kohane. 2017. "External triggering and triggered targeting strategies for drug delivery." *Nat. Rev. Mater* 2:17020.

Wei, Y., L. Quan, C. Zhou, and Q. Zhan. 2018. "Factors relating to the biodistribution & clearance of nanoparticles & their effects on *in vivo* application." *Nanomedicine* 13 (12):1495–512.

Wilhelm, S., A. J. Tavares, Q. Dai, S. Ohta, J. Audet, H. F. Dvorak, and W. C. W. Chan. 2016. "Analysis of nanoparticle delivery to tumours." *Nat Rev Mater* 1 (5):16014.

Xu, W., K. Kattel, J. Y. Park, Y. Chang, T. J. Kim, and G. H. Lee. 2012. "Paramagnetic nanoparticle T1 and T2 MRI contrast agents." *Phys Chem Chem Phys* 14 (37):12687–700.

Yan, L., A. Amirshaghaghi, D. Huang, J. Miller, J. M. Stein, T. M. Busch, Z. Cheng, and A. Tsourkas. 2018. "Protoporphyrin IX (PpIX)-coated superparamagnetic iron oxide nanoparticle (SPION) nanoclusters for magnetic resonance imaging and photodynamic therapy." *Adv Funct Mater* 28 (16):1707030.

Yanagioka, M., H. Kurita, T. Yamaguchi, and S.-I. Nakao. 2003. "Development of a molecular recognition separation membrane using cyclodextrin complexation controlled by thermosensitive polymer chains." *Ind Eng Chem Res* 42 (2):380–5.

Yang, J., P. Zhang, L. Tang, P. Sun, W. Liu, P. Sun, A. Zuo, and D. Liang. 2010. "Temperature-tuned DNA condensation and gene transfection by PEI-g-(PMEO2MA-b-PHEMA) copolymer-based nonviral vectors." *Biomaterials* 31 (1):144–55.

Yang, Z., J. Song, Y. Dai, J. Chen, F. Wang, L. Lin, Y. Liu et al. 2017. "Self-assembly of semiconducting-plasmonic gold nanoparticles with enhanced optical property for photoacoustic imaging and photothermal therapy." *Theranostics* 7 (8):2177–85.

Ye, Y., and X. Chen. 2011. "Integrin targeting for tumor optical imaging." *Theranostics* 1:102–26.

Zeng, Z., Z. Peng, L. Chen, and Y. Chen. 2014. "Facile fabrication of thermally responsive pluronic F127-based nanocapsules for controlled release of doxorubicin hydrochloride." *Colloid Polym Sci* 292 (7):1521–30.

Zhang, Q., C. Weber, U. S. Schubert, and R. Hoogenboom. 2017. "Thermoresponsive polymers with lower critical solution temperature: from fundamental aspects and measuring techniques to recommended turbidimetry conditions." *Mater Horiz* 4 (2):109–16.

Zhang, Y., J. Yu, H. N. Bomba, Y. Zhu, and Z. Gu. 2016. "Mechanical force-triggered drug delivery." *Chem Rev* 116 (19):12536–63.

Zheng, M., T. Jiang, W. Yang, Y. Zou, H. Wu, X. Liu, F. Zhu et al. 2019. "The siRNAsome: a cation-free and versatile nanostructure for siRNA and drug co-delivery." 58 (15):4938–42.

Zhong, J., S. Yang, L. Wen, and D. Xing. 2016. "Imaging-guided photoacoustic drug release and synergistic chemo-photoacoustic therapy with paclitaxel-containing nanoparticles." *J Control Release* 226:77–87.

4 C₆₀-Fullerenes as an Emerging Cargo Carrier for the Delivery of Anti-Neoplastic Agents
Promises and Challenges

Nagarani Thotakura and Kaisar Raza
Central University of Rajasthan

CONTENTS

4.1 Introduction 49
4.2 Synthetic Methods of C₆₀Fs 50
    4.2.1 Arc Vaporisation Technique 51
    4.2.2 Laser Ablation...51
    4.2.3 Hydrocarbon Combustion 51
    4.2.4 Other Methods 51
4.5 Functionalisation of C₆₀Fs 51
    4.5.1 Nucleophilic Addition Reactions 52
        4.5.1.1 Bingel Reactions 52
        4.5.1.2 Addition of Ylides 52
    4.5.2 Cycloaddition Reactions 52
    4.5.3 Miscellaneous Functionalisations...52
4.6 Medical Applications of C₆₀Fs...53
    4.6.1 Antioxidant and Neuroprotective Agent 53
    4.6.2 Antiviral Agent..53
    4.6.3 Photodynamic Therapy...53
    4.6.4 Anti-Inflammatory Agent 54
    4.6.5 Antitumour Agent...54
4.7 C₆₀Fs as Drug Delivery Carriers to Neoplastic Cells 54
4.8 Toxicity Profile and the Challenges of C₆₀Fs..55
4.9 Conclusions...56
Conflict of Interest 56
References...56

4.1 INTRODUCTION

Delivery of drug molecules of more significant size generally poses a considerable challenge such as poor solubility, which indirectly affects the absorption and bioavailability of the drug (Khadka et al. 2014). This also results in *in vivo* instability, adverse effects and issues with targeted drug delivery (Williams et al. 2013). The advent of nanotechnology has played a vital role in the development of novel drug delivery systems (NDDS), which possess drug targeting potential along with the controlled drug release (Lam et al. 2017). The carriers employed for the NDDS can be either organic or inorganic. Various nanocarriers such as nanoparticles, liposomes, polymeric micelles and mixed micelles have been used for the drug delivery process (Din et al. 2017). On the other hand, carbon-based nanocarriers such as graphene oxide (GO), carbon nanotubes (CNTs) and C₆₀-fullerenes (C₆₀Fs) shown in Figure 4.1 have also been explored for better drug delivery. Table 4.1 shows the differences between these three carbon nanocarriers.

C₆₀Fs belong to the family where the molecule is purely made up of carbon atoms. There are several molecules where the number of carbon atoms varies from 36 to many thousands (Wudl, 2002). In 1985, another allotrope of the carbon family, namely fullerenes, was discovered. They are made of hexagonal and pentagonal rings containing sp² hybridised carbon atoms at each corner of the ring (Kumar and Kumbhat 2016). Due to the strain in each pentagon, there is curvature or out-of-plane deformation near the pentagon. This curvature results in the formation of characteristic cage property of fullerenes. The stable form can be achieved by

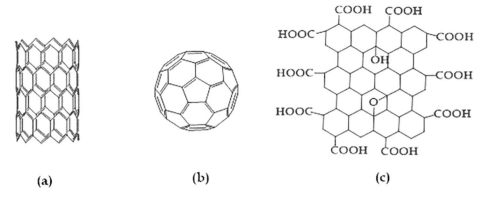

FIGURE 4.1 Basic structures of (a) CNTs, (b) C_{60}-fullerenes and (c) GO.

TABLE 4.1
Comparison of C_{60}Fs, CNTs and GO

Properties	C_{60}Fs	CNTs	GO
Crystal form	Tetragonal	Icosahedral	Hexagonal
Dimension	Non-dimensional	One-dimensional	Two-dimensional
Surface area (m²/g)	80–90	Approx. 1,300	Approx. 1,500
Density (g/cm³)	1.72	Greater than 1	Greater than 1
Optical property	Non-linear	Depends on the structure	97.7% of optical transmittance
Electronic property	Insulator	Metallic and semiconducting	Semi-metal, zero-gap semiconductor
Tenacity	Elastic	Flexible elastic	Flexible elastic

the implementation of the isolated pentagon rule, which means no two pentagons should be adjacent to each other (Schwerdtfeger, Wirz, and Avery 2015). In any type of fullerenes, the number of pentagons is limited to 13, which results in the achievement of the curvature enough for the formation of perfect football shape. However, the hexagon ring count may vary based on the carbon count in the type of fullerene. This variation is dependent on equation 4.1 shown below. In a molecule of C_{60}F, a total of 60 carbon atoms are arranged within 20 hexagons and 12 pentagons. The cage- or ring-like structure entirely depends on the arrangement of the pentagonal rings. These rings are distributed as per the isolated pentagon rule, resulting in the formation of the sphere, as mostly observed in the C_{60} isomers (Tan et al. 2009).

$$\text{No. of Hexagons} = \frac{(\text{Carbon atoms} - 20)}{2} \quad (4.1)$$

The two well-known isomers of fullerenes include C_{60} and C_{70}, respectively. C_{60}Fs are also known as buckminsterfullerenes. This is named after the eminent scientist R. Buckminster Fuller, as he designed the geodesic domes that have structural resemblance with a molecule of C_{60}. These two molecules can attain the isolated pentagon rule only in their single configuration. These configurations are known to be highly stable and obtained for usage (Kroto et al. 1985).

These fullerenes having the carbon atoms from 20 to 300 are identified to date. The fullerenes that contain the carbon atoms less than 300 exist as the single-shelled structures. These are also known as buckyballs (Stankevich, Nikerov, and Bochvar 1984). If metals such as lanthanum, cerium and xanthium are encapsulated within these individual shells, they are known as metallofullerenes or endohedral fullerenes (Thakral and Mehta 2006). If the carbon number is more than 300, they can be either single- or multi-shelled, named as giant fullerenes. The multi-shelled structures are known as the onions as they resemble the daily used onions of the kitchen. Giant fullerenes can form the nanotubes, which are named as buckytubes. These tubes also have a similar structural backbone as fullerenes (Georgakilas et al. 2015).

These C_{60}Fs are highly stable and lipophilic, which is vital for the smooth crossing of the cell membranes. For the delivery of various drug molecules or drugs with the biomarker or drug with targeting agents, they are multi-functionalised (Montellano et al. 2011). In comparison with other carbon nanomaterials, C_{60}Fs can be selective due to advantages in the point of their chemical reactivity and the loading potential along with the controlled pattern of the drug release (Mendes et al. 2013). By embellishing the multifunctional uses of these cage-like structures, there can be better development of a derivative with a realistic biomedical application (Patel, Singh, and Kim 2019).

4.2 SYNTHETIC METHODS OF C_{60}FS

There are various methods for the synthesis of C_{60}-fullerenes. They are mentioned in Figure 4.2.

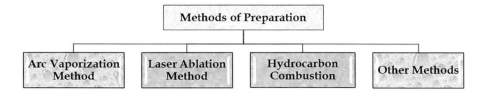

FIGURE 4.2 Pictorial representation of multiple methods employed for the synthesis of fullerenes.

4.2.1 Arc Vaporisation Technique

The biomedical applications of fullerenes were revolutionised by the discovery that pure resistive vaporisation of graphite rods could produce fullerenes in substantial yield (Weston and Murthy 1996). This procedure, which is often termed the Kratschmer–Huffman method, is a simple and inexpensive help in preparing huge amounts of carbon soot, which contains fullerenes. In this method, two high-purity graphite rods are clipped together, and high current is passed through the chamber along with helium gas to the pressure of 150–250 torr (Kyesmen, Onoja, and Amah 2016). As water and oxygen are the two materials that extensively inhibit the preparative method, the most important thing to be done is clearing the oxygen present in the chamber with the help of a vacuum and filling it with pure helium. Electrodes are positioned so that the carbon rods are just touching, and then, vaporisation is initialised through the movement of high current between two rods. The crude carbon product or soot formed by this vaporisation method is collected onto the water-cooler inner surface and is readily scratched off from the walls by using a stiff brush. This soot contains a mixture of carbon products, including $C_{60}Fs$ and giant fullerenes belonging to their family (Churilov 2008).

4.2.2 Laser Ablation

The next technique, which is very useful and powerful for preparing fullerene clusters, involves laser ablation of graphite in a helium atmosphere (Buseck 2002). Traditionally, laser ablation has to be considered as the foremost technique, which helped in the preparation of fullerene clusters in the gas phase, because the dense carbon plasma created during laser vaporisation cools too quickly (Ying et al. 1996). The rapidly cooling plasma does not provide sufficient time for the growth of carbon fragments to rearrange into stable closed fullerene structures. In this, graphite is ablated within a high-temperature furnace. By carrying out ablation at an elevated temperature, the plasma cools very slowly, resulting in the growth of carbon clusters. They even have sufficient time to rearrange (anneal) into stable fullerenes. It is observed that ablation at low temperatures reduces the yield of fullerenes. So, to avoid this reduction in the yield, the temperature of the furnace is maintained not less than 1,200°C (Itagaki et al. 2019).

4.2.3 Hydrocarbon Combustion

The production of fullerenes from hydrocarbon combustion was first reported by Howard and co-workers in 1991. Notably, the combustion of benzene or any hydrocarbon–oxygen mixtures around 1,200°C in laminar-flow flames produced significant quantities of fullerenes by the well-known process named pyrolysis (Mckinnon, Bell, and Barkley 1992).

4.2.4 Other Methods

Several other techniques, including inductively coupled RF evaporation of graphite targets, electron beam evaporation, low-pressure helium sputtering and microwave technologies, are being used for the preparation of fullerenes and other carbon products (Mojica, Alonso, and Méndez 2013a). Of all the methods, it was observed that only the combustion technique is sufficiently developed to be considered as a general method for the production of macroscopic quantities of fullerenes (Nimibofa et al. 2018).

4.5 FUNCTIONALISATION OF $C_{60}Fs$

One of the significant challenges faced in the usage of fullerenes is their low solubility in most of the commonly employed solvent systems. This indirectly affects the processability by a substantial decrease in the usage graph (Schur et al. 2008). There is a need to distract various solvent–solvent interactions for increasing the solvation of $C_{60}Fs$. But these interactions can't be that easily disrupted as there is no permanent dipole moment along with the rigid geometric structure of $C_{60}Fs$ (Chaban, MacIel, and Fileti 2014). Buckyballs are almost insoluble in polar and dipolar aprotic solvents such as acetonitrile, dimethyl sulfoxide (DMSO), methanol and tetrahydrofuran (THF). It shows sparingly soluble properties in alkanes with low molecular weight. As the non-polar nature increases, the solubility of $C_{60}Fs$ increases. In tetrachloroethylene and tetrachloroethane, the solubility is substantially high in comparison with alkanes (Marcus et al. 2001). Due to the great electron-accepting nature of the $C_{60}Fs$, they form charge transfer complexes in the solvents that have electron-donating property. Some examples of these electron donor solvents include anilines and tertiary amines (Sivaraman et al. 1992). The solubility of $C_{60}Fs$ in water and biological fluids is negligible, which results in the development of various functionalised derivatives of fullerenes through different chemical reactions (Lahir 2017), which are discussed below.

4.5.1 Nucleophilic Addition Reactions

This type of reaction includes Bingel-type reactions where α-carbanions are involved. Another well-known example is the addition of ylides, which are discussed in detail.

4.5.1.1 Bingel Reactions

In this reaction, C_{60}Fs are treated with diethyl bromomalonate in the presence of sodium hydride. The mechanism involved in the reaction is shown in Figure 4.3. Initially, α-halo carbanion formed by the deprotonation of the α-bromomalonate is added to the anionic fullerene. Later, the bromide group is displaced by the nucleophilic substation resulting in the methanofullerenes (Thong et al. 2016). In this reaction, along with sodium hydride, 1,8-diazabicyclo [5.4.0] undec-7-ene (DBU) can also be used as a catalyst. Both catalysts are bases with low nucleophilicity, preventing the additions over the surface of fullerenes. This reaction occurs exclusively at 6-6 bonds of the fullerene (Li et al. 2006). When the attachment of the complex structures to the core of fullerene is planned, this bromomalonate derivative precursor can reduce the yield of the final product (Yan et al. 2015). Modifications are done in the reaction such as preparation of the α-halomalonate from the malonate source along with the use of a non-nucleophilic base. Another change involves the combination of DBU and CBr_4 as catalyst and source of bromine to form α-bromomalonates. These modified reactions are performed under a mild environment with the involvement of different functional groups. The methanofullerenes obtained from this reaction are highly stable with a rigid cyclopropane ring and distinctly directed ester groups. Bingel reaction can be applied for the synthesis of a large variety of methanofullerenes such as 3-bromopenta-1,4-dienes, α-bromocyanoacetates, chloroketones and α-bromonitromethanes (Mateo-Alonso, Bonifazi, and Prato 2006).

Even though there are various advantages in the usage of Bingel cyclopropanation reaction, its use is restricted due to the subtle nature of the fullerene-fused di(alkoxycarbonyl) cyclopropane ring to reductive conditions. It is also reported that the obtained final product is unstable in a state like reductive electrochemistry (Nuretdinov et al. 2000). This results in the retro-Bingel reaction. This reaction removes the malonate resulting in the naïve fullerenes and the malonates used for Bingel reaction. These retro-reactions are even catalysed in the presence of amalgamated magnesium (Moonen et al. 2000). A modified Bingel reaction is preferred to synthesise methanofullerene derivatives to avoid the challenges as mentioned above (Yanilkin et al. 2002).

4.5.1.2 Addition of Ylides

When C_{60}Fs react with phosphonium ylides, it yields in the methanofullerenes, as shown in Figure 4.4. This reaction mechanism involves two steps. The first step is the nucleophilic addition at the carbanionic centre. This step is followed by the intramolecular nucleophilic substitution reactions (SN1), resulting in the formation of the cyclopentane ring. The yield of the reaction generally ranges from 15% to 25%. The same reaction can be carried out using sulphonium ylides, with a good yield of 30%–85% (Figure 4.4) (Biglova and Mustafin 2019).

4.5.2 Cycloaddition Reactions

C_{60}Fs undergo cycloaddition reactions due to their behaviour as 1,3-dipolarophiles and electron-deficient dienophiles (Sliwa 1995). The subtypes of these cycloaddition reactions include [1+2], [2+2], [3+2] and [4+2]. All these reactions take place at 6-6 ring junctions of fullerene with double bonds. There are a large number of compounds that can be synthesised using these reactions (Prato et al. 1993). The addition of carbenes and nitrenes results in methanofullerenes and fulleroaziridines [60]. Fullerene oligomers are also synthesised using these reactions (Mojica, Méndez, and Alonso 2013b).

4.5.3 Miscellaneous Functionalisations

Furthermore, fullerenes are encapsulated into unique molecules such as calixarenes, cyclodextrins, polyvinylpyrrolidone and various nanocarriers such as liposomes and micelles. The combination of lipid membranes and C_{60}Fs showed exciting aspects for the controlled delivery of drugs in the biological systems (Goodarzi et al. 2017). Suspensions of C_{60}Fs are prepared using the co-solvent method. In this method, initially, fullerenes are dissolved into benzene or toluene, which was poured into THF. This mixture is added into the acetone drop-wise, followed by water. All

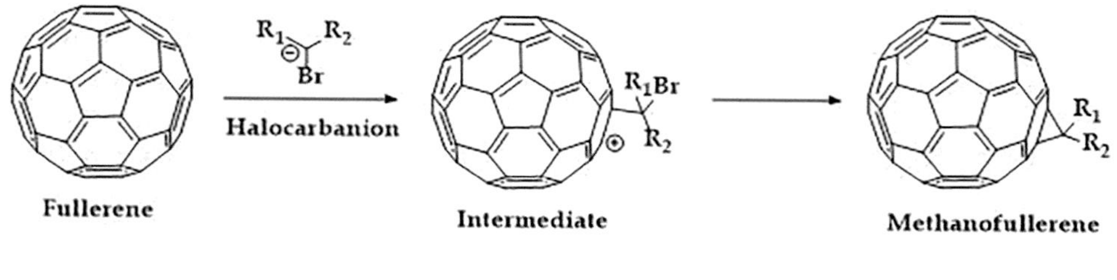

FIGURE 4.3 Pictorial representation of the Bingel reaction.

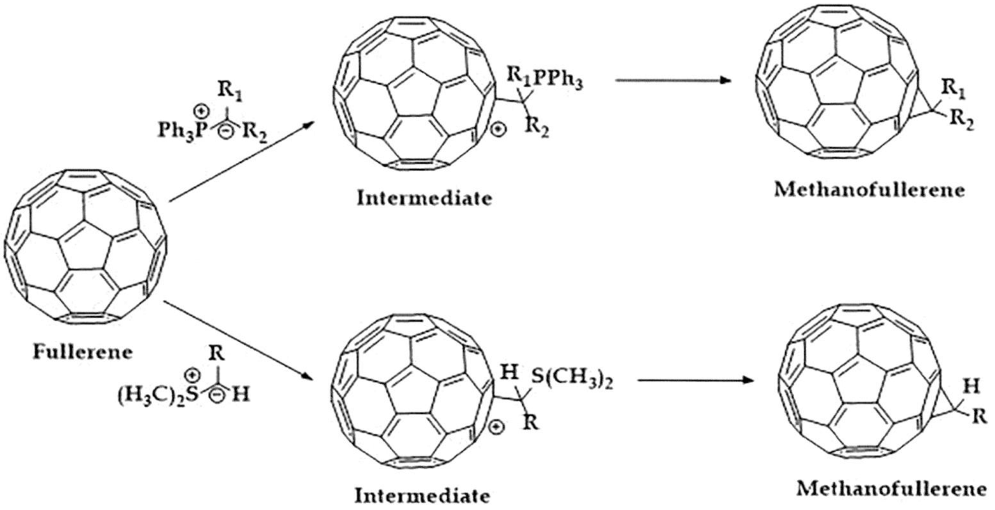

FIGURE 4.4 Scheme showing the nucleophilic addition of (a) phosphonium and (b) sulphonium ylides continued by the nucleophilic substitution.

the organic solvents are evaporated, resulting in the aqueous suspensions of C_{60}Fs (Ma and Bouchard 2009). Other than these, certain chemical functionalisations such as the addition of amino acids and polymers, hydroxylation and carboxylation are also observed to improvise the solubility of these carbon moieties (Kumar and Raza 2017).

4.6 MEDICAL APPLICATIONS OF C_{60}Fs

Both the naive fullerenes and their derivatives are reported for their various pharmacological effects. Various biomedical applications of C_{60}Fs are discussed in detail. Table 4.2 shows the marketed formulations of C_{60}Fs (Bakry et al. 2007).

4.6.1 Antioxidant and Neuroprotective Agent

As fullerenes has ability to uptake six electrons in the solution form. Scientists proved that the water-soluble fullerene derivatives are capable of the neuroprotective action by studying in an amyloid-β (Aβ) rat model of Alzheimer's disease (AD) (Vorobyov et al. 2015). They suggested the mediation through the pre-synaptic dopamine receptors. Reports are showing that the polar derivatives of C_{60}Fs can show the effective pharmacological response in various chronic and acute neurodegenerative diseases (Dugan et al. 1997).

4.6.2 Antiviral Agent

Several reports are showing the antiviral potential of C_{60}Fs and their derivatives. The antioxidant potential and the unique molecular design of fullerenes are responsible for the antiviral activity (Sijbesma et al. 1993). Fullerenes are capable of making complexes with HIV protease (HIV-P), making it inactive (Friedman et al. 1993). Derivatives of fullerenes also show antiviral properties when the substituents are at the trans-2 position with positive charges near to it (Marchesan et al. 2005). Falynskova and co-workers synthesised fullerene-(tris-aminocaproic acid) hydrate (FTACAH) and reported it to be active against the respiratory syncytial virus (RSV) (Falynskova et al. 2014). Shoji and his team prepared 12 derivatives and screened them for inhibiting endonucleases by an *in vitro* PA endonuclease inhibition assay. They have performed docking simulation analysis of C_{60}F derivatives and PA endonucleases, which proved the binding of fullerene derivatives to the active site of the endonuclease. It is also known that C_{60}Fs can be used for drug delivery in the treatment of influenza (Shoji et al. 2013).

4.6.3 Photodynamic Therapy

C_{60}Fs are capable of production of reactive oxygen species (ROS) in the biological system, which can be the reason for the development of various physiological disorders (Mroz

TABLE 4.2
Marketed Formulations of C_{60}Fs

Product Name	Manufacturer	Applicability
NOVA C_{60}	NOVA C_{60} Skin Solutions	Anti-wrinkling, anti-ageing and anti-inflammatory effects
LipoFullerene™	Vitamin C_{60} Bioresearch Corporation	Anti-wrinkling and anti-acne effects
Vitamin C_{60} Radical Sponge®	Vitamin C_{60} Bioresearch Corporation	Sunscreen activity along with skin-whitening nature

et al. 2007). Fullerenes have right π-electron conjugation resulting in the absorbance of the visible radiations with a long triplet yield (Yin et al. 2014). This property raised the role of fullerenes in photodynamic therapy (PDT). PDT involves the use of photosensitisers and light sources to generate ROS, which kills tumour cells (Sharma, Chiang, and Hamblin 2011). Based on the nature of the derivatisation of fullerene, molecules can inactivate the microorganisms and tumour cells effectively (Rašović 2017). Under appropriate conditions, $C_{60}Fs$ can be more effective photosensitisers than the marketed ones that are used for the management of certain diseases (Partha and Conyers 2009).

4.6.4 Anti-Inflammatory Agent

Inflammation caused due to the ROS production resulting in some allergic conditions can be controlled by using the water-soluble functionalised fullerenes (Roursgaard et al. 2008). Its anti-inflammatory effects against atopic dermatitis were verified in various animal models, both subcutaneously and epicutaneously. They found that this approach can be a better substitute for treating allergies and inflammatory diseases (Shershakova et al. 2016).

4.6.5 Antitumour Agent

$C_{60}Fs$ are well known for their anti-neoplastic and anti-apoptotic activity. Researchers proved that the carboxylic acid derivatives of fullerenes are useful in the prevention of apoptosis of hepatic tumour cells by the neutralisation of ROS produced by transforming growth factor (TGF-b) (Roursgaard et al. 2008). Various derivatives of fullerenes possess anticancer potential (Mashino et al. 2003). Derivatives such as fullerenols (FLUs) (Zhu et al. 2008), multi-hydroxylated endohedral metallofullerenol $[Gd@C_{82}(OH)_{22}]_n$ (Meng et al. 2010), epoxidised estradiol C_{60} conjugate (Pérez-Manríquez et al. 2013) and C_{60}-bis(N,N-dimethylpyrrolidinium iodide) (Raoof et al. 2012) are reported for the *in vitro* and *in vivo* anticancer activity.

4.7 $C_{60}FS$ AS DRUG DELIVERY CARRIERS TO NEOPLASTIC CELLS

Cancer is known as the frenzied proliferation of body cells. This may be due to various reasons such as exposure to toxic chemicals, genetic changes in the human biological systems, and ionising radiation. Multiple drugs have been used for cancer chemotherapy, but the main challenge in the usage of these drugs in chemotherapy is their side effects. This can be overlooked by targeted drug delivery. Another problem faced in the treatment process is drug resistance, which is reduced by the use of multiple drugs. This further increased the side effects as the drug number and the concentrations are rising within the biological system (Blackadar 2016).

Till now, there are reports of various nanoparticulate systems developed for the site-specific delivery of chemotherapeutic drugs (Thotakura et al. 2013; Bhatia et al. 2012, 2013; Raza, Kumar, Kiran et al. 2016; Yadav et al. 2016; P. Kumar, Raza et al. 2016; Thotakura et al. 2017; Raza, Kumar, Misra et al. 2016; P. Kumar, Sharma et al. 2016; Thakur et al. 2016; Madhwi et al. 2017; Singh et al. 2017; Thotakura et al. 2017a; P. Kumar, Kumar, et al. 2017; P. Kumar, Sharma et al. 2017; M. Kumar, Sharma, Misra et al. 2018; Thotakura et al. 2019; Singh et al. 2019; Harsha et al. 2019; Thotakura et al. 2017b). Among them, some, such as Doxil®, a liposomal injection of doxorubicin, are available on the market (Duggan et al. 2011). One of the nanoparticulate systems includes $C_{60}Fs$, which possess some strong structural properties. As these cargo vehicles are capable of carrying multiple drugs, they can avoid the drug resistance along with the targeted action. This helps in increasing the efficacy with a reduction in side effects (Goodarzi et al. 2017). Due to the cardiotoxicity, doxorubicin is conjugated with fullerenes with 100% drug release at pH 5. These results prove that the conjugation helps in achieving the drug levels at the target site along with decreased challenges of naïve drug. The hydrophilic drug, doxorubicin, is conjugated with the hydrophobic carrier, $C_{60}Fs$, using ethylene glycol as a linker. The final conjugate is found to be having water-soluble properties. In another report, scientists used buckysomes to conceal the hydrophilic surface of paclitaxel. This conjugate helped in enhancing the absorption of the drug in the biological system (Lu et al. 2009).

Based on the 'on–off' drug delivery technique, a novel conjugate is developed. In this, the conjugate of doxorubicin and fullerene is coated with the hydrophilic agent to improve the solubility. This novel drug conjugate is found to be highly stable at the physiological conditions, which are the off state. Later, when the on state is activated, the fullerenes, antioxidant nature and the drug's anti-neoplastic nature showed the pharmacological action in two different pathways. The first step is the generation of the ROS, which results in cell death. The second pathway includes the drug release at the target site. Among various nanocarriers, fullerenes and their derivatives can help in better drug delivery with safety and targeted nature (Kazemzadeh and Mozafari 2019). These cargo vehicles can be a precious invention in effective chemotherapy and PDT if proper evaluation studies are performed. Some of the research ideas of the 'Fullerenes and their derivatives in cancer chemotherapy' of Raza and his team are discussed in the below paragraphs.

Kumar et al. (2018) prepared monomethyl fumarate-conjugated lysine-tagged $C_{60}F$ (CF-LYS-TEG-MMF) for the delivery to brain cancer cells via the Prato reaction. This conjugate was found to be erythrocyte biocompatible with increased cytotoxic action against neuroblastoma cell lines. The drug release from the nanoconjugate was pH-dependent with minimal amount of drug at the normal physiological pH of 7.4. This helped in the release of the drug at the specific site, i.e. cancer cells with lower pH range. Through pharmacokinetic study in rodents, they observed that there is significant enhancement in the bioavailability of drug at the central compartment, plasma concentrations and bio-residence of the drug along with the reduced rate of clearance. All these results in combination helped in proving the role of CF-LYS-TEG-MMF in the efficient drug delivery to the brain, with reduced dose and side effects (M. Kumar, Sharma, Kumar et al. 2018).

Thotakura et al. (2018) synthesised docetaxel-conjugated aspartic acid-derivatised hydroxylated fullerenes. This study was planned in order to deliver docetaxel to the cancer cells with improved efficacy. The conjugate was in nanometric range, which was supported by the micromeritics and zeta potential studies. Pharmacokinetic evaluation of the fullerene conjugate resulted in the enhanced bioavailability and residence time in the biological system. They reported that the high plasma profile and longer durations of the drug through the nanoconjugate can fairly improve the access at the site of action along the dose reduction. These are most desired in the chemotherapy, which can be achieved by aspartic acid-linked FLUs (Thotakura et al. 2018).

Bahuguna et al. (2017) synthesised water-soluble FLUs and used them for the delivery of methotrexate (MTX). This group reported that the water-soluble FLUs synthesis was followed by the conjugation of MTX using ester linkage, which is stable at the normal physiological pH and hydrolysable at the intracellular level. Low protein binding was observed along with increased drug loading. The conjugate showed pH-dependent drug release and reduced IC_{50} values indicating the increased efficacy, in comparison with the naïve drug. The results of cell viability were further confirmed by the confocal laser scanning microscopy, indicating the penetration of the nanoconjugate into the cell cytoplasm as well as nucleoplasm. FLU-MTX offered enhanced availability, residence time and half-life with the reduced rate of elimination in rodents (Bahuguna et al. 2018).

Joshi et al. (2017) employed amination for the functionalisation of fullerenes which are further conjugated with MTX. C_{60}Fs underwent 1,3-cycloaddition reaction in the presence of glycine and paraformaldehyde. More amount of drug release was found at cancerous cell pH, rather than the physiological pH. The nanoconjugate was found to be haemocompatible with enhanced cytotoxicity. The cellular uptake of the conjugate supports the cell viability assay. Even the pharmacokinetic study proved the availability, and the half-life of the drug is increased with reduced clearance. They reported that these observations accustomed the prospective of fullerenes to the delivery of chemotherapeutics more safely and effectively (Joshi et al. 2017).

Misra et al. (2017) synthesised fullerenes that are tethered to glycine for the delivery of docetaxel. They characterised the synthesised conjugates using FT-IR and NMR. They found that the conjugate was in nanometric size range with the ability to release the drug at the cancer cell due to the pH-dependent drug release profile of the drug from the conjugate. There was substantial enhancement in the cytotoxicity profile against MDA-MB-231 cells. Even the developed nanoconjugate system can be tolerated by the erythrocytes. The internalisation of the conjugate into the cytoplasm and the nucleus was proved by the use of confocal laser scanning microphotographs. Pharmacokinetic studies also showed the desirable results along with the low protein binding of the drug. In conclusion, they mention that 'The glycinated fullerenes can serve as promising "cargo vehicles" for the delivery of anti-cancer drugs safely and effectively' (Misra, Thotakura et al. 2017).

Misra, Kumar et al. (2017) prepared glycinated fullerenes for intracellular delivery of tamoxifen with improved anticancer activity and pharmacokinetics. In this study, C_{60}Fs were glycinated, and with the linker, N-desmethyl tamoxifen was conjugated. The present findings indicate the superiority of glycine-tethered N-TAM over the plain drug, not only in terms of safety and efficacy but also on the pharmacokinetic fronts. The in vitro and preclinical data unequivocally demonstrate the promises of such surface-decorated C_{60} nanocarriers in enhancing the desirable outcomes in cancer chemotherapy. These solvent- and surfactant-free nanoconstructs offer a new future in the domain of drug delivery, which is not yet thoroughly explored. There should be more efforts to explore the potential of amino acid-tagged C_{60}F derivatives for various anticancer agents' site-specific delivery. These systems offer a poisonous bait to the cancer cells, as the need for an amino acid is tremendous in the overgrowing cancerous cells. The amino acid-tagged anticancer drug-loaded nanocarriers get quickly engulfed by cancer cells and offer a considerable promise in cancer targeting (Misra, Kumar et al. 2017).

Raza et al. (2015) synthesised carboxylated fullerenes by the reductive oxidation followed by the conjugation of docetaxel. They mentioned that the use of this nanoconjugate system helped in achieving the enhanced cytotoxicity along with the improvised pharmacokinetic profile. Blood compatibility of the systems is high with the demand of the further exploration in the suitable animal models to draw a hypothesis of its use for mankind. They stated that 'However, this approach offers a promise which can be extrapolated to other anticancer agents with similar challenges' (Raza et al. 2015).

4.8 TOXICITY PROFILE AND THE CHALLENGES OF C_{60}FS

Along with efficacy and targeted activity towards the diseased site, there are specific issues to be handled. Along with the theranostic properties, it was reported that C_{60}Fs possess tissue toxicity, which should be carefully addressed before using these carriers for the futuristic clinical applications (Yuan et al., 2019). Distribution and degradation of fullerenes within the biological systems are correlated with their biocompatibility issues. Through in vivo studies, it was found that these materials are degradable based on the physicochemical properties and their type (Zhang et al. 2018). In most cases, it was noticed that these carriers are excreted from the body through the renal clearance without the degradation. The degradation products of these carriers are carbon species (Bhattacharya et al. 2016). However, various reports mentioned that toxicity of using these carriers is due to the trace elements used during their synthesis. To overcome this toxicity issue, the purity of these carriers has to be increased. The surface properties of these nanocarriers determine the dispersibility and biocompatibility within the biological systems (Cha et al. 2013). Modification of $C_{60}Fs$ by the functionalisation results in the enhancement of the dispersibility in the aqueous solvents, which directly relates to the biocompatibility of these carriers. Another vital factor to be considered during the biological use of these vehicles is their biodistribution (Magrez et al. 2006). In the recent development of technology, scientists used label-free mass spectroscopy imaging for the detection of distribution of these carbon nanocarriers in the rodent models employing intrinsic carbon cluster fingerprint signal (Madannejad et al. 2019). However, this method could not determine the same if these particles were administered *in vivo*. Henceforth, radiolabelling of carriers was used for the determination of biodistribution (Wang et al. 2013).

The first water-soluble fullerenes synthesised were polyhydroxylated gadolinium metallofullerene nanocrystals (GFNCs). After the *in vivo* evaluation studies of GFNCs, it was found that they can be easily excreted from the body within a few days reducing the toxicity (Guan et al. 2016). Scientists continued this study further to know the long-term toxic effects using various concentrations of GFNCs (Sitharaman et al. 2004). On functionalisation of these GFNCs with PEG, through histopathology and biochemical parameter analysis, it was found that there was a significant decrease in the low term toxicity (Chen et al. 2012). Biodistribution studies were performed in rats using carboxylated fullerenes, which are ^{14}C radiolabelled. After administration through the intravenous route, the nanocarrier reached all the significant organs without any toxicity, but by injecting the same formulation through the intraperitoneal route, tissue toxicity was observed. It was concluded that the administration of a high amount of fullerenes could cause inflammatory issues at a particular site, but dilution of concentration through mixing with the bloodstream can minimise these conditions (Wang et al. 2016). As fullerenes are found to be lipophilic, there may be a chance of a reduced rate of excretion, which leads to the accumulation in the internal organs, resulting in the long-term toxicity (Zhao, Striolo, and Cummings 2005).

There is no proper conclusion regarding the toxicity and biodistribution of $C_{60}Fs$, even after performing numerous studies. We can notice that the toxicity and biocompatibility issues mainly depend on the physicochemical properties such as size, shape and functional groups on the surface of cargo vehicles. Even the biological interactions, the formation of the corona and aggregation within the organs affect these parameters. Therefore, there is a need for a complete investigation of *in vitro* and *in vivo* parameters such as planning of study in a large group of animals and determination of mechanisms involved in the biological interactions (Khan et al., 2017). These all factors helps in the irradiating theranostic properties of carbon-based nanocarriers, which results in better clinical applications to the future generation.

4.9 CONCLUSIONS

Even though the cosmeceutical and pharmaceutical applications of carbon nanocarriers in human life have been from time immemorial, the evidence is generated recently, proving their advantages in the treatment of various diseased conditions. Among them, $C_{60}Fs$ gained significant attention due to their potential in the delivery of the drug and genes. This helped in increasing the efficacy, pharmacokinetic potential and bioavailability of the anticancer agents. But the biocompatibility and toxicity of fullerenes are the major challenges in drug delivery. However, the precaution 'High doses make poison' should be taken in the usage of nanomaterials, even the carbon-based ones. The preclinical evidence of the $C_{60}Fs$' utilisation in drug delivery is encouraging, which is further boosted by the availability of the fullerene-based marketed products. If these cargo vehicles are explored, there will be further improvement in the clinical aspects which could benefit in developing patient-centric nanomedicine.

CONFLICT OF INTEREST

The authors confirm that this chapter content has no conflict of interest.

REFERENCES

Bahuguna, Shradha, Manish Kumar, Gajanand Sharma, Rajendra Kumar, Bhupinder Singh, and Kaisar Raza. 2018. "Fullerenol-Based Intracellular Delivery of Methotrexate: A Water-Soluble Nanoconjugate for Enhanced Cytotoxicity and Improved Pharmacokinetics." *AAPS PharmSciTech* 19 (3): 1084–92. doi:10.1208/s12249-017-0920-0.

Bakry, Rania, Rainer M. Vallant, Muhammad Najam-ul-Haq, Matthias Rainer, Zoltan Szabo, Christian W. Huck, and Günther K. Bonn. 2007. "Medicinal Applications of Fullerenes." *International Journal of Nanomedicine*. Dove Press.

Bhatia, Amit, Bhupinder Singh, Kaisar Raza, Anshuman Shukla, Basant Amarji, and Om Prakash Katare. 2012. "Tamoxifen-Loaded Novel Liposomal Formulations: Evaluation of Anticancer Activity on DMBA-TPA Induced Mouse Skin Carcinogenesis." *Journal of Drug Targeting* 20 (6): 544–50. doi:10.3109/1061186X.2012.694887.

Bhatia, Amit, Bhupinder Singh, Kaisar Raza, Sheetu Wadhwa, and Om Prakash Katare. 2013. "Tamoxifen-Loaded Lecithin Organogel (LO) for Topical Application: Development, Optimization and Characterization." *International Journal of Pharmaceutics* 444 (1–2): 47–59. doi:10.1016/j.ijpharm.2013.01.029.

Bhattacharya, Kunal, Sourav P. Mukherjee, Audrey Gallud, Seth C. Burkert, Silvia Bistarelli, Stefano Bellucci, Massimo Bottini, Alexander Star, and Bengt Fadeel. 2016. "Biological Interactions of Carbon-Based Nanomaterials: From Coronation to Degradation." *Nanomedicine: Nanotechnology, Biology, and Medicine* 2 (4). Elsevier Inc.: 639. doi:10.1016/j.nano.2015.11.011.

Biglova, Yulya N., and Akhat G. Mustafin. 2019. "Nucleophilic Cyclopropanation of [60]Fullerene by the Addition-Elimination Mechanism." *RSC Advances*. Royal Society of Chemistry. doi:10.1039/c9ra04036f.

Blackadar, Clarke Brian. 2016. "Historical Review of the Causes of Cancer." *World Journal of Clinical Oncology*. Baishideng Publishing Group Co., Limited. doi:10.5306/wjco.v7.i1.54.

Buseck, Peter R. 2002. "Geological Fullerenes: Review and Analysis." *Earth and Planetary Science Letters* 203 (3–4). Elsevier: 781–92. doi:10.1016/S0012-821X(02)00819-1.

Cha, Chaenyung, Su Ryon Shin, Nasim Annabi, Mehmet R. Dokmeci, and Ali Khademhosseini. 2013. "Carbon-Based Nanomaterials: Multifunctional Materials for Biomedical Engineering." *ACS Nano*. doi:10.1021/nn401196a.

Chaban, Vitaly V., Cleiton MacIel, and Eudes Eterno Fileti. 2014. "Solvent Polarity Considerations Are Unable to Describe Fullerene Solvation Behavior." *Journal of Physical Chemistry B* 118 (12). American Chemical Society: 3378–84. doi:10.1021/jp4116639.

Chen, Zhiyun, Lijing Ma, Ying Liu, and Chunying Chen. 2012. "Applications of Functionalized Fullerenes in Tumor Theranostics." *Theranostics*. doi:10.7150/thno.3509.

Churilov, G. Churilov. 2008. "Synthesis of Fullerenes and Other Nanomaterials in Arc Discharge." *Fullerenes Nanotubes and Carbon Nanostructures* 16:395–403. doi:10.1080/15363830802281641.

Din, Fakhar Ud, Waqar Aman, Izhar Ullah, Omer Salman Qureshi, Omer Mustapha, Shumaila Shafique, and Alam Zeb. 2017. "Effective Use of Nanocarriers as Drug Delivery Systems for the Treatment of Selected Tumors." *International Journal of Nanomedicine*. Dove Medical Press Ltd. doi:10.2147/IJN.S146315.

Dugan, Laura L., Dorothy M. Turetsky, Cheng Du, Doug Lobner, Mark Wheeler, C. Robert Almli, Clifton K.F. Shen, Tien Yau Luh, Dennis W. Choi, and Tien Sung Lin. 1997. "Carboxyfullerenes as Neuroprotective Agents." *Proceedings of the National Academy of Sciences of the United States of America* 94 (17): 9434–39. doi:10.1073/pnas.94.17.9434.

Duggan, Sean T., Gillian M. Keating, G. Ferrandina, J. P. Kesterson, D. Lorusso, and F. Muggia. 2011. "Pegylated Liposomal Doxorubicin: A Review of Its Use in Metastatic Breast Cancer, Ovarian Cancer, Multiple Myeloma and AIDS-Related Kaposis Sarcoma." *Drugs*. Springer International Publishing. doi:10.2165/11207510-000000000-00000.

Falynskova, I. N., Ionova, K. S., Dedova, A. V., Leneva, I. A., Makhmudova, N. R., and Rasnetsov. L. D. 2014. "Antiviral Activity of Fullerene-(Tris-Aminocaproic Acid) Hydrate against Respiratory Syncytial Virus in HEp-2 Cell Culture." *Pharmaceutical Chemistry Journal* 48 (2). Springer New York LLC: 85–88. doi:10.1007/s11094-014-1053-3.

Friedman, Simon H., Dianne L. DeCamp, George L. Kenyon, Rint P. Sijbesma, Gordana Srdanov, and Fred Wudl. 1993. "Inhibition of the HIV-1 Protease by Fullerene Derivatives: Model Building Studies and Experimental Verification." *Journal of the American Chemical Society* 115 (15). American Chemical Society: 6506–9. doi:10.1021/ja00068a005.

Georgakilas, Vasilios, Jason A. Perman, Jiri Tucek, and Radek Zboril. 2015. "Broad Family of Carbon Nanoallotropes: Classification, Chemistry, and Applications of Fullerenes, Carbon Dots, Nanotubes, Graphene, Nanodiamonds, and Combined Superstructures." *Chemical Reviews*. American Chemical Society. doi:10.1021/cr500304f.

Goodarzi, Saba, Tatiana Da Ros, João Conde, Farshid Sefat, and Masoud Mozafari. 2017. "Fullerene: Biomedical Engineers Get to Revisit an Old Friend." *Materials Today*. Elsevier B.V. doi:10.1016/j.mattod.2017.03.017.

Guan, Mirong, Jie Li, Qingyan Jia, Jiechao Ge, Daiqin Chen, Yue Zhou, Pengfei Wang et al. 2016. "A Versatile and Clearable Nanocarbon Theranostic Based on Carbon Dots and Gadolinium Metallofullerene Nanocrystals." *Advanced Healthcare Materials* 5 (17). Wiley-VCH Verlag: 2283–94. doi:10.1002/adhm.201600402.

Harsha, P. J., Nagarani Thotakura, Manish Kumar, Saurabh Sharma, Anupama Mittal, Rajneet Kaur Khurana, Bhupinder Singh, Poonam Negi, and Kaisar Raza. 2019. "A Novel PEGylated Carbon Nanotube Conjugated Mangiferin: An Explorative Nanomedicine for Brain Cancer Cells." *Journal of Drug Delivery Science and Technology* 53 (October). Elsevier BV: 101186. doi:10.1016/j.jddst.2019.101186.

Itagaki, H., Fujiwara, Y., Minowa, Y., kehara, Y., Kaneko, T., Okazaki, T., Iizumi, Y., Kim, J., and Sakakita, H. 2019. "Synthesis of Endohedral-Fullerenes Using Laser Ablation Plasma from Solid Material and Vaporized Fullerenes." *AIP Advances* 9 (7). American Institute of Physics Inc.: 075324-1–075324-27. doi:10.1063/1.5100980.

Joshi, Mayank, Pramod Kumar, Rajendra Kumar, Gajanand Sharma, Bhupinder Singh, Om Prakash Katare, and Kaisar Raza. 2017. "Aminated Carbon-Based 'Cargo Vehicles' for Improved Delivery of Methotrexate to Breast Cancer Cells." *Materials Science and Engineering C* 75: 1376–88. doi:10.1016/j.msec.2017.03.057.

Kazemzadeh, Houman, and Masoud Mozafari. 2019. "Fullerene-Based Delivery Systems." *Drug Discovery Today*. Elsevier Ltd. doi:10.1016/j.drudis.2019.01.013.

Khadka, Prakash, Jieun Ro, Hyeongmin Kim, Iksoo Kim, Jeong Tae Kim, Hyunil Kim, Jae Min Cho, Gyiae Yun, and Jaehwi Lee. 2014. "Pharmaceutical Particle Technologies: An Approach to Improve Drug Solubility, Dissolution and Bioavailability." *Asian Journal of Pharmaceutical Sciences*. Shenyang Pharmaceutical University. doi:10.1016/j.ajps.2014.05.005.

Kroto, Harold W., James R. Heath, Sean C. O'Brien, Robert F. Curl, and Richard E. Smalley. 1985. "C60: Buckminsterfullerene." *Nature* 318 (6042): 162–63. doi:10.1038/318162a0.

Kumar, Manish, and Kaisar Raza. 2017. "C60-Fullerenes as Drug Delivery Carriers for Anticancer Agents: Promises and Hurdles." *Pharmaceutical Nanotechnology* 5 (3). Bentham Science Publishers Ltd.: 169–79. doi:10.2174/2211738505666170301142232.

Kumar, Narendra, and Sunita Kumbhat. 2016. *Essentials in Nanoscience and Nanotechnology*. John Wiley & Sons, Inc. https://books.google.co.in/books?id=jVTWCwAAQBAJ&pg=PA192&lpg=PA192&dq=In+1985,+another+allotrope+of+the+carbon+family,+namely,+fullerenes,+was+discovered.+They+are+made+of+hexagonal&source=bl&ots=npIbo-M6rt&sig=ACfU3U3KNQcT09aWGX5o1BXioZF9qh0IOQ&hl=en&sa.

Kumar, Manish, Gajanand Sharma, Rajendra Kumar, Bhupinder Singh, Om Prakash Katare, and Kaisar Raza. 2018. "Lysine-Based C60-Fullerene Nanoconjugates for Monomethyl Fumarate Delivery: A Novel Nanomedicine for Brain

Cancer Cells." *ACS Biomaterials Science and Engineering* 4 (6). American Chemical Society: 2134–42. doi:10.1021/acsbiomaterials.7b01031.

Kumar, Manish, Gajanand Sharma, Charu Misra, Rajendra Kumar, Bhupinder Singh, Om Prakash Katare, and Kaisar Raza. 2018. "N-Desmethyl Tamoxifen and Quercetin-Loaded Multiwalled CNTs: A Synergistic Approach to Overcome MDR in Cancer Cells." *Materials Science and Engineering C* 89 (August). Elsevier Ltd: 274–82. doi:10.1016/j.msec.2018.03.033.

Kumar, Pramod, Rajendra Kumar, Bhupinder Singh, Ruchi Malik, Gajanand Sharma, Deepak Chitkara, O. P. Katare, and Kaisar Raza. 2017. "Biocompatible Phospholipid-Based Mixed Micelles for Tamoxifen Delivery: Promising Evidences from *In-Vitro* Anticancer Activity and Dermatokinetic Studies." *AAPS PharmSciTech* 18 (6). Springer New York LLC: 2037–44. doi:10.1208/s12249-016-0681-1.

Kumar, Pramod, Kaisar Raza, Lokesh Kaushik, Ruchi Malik, Shweta Arora, and Om Prakash Katare. 2016. "Role of Colloidal Drug Delivery Carriers in Taxane-Mediated Chemotherapy: A Review." *Current Pharmaceutical Design* 22. Bentham Science Publishers: 5127–43.

Kumar, Pramod, Gajanand Sharma, Rajendra Kumar, Ruchi Malik, Bhupinder Singh, Om Prakash Katare, and Kaisar Raza. 2016. "Promises of a Biocompatible Nanocarrier in Improved Brain Delivery of Quercetin: Biochemical, Pharmacokinetic and Biodistribution Evidences." *International Journal of Pharmaceutics* 515 (1–2). Elsevier B.V.: 307–14. doi:10.1016/j.ijpharm.2016.10.024.

Kumar, Pramod, Gajanand Sharma, Rajendra Kumar, Ruchi Malik, Bhupinder Singh, Om Prakash Katare, and Kaisar Raza. 2017. "Stearic Acid Based, Systematically Designed Oral Lipid Nanoparticles for Enhanced Brain Delivery of Dimethyl Fumarate." *Nanomedicine* 12 (23). Future Medicine Ltd.: 2607–21. doi:10.2217/nnm-2017-0082.

Kyesmen, Pannan Isa, Audu Onoja, and Alexander Nwabueze Amah. 2016. "Fullerenes Synthesis by Combined Resistive Heating and Arc Discharge Techniques." *SpringerPlus* 5 (1). SpringerOpen. doi:10.1186/s40064-016-2994-7.

Lahir, Yogendrakumar. 2017. "Impacts of Fullerene on Biological Systems." *Clinical Immunology, Endocrine & Metabolic Drugs* 4 (1). Bentham Science Publishers Ltd.: 47–58. doi:10.2174/2212707004666171113151624.

Lam, Pik Ling, Wai Yeung Wong, Zhaoxiang Bian, Chung Hin Chui, and Roberto Gambari. 2017. "Recent Advances in Green Nanoparticulate Systems for Drug Delivery: Efficient Delivery and Safety Concern." *Nanomedicine*. Future Medicine Ltd. doi:10.2217/nnm-2016-0305.

Li, Huaping, Sk Anwarul Haque, Alex Kitaygorodskiy, Mohammed J. Meziani, Maria Torres-Castillo, and Ya Ping Sun. 2006. "Alternatively Modified Bingel Reaction for Efficient Syntheses of C 60 Hexakis-Adducts." *Organic Letters* 8 (24). American Chemical Society: 5641–43. doi:10.1021/ol062391d.

Lu, Fushen, Sk Anwarul Haque, Sheng Tao Yang, Pengju G. Luo, Lingrong Gu, Alex Kitaygorodskiy, Huaping Li, Sebastian Lacher, and Ya Ping Sun. 2009. "Aqueous Compatible Fullerene-Doxorubicin Conjugates." *Journal of Physical Chemistry C* 113 (41). NIH Public Access: 17768–73. doi:10.1021/jp906750z.

Ma, Xin, and Dermont Bouchard. 2009. "Formation of Aqueous Suspensions of Fullerenes." *Environmental Science and Technology* 43 (2). American Chemical Society: 330–36. doi:10.1021/es801833p.

Madannejad, Rasoul, Nahid Shoaie, Fatemeh Jahanpeyma, Mohammad Hasan Darvishi, Mostafa Azimzadeh, and Hamidreza Javadi. 2019. "Toxicity of Carbon-Based Nanomaterials: Reviewing Recent Reports in Medical and Biological Systems." *Chemico-Biological Interactions*. Elsevier Ireland Ltd. doi:10.1016/j.cbi.2019.04.036.

Madhwi, Rajendra Kumar, Pramod Kumar, Bhupinder Singh, Gajanand Sharma, Om Prakash Katare, and Kaisar Raza. 2017. "*In Vivo* Pharmacokinetic Studies and Intracellular Delivery of Methotrexate by Means of Glycine-Tethered PLGA-Based Polymeric Micelles." *International Journal of Pharmaceutics* 519 (1–2): 138–44. doi:10.1016/j.ijpharm.2017.01.021.

Magrez, Arnaud, Sandor Kasas, Valérie Salicio, Nathalie Pasquier, Jin Won Seo, Marco Celio, Stefan Catsicas, Beat Schwaller, and László Forró. 2006. "Cellular Toxicity of Carbon-Based Nanomaterials." *Nano Letters* 6 (6): 1121–25. doi:10.1021/nl060162e.

Marchesan, Silvia, Tatiana Da Ros, Giampiero Spalluto, Jan Balzarini, and Maurizio Prato. 2005. "Anti-HIV Properties of Cationic Fullerene Derivatives." *Bioorganic and Medicinal Chemistry Letters* 15 (15): 3615–18. doi:10.1016/j.bmcl.2005.05.069.

Marcus, Yizhak, Allan L. Smith, Korobov, M. V., Mirakyan, A. L., Avramenko, N. V., and Stukalin, E. B. 2001. "Solubility of C60 Fullerene." *Journal of Physical Chemistry B* 105 (13). American Chemical Society: 2499–2506. doi:10.1021/jp0023720.

Mashino, Tadahiko, Dai Nishikawa, Kyoko Takahashi, Noriko Usui, Takao Yamori, Masako Seki, Toyoshige Endo, and Masataka Mochizuki. 2003. "Antibacterial and Antiproliferative Activity of Cationic Fullerene Derivatives." *Bioorganic and Medicinal Chemistry Letters* 13 (24). Elsevier Ltd: 4395–97. doi:10.1016/j.bmcl.2003.09.040.

Mateo-Alonso, Aurelio, Davide Bonifazi, and Maurizio Prato. 2006. "Functionalization and Applications of C_{60} Fullerene." In *Recent Developments in Chemistry, Physics, Materials Science and Device Applications*, 155–89. Elsevier. doi:10.1016/B978-044451855-2/50010-3.

Mckinnon, J. Thomas, William L. Bell, and Robert M. Barkley. 1992. "Combustion Synthesis of Fullerenes." *Combustion and Flame* 88 (1). Elsevier: 102–12. doi:10.1016/0010-2180(92)90010-M.

Mendes, Rafael G., Alicja Bachmatiuk, Bernd Büchner, Gianaurelio Cuniberti, and Mark H. Rümmeli. 2013. "Carbon Nanostructures as Multi-Functional Drug Delivery Platforms." *Journal of Materials Chemistry B* 1 (4). Royal Society of Chemistry: 401–28. doi:10.1039/c2tb00085g.

Meng, Huan, Gengmei Xing, Baoyun Sun, Feng Zhao, Hao Lei, Wei Li, Yan Song et al. 2010. "Potent Angiogenesis Inhibition by the Particulate Form of Fullerene Derivatives." *ACS Nano* 4 (5). American Chemical Society: 2773–83. doi:10.1021/nn100448z.

Misra, Charu, Manish Kumar, Gajanand Sharma, Rajendra Kumar, Bhupinder Singh, Om Prakash Katare, and Kaisar Raza. 2017. "Glycinated Fullerenes for Tamoxifen Intracellular Delivery with Improved Anticancer Activity and Pharmacokinetics." *Nanomedicine* 12 (9): 1011–23. doi:10.2217/nnm-2016-0432.

Misra, Charu, Nagarani Thotakura, Rajendra Kumar, Bhupinder Singh, Gajanand Sharma, Om Prakash Katare, and Kaisar Raza. 2017. "Improved Cellular Uptake, Enhanced Efficacy and Promising Pharmacokinetic Profile of Docetaxel Employing Glycine-Tethered C_{60}-Fullerenes." *Materials Science and Engineering: C* 76 (July): 501–8. doi:10.1016/j.msec.2017.03.073.

Mojica, Martha, Julio A. Alonso, and Francisco Méndez. 2013a. "Synthesis of Fullerenes." *Journal of Physical Organic Chemistry* 26 (7). John Wiley & Sons, Ltd: 526–39. doi:10.1002/poc.3121.

Mojica, Martha, Francisco Méndez, and Julio A. Alonso. 2013b. "Growth of Fullerene Fragments Using the Diels-Alder Cycloaddition Reaction: First Step towards a C60 Synthesis by Dimerization." *Molecules* 18 (2). Multidisciplinary Digital Publishing Institute (MDPI): 2243–54. doi:10.3390/molecules18022243.

Montellano, Alejandro, Tatiana Da Ros, Alberto Bianco, and Maurizio Prato. 2011. "Fullerene C_{60} as a Multifunctional System for Drug and Gene Delivery." *Nanoscale* 3 (10). The Royal Society of Chemistry: 4035–41. doi:10.1039/c1nr10783f.

Moonen, Nicolle N. P., Carlo Thilgen, Luis Echegoyen, and François Diederich. 2000. "The Chemical Retro-Bingel Reaction: Selective Removal of Bis(Alkoxycarbonyl) Methano Addends from C_{60} and C_{70} with Amalgamated Magnesium." *Chemical Communications*, no. 5 (March). Royal Society of Chemistry: 335–36. doi:10.1039/a909704j.

Mroz, Pawel, Anna Pawlak, Minahil Satti, Haeryeon Lee, Tim Wharton, Hariprasad Gali, Tadeusz Sarna, and Michael R. Hamblin. 2007. "Functionalized Fullerenes Mediate Photodynamic Killing of Cancer Cells: Type I versus Type II Photochemical Mechanism." *Free Radical Biology and Medicine* 43 (5). Free Radic Biol Med: 711–19. doi:10.1016/j.freeradbiomed.2007.05.005.

Nimibofa, Ayawei, Ebelegi Augustus Newton, Abasi Yameso Cyprain, and Wankasi Donbebe. 2018. "Fullerenes: Synthesis and Applications." *Journal of Materials Science Research* 7 (3). Canadian Center of Science and Education: 22. doi:10.5539/jmsr.v7n3p22.

Nuretdinov, I. A., V. V. Yanilkin, V. P. Gubskaya, N. I. Maksimyuk, and Lucia Sh. Berezhnaya. 2000. "Electrochemical Reduction of Some Methanofullerenes. On the Mechanism of the Retro-Bingel Reaction." *Russian Chemical Bulletin* 49 (3). Kluwer Academic/Plenum Publishers: 427–30. doi:10.1007/BF02494770.

Partha, Ranga, and Jodie L. Conyers. 2009. "Biomedical Applications of Functionalized Fullerene-Based Nanomaterials." *International Journal of Nanomedicine*. Dove Press. doi:10.2147/ijn.s5964.

Patel, Kapil D., Rajendra K. Singh, and Hae Won Kim. 2019. "Carbon-Based Nanomaterials as an Emerging Platform for Theranostics." *Materials Horizons*. Royal Society of Chemistry. doi:10.1039/c8mh00966j.

Pérez-Manríquez, Liliana, Estrella Ramos, Eduardo Rangel, and Roberto Salcedo. 2013. "Interaction between Epoxidised Estradiol and Fullerene (C_{60}): Possible Anticancer Activity." *Molecular Simulation* 39 (8). Taylor & Francis Group: 612–20. doi:10.1080/08927022.2012.758845.

Prato, M., Suzuki, T., Foroudian, H., Li, Q., Khemani, K., Wudl, F., Leonetti, J., et al. 1993. "[3+2] and [4+2] Cycloadditions of C60." *Journal of the American Chemical Society* 115 (4). American Chemical Society: 1594–95. doi:10.1021/ja00057a065.

Raoof, Mustafa, Yuri Mackeyev, Matthew A. Cheney, Lon J. Wilson, and Steven A. Curley. 2012. "Internalization of C60 Fullerenes into Cancer Cells with Accumulation in the Nucleus via the Nuclear Pore Complex." *Biomaterials* 33 (10). Biomaterials: 2952–60. doi:10.1016/j.biomaterials.2011.12.043.

Rašović, I. 2017. "Water-Soluble Fullerenes for Medical Applications." *Materials Science and Technology* 33 (7). Taylor and Francis Ltd.: 777–94. doi:10.1080/02670836.2016.1198114.

Raza, Kaisar, Dinesh Kumar, Chanchal Kiran, Manish Kumar, Santosh Kumar Guru, Pramod Kumar, Shweta Arora, Gajanand Sharma, Shashi Bhushan, and O. P. Katare. 2016. "Conjugation of Docetaxel with Multiwalled Carbon Nanotubes and Codelivery with Piperine: Implications on Pharmacokinetic Profile and Anticancer Activity." *Molecular Pharmaceutics* 13 (7). American Chemical Society: 2423–32. doi:10.1021/acs.molpharmaceut.6b00183.

Raza, Kaisar, Nitesh Kumar, Charu Misra, Lokesh Kaushik, Santosh Kumar Guru, Pramod Kumar, Ruchi Malik, Shashi Bhushan, and Om Prakash Katare. 2016. "Dextran-PLGA-Loaded Docetaxel Micelles with Enhanced Cytotoxicity and Better Pharmacokinetic Profile." *International Journal of Biological Macromolecules* 88 (July). Elsevier B.V.: 206–12. doi:10.1016/j.ijbiomac.2016.03.064.

Raza, Kaisar, Nagarani Thotakura, Pramod Kumar, Mayank Joshi, Shashi Bhushan, Amit Bhatia, Vipin Kumar et al. 2015. "C_{60}-Fullerenes for Delivery of Docetaxel to Breast Cancer Cells: A Promising Approach for Enhanced Efficacy and Better Pharmacokinetic Profile." *International Journal of Pharmaceutics* 495 (1). Elsevier: 551–59. doi:10.1016/j.ijpharm.2015.09.016.

Roursgaard, Martin, Steen S. Poulsen, Christopher L. Kepley, Maria Hammer, Gunnar D. Nielsen, and Søren T. Larsen. 2008. "Polyhydroxylated C_{60} Fullerene (Fullerenol) Attenuates Neutrophilic Lung Inflammation in Mice." *Basic and Clinical Pharmacology and Toxicology* 103 (4): 386–88. doi:10.1111/j.1742-7843.2008.00315.x.

Schur, D. V., Yu Zaginaichenko, S., Zolotarenko, A. D., and Veziroglu, T. N. 2008. "Solubility and Transformation of Fullerene C_{60} Molecule." In *NATO Science for Peace and Security Series C: Environmental Security*, PartF2: 85–95. Springer Verlag. doi:10.1007/978-1-4020-8898-8_7.

Schwerdtfeger, Peter, Lukas N. Wirz, and James Avery. 2015. "The Topology of Fullerenes." *Wiley Interdisciplinary Reviews: Computational Molecular Science* 5 (1). Blackwell Publishing Inc.: 96–145. doi:10.1002/wcms.1207.

Sharma, Sulbha K., Long Y. Chiang, and Michael R. Hamblin. 2011. "Photodynamic Therapy with Fullerenes *In Vivo*: Reality or a Dream?" *Nanomedicine (London, England)* 6 (10): 1813–25. doi:10.2217/nnm.11.144.

Shershakova, Nadezda, Elena Baraboshkina, Sergey Andreev, Daria Purgina, Irina Struchkova, Oleg Kamyshnikov, Alexandra Nikonova, and Musa Khaitov. 2016. "Anti-Inflammatory Effect of Fullerene C_{60} in a Mice Model of Atopic Dermatitis." *Journal of Nanobiotechnology* 14 (1). BioMed Central Ltd. doi:10.1186/s12951-016-0159-z.

Shoji, Masaki, Etsuhisa Takahashi, Dai Hatakeyama, Yuma Iwai, Yuka Morita, Riku Shirayama, Noriko Echigo et al. 2013. "Anti-Influenza Activity of C60 Fullerene Derivatives." *PLoS One* 8 (6). Public Library of Science. doi:10.1371/journal.pone.0066337.

Sijbesma, R., G. Srdanov, F. Wudl, J. A. Castoro, Charles Wilkins, Simon H. Friedman, Diane L. DeCamp, and George L. Kenyon. 1993. "Synthesis of a Fullerene Derivative for the Inhibition of HIV Enzymes." *Journal of the American Chemical Society* 115 (15). American Chemical Society: 6510–12. doi:10.1021/ja00068a006.

Singh, Anupama, Nagarani Thotakura, Rajendra Kumar, Bhupinder Singh, Gajanand Sharma, Om Prakash Katare, and Kaisar Raza. 2017. "PLGA-Soya Lecithin Based Micelles for Enhanced Delivery of Methotrexate: Cellular Uptake, Cytotoxic and Pharmacokinetic Evidences." *International Journal of Biological Macromolecules* 95 (February). Elsevier B.V.: 750–56. doi:10.1016/j.ijbiomac.2016.11.111.

Singh, Anupama, Nagarani Thotakura, Bhupinder Singh, Shikha Lohan, Poonam Negi, Deepak Chitkara, and Kaisar Raza. 2019. "Delivery of Docetaxel to Brain Employing Piperine-Tagged PLGA-Aspartic Acid Polymeric Micelles: Improved Cytotoxic and Pharmacokinetic Profiles." *AAPS PharmSciTech* 20 (6). Springer New York LLC. doi:10.1208/s12249-019-1426-8.

Sitharaman, Balaji, Robert D. Bolskar, Irene Rusakova, and Lon J. Wilson. 2004. "Gd@C_{60}[C(COOH)2]$_{10}$ and Gd@C_{60}(OH)$_x$: Nanoscale Aggregation Studies of Two Metallofullerene MRI Contrast Agents in Aqueous Solution." *Nano Letters* 4 (12): 2373–78. doi:10.1021/nl0485713.

Sivaraman, N., Dhamodaran, R., Kaliappan, I., Srinivasan, T. G., Vasudeva Rao, P. R., and Mathews, C. K. 1992. "Solubility of C_{60} in Organic Solvents." *Journal of Organic Chemistry* 57 (22). American Chemical Society: 6077–79. doi:10.1021/jo00048a056.

Sliwa, Wanda. 1995. "Cycloaddition Reactions of Fullerenes." *Fullerene Science and Technology* 3 (3). Taylor & Francis Group: 243–81. doi:10.1080/153638X9508543782.

Stankevich, Ivan V., Mikhail Vasil'evich Nikerov, and Dmitrii Anatol'evich Bochvar. 1984. "The Structural Chemistry of Crystalline Carbon: Geometry, Stability, and Electronic Spectrum." *Russian Chemical Reviews* 53 (7). Turpion-Moscow Limited: 640–55. doi:10.1070/rc1984v053n07abeh003084.

Tan, Yuan Zhi, Su Yuan Xie, Rong Bin Huang, and Lan Sun Zheng. 2009. "The Stabilization of Fused-Pentagon Fullerene Molecules." *Nature Chemistry*. Nature Publishing Group. doi:10.1038/nchem.329.

Thakral, Seema, and Mehta, R.. 2006. "Fullerenes: An Introduction and Overview of Their Biological Properties." *Indian Journal of Pharmaceutical Sciences*. Medknow Publications and Media Pvt. Ltd. doi:10.4103/0250-474X.22957.

Thakur, Chanchal Kiran, Nagarani Thotakura, Rajendra Kumar, Pramod Kumar, Bhupinder Singh, Deepak Chitkara, and Kaisar Raza. 2016a. "Chitosan-Modified PLGA Polymeric Nanocarriers with Better Delivery Potential for Tamoxifen." *International Journal of Biological Macromolecules* 93 (December). Elsevier B.V.: 381–89. doi:10.1016/j.ijbiomac.2016.08.080.

Thakur, Chanchal Kiran, Nagarani Thotakura, Rajendra Kumar, Pramod Kumar, Bhupinder Singh, Deepak Chitkara, and Kaisar Raza. 2016b. "Chitosan-Modified PLGA Polymeric Nanocarriers with Better Delivery Potential for Tamoxifen." *International Journal of Biological Macromolecules* 93 (December). Elsevier: 381–89. doi:10.1016/J.IJBIOMAC.2016.08.080.

Thong, Nguyen Minh, Thi Chinh Ngo, Duy Quang Dao, Tran Duong, Quoc Tri Tran, and Pham Cam Nam. 2016. "Functionalization of Fullerene via the Bingel Reaction with α-Chlorocarbanions: An ONIOM Approach." *Journal of Molecular Modeling* 22 (5). Springer Verlag. doi:10.1007/s00894-016-2981-5.

Thotakura, Nagarani, Mukesh Dadarwal, Pramod Kumar, Gajanand Sharma, Santosh Kumar Guru, Shashi Bhushan, Kaisar Raza, and Om Prakash Katare. 2017. "Chitosan-Stearic Acid Based Polymeric Micelles for the Effective Delivery of Tamoxifen: Cytotoxic and Pharmacokinetic Evaluation." *AAPS PharmSciTech* 18 (3). Springer New York LLC: 759–68. doi:10.1208/s12249-016-0563-6.

Thotakura, Nagarani, Mukesh Dadarwal, Rajendra Kumar, Bhupinder Singh, Gajanand Sharma, Pramod Kumar, Om Prakash Katare, and Kaisar Raza. 2017a. "Chitosan-Palmitic Acid Based Polymeric Micelles as Promising Carrier for Circumventing Pharmacokinetic and Drug Delivery Concerns of Tamoxifen." *International Journal of Biological Macromolecules* 102 (September). Elsevier B.V.: 1220–25. doi:10.1016/j.ijbiomac.2017.05.016.

Thotakura, Nagarani, Mukesh Dadarwal, Rajendra Kumar, Bhupinder Singh, Gajanand Sharma, Pramod Kumar, Om Prakash Katare, and Kaisar Raza. 2017b. "Chitosan-Palmitic Acid Based Polymeric Micelles as Promising Carrier for Circumventing Pharmacokinetic and Drug Delivery Concerns of Tamoxifen." *International Journal of Biological Macromolecules* 102 (September). Elsevier: 1220–25. doi:10.1016/J.IJBIOMAC.2017.05.016.

Thotakura, Nagarani, Gajanand Sharma, Bhupinder Singh, Vipin Kumar, and Kaisar Raza. 2018. "Aspartic Acid Derivatized Hydroxylated Fullerenes as Drug Delivery Vehicles for Docetaxel: An Explorative Study." *Artificial Cells, Nanomedicine and Biotechnology* 46 (8): 1763–72. doi:10.1080/21691401.2017.1392314.

Thotakura, Nagarani, Saurabh Sharma, Rajneet Kaur Khurana, Penke Vijaya Babu, Deepak Chitkara, Vipin Kumar, Bhupinder Singh et al. 2013. "Tamoxifen-Loaded Poly(L-Lactide) Nanoparticles: Development, Characterization and In Vitro Evaluation of Cytotoxicity." *Aaps Pharmscitech* 60 (March): 135–42. doi:10.1016/j.msec.2015.11.019.

Thotakura, Nagarani, Saurabh Sharma, Rajneet Kaur Khurana, Penke Vijaya Babu, Deepak Chitkara, Vipin Kumar, Bhupinder Singh, and Kaisar Raza. 2019. "Aspartic Acid Tagged Carbon Nanotubols as a Tool to Deliver Docetaxel to Breast Cancer Cells: Reduced Hemotoxicity with Improved Cytotoxicity." *Toxicology in Vitro* 59 (September). Elsevier Ltd: 126–34. doi:10.1016/j.tiv.2019.04.012.

Vorobyov, Vasily, Vladimir Kaptsov, Rita Gordon, Ekaterina Makarova, Igor Podolski, and Frank Sengpiel. 2015. "Neuroprotective Effects of Hydrated Fullerene C_{60}: Cortical and Hippocampal EEG Interplay in an Amyloid-Infused Rat Model of Alzheimer's Disease." *Journal of Alzheimer's Disease* 45 (1). IOS Press: 217–33. doi:10.3233/JAD-142469.

Wang, Chenglong, Yitong Bai, Hongliang Li, Rong Liao, Jiaxin Li, Han Zhang, Xian Zhang, Sujuan Zhang, Sheng Tao Yang, and Xue Ling Chang. 2016. "Surface Modification-Mediated Biodistribution of 13C-Fullerene C60 in Vivo." *Particle and Fibre Toxicology* 13 (1). BioMed Central Ltd.: 1–14. doi:10.1186/s12989-016-0126-8.

Wang, Haifang, Sheng Tao Yang, Aoneng Cao, and Yuanfang Liu. 2013. "Quantification of Carbon Nanomaterials in Vivo." *Accounts of Chemical Research* 46 (3): 750–60. doi:10.1021/ar200335j.

Weston, A., and Murthy, M. 1996. "Synthesis of Fullerenes: An Effort to Optimize Process Parameters." *Carbon* 34 (10). Elsevier Ltd: 1267–74. doi:10.1016/0008-6223(96)00084-X.

Williams, Hywel D., Natalie L. Trevaskis, Susan A. Charman, Ravi M. Shanker, William N. Charman, Colin W. Pouton, and Christopher J. H. Porter. 2013. "Strategies to Address Low Drug Solubility in Discovery and Development." *Pharmacological Reviews*. American Society for Pharmacology and Experimental Therapy. doi:10.1124/pr.112.005660.

Wudl, Fred. 2002. "Fullerene Materials." *Journal of Materials Chemistry*. The Royal Society of Chemistry. doi:10.1039/b201196d.

Yadav, Harsh, Pramod Kumar, Vikas Sharma, Gajanand Sharma, Kaisar Raza, and Om Prakash Katare. 2016. "Enhanced Efficacy and a Better Pharmacokinetic Profile of Tamoxifen Employing Polymeric Micelles." *RSC Advances* 6 (58). Royal Society of Chemistry: 53351–57. doi:10.1039/c6ra10874a.

Yan, Weibo, Stefan M. Seifermann, Philippe Pierrat, and Stefan Bräse. 2015. "Synthesis of Highly Functionalized C60 Fullerene Derivatives and Their Applications in Material and Life Sciences." *Organic and Biomolecular Chemistry*. Royal Society of Chemistry. doi:10.1039/c4ob01663g.

Yanilkin, V. V., Nastapova, N. V., Gubskaya, V. P., Morozov, V. I., Lucia Sh. Berezhnaya, and Nuretdinov, I. A. Nuretdinov. 2002. "Retro-Bingel Reaction in the Electrochemical Reduction of Bis(Dialkoxyphosphoryl) Methanofullerenes." *Russian Chemical Bulletin* 51 (1): 72–77. doi:10.1023/A:1015005628871.

Yin, Rui, Min Wang, Ying Ying Huang, Huang Chiao Huang, Pinar Avci, Long Y. Chiang, and Michael R. Hamblin. 2014. "Photodynamic Therapy with Decacationic [60]Fullerene Monoadducts: Effect of a Light Absorbing Electron-Donor Antenna and Micellar Formulation." *Nanomedicine: Nanotechnology, Biology, and Medicine* 10 (4). Elsevier Inc.: 795–808. doi:10.1016/j.nano.2013.11.014.

Ying, Z. C., Hettich, R. L., Compton, R. N., and Haufler, R. E. 1996. "Synthesis of Nitrogen-Doped Fullerenes by Laser Ablation." *Journal of Physics B: Atomic, Molecular and Optical Physics* 29 (21). IOP Publishing: 4935. doi:10.1088/0953-4075/29/21/007.

Yuan, Xia, Xiangxian Zhang, Lu Sun, Yuquan Wei, and Xiawei Wei. 2019. "Cellular Toxicity and Immunological Effects of Carbon-Based Nanomaterials." *Particle and Fibre Toxicology*. BioMed Central Ltd. doi:10.1186/s12989-019-0299-z.

Zhang, Yanyan, Minghao Wu, Mingjie Wu, Jingyi Zhu, and Xuening Zhang. 2018. "Multifunctional Carbon-Based Nanomaterials: Applications in Biomolecular Imaging and Therapy." *ACS Omega*. American Chemical Society. doi:10.1021/acsomega.8b01071.

Zhao, Xiongce, Alberto Striolo, and Peter T. Cummings. 2005. "C_{60} Binds to and Deforms Nucleotides." *Biophysical Journal* 89 (6). Biophysical Society: 3856–62. doi:10.1529/biophysj.105.064410.

Zhu, Jiadan, Zhiqiang Ji, Jing Wang, Ronghua Sun, Xiang Zhang, Yang Gao, Hongfang Sun et al. 2008. "Tumor-Inhibitory Effect and Immunomodulatory Activity of Fullerol $C_{60}(OH)_x$." *Small* 4 (8). John Wiley & Sons, Ltd: 1168–75. doi:10.1002/smll.200701219.

Section B

Bioactive Delivery Systems

5 Pharmaceutical and Biomedical Applications of Multifunctional Quantum Dots

Devesh Kapoor
Dr. Dayaram Patel Pharmacy College

Swapnil Sharma, Kanika Verma, and Akansha Bisth
Banasthali Vidyapith

Yashu Chourasiya
Smriti College of Pharmaceutical Education

Mayank Sharma and Rahul Maheshwari
SVKM's NMIMS

CONTENTS

5.1 Introduction 66
5.2 History of QDs 66
5.3 Types of QDs 66
5.4 Synthesis of QDs 67
    5.4.1 Organic-Phase Method/Organometallic Chemistry Method 67
    5.4.2 Water-Phase Method/Aqueous Solution Method 68
    5.4.3 Hydrothermal and Microwave-Assisted Irradiation Methods 68
    5.4.4 Laser Ablation Techniques for QDs 68
    5.4.5 Molecular Beam Epitaxy (MBE) and Nanopatterning for QDs 68
5.5 Properties of QDs 69
5.6 Biomedical Applications of QDs 69
    5.6.1 *In Vivo* Cell Imaging 70
        5.6.1.1 Synaptic Neurotransmission 72
        5.6.1.2 Single Protein Tracking 72
        5.6.1.3 Cell Tracking and Migration 72
    5.6.2 *In Vitro/Ex Vivo* Cell Imaging 73
    5.6.3 Tissue Imaging 73
    5.6.4 Diagnostic Tool for Detection of Diseases 73
    5.6.5 Development of Diagnostic Test Systems 74
    5.6.6 Biosensors/Biomarkers for Detection of Mutation, Multiplexed Target and miRNA Detection 74
    5.6.7 Drug Delivery/Carrier for Treatment 75
        5.6.7.1 Ocular Diseases 75
        5.6.7.2 Cardiovascular Diseases 75
        5.6.7.3 Neurological Disorders 76
        5.6.7.4 Hepatic Diseases 76
        5.6.7.5 Antibiotic-Resistant Infection 76
        5.6.7.6 Tumours 76
        5.6.7.7 Renal Diseases 78
        5.6.7.8 Others 78
    5.6.8 Cell Labelling 78
5.7 Conclusion 79
References 79

DOI: 10.1201/9781003043164-5

5.1 INTRODUCTION

Conventional formulations developed by pharmaceutical industries are in abundance, though with limited efficacy, poor permeation, decreased bioavailability and toxicity (Uehara et al. 2010). Nanotechnology is a tool for delivering drugs at specific target sites using intelligent and smart nanocarriers having well-defined sizes and shapes (Lalu et al. 2017; Tekade et al. 2017). Nanostructured materials possess the ability to bridge the gap between molecular and bulk levels and therefore create new avenues for a wide range of applications in biology, electronics and optoelectronics (Maheshwari et al. 2015; Sharma et al. 2015; Maheshwari et al. 2018; Moondra et al. 2018). Based on particle size, these nanostructures can be categorised as zero-dimensional or quantum dots (QDs), one-dimensional or quantum wires, and two-dimensional or quantum wells (Bera et al. 2010).

QDs, also known as semiconductor nanocrystals, pertain to unique electronic and optical properties, including multiplexed capabilities, long-term photostability, high signal brightness, simultaneous excitation of multiple fluorescence colours and size-tunable light emission (Jin et al. 2011). QDs are nanometre-sized semiconductor structures with dimensions smaller than the de Broglie wavelength (Mandal and Chakrabarti 2017). Nanometre size increases the particle surface area-to-volume ratio, which further enables surface modifications to ameliorate reactivity, solubility and biocompatibility (Wagner et al. 2019). QDs are proven to be integrated with a range of applications in biomedical sciences, including fluorescent assay for drug discovery, bioimaging, detection of disease, intracellular reporting and protein tracking (Rosenthal et al. 2011). Moreover, these nanometre-sized QDs also overcome severe toxicity, decrease effective dose and increase sensitivity (Wagner et al. 2019).

5.2 HISTORY OF QDs

At the end of 1970, during the crisis of petroleum, researchers aimed at discovering alternatives for solar energy conversion. This provokes the investigators to synthesise semiconductor crystals in solution and screens their optoelectronic properties. Meanwhile, they observed blueshift with a decrease in nanocrystal size and explained quantum confinement effect. Typically, the studies on QDs were initiated in physics and then emerged through medical and technical fields. Of note, quantum oscillator, transistor, multispectral fluorescent dye imaging, filters, detectors, data analysis technique and QD light imaging device are some inventions in the field of physics (Bera et al. 2010; Efros and Nesbitt 2016). Interestingly, the outstanding and unique properties of QDs make them novel drug delivery and targeting approaches (Figures 5.1 and 5.2).

5.3 TYPES OF QDs

These are nanoscale man-fabricated crystals that can convert light spectrum into diverse colours. According to the size of these QDs, every dot emits a different colour. QDs can be classified into distinct types based on their composition and structure, such as core-type QDs, core-shell QDs and alloyed QDs (Liu and Su 2014). A typical structure of QDs is presented in Figure 5.3.

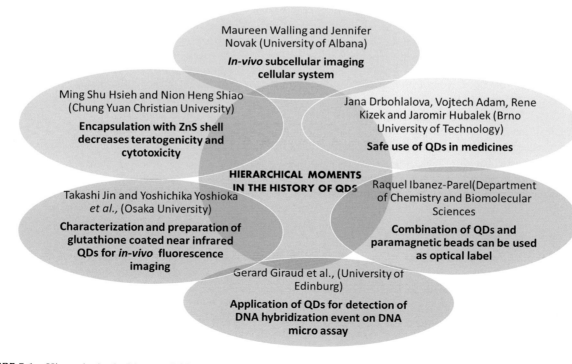

FIGURE 5.1 Hierarchy in the history of QDs. Hierarchical moments marked in the journey of development of QDs by global universities including its application in medicine as imaging system, diagnostic tools and optical labelling.

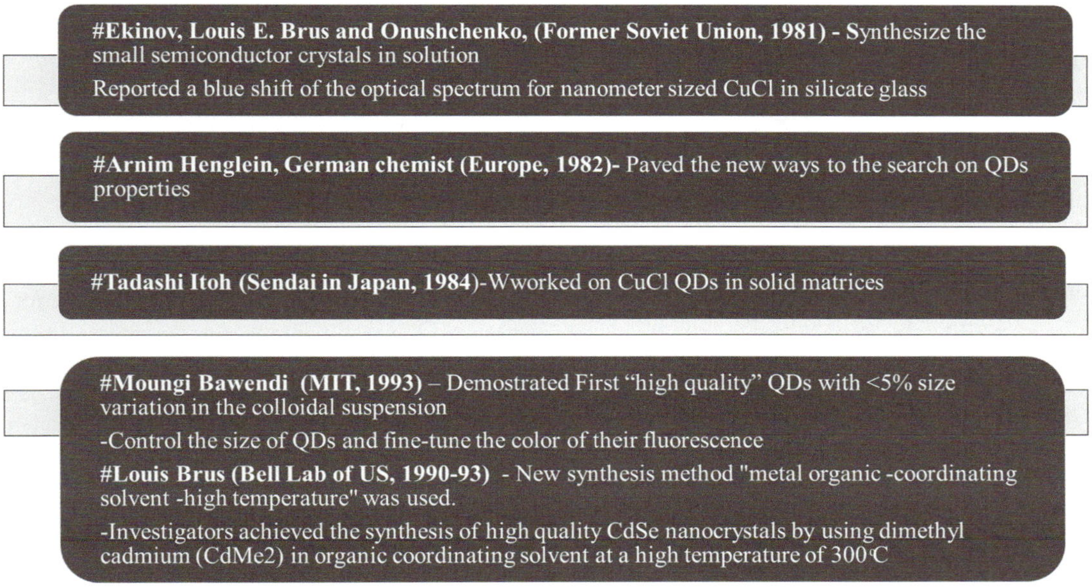

#Ekinov, Louis E. Brus and Onushchenko, (Former Soviet Union, 1981) - Synthesize the small semiconductor crystals in solution
Reported a blue shift of the optical spectrum for nanometer sized CuCl in silicate glass
#Arnim Henglein, German chemist (Europe, 1982)- Paved the new ways to the search on QDs properties
#Tadashi Itoh (Sendai in Japan, 1984)-Wworked on CuCl QDs in solid matrices
#Moungi Bawendi (MIT, 1993) – Demostrated First "high quality" QDs with <5% size variation in the colloidal suspension
-Control the size of QDs and fine-tune the color of their fluorescence
#Louis Brus (Bell Lab of US, 1990-93) - New synthesis method "metal organic -coordinating solvent -high temperature" was used.
-Investigators achieved the synthesis of high quality CdSe nanocrystals by using dimethyl cadmium (CdMe2) in organic coordinating solvent at a high temperature of 300°C

FIGURE 5.2 Historical representation of QDs. The figure highlights the contribution of eminent scientists globally for historical development of QDs possessing outstanding and unique properties that make them a novel drug delivery material in targeting and delivery.

Shell
Functionaliser
Core
QD
Ampiphillic polymer
Target molecule

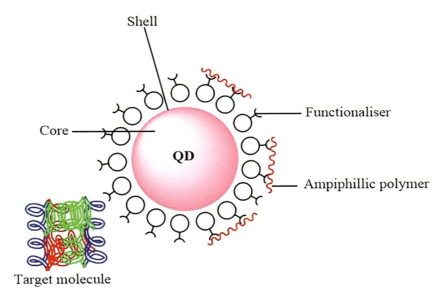

FIGURE 5.3 Structure of a typical QD. Diagrammatic representation of a basic QD with core (semiconductor such as CdSe or CdTe), shell (such as ZnS), amphiphilic polymers (such as PEG) and target molecules such as peptides or antibodies.

5.4 SYNTHESIS OF QDs

5.4.1 Organic-Phase Method/Organometallic Chemistry Method

The organometallic chemistry method is considered as the most crucial method for the synthesis of regular and uniform core-structured monodisperse QDs with high quantum yield in non-polar organic solvents (Aswathi et al. 2018). Different sizes of QDs can be obtained by varying temperature and reaction time conditions. In general, bis(trimethylsilyl)selenium ($(TMS)_2Se$) and Me_2Cd are two profusely used organometallic precursors. Monodisperse CdSe can be obtained based on the pyrolysis of organometallic reagents by injection into a hot coordinating solvent between 250°C and 300°C. The adsorption of ligand such as tri-n-octylphosphine oxide (TOPO) leads to annealing of cores in coordinating solvents (Jin et al. 2011).

5.4.2 Water-Phase Method/Aqueous Solution Method

The aqueous solution method is a cost-effective and eco-friendly procedure of QD synthesis. The technique has direct applications in biological research without the involvement of the ligand exchange procedure. In general, glutathione (GSH), 3-mercaptopropionic acid (3-MPA) or other hydrosulfyl-containing materials are the commonly used ligands for the production of CdTe QDs in aqueous solution. In addition to this, ionic perchlorates such as Al_2Te_3 and $Cd(ClO)_4 \cdot 6H_2O$ are used as the precursors. Thiol-capped CdTe QDs were the first synthesised aqueous dispersed QDs; however, they showed low quantum yields, poor stability and broad size distribution when compared with the QDs produced by the organometallic method (Jin et al. 2011; Aswathi et al. 2018).

5.4.3 Hydrothermal and Microwave-Assisted Irradiation Methods

Hydrothermal and microwave-assisted irradiation methods are used to produce QDs with high quantum yields and narrow size distribution. Moreover, these methods also reduce the reaction time and surface defects generated during the growth process of QDs due to high pressure. In brief, all reaction reagents are heated at high temperatures up to supercritical temperature into the hermetic container. In the microwave-assisted irradiation method, microwave irradiation is considered as a heating source which aids in optimising synthesis conditions. An increase in heat from this system produces homogenous QDs with high yield (17%) (Jin et al. 2011; Aswathi et al. 2018). Different synthesis methods of QDs are presented in Figure 5.4.

5.4.4 Laser Ablation Techniques for QDs

This method is reported as a clean technique for the fabrication of QDs due to less waste production. In this method, nanoparticles are fabricated by employing high-energetic laser light on metals or crystals. Anikin et al. used laser of Cu to obtain QDs of CdS/ZnSe in open air by ablating CdS/ZnSe under a slim liquid layer just above the semiconductor surface. They employed different liquid mediums such as isobutanol, diethylene glycol, ethanol and dimethyl sulfoxide (Horoz et al. 2012).

5.4.5 Molecular Beam Epitaxy (MBE) and Nanopatterning for QDs

The MBE is an epitaxial growth technology that relies on the interaction of molecules of thermal energy and beams of atoms on a heated crystalline surface along with ultra-high-vacuum conditions. One of the latest methods for QD fabrication is droplet epitaxy (DE). This method is fully

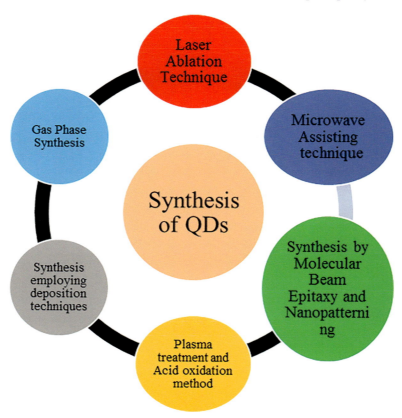

FIGURE 5.4 Various approaches for the synthesis of QDs. Diagrammatic representation of multiple synthesis approaches considered during the development of QDs.

compatible with MBE technology and produces various zero-dimensional quantum structures such as double ring-like QDs and inverse QDs. Claro et al. (2019) reported the development of self-assembled Bi_2Se_3 QDs by MBE on GaAs substrates employing the droplet epitaxy technique (DET) (Claro et al. 2019).

5.5 PROPERTIES OF QDs

Structurally, QDs are composed of semiconductor core (groups III–V and II–VI), enclosed within a shell formed of another semiconductor material (Figure 5.3). Notably, QDs have five unique properties that give them their distinct capabilities. Firstly, the diameter of QDs ranges from 2 to 10 nm comprising approximately 200–10,000 atoms. Secondly, QDs discern themselves with other fluorescent proteins and organic dyes in offering unique electronic and optical properties for targeted drug delivery and imaging (Bajwa et al. 2016).

QDs are exceptionally bright due to high fluorescent quantum yields and possess large absorption extinction coefficients. Thirdly, electronic and optical properties include composition- and size-tunable light, narrow and Gaussian emission spectra, signal brightness, multiple fluorescent colour excitation, and resistance to photobleaching, which enables extended dynamic imaging. Fourth, QDs blinking, it is interesting that different coloured QDs can be excited simultaneously with minimal spectral overlapping, with a single light source, which provides significant advantages for multiplexed detection of target molecules (Peng and Li 2010; Rosenthal et al. 2011). Despite the hydrophobic nature of QDs, the large surface area required for attaching biofunctional molecules necessitates solubilisation of QDs. Fifth, they are photochemically robust and inorganic (Rosenthal et al. 2011). The well-understood and well-documented property is an inverse relationship that exists between energy band gap and nanocrystal size. In simpler words, with the increase in the energy band gap, the nanocrystal size decreases and the corresponding emission/excitation wavelength also decreases. This concept is known as the quantum size effect (Wagner et al. 2019).

Advantages and properties of QDs in biological and chemical research concerning their spectral overlapping, low signal intensity and photobleaching distinguish them significantly from green fluorescent proteins and traditional fluorescent organic dyes (Jin et al. 2011). Also, QDs resist photobleaching for longer fluorescent durations and show minimal cytotoxicity as compared to an organic fluorescent dye. Due to reduced photobleaching tendencies, QDs can preserve and image the number of samples a number of times (Amaral et al. 2020). Such properties of QDs considered being potential candidates as labels and luminescent probes in biological applications ranging from biological imaging, diagnosis of disease to molecular histopathology. Numerous studies have also been reported on the use of QDs for *in vitro* and *in vivo* imaging (Jin et al. 2011).

5.6 BIOMEDICAL APPLICATIONS OF QDs

For the most promising application of QDs in biological imaging, several advanced studies have focused on developing fluorescent probes. QDs possess the ability to play a significant role in cell biology, *in vivo* and *in vitro*, and have even replaced a lot of fluorescent dye molecules when coupled with biological molecules, as shown in Table 5.1 (Jin et al. 2011; Zhao and Zeng 2015).

TABLE 5.1
Applications of QDs in Molecular, Cell, Tissue and Animal Model Cancer Imaging

Category	Components	Findings	Future Needs	References
In vivo cell imaging	Fibrous phosphorus quantum dots (FPQDs)	FPQDs were used for live imaging using simple fluorescent microscopy with 4′,6-diamidino-2-phenylindole (DAPI) filter. Bioimages of human adenocarcinoma cells were acquired	–	Amaral et al. (2020)
	Streptavidin-conjugated green QDs (QD525), Tetrazine-conjugated red QD (QD625) and transcription activator-like effector (TALE)-labelled QDs	Single-genome loci were imaged *in vivo* by combining QDs with TALEs labelling technique	Colocalisation microscopy can then be employed to examine the bound QD-TALEs. The ultrasensitive analysis may contribute to the study of chronic HIV-1 infections, as well as virus detection and treatments at the live-cell level	Ma et al. (2017)
	Biocompatible heavy metal-free/cadmium-free QD nanoparticles (bio CFQD® nanoparticles)	Effectiveness and applicability of novel biocompatible water-soluble indium-based QDs for *in vivo* axillary lymphatic mapping and lymph nodes in a prostate cancer mouse model	Due to the absence of toxic elements in bio CFQD® nanoparticles, they may further be used in biomedical applications, but after additional work to study long-term toxicity	Yaghini et al. (2016)

(Continued)

TABLE 5.1 (*Continued*)
Applications of QDs in Molecular, Cell, Tissue and Animal Model Cancer Imaging

Category	Components	Findings	Future Needs	References
In vitro imaging	PEG-functionalised QDs	Non-aggregated QDs were internalised by microglia *in vivo* and in slices, irrespective of the brain-relevant platform and surface chemistry. Thus, providing a brain microenvironment allows the development of QD-based imaging probes that enables targeting an interested region in the CNS	Results guided to further engineer candidate QD-based imaging probes for neurological applications	Zhang et al. (2019)
	QD-conjugated probes	Cell clone formation assay based on QD molecular imaging provided a novel method to understand the proliferative feature and morphological characteristics of cancer cells providing an insight onto tumour biology	There is theoretical feasibility and technical possibility to develop a differentiation strategy which may allow for the control and coordination of cancer cells	Geng et al. (2016)
Drug delivery	GQD-PEG-BFG-Pyr-RF and GQD-Pyr-RF	GQD-PEG-BFG-Pyr-RF possesses less cytotoxic effect, and subsequently, its effect on three cancer cell lines was compared with the effect produced by anti-cancer drug (GQD-Pyr-RF) which was similar to nanovector lacking targeted riboflavin ligand (GQD-PEG-BFG)	The therapy opens new possibilities for poorly water-soluble anti-cancer drugs	Iannazzo et al. (2019)
	FA-PEG-cGQDs-MTN nanosystem	The nanodelivery system showed low systemic toxicity, improved anti-tumour ability and targeting capacity in a live animal without the presence of systemic adverse effects	Despite the development of such a nanosystem, clinical settings should be carefully considered by using clinical tumour models. Also, mimetic membranes should be used to improve the immune evasion capabilities	Li et al. (2019)
	Carboxylated PEGylated QDs and QD-encoded microcapsules	The conjugate of a monoclonal antibody targeting HER2 and microcapsule provided specific and sensitive antibody-mediated binding of microcapsules with living cancer cells	QD-encoded microcapsules can be biocompatible fluorescent agents for live-cell targeting, and serve as the basic platform for further development of targeted systems for diagnosis and therapy of a variety of tumour entities	Nifontova et al. (2019)
Cell labelling	Qdot® 625 ITK™ carboxyl QD	The carboxylated QDs can used as an effective and non-specific dye for labelling bone marrow MSCs	Results showed promising future in tissue engineering, anti-cancer drug delivery, cellular therapy and fundamental stem cell biology	Kundrotas et al. (2019)
Biosensors	Fe_3O_4/CdSe composite QD	For designing magnetic Fe_3O_4/CdSe QD-based electrochemiluminescence for assay of cancer cells using cyclic amplification technology	The design and development of magnetic QD-based ECL have the potential in clinical applications	Jie et al. (2013)

5.6.1 IN VIVO CELL IMAGING

In recent years, applications of QDs have been noticeable in the field of cellular biology and living cells (Table 5.1). Weng et al. aimed at incorporating exclusively luminescent QDs in immunoliposomes for imaging, diagnosis and treatment of cancer. They synthesised QD-conjugated immunoliposome-based nanoparticles by inserting anti-HER2 scFv in HER2-overexpressing MCF-7/HER2 and SK-BR-3 cells. They showed that QD-conjugated immunoliposome-based nanoparticles increase the circulation of QDs in athymic mice. Also, plasma $t_{1/2}$ was found to be ~2.9 h when compared with free QDs with $t_{1/2}$ < 10 min. *In vivo* fluorescent imaging was used for confirming localisation of QD-conjugated immunoliposome-based nanoparticles in MCF-7/HER2 xenograft models (Weng et al. 2008).

QDs have a unique property that with increasing separation between emission and excitation wavelength, the absorbance of QDs also increases. Based on this, much of enthusiasm is generated for applying QDs in *in vivo* stems, since photon yield should be proportional to the integral of the broadband absorption. Lim et al. hypothesised fluorescent QDs to be excellent contrast agents for biomedical imaging and assays. Based on a validated mathematical model, they explored the effects of tissue thickness, water-to-haemoglobin ratio, the wavelength dependence of scattering, tissue scatter and tissue absorbance on the performance of QDs. In conclusion, the excitation wavelength remains constrained and should be selected carefully based on the particular application of QD, when embedded *in vivo*. Near-infrared QD was produced and optimised for imaging surface vasculature with a silicon CCD camera and white light excitation. They were also used for imaging coronary vasculature *in vivo* and may have applications in designing fluorescent QD contrast agents optimised for specific biomedical applications (Lim et al. 2003).

Even after specific selection of emission and excitation wavelengths, many factors such as pharmacokinetics, toxicity, photostability, chemical stability and quantum yield remain unanswerable for their impact on QDs for biomedical application. Since, no reports were published for screening effect of QDs on such factors, after its *in vivo* administration into tumours to map sentinel nodes. Ballou et al. used oligomeric phosphines for capping QDs, which preserved photostability for at least initial contact hours of QDs with plasma. This paramount imparted aqueous stability, solubility, maximised quantum yield and minimised non-specific tissue interactions. They also investigated that retention and binding in the lymph node are affected by changing the surface charge of PEG-conjugated QDs (Ballou et al. 2007).

Amaral et al. developed a facile solution-based method for synthesis of fibrous phosphorus QDs with a height of 2.7 ± 1.3 nm and an average size of 3.8 ± 0.9 nm. Remarkable stability was obtained along with fluorescence properties in the indigo-blue region. Further, these fibrous phosphorus QDs were used as fluorescent labels in live bioimaging of human adenocarcinoma cells (Amaral et al. 2020).

Small peptides known as cell-penetrating peptides (CPPs) are able to traverse cell membrane and deliver a variety of molecules within a living cell. Such agents tend to deliver QDs across the membrane with minimal toxic and non-specific absorption effect. Liu et al. designed polyethylene glycol lipid-coated carboxyl-functionalised, monodisperse and water-soluble indium phosphide/zinc sulphide (InS/ZnS) QDs. They characterised cellular internalisation and physicochemical properties of CPP/QInP complexes and carboxyl- and PEG-bifunctionalised indium phosphide (InP)/ZnS QDs (QInP). These CPP/QInP complexes were efficient in delivering QInP in human A549 cells. This was evident when QInP (<1 M) and CPP/QInP complexes (500 nM) did not affect cell viability. Conclusively, they reported that PEGylated and carboxylated biofunctionalised QInP were biocompatible nanoparticles with potential applications in drug delivery and bioimaging studies (Liu et al. 2013).

Ma et al. designed a strategy for live-cell imaging of single genomic loci associated with cellular functions, pathogenic infections or genetic diseases. They combined the QD labelling technique with transcription activator-like effectors (TALEs) which specifically targeted HIV-1 pro-viral DNA sequences. Besides, two bio-orthogonal ligation reactions were used for labelling with varied colours. The first TALE was well labelled with tetrazine-conjugated red QD via the Diels–Alder cycloaddition and fused to short LplA acceptor peptide. The second TALE was labelled with streptavidin-conjugated green QDs (QD525) and merged with AP tag. Thus, TALEs labelled with QDs (QD-TALE) entered the cell nucleus via nuclear localisation sequence of TALE, identified single HIV-1 pro-viral loci by providing fluorescence and mapped in live U1 cells (Ma et al. 2017).

Yaghini et al. evaluated novel biocompatible heavy metal-free/cadmium-free QD nanoparticles (bio CFQD® nanoparticles) with good photoluminescence quantum yield. They evaluated the potential for mapping the lymph node by *ex vivo* imaging of regional lymph nodes after subcutaneous administration in rat paw. QDs were shown to accumulate selectively and quickly in thoracic and axillary regional lymph nodes according to chemical extraction and photoluminescent imaging methods (Yaghini et al. 2016).

Moreover, Zhao et al. proposed utilisation of QDs as spectra; converters that convert Cerenkov luminescence-blue emission to near-infrared light that is less absorbed or scattered *in vivo*. Cerenkov luminescence is an imaging modality that uses light produced during radioactive decay of clinically used isotopes. They showed that experimentation related to tissue phantom increases transmission intensity and penetration depth for Cerenkov luminescence in the presence of near-infrared QDs. They also developed three types of ^{89}Zr dual-labelled nanoparticles and near-infrared QDs based on polymeric nanoplatforms, nanoemulsions and lipid micelles, which enable co-delivery of radionuclide and QDs for maximised spectral conversion efficiency. Finally, they showed applications of self-illuminating nanoparticles for imaging tumours and lymph nodes in a prostate cancer mouse model (Zhao et al. 2017).

Apart from several advantages such as tunable emission and excitation spectra, low toxicity, surface functionality, photostability and photoluminescence required for *in vivo* imaging, no systemic evaluation of QDs in the brain microenvironment has been done yet. Zhang et al. investigated core-shell CdSe/CdS QDs' cellular uptake, colloidal uptake and *in vivo*, *ex vivo* and *in vitro* toxicity in the brain. They targeted red-emitting aqueous dispersible QDs with three surface functionalities: PEG-5000k-hydroxyl (PEG-OH), PEG-5000k-methoxy (PEG-OMe) and 3-MPA. They found that surface functionality plays an important role and is a dependable factor for QDs. PEG-conjugated QDs were protected from aggregation in neurophysiologically relevant tissues and fluids, allowing greater penetration. The behaviour of QDs differed in cultured slices as compared to monolayer cultures. Upon systemic administration, non-aggregated QDs were internalised by microglia *in vivo* and in slices, irrespective of brain-relevant platform and surface chemistry. Thus, providing brain microenvironment allows the development of QD-based imaging probes that enables targeting an interested region in the CNS (Zhang et al. 2019).

5.6.1.1 Synaptic Neurotransmission

Synapses are the medium for neurons to communicate, which shows enrichment for specialised receptors. Ehlers et al. reported that the mobility of GluR1 (glutaminergic receptor) was restricted to subregions of post-synaptic membrane. Such GluR1 mobility defines a new input-specific mechanism for AMPA receptor regulation in abundance (Ehlers et al. 2007).

Murphy-Royal et al. found that glutamate transporter (GLT-1) is enormously available on rat astrocyte. Surface modification of GLT-1 was sensitive to glial and neuronal activities and was decreased in the vicinity of glutamatergic synapses, leading to retention of transporter receptors. They gave first evidence for the physiological role of GLT-1 in shaping synaptic transmission, as improper diffusion of GLT-1 membrane via cross-linking in *in vivo* and *in vitro*. It increases time course for synaptic glutamate transmission (Murphy-Royal et al. 2015).

Extracellular space separates brain cells from each other and is vital as it provides a medium for drug and drug delivery vectors. Glia and neurons to access nutrients and chemical signalling necessitate diffusion within extracellular space. Thorne and Nicholson showed that water-soluble QDs and dextran with Stokes–Einstein diameter diffuse within the extracellular space of rat neocortex using integrative optical imaging. They were able to measure the width of extracellular space, i.e. 38–42 nm. The results improved modelling of neurotransmitter spread after ectopic release and spillover and established size limits for the diffusion of drug delivery vectors such as nanoparticles, liposomes and viruses in brain extracellular space (Thorne and Nicholson 2006).

Interestingly, knowledge regarding the application of QDs in presynaptic terminal has always been limited compared to diffusive behaviours of post-synaptic receptors. In general, disruption of actin induces a reduction in the diffusive behaviour of synaptic vesicle at synapse, while disruption of microtubules only decreases extrasynaptic mobility. Also, a significant increase in inter-boutonal and synaptic trafficking was produced by glycine-induced synaptic potentiation, which was actin- and NMDA receptor-dependent, while NMDA-induced synaptic depression reduced the mobility of synaptic vesicles at synapses. Results showed that SynaptopHluorin (sPH)-AP-QDs revealed unobserved trafficking properties of synaptic vesicles around synapses (Lee et al. 2012). Mansson et al. demonstrated novel application of streptavidin-coated CdSe QD-labelled isolated actin filament with preserved actomyosin functions. They evidenced that labelling covers both cross-linking of filaments and cargo (enzymes) transportation. Such photostable and bright QDs facilitate filament tracking and cargo detection for extended periods (Månsson et al. 2004).

Modi et al. demonstrated the use of high and small specific amine-modified QD-conjugated recombinant single-domain antibody fragment (VHH fragment) against green fluorescent protein to deliver information on diffusion of adhesive molecules at neurotransmitter receptor and growth cone at synapses. The results revealed that QD nanobodies were able to quantify the dynamics of neurotransmitter receptors at both inhibitory and excitatory synapses in *ex vivo* rat brain slices as well as primary neuronal cells. They demonstrated a strategy for multiple imaging of targeted/tagged proteins to monitor simultaneous behaviour, transport, diffusion and clustering (Modi et al. 2018). Caglar et al. developed a tool for analysing response of photoluminescence *in vivo* under AC and DC voltage changes and highlighted their imaging potential. They screened InP/ZnS and CdSe/CdS QDs to develop characteristics of PL/voltage on a chip. Such measurements with neuronal cells showed that QDs were used to track voltage changes sensitively. Also, CdSe/CdS QDs with more significant photoluminescence effect on depolarisation of membrane have lower cytotoxicity, which makes InP/ZnS more suitable for sensing in living (Caglar et al. 2019). In addition to this, Efros et al. examined the role of semiconductor trioctylphosphine oxide-coated QDs incorporated in liposome in addressing challenges of real-time optical voltage imaging (Efros et al. 2018).

5.6.1.2 Single Protein Tracking

QD single-particle tracking is a novel super-resolution imaging technique that utilises semiconductor nanocrystal QDs as a powerful tool for protein and lipid behaviour analysis in the membrane and as fluorescent probes (Bannai et al. 2020). However, targeting single-particle in rat hippocampus neuron with pH-sensitive QD probe reports movements of receptors on the surface. Taylor et al. evaluated a subpopulation of neuronal EphB2 receptors, which directed motion between the plasma membrane and synapses (Taylor et al. 2018). Varela et al. targeted dopamine receptors with functionalised QDs and performed single-molecule tracking *in vivo*. They also proposed a novel way to delocalised and non-inflammatory ways to deliver nanoparticles *in vivo* in the brain, which allowed to track and label genetically engineered surface of dopamine receptors in neurons, revealing regulation of activity and inherent behaviour in pathological and physiological animals (Varela et al. 2016).

5.6.1.3 Cell Tracking and Migration

Jayagopal et al. demonstrated a technique for *ex vivo* imaging of cellular recruitment in atherogenesis which utilises colour-code cell types to QDs within lesion area. It's well known that atherosclerosis progression is associated with infiltration of leucocytes within lesions. They coated QDs with fluorescently labelled immunomagnetically isolated macrophages/monocytes and T lymphocytes to maurocalcine (CPPs). QD–maurocalcine bioconjugates efficiently labelled both cell types with preserved cell viability and efficiency and did not disturb native leucocyte functionality in endothelial adhesion assay and cytokine release. They reinfused QDs–macrophages/monocytes and T lymphocytes in the ApoE-deficient mouse model of atherosclerosis and further observed that within two days of injection, the QD-labelled cells were visible in atherosclerotic plaque.

High signal-to-noise (S/N) ratio imaging of multiple biomarkers and cell types within the same specimen was enabled by this method of tracking leucocytes in lesions. Further, it also possesses great applicability in investigating the role of distinct circulating leucocyte subsets in the development and progression of plaques (Jayagopal et al. 2009)

Bilen et al. evaluated the feasibility of time-resolved fluorescence spectroscopy (TRFS) and scanning acoustic microscopy (SAM) in the characterisation of atherosclerotic plaque. They performed dual-modality imaging of human carotid atherosclerotic plaque, where they showed that acoustic impedance values were statistically lower in collagen-rich regions compared to calcified areas. They involved CdTe/CdS QDs for atherosclerotic plaque imaging using TRFS and showed a difference in fluorescence lifetime values of QDs in both the regions. Where TRFS provides information regarding the molecular environment of plaque, SAM highlighted the mechanism and structural information of the plaque (Bilen et al. 2018).

5.6.2 In Vitro/Ex Vivo Cell Imaging

In addition to their usage for *in vivo* imaging at the cellular and molecular levels, QDs have also been used widely as *in vitro* imaging agents (Table 5.1). Near-infrared QDs are an emerging novel class of fluorescent labels with strong tissue penetrability, sufficient electron density, excellent fluorescent stability and high fluorescent intensity. Such QDs possess the potential for *in vivo* imaging, early cancer diagnosis and high-resolution electron microscopy. Brunetti et al. constructed NT4 cancer-selective tetra-branched peptides conjugated with near-infrared QDs and functionalised them with amine-derived PEG. They also observed its promising role in imaging and targeting tumour cells in *in vivo* HT29 xenografted mice and *in vitro* HT29 cancer cells. The formulated near-infrared QDs with NT4 cancer-selective tetra-branched peptides were effective cancer theranostics as compared with the well-established high cancer selectivity. The PEG-coated QDs produced desired bioactivity with biocompatibility, stability, improved water solubility, tumour retention and reduced aggregation and systemic toxicity (Brunetti et al. 2018).

On the other side, nanometre-sized luminescent semiconductor QDs have also been involved as therapeutic and imaging agents in numerous diseases. Zhang et al. investigated cellular uptake, colloidal stability and toxicity of QDs in *in vivo*, *in vitro* and *ex vivo* environments of the brain. They observed that the behaviour of QDs is dependent on surface functionality and its treatment with cultured organotypic whole-hemisphere slices leads to increased metallothionein levels and dose-dependent toxicity. No change was obtained in the expression of mRNA in inflammatory cytokines or oxidative stress markers. PEG coating over the surface of QDs provided protection from aggregation in neurophysiological tissues and fluids. Notably, the brain microenvironment enables the development of QD-based imaging probes capable of targeting regions in the central nervous system and alters cellular interactions and localisation based on intended outcome (Zhang et al. 2019).

Geng et al. investigated the behaviour of clonal growth and analysed proliferation characteristics of varied cancer cells including SGC7901 human gastric cancer cell line, SW480 human colon cancer cell line and MCF 7 human breast cancer cell line. *In vitro* progression and tumour development were stimulated by cell clone formation assay. This was analysed using pan-CK and proliferating cell nuclear antigen Ki67, marked by different QD-conjugated probes. Parameters such as distribution in clone, Ki67 expression, discrete tendency, cell morphology and clone formation rate were investigated using QD-based molecular-targeted imaging. All three cell lines showed significant expression of Ki67 in clones and asymmetric growth behaviour. As a result, cell clone formation assay based on QD molecular imaging provided a novel method to understand the proliferative feature and morphological characteristics of cancer cells, providing an insight into tumour biology. This suggested that there is theoretical feasibility and technical possibility to develop a differentiation strategy which may allow for the control and coordination of cancer cells (Geng XF et al. 2016).

5.6.3 Tissue Imaging

Near-infrared fluorescence provides several advantages for *in vivo* and tissue imaging (Cassette et al. 2013). Optical imaging in preclinical and clinical settings provides enriched biological information, particularly when coupled with the targeted nanoparticles. Ryan et al. used clinical X-ray system to map the distribution of CdTe QD in mice to demonstrate excitation of X-rays of QDs emitting in near-infrared. They elicited near-infrared signals from the deep organ with short durable and tolerable radiations to permit *in vivo* applications. Notably, the application of keV X-rays to produce emission from tissue and QDs presents a novel bioimaging technology (Ryan et al. 2019).

5.6.4 Diagnostic Tool for Detection of Diseases

Tumour heterogeneity is one of the challenging and most important problems not only in understanding the mechanism of cancer but also in developing therapeutics to eradicate cancer cells. Liu et al. demonstrated the use of wavelength-resolved imaging and multiplexed QD–antibody conjugates for molecular mapping of human prostate cancer tissue. Multiplexed QD mapping provides morphological and molecular information for a clinical diagnostic application that is unavailable from traditional profiling and staining methods. They showed detection and characterisation of single and prostate gland cancer cells by using a panel of four protein biomarkers (α-methyl acyl-CoA racemase, p63, high molecular weight cytokeratin and E-cadherin). The results revealed the presence of tumour heterogeneity at the developmental, cellular and molecular levels, which permits direct visualisation of prostate undergoing structural transitions up to malignancy of cells (Liu et al. 2010).

Further, Shi et al. developed a biocompatible multifunctional GQD-coated, high-luminescence magnetic nanoplatform for the selective separation and diagnosis of glypican-3 (GPC3)-expressed HepG2 liver circulating tumour cells from infected blood. Experimental results indicated that, because of the presence of a two-photon absorption cross section (40530 GM), an anti-GPC3 antibody-attached GOQD-coated magnetic nanoplatform could be incorporated as a two-photon luminescence platform for bright and selective imaging of HepG2 tumour cells in biological transparency window (960 nm). These results were evident from SK-BR-3 breast cancer cells and non-targeted GPC3(−) cells that showed two-photon imaging and are high selective for HepG2 hepatocellular carcinoma (HCC) tumour cells (Shi et al. 2015). Also, Morales-Narvaez et al. compared the potential of fluorescent dye Alexa 647 and CdSe/ZnS QDs as a reporter in sandwich immunocomplex microarray assay to detect apolipoprotein E. Although the performance of QDs varied as a function of excitation wavelength, they were proved as efficient reporters in microarrays. At 532 nm, QD microarray provided a limit of detection of ~62 pg/mL; however, at 633 nm, it provided a limit of detection of ~247 pg/mL. This was seven times more than that of ELISA and five times more than that of Alexa microarray. At last, human serum samples were also assessed, which gave high acceptability and sensitivity. Thus, the approach could be extended to the multiplexed detection of apolipoprotein and other Alzheimer-related biomarkers (Morales-Narváez et al. 2012).

5.6.5 Development of Diagnostic Test Systems

Samuel et al. studied the positively charged and negatively charged CdTe QDs' effect on human platelet functions in the absence or presence of plasma. They investigated interactions of QDs with platelet using transmission electron microscopy, atomic force microscopy and immunofluorescence. Also, the QD–platelet effects were screened using gelatin zymography, flow cytometry, quartz crystal microbalance with dissipation, and light aggregometry. They showed that binding of QDs with platelet plasma membrane was due to matrix metalloproteinase-2 release and upregulation of glycoprotein IIB/IIIa and P-selectin receptors. Further, the mechanism of the functional response of platelets to ultra-small QDs was unravelled for the first time. They reported that QDs can stimulate platelet aggregation in both underflow and no-flow conditions (Samuel et al. 2015). Kim et al. developed a clinical validation of QD barcode-based technology for diagnosing hepatitis B virus. Also, involvement of multiplexed QD barcode for the detection of multiple regions of the viral genome was reported to improve clinical sensitivity (Kim et al. 2016). Zhang et al. developed a sensitive and simple method for the detection of infectious disease markers that conjugate the dot-blot immunoassay with reporter as CdSe/ZnSe/ZnS core/shell/shell QD nanobeads prepared using oil-in-water emulsion evaporation technique. They detected proteins in a step test by developing QD nanobeads as a signal indicator, as low as 78 pg hepatitis surface antigen (Zhang et al. 2014).

5.6.6 Biosensors/Biomarkers for Detection of Mutation, Multiplexed Target and miRNA Detection

Since the last few decades, one of the greatest achievements in nanomaterials has been the development of biosensors. Biosensors are devices containing biological sensing elements either integrated or connected in transducers. After recognition of specific molecules in the body on the basis of structure including receptor hormone, enzyme–substrate, antibody and antigen, biosensors exhibit their mechanism of action (Kricka 1988; Buch and Rechnitz 1989; Zhang et al. 2009). To produce highly efficient biosensors, the substrate selected for sensing material dispersion is a prerequisite. Different types of nanomaterials, including golden nanoparticles, carbon nanotubes, magnetic nanoparticles and QDs, can be applied as biosensors. Involving QDs with their unique properties has provided novel strategies for quantification and identification of biologically relevant molecules (Matea et al. 2017).

Chan and Nie were the first to publish reports of ZnS-CdSe QDs for ultrasensitive non-isotropic detection (Chan and Nie 1998). Johri-Ahar et al. presented a novel nanostructured immunosensor for early detection of cancer antigen 125 (CA-125) serum biomarker in patients with ovarian cancer. The immunosensor was composed of gold electrode-modified mercaptopropionic acid (MPA) and conjugated with anti-CA-125 monoclonal antibody (mAb), CdSe QDs and silica-coated gold nanoparticles (AuNP-SiO$_2$). This conjugation resulted in the sensitive detection of CA-125 down to 0.0016 u/mL. They characterised the engineered MPA-AuNP-SiO2-QD-mAb immunosensor using different spectral techniques (Johari-Ahar et al. 2015). Yang et al. used spectroscopic fluorescence techniques to propose a sensitive sensor for probing the interaction of proteins with clofazimine (an effective drug against multidrug-resistant breast cancer and tuberculosis). They developed CdZnSeS/ZnS alloyed core/thick-shell QDs as energy donors Förster resonance energy transfer (FRET) applications. These QDs were further coated with multifunctional polymers containing dihydrolipoic acid through a direct ligand exchange method and functionalised with cyanine-3-labelled human serum albumin (Yang et al. 2016).

Andreadou et al. established a Leishmania-specific surface antigen detection method based on a combination of CdSe QDs and magnetic beads with a low limit of detection (3125 ng/μL) and high specificity for Leishmania DNA. Thus, the principle behind the method is that the detection was performed by QDs and analytes were isolated from solution with the help of magnetic beads (Andreadou et al. 2016). In another study, core/multi-shell QDs (CdSe@ZnS@

CdS@ZnS) produce 100% quantum yield with photoluminescence as high as obtained by layer-by-layer deposition. The coating was performed layer by layer, where each layer deposited onto CdSe core had some specific work. The first ZnS shell acted as a potential barrier for electron–hole pairs, the ZnS shell acted as an outer shell or potential barrier, and the CdSe shell acted as a separator between two ZnS shells. They characterised two-photon properties of QDs with varied sizes. Thus, the obtained large two-photon excitation acts cross-sectionally and makes QDs an efficient photoluminescent material for multiphoton microscopy (Linkov et al. 2016). Geng et al. proposed CdSe-ZnS-conjugated haem nanoprobes in an *in vivo* and *in vitro* experimental set-up to investigate the haem iron absorption complex. Their results presented that such conjugation is suitable for tracing ham iron absorption by both endocytosis and active transport haemoglobin carrier protein-1 pathways involved in haem uptake in Caco-2 cells (Geng L et al. 2016).

5.6.7 Drug Delivery/Carrier for Treatment

Over the last decade, researchers have paid great attention to the multimodal drug delivery system. Due to unique physicochemical properties, QDs are considered as an efficient tool in theranostic applications also (Zayed et al. 2019). The term theranostics can be defined as ongoing efforts to develop individualised and more specific therapies for various diseases and to combine therapeutics and diagnostics into a single agent. Advanced drug delivery systems have been designed as strategies to resolve many complications associated with conventional formulations and provide site-specific targeted delivery of a drug (Singh and Bajwa 2016). Further strategies to increase biocompatibility and reduce toxicity have also been developed through hybridisation with lipids, polysaccharides, protein or polymers, offering tumour targeting and enhanced bioavailability (Zayed et al. 2019).

5.6.7.1 Ocular Diseases

QDs have always been suitable for multi- and single-colour bioimaging of biomolecules. QDs with near-infrared emission have been developed for *in vivo* imaging due to their absorption and scattering ability of incident light, yielding good optical signals and tissue penetration. In ophthalmology, an autofluorescence may be caused due to visible light from ocular structures leading to a reduction in the contrast of ocular fluorescence. However, QDs have the ability to offer both near-infrared and visible emission of the electromagnetic spectrum without disturbing autofluorescence and also offer a range of applications such as labelling and bioimaging in ophthalmology (Sarwat et al. 2019). Inefficient penetration across the plasma membrane has always hampered the delivery of gene, therapeutic molecule and drug to ocular tissues. Johnson et al. used a novel peptide with protein transduction properties for ocular delivery of small and large molecules across plasma membrane *in vivo* and *in vitro*. The peptide was able to enter the cell in a temperature-dependent manner within 5 min. Also, they reported that the peptide has a tendency to compact and deliver plasmid DNA. Small interfering RNA duplexes to cell using the peptide allowed silencing and achieving transgene expression in more than 50% human embryonic retinoblasts. They evidenced that the peptide for ocular delivery entered neural retina and localised to the ganglionic cell, photoreceptors and retinal pigment epithelium.

Moreover, peptides were also able to enter dura of optic nerve, choroid, sclera and corneal epithelium via topical application. The peptide for ocular delivery functions as bacteriostatic, which enables it to be a carrier of molecules to post-mitotic neural ocular tissue (Johnson et al. 2008). Further, Olsan et al. administered single intravitreal injection of biotin-conjugated QDs composed of Cd/Se coated with a thin layer of ZnS in the Royal Chemical Society rat model of progressive photoreceptor degeneration. Over weeks of post-procedure, they observed a gradual reduction in the amplitude of electroretinogram recording. They reported the effectiveness of intravitreal injection of photoactive QDs as a technology in progressive retinal degenerations followed by increased retinal electrical activity (Olson et al. 2012). In line with this, Olsan et al. also performed another experiment where they administered a single intravenous injection of biotin-conjugated photoactive QDs in the Royal Chemical Society rat model of progressive photoreceptor degeneration. Over 6 weeks of post-procedure, they observed a gradual reduction in the amplitude of electroretinogram recording. They used photoactive QDs as carriers to deliver localised electrical stimulation in retinal degeneration. The technology also possesses a tendency to overcome limitations associated with presently available treatment for blinding retinal disease. First, QD was able to deliver more precisely at the cellular level. Second, the ability of QD to conjugate allows precise targeting of different cells within retina. Third, this strategy can be employed as preventive rather than reversing a loss in pronounced photoreceptor degeneration. Fourth, therapy can be delivered with a needle rather than significant eye surgery (Olson and Mandava 2014).

5.6.7.2 Cardiovascular Diseases

Sun et al. designed trifunctional simian virus (SV40)-based nanoparticles for *in vivo* targeting and imaging of atherosclerotic plaque. Further, they encapsulated near-infrared QDs in nanoparticles and also incorporated Hirulog, an anticoagulant drug. They imaged fluorescent atherosclerotic plaque non-invasively in live ApoE-deficient mice. Targeted SV40-based nanoparticles were also able to deliver Hirulog to an atherosclerotic plaque region efficiently. The SV40-based nanoparticles encapsulating QDs showed remarkable optical properties for *in vivo* imaging. Also, early, developmental and late stages of atherosclerosis were quickly targeted and imaged in a live animal using targeting peptides for fibrin, macrophage and vascular cell adhesion molecule-1. Thus, a multifunctional and multivalent SV40-based nanoparticle was developed suitable for drug delivery, molecular targeting and *in vivo* imaging (Sun et al. 2016).

5.6.7.3 Neurological Disorders

Crossing the blood–brain barrier for the delivery of drugs and active agents from blood to the central nervous system is one of the challenging issues faced by scientists. The development of functionalised bioconjugated QDs to address this challenge for both treatment and diagnosis of brain diseases, such as HIV-associated encephalopathy or Alzheimer's disease, had been investigated in nanomedicine (Xu et al. 2013).

Medina et al. designed a novel approach for barcoding nanoparticles composed of poly(lactic-co-glycolic acid) (PLGA) with spectrally defined QDs to allow direct fluorescent detection of nanoparticle fate with subcellular resolution. Also, the biophysical properties of nanoparticles or their interaction with other cells are unaffected by QD labelling. *In vivo* imaging allowed simultaneous visualisation of the interaction of bEnd.3 cells with the targeted nanoparticle, confirming that surface modification with TAT, a CPP, increases their biophysical association with cell surfaces. Further, this modification with the CPP also facilitates brain-specific delivery specifically to brain vasculature along with tracking and imaging (Medina et al. 2017).

Paris-Robidas et al. conjugated mAbs (Ri7) with QDs targeting murine transferrin receptor. They were the first to confirm specific transferrin receptor-mediated endocytosis of QD–Ri7 in bEnd5 and N2A cells. As compared to control, intravenously injected Ri7-QD showed a fourfold higher volume of distribution in brain tissues. Most QDs within brain capillary endothelial cells were seen in small vesicles, with a small portion in multivesicular bodies and tubular structures. Parenchymal penetration of QD–Ri7 was shallow and even comparable to IgG. Thus, results showed that the QD–Ri7 complex undergoes endocytosis by brain capillary endothelial cells if administered systemically (Paris-Robidas et al. 2016).

5.6.7.4 Hepatic Diseases

Hunt et al. showed targeted delivery of Ag_2S QDs to hepatocytes *in vitro* and *in vivo* or liver sinusoidal endothelial cells (LSECs) and rapid absorption across small intestine after oral administration. The QDs were radiolabelled with ^{14}C-metformin, a fluorescent tag or 3H-oleic acid within a drug binding site. They also included three biopolymer shell coatings, namely heparin, gelatin and formaldehyde-treated serum albumin (FSA). The passage across the small intestine into mesenteric veins was mediated via micropinocytosis and clathrin endocytosis. Two hours post-ingestion, the bioavailability of ^{14}C-metformin was increased fivefold by conjugation with QD-FSA, whereas uptake of metformin to LSECs was improved 50-fold by using such QDs. Endocytosis of QDs by SK-Hep1 cells was via QD-conjugated caveolae and clathrin-mediated pathways taken up into lysosomes (Hunt et al. 2020).

Shi et al. produced HCC-targeted delivery vehicles by covalently coupling 5-fluorouracil acetic acid (FUA) and folic acid on the surface of ZnCdSe/ZnS QDs (Folic acid-QDs-FUA). The prepared complex was characterised using different spectral techniques. The drug-loading content, zeta potential and average particle size were 36.85% ± 1.61%, −13.3 mV and 220.28 nm, respectively. Folic acid-QDs-FUA showed reduced cytotoxicity and targeted the HCC (HepG2 and SMMC-7721) more easily. *In vivo* experiments showed that mice treated with Folic acid-QDs-FUA produced superior tumour suppression. Thus, Folic acid-QDs-FUA can be used for improving the efficacy of 5-FU and tumour targeting with limited toxicity (Shi et al. 2018).

5.6.7.5 Antibiotic-Resistant Infection

Sarkar et al. prepared highly luminescent carbon QDs from aloe vera leaves using the carbonisation pathway. The prepared QDs were characterised through different spectral techniques, and functional carbon QDs were screened to be non-cytotoxic. Also, a cytotoxicity study of prepared carbon QDs was performed in HepG2, A549 and HeLa cell lines. The high aqueous dispersibility, less cytotoxicity and biocompatibility of synthesised carbon QDs were aimed to design carbon QDs tailored calcium alginate (CA) hydrogen films with goal to controlled delivery of vancomycin in the gastrointestinal tract. The drug-loading capacity of CA/carbon QDs was increased to 89% as compared to CA/carbon QDs with β-cyclodextrin (β-CD), which increased to 96%. Thus, at pH 1.5, a lower release rate (56% in 120 h) and high drug-loading capacity of CA/carbon QDs with β-CD can be utilised for drug delivery (Sarkar et al. 2017).

Wansapura et al. developed chitin–CdTe QD hybrid films with an antibacterial effect against Pseudomonas aeruginosa and Staphylococcus aureus by combining CdTe QDs with chitin using the facile aqueous synthesis route. They characterised chitin–CdTe QD using X-ray diffraction analysis, thermogravimetric analysis, Fourier transform infrared spectroscopy, energy-dispersive X-ray spectroscopy and high-resolution field emission scanning electron microscopy. They also reported that an efficient antibacterial activity was shown by chitin–CdTe QD films against Gram-negative and Gram-positive bacteria. Conclusively, they evidenced that chitin–CdTe QD films might be desirable antibacterial agents for a variety of biomedical applications, including implants, ophthalmology, packaging, drug delivery systems, burn treatments and wound dressings (Wansapura et al. 2017).

5.6.7.6 Tumours

At times, the direct intracellular delivery of peptides or proteins has remained a challenge due to the bioavailability barrier across the membrane. Several pharmacological agents were discovered using the traditional approach to modulate protein functions and deliver across the cell. However, these specific molecules and drugs possess some limitations, such as poor tissue distribution, toxicity, unwanted side effects and target specificity. Likewise, over the past decades, tremendous advances have been made for the development of molecular techniques for gene delivery and protein expressions, but they have been of surprisingly little benefit for the management of genetic disorders. Apart from these, *in vivo* gene therapy approaches based on adenoviral vectors are also associated with toxicity and lack of specificity (Wadia and Dowdy 2005). Based on cell-penetrating motif of an anti-cancer peptide, i.e. buforin IIb, Lim et al. developed CPPs with cancer cell specificity. They found a 17-amino acid peptide (BR2) to be a cancer-specific derivative among all derivatives, without any cytotoxicity. BR2 entered the cell through lipid-mediated macropinocytosis after specifically targeting cancerous cells via interactions with gangliosides. Besides, BR2 was reported to translocate efficiently as compared to the CPP TAT (49-57). BR2 fused with single-chain variable fragment (scFv) directed towards K-ras-mutated HCT116 cells to demonstrate ability of BR2 as specific cancer cell carrier. It was reported that BR2-fused scFv shows a higher degree of apoptosis when compared with TAT-fused scFv. They concluded that the CPP BR2 possesses great potential to be used as a drug delivery carrier with cancer cell specificity (Lim et al. 2013).

Iannazzo et al. synthesised graphene QD (GQD)-based biocompatible and fluorescent nanovector for targeted delivery of benzofuran structure (BFG)-derived anti-cancer drug bearing targeting ligand riboflavin. The highly water-dispersible nanoparticles were covalently linked to the anti-cancer drug via cleavage PEG linker, while the ligand riboflavin (RF) was conjugated with graphene QD by π–π interaction using a pyrene (Pyr) linker. They showed that the designed drug delivery system (GQD-PEG-BFG-Pyr-RF) has less cytotoxic effect, and subsequently, its impact on three cancer cell lines was compared with the effect produced by the anti-cancer drug (GQD-Pyr-RF) which was similar to nanovector lacking targeted riboflavin ligand (GQD-PEG-BFG) (Iannazzo et al. 2019).

Nifontova et al. screened the cytotoxicity of QD-encoded microcapsules and validated an approach for the activation of microcapsule's surface for functionalisation with trastuzumab. Trastuzumab is a clinically available humanised mAb for the treatment of breast cancer-targeting human epidermal growth factor receptor 2 (HER2). They reported that polymer shell-encapsulated QDs synthesised using the layer-by-layer deposition method produce a highly fluorescent polyelectrolyte microcapsule with biocompatibility and homogenous size distribution. Activation of carbodiimide surface evidenced that QD-encapsulated microcapsules show optical characteristics and optimal dispersion before conjugation with antibody. Moreover, the conjugate of mAb targeting HER2 and microcapsule provided specific and sensitive antibody-mediated binding of microcapsule with living cancer cells (Nifontova et al. 2019).

Li et al. formulated a smart nanosystem for reducing side effects and improving the therapeutic efficiency of mitoxantrone (MTN) by cross-linking carboxylated GQDs (cGQDs) with NH2-PEG-NH2 and folic acid modification. The novel drug delivery system showed a remarkable drug-loading capacity (40.1%) and entrapment efficiency (97.5%). Morphological studies, i.e. cell imaging, showed that the nanosystem enters human cervical cells via a micropinocytosis-dependent pathway. Further, the nanodelivery system showed low systemic toxicity, improved anti-tumour ability and targeting capacity in a live animal without having any systemic adverse effects (Li et al. 2019).

Khodadadei et al. synthesised blue-fluorescent nitrogen-doped GQDs using the hydrothermal method with urea as a nitrogen source and citric acid as a carbon source. Fourier transform infrared resonance spectroscopy confirmed the existence of nitrogen-doped GQDs. Further, methotrexate (MTX), an anti-cancer drug, was loaded within nitrogen-doped GQDs to prepare MTX–nitrogen-doped GQDs as a delivery system. Successful loading of MTX in nitrogen-doped GQDs was indicated by the presence of a strong π-π stacking interaction between nitrogen-doped GQDs and MTX as confirmed by UV spectroscopy and Fourier transform infrared resonance spectroscopy. Also, the *in vitro* cytotoxicity study on MCF-7 breast cancer cell line showed that MTX–nitrogen-doped GQDs are toxic than drug-free nitrogen-doped GQD nanocarriers (Khodadadei et al. 2017).

Olerile et al. screened the potential of a co-loaded [QDs (CdTe/CdS/ZnS) and paclitaxel] nanostructured lipid carriers (NLC) as a parenteral multifunctional delivery system. The co-loaded NLC was formed by the low-temperature solidification and emulsion evaporation methods using soya phosphatidylcholine, oleic acid and glyceryl monostearate as lipid matrix. In characterising co-loaded NLC, the zeta potential, polydispersity index and particle size were found, -0.22 ± 0.03 mV, 0.17 ± 0.04 mV and 115.93 ± 1.61 nm, respectively. A higher drug-loading capacity of $4.68\% \pm 0.04\%$ and entrapment efficiency of $80.70\% \pm 2.11\%$ were recorded with a spheroid-like shape and smooth surface. The tumour growth inhibition rate and IC_{50} were found to be 77.85% and 1.05 ± 0.58 M, respectively. The *ex vivo* and *in vivo* results of co-loaded NLC indicated its capability to target and detect H22 tumour. Thus, the co-loaded NLC formulation was qualified as a parenteral drug delivery system for cancer theragnostic (Olerile et al. 2017).

Li et al. for the first time explored mitochondria as a delivery system of doxorubicin (Dox) and carbon QDs for *in vivo* imaging. To achieve *in vivo* imaging, they prepared near-fluorescent carbon QD (DyLight 680) and loaded it in the mitochondria (Mito-CQD-Dy680). As a carrier, mitochondrion was found to be compatible with carbon QD and also preserves the optical properties of carbon QD. Moreover, the system also improves the biodistribution of carbon QD and increases its retention time after intravenous injection. An enhanced therapeutic effect was shown by mitochondrion loaded with Dox as compared to free Dox. Thus, the mitochondria-based aircraft system possesses high potential for drug delivery and bioimaging against cancer and other diseases (Li et al. 2018).

5.6.7.7 Renal Diseases

Renal disease has become a prevalent problem to public health for which the application of kidney-related drug delivery system has profound transformative potential. Such drug delivery system enhances the efficacy and reduces undesired side effects of the potent drug for treating renal diseases (Liu et al. 2019).

5.6.7.8 Others

Omidi et al. produced vascular endothelial growth factor (VEGF)-loaded PLGA carbon QD microspheres using microfluidic platforms. VEGF was PEGylated to prevent structural stability and protein functional stability during the encapsulation process. The produced microsphere was highly monodispersed and intact. The loading efficiency of PEGylated VEGF in microparticles varied from 51% to 69%, and >90% of PEGylated VEGF was even released within 28 days, which was monitored by carbon QDs (Omidi et al. 2019).

5.6.8 Cell Labelling

The optical properties of QDs, precisely the wavelength of their fluorescence, depend on their size. Typically, such small size provides no direct molecular or structural observations from microscopy and needs to be labelled with a marker to be observed. It is well known that fluorescent labelling is one of the modest techniques involved in cell biology. Since QDs have unique and constant optical properties, they were utilised for the purpose of cell labelling. Interestingly, QDs have the ability to tag intracellular and multi-inner components concurrently in living cells for varied time intervals. They can be attached to a molecule like an antibody that will bind the target, or they can bind the target to be visualised. Different coloured QDs have the ability to label varied cellular components, which can be easily visualised *in vivo* or with fluorescence microscopy (Parak et al. 2005; Bajwa et al. 2016). For instance, ganglion was labelled by the conjugate of biotinylated cholera toxin B with QD–avidin for animal bioimaging, lignin and cellulose were labelled with CdSe QDs for plant bioimaging (Cheki et al. 2013), and anti-E. coli 0157-coated and streptavidin-coated were used for measuring bacterial cell in prokaryotic imaging and (Bajwa et al. 2016).

Further, the light stability of QD markers makes long-term tracking of biological molecules possible, using the labelling technique. Monitoring interactions among and within the living cells as they differentiate after growing is crucial for understanding the development of organisms. In reference to this, fluorescence microscopy has been the most widely used approach for non-invasive and high-resolution *in vivo* imaging and organic fluorophores have been the commonly used tag in fluorescence microscopy but with few limitations. Fluorescent QDs an inorganic fluorescent nanocrystal emerged to overcome the barriers and function as useful alternative for multicolour imaging. Jaiswal et al. used the mixed-surface self-assembly approach and positively charged adapters to prepare protein-conjugated QDs coated with negatively charged dihydroxylipoic acid. For instance, naturally charged avidin protein allows stable conjugation of QDs to antibodies or ligands that can be biotinylated, whereas the involvement of protein fused to oligohistidine peptide or positively charged leucine zipper peptide obviates the need for biotinylating the target molecule (Jaiswal et al. 2004).

Valley et al. involved two-coloured single QD tracking to analyse kinetics of receptor dimerisation on living. They showed that mutants are capable of forming stable ligand-independent dimers as compared to wild-type epidermal growth factor receptor (EGFR/erbB1/Her1). Several EGFR mutants have proved that ligand-independent signals are commonly available in non-small-cell lung cancers, including exon 19 deletion and kinase domain mutations L858R. Measurements of live-cell FRET showed that mutations of L858R kinase modify the ectodomain structure such that the unliganded mutant EGFR adopts a dimerisation-competent and extends confirmation. Such results support the model considering that dysregulated activity of non-small-cell lung cancer-associated kinase mutants is driven by the interaction of both extracellular domains and kinase that leads to dimerisation (Valley et al. 2015).

Despite having attractive properties such as photostability, brightness and the ability to excite different QDs with single wavelength over fluorescent proteins and dyes, only a few QD conjugates are commercially available for labelling cellular targets. Francis et al. determined the specificity of commercially available QD625 conjugated with a secondary antibody against primary IgG antibody. Simultaneously, the antigen was labelled with fluorescent dye coupled to the secondary antibody and QD 525-conjugated secondary antibody. They observed that fluorescent dye-coupled secondary antibody labelled all intended targeted proteins, but the QD-conjugated secondary antibody, specifically anti-tubulin QD625 conjugate, was able to bind quickly to some of the cytosol-based protein targets only. This labelling corresponded to steric hindrance associated with the size of the QD-conjugated secondary antibody (Francis et al. 2017).

Tracking human mesenchymal stem cells (MSCs) in living cells has remained a critical component of the evaluation of efficacy and safety of therapeutic cell products. Thus, the cell needs to be labelled with agents to allow visualisation of migration of MSCs. Kundrotas et al. designed a study to evaluate the cytotoxicity, uptake dynamics and extracellular and subcellular distribution of non-targeted carboxylated QDs in the bone marrow. They revealed that no negative impact is produced by QD on the viability of MSCs throughout the experimental period in cells. However, the presence of lipid droplets was observed in some MSCs. At low cell growing densities, QDs distribute within the MSC cytoplasm already after 1 h of incubation, reaching saturation after 6 h. Also, QDs prominently mark MSC long filopodia-like structures attaching neighbouring cells, whereas at high-cell-density cultivation, QD distribution was prominent in the extracellular matrix of MSCs.

Interestingly, the average photoluminescence time of distribution of QDs in the extracellular matrix was more than a lifetime of QD entrapped in endosomes. In conclusion, they stated that carboxylated QDs could be used as an effective and non-specific dye for labelling bone marrow MSCs (Kundrotas et al. 2019).

5.7 CONCLUSION

In recent years, QDs have attracted fascinating attention as a promising and most valuable candidate in areas of imaging, sensor, labelling and drug delivery. Their favourable characteristics such as good biocompatibility, cost-effectiveness and low toxicity emphasise their use as a theranostic agent for cancer and as a drug delivering agent in various diseases affecting systems such as renal, hepatic, neurological, cardiovascular and ocular. Further, the advancements in the field of QDs have reduced their cytotoxicity, expanded the scope of their biomedical applications and rendered QDs as the crucial device in the research of cellular processes such as receptor trafficking, uptake and intracellular delivery. Also, a part has shown a significant breakthrough in the field for antibiotic-resistant infections, microbial infections and detection of antigens. Owing to the initial success of QDs and their conjugates employed in biomedical applications, it is convincing that future research will continue to focus on the identification of various tumours with minimal side effects and evaluate the mechanisms related to drug and disease. Different ongoing research on QDs targets achieving a selective approach for labelling, better fluorescence and higher biosafety. Also, vigorous research is focusing on studying the effects of interference of QDs with normal physiology. To see the realistic translation of QDs into clinical applications, several issues still need to be addressed, such as environmental impact, synthesis protocol scalability, body clearance and overall toxicity.

REFERENCES

Amaral PE, Hall Jr DC, Pai R, Król JE, Kalra V, Ehrlich GD, Ji H-F. 2020. Fibrous phosphorus quantum dots for cell imaging. *ACS Applied Nano Materials*. 3(1): 752–759.

Andreadou M, Liandris E, Gazouli M, Mataragka A, Tachtsidis I, Goutas N, Vlachodimitropoulos D, Ikonomopoulos J. 2016. Detection of Leishmania-specific DNA and surface antigens using a combination of functionalized magnetic beads and cadmium selenite quantum dots. *Journal of Microbiological Methods*. 123:62–67.

Aswathi M, Ajitha A, Akhina H, Lovely M, Thomas S. 2018. Quantum dots: a promising tool for biomedical application. *JSM Nanotechnology & Nanomedicine*. 6:1066.

Bajwa N, Mehra NK, Jain K, Jain NK. 2016. Pharmaceutical and biomedical applications of quantum dots. *Artificial Cells, Nanomedicine, and Biotechnology*. 44(3):758–768.

Ballou B, Ernst LA, Andreko S, Harper T, Fitzpatrick JA, Waggoner AS, Bruchez MP. 2007. Sentinel lymph node imaging using quantum dots in mouse tumor models. *Bioconjugate Chemistry*. 18(2):389–396.

Bannai H, Inoue T, Hirose M, Niwa F, Mikoshiba K. 2020. Synaptic function and neuropathological disease revealed by quantum dot-single-particle tracking. In: Yamamoto N, Okada Y, editors. *Single Molecule Microscopy in Neurobiology*. Springer: New York; pp. 131–155.

Bera D, Qian L, Tseng T-K, Holloway PH. 2010. Quantum dots and their multimodal applications: a review. *Materials*. 3(4):2260–2345.

Bilen B, Gokbulut B, Kafa U, Heves E, Inci MN, Unlu MB. 2018. Scanning acoustic microscopy and time-resolved fluorescence spectroscopy for characterization of atherosclerotic plaques. *Scientific Reports*. 8(1):1–11.

Brunetti J, Riolo G, Gentile M, Bernini A, Paccagnini E, Falciani C, Lozzi L, Scali S, Depau L, Pini A. 2018. Near-infrared quantum dots labelled with a tumor selective tetrabranched peptide for *in vivo* imaging. *Journal of Nanobiotechnology*. 16(1):1–10.

Buch R, Rechnitz G. 1989. Intact chemoreceptor-based biosensors: responses and analytical limits. *Biosensors*. 4(4):215–230.

Caglar M, Pandya R, Xiao J, Foster SK, Divitini G, Chen RY, Greenham NC, Franze K, Rao A, Keyser UF. 2019. All-optical detection of neuronal membrane depolarization in live cells using colloidal quantum dots. *Nano Letters*. 19(12):8539–8549.

Cassette E, Helle M, Bezdetnaya L, Marchal F, Dubertret B, Pons T. 2013. Design of new quantum dot materials for deep tissue infrared imaging. *Advanced Drug Delivery Reviews*. 65(5):719–731.

Chan WC, Nie S. 1998. Quantum dot bioconjugates for ultrasensitive nonisotopic detection. *Science*. 281(5385):2016–2018.

Cheki M, Moslehi M, Assadi M. 2013. Marvelous applications of quantum dots. *European Review for Medical and Pharmacological Sciences*. 17(9):1141–1148.

Claro MS, Levy I, Gangopadhyay A, Smith DJ, Tamargo MC. 2019. Self-assembled Bismuth Selenide (Bi_2Se_3) quantum dots grown by molecular beam epitaxy. *Scientific Reports*. 9(1):1–7.

Efros AL, Delehanty JB, Huston AL, Medintz IL, Barbic M, Harris TD. 2018. Evaluating the potential of using quantum dots for monitoring electrical signals in neurons. *Nature Nanotechnology*. 13(4):278–288.

Efros AL, Nesbitt DJ. 2016. Origin and control of blinking in quantum dots. *Nature Nanotechnology*. 11(8):661.

Ehlers MD, Heine M, Groc L, Lee M-C, Choquet D. 2007. Diffusional trapping of GluR1 AMPA receptors by input-specific synaptic activity. *Neuron*. 54(3):447–460.

Francis JE, Mason D, Lévy R. 2017. Evaluation of quantum dot conjugated antibodies for immunofluorescent labelling of cellular targets. *Beilstein Journal of Nanotechnology*. 8(1):1238–1249.

Geng L, Duan X, Wang Y, Zhao Y, Gao G, Liu D, Chang Y-Z, Yu P. 2016. Quantum dots-hemin: preparation and application in the absorption of heme iron. *Nanomedicine: Nanotechnology, Biology and Medicine*. 12(7):1747–1755.

Geng XF, Fang M, Liu SP, Li Y. 2016. Quantum dot-based molecular imaging of cancer cell growth using a clone formation assay. *Molecular Medicine Reports*. 14(4):3007–3012.

Horoz S, Lu L, Dai Q, Chen J, Yakami B, Pikal J, Wang W, Tang J. 2012. CdSe quantum dots synthesized by laser ablation in water and their photovoltaic applications. *Applied Physics Letters*. 101(22):223902.

Hunt NJ, Lockwood GP, Le Couteur FH, McCourt PA, Singla N, Kang SWS, Burgess A, Kuncic Z, Le Couteur DG, Cogger VC. 2020. Rapid intestinal uptake and targeted delivery to the liver endothelium using orally administered silver sulfide quantum dots. *ACS Nano.* 14(2): 1492–1507.

Iannazzo D, Pistone A, Celesti C, Triolo C, Patané S, Giofré SV, Romeo R, Ziccarelli I, Mancuso R, Gabriele B. 2019. A smart nanovector for cancer targeted drug delivery based on graphene quantum dots. *Nanomaterials.* 9(2):282.

Jaiswal JK, Goldman ER, Mattoussi H, Simon SM. 2004. Use of quantum dots for live cell imaging. *Nature Methods.* 1(1):73–78.

Jayagopal A, Su YR, Blakemore JL, Linton MF, Fazio S, Haselton FR. 2009. Quantum dot mediated imaging of atherosclerosis. *Nanotechnology.* 20(16):165102.

Jie G, Zhao Y, Niu S. 2013. Amplified electrochemiluminescence detection of cancer cells using a new bifunctional quantum dot as signal probe. *Biosensors and Bioelectronics.* 50:368–372.

Jin S, Hu Y, Gu Z, Liu L, Wu H-C. 2011. Application of quantum dots in biological imaging. *Journal of Nanomaterials.* 2011, Article ID 834139.

Johari-Ahar M, Rashidi M, Barar J, Aghaie M, Mohammadnejad D, Ramazani A, Karami P, Coukos G, Omidi Y. 2015. An ultra-sensitive impedimetric immunosensor for detection of the serum oncomarker CA-125 in ovarian cancer patients. *Nanoscale.* 7(8):3768–3779.

Johnson LN, Cashman SM, Kumar-Singh R. 2008. Cell-penetrating peptide for enhanced delivery of nucleic acids and drugs to ocular tissues including retina and cornea. *Molecular Therapy.* 16(1):107–114.

Khodadadei F, Safarian S, Ghanbari N. 2017. Methotrexate-loaded nitrogen-doped graphene quantum dots nanocarriers as an efficient anticancer drug delivery system. *Materials Science and Engineering: C.* 79:280–285.

Kim J, Biondi MJ, Feld JJ, Chan WC. 2016. Clinical validation of quantum dot barcode diagnostic technology. *ACS Nano.* 10(4):4742–4753.

Kricka L. 1988. Molecular and ionic recognition by biological systems. In: Edmonds TE, editor. *Chemical Sensors.* Blackie and Sons: Glasgow; pp. 3–14.

Kundrotas G, Karabanovas V, Pleckaitis M, Juraleviciute M, Steponkiene S, Gudleviciene Z, Rotomskis R. 2019. Uptake and distribution of carboxylated quantum dots in human mesenchymal stem cells: cell growing density matters. *Journal of nanobiotechnology.* 17(1):39.

Lalu L, Tambe V, Pradhan D, Nayak K, Bagchi S, Maheshwari R, Kalia K, Tekade RK. 2017. Novel nanosystems for the treatment of ocular inflammation: current paradigms and future research directions. *Journal of Controlled Release.* 268:19–39.

Lee S, Jung KJ, Jung HS, Chang S. 2012. Dynamics of multiple trafficking behaviors of individual synaptic vesicles revealed by quantum-dot based presynaptic probe. *PLoS One.* 7(5):e38045.

Li W-Q, Wang Z, Hao S, Sun L, Nisic M, Cheng G, Zhu C, Wan Y, Ha L, Zheng S-Y. 2018. Mitochondria-based aircraft carrier enhances *in vivo* imaging of carbon quantum dots and delivery of anticancer drug. *Nanoscale.* 10(8):3744–3752.

Li Z, Fan J, Tong C, Zhou H, Wang W, Li B, Liu B, Wang W. 2019. A smart drug-delivery nanosystem based on carboxylated graphene quantum dots for tumor-targeted chemotherapy. *Nanomedicine.* 14(15):2011–2025.

Lim KJ, Sung BH, Shin JR, Lee YW, Yang KS, Kim SC. 2013. A cancer specific cell-penetrating peptide, BR2, for the efficient delivery of an scFv into cancer cells. *PloS one.* 8(6):e66084.

Lim YT, Kim S, Nakayama A, Stott NE, Bawendi MG, Frangioni JV. 2003. Selection of quantum dot wavelengths for biomedical assays and imaging. *Molecular Imaging.* 2(1):15353500200302163.

Linkov P, Krivenkov V, Nabiev I, Samokhvalov P. 2016. High quantum yield CdSe/ZnS/CdS/ZnS multishell quantum dots for biosensing and optoelectronic applications. *Materials Today: Proceedings.* 3(2):104–108.

Liu BR, Winiarz JG, Moon J-S, Lo S-Y, Huang Y-W, Aronstam RS, Lee H-J. 2013. Synthesis, characterization and applications of carboxylated and polyethylene-glycolated bifunctionalized InP/ZnS quantum dots in cellular internalization mediated by cell-penetrating peptides. *Colloids and Surfaces B: Biointerfaces.* 111:162–170.

Liu CP, Hu Y, Lin JC, Fu HL, Lim LY, Yuan ZX. 2019. Targeting strategies for drug delivery to the kidney: from renal glomeruli to tubules. *Medicinal Research Reviews.* 39(2):561–578.

Liu J, Lau SK, Varma VA, Moffitt RA, Caldwell M, Liu T, Young AN, Petros JA, Osunkoya AO, Krogstad T. 2010. Molecular mapping of tumor heterogeneity on clinical tissue specimens with multiplexed quantum dots. *ACS Nano.* 4(5):2755–2765.

Liu S, Su X. 2014. The synthesis and application of I–III–VI type quantum dots. *RSC Advances.* 4(82):43415–43428.

Ma Y, Wang M, Li W, Zhang Z, Zhang X, Tan T, Zhang X-E, Cui Z. 2017. Live cell imaging of single genomic loci with quantum dot-labeled TALEs. *Nature Communications.* 8(1):1–8.

Maheshwari R, Chourasiya Y, Bandopadhyay S, Katiyar PK, Sharma P, Deb PK, Tekade RK. 2018. Chapter 1 – Levels of solid state properties: role of different levels during pharmaceutical product development. In: Tekade RK, editor. *Dosage Form Design Parameters.* Academic Press: London; pp. 1–30.

Maheshwari R, Tekade M, Sharma PA, Tekade RK. 2015. Nanocarriers assisted siRNA gene therapy for the management of cardiovascular disorders. *Current Pharmaceutical Design.* 21(30):4427–4440.

Mandal A, Chakrabarti S. 2017. *Impact of Ion Implantation on Quantum Dot Heterostructures and Devices.* Springer: Singapore.

Månsson A, Sundberg M, Balaz M, Bunk R, Nicholls IA, Omling P, Tågerud S, Montelius L. 2004. In vitro sliding of actin filaments labelled with single quantum dots. *Biochemical and Biophysical Research Communications.* 314(2):529–534.

Matea CT, Mocan T, Tabaran F, Pop T, Mosteanu O, Puia C, Iancu C, Mocan L. 2017. Quantum dots in imaging, drug delivery and sensor applications. *International Journal of Nanomedicine.* 12:5421.

Medina DX, Householder KT, Ceton R, Kovalik T, Heffernan JM, Shankar RV, Bowser RP, Wechsler-Reya RJ, Sirianni RW. 2017. Optical barcoding of PLGA for multispectral analysis of nanoparticle fate *in vivo*. *Journal of Controlled Release.* 253:172–182.

Modi S, Higgs NF, Sheehan D, Griffin LD, Kittler JT. 2018. Quantum dot conjugated nanobodies for multiplex imaging of protein dynamics at synapses. *Nanoscale.* 10(21):10241–10249.

Moondra S, Maheshwari R, Taneja N, Tekade M, Tekadle RK. 2018. Chapter 6 – Bulk level properties and its role in formulation development and processing. In: Tekade RK, editor. *Dosage Form Design Parameters*. Academic Press: London; pp. 221–256.

Morales-Narváez E, Montón H, Fomicheva A, Merkoçi A. 2012. Signal enhancement in antibody microarrays using quantum dots nanocrystals: application to potential Alzheimer's disease biomarker screening. *Analytical Chemistry*. 84(15):6821–6827.

Murphy-Royal C, Dupuis JP, Varela JA, Panatier A, Pinson B, Baufreton J, Groc L, Oliet SH. 2015. Surface diffusion of astrocytic glutamate transporters shapes synaptic transmission. *Nature neuroscience*. 18(2):219–226.

Nifontova G, Ramos-Gomes F, Baryshnikova M, Alves F, Nabiev I, Sukhanova A. 2019. Cancer cell targeting with functionalized quantum dot-encoded polyelectrolyte microcapsules. *Frontiers in Chemistry*. 7:34.

Olerile LD, Liu Y, Zhang B, Wang T, Mu S, Zhang J, Selotlegeng L, Zhang N. 2017. Near-infrared mediated quantum dots and paclitaxel co-loaded nanostructured lipid carriers for cancer theragnostic. *Colloids and Surfaces B: Biointerfaces*. 150:121–130.

Olson J, Mandava N. 2014. Method for stimulating retinal response using photoactive devices. Google Patents.

Olson JL, Velez-Montoya R, Mandava N, Stoldt CR. 2012. Intravitreal silicon-based quantum dots as neuroprotective factors in a model of retinal photoreceptor degeneration. *Investigative Ophthalmology & Visual Science*. 53(9):5713–5721.

Omidi M, Hashemi M, Tayebi L. 2019. Microfluidic synthesis of PLGA/carbon quantum dot microspheres for vascular endothelial growth factor delivery. *RSC Advances*. 9(57):33246–33256.

Parak WJ, Pellegrino T, Plank C. 2005. Labelling of cells with quantum dots. *Nanotechnology*. 16(2):R9.

Paris-Robidas S, Brouard D, Emond V, Parent M, Calon F. 2016. Internalization of targeted quantum dots by brain capillary endothelial cells *in vivo*. *Journal of Cerebral Blood Flow & Metabolism*. 36(4):731–742.

Peng C-W, Li Y. 2010. Application of quantum dots-based biotechnology in cancer diagnosis: current status and future perspectives. *Journal of Nanomaterials*. 2010, Article ID 676839.

Rosenthal SJ, Chang JC, Kovtun O, McBride JR, Tomlinson ID. 2011. Biocompatible quantum dots for biological applications. *Chemistry & Biology*. 18(1):10–24.

Ryan SG, Butler MN, Adeyemi SS, Kalber T, Patrick PS, Thin MZ, Harrison IF, Stuckey DJ, Pule M, Lythgoe MF. 2019. imaging of X-ray-excited emissions from quantum dots and biological tissue in whole mouse. *Scientific Reports*. 9(1):1–10.

Samuel SP, Santos-Martinez MJ, Medina C, Jain N, Radomski MW, Prina-Mello A, Volkov Y. 2015. CdTe quantum dots induce activation of human platelets: implications for nanoparticle hemocompatibility. *International Journal of Nanomedicine*. 10:2723.

Sarkar N, Sahoo G, Das R, Prusty G, Swain SK. 2017. Carbon quantum dot tailored calcium alginate hydrogel for pH responsive controlled delivery of vancomycin. *European Journal of Pharmaceutical Sciences*. 109:359–371.

Sarwat S, Stapleton F, Willcox M, Roy M. 2019. Quantum dots in ophthalmology: a literature review. *Current Eye Research*. 44(10):1037–1046.

Sharma PA, Maheshwari R, Tekade M, Tekade RK. 2015. Nanomaterial based approaches for the diagnosis and therapy of cardiovascular diseases. *Current Pharmaceutical Design*. 21(30):4465–4478.

Shi X, He D, Tang G, Tang Q, Xiong R, Ouyang H, Yu C-y. 2018. Fabrication and characterization of a folic acid-bound 5-fluorouracil loaded quantum dot system for hepatocellular carcinoma targeted therapy. *RSC Advances*. 8(35):19868–19878.

Shi Y, Pramanik A, Tchounwou C, Pedraza F, Crouch RA, Chavva SR, Vangara A, Sinha SS, Jones S, Sardar D. 2015. Multifunctional biocompatible graphene oxide quantum dots decorated magnetic nanoplatform for efficient capture and two-photon imaging of rare tumor cells. *ACS Applied Materials & Interfaces*. 7(20):10935–10943.

Singh PA, Bajwa N. 2016. Nano technical trends for cancer treatment: quantum dots a smart drug delivery system. *International Journal of Pharmaceutical Sciences and Research*. 7(4):1360.

Sun X, Li W, Zhang X, Qi M, Zhang Z, Zhang X-E, Cui Z. 2016. *In vivo* targeting and imaging of atherosclerosis using multifunctional virus-like particles of simian virus 40. *Nano Letters*. 16(10):6164–6171.

Taylor RD, Heine M, Emptage NJ, Andreae LC. 2018. Neuronal receptors display cytoskeleton-independent directed motion on the plasma membrane. *Iscience*. 10:234–244.

Tekade RK, Maheshwari R, Soni N, Tekade M, Chougule MB. 2017. Chapter 1 – Nanotechnology for the development of nanomedicine A2 – Mishra, Vijay. In: Kesharwani P, Amin MCIM, Iyer A, editors. *Nanotechnology-Based Approaches for Targeting and Delivery of Drugs and Genes*. Academic Press: Amsterdam; pp. 3–61.

Thorne RG, Nicholson C. 2006. In vivo diffusion analysis with quantum dots and dextrans predicts the width of brain extracellular space. *Proceedings of the National Academy of Sciences*. 103(14):5567–5572.

Uehara T, Ishii D, Uemura T, Suzuki H, Kanei T, Takagi K, Takama M, Murakami M, Akizawa H, Arano Y. 2010. γ-glutamyl PAMAM dendrimer as versatile precursor for dendrimer-based targeting devices. *Bioconjugate Chemistry*. 21(1):175–181.

Valley CC, Arndt-Jovin DJ, Karedla N, Steinkamp MP, Chizhik AI, Hlavacek WS, Wilson BS, Lidke KA, Lidke DS. 2015. Enhanced dimerization drives ligand-independent activity of mutant epidermal growth factor receptor in lung cancer. *Molecular Biology of the Cell*. 26(22):4087–4099.

Varela JA, Dupuis JP, Etchepare L, Espana A, Cognet L, Groc L. 2016. Targeting neurotransmitter receptors with nanoparticles *in vivo* allows single-molecule tracking in acute brain slices. *Nature Communications*. 7(1):1–10.

Wadia JS, Dowdy SF. 2005. Transmembrane delivery of protein and peptide drugs by TAT-mediated transduction in the treatment of cancer. *Advanced Drug Delivery Reviews*. 57(4):579–596.

Wagner AM, Knipe JM, Orive G, Peppas NA. 2019. Quantum dots in biomedical applications. *Acta Biomaterialia*. 94:44–63.

Wansapura PT, Dassanayake RS, Hamood A, Tran P, Moussa H, Abidi N. 2017. Preparation of chitin-CdTe quantum dots films and antibacterial effect on *Staphylococcus aureus* and *Pseudomonas aeruginosa*. *Journal of Applied Polymer Science*. 134(22):44904.

Weng KC, Noble CO, Papahadjopoulos-Sternberg B, Chen FF, Drummond DC, Kirpotin DB, Wang D, Hom YK, Hann B, Park JW. 2008. Targeted tumor cell internalization and imaging of multifunctional quantum dot-conjugated immunoliposomes *in vitro* and *in vivo*. *Nano Letters.* 8(9):2851–2857.

Xu G, Mahajan S, Roy I, Yong K-T. 2013. Theranostic quantum dots for crossing blood–brain barrier *in vitro* and providing therapy of HIV-associated encephalopathy. *Frontiers in Pharmacology.* 4:140.

Yaghini E, Turner HD, Le Marois AM, Suhling K, Naasani I, MacRobert AJ. 2016. *In vivo* biodistribution studies and *ex vivo* lymph node imaging using heavy metal-free quantum dots. *Biomaterials.* 104:182–191.

Yang HY, Fu Y, Jang M-S, Li Y, Lee JH, Chae H, Lee DS. 2016. Multifunctional polymer ligand interface CdZnSeS/ZnS quantum Dot/Cy3-labeled protein pairs as sensitive FRET sensors. *ACS Applied Materials & Interfaces.* 8(51):35021–35032.

Zayed DG, AbdElhamid AS, Freag MS, Elzoghby AO. 2019. Hybrid quantum dot-based theranostic nanomedicines for tumor-targeted drug delivery and cancer imaging. *Future Medicine.*

Zhang M, Bishop BP, Thompson NL, Hildahl K, Dang B, Mironchuk O, Chen N, Aoki R, Holmberg VC, Nance E. 2019. Quantum dot cellular uptake and toxicity in the developing brain: implications for use as imaging probes. *Nanoscale Advances.* 1(9):3424–3442.

Zhang P, Lu H, Chen J, Han H, Ma W. 2014. Simple and sensitive detection of HBsAg by using a quantum dots nanobeads based dot-blot immunoassay. *Theranostics.* 4(3):307.

Zhang X, Guo Q, Cui D. 2009. Recent advances in nanotechnology applied to biosensors. *Sensors.* 9(2):1033–1053.

Zhao M-X, Zeng E-Z. 2015. Application of functional quantum dot nanoparticles as fluorescence probes in cell labeling and tumor diagnostic imaging. *Nanoscale Research Letters.* 10(1):1–9.

Zhao Y, Shaffer TM, Das S, Pérez-Medina C, Mulder WJ, Grimm J. 2017. Near-infrared quantum dot and 89Zr dual-labeled nanoparticles for *in vivo* Cerenkov imaging. *Bioconjugate Chemistry.* 28(2):600–608.

6 PLGA-Based Micro- and Nano-particles
From Lab to Market

Akash Chaurasiya, Parameswar Patra, and Pranathi Thathireddy
Birla Institute of Technology and Science

Amruta Gorajiya
Amneal Pharmaceuticals

CONTENTS

Abbreviations 83
6.1 Introduction 83
6.2 Physicochemical Properties of PLGA 84
    6.2.1 Molecular Weight 84
    6.2.2 Solubility 84
    6.2.3 Polymer Erosion 84
    6.2.4 PLA to PGA Content 84
    6.2.5 Glass Transition Temperature (Tg) 84
6.3 Advantages and Limitations of PLGA 85
6.4 PLGA Application 85
    6.4.1 Controlled and Sustained Drug Delivery 85
    6.4.2 Drug Stabilisation 86
    6.4.3 Targeted Drug Delivery 86
    6.4.4 PLGA-Based Carrier System for Remotely Stimulated Cancer Therapy 87
6.5 PLGA-Based Commercially Available Products 89
6.6 Challenges for PLGA-Based Drug Delivery Systems – Clinical and Commercial Success 90
    6.6.1 Scale-Up and Large-Scale Production 90
    6.6.2 Biocompatibility and Safety Challenges 91
    6.6.3 Intellectual Property (IP) 92
    6.6.4 Regulatory Requirements 92
6.7 Conclusion and Future Perspectives for PLGA as Carrier System 92
References 92

ABBREVIATIONS

DNA	Deoxyribonucleic acid
EMA	European Medicines Agency
GMP	Good Manufacturing Practices
IP	Intellectual property
LAR	Long-acting release
LHRH	Luteinizing hormone-releasing hormone
MP	Microparticles
NP	Nanoparticle
PEG	Polyethylene glycol
PGA	Polyglycolic acid
PLA	Polylactic acid
PLGA	Poly(lactic-co-glycolic) acid
PMDA	Pharmaceuticals and Medical Devices Agency
PVA	Polyvinyl alcohol
RNA	Ribonucleic acid
T_g	Glass transition temperature
TGA	Therapeutic Goods Administration
USFDA	United States Food and Drug Administration

6.1 INTRODUCTION

The application of polymers in pharmaceutical products offers valuable tools for the treatment of diseases. Therefore, a wide range of non-degradable and biodegradable polymers have been developed and investigated for the drug delivery aspect. Non-degradable polymers have shown limited use due to the toxicological aspects and therefore stimulate the foundations for the invention and usage of biodegradable polymers in pharmaceutical applications. Among the various biodegradable polymers, poly(lactic-co-glycolic) acid (PLGA) gains major attention in commercial usage due

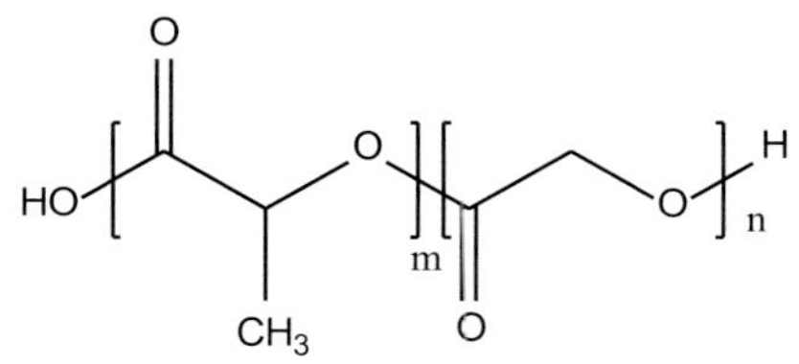

FIGURE 6.1 Structure of PLGA

to its unique characteristics and several advantages (Mir et al. 2017). PLGA is a copolymer of polylactic acid (PLA)/polyglycolic acid (PGA) and, by virtue of its biocompatibility, biodegradability and non-toxicity characteristics, has attracted many scientists to investigate its potential in drug delivery (Figure 6.1). Apart from this, PLGA, with varied ratios of PLA and PGA, exhibits a wide range of erosion rates and thereby provides flexibility for product development scientists to develop controlled and targeted drug delivery systems as per intended usage. PLGA is well approved as a functional excipient by regulatory agencies (USFDA, EMA, etc.) across the globe and therefore has been used in various clinically successful formulations such as microspheres, nanoparticulates and implants (Kapoor et al. 2015).

Currently, PLGA-based carrier system has been under extensive research for product development of various small, large and other macromolecules for several disease treatments, notably cancer, cardiovascular, neurodegenerative and inflammatory, along with theranostic applications. Despite the fact that PLGA has proven its application in pharmaceutical space and controlled drug delivery, it also becomes important to consider various factors to develop suitable formulation for intended application. In this chapter, we will address an overview of possible applications, limitations and future perspectives for usages of PLGA in drug delivery for commercial use.

6.2 PHYSICOCHEMICAL PROPERTIES OF PLGA

Physicochemical properties of polymers are important in the design and development of any formulation to impart desired characteristics. Particularly with polymers such as PLGA, which can be modified in many ways to attain desirable characteristics of the formulation, a thorough understanding of the physicochemical properties would aid in designing a better formulation. Some of the important physicochemical properties that play an important role in the design and development of PLGA-based carrier systems are molecular weight, hydrophilic or hydrophobic nature, crystallinity and glass transition temperature (T_g) (Rana et al. 2013).

6.2.1 MOLECULAR WEIGHT

The variation in molecular weight of PLGA can influence chemical and physical properties of the carrier system and regulate the polymer degradation and drug-release rate. Higher molecular weight PLGA has long polymer chain which in turn takes longer time for degradation and thereby useful for design of extended-release formulations (Sharma et al. 2016), whereas lower molecular weight PLGA shows a higher release rate due to its faster degradation and looser structure.

6.2.2 SOLUBILITY

PLGA is soluble in a wide range of non-aqueous solvents such as tetrahydrofuran, dichloromethane and dimethyl sulfoxide. PLGA undergoes hydrolysis in presence of water due to ester linkages. Compared to PGA, PLA is more hydrophobic due to the presence of methyl side groups, and therefore, lactide-rich polymers degrade slowly (Schliecker et al. 2003).

6.2.3 POLYMER EROSION

Erosion of PLGA occurs due to hydrolytic degradation through the acidic end groups present in polymer chains. Understanding of hydrolytic degradation is important as it leads to a change in physical and chemical properties of PLGA which thereby can affect the drug leakage or release polymeric carrier system (Engineer et al. 2011). When polymeric carrier system (particle) came in contact with water, void spaces are created within the particle, which thereby leads to a marked increase in porosity and encourages diffusion of the drug outwards. The influence of functional end group on polymer chain is also reported, as ester end groups degrade slower than carboxylic acid end groups (Gentile et al. 2014).

6.2.4 PLA TO PGA CONTENT

The PLA/PGA content in PLGA copolymer plays an important role in the design of carrier systems for intended application. The presence of additional methyl side groups in PLA makes it more hydrophobic than PGA; hence, lactide-rich PLGA copolymers are less hydrophilic, absorb less water and subsequently degrade more slowly (Lu et al. 2000). Additionally, crystallinity also plays a role in hydration of PLGA copolymer due to more water penetration as compared to amorphous materials. Therefore, PGA, being hydrophilic and crystalline in nature, increases the wettability of PLGA copolymer, and thus, higher PGA content shows rapid degradation and drug release (Makadia et al. 2011). Various forms of PLGA based on PLA/PGA content such as 85:15, 75:25, 65:35 and 50:50 are commercially available and provide flexibility to product development scientists for the design of PLGA-based formulation with desirable drug-release behaviour (D'souza et al. 2014).

6.2.5 GLASS TRANSITION TEMPERATURE (T_G)

Higher T_g values of polymer help in maintaining a rigid, glassy structure of carrier system, which thereby can be useful in designing extended-release formulations with slow drug release (Gentile et al. 2014). Polymers with low T_g values (near human body temperature) exist in fluid form and easily degrade and facilitate faster drug release (Passerini et al. 2001; D'souza et al. 2014). PLA/PGA compositions have a direct effect on T_g and crystallinity of PLGA, which further influences degradation and drug release. The T_g value of PGA and PLA is reported to be in the range from 35°C to 40°C from 55°C to 60°C, respectively; therefore, as the glycolic content is increased, the T_g value of PLGA is decreased.

6.3 ADVANTAGES AND LIMITATIONS OF PLGA

PLGA possesses unique characteristics of biodegradability, biocompatibility and non-immunogenicity; therefore, it is successfully approved in more than 20 marketed products across the globe (Kapoor et al. 2015). The success of this polymer is depicted from the fact that PLGA is widely investigated for delivery of wide range of drugs and commercially successful also. Based on reported results and benefits, here are some of the important advantages of PLGA as a carrier system:

- Different grades and molecular weights of PLGA offer great flexibility to formulation scientists to develop products of desired drug-release characteristics (Mir et al. 2017).
- PLGA can be used for entrapment of both categories of drugs, i.e. hydrophilic and hydrophobic, which thereby extended its application as potential carrier system for a broad range of drug molecules.
- Sustained release of drug through PLGA matrix offers a reduction in the frequency of administration and maintains required drug concentration in plasma and possible dose reduction, which thereby improve the safety and efficacy of the drug along with patient compliance (Samani et al. 2015).
- The metabolic degradation of PLGA leads to formation of non-toxic metabolic products, i.e. lactic acid and glycolic acid, which easily get eliminated from the body, and therefore, PLGA offers the least toxicity among other polymers (Silva et al. 2016).
- PLGA remains stable during the lyophilisation process and therefore offers the possibility to develop a stable lyophilised products for thermolabile drugs (Samani et al. 2015).
- Various biomolecules and ligands can be coupled with PLGA (like PEG), which thereby can be used to manufacture surface-modified nanoparticles (NPs) for site-specific delivery of drugs (Muthu et al. 2009).
- PLGA offers a great possibility for entrapment and delivery of large molecules and therefore is under development for biological products such as vaccines, DNA, peptides and proteins (Muthu et al. 2009).

Despite multiple advantages, PLGA possesses certain limitations that somehow restrict its application as carrier system in some areas:

- Poor osteoconductivity PLGA limits its use in bone tissue engineering (Gentile et al. 2014).
- Usage of a large amount of surfactant such as polyvinyl alcohol (PVA) compromises the biocompatibility behaviour of the PLGA carrier system.
- Complex manufacturing process for PLGA-based carrier systems (microspheres, NPs, etc.) needs highly trained manpower and efficient supervision (Bhatia et al. 2013).
- Improper design and development of PLGA carrier system can cause issues related to burst release and instability of therapeutic moieties.
- For large molecule-based products, adsorption of protein or peptide to PLGA can cause loss of activity (Bhatia et al. 2013).
- Marketing approval of PLGA-based products needs extensive characterisation of developed formulation and a good understanding of regulatory science.

6.4 PLGA APPLICATION

PLGA is one of the most extensively used polymers in the design and development of drug delivery systems for biomedical applications owing to its biodegradability, biosafety, biocompatibility, properties, tumour embolisation and specificity, versatility in formulation and functionalisation (Fuchs et al. 2017; Pandita et al. 2015; Porcu et al. 2017). Several scientists have investigated the potential application of PLGA in various domains of drug delivery and reported highly encouraging results (Zhang et al. 2018; Mirakabad et al. 2014). Based on the outcome of various studies done so far, there are some important applications of PLGA as a carrier system for drug delivery.

6.4.1 Controlled and Sustained Drug Delivery

The PLGA-based carrier system provides controlled release of drug in various physiological conditions. Various grades of PLGA with different PLA/PGA ratios and molecular weights offer flexibility to develop a formulation of desired release characteristics. The release of drug from PLGA matrix is attributed to the diffusion of water inside the matrix followed by swelling and erosion of the carrier system. The mechanism of controlled release of drugs by PLGA-based carrier system, as represented in Figure 6.2, can be summarised as (i) diffusion (water-filled pores and polymer matrix), (ii) osmotic impelling and (iii) degradation/erosion (surface and bulk) (Nair et al. 2019; Chen et al. 2020; Fredenberg et al. 2011).

The diffusion of the drug occurs through intact polymer matrix into pores filled with water. Water uptake causes swelling of polymeric chains, inducing the formation of

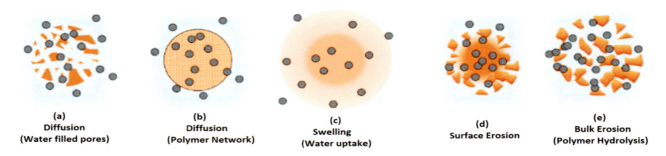

FIGURE 6.2 Drug release mechanisms in PLGA particles.

new pores and/or osmotic pressure. During swelling, the volume rises, the effective diffusion coefficient of the drug is increased, and more molecules enter the aqueous part. Surface erosion occurs when polymers start degrading from the exterior part towards the interior. Finally, bulk erosion occurs when water penetrates the bulk of the polymer, which results in homogeneous degradation of the entire matrix (Uhrich et al. 1999; Park et al. 2019; Kamaly et al. 2016; Hines et al. 2013).

6.4.2 Drug Stabilisation

Many small and large molecules possess serious stability issues, as they are prone to various degradation pathways such as hydrolytic, oxidative, thermal, etc. Development of stable pharmaceutical products is always a key consideration for formulation scientists intending for commercial applications. PLGA-based carrier systems are well reported for the protection of drugs against enzymatic and chemical degradation (Qiu et al. 2006; Lee et al. 2018). PLGA provides the unique possibility of formulating suitable delivery system which can protect therapeutic moieties from degradation pathways and provide inherent stability for longer duration of time. Therefore, PLGA-based drug products can be a successful strategy with the intention of developing stable pharmaceutical products.

6.4.3 Targeted Drug Delivery

Clinical and commercial success of PLGA in sustained drug delivery and stabilisation of therapeutic moieties has attracted scientists and pharmaceutical companies across the globe to investigate its potential in targeted delivery of drugs and gene delivery as well. PLGA-based NPs can be smartly engineered for targeted and site-specific delivery of therapeutic moieties. Structural changes in PLGA are feasible by virtue of that drug can be entrapped in structurally modified PLGA NPs and can be delivered directly to the site of action. Specific targeting of PLGA NPs allows for their differential spatial localisation within the body, minimising the therapeutic drug payloads' off-target adverse effects. Surface modification of PLGA facilitates active and passive targeting of therapeutic moieties (Hines et al. 2013). Passive targeting refers to the preferential accumulation of NPs (bearing no affinity ligands) at active sites and is directly related to the characteristic bio-physicochemical properties of the NPs (size, shape, charge, inherent polymer properties, etc.). PLGA NPs possessing high stability, flexibility and tunable prolonged blood circulation time are ideal to use. Active targeting refers to incorporation of affinity ligands on the surface of the nanocarriers, for binding to antigens/receptors overexpressed in diseased cells or to the extracellular matrix proteins (Mirakabad et al. 2014).

A new series of six-arm star-shaped poly(lactic-co-glycolic acid) (6-s-PLGA) was synthesised by ring-opening polymerisation (Qiu et al. 2006). Paclitaxel-loaded six-arm star-shaped poly(lactic-co-glycolic acid) nanoparticles (6-s-PLGA-PTX-NPs) were prepared under the conditions optimised by the orthogonal testing. The encapsulation efficiency of the 6-s-PLGA-PTX-NPs was higher than that of the L-PLGA-PTX-NPs. In terms of *in vitro* release of NPs, paclitaxel (PTX) was released more slowly and more steadily from 6-s-PLGA than from linear PLGA. The polymer has good biocompatibility and the 6-s-PTX-PLGA-NPs exhibited a smaller size distribution and higher drug-loading capacity than the L-PTX-PLGA-NPs. The results suggest that 6-s-PLGA may be promising for application in PTX delivery to enhance sustained antiproliferative therapy (Chen et al. 2013).

A recent study showed that biodegradable poly(lactic-co-glycolic acid)/polyethylene glycol nanoparticles (PLGA-PEG NPs) are efficient vehicles for the therapeutic delivery of microRNAs. These PLGA/PEI NPs efficiently bind with pEGFP and pβ-gal plasmid DNA, and the formed PLGA/PEI/DNA complexes displayed higher transfection efficiency with lower toxicity compared with PEI polyplexes (Devulapally et al. 2015). These PEGylated PLGA/PEI copolymer NPs that are efficiently complexed with pcD-NATK-NTR plasmid DNA express TK-NTR dual therapeutic reporter genes when US-MB-mediated sonoporation is applied to TNBC tumours in small animals (Shau et al. 2012; Ortega et al. 2016). *In vivo* analysis indicated that PEGylated PLGA/PEG-NPs can successfully deliver the GDEPT gene into tumours in mice as indicated by

CytoCy5S imaging. The animals delivered with GDEPT genes and prodrugs showed a significant reduction in tumour size (2.3-fold) compared with untreated control mice (Devulapally et al. 2018).

A highly stable luteinizing hormone-releasing hormone (LHRH)-conjugated PEGylated PLGA NPs were developed for the successful treatment of prostate cancers (Danhier et al. 2012). The docetaxel (DTX)/PLGA-LHRH micelles possessed a uniform spherical shape with an average diameter of ~170 nm. The micelles exhibited a controlled drug release for up to 96 h which can minimise the non-specific systemic spread of toxic drugs during circulation while maximising the efficiency of tumour-targeted drug delivery. The LHRH-conjugated micelles showed enhanced cellular uptake and exhibited significantly higher cytotoxicity against LNCaP cancer cells (Prabaharan et al. 2009; Cao et al. 2016).

6.4.4 PLGA-Based Carrier System for Remotely Stimulated Cancer Therapy

Stimuli-responsive PLGA-based carrier systems can be engineered to release their therapeutic payload on 'signal' according to specific stimuli triggered via chemical, biochemical or physical means. This stimulated release sequentially causes changes in the nanocarrier/microcarrier structure or chemistry, leading to the release of the therapeutic payload in a biological environment, as represented in Figure 6.3. These PLGA-based delivery systems have been designed to minimise systemic toxicities and unfavourable drug–plasma interactions, minimise dose and treat disease more efficiently (Han et al. 2016). Various studies have been conducted to investigate the potential of PLGA-based carrier systems for remotely stimulated cancer therapy (Table 6.1).

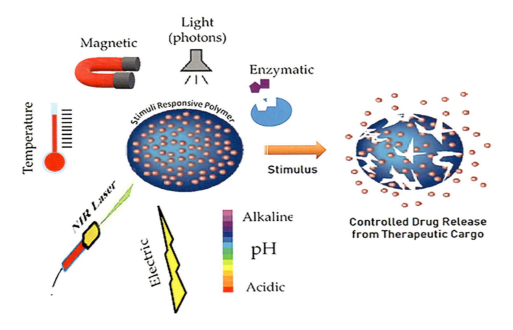

FIGURE 6.3 Exogenously stimulated drug release by different stimuli.

TABLE 6.1
Recent Applications of PLGA-Based Products for Remotely Stimulated Cancer Therapy

Remote Stimulation	Formulation	Anticancer Agent	Cell Culture	Therapeutic Outcome	References
pH	PLGA-NaHCO$_3^-$ hollow microspheres	Rapamycin	MCF-7 cells	Rapamycin (an inhibitor of mTOR) inhibits cell growth and promotes autophagy. Combined CQ–NaHCO3-loaded HMs and rapamycin–NaHCO3-loaded HMs efficiently induce cancer cell death *In vitro*: MCF-7 cell survival rate markedly decreased *In vivo*: enhanced cell death induced by CQ or CQ–NaHCO$_3^-$-loaded PLGA microspheres	Liang et al. (2015)

(Continued)

TABLE 6.1 (*Continued*)
Recent Applications of PLGA-Based Products for Remotely Stimulated Cancer Therapy

Remote Stimulation	Formulation	Anticancer Agent	Cell Culture	Therapeutic Outcome	References
	PLGA/no quantum dots	ZnO	A549 cells	ZnO quantum dots encapsulated with PLGA nanoparticles are pH responsive and easily target cancerous cells based on the acidic tumour microenvironment *In vitro*: induced apoptosis in the metastatic A549 cancer cells	Kim et al. (2020)
	Lipid hybrid PLGA nanoparticles	Doxorubicin	MDA-MB-231 cells, A549 cells, KB cells	pH-triggered surface charge-inverted lipid hybrid PLGA nanoparticles significantly suppressed the growth of tumour in tumour-bearing nude mice and reduced the systemic toxicity of DOX *In vitro*: enhanced escape of DOX from lysosome and delivered more amount of DOX to the nucleus *In vivo*: increased the efficacy of DOX in tumour-bearing mice	Du et al. (2016)
	PLGA-FeO nanoparticles	Sorafenib	McA-RH7777 hepatoma cells	pH-triggered drug-eluting nanocomposite (pH-DEN) made up of PLGA, iron oxide nanocubes, pH-responsive synthetic peptides with lipid tails *In vitro*: efficiently localised to the targeted tumour regions *In vivo*: sorafenib release from IA-infused sorafenib–pH-DENs into hepatocellular carcinoma tumour and significant tumour growth inhibition with anti-angiogenic effects	Park et al. (2016)
	PLGA(UCNPs/DOX) nanocapsules	Doxorubicin	H460 cancer cells	PLGA-upconversion nanoparticles (UCNPs) used as theranostic agents for cancer cell imaging and cellular cytotoxic effect *In vitro*: Dox releases on the 16th day about 35.7% and 14.58% at pH 5.0 and 7.4, respectively	Zhao et al. (2017)
Near infrared	PLGA-PTX/HAuNS microspheres	Paclitaxel	MDA-MB-231 cells and U87 cells	PTX release from microsphere is controlled by NIR laser power. Intratumoural injection of microsphere followed by NIR irradiation results in a significant delay in tumour growth *In vitro*: antitumour effect of PTX/HAuNS microspheres was enhanced by irradiated NIR laser *In vivo*: laser irradiation caused burning of tissue, but it gradually healed and became scars. No microscopic tumour cells were found in the scar tissues	You et al. (2010)
	GNPs-pD-PTX-PLGA microspheres	Paclitaxel	PANC-1 cells	Polydopamine (pD) used as a linker to stick to GNPs to the surface of PTX-loaded PLGA microspheres. NIR showed more ROS generation and a reduction in the expression levels of antioxidant enzyme (SOD2 and catalase) *In vitro*: less than 70% of PTX in 50 days, without any significant initial burst release. The GNPs-pD-PTX-PLGAMs also showed a concentration- and time-dependent rise in temperature	Banstola et al. (2019)
Laser	GNP-PLGA-PFH microparticles	Perfluorohexane liquid	MDA-MB-231 cells	Laser irradiation after percutaneous injection of the particles triggered gas expulsion *In vitro*: after irradiation with laser vaporisation occurs, the bubble expanded within the cell, and the cell was destroyed by the expanding bubble *In vivo*: After treatment, significant voids were observed in the ultrasound images and attributed to tissue injury and necrosis, and blood flow reduced within the lymph node	Sun et al. (2014)
	ANG/GS/PLGA/DTX nanoparticles	Docetaxel	U87MG cells	Angiopep-2 attached on gold nanoshell through Au-S bond, high active targeting capability and much higher antitumour efficiency without palpable toxic effects *In vitro*: 17.0% increased cumulative drug release compared to that of ANG/GS/PLGA/DTX NPs without irradiation *In vivo*: laser irradiation caused a severe apoptosis (about 39.6%); no difference in late apoptosis percentage	Hao et al. (2015)

(*Continued*)

TABLE 6.1 (*Continued*)
Recent Applications of PLGA-Based Products for Remotely Stimulated Cancer Therapy

Remote Stimulation	Formulation	Anticancer Agent	Cell Culture	Therapeutic Outcome	References
Redox	Bi(mPEG-PLGA)-Se2 micelles	Doxorubicin		DCC/DMAP as coupling agent, drug release in cytosol and nuclei of cancer cells, pronounced antitumour activities against HeLa cells *In vitro*: after 72 h, approximately 72% and 69% of drug cargo were released in the presence of GSH and H2O2, respectively *In vivo*: no any intrinsic toxicities towards HaCaT and HeLa cells. DOX-loaded PLGA micelle significantly decreased the cancer cell load within 24 h at 7.5 µg/mL drug concentration	Birhan et al. (2019)
Temperature	TMZ-PLGA nanoparticles	Temozolomide	RG2 rat glioma cells	Thermo-reversible hydrogel system comprising Pluronic® F-127, efficiently taken up by the glioblastoma cells *In vitro*: increased the apparent half-life of TMZ to 18 h, 10% drug immediate release in the first 12 h; plateau reached in the next 60 days. Prodrug passed the BBB and converted to the active substance	Sayıner et al. (2020)
Magnetic	PLGA-MP nanoparticles	Methotrexate (MTX)	SK-BR-3	Magnetisation saturation is 43 emu/g in nanoparticles and showed similar cell cytotoxicity in comparison with MTX solution *In vitro*: particles are localised in the desired target site; they mostly affect malignant cells and reduce the side effects of MTX. Cell viability was higher for MTX NP than the MTX solution	Vakilinezhad et al. (2018)

6.5 PLGA-BASED COMMERCIALLY AVAILABLE PRODUCTS

PLGA application as carrier system for drug delivery has been increasing, and a large number of products are approved across the globe by various regulatory agencies (such as USFDA and EMA). The first-ever approved PLGA product dating back to 1986 is Decapeptyl® SR with triptorelin acetate as API for the treatment of prostate cancer via the intramuscular (i.m.) route. In the year 1989, Lupron® Depot was approved containing leuprolide acetate as active ingredient which is a synthetic analogue of naturally occurring gonadotropin-releasing hormone. Lupron Depot is a PLGA microsphere-based product approved by the USFDA, and it created a benchmark in complex manufacturing process. Because of critical multistep manufacturing process, no generic version of Lupron Depot is approved by the USFDA till date. Recently, Lupron® Depot PED was approved by the USFDA in 2011 for the treatment of central precocious puberty. Eligard® Kit is another product from the same category of leuprolide acetate-based formulation approved for subcutaneous (s.c.) delivery in various strengths (7.5 mg/vial, 22.5 mg/vial, 30 mg/vial and 45 mg/vial) with different grades of PLGA in each pack size. It was approved in 2002. Lupaneta® Pack approved in 2012 contains leuprolide acetate and norethindrone acetate for the treatment of endometriosis via i.m. administration. Zoladex® is another product containing gonadotropin-releasing hormone injectable implant approved by the USFDA in 1996 for the treatment of prostate and breast cancer.

Sandostatin® LAR containing octreotide (peptide) is a long-acting microsphere formulation approved in 1997 for the treatment of acromegaly and diarrhoea with carcinoid syndrome via the i.m. route. Signifor® LAR (pasireotide pamoate) is another product approved in 2014 for treating acromegaly and Cushing's disease. While Sandostatin® LAR is considered the first-generation somatostatin receptor (SR) analogue product, Signifor® is considered the second-generation SR analogue. In 1998, Atridox® containing doxycycline hyclate was approved for periodontal disease through the sub-gingival route. It is a biodegradable product that comes as two parts, one syringe with API and the other with Atrigel delivery system which when mixed and placed in periodontal pockets forms a gel and releases drug in a sustained manner. For a similar indication, Arestin® containing minocycline HCl sustained-release microspheres was also approved in 2001.

Risperdal® Consta (risperidone microspheres) is a benchmark PLGA-based product approved in 2003 by the USFDA for the treatment of schizophrenia and bipolar disorder. Vivitrol®, PLGA microspheres encapsulating naltrexone, was approved in 2006 for treating alcohol dependence and later added new indication as opioid dependence in 2010. Sublocade®, approved in 2017, containing buprenorphine is another product of PLGA, once in a month injectable implant administered by the subcutaneous route. Bydureon® (exenatide) microspheres were prepared by Medisorb™ microsphere technology for sustained release of exenatide and approved in 2012 for the treatment of type 2 diabetes mellitus via s.c. administration. Ozurdex®

TABLE 6.2
Commercially Available PLGA-Based Products

Product Name	Active Pharmaceutical Ingredient (API)	Primary Indication	Formulation Characteristics	Year of Approval	References
Decapeptyl® SR	Triptorelin	Precocious puberty	Sustained-release injectable microspheres	1986	Decapeptyl SMC
Prostap® SR	Leuprorelin acetate	Prostate cancer in men and breast cancer in pre- and perimenopausal women	Prolonged-release suspension for injection	1996–1997	Prostap SR SMC
Lupron® Depot	Leuprolide	Endometriosis	PLGA microspheres	2011	Lupron Depot PI
Lupron® Depot PED	Leuprolide	Central precocious puberty	PLGA microspheres	2011	Lupron Depot PED PI
Eligard®	Leuprolide acetate	Prostate cancer	In situ forming injectable depot	2011	Eligard PI
Lupaneta® Pack	Leuprolide acetate and norethindrone	Endometriosis	PLGA microparticle depot suspension	2012	Lupaneta Pack PI
Sandostatin® LAR	Octreotide	Acromegaly and symptoms of neuroendocrine tumours	PLGA microparticles	1997	Sandostatin LAR Depot PI
Arestin®	Minocycline HCl	Adult periodontitis	PLGA microspheres	2001	Arestin PI
Atridox®	Doxycycline hyclate	Chronic adult periodontitis	PLA depot	1998	Atridox PI
Risperdal® Consta®	Risperidone	Schizophrenia and bipolar disorder	Extended-release microspheres	2003	Risperdal Consta PI
Perseris®	Risperidone	Schizophrenia	In situ forming PLGA gel	2018	Perseris PI
Suprefact® Depot	Buserelin	Advanced carcinoma of prostate gland	PLGA microspheres	–	Suprefact Depot PI
Zoladex®	Goserelin	Prostate cancer and breast cancer	PLGA rods	1996	Zoladex PI
Vivitrol®	Naltrexone	Alcohol dependence	Extended-release PLGA microspheres	2006	Vivitrol PI
Sublocade®	Buprenorphine	Opioid use disorder	PLGA in situ gel	2017	Sublocade PI
Somatuline® Depot	Lanreotide	Acromegaly	Sustained-release PLGA microspheres	2007	Somatuline Depot PI
Trelstar® Depot	Triptorelin pamoate	Advanced prostate cancer	Lyophilised PLGA microparticles	2000	Trelstar Depot PI
Bydureon®	Exenatide	Type 2 diabetes	PLGA microspheres	2012	Bydureon PI
Signifor® LAR	Pasireotide pamoate	Acromegaly and Cushing's disease	PLGA microspheres	2012	Signifor LAR PI
Ozurdex®	Dexamethasone	Macular oedema	PLGA solid implant	2009	Ozurdex PI
Sinuva®	Mometasone furoate	Chronic rhinitis	PLGA solid implant	2017	Sinuva PI

implant containing 0.7 mg (700 µg) dexamethasone in the NOVADUR™ solid polymer drug delivery system was approved in 2009 for the treatment of diabetic macular oedema. The NOVADUR™ system contains PLGA polymer matrix for administration via ocular intravitreal implant. Triptodur® (triptorelin pamoate) was approved in 2017 for the treatment of paediatric patients 2 years and older with central precocious puberty via i.m. administration. A consolidated list of commercially available PLGA-based products is presented in Table 6.2.

6.6 CHALLENGES FOR PLGA-BASED DRUG DELIVERY SYSTEMS – CLINICAL AND COMMERCIAL SUCCESS

Commercial success of PLGA-based products depends on various aspects and thus creates multiple challenges for pharmaceutical researchers intending for commercial applications.

The development of PLGA-based products is not only related to optimisation, drug loading and drug release, but focus on scale-up, manufacturing and batch-to-batch reproducibility, toxicological aspects, regulatory considerations and intellectual issues is also needed (Narang et al. 2013; Hafner et al. 2014). The complexity involved in multistep manufacturing process creates a major challenge in manufacturing and approval of these products. In this section, various challenges associated with PLGA product development and manufacturing are discussed. Various key challenges and obstacles identified during product development, evaluation and clinical studies are summarised in Table 6.3.

6.6.1 Scale-Up and Large-Scale Production

PLGA-based formulations possess complex structural and physicochemical properties that play a significant role in commercial success of these products. The

TABLE 6.3
Key Considerations and Obstacles of Development of PLGA-Based Products

Key Considerations	Major Obstacles
Pharmaceutical Formulation Design	
• Route of administration • Reduced complexity in formulation design • Final dosage form for human use • Biocompatibility and biodegradability • Pharmaceutical stability (physical and chemical)	• Scale-up/large-scale production as per GMP, e.g. reproducibility, infrastructure, techniques, expertise and cost • Extensive characterisation, e.g. size and polydispersity, morphology, charge, encapsulation, surface modifications, purity and stability
Preclinical Evaluation	
• Need for validated and standardised assays for early detection of toxicity • Development of suitable animal models • Adequate understanding of *in vivo* behaviour	• Conducting specialised toxicology studies • Interaction of the product with tissues and cells • Adequate structural stability of the product in physiological conditions • Accumulation in target organs/tissues/cells
Clinical Evaluation for Commercialisation	
• Simplification of development pathways from invention to commercialisation • Evaluation of safety and efficacy in humans • Optimal clinical trial design	• Lack of clear regulatory guidelines specific for PLGA-based products • Complexity associated with IP landscape for these products • Limited understanding of the biological interaction of these products

development, characterisation and validation of these complex products can be challenging due to various optimisation parameters such as size distribution, morphology, charge, purity, drug encapsulation efficiency, coating efficiency and density of conjugated ligand(s) (Teli et al. 2010). Additionally, batch-to-batch variation of the finished product at a large manufacturing scale can lead to significant changes in physicochemical properties, pharmacokinetic parameters and/or pharmacodynamic interactions which thereby can affect its safety and efficacy (Tinkle et al. 2014; Barz et al. 2015). Proper development of manufacturing process is an essential element for clinical translation of PLGA-based products so that they can be easily scaled up with potential manufacturability and batch-to-batch reproducibility (Grainger 2013). These PLGA-based products are manufactured by multistep process, and therefore, a good understanding of critical process parameters is essential for the production of these products with desired properties as per set specification (Teli et al. 2010). In general, PLGA-based complex product manufacturing is challenged by potential issues such as the following:

- quality issues
- scalability complexities
- incomplete purification from contaminants (by-products and starting materials)
- high material and manufacturing costs
- low production yield
- insufficient batch-to-batch reproducibility, consistency and storage stability of the finished product
- lack of infrastructure facility and/or in-house expertise and trained manpower
- chemical instability or denaturation of the encapsulated compound during the manufacturing process
- complex characterisation/analytical method

6.6.2 Biocompatibility and Safety Challenges

In-depth toxicological evaluation is essential for these PLGA-based complex products during clinical studies to determine the overall safety for human use (Nystrom et al. 2012). During the early phase of drug product development (starting from the preclinical stage), all excipients including developed product should undergo specific toxicological studies related to intended usage. These studies will be the basis for performing toxicological studies during clinical trials to understand the toxicity of developed PLGA complex products (Dobrovolskaia et al. 2013; Tinkle et al. 2014). Any change in drug substance and drug product during the development process can have an impact on its safety and efficacy. As per regulatory expectations, in the case of any of the following changes, sufficient studies are needed to address all toxicological concerns:

- changes in the synthesis procedure or reagents used for the synthesis drug substance, product or formulation
- changes in the existing manufacturing method (e.g. chemical synthesis, fermentation or derivation from a natural source)
- changes in the raw material source
- changes in the sterilisation method of the drug substance or drug product
- changes in the route of administration
- changes in the constitution and dosage form of the product

- changes in the drug product manufacturing process
- changes in the drug product container closure system

If any of these changes are needed, the product has to undergo specific studies to evaluate product equivalency, quality and safety along with other comparison testing to understand the impact of modification.

6.6.3 Intellectual Property (IP)

All these PLGA-based complex products are developed and manufactured using a specific platform technology; therefore, proper landscaping of IP information is very critical. Patent filing provides opportunities to inventor/sponsor to protect their IP and leverage their developments over the marketplace. Simultaneously, it also restricts others to adopt the developed technology related to these complex products. Therefore, proper landscaping is needed before adopting any strategies related to development, manufacturing and characterisation processes (Satalkar et al. 2016). The patent reviewer must have expertise and training concerning the emerging fields of polymer (PLGA)-based products for therapeutic interventions. Because of complexity in IP landscaping and lack of information, the situation becomes highly complicated and causes a roadblock for development, manufacturing or marketing approval of these complex products. The complexities of IP associated with PLGA-based products can lead to costly litigation and halt commercialisation efforts of drug products (Tinkle et al. 2014).

6.6.4 Regulatory Requirements

Commercialisation of these complex products depends upon appropriate submission of all documents/dossiers related to development, manufacturing and preclinical/clinical studies, as per regulatory agencies' (USFDA, EMA, TGA, PMDA, etc.) expectations and guidelines (Gaspar 2007; Tinkle et al. 2014; Sainz et al. 2015). The deficiency of clear regulatory and safety guidelines limits the development and approval of these products in a timely manner. Many regulatory agencies (such as USFDA and EMA) are in the process of building special guidance for submission and approval of these products, but still at many places the approval is regulated through a conventional framework guided by some of the significant regulatory authorities worldwide. Due to the significant increase in polymer-/lipid-based products in the last few years, it becomes highly important for regulatory agencies to draft specific guidance for all stages of development and manufacturing. In recent times, the USFDA is in the process of providing product-specific guidance for these complex products which helps the applicant to generate data as per the regulatory agency expectations. Apart from this, the USFDA, as per the GDUFA programme, provided commitment on building a regulatory framework for complex product development and approval.

6.7 CONCLUSION AND FUTURE PERSPECTIVES FOR PLGA AS CARRIER SYSTEM

In the last few decades, PLGA-based products have drawn attention as a flexible carrier system for a broad range of therapeutics. The potency and flexibility of these products showed promising results in drug delivery for various functions such as controlled release, targeting and imaging. The major drawback associated with PLGA-based therapeutics is complex manufacturing processes which limits the conversion of clinically effective formulation into commercially viable product. To mitigate this limitation, there is a need to focus on simplifying and streamlining the manufacturing process so that more clinical applications can come into reality. The regulatory agencies should also focus aggressively in a scientific manner on developing specific guidance and more clarity on these products for successful regulatory submission. The potential of PLGA to be used as a nanoparticulate system provides a new era in the world of nanomedicine with great possibilities. From the last many years, scientists across the globe are investigating the potential of PLGA NPs for site-specific delivery of drugs, proteins, peptides, etc. The information generated so far can be helpful in focusing on clinical and commercial application of nanomedicine for treatment, diagnostic and imaging purposes. There is a need to establish the effect of PLGA-based products on immune system response, which thereby would be helpful in understanding PLGA-mediated immunotherapeutic approaches for the treatment of various disorders such as cancer. More research in the field of toxicological aspects is needed for the development of safe products for clinical applications. The application of artificial intelligence in design, development and probable predictions will be of great help for scientific community and will definitely help to turn concept into reality in a timely manner.

REFERENCES

Arestin® Package Insert (PI). https://www.accessdata.fda.gov/drugsatfda_docs/label/2017/050781s020lbl.pdf.

Atridox Package Insert (PI). https://www.accessdata.fda.gov/drugsatfda_docs/label/2011/050751s015lbl.pdf.

Banstola, A., Pham, T.T., Jeong, J.H. et al. 2019. Polydopamine-tailored paclitaxel-loaded polymeric microspheres with adhered NIR-controllable gold nanoparticles for chemo-phototherapy of pancreatic cancer. *Drug Deliv* 26(1): 629–640.

Barz, M., Luxenhofer, R., Schillmeier, M. 2015. Quo vadis nanomedicine? *Nanomedicine* 10(20): 3089–3091.

Bhatia, A., Singh, B., Raza, K. et al. 2013. Tamoxifen-loaded lecithin organogel (LO) for topical application: development, optimization and characterization. *International Journal of Pharmaceutics* 444 (1–2): 47–59.

Birhan, Y.S., Hailemeskel, B.Z., Mekonnen, T.W. et al. 2019. Fabrication of redox-responsive Bi (mPEG-PLGA)-Se$_2$ micelles for doxorubicin delivery. *International Journal of Pharmaceutics* 567: 118486.

Bydureon® Package Insert (PI). https://www.accessdata.fda.gov/drugsatfda_docs/label/2020/022200s030lbl.pdf

Cao, L.B., Zeng, S., Zhao, W. 2016. Highly stable PEGylated poly (lactic-co-glycolic acid) (PLGA) nanoparticles for the effective delivery of docetaxel in prostate cancers. *Nanoscale Research Letters* 11 (1): 305.

Chen, X., Song, L., Li, X. et al. 2020. Co-delivery of hydrophilic/hydrophobic drugs by multifunctional yolk-shell nanoparticles for hepatocellular carcinoma theranostics. *Chemical Engineering Journal* 389: 124416.

Chen, Y., Yang, Z., Liu, C. et al. 2013. Synthesis, characterization, and evaluation of paclitaxel loaded in six-arm star-shaped poly (lactic-co-glycolic acid). *International Journal of Nanomedicine* 8: 4315–4326.

Danhier, F., Ansorena, E., Silva, J.M. et al. 2012. PLGA-based nanoparticles: an overview of biomedical applications. *Journal of Controlled Release* 161 (2): 505–522.

Decapeptyl SMC. https://www.medicines.org.uk/emc/product/780/smpc.

Devulapally, R., Lee, T., Shah, A. et al. 2018. Ultrasound-guided delivery of thymidine kinase-nitroreductase dual therapeutic genes by PEGylated-PLGA/PIE nanoparticles for enhanced triple negative breast cancer therapy. *Nanomedicine (London)* 13 (9): 1051–1066.

Devulapally, R., Sekar, N.M., Sekar, T.V. et al. 2015. Polymer nanoparticles mediated co delivery of antimiR-10b and antimiR-21 for achieving triple negative breast cancer therapy. *ACS Nano* 9 (3): 2290–2302.

Dobrovolskaia, M.A., McNeil S.E. 2013. Understanding the correlation between *in vitro* and *in vivo* immunotoxicity tests for nanomedicines. *Journal of Controlled Release* 172(2): 456–466.

D'Souza, S., Rossella, D., Patrick, P. D. et al. 2014. Effect of hydration on physichochemical properties of end-capped PLGA. *Advances in Biomaterials* 2014: 9.

Du, J.B., Cheng, Y., Teng, Z.H. et al. 2016. pH-Triggered surface charge reversed nanoparticle with active targeting to enhance the antitumor activity of doxorubicin. *Molecular Pharmaceutics* 13 (5): 1711–1722.

Eligard Package Insert (PI). https://www.accessdata.fda.gov/drugsatfda_docs/label/2019/021343s039,021379s041,021488s036,021731s037lbl.pdf.

Engineer, C., Parikh, K.J., Raval, A. et al. 2011. Review on hydrolytic degradation behavior of biodegradable polymers from controlled drug delivery system. *Trends in Biomaterials and Artificial Organs* 25 (2): 79–85.

Fredenberg, S., Marie, W., Reslow, M. et al. 2011. Pore formation and pore closure in poly (D,L-lactide-co-glycolide) films. *Journal of Controlled Release* 150 (2): 142–149.

Fuchs, K., Duran, R., Denys, A. et al. 2017. Drug-eluting embolic microspheres for local drug delivery – state of the art. *Journal of Controlled Release* 262: 127–138.

Gaspar, R. 2007. Regulatory issues surrounding nanomedicine: setting the scene for the next generation of nanopharmaceuticals. *Nanomedicine* 2 (2): 143–147.

Gentile, P., Chiono, V., Carmagnola, I. et al. 2014. An overview of poly (lactic-co-glycolic) acid (PLGA)-based biomaterials for bone tissue engineering. *International Journal of Molecular Sciences*. 15 (3): 3640–3659.

Grainger, D.W. 2013. Connecting drug delivery reality to smart materials design. *International Journal of Pharmaceutics* 454 (1): 521–524.

Hafner, A., Lovrić, J., Lakos, G.P. et al. 2014. Nanotherapeutics in the EU: an overview on current state and future directions. *International Journal of Nanomedicine* 9: 1005–1023.

Han, F.Y., Thurecht, K.J., Whittaker, A.K. et al. 2016. Bioerodable PLGA-based microparticles for producing sustained-release drug formulations and strategies for improving drug loading. *Frontiers in Pharmacology* 7: 185.

Hao, Y., Zhang, B., Zheng, C. et al. 2015. The tumor-targeting core-shell structured DTX-loaded PLGA@Au nanoparticles for chemo-photothermal therapy and X-ray imaging. *Journal of Controlled Release* 220: 545–555.

Hines, D.J., Kaplan, D.L. 2013. Poly (lactic-co-glycolic) acid-controlled-release systems: experimental and modeling insights. *Critical Reviews in Therapeutic Drug Carrier Systems* 30 (3): 257–276.

Kamaly, N. Yameen, B., Wu, J. et al. 2016. Degradable controlled-release polymers and polymeric nanoparticles: mechanisms of controlling drug release. *Chemical Reviews* 116 (4): 2602–2663.

Kapoor, N.D., Bhatia, A., Ripandeep, K., Ruchi, S. et al. 2015. PLGA: a unique polymer for drug delivery. *Theraputic Delivery* 6 (1): 41–58.

Kim, H., Meghani, N., Park, M. et al. 2020. Electrohydrodynamically atomized pH-responsive PLGA/ZnO quantum dots for local delivery in lung cancer. *Macromolecular Research* 28 (4): 407–414.

Lee, P.W., Pokorski, J.K. 2018. Poly(lactic-co-glycolic acid) devices: production and applications for sustained protein delivery. *Wiley Interdisciplinary Reviews Nanomedicine and Nanobiotechnology* 10:1002.

Liang, X., Yang, Y., Wang, L. et al. 2015. pH-Triggered burst intracellular release from hollow microspheres to induce autophagic cancer cell death. *Journal of Materials Chemistry B* 3 (48): 9383–9396.

Lu, L., Susan, J.P., Michelle, D.L. et al. 2000. *In vitro* and *in vivo* degradation of porous poly (DL-lactic-co-glycolic acid) foams. *Biomaterials* 21 (18): 1837–1845.

Lupaneta Pack Package Insert (PI). https://www.accessdata.fda.gov/drugsatfda_docs/label/2013/203696s001lbl.pdf.

Lupron Depot Package Insert (PI). https://www.accessdata.fda.gov/drugsatfda_docs/label/2014/020517s036_019732s041lbl.pdf.

Lupron Depot-PED Package Insert (PI). https://www.accessdata.fda.gov/drugsatfda_docs/label/2020/020263s046lbl.pdf.

Makadia, H.K., Siegel, S.J., 2011. Poly lactic-co-glycolic acid (PLGA) as biodegradable controlled drug delivery carrier. *Polymers* 3 (3): 1377–1397.

Mir, M., Ahmed, N., Rehman, A., 2017. Recent applications of PLGA based nanostructures in drug delivery. *Colloids and Surfaces B: Biointerfaces* 159: 217–231.

Mirakabad, S.T., F., Nejati, K., Akbarzadeh, A. et al. 2014. PLGA-based nanoparticles as cancer drug delivery systems. *Asian Pacific Journal of Cancer Prevention* 15 (2): 517–535.

Muthu, M.S., Manoj, K.R., Mishra A. et al. 2009. PLGA nanoparticle formulations of risperidone: preparation and neuropharmacological evaluation. *Nanomedicine: Nanotechnology, Biology and Medicine* 5 (30): 323–333.

Nair, A.B. Sreeharsha, N., Al-Dhubiab B.E. et al. 2019. HPMC- and PLGA-based nanoparticles for the mucoadhesive delivery of sitagliptin: optimization and *in vivo* evaluation in rats. *Materials (Basel)* 12 (24): 4239.

Narang A.S., Chang R.K., Narang M.A. et al. 2013. Pharmaceutical development and regulatory considerations for nanoparticles and nanoparticulate drug delivery systems. *Journal of Pharmaceutical Sciences*. 102 (11): 3867–3882.

Nystrom, A.M., Fadeel B. 2012. Safety assessment of nanomaterials: implications for nanomedicine. *Journal of Controlled Release* 161 (2): 403–408.

Ortega, M., Giron, MD., Salto, R. et al. 2016. Polyethyleneimine-coated gold nanoparticles: straightforward preparation of efficient DNA delivery nanocarriers. *Chemistry: An Asian Journal* 11 (23): 3365–3375.

Ozurdex Package Insert (PI). https://www.accessdata.fda.gov/drugsatfda_docs/label/2018/022315s012lbl.pdf.

Pandita, D., Kumar, S., Lather, V. 2015. Hybrid poly(lactic-co-glycolic acid) nanoparticles: design and delivery prospectives. *Drug Discov Today* 20 (1): 95–104.

Park, K., Skidmore, S., Hadar, J. et al. 2019. Injectable, long-acting PLGA formulations: analyzing PLGA and understanding microparticle formation. *Journal of Controlled Release* 304: 125–134.

Park, W., Chen, J., Cho, S. et al. 2016. Acidic pH-triggered drug-eluting nanocomposites for magnetic resonance imaging-monitored intra-arterial drug delivery to hepatocellular carcinoma. *ACS Applied Materials & Interfaces* 8 (20): 12711–12719.

Passerini, N., Craig D.Q. 2001. An investigation into the effects of residual water on the glass transition temperature of polylactide microspheres using modulated temperature DSC. *Journal of Controlled Release* 73 (1): 111–115.

Perseris Package Insert (PI). https://www.accessdata.fda.gov/drugsatfda_docs/label/2020/210655s004lbl.pdf.

Porcu, E.P. Salis, A., Rassu, G. et al. 2017. Engineered polymeric microspheres obtained by multi-step method as potential systems for transarterial embolization and intraoperative imaging of HCC: preliminary evaluation. *European Journal of Pharmaceutics and Biopharmaceutics* 117: 160–167.

Prabaharan, M., Grailer, J., Pilla, S. et al. 2009. Folate-conjugated amphiphilic hyperbranched block copolymers based on Boltorn H40, poly (L-lactide) and poly (ethylene glycol) for tumor-targeted drug delivery. *Biomaterials* 30 (16): 3009–3019.

Prostap SR. https://www.medicines.org.uk/emc/product/4650/smpc.

Qiu, L.Y., Bae, Y.H. 2006. Polymer architecture and drug delivery. *Pharmaceutical Research* 23 (1): 1–30.

Rana, S., Chaudhary, H., Kholi, K. et al. 2013. Effect of physicochemical properties on biodegradable polymers on nano drug delivery. *Polymer Reviews* 53 (4): 546–567.

Risperdal Consta® Package Insert (PI). https://www.accessdata.fda.gov/drugsatfda_docs/label/2020/021346s063lbl.pdf.

Sainz, V., Matos, A. I., Peres, C.Z.E., et al. 2015. Regulatory aspects on nanomedicine. *Biochemical and Biophysical Research Communications* 468 (3): 504–510.

Samani, S.M., Taghipour, B., 2015. PLGA micro and nanoparticles in delivery of peptides and proteins; problems and approaches. *Pharmaceutical Development and Technology* 20 (4): 385–393.

Sandostatin LAR® Depot Package Insert (PI). https://www.accessdata.fda.gov/drugsatfda_docs/label/2008/021008s021lbl.pdf.

Satalkar, P., Elger B.S., Shaw, D.M. et al. 2016. Defining nano, nanotechnology and nanomedicine: why should it matter? *Science and Engineering Ethics* 22: 1255–1276.

Sayıner, Ö., Arısoy, S., Comoglu, T. et al. 2020. Development and *in vitro* evaluation of temozolomide-loaded PLGA nanoparticles in a thermoreversible hydrogel system for local administration in glioblastoma multiforme. *Journal of Drug Delivery Science and Technology* 57: 101627.

Schliecker, G., Carsten S., Stefan F. et al. 2003. Characterization of a homologous series of D, L-lactic acid oligomers; a mechanistic study on the degradation kinetics *in vitro*. *Biomaterials* 24 (21): 3835–3844.

Sharma, S., Parmar, A., Kori, S. et al. 2016. PLGA-based nanoparticles: a new paradigm in biomedical applications. *TrAC Trends in Analytical Chemistry* 8: 30–40.

Shau, M.D., Shih, M.F., Lin., C.C., et al. 2012. A one-step process in preparation of cationic nanoparticles with poly(lactide-co-glycolide)-containing polyethylenimine gives efficient gene delivery. *European Journal of Pharmaceutical Sciences* 46 (5): 522–529.

Signifor LAR Package Insert (PI). https://www.accessdata.fda.gov/drugsatfda_docs/label/2020/203255s008lbl.pdf.

Silva, A.L., Soema, P.C., Slutter B. et al. 2016. PLGA particulate delivery systems for subunit vaccines: linking particle properties to immunogenicity. *Human Vaccines & Immunotherapeutics* 12 (4): 1056–1069.

Sinuva Package Insert (PI). https://www.accessdata.fda.gov/drugsatfda_docs/label/2020/209310s003lbl.pdf.

Somatuline® Depot Package Insert (PI). https://www.accessdata.fda.gov/drugsatfda_docs/label/2019/022074s024lbl.pdf.

Sublocade Package Insert (PI). https://www.accessdata.fda.gov/drugsatfda_docs/label/2020/209819s012lbl.pdf.

Sun, Y., Wang, Y., Niu, C. et al. 2014. Laser-activatible PLGA microparticles for image-guided cancer therapy *in vivo*. *Advanced Functional Materials* 24 (48): 7674–7680.

Suprefact® Depot Package Insert (PI). http://products.sanofi.ca/en/suprefact-depot.pdf.

Teli M.K., Mutalik, S., Rajanikant, G.K. 2010. Nanotechnology and nanomedicine: going small means aiming big. *Current Pharmaceutical Design*. 16 (16): 1882–1892.

Tinkle, S., Bawa, R., Barenholz Y.C. et al. 2014. Nanomedicines: addressing the scientific and regulatory gap. *Annals of the New York Academy of Sciences*. 13: 35–56.

Trelstar LA Package Insert (PI). https://www.accessdata.fda.gov/drugsatfda_docs/nda/2001/21-288_Trelstar_prntlbl.pdf.

Trelstar Package Insert (PI). https://www.accessdata.fda.gov/drugsatfda_docs/label/2018/020715s040,021288s035,022427s015lbl.pdf.

Triptodur Package Insert (PI). https://www.accessdata.fda.gov/drugsatfda_docs/label/2017/208956s000lbl.pdf.

Uhrich, K.E. Cannizzaro, S.M., Langer, R.S. et al. 1999. Polymeric systems for controlled drug release. *Chemical Reviews* 99 (11): 3181–3198.

Vakilinezhad, M.A., Alipour, S., Montaseri, H. 2018. Fabrication and *in vitro* evaluation of magnetic PLGA nanoparticles as a potential methotrexate delivery system for breast cancer. *Journal of Drug Delivery Science and Technology* 44: 467–474.

Vivitrol Package Insert (PI). https://www.accessdata.fda.gov/drugsatfda_docs/label/2020/021897s049lbl.pdf.

You, J., Shao, R., Wei, X. et al. 2010. Near-infrared light triggers release of paclitaxel from biodegradable microspheres: photothermal effect and enhanced antitumor activity. *Small* 6 (9): 1022–1031.

Zhang, Z., Wang, X., Li, B. et al. 2018. Development of a novel morphological paclitaxel-loaded PLGA microsphere for effective cancer therapy: *in vitro* and *in vivo* evaluations. *Drug Delivery*. 25 (1): 166–177.

Zhao, J., Yang, H., Li, J. et al. 2017. Fabrication of pH-responsive PLGA (UCNPs/DOX) nanocapsules with up conversion luminescence for drug delivery. *Scientific Reports* 7 (1): 18014.

Zoladex® Package Insert (PI). https://www.accessdata.fda.gov/drugsatfda_docs/label/2015/019726s059,020578s037lbl.pdf.

# 7 Targeted Lipid-Based Nanoparticles for Nucleic Acid Delivery in Cancer Therapy

Prashant Upadhaya, Swapnil Talkar, Vinod Ghodake, Namrata Kadwadkar, Anjali Pandya, Sreeranjini Pulakkat, and Vandana B. Patravale
Institute of Chemical Technology

CONTENTS

Abbreviations 95
7.1 Introduction 95
7.2 Lipid Nanosystems for Nucleic Acid Delivery in Cancer 97
    7.2.1 Lipoplexes (Liposome + Nucleic Acid) 97
    7.2.2 Solid-Lipid Nanoparticles (SLNs) 99
    7.2.3 Nanostructured Lipid Carriers (NLCs) 101
    7.2.4 Stable Nucleic Acid-Lipid Particles (SNALPs) 101
    7.2.5 High-Density Lipoproteins (HDL) 102
    7.2.6 Lipidoids 104
7.3 Clinical Trials Involving Lipid Nanosystems for Nucleic Acid Delivery in Cancer 105
7.4 Conclusion and Future Prospects 108
References 108

ABBREVIATIONS

CPC:	Cetylpyridinium chloride
CTAB:	Cetrimide
DOTAP:	N-[1-(2,3-Dioleoyloxy)propyl]-N,N,N-trimethyl ammonium chloride
DDAB:	Dimethyl dioctadecyl ammonium bromide
cSLNs:	Cationic solid-lipid nanoparticles
PcSLN:	Paclitaxel-loaded cSLN
TC:	Tricaprin
DOPE:	Dioleoyl phosphatidyl ethanolamine
NLC:	Nanostructured lipid carriers
PNLCs:	Polycationic nanostructured lipid carriers
SNALPs:	Stable Nucleic Acid Lipid Particles

7.1 INTRODUCTION

One of the prime anomalies associated with cancer is the change in the expression of significant proteins belonging to the cellular signalling pathways. As a result, numerous proteins involved in cellular growth, including growth factors, intracellular mediators, transcription factors etc., have been found altered via several mechanisms such as enhanced expression, mutation, chimeric protein generation etc. In order to fight cancer, it is of prime importance to find explicit ligands with the potential to detect and measure the change in protein expression. These ligands are required to not only discriminate amid the oncogenic and non-oncogenic forms of protein in the signalling pathway but also enumerate the level of expression of the same both *in vitro* and *in vivo* (Cerchia et al. 2002). Thus, the use of nucleic acids such as plasmids, siRNA, oligonucleotides, antisense nucleotides etc. as therapeutics has gained an important future direction in the field of molecular medicine. Nucleic acid therapy generally involves the delivery of nucleic acids which act at the molecular genetic level to induce or suppress a specific gene. The plasmid DNA vectors are generally used for the intra-nuclear delivery either to swap or to substitute a specific genetic function; similarly, intra-cytoplasmatic transfer of the siRNA results in falling of the endogenous genes in a sequence-specific manner (Wagner, Kircheis, and Walker 2004).

The nucleic acids encompass the nucleobases such as adenine, guanine, cytosine and thymine in DNA and uracil in the RNA, which form the double helix strands by means of hydrogen bonding in the complementary bases. These organisational structures have programmable self-assembly and organisational features, thereby opening multiple avenues for nucleic acid applications on the therapeutic level besides their basic function of biopolymer and genetic information storage units (Jin et al. 2020). The mechanisms of therapeutic effects of nucleic acids are therefore diverse and include neoangiogenesis (Filleur et al. 2003; Gunther, Wagner, and Ogris 2005), apoptosis induction

DOI: 10.1201/9781003043164-7

(Shir et al. 2005), immunostimulatory response activation (Heller et al. 2006), tumour proliferation reduction (Rait et al. 2002; Aigner et al. 2002) and replacement or deletion of genes (Ndoye et al. 2004).

Nucleic acid therapies have been anticipated to treat extreme grave diseases requiring systemic administration of the gene to the target cells which are affected (De Jesus and Zuhorn 2015). In order to accomplish the desired therapeutic effects, the nucleic acids are required to cross multiple biological barriers and membranes such as the plasmatic and the endosomal membranes. Further, the delivery challenges include the endosomal escape, cytosolic delivery or nuclear delivery, which can be taken care off with the help of viral vectors or ligand-grafted non-viral vectors. Other than the biological barriers, another potential barrier to be taken into account in the delivery of the nucleic acids to the targeted cells are the nucleases such as RNAses and DNAses in the biological fluids and intracellular compartments (Figure 7.1) (Pérez-Martínez et al. 2011; Zuhorn, Engberts, and Hoekstra 2007; Wiethoff and Middaugh 2003). To confront the aforementioned barriers, numerous gene delivery devices/vectors have been investigated to effectively protect the nucleic acids from being degraded while in the gastrointestinal tract, bloodstream etc. thereby facilitating competent translocation across barriers and membranes.

Generally, the gene delivery vectors are of two types: viral and non-viral vectors; each class exhibits its explicit advantages and disadvantages. Viral delivery systems such as adeno-associated viruses, alpha virus, flavivirus, herpes simplex virus, measles virus, rhabdovirus, retrovirus, lentivirus, poxvirus, picornavirus, Newcastle disease virus etc. have been extensively explored for gene delivery, and promising *in vivo* results have been offered by the same (Mingozzi et al. 2009; Manno et al. 2006; Lundstrom 2018).

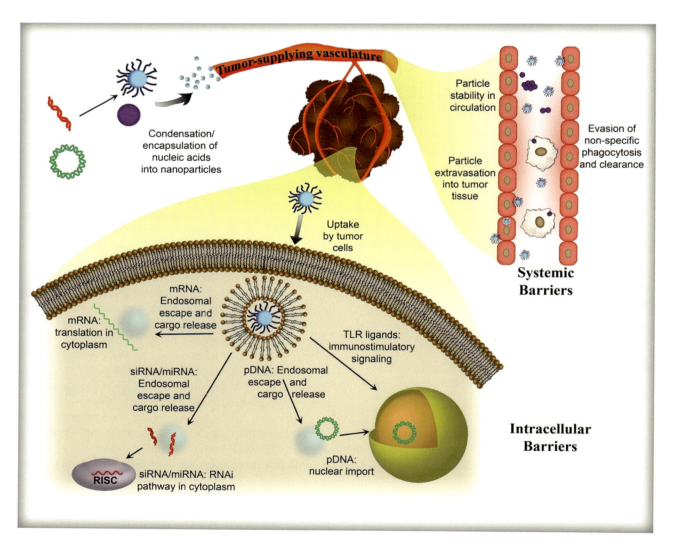

FIGURE 7.1 Various factors to be considered while designing targeted *in vivo* nucleic acid delivery vectors. Nucleic acid complexation, transportation across blood vessels and the tumour microenvironment, efficient cellular uptake and internalisation, followed by endosomal escape, nucleic acid dissociation and cytosolic release. (Reproduced with permission from Vaughan, Green, and Tzeng (2020b).)

However, there have been several serious setbacks owing to the medical complications and potential side effects such as induction of immune response and inflammatory response in the host. Additionally, viral mutagenesis and limitations in production at the large scale render viral vectors less feasible. For the above reasons, non-viral delivery has been focused upon recently as it serves as a potentially less hazardous and an efficient and expedient alternative (Goyal et al. 2011; Wei et al. 2011; Lobovkina et al. 2011).

Worldwide, researchers have developed a variety of non-viral delivery/vectors, including inorganic, lipid and polymer-assembled nanoparticles such as cationic lipids (Rehman, Zuhorn, and Hoekstra 2013; Hope 2014), cationic polysaccharide particles (Saranya et al. 2011), cyclodextrins (Chaturvedi et al. 2011), polymeric nanoparticles (Laga et al. 2012), metallic nanoparticles (Sokolova and Epple 2008), solid-lipid nanoparticles (SLNs) (Bondì and Craparo 2010), liposomes (Elouahabi and Ruysschaert 2005; Ruozi et al. 2003; Zhou, Liu, and Liang 2004), cell-penetrating peptides (Bolhassani 2011), micelles (Kataoka et al. 2005), dendrimers (Palmerston Mendes, Pan, and Torchilin 2017) etc. The aforementioned systems (lipidic systems) have showed convincible projections; however, less attention has been on the same (H du Plessis et al. 2014; Hayes et al. 2006; Li and Szoka 2007). These aforementioned vectors need to achieve tumour targeting at several levels such as (i) targeted delivery of the gene via physical targeting such as active ligand-based targeting, passive targeting via prolonged systemic circulation etc. (Mendelsohn 2002; Yarden 2001; Maeda 2001); (ii) transductional targeting by preferred intracellular targeting into the cell nucleus, which includes the use of vectors transducing dividing cells only, nuclear uptake specific to a certain cell type, light-directed releasing vectors etc. (Vacik et al. 1999; Brunner et al. 2000; Rainov 2000); (iii) transcriptional targeting with the aid of expression cassette applying to tumour/tissue-specific promoter/enhancers, hypoxia/radiation-responsive elements, viral vectors with tumour discerning replication etc. (Rodriguez et al. 1997; Hallahan et al. 1995; Dachs et al. 1997); and (iv) biological anticancer effect by delivering nucleic acids exploiting unique properties of tumours such as angiogenesis inhibition, interference in metastasis, antitumour immune response induction etc. (Nemunaitis et al. 2000).

The lipidic nanoparticles aid in entrapping the nucleic acids in the matrix/core and protecting from the harsh environment of enzymes and pH, targeting to the desired cells with the aid of ligand grafting, and controlled release of the components. Additionally, properties of the lipidic systems such as ease of reproducibility, economic feasibility, long-term stability etc. present added advantages to the fabricated delivery systems for nucleic acids. One of the most significant recent advances in the field of lipid-based nucleic acid delivery is epitomised by the first-ever FDA and European Commission-approved RNAi therapy, Patisiran (Onpattro™), produced by Alnylam Pharmaceutical Corporation. Further, the gene silencing technology is constantly being worked upon and the RNAi mechanism has been explored a lot in recent years (Ledford 2018; Fire et al. 1998). The following sections describe some of the recent strategies employing lipidic nanosystems for targeted delivery of nucleic acids in cancer therapy.

7.2 LIPID NANOSYSTEMS FOR NUCLEIC ACID DELIVERY IN CANCER

7.2.1 Lipoplexes (Liposome + Nucleic Acid)

Lipoplexes are non-viral vectors mostly designed to deliver nucleic acids like DNA, RNA, mRNA and siRNA. Lipoplexes are considered for nucleic acid delivery because of their provision to deliver nucleic acids of various types and sizes. They usually comprise one or more lipids, either anionic, neutral or cationic lipids, and the nucleic acid (Rafael et al. 2015). However, the negative charge of nucleic acid and cell membrane makes the cationic lipid to be the most attractive candidate for gene delivery. Lipoplexes need to have specific properties for the delivery of nucleic acids such as non-immunogenicity, less antigenicity, better physical stability, better pharmacokinetic characteristics (which ensure the delivery at the desired site) and the ability to deliver intact cargo at the disease site, in addition to being safe and well tolerated (MacLachlan 2007).

A typical liposomal carrier consists of a cationic lipid, helper lipid and an active targeting moiety. Cationic lipids are amphiphilic molecules made up of a positive charged polar head group, and one or more hydrophobic chain(s) (chain lengths vary commonly between C8:0 and C18:1). The head group defines the characteristics of the system and is responsible for the transfection efficiency of the lipoplexes. Therefore, variations in the head groups are carried out during the synthesis of synthetic cationic lipids. Similarly, the structural modifications can also be carried out in the head group as well as the hydrophobic group for the modulation of transfection efficiency and cytotoxicity (Rafael et al. 2015). It provides a positive charge to the system, which in turn interacts with the negatively charged phosphate backbone of nucleic acids, allowing them to undergo condensation into a more compact organisation (Mahato, Rolland, and Tomlinson 1997).

The phase transition temperature of a lipid is the temperature at which the lipid phase converts into the gel phase and hydrocarbon chains are fully extended in an orderly fashion. This temperature is affected by various factors such as hydrocarbon chain length, degree of unsaturation, head group charge and species. The most influencing factor is the hydrocarbon chain length; as length increases, van der Waals interaction increases, which requires more energy to become disordered. Similarly, the introduction of a double bond puts a kink in the chain, which requires much less energy to become disordered (Koynova and Tenchov 2013). Above phase transition temperature, lipid becomes highly fusogenic (self-organised into bilayer that can form stable liposomes). This fusogenicity increases when cationic lipid is less saturated because of which liposomes with highly saturated cationic lipids show the highest *in vitro* transfection efficiency, whereas liposomes with highly unsaturated cationic lipid show the highest *in vitro* as well as *in vivo* transfection efficiency (MacLachlan 2007). Helper lipids also play an important role in fusogenicity in addition to the cationic lipid. This fusogenicity contributed by the cationic lipid and helper lipids enables liposomes in fusing with the cell membrane of a target cell. For example, DOPE (dioleoyl phosphatidyl ethanolamine) increases the transfection efficiency by 5–10 times when used along with DOTAP (*N*-[1-(2,3-dioleoyloxy)propyl]-*N*,*N*,*N*-trimethyl ammonium chloride) (cationic lipid) (Hui et al. 1996). DOPE causes the destabilisation of lipid bilayer after fusion, similar to that of cationic lipid. Some cationic lipids efficiently act as a fusogenic agent alone or in the presence of non-fusogenic lipid, i.e. cholesterol, but this kind of systems demonstrate less transfection efficacy when it comes to *in vivo* applications (MacLachlan 2007).

Common techniques employed for the encapsulation of nucleic acid into the liposomes include simple mixing method (mixing of nucleic acid with the liposomes), ethanol drop method (dropwise addition of ethanolic lipid solution into the aqueous solution of nucleic acid), ethanol-destabilised liposome method (preformed liposomes are destabilised by the controlled addition of ethanol) and reverse-phase evaporation method (chloroform-lipid solution and aqueous solution of nucleic acid are added simultaneously, then excess of aqueous chloroform solution along with methanol is added, followed by drying using rotary evaporator and rehydration of the resulting film) (MacLachlan 2007).

Active targeting of the delivery system helps to significantly increase the payload of the drug at the disease site. The main idea is to increase intracellular delivery and therapeutic index of a drug/therapeutic agent by the accumulation and retention of it by attaching cell-oriented or targeted specialised moieties/ligands. This can be achieved through the use of antibodies against highly expressed cell surface molecules (Leonetti et al. 1990), ligands against cell surface receptors (Hara et al. 1995), glycoproteins or glycolipids (Sasaki et al. 1994), vitamins (Lee and Huang 1996), etc. The main advantage of active targeting is that the targeted system does not rapidly clear from the systemic circulation as in the case of passive administration. Hence, many of the researchers are considering active targeting approaches.

Types of lipoplexes are as follows:

A. **siRNA lipoplexes**: siRNA is a synthetic 23-to 25-nucleotide-long double-stranded RNA that enables gene silencing and down-regulation of a specific gene expression by the mechanism of RNA interference (RNAi) (Dana et al. 2017). Nowadays, siRNA is emerging as an innovative drug for the treatment of a host of diseases, including cancer, because it targets the mRNA with fewer side effects (Oh and Park 2009). Luciferase siRNA was effectively transfected with the help of cationic liposomes, which led to high knockdown efficiency (Barichello et al. 2012).

B. **miRNA lipoplexes**: miRNA or micro RNA is short, i.e. 22-nucleotide-long, biologically conserved non-coding RNA molecule found indigenously in plants, animals and some viruses. They involve mostly in gene regulation through the gene silencing process, which is the same as RNAi. They offer a powerful tool to the researchers for gene manipulation. Unlike in siRNA, they have less complementarity with target miRNA but that is sufficient for its degradation or down-regulation (Wahid et al. 2010). Wu et al. demonstrated that miR-133b (a tumour suppressor) is efficiently delivered using cationic lipoplexes for the treatment of non-small-cell lung cancer (NSCLC) (Wu et al. 2011).

C. **DNA (plasmid DNA) lipoplexes**: DNA (the wild type of gene) is used for the restoration of loss of function of the gene, whereas plasmid DNA is used for the continuous expression of that specific gene. This approach is utilised in cancer for replacing the mutated gene with a normal one (using DNA) or the introduction of the new genes to fight against the disease (using plasmid DNA) (Talkar and Patravale 2020). For instance, Charoensit et al. showed that ATRA-cationic liposome/IL-12 plasmid DNA (pDNA) complexes efficiently produced the storm T cells and natural killer cells in lung metastasis of mice. Interleukin 12 (IL-12) is a pro-inflammatory antitumour cytokine that stimulates helper T cell and natural killer cells (Charoensit et al. 2010).

D. **Oligonucleotide lipoplexes**: Oligonucleotide is a short stretch of DNA or RNA. Alteration in gene expression has been persuaded by complementarity binding of the oligonucleotide to the target mRNA or gene in cancer treatment. Koh et al. developed lipoplexes containing Bcl-2 antisense deoxy-oligonucleotide and demonstrated greater down-regulation of respective genes in leukaemia cells (Koh et al. 2010).

The use of lipoplexes for nucleic acid delivery has been discussed in detail in several reviews (Singh, Trivedi, and Jain 2018; Kulkarni et al. 2018; Kulkarni, Cullis, and van der Meel 2018; C. Chen, Yang, and Tang 2018; Pal Singh et al. 2019; Xiao et al. 2019). Though lipoplexes have a great potential to incorporate nucleic acids, *in vivo* delivery at a specific target site remains a major challenge for the researchers (Shim et al. 2013). Hence, the following section focusses on the efforts to modify the lipoplexes surface using peptides, antibodies, ligands, receptors and affibodies, etc.

Jiang et al. developed lipoplexes where cationic liposome surface was actively modified with RGD (arginine-glycine-aspartic acid) peptide (which binds to the integrin) for the delivery of siRNA against P-glycoprotein for the multidrug-resistant tumours. The study showed that there was a huge accumulation of siRNA at targeted cells and there was also the down-regulation of that gene (Jiang et al. 2010). Folic acid (FA) was conjugated by Duarte et al. to the cationic liposome by electrostatic association for the delivery of HSV-tk/GCV suicidal gene. It was concluded that FA-conjugated lipoplexes showed greater antitumour activity than the plain lipoplexes, in both SCC-VII and TSA cell lines. Further, the results were corroborated with a drastic inhibition of both cell lines because of more arresting of the cell cycle at S phase in FA-lipoplexes than the plain one (Duarte, Faneca, and Lima 2012). Nakase et al. developed a cationic liposomes coupled with transferrin receptors (highly expressed on cancer cells) for the delivery of the plasmid pcGS*p53* carrying p53 gene for osteosarcoma treatment. The plasmid pcGS*p53* contained wild-type p53 DNA under the control of a cytomegalovirus promoter. Because of transferrin, transfection efficacy was increased by 3.7 times, and p53 expression was increased by 1.4 times higher than the unmodified liposomes (Nakase et al. 2005).

Table 7.1 summarises more studies about active targeting of nucleic acid-lipoplexes.

Even after decades of research in the field of lipoplexes for nucleic acid delivery, they still face a few limitations, for instance their low stability in the systemic circulation owing to their cationic charge. Further, the cationic lipid components interact with serum proteins and may potentially disrupt the integrity of the delivery vehicle or form aggregates limiting cellular uptake. It is only a matter of time that nucleic acids may emerge as new class of therapeutics considering several clinical trials involving lipoplexes are ongoing and their success will depend on the constant efforts to develop safe, stable, effective targeted cationic lipid-based carriers.

7.2.2 SOLID-LIPID NANOPARTICLES (SLNs)

In recent years, lipidic nanoparticles have been extensively explored in the quest for developing new, safe and effective non-viral dosage forms anticipated for gene delivery (Ekambaram, Sathali, and Priyanka 2012; Bunjes 2011). Lipidic nanoparticles are categorised as SLNs and nanostructured lipid carriers (NLCs). SLNs are the first-generation nanoparticles prepared with lipids and stabilised by emulsifier. These lipids are solid at room temperature. The size of the SLNs is in submicron (less than 1,000 nm) range (Yoon, Woo, and Yoon 2013), and they show various advantages such as protection of drug against chemical degradation, feasibility of large-scale productions using high-pressure homogenisation technique, biocompatibility and biodegradability. However, SLNs do encompass a few disadvantages: the loading capacity of drug in SLNs is low because of their perfect crystalline structure. During the storage, expulsion of drug is observed due to the crystallisation process. Another problem is the initial burst release of drug. To overcome the limitations of SLNs, second generation of lipid-based nanocarriers, i.e. NLCs, were formulated, which are made from a mixture of solid and liquid lipids. In NLCs, because of imperfect crystal structure, higher amount of drug can be loaded. The crystallisation of lipids will not happen during the manufacturing and storage periods and liquid lipid expulsion of drug is not observed.

TABLE 7.1
Examples of Active Targeting of Lipoplexes for Nucleic Acid Delivery

Gene	Targeting Ligand	Types of Cancer	Significance (as Compared to Non-Targeted Lipoplexes) and References
P53 and PTEN DNA	Transferrin	Prostate cancer	Increase in growth inhibition of tumour as high accumulation of p53 at tumour site (Seki et al. 2002)
Anti-HER-2 antisense oligonucleotide	Folate	Breast cancer	Notable inhibition of growth in aggressive breast cancer (Rait et al. 2002)
Eglin 5 (EG5) siRNA	Atherosclerotic plaque-specific peptide-1 (AP-1)	Solid cancers	Induced strong antitumour activity (Luo et al. 2019)
Apolipoprotein B (ApoB)- siRNA	Chondroitin sulphate, poly-l-glutamic acid and poly-aspartic acid	Liver cancer	Significant suppression of ApoB in the liver (Hattori et al. 2014)
Luciferase siRNA	Hyaluronic acid	Metastatic lung cancer	Inhibition of the expression of luciferase gene (Leite Nascimento et al. 2016)
hTERT promoter-regulated plasmid that encodes a matrix protein (MP)	Folate alpha	Ovarian cancer	Significantly inhibited the growth of tumours and increased the survival of mice (He et al. 2016)
Plasmid-pCMVLuc	Cholesteryl-3β-N-(4-amino phenyl-β-D-galactopyranos-yl) carbamate	Hepato-carcinoma	Sixfold increase in accumulation of luciferase in HEPG2 cells (Narainpersad, Singh, and Ariatti 2012)
Co-delivery of thioredoxin 1 shRNA and doxorubicin	Folic acid receptor	Hepatocellular carcinoma	Induced apoptosis in HCC cell line more efficiently than co-delivery of control shRNA and doxorubicin (Li et al. 2014)

NLCs can also increase drug solubility in lipid matrix and show more controllable release profiles in comparison with SLNs. Lipidic nanoparticle-based delivery for nucleic acid usually depends on the use of (ionisable) polycations, which provide an efficient binding site for negatively charged nucleic acids via electrostatic interactions. On the basis of this principle, a great variety of vectors have been designed for nucleic acid delivery (Ghasemiyeh and Mohammadi-samani 2018).

Since the last two decades, cationic SLNs have been prepared and used as an effective tool for gene delivery for cancer treatment (Olbrich et al. 2001). Few studies have successfully reported the delivery of combination of drug and gene together entrapped in lipidic nanoparticles. Furthermore, the surface of nanoparticles can be readily modified by specific ligands; thus, targeted delivery of gene to specific organs, cell and tissue is possible (Mechanisms, Jesus, and Zuhorn 2015). To form a complex of the negatively charged nucleic acids and SLNs, lipidic nanoparticles are formulated by incorporating a positively charged co-surfactant in their formulation, for example N,N-di-(b-stearoylethyl)-N,Ndimethyl-ammonium chloride, benzalkonium chloride (BA), cetylpyridinium chloride (CPC), cetrimide (CTAB), DOTAP or dimethyl dioctadecyl ammonium bromide (DDAB) (Manjunath, Reddy, and Venkateswarlu 2005). Lipidic nanoparticles are produced by different methods such as high-pressure homogenisation (hot homogenisation, cold homogenisation and ultrasound or high-speed homogenisation method), solvent emulsification, microemulsion method, spray drying and double emulsion method. Among these methods, microemulsion, homogenisation and solvent evaporation methods are commonly used to prepare cationic SLNs (Ezzati et al. 2015).

Yu et al. developed cationic SLNs (cSLNs) for co-delivery of paclitaxel and siRNA using 1,2-dioleoyl-sn-glycero-3-ethylphosphocholine lipid with the help of emulsification solidification method. The developed SLNs were characterised for particle size, zeta potential and gel retardation. The mean particle size and zeta potential were 140.4 ± 12.9 nm and 43.9 ± 2.7 mV, respectively. The gel retardation study confirmed the electrostatic complex formation of cSLN with siRNAs, and the particle size of the complex was found to be 183.1 ± 12.0 nm. Cellular uptake study revealed the enhanced transfection of siRNA using paclitaxel-loaded cSLN (PcSLN). Further, in vitro and in vivo studies of formulation performed in comparison with the plain drug and siRNA showed better efficacy of formulation. The in vitro studies in KB cells showed the lowest survival of cancer cells after co-treatment of siMCL1 and PTX using PcSLN ($38.8\% \pm 1.5\%$) as compared to free PTX ($90.9\% \pm 2.8\%$) or naked siMCL1 ($91.9\% \pm 7.6\%$). In vivo studies show that co-delivery of siMCL1 with PTX using PcSLN significantly inhibited the growth of tumour tissues as compared to other treatment groups. On day 16 after inoculation, the average tumour volume of control group treated with saline solution was 1373.1 ± 241.0 mm, the second group treated intratumorally with siMCL1 complexed to empty cSLN and free PTX showed an average tumour volume of 714.2 ± 230.0 mm, while the intratumoral co-delivery of PTX and siMCL1 using PcSLN resulted in a significant reduction in tumour volume to 172.0 ± 73.7 mm (Yu et al. 2012).

In another study, Choi et al. prepared novel cationic SLNs for enhanced p53 gene transfer to lung cancer cells. The SLNs were prepared by mixing tricaprin (TC) as a core, [N-(N_0, N_0-dimethylaminoethane) carbamoyl] cholesterol (DC-Chol), DOPE and Tween-80 in various ratios. The average particle size was observed around 100 nm, and zeta potential was in between 8 and 13 mv for different SLNs. The transfection efficiency of SLN1 prepared in the ratio of (TC:DC-Chol:DOPE:Tween 80 = 0.3:0.3:0.3:1) was the highest among all the SLN formulations, even higher than that of commercially available Lipofectin. PCR analysis indicated that the prepared SLNs stabilised the plasmid DNA in serum in a manner comparable to that of commercially available Lipofectin®. Furthermore, novel cationic SLNs prolonged the expression of mRNA of the plasmid DNA in various organs for up to 5 days. The pp53-EGFP/SLNs complexes inhibited the cell growth, thereby indicating that these complexes were efficient not only in increasing the expression of p53 proteins, but also in inhibiting the growth of lung cancer cells. The cells treated with pp53-EGFP alone showed no significant difference compared with the untreated group. The SLN-mediated transfer of the p53 gene apparently restored a functional apoptotic pathway. Both SLNs and Lipofectin® complexes with pp53-EGFP showed higher levels of apoptosis compared with the control and pp53-EGFP alone (Hee et al. 2008).

Jin et al. studied the delivery of c-Met siRNA to glioblastoma using cationic SLNs prepared by using the modified emulsification-solvent evaporation method. siRNA targeting humanc-Met(sense; 5′-GUGCAGUAUCCUCUGACAGUU-(CH2)6NH$_2$–3′,antisense; 5′-CUGUCAGAGGAUACUGC ACUU-3′) conjugated with PEG through a disulphide bond and non-targeting (NT) siRNA (sense; 5′-GUUCAGCGUGUCCGGCGAGTT-(CH2)6-NH$_2$–3′, antisense; 5′-CUCGCCGGACACGCU GCUGAACTT-3′), both modified with 3′-hexylamine on the sense strand siRNA, was incubated with SLN at room temperature for 15 min in deionised water. The mean diameter of SLN measured by laser light scattering was found to be 117.4 ± 11.7 nm, and the zeta potential value was 37.3 ± 2.3 mV. In the proliferation assay, treatment with the c-Met siRNA-PEG/SLN complex significantly reduced tumour cell proliferation in a dose-dependent manner; 20 and 40 nM groups showed 13.1% and 23.4% tumour cell proliferation reduction, respectively, compared with NT siRNA-PEG/SLN complex group. These results indicated that the prepared c-Met siRNA-PEG/SLN complex specifically reduced the proliferation of glioblastoma cells by controlling the expression of c-Met in vitro. In the in vivo study, after 2 weeks of implantation of tumour cells, four groups ($n = 7$ for each group) were administered intravenous injection of SLN alone (control), 0.125, 0.5, or 2 mg/kg of c-Met siRNA-PEG/SLN complex three times once a week. The tumour volumes were measured on the day of last c-Met siRNA-PEG/SLN complex injection. The group treated with c-Met siRNA-PEG/SLN complex significantly inhibited U-87MG tumour growth in a dose-dependent manner; 0.125, 0.5 and 2 mg/kg groups showed 50%, 62% and 91% tumour volume reduction, respectively, compared with the control group (*P<0.05 and **P<0.01 vs. control). Immunohistochemical staining of the tumour sections demonstrated that there was a down-regulation of c-Met with c-Met siRNA-PEG/SLN complex administration (Jin et al. 2011).

Although there have been reports of cationic SLNs being investigated as transfection agents, there is a lack of knowledge as how specifically SLNs accommodate the nucleic acid therapeutics and their downstream intracellular processing. Further investigations on structure-function relationships of cationic SLNs incorporating nucleic acids and their interaction with the target membranes will help in translating these preclinical studies involving SLNs into clinical applications.

7.2.3 Nanostructured Lipid Carriers (NLCs)

Currently, there are three major types of NLCs: cationic NLC, neutral NLC and targeting moiety modified NLCs, employed for tumour gene therapy. NLCs have bright prospects for many clinical applications because of their superior biocompatibility, high biodegradability and low immunogenicity. Cationic NLCs are positively charged lipid vesicles and can form complex with negatively charged substances, including proteins, polypeptides, oligonucleotides, RNAs and DNAs. The composition of most cationic NLCs includes a cationic head, a hydrophobic hydrocarbon backbone and a linker region. In recent years, neutral NLCs have also garnered attention as a novel carrier for the delivery of miRNAs. Neutral NLCs do not possess charge on the lipid and thus aid in overcoming the disadvantages that can be attributed to surface charge. For example, neutral NLCs are not filtered by the liver, avoid adhering to endothelium or taken up by macrophages as they do not form aggregates in biofluids. Though delivery of nucleic acids by using both cationic NLCs and neutral NLCs was possible, the target efficiency of these delivery systems was still lacking *in vivo*. Therefore, surface-modified NLCs have been the centre of attention of current research works. To avoid uptake of NLCs by macrophages, they are surface-coated with biocompatible polymers such as PEG, which significantly improved the stability and half-life of NLCs. Moreover, the ligands of specific cells can be conjugated to the surface of NLCs to increase the concentrations of miRNA/NLC complex in the target tissues (Wang et al. 2018).

In one of the earlier studies, Zhang et al. developed novel polycationic NLCs (PNLCs) with triolein for efficient gene delivery. PNLCs were prepared using the emulsion-solvent evaporation method, and their gene transfer properties were evaluated in human lung adeno carcinoma cell line SPC-A1 and Chinese Hamster ovary (CHO) cells. Significant enhancement of transfection efficiency of PNLCs, comparable to that of Lipofectamine TM2000, was observed after the addition of triolein to the PNLC formulation. In the presence of 10% serum, the transfection efficiency of the optimised PNLC formulation was not significantly changed in either cell line, whereas that of Lipofectamine TM2000 was greatly decreased in both (Zhang et al. 2008).

Chen et al. prepared NLCs of temozolomide (TMZ) and gene for gliomatosis cerebri combination therapy. NLCs were prepared by solvent diffusion method, and the TMZ/DNA-NLCs had a particle size of 179 nm with a zeta potential of +23 mV. The *in vivo* transfection efficacy studies in tumour-bearing BALB/c nude mice after 48 and 72 h of administering TMZ/DNA-NLCs, DNA-NLCs, and TMZ-NLCs, Lipofectamine 2000 /DNA and naked DNA showed significantly higher transfection efficiency for TMZ/DNA-NLCs and DNA/NLCs as compared to other formulations (p < 0.05). The *in vivo* antitumour efficacy of formulation and free drug was evaluated in U87MG solid tumours in mice. The TMZ/DNA-NLCs group showed a significant reduction of tumour volume (588 mm^3) when compared with control group treated with saline solution (1,605 mm^3) and free TMZ-treated groups (857 mm^3) after a 15-day treatment. These results indicate that tumour growth was significantly inhibited by NLCs formulations. The rate of inhibition of tumour for treatment with TMZ/DNA-NLCs was 3.3 times higher than that for free TMZ solution (Chen et al. 2015).

Han et al. prepared transferrin-modified NLCs as a multifunctional nanomedicine for co-delivery of green fluorescence protein plasmid (pEGFP) and doxorubicin. In this study, the authors developed pEGFP and DOX-loaded NLCs and coated by transferrin-containing ligands. The particle and zeta potential of transferrin-modified DOX and pEGFP co-encapsulated NLCs (T-NLC) were 157 nm and +19 mV, respectively, with 82% gene loading. The developed NLC formulations showed 80%–100% *in vitro* cell viability compared with the control. Further, transferrin-modified NLCs showed higher *in vitro* and *in vivo* transfection in A549 cells, and subsequently enhanced antitumour activity in tumour-bearing mice as compared to naked pEGFP and unmodified NLCs (Han et al. 2014). Thus, it is evident that active targeting of lipid nanoparticles can significantly enhance both *in vitro* and *in vivo* transfection efficiency and improve the efficacy of gene therapy for various cancers. However, the research in this direction is limited to preclinical studies and warrants a detailed structure-function analysis of these formulations before proceeding to clinical evaluations.

7.2.4 Stable Nucleic Acid-Lipid Particles (SNALPs)

SNALPs constitute a class of lipid carriers with a high transition temperature phospholipid, and an ionisable cationic phospholipid. They also have a diffusible coating of polyethylene glycol-lipid (PEG-lipid) conjugate to stabilise and provide a neutral and hydrophilic exterior. Although SNALPs have neutral surface charge, positively charged inner membranes aid in efficient encapsulation of negatively charged nucleic acid drugs. First developed by Semple and colleagues in 2001, SNALPS showed higher encapsulation efficiency and stability in biological fluids and the conventional cationic liposomes (Semple et al. 2001). They rely on EPR effect due to prolonged circulation time in the blood and accumulate at the sites of vascular leakage, especially at tumour sites. Upon reaching the tumour site, they are easily endocytosed by cancer cells and deliver siRNA intracellularly. In addition to cancer, SNALPs have been explored for the treatment of dyslipidaemia, hypercholesterolaemia (Zimmermann et al. 2006), hepatitis B viral infection (Morrissey et al. 2005), Ebola (Geisbert et al. 2006) etc.

In cancer therapy, therapeutic efficacy of knockdown of CSN5 gene that plays a crucial role in cell proliferation and senescence was studied by administering intravenous injections of either β-galactosidase-targeted siRNA/SNALP or CSN5-targeted siRNA/SNALP in mouse xenograft model of hepatocarcinoma. Further, the use of SNALPs to deliver the siRNA aided in reducing the dosage to 2 mg/kg, a tenfold reduction as compared to the standard dose of siRNA without a carrier. The hepatic tumour growth was effectively inhibited by systemic delivery of CSN5 siRNA by SNALPs, while tumours in β-galactosidase-targeted siRNA/SNALP group grew rapidly in the liver parenchyma (Lee et al. 2011). SNALPs were used for delivering miRNA too. miRNA 119b-5p, which plays a role in the impairment of cancer stem cells, was efficiently encapsulated into high transition temperature phospholipid disteroylphosphatidylcholine, DODAP-based SNALPs, and found to significantly down-regulate the expression of cancer stem cell markers in several tumorigenic cell lines, including colon (HT29, CaCo-2 and SW480), breast (MDA-MB231T, MCF-7), prostate (PC-3), glioblastoma (U-87) and medulloblastoma (Daoy, ONS-76 and UW-228) (Antonellis et al. 2013). The same SNALP formulation was used for encapsulating tumour suppressor miRNA (miRNA-34a). The developed SNALP-miRNA-34a complexes of 150 nm, negative zeta potential and an encapsulation efficiency of about 83% demonstrated a significant cell growth inhibition in a multiple myeloma cell line (SKMM-1) after 48 h of treatment. Further *in vivo* evaluations showed tumour growth inhibition via apoptosis in the tested mice model (Di Martino et al. 2014).

Judge et al. developed chemically modified siRNA targeting the cell cycle proteins polo-like kinase 1 (PLK1) and kinesin spindle protein (KSP), and formulated them in SNALPS. A single intravenous administration was found to induce extensive mitotic disruption and tumour cell apoptosis due to target gene silencing in both hepatic and subcutaneous tumour models in mice (Judge et al. 2009). Following this, clinical trials of SNALPs encapsulating siRNA targeting Plk1 (TKM080301, Tekmira Pharmaceuticals Corporation, Canada) were started. Phase I/II studies in adult patients with advanced solid tumours or lymphomas (Northfelt et al. 2013), neuroendocrine tumours or adrenocortical carcinoma (Ramanathan et al. 2014) were well tolerated and showed promising evidence of anti-tumour effect, while in case of advanced hepatocellular carcinoma (HCC), although it showed a favourable toxicity profile, improvements in overall survival were limited and did not support larger randomised trials evaluating it as a single agent (El Dika et al. 2019). ALN-VSP02, another SNALP formulation incorporating siRNAs targeting dual targets KSP and vascular endothelial growth factor (VEGF), was developed by Alnylam Pharmaceuticals, USA. In a Phase I clinical trial of ALN-VSP02 in patients with advanced solid tumours involving liver, the drug was well tolerated and there was evidence of residual siRNAs and mRNAs cleavage products in the liver and antitumour effects in hepatic and extrahepatic sites of disease, including a complete response (Tabernero et al. 2013). Although two other clinical trials involving SNALPs for delivering an siRNA to inhibit the overexpressed proto-oncogene c-Myc were initiated by Dicerna Pharmaceuticals in 2014, they were terminated as the preliminary results were not convincing for the company to continue with further development (Tolcher et al. 2015). To date, SNALP formulations have been the most successful lipid-based nanocarriers in clinical trials and have led to the development of the first RNAi therapeutic in the market, Patisiran that silences the gene associated with the rare disease hereditary transthyretin amyloidosis (Adams et al. 2018). Although SNALP formulations can be easily designed and customised for the treatment of different cancers, toxicity issues related to their cationic lipid-based structure still need to be probed carefully before their clinical use.

7.2.5 High-Density Lipoproteins (HDL)

Endogenous HDL is involved in the reverse cholesterol transport to deliver excess cholesterol to the liver. In addition, they are also known to transport proteins, hormones, protease inhibitors, immune function mediators, vitamins and miRNAs to target cells or tissues. Typically, the HDL particle consists of a central core of esterified cholesterol, surrounded by a phospholipid monolayer, free cholesterol and apolipoproteins (Shah et al. 2013). Owing to their biocompatibility and receptor-mediated uptake via scavenger receptor type B1 (SR-B1) receptors, they have been explored as carriers for lipophilic compounds as well as nucleic acids. The use of HDL for nucleic acid delivery was driven by the information that lipid-modified siRNAs bind to lipoproteins (HDL and LDL) in the serum after systemic administration and get delivered to tissues that express receptors for these specific lipoproteins. Further, HDL was found to bind and stabilise nucleic acids and deliver directly to the cytosol of the cells that express SR-B1 (Wolfrum et al. 2007; Vickers et al. 2011). However, source-dependent inconsistency in the composition of endogenous HDL and the risk of harbouring infectious agents limited their widespread use.

Therefore, synthetic or reconstituted HDL (rHDL) nanoparticles are now being developed for drug/gene delivery. Although rHDL were initially developed to understand the structure and functions of endogenous HDL, their inherent biocompatibility, amphipathic nature and structural similarity to HDL aid in receptor-ligand interactions and specific delivery of lipophilic drugs to some degree. Further target specificity can be tailored by appending other targeting moieties to the apolipoprotein A-I.

Several tumour cells overexpress SR-B1 receptors to scavenge HDL particles to maintain their high growth. This characteristic has been exploited for the targeted delivery of rHDL nanoparticles incorporating siRNA. For instance, Yang et al. prepared HDL-mimicking peptide-phospholipid scaffold (HPPS), the central components of which were phospholipids, cholesteryl oleate and amphipathic α-helical peptides that mimic apolipoprotein A-I. Cholesterol-modified bcl-2 siRNA was intercalated into the phospholipid monolayer of HPPS to form stable nanoparticles, which then selectively delivered the cholesterol-modified siRNA into the cytosol of treated KB cells, resulting in a dose-dependent decrease in Bcl-2 protein expression (Yang et al. 2011). In another study, Shahzad et al. demonstrated the delivery of siRNA via rHDL nanoparticles and effective silencing of signal transducer and activator of transcription 3 (STAT3) and focal adhesion kinase (FAK) proteins acting as key factor to cancer growth and metastasis in orthotopic mouse models of ovarian and colorectal cancer (Shahzad et al. 2011). Ding et al. prepared fluorescently tagged cholesterol-conjugated siRNA-loaded rHDL nanoparticles using thin-film dispersion method and studied their uptake in HCC cell line, HepG2. It was found that rHDL-mediated transfection enabled efficient uptake and specific cytoplasmic delivery of the cholesterol-tagged siRNA and subsequent growth inhibition and decrease of Pokemon and Bcl-2 protein expression in HepG2 cells. Further *in vivo* studies in nude mice revealed a significant accumulation in tumour tissue and subsequent tumour growth inhibition and decreased Pokemon and Bcl-2 protein expression after i.v. administration in comparison with lipoplexes loaded with the same cargo (Ding et al. 2012).

Another set of interesting studies involved the development of gold nanoparticle templated HDL biomimetics for DNA- or RNA-based antisense therapy to achieve the knockdown of a target gene expression and further downstream effects in treated cells. The gold nanoparticle templates aided in controlling the size, shape and surface chemistry of the developed HDL biomimetics. For instance, Thaxton and colleagues developed HDL-gold nanoparticles (HDL-Au) by mixing and overnight stirring of aqueous suspension of citrate-stabilised gold nanoparticles and purified APOA1, followed by the addition of disulphide-functionalised lipid, 1,2-dipalmitoyl-sn-glycero-3-phosphoethanolamine-*N*-[3-(2-pyridyldithio) propionate], and amine-functionalised lipid, DPPC in the ratio 1:1 (Thaxton et al. 2009). Further, they utilised these HDL-Au nanoparticle platforms to adsorb cholesterated-DNA and analysed these conjugates to regulate target RNA, miR-210, in cultured PC-3 prostate cancer cells (McMahon et al. 2011). The same platform system was used to deliver cholesteryl-modified anti-angiogenic RNAi to endothelial cells for the knockdown of VEGFR2 expression. The schematic diagram representing the preparation of HDL-Au nanocarrier for cholesterylated oligonucleotide delivery is given in Figure 7.2. Efficient delivery of RNAi by these biomimetic HDL nanoparticles resulted in the modulation of VEGF responses *in vitro* and inhibition of angiogenesis *in vivo* after systemic administration in a mouse non-small-cell lung carcinoma model (Tripathy et al. 2014).

To avoid concerns about off-target toxicity towards other SR-B1-expressing organs like the liver and adrenal gland, $\alpha_v\beta_3$-integrin-specific cyclic-RGDyk peptide

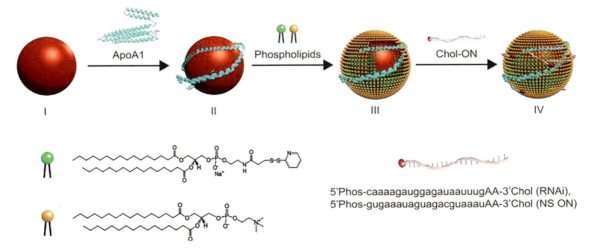

FIGURE 7.2 Scheme for the preparation HDL-Au nanocarrier for cholesterylated oligonucleotide delivery. (I) Gold nanoparticles (5 nm, red spheres) were surface-functionalised with apolipoprotein A1 (blue) and two phospholipids [DPPC (yellow) and 1,2-dipalmitoyl-sn-glycero-3-phosphoethanolamine-*N*-[3-(2-pyridyldithio) propionate (green)] to form HDL-Au nanoparticles, followed by the addition of cholesterylated oligonucleotides. (Reproduced with permission from Tripathy et al. (2014).)

was conjugated to the apolipoprotein A-I of native HDLs to confer them with selective capacity to reroute to the tumour receptors. Further, cholesterol siRNA-incorporated peptide-modified HDL nanoparticles were prepared, and they exhibited efficient siRNA protection in systemic circulation, favourable proton sponge effect and rapid endolysosomal escape. To demonstrate targeted gene silencing therapy, Pokemon-specific siRNA was incorporated and upon administration, extensive tumour growth inhibition, improvement in the survival period, and decreased pokemon expression were observed (Ding et al. 2017). HDL nanosystems have been explored for combination therapy too. Synthetic HDL nanodiscs were explored for co-delivery of a hydrophobic anticancer drug, docetaxel and CpG oligodeoxynucleotide to glioblastoma (GBM) tumour microenvironment. When administered along with radiation, the HDL mimicking nanodiscs induced tumour regression and elicited anti-GBM immunological memory, preventing the development of tumour upon tumour rechallenge in treated mice (Kadiyala et al. 2019). The same platform was used for the delivery of docetaxel along with Toll-like receptor 9 (TLR9) agonist CpG oligonucleotide for chemoimmunotherapy of colon adenocarcinoma. Intratumoral administration of this nanosystem in MC-38 tumour-bearing mice showed a significant improvement in overall survival compared to chemotherapy alone (Scheetz et al. 2020).

Thus, nanosystems based on endogenous HDL and HDL biomimetics have proved to be potential candidates for targeted drug and/or nucleic acid delivery to tumour cells. However, limited availability of the protein component apolipoprotein A-I has hampered the scale-up and commercial development of HDL drug delivery vehicles. Another challenge is to identify the ideal platform candidate to formulate the drug from among the limitless combinations of molecules in each synthetic HDL. Finally, there is a need to ascertain specificity of these biomimetics and avoid safety concerns to off-target cells. Although the studies so far have not raised any concerns regarding the safety of HDL, continued research and refinement of techniques are key to the clinical translation of HDL nanoparticles.

7.2.6 Lipidoids

Lipidoid particles are lipid-like delivery molecules that were developed to improve SNALP formulations and their development involved concepts of combinatorial synthesis and high-throughput screening. Combinatorial synthesis aids in having diverse chemical structures in lipids, and high-throughput screening can be employed for the rapid determination of structure-activity relationships and identification of best suited lipids for the delivery of a particular cargo. Further along with helper lipids like DOPE, cholesterol, formulations of the selected lipids can be employed for nucleic acid delivery. Michael addition of amines to alkyl acrylamides or acrylates, alkylation of amines, epoxide and amine ring-opening reaction, and click chemistry methods have been employed so far for the synthesis of lipids (Altınoglu, Wang, and Xu 2015). One of the initial studies involving lipidoids was carried out by Lenssen et al. in which cationic lipids based on 3-methylamino-1,2-dihydroxypropane as polar, cationic lipid part were synthesised and screened to identify a better transfection agent. The transfection efficacy of the identified candidate named KL 1-14 was further tested in the cell lines MDA-MB-468, MCF-7, MDCK-C7, and primary dendritic cells (DC), and KL-1-14/Chol in the ratio 1:0.6 showed two to four times higher transfection efficacy than DOTAP (Lenssen et al. 2002). The pioneering research in this field by Akinc et al. involved the development of a large library of 1,200 lipidoids to identify a promising candidate for siRNA delivery. The identified lipidoid, 98N12-5, demonstrated specific silencing of endogenous gene transcripts using siRNA or miRNA in the hepatocytes of mice, rats and nonhuman primates at much lower doses of siRNA than those required by the original SNALP formulation (Altınoglu, Wang, and Xu 2015; Akinc et al. 2008). Huang et al. then used the same lipidoid, 98N12-5 to formulate claudin-3 (CLDN3) siRNA and evaluated its efficacy to suppress ovarian cancer. Intratumoral and intraperitoneal injections of lipidoid/CLDN3 siRNA in tumour-bearing mice resulted in dramatic silencing of CLDN3, enhanced apoptosis and reduction in cell proliferation and tumour burden (Huang et al. 2009). In similar lines, Qiaobing Xu and coworkers developed a series of lipidoids, including bioreducible lipidoids based on Michael addition of aliphatic amines and acrylate for the delivery of DNA (Sun et al. 2012), mRNA (Wang et al. 2012) and siRNA (Wang et al. 2014), and demonstrated their ability to transfect fibroblasts and cancerous cell lines. They also demonstrated *in vitro* degradation of bioreducible lipidoids using an intracellular siRNA trafficking study and noted that a highly reductive intracellular environment enhances siRNA release and gene knockdown efficiency (Wang et al. 2014). In another study, Ding et al. developed negative lipidoid nanoparticles based on 98N12-5(1), mPEG2000-C12/C14 glyceride and cholesterol, for delivering siRNA for effectively silencing a proliferation-inducing ligand (APRIL) that stimulates tumour growth and is overexpressed in colorectal cancer. Local enema delivery of these lipidoids silenced APRIL *in vitro* and *in vivo*, thereby resulting in the suppression of proliferation, metastasis and apoptosis-related cytokine expression in colorectal cancer cells (Ding et al. 2013, 2014). Dong et al. attempted co-delivery of siRNA and plasmid DNA using lipidoids, and utilised firefly luciferase pDNA and Tie2 (receptor tyrosine kinase expressed in many organs) siRNA to demonstrate significant luciferase expression (pDNA) and reduction in Tie2 expression (siRNA) in the lung, liver and spleen after intravenous injections in mice (Dong et al. 2014). In another study, silencing of HoxA1 gene, responsible for early mammary cancer progression, was silenced *in vivo* via intraductal delivery of siRNA-incorporated lipidoids. Upon the treatment of transgenic mice with early-stage breast cancer, significant suppression of mammary epithelial cell proliferation and tumour incidence were observed, indicating that this strategy could be employed to prevent breast cancer progression (Brock et al. 2014). Recently, a series of lipidoids based on neurotransmitter derivatives that could cross BBB were developed through Michael addition between the primary amine of the neurotransmitters and acrylate-containing hydrophobic tails. These lipidoids when introduced into otherwise impermeable lipid nanoformulations enabled their BBB permeability. Small molecules (amphotericin B), macromolecules like antisense oligonucleotides (Tau-ASOs) and proteins [green fluorescent protein (GFP)-Cre] could be delivered into mouse brain after intravenous injection of this nanosystem, thereby establishing its utility in treating CNS tumours and neurodegenerative disorders (Ma et al. 2020). Although the therapeutic efficacy of lipidoids and SNALPs has been established in a number of *in vivo* studies, the chances of splenomegaly and immunostimulatory effects have also been reported. Some of the clinical trials involving SNALPS have been terminated owing to influenza-like syndrome in treated patients and requirement for premedication with corticosteroids. Citing these reasons, Groot et al. developed and studied the *in silico* and *in vitro* immunogenicity of lipidoid-polymer hybrid nanosystems in comparison with lipoplexes and SNALPS. Although these hybrid systems were found to be less immunogenic, further *in vivo* safety and efficacy studies at higher doses and longer exposure times are warranted before proceeding with clinical evaluations (Groot et al. 2018).

7.3 CLINICAL TRIALS INVOLVING LIPID NANOSYSTEMS FOR NUCLEIC ACID DELIVERY IN CANCER

As discussed in the previous sections, we are aware of various lipid-based nanoparticle platforms that are used for nucleic acid-based therapeutics and their classification. The delivery methods for siRNAs, ASOs and mRNAs have shown a marked progress due to the usage of lipid carriers such as liposomes and lipoplexes, LNPs (SLNs and NLCs) and SNALPs (Patravale, Dandekar, and Jain 2012). Over more than a decade of research and development in cancer therapy, the gene delivery is finally turning into clinical reality. The use of lipids as a carrier for the delivery of precious cargo has advanced to a point where it enables competent stability and immune elusion, meanwhile imparting efficacy and specificity. Preclinical efficacy and successful translation of new nanosized therapy to become FDA-approved treatment is an expedition facing many failures. Nevertheless, the spike in the number of clinical trials for nucleic acid nanotherapy especially in Phase I and Phase II tells a tale that the research is going on the precise route to achievement.

Liposomes, lipoplexes and LNP-based nanodesigns with the ability to localise in the liver by means of accumulation in Kupffer cells or hepatocytes and deliver RNAi cargos owing to their diameters of ≤100 nm, have successfully reached Phase II. Liver is the fundamental organ for metabolism, detoxification and factor production; hepatocyte-targeted nucleic acid therapies impact liver-specific diseases, such as hepatic cancer, hypercholesterolemia and viral hepatitis, along with systemic diseases such as familial amyloidosis and hyperoxaluria that express culprit mutant genes in the liver, including TTR and glycolate oxidase, respectively. Examples of a few clinical trials involving lipid nanocarriers for cancer therapy are cited in Table 7.2. Some of the important clinical studies have also been discussed in the following section.

A Phase 1b/2, multicentre, dose escalation trial was conducted to determine the safety, tolerance, maximum tolerated dose (MTD) and recommended Phase 2 dose of DCR-MYC, an LNP-formulated siRNA oligonucleotide in HCC patients. The formulation being investigated was a stable lipid particle suspension that targets the oncogene MYC and thereby inhibits cancer cell growth. MYC oncogene activation is important to the growth of many hematologic and solid tumour malignancies. The target group of the study included patients who were identified as sorafenib-refractory and sorafenib-intolerant regardless of dose reduction. The group of patients also included the cases in which no other suitable therapy was available and effective including sorafenib. DCR-MYC was administered as IV infusion across five dose levels (0.1, 0.125, 0.156, 0.2 and 0.3 mg/kg) for a median age of 58 years to determine MTD. The patients did not show significant treatment-related adverse effects, including fatigue, nausea and infusion reactions, except a subject with primitive neuroectodermal tumour who experienced dose-limiting toxicity (DLT) and fatigue at a dose of 0.1 mg/kg. All the subjects experienced a complete metabolic response (based on FDG-PET) post cycle 1, which was sustained for more than 8 months without further treatment and multiple subjects indicated tumour shrinkage. Serious adverse events (SAEs) covering a range of cardiac disorders, general disorders, hepatobiliary disorders, nervous system disorders, respiratory, thoracic and mediastinal disorders and vascular disorders were also zero, indicating that DCR-MYC was well tolerated. Thus, the data from clinical trials NCT02110563 and NCT02314052 indicate promising initial clinical and metabolic responses across five dose levels of LNP DRC-MYC formulation and support early validation of MYC as a therapeutic target (Tolcher et al. 2015).

Another lipid-based formulation, Atu027, was investigated in an open-label, single-centre, dose-determining Phase I study in subjects with advanced solid cancer. Atu027 is a liposomal formulation of AtuPLEX (Santel et al. 2006) containing chemically stabilised siRNA, which silences the expression of protein kinase N3 (Leenders et al. 2004). Atu027 has four components: 23-mer blunt-ended chemically stabilised siRNA, AtuRNA 23R/H19, a cationic lipid (AtuFECT01), a neutral helper lipid, and pegylated lipid, which become lipoplexed in isotonic sucrose carrier. It is in the form of a sterile lyophilised concentrate which is used for infusion (Schultheis et al. 2014). The NCT00938574 trial enrolled 34 patients with advanced solid tumours

TABLE 7.2
Lipid Nanocarriers in Clinical Trials for Cancer Therapy

Name	Treatment	Genetic/Protein Target	Delivery Vehicle	Administration Method	Disease	ClinicalTrials.gov Identifier/Status
siRNA-EphA2-DOPC	siRNA	EphA2	Liposome	Intravenous infusion	Advanced malignant solid neoplasm	NCT01591356/ phase I
Atu027	siRNA	PKN3	Liposome	Intravenous infusion	Advanced solid tumours	NCT00938574/ phase I
		siRNA vs. PKN3	Liposome		Pancreatic cancer	NCT01808638/ completed
		siRNA vs. PKN3	Liposome		Pancreatic cancer	NCT01808638/ completed
		siRNA vs. PKN3	Liposome		Solid tumour	NCT00938574/ completed
EphA2	siRNA	siRNA vs. EphA2	Liposome		Solid tumour	NCT01591356/ phase I
Lipo-MERIT	mRNA	Tumour-associated antigens	mRNA-Lipoplex	Intravenous infusion	Melanoma	NCT02410733/ phase I
TNBC-MERIT	mRNA	Tumour-associated antigens	mRNA-Lipoplex	Intravenous infusion	Triple-negative breast cancer	NCT02316457/ phase I
DCR-MYC	siRNA	MYC	LNP	Intravenous infusion	Solid cancer, multiple myeloma, non-Hodgkins lymphoma, pancreatic neuroendocrine tumours, PNET NHL	NCT02110563/ phase I
					Hepatocellular carcinoma	NCT02314052/ phase I/II
LErafAON-ETU	ASO	C-raf	LNP	Intravenous infusion	Advanced cancer	NCT00100672/ phase I
Prexigebersen (BP1001)	ASO	Grb2	LNP	Intravenous infusion	Myeloid leukaemia	NCT02781883/ phase II

who received ten escalating doses of Atu027 to determine the MTD and DLT by intravenous infusions twice per week during a 28-day cycle. The treatment response was monitored by computed tomography/magnetic resonance imaging at baseline, at the end of treatment (EoT) and at the final follow-up (EoS), and was assessed according to RECIST. Twenty-five subjects successfully completed the full treatment regimen, while nine patients withdrew prior to completion due to progression of the condition, SAE incidence, death or loss during follow-up. Pharmacokinetic analysis was done by ELISA method that was adjusted to quantify single-strand RNA oligonucleotide. The plasma pharmacokinetic profile of AtuRNA 027/23H siRNA antisense double-strand exhibited mean siRNA plasma levels in human patients after single treatment itself for the indicated dose levels: 0.001, 0.003, 0.009, 0.018, 0.036, 0.072, 0.120, 0.180, 0.253 and 0.336 mg/kg. In all the subjects, an equivalent maximal siRNA plasma level was achieved after administering Atu027 at 0.180 mg/kg impeding the success of *PKN3* target gene inhibition in patients treated with this dose or higher doses. Atu027 was well tolerated up to dose levels of 0.336 mg/kg; most adverse events were low-grade toxicities (grade 1 or 2), and no MTD was reached by escalation method.

The Atu027 siRNA lipoplex delivers siRNAs ingenuously into the cytoplasm of vascular endothelial cells *in vivo* for sequence-specific cleavage of the cognate PKN3-mRNA by an RNAi mode of action. As the system contains particulate compounds, some transient responses of the innate immune system such as complement activation and cytokine elevation were stated for Atu027; these changes had no clinical significance, and hence, no immunosuppressive premedication was required. Atu027 is not indicated for particular malignancy, and hence, its therapeutic application in various tumours is plausible.

Utilisation of DoE-based optimisation to establish relation between LNP material composition, its characteristics and its impact on performance will aid into tailoring reproducible scalable LNPs (Patravale, Disouza, and Rustomjee 2016). The ability for reproducible and scalable manufacturing of LNPs is a very important criterion for clinical development. Various clinical trials have made it clear that formulations need to be optimised for each type of nucleic acid payload and are certainly not interchangeable. LNPs are a versatile platform for unlocking the therapeutic potential of several types of nucleic acid-based therapeutics (Evers et al. 2018). The initiation of SNALPs since 2001 by Semple and colleagues has gone under various researches and bore its fruits in 2018 when FDA approved marketing of the first RNAi therapy, Patisiran (ALN-TTR02) for TransThyRetin (TTR)-mediated amyloidosis. Patisiran specifically binds to a genetically conserved sequence in the 3′ untranslated region (3′UTR) of mutant and wild-type TTR mRNA. It is available as a single-dose glass vial containing a sterile, preservative-free, white to off-white, opalescent, homogeneous solution for intravenous infusion (Adams et al. 2018; Zhang, Goel, and Robbie 2020). The structural formula is as given in Figure 7.3.

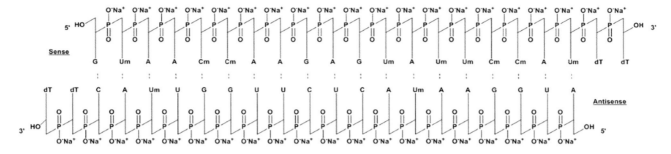

FIGURE 7.3 Structural formula of Patisiran (ALN-TTR02). ('Onpattro-Epar-Public-Assessment-Report_.Pdf' n.d.)

SNALP systems for nucleic acid delivery that have undergone clinical trials at varied phases are mentioned in Table 7.3; a few have been terminated or put on hold due to varied reasons of safety, toxicity or efficacy in Phase II of clinical trials. An open-label, multicentre, Phase I, dose escalation study with Phase II expansion cohort was conducted to determine the safety, pharmacokinetics and antitumour efficacy of TKM-080301 administered intravenously in advanced HCC patients. Preclinical studies with TKM-080301 had shown a robust activity in subcutaneous and orthotopic liver tumour models in mice based on EPR effect within the tumours. The lipidic-based composition not only protects the siRNA but also prevents rapid clearance and effective intracellular uptake by the PKL1 target (Semple et al. 2011). The Phase I study enrolled 43 patients, and the initiating dose of TKM-080301 was 0.3 mg/kg. As two subjects showed grade 4 thrombocytopenia, MTD was delignated to 0.6 mg/kg from 0.75 mg/kg. The safety review panel reported that 90% of the adverse events reported by the subjects were mild to moderate in severity. Out of the enrolled 43 subjects, 6 showed SAEs but none of SAEs were severe and were declared dubious, unexpected and treatment related by the investigators. Thirty-nine of the patients had evaluable post-baseline scan, whereas four patients did not have any evaluable post-baseline scan. Clinical trial NCT02191878 suggested that treatment with TKM-080301 was mostly safe and accepted in subjects with HCC. Barring two subjects, the remainder patients showed and reported at least one adverse effect. Due to an on-target siRNA effect of PLK1 inhibition, it caused a reduction in spleen volume in three subjects only. Treatment-emergent adverse effects (TEAEs) reported included were pyrexia (20 subjects; 46.5%); fatigue (16 subjects; 37.2%); chills (15 subjects; 34.9%); peripheral oedema (8 subjects; 18.6%); non-cardiac chest pain (3 subjects; 7.0%); and malaise, oedema and pain in two subjects (4.7%) each. Infusion-related TEAEs were observed in a total of 17 subjects (39.5%). A total of 22 subjects died. Three fatal TEAES during the study were evaluated and reported as unlikely or not related to TKM-080301. In conclusion, TKM-080301 was safe and well tolerated but did not establish clinically meaningful antitumour activity, the secondary endpoint of this trial (El Dika et al. 2019).

A wide range of cancer subtypes can be treated with nucleic acid-based lipidic delivery systems. Systematic engineering of the lipidic systems to yield the nanotherapies is the lesson what we learn from the data of clinical trials conducted till date. Before going ahead, a thorough review of safety failures in Phase II/III clinical trials is critical to the development of nanomedicines, in addition to a review

TABLE 7.3
SNALPS in Clinical Trials for Cancer Therapy

Name	Treatment	Genetic/Protein Target	Delivery Vehicle	Administration Method	Disease	ClinicalTrials.gov Identifier/Phase
TKM 080301	siRNA	PLK1	SNALP	Liver injection	Colorectal cancer with hepatic metastases, pancreatic cancer with hepatic metastases, gastric cancer with hepatic metastases, breast cancer with hepatic metastases, ovarian cancer with hepatic metastases	NCT01437007/ completed
				Intravenous infusion	Liver cancer	NCT02191878/ completed
				Intravenous infusion	Adrenocortical carcinoma	NCT01262235/ completed
ALN-VSP02	siRNA	VEGFR1	SNALP	Intravenous infusion	Solid tumours	NCT01158079/ phase I
						NCT00882180/ phase I

SNALP, stabilised nucleic acid lipid nanoparticle; VEGFR1, VEGF receptor-1.

of efficacy failures in Phase II/III clinical trials. The critical Phase II/III studies for nucleic acid nanotherapies suggest that the liver's role as an immunologically complex organ needs to be factored in at the design stage (Herrera et al. 2018). The inconsistency in the late-stage clinical trial failures indicates a discrepancy of preclinical and Phase I safety studies. The main reason for the difference is the use of population which is healthy in Phase I and preclinical, whereas the phase II/III includes the population with the disease. To overcome this problem, preclinical testing and characterisation of nanoparticle pharmacological properties (biodistribution, clearance, efficacy, toxicity, off-target effects, etc.) should be performed in simulated systems, including age as one of the factors and the use of biomarkers relevant to the disease patients to aid in bridging the gap between failure to success.

7.4 CONCLUSION AND FUTURE PROSPECTS

Delivery of nucleic acids has been a topic of extensive research because of their ability to function at the genetic level. DNA, mRNA, siRNA, miRNA and immunostimulatory nucleic acids are some of the major classes of nucleic acid-based molecules used for cancer therapy (Vaughan, Green, and Tzeng 2020a). The ease of fabrication and biocompatibility of lipid-based delivery systems have paved the way for their wide acceptance for nucleic acid delivery. However, the lipid-based nanoparticulate systems also encompass certain drawbacks like endosomal escape leading to toxicity, immunogenicity and unwanted biointeractions. The fact that over 30 RNA-based therapeutics have been exposed to clinical trials in the last decade and only three among them have reached the market signifies the level of challenges encountered for successful delivery of nucleic acids (Samaridou, Heyes, and Lutwyche 2020).

Overcoming these drawbacks is something which is being constantly studied owing to the future prospects of these systems for nucleic acid delivery, especially in morbid conditions like cancer. For example, ligand-targeted delivery of cationic lipids is on the exploration stage and the clinical applicability is still awaited (Wahane et al. 2020). A major cause of cancer therapy failures lies with the metastasis phenomenon, which complicates the physiological conditions to an irreversible extent. The EPR effect has been heterogeneous among various tumour types, and hence, its detailed investigation has been difficult and a major cause of hurdles in clinical trials (Jin et al. 2020). Nonetheless, these complex clinical trials are slowly becoming a reality, and an expansion of research in this direction will open new avenues for bench to bed transformations of novel cancer therapies.

A rise in the popularity of non-viral-based strategies is attributed to the need for the development of next-generation gene therapeutics. Lipid-based nanoplatforms, in particular, are deemed to be the most promising clinically based on their success with improving safety and efficacy of drugs like paclitaxel (Lipusu® – paclitaxel liposomes) and vincristine (Marqibo® – vincristine sulphate liposomes) (Yang and Merlin 2020). Lipidic nanosystems offer versatility in their tailoring ability with respect to design, particle size, morphology, surface modifications, targeting, controlled release and dose reduction. The flexibility of these systems in combination with the growing understanding of genetics and cancer pathophysiology would prove to be highly impactful in near future. With the scientific community inclining towards personalised medicines in the current decade, fabrication of delivery systems to cater to such individualistic needs would revolutionise the field cancer therapy. A continuous progress in the development of lipid-based nanotherapeutics has given hopes that if the potency and tolerability of these systems can be desirably catered to, in future, they can overcome major obstacles present today. Cancer can be described as the disease of unknowns and will always demand advancements to cater to the physiological hurdles. Optimum utilisation of resources, both from the fields of drug delivery science and cancer pathophysiology, is the way forward in order to achieve long-term translational (and clinical) progress.

REFERENCES

Adams, David, Alejandra Gonzalez-Duarte, William D. O'Riordan, Chih-Chao Yang, Mitsuharu Ueda, Arnt V. Kristen, Ivailo Tournev, et al. 2018. "Patisiran, an RNAi Therapeutic, for Hereditary Transthyretin Amyloidosis." *New England Journal of Medicine* 379 (1): 11–21. doi:10.1056/NEJMoa1716153.

Aigner, Achim, Dagmar Fischer, Thomas Merdan, Carola Brus, Thomas Kissel, and Frank Czubayko. 2002. "Delivery of Unmodified Bioactive Ribozymes by an RNA-Stabilizing Polyethylenimine (LMW-PEI) Efficiently Down-Regulates Gene Expression." *Gene Therapy* 9 (24): 1700–7

Akinc, Akin, Andreas Zumbuehl, Michael Goldberg, Elizaveta S Leshchiner, Valentina Busini, Naushad Hossain, Sergio A Bacallado, et al. 2008. "A Combinatorial Library of Lipid-Like Materials for Delivery of RNAi Therapeutics." *Nature Biotechnology* 26 (5): 561–69. doi:10.1038/nbt1402.

Altınoglu, Sarah, Ming Wang, and Qiaobing Xu. 2015. "Combinatorial Library Strategies for Synthesis of Cationic Lipid-Like Nanoparticles and their Potential Medical Applications." *Nanomedicine (London, England)* 10 (4): 643–57. doi:10.2217/nnm.14.192.

Antonellis, Pasqualino de, Lucia Liguori, Annarita Falanga, Marianeve Carotenuto, Veronica Ferrucci, Immacolata Andolfo, Federica Marinaro, et al. 2013. "MicroRNA 199b-5p Delivery through Stable Nucleic Acid Lipid Particles (SNALPs) in Tumorigenic Cell Lines." *Naunyn-Schmiedeberg's Archives of Pharmacology* 386 (4): 287–302. doi:10.1007/s00210-013-0837-4.

Barichello, José Mario, Shinji Kizuki, Tatsuaki Tagami, Luiz Alberto Lira Soares, Tatsuhiro Ishida, Hiroshi Kikuchi, and Hiroshi Kiwada. 2012. "Agitation during Lipoplex Formation Harmonizes the Interaction of SiRNA to Cationic Liposomes." *International Journal of Pharmaceutics* 430 (1–2): 359–65. doi:10.1016/j.ijpharm.2012.04.006.

Bolhassani, Azam. 2011. "Potential Efficacy of Cell-Penetrating Peptides for Nucleic Acid and Drug Delivery in Cancer." *Biochimica et Biophysica Acta (BBA)-Reviews on Cancer* 1816 (2): 232–246.

Bondì, Maria Luisa, and Emanuela Fabiola Craparo. 2010. "Solid Lipid Nanoparticles for Applications in Gene Therapy: A Review of the State of the Art." *Expert Opinion on Drug Delivery* 7 (1): 7–18.

Brock, Amy, Silva Krause, Hu Li, Marek Kowalski, Michael S. Goldberg, James J. Collins, and Donald E. Ingber. 2014. "Silencing *HoxA1* by Intraductal Injection of SiRNA Lipidoid Nanoparticles Prevents Mammary Tumor Progression in Mice." *Science Translational Medicine* 6 (217): 217ra2. doi:10.1126/scitranslmed.3007048.

Brunner, Sylvia, Thomas Sauer, Sebastian Carotta, Matt Cotten, Mediyha Saltik, and Ernst Wagner. 2000. "Cell Cycle Dependence of Gene Transfer by Lipoplex, Polyplex and Recombinant Adenovirus." *Gene Therapy* 7 (5): 401–7.

Bunjes, Heike. 2011. "Current Opinion in Colloid & Interface Science Structural Properties of Solid Lipid Based Colloidal Drug Delivery Systems." *Current Opinion in Colloid & Interface Science* 16 (5): 405–11. doi:10.1016/j.cocis.2011.06.007.

Cerchia, Laura, Jörg Hamm, Domenico Libri, Bertrand Tavitian, and Vittorio De Franciscis. 2002. "Nucleic Acid Aptamers in Cancer Medicine." *FEBS Letters* 528 (1–3): 12–16.

Charoensit, Pensri, Shigeru Kawakami, Yuriko Higuchi, Fumiyoshi Yamashita, and Mitsuru Hashida. 2010. "Enhanced Growth Inhibition of Metastatic Lung Tumors by Intravenous Injection of ATRA-Cationic Liposome/IL-12 PDNA Complexes in Mice." *Cancer Gene Therapy* 17 (7): 512–22. doi:10.1038/cgt.2010.12.

Chaturvedi, Kiran, Kuntal Ganguly, Anandrao R. Kulkarni, Venkatarao H. Kulkarni, Mallikarjuna N. Nadagouda, Walter E. Rudzinski, and Tejraj M. Aminabhavi. 2011. "Cyclodextrin-Based SiRNA Delivery Nanocarriers: A State-of-the-Art Review." *Expert Opinion on Drug Delivery* 8(11): 1455–68.

Chen, Changmai, Zhenjun Yang, and Xinjing Tang. 2018. "Chemical Modifications of Nucleic Acid Drugs and their Delivery Systems for Gene-Based Therapy." *Medicinal Research Reviews* 38 (3): 829–69. doi:10.1002/med.21479.

Chen, Zhihua, Xianliang Lai, Shuxin Song, Xingen Zhu, and Jianming Zhu. 2015. "Nanostructured Lipid Carriers Based Temozolomide and Gene Co-Encapsulated Nanomedicine for Gliomatosis Cerebri Combination Therapy." *Drug Delivery* 00 (00): 1–5. doi:10.3109/10717544.2015.1038857.

Dachs, Gabi U., Adam V. Patterson, John D. Firth, Peter J. Ratcliffe, K. M. Stuart Townsend, Ian J. Stratford, and Adrian L. Harris. 1997. "Targeting Gene Expression to Hypoxic Tumor Cells." *Nature Medicine* 3(5): 515–20.

Dana, Hassan, Ghanbar Mahmoodi Chalbatani, Habibollah Mahmoodzadeh, Rezvan Karimloo, Omid Rezaiean, Amirreza Moradzadeh, Narges Mehmandoost, et al. 2017. "Molecular Mechanisms and Biological Functions of SiRNA." *International Journal of Biomedical Science: IJBS* 13(2): 48–57.

De Jesus, Marcelo B., and Inge S. Zuhorn. 2015. "Solid Lipid Nanoparticles as Nucleic Acid Delivery System: Properties and Molecular Mechanisms." *Journal of Controlled Release* 201: 1–13. doi:10.1016/j.jconrel.2015.01.010.

Di Martino, Maria Teresa, Virginia Campani, Gabriella Misso, Maria Eugenia Gallo Cantafio, Annamaria Gullà, Umberto Foresta, Pietro Hiram Guzzi, et al. 2014. "In Vivo Activity of MiR-34a Mimics Delivered by Stable Nucleic Acid Lipid Particles (SNALPs) against Multiple Myeloma." *PloS One* 9 (2): e90005. doi:10.1371/journal.pone.0090005.

Ding, Weifeng, Feng Wang, Jianfeng Zhang, Yibing Guo, Shaoqing Ju, and Huimin Wang. 2013. "A Novel Local Anti-Colorectal Cancer Drug Delivery System: Negative Lipidoid Nanoparticles with a Passive Target via a Size-Dependent Pattern." *Nanotechnology* 24 (37): 375101. doi:10.1088/0957-4484/24/37/375101.

Ding, Weifeng, Guihua Wang, Keke Shao, Feng Wang, Hua Huang, Shaoqing Ju, Hui Cong, and Huimin Wang. 2014. "Amelioration of Colorectal Cancer Using Negative Lipidoid Nanoparticles to Encapsulate SiRNA against APRIL by Enema Delivery Mode." *Pathology & Oncology Research* 20 (4): 953–64. doi:10.1007/s12253-014-9779-5.

Ding, Yang, Yue Han, Ruoning Wang, Yazhe Wang, Cheng Chi, Ziqiang Zhao, Huaqing Zhang, Wei Wang, Lifang Yin, and Jianping Zhou. 2017. "Rerouting Native HDL to Predetermined Receptors for Improved Tumor-Targeted Gene Silencing Therapy." *ACS Applied Materials & Interfaces* 9 (36): 30488–501. doi:10.1021/acsami.7b10047.

Ding, Yang, Wei Wang, Meiqing Feng, Yu Wang, Jianping Zhou, Xuefang Ding, Xin Zhou, Congyan Liu, Ruoning Wang, and Qiang Zhang. 2012. "A Biomimetic Nanovector-Mediated Targeted Cholesterol-Conjugated SiRNA Delivery for Tumor Gene Therapy." *Biomaterials* 33 (34): 8893–905. doi:10.1016/j.biomaterials.2012.08.057.

Dong, Yizhou, Ahmed A. Eltoukhy, Christopher A. Alabi, Omar F. Khan, Omid Veiseh, J. Robert Dorkin, Sasilada Sirirungruang, et al. 2014. "Lipid-Like Nanomaterials for Simultaneous Gene Expression and Silencing *In Vivo*." *Advanced Healthcare Materials* 3 (9): 1392–97. doi:10.1002/adhm.201400054.

Duarte, Sónia, Henrique Faneca, and Maria C. Pedroso De Lima. 2012. "Folate-Associated Lipoplexes Mediate Efficient Gene Delivery and Potent Antitumoral Activity *In Vitro* and *In Vivo*." *International Journal of Pharmaceutics* 423 (2): 365–77. doi:10.1016/j.ijpharm.2011.12.035.

Ekambaram, Prabhakaran, A. Abdul Hasan Sathali, and Karunanidhi Priyanka. 2012. "Solid Lipid Nanoparticles: A Review." *Reviews and Chemical Communications* 2 (1): 80–102.

El Dika, Imane, Ho Yeong Lim, Wei Peng Yong, Chia-Chi Lin, Jung-Hwan Yoon, Manuel Modiano, Bradley Freilich, et al. 2019. "An Open-Label, Multicenter, Phase I, Dose Escalation Study with Phase II Expansion Cohort to Determine the Safety, Pharmacokinetics, and Preliminary Antitumor Activity of Intravenous TKM-080301 in Subjects with Advanced Hepatocellular Carcinoma." *The Oncologist* 24 (6): 747. doi:10.1634/theoncologist.2018-0838.

Elouahabi, Abdelatif, and Jean-Marie Ruysschaert. 2005. "Formation and Intracellular Trafficking of Lipoplexes and Polyplexes." *Molecular Therapy* 11(3): 336–47.

Evers, Martijn J. W., Jayesh A. Kulkarni, Roy van der Meel, Pieter R. Cullis, Pieter Vader, and Raymond M. Schiffelers. 2018. "State-of-the-Art Design and Rapid-Mixing Production Techniques of Lipid Nanoparticles for Nucleic Acid Delivery." *Small Methods* 2 (9): 1700375. doi:10.1002/smtd.201700375.

Ezzati, Jafar, Nazhad Dolatabadi, Hadi Valizadeh, and Hamed Hamishehkar. 2015. "Solid Lipid Nanoparticles as Efficient Drug and Gene Delivery Systems: Recent Breakthroughs." *Advanced Pharmaceutical Bulletin* 5 (2): 151–59. doi:10.15171/apb.2015.022.

Filleur, Stéphanie, Aurélie Courtin, Slimane Ait-Si-Ali, Julien Guglielmi, Carole Merle, Annick Harel-Bellan, Philippe Clézardin, and Florence Cabon. 2003. "SiRNA-Mediated Inhibition of Vascular Endothelial Growth Factor

Severely Limits Tumor Resistance to Antiangiogenic Thrombospondin-1 and Slows Tumor Vascularization and Growth." *Cancer Research* 63 (14): 3919–22.

Fire, Andrew, SiQun Xu, Mary K. Montgomery, Steven A. Kostas, Samuel E. Driver, and Craig C. Mello. 1998. "Potent and Specific Genetic Interference by Double-Stranded RNA in *Caenorhabditis elegans*." *Nature* 391 (6669): 806–11.

Geisbert, Thomas W., Lisa E. Hensley, Elliott Kagan, Erik Zhaoying Yu, Joan B. Geisbert, Kathleen Daddario-DiCaprio, Elizabeth A. Fritz, et al. 2006. "Postexposure Protection of Guinea Pigs against a Lethal Ebola Virus Challenge is Conferred by RNA Interference." *The Journal of Infectious Diseases* 193 (12): 1650–57. doi:10.1086/504267.

Ghasemiyeh, Parisa, and Soliman Mohammadi-samani. 2018. "Solid Lipid Nanoparticles and Nanostructured Lipid Carriers as Novel Drug Delivery Systems: Applications, Advantages and Disadvantages" *Research in Pharmaceutical Sciences* 13 (February): 288–303.

Goyal, Ritu, Sushil Kumar Tripathi, Shilpa Tyagi, Kristipati Ravi Ram, Kausar Mahmood Ansari, Yogeshwar Shukla, Debapratim Kar Chowdhuri, Pradeep Kumar, and Kailash Chand Gupta. 2011. "Gellan Gum Blended PEI Nanocomposites as Gene Delivery Agents: Evidences from In Vitro and In Vivo Studies." *European Journal of Pharmaceutics and Biopharmaceutics* 79(1): 3–14.

Groot, Anne Marit de, Kaushik Thanki, Monique Gangloff, Emily Falkenberg, Xianghui Zeng, Djai C. J. van Bijnen, Willem van Eden, et al. 2018. "Immunogenicity Testing of Lipidoids *In Vitro* and *In Silico*: Modulating Lipidoid-Mediated TLR4 Activation by Nanoparticle Design." *Molecular Therapy – Nucleic Acids* 11 (June): 159–69. doi:10.1016/j.omtn.2018.02.003.

Gunther, Michael, Ernst Wagner, and Manfred Ogris. 2005. "Specific Targets in Tumor Tissue for the Delivery of Therapeutic Genes." *Current Medicinal Chemistry-Anti-Cancer Agents* 5 (2): 157–71.

H du Plessis, Lissinda, Etienne B. Marais, Faruq Mohammed, and Awie F. Kotze. 2014. "Applications of Lipid Based Formulation Technologies in the Delivery of Biotechnology-Based Therapeutics." *Current Pharmaceutical Biotechnology* 15 (7): 659–72.

Hallahan, Dennis E., Helena J. Mauceri, Lisa P. Seung, Edward J. Dunphy, Jeffrey D. Wayne, Nader N. Hanna, Alicia Toledano, Samuel Hellman, Donald W. Kufe, and Ralph R. Weichselbaum. 1995. "Spatial and Temporal Control of Gene Therapy Using Ionizing Radiation." *Nature Medicine* 1 (8): 786–91.

Han, Yiqun, Ye Li, Peng Zhang, Jiping Sun, Xianzhen Li, Xin Sun, and Fansheng Kong. 2014. "Nanostructured Lipid Carriers as Novel Drug Delivery System for Lung Cancer Gene Therapy" 7450: 1–5. doi:10.3109/10837450.2014.996 900.

Hara, Toshifumi, Yukihiko Aramaki, Shinako Takada, Katsuro Koike, and Seishi Tsuchiya. 1995. "Receptor-Mediated Transfer of PSV2CAT DNA to a Human Hepatoblastoma Cell Line HepG2 Using Asialofetuin-Labeled Cationic Liposomes." *Gene* 159 (2): 167–74. doi:10.1016/0378-1119(95)00100-K.

Hattori, Yoshiyuki, Ayako Nakamura, Shohei Arai, Mayu Nishigaki, Hiroyuki Ohkura, Kumi Kawano, Yoshie Maitani, and Etsuo Yonemochi. 2014. "In Vivo SiRNA Delivery System for Targeting to the Liver by Poly-l-Glutamic Acid-Coated Lipoplex." *Results in Pharma Sciences* 4 (January): 1–7. doi:10.1016/j.rinphs.2014.01.001.

Hayes, Mark, Daryl Drummond, Dmitri Kirpotin, Wei Wen Zheng, Charles Noble, John Park, James D. Marks, Christopher Benz, and Kim Ke Hong. 2006. "Genospheres: Self-Assembling Nucleic Acid-Lipid Nanoparticles Suitable for Targeted Gene Delivery." *Gene Therapy* 13 (7): 646–51

He, Zhi Yao, Feng Deng, Xia Wei Wei, Cui Cui Ma, Min Luo, Ping Zhang, Ya Xiong Sang, et al. 2016. "Ovarian Cancer Treatment with a Tumor-Targeting and Gene Expression-Controllable Lipoplex." *Scientific Reports* 6 (1): 1–13. doi:10.1038/srep23764.

Hee, Sung, Su-Eon Jin, Mi-Kyung Lee, Soo-Jeong Lim, Jeong-Sook Park, Byung-Gyu Kim, Woong Shick, and Chong-Kook Kim. 2008. "Novel Cationic Solid Lipid Nanoparticles Enhanced P53 Gene Transfer to Lung Cancer Cells." *European Journal of Pharmaceutics and Biopharmaceutics* 68: 545–54. doi:10.1016/j.ejpb.2007.07.011.

Heller, Loree, Kathleen Merkler, Jeffrey Westover, Yolmari Cruz, Domenico Coppola, Kaaron Benson, Adil Daud, and Richard Heller. 2006. "Evaluation of Toxicity Following Electrically Mediated Interleukin-12 Gene Delivery in a B16 Mouse Melanoma Model." *Clinical Cancer Research* 12 (10): 3177–83.

Herrera, Victoria L. M., Aaron H. Colby, Nelson Ruiz-Opazo, David G. Coleman, and Mark W. Grinstaff. 2018. "Nucleic Acid Nanomedicines in Phase II/III Clinical Trials: Translation of Nucleic Acid Therapies for Reprogramming Cells." *Nanomedicine (London, England)* 13 (16): 2083–98. doi:10.2217/nnm-2018-0122.

Hope, Michael J. 2014. "Enhancing SiRNA Delivery by Employing Lipid Nanoparticles." *Therapeutic Delivery* 5 (6): 663–73.

Huang, Yu-Hung, Yunhua Bao, Weidan Peng, Michael Goldberg, Kevin Love, David A. Bumcrot, Geoffrey Cole, Robert Langer, Daniel G. Anderson, and Janet A. Sawicki. 2009. "Claudin-3 Gene Silencing with SiRNA Suppresses Ovarian Tumor Growth and Metastasis." *Proceedings of the National Academy of Sciences* 106 (9): 3426. doi:10.1073/pnas.0813348106.

Hui, Sek Wen, Marek Langner, Ya-Li Zhao, Patrick Ross, Edward Hurley, and Karen Chan. 1996. "The Role of Helper Lipids in Cationic Liposome-Mediated Gene Transfer." *Biophysical Journal* 71 (2): 590–9.

Jiang, Juan, Shi Jin Yang, Jian Cheng Wang, Li Juan Yang, Zhen Zhong Xu, Ting Yang, Xiao Yan Liu, and Qiang Zhang. 2010. "Sequential Treatment of Drug-Resistant Tumors with RGD-Modified Liposomes Containing SiRNA or Doxorubicin." *European Journal of Pharmaceutics and Biopharmaceutics* 76 (2): 170–78. doi:10.1016/j.ejpb.2010.06.011.

Jin, Jun-O., Gyurin Kim, Juyoung Hwang, Kyung Ho Han, Minseok Kwak, and Peter C. W. Lee. 2020. "Nucleic Acid Nanotechnology for Cancer Treatment." *Biochimica et Biophysica Acta – Reviews on Cancer*: 188377. doi:10.1016/j.bbcan.2020.188377.

Jin, Juyoun, Ki Hyun Bae, Heekyoung Yang, Se Jeong Lee, Hyein Kim, Yonghyun Kim, Kyeung Min Joo, Soo Won Seo, Tae Gwan Park, and Do-Hyun Nam. 2011. "*In Vivo* Specific Delivery of C-Met SiRNA to Glioblastoma Using Cationic Solid Lipid Nanoparticles." *Bioconjugate Chemistry* 22 (12): 2568–72.

Judge, Adam D., Marjorie Robbins, Iran Tavakoli, Jasna Levi, Lina Hu, Anna Fronda, Ellen Ambegia, Kevin McClintock, and Ian MacLachlan. 2009. "Confirming the RNAi-Mediated Mechanism of Action of SiRNA-Based Cancer Therapeutics in Mice." *The Journal of Clinical Investigation* 119 (3): 661–73. doi:10.1172/JCI37515.

Kadiyala, Padma, Dan Li, Fernando M. Nuñez, David Altshuler, Robert Doherty, Rui Kuai, Minzhi Yu, et al. 2019. "High-Density Lipoprotein-Mimicking Nanodiscs for Chemo-Immunotherapy against Glioblastoma Multiforme." *ACS Nano* 13 (2): 1365–84. doi:10.1021/acsnano.8b06842.

Kataoka, Kazunori, Keiji Itaka, Nobuhiro Nishiyama, Yuichi Yamasaki, Motoi Oishi, and Yukio Nagasaki. 2005. "Smart Polymeric Micelles as Nanocarriers for Oligonucleotides and SiRNA Delivery." In *Nucleic Acids Symposium Series*, 49: 17–18, Oxford: Oxford University Press.

Koh, Chee Guan, Xulang Zhang, Shujun Liu, Sharon Golan, Bo Yu, Xiaojuan Yang, Jingjiao Guan, et al. 2010. "Delivery of Antisense Oligodeoxyribonucleotide Lipopolyplex Nanoparticles Assembled by Microfluidic Hydrodynamic Focusing." *Journal of Controlled Release* 141 (1): 62–69. doi:10.1016/j.jconrel.2009.08.019.

Koynova, Rumiana, and Boris Tenchov. 2013. "Phase Transitions and Phase Behavior of Lipids." In *Encyclopedia of Biophysics*, edited by Gordon C. K. Roberts, 1841–54. Berlin, Heidelberg: Springer. doi:10.1007/978-3-642-16712-6_542.

Kulkarni, Jayesh A., Pieter R. Cullis, and Roy van der Meel. 2018. "Lipid Nanoparticles Enabling Gene Therapies: From Concepts to Clinical Utility." *Nucleic Acid Therapeutics* 28 (3): 146–57. doi:10.1089/nat.2018.0721.

Kulkarni, Jayesh A., Maria M. Darjuan, Joanne E. Mercer, Sam Chen, Roy van der Meel, Jenifer L. Thewalt, Yuen Yi C. Tam, and Pieter R. Cullis. 2018. "On the Formation and Morphology of Lipid Nanoparticles Containing Ionizable Cationic Lipids and SiRNA." *ACS Nano* 12 (5): 4787–95. doi:10.1021/acsnano.8b01516.

Laga, Richard, Robert Carlisle, Mark Tangney, Karel Ulbrich, and Len W. Seymour. 2012. "Polymer Coatings for Delivery of Nucleic Acid Therapeutics." *Journal of Controlled Release* 161(2): 537–53.

Ledford, Heidi. 2018. "Gene-Silencing Technology Gets First Drug Approval after 20-Year Wait." *Nature* 560 (7718): 291–93.

Lee, Robert J., and Leaf Huang. 1996. "Folate-Targeted, Anionic Liposome-Entrapped Polylysine-Condensed DNA for Tumor Cell-Specific Gene Transfer." *Journal of Biological Chemistry* 271 (14): 8481–87. doi:10.1074/jbc.271.14.8481.

Lee, Yun-Han, Adam Judge, Daekwan Seo, Mitsuteru Kitade, Luis Gómez-Quiroz, Tsuyoshi Ishikawa, Jesper Bøje Andersen, et al. 2011. "Molecular Targeting of CSN5 in Human Hepatocellular Carcinoma: A Mechanism of Therapeutic Response." *Oncogene* 30 (40): 4175–84. doi:10.1038/onc.2011.126.

Leenders, Frauke, Kristin Möpert, Anett Schmiedeknecht, Ansgar Santel, Frank Czauderna, Manuela Aleku, Silke Penschuck, et al. 2004. "PKN3 is Required for Malignant Prostate Cell Growth Downstream of Activated PI 3-Kinase." *The EMBO Journal* 23 (16): 3303–13. doi:10.1038/sj.emboj.7600345.

Leite Nascimento, Thais, Hervé Hillaireau, Juliette Vergnaud, Melania Rivano, Claudine Deloménie, Delphine Courilleau, Silvia Arpicco, Jung Soo Suk, Justin Hanes, and Elias Fattal. 2016. "Hyaluronic Acid-Conjugated Lipoplexes for Targeted Delivery of SiRNA in a Murine Metastatic Lung Cancer Model." *International Journal of Pharmaceutics* 514 (1): 103–11. doi:10.1016/j.ijpharm.2016.06.125.

Lenssen, Karl, Peter Jantscheff, Günter von Kiedrowski, and Ulrich Massing. 2002. "Combinatorial Synthesis of New Cationic Lipids and High-Throughput Screening of their Transfection Properties." *ChemBioChem* 3 (9): 852–58. doi:10.1002/1439-7633(20020902)3:9<852::AID-CBIC852>3.0.CO;2-A.

Leonetti, Jean-Paul, Patrick Machyt, Genevitve Degols, Bernard Lebleu, and Lee Lesermant. 1990. "Antibody-Targeted Liposomes Containing Oligodeoxyribonucleotides Complementary to Viral RNA Selectively Inhibit Viral Replication (Antisense Oligonucleotides/Vesicular Stomatitis Virus/Antiviral Therapy/Intracellular Delivery)." *Proceedings of the National Academy of Sciences of the United States of America* 87 (7): 2448–51.

Li, Weijun, and Francis C. Szoka. 2007. "Lipid-Based Nanoparticles for Nucleic Acid Delivery." *Pharmaceutical Research* 24 (3): 438–49.

Li, Wenjie, Jing Shi, Chun Zhang, Min Li, Lu Gan, Huibi Xu, and Xiangliang Yang. 2014. "Co-Delivery of Thioredoxin 1 ShRNA and Doxorubicin by Folate-Targeted Gemini Surfactant-Based Cationic Liposomes to Sensitize Hepatocellular Carcinoma Cells." *Journal of Materials Chemistry B* 2 (30): 4901–10. doi:10.1039/C4TB00502C.

Lobovkina, Tatsiana, Gunilla B. Jacobson, Emilio Gonzalez-Gonzalez, Robyn P. Hickerson, Devin Leake, Roger L. Kaspar, Christopher H. Contag, and Richard N. Zare. 2011. "*In Vivo* Sustained Release of SiRNA from Solid Lipid Nanoparticles." *ACS Nano* 5(12): 9977–83.

Lundstrom, Kenneth. 2018. "Viral Vectors in Gene Therapy." *Diseases* 6 (2): 42.

Luo, Jie, Miriam Höhn, Sören Reinhard, Dominik M. Loy, Philipp Michael Klein, and Ernst Wagner. 2019. "IL4-Receptor-Targeted Dual Antitumoral Apoptotic Peptide – SiRNA Conjugate Lipoplexes." *Advanced Functional Materials* 29 (25): 1900697. doi:10.1002/adfm.201900697.

Ma, Feihe, Liu Yang, Zhuorui Sun, Jinjin Chen, Xuehui Rui, Zachary Glass, and Qiaobing Xu. 2020. "Neurotransmitter-Derived Lipidoids (NT-Lipidoids) for Enhanced Brain Delivery through Intravenous Injection." *Science Advances* 6 (30): eabb4429. doi:10.1126/sciadv.abb4429.

MacLachlan, Ian. 2007. "Liposomal Formulations for Nucleic Acid Delivery." In *Antisense Drug Technology: Principles, Strategies, and Applications, Second Edition*, edited by Stanley T. Crooke, 237–70. Boca Raton, FL: CRC Press. https://doi.org/10.1201/9780849387951.

Maeda, Hiroshi. 2001. "The Enhanced Permeability and Retention (EPR) Effect in Tumor Vasculature: The Key Role of Tumor-Selective Macromolecular Drug Targeting." *Advances in Enzyme Regulation* 41: 189–207.

Mahato, Ram I., Alain Rolland, and Eric Tomlinson. 1997. "Cationic Lipid-Based Gene Delivery Systems: Pharmaceutical Perspectives." *Pharmaceutical Research*. doi:10.1023/A:1012187414126.

Manjunath, K., J. Suresh Reddy, and V. Venkateswarlu. 2005. "Solid Lipid Nanoparticles as Drug Delivery Systems." *Methods and Findings in Experimental and Clinical Pharmacology* 27 (2): 1–20. doi:10.1358/mf.2005.27.2.876286.

Manno, Catherine S., Glenn F. Pierce, Valder R. Arruda, Bertil Glader, Margaret Ragni, John JE Rasko, Margareth C. Ozelo, Keith Hoots, Philip Blatt, and Barbara Konkle. 2006. "Successful Transduction of Liver in Hemophilia by AAV-Factor IX and Limitations Imposed by the Host Immune Response." *Nature Medicine* 12(3): 342–47.

McMahon, Kaylin M, R Kannan Mutharasan, Sushant Tripathy, Dorina Veliceasa, Mariana Bobeica, Dale K Shumaker, Andrea J Luthi, et al. 2011. "Biomimetic High Density Lipoprotein Nanoparticles for Nucleic Acid Delivery." *Nano Letters* 11 (3): 1208–14. doi:10.1021/nl1041947.

Mendelsohn, John. 2002. "Targeting the Epidermal Growth Factor Receptor for Cancer Therapy." *Journal of Clinical Oncology* 20 (18): 1s–13s.

Mingozzi, Federico, Janneke J. Meulenberg, Daniel J. Hui, Etiena Basner-Tschakarjan, Nicole C. Hasbrouck, Shyrie A. Edmonson, Natalie A. Hutnick, Michael R. Betts, John J. Kastelein, and Erik S. Stroes. 2009. "AAV-1–Mediated Gene Transfer to Skeletal Muscle in Humans Results in Dose-Dependent Activation of Capsid-Specific T Cells." *Blood* 114 (10): 2077–86.

Morrissey, David V., Jennifer A. Lockridge, Lucinda Shaw, Karin Blanchard, Kristi Jensen, Wendy Breen, Kimberly Hartsough, et al. 2005. "Potent and Persistent *In Vivo* Anti-HBV Activity of Chemically Modified SiRNAs." *Nature Biotechnology* 23 (8): 1002–7. doi:10.1038/nbt1122.

Nakase, Minoru, Madoka Inui, Kenya Okumura, Takahiko Kamei, Shinnosuke Nakamura, and Toshiro Tagawa. 2005. "P53 Gene Therapy of Human Osteocarcinoma Using a Transferrin-Modified Cationic Liposome." *Molecular Cancer Therapeutics* 4 (4): 625–31. doi:10.1158/1535-7163.MCT-04-0196.

Narainpersad, Nicolisha, Moganavelli Singh, and Mario Ariatti. 2012. "Novel Neo Glycolipid: Formulation into Pegylated Cationic Liposomes and Targeting of DNA Lipoplexes to the Hepatocyte-Derived Cell Line HepG2." *Nucleosides, Nucleotides and Nucleic Acids* 31 (3): 206–23. doi:10.1080/15257770.2011.649331.

Ndoye, Alioune, Jean-Louis Merlin, Agnès Leroux, Gilles Dolivet, Patrick Erbacher, Jean-Paul Behr, Kristian Berg, and François Guillemin. 2004. "Enhanced Gene Transfer and Cell Death Following P53 Gene Transfer Using Photochemical Internalisation of Glucosylated PEI-DNA Complexes." *The Journal of Gene Medicine: A Cross-Disciplinary Journal for Research on the Science of Gene Transfer and Its Clinical Applications* 6(8): 884–94.

Nemunaitis, John, Stephen G. Swisher, Therese Timmons, Dee Connors, Michael Mack, Lesah Doerksen, David Weill, Juliette Wait, David Lawrence, Bonnie Kemp et al. 2000. "Adenovirus-Mediated P53 Gene Transfer in Sequence with Cisplatin to Tumors of Patients with Non-Small-Cell Lung Cancer." *Journal of Clinical Oncology* 18 (3): 609–622.

Northfelt, Donald W., Solomon I. Hamburg, Mitesh J. Borad, Mahesh Seetharam, Kelly Kevelin Curtis, Peter Lee, Brynne Crowell, et al. 2013. "A Phase I Dose-Escalation Study of TKM-080301, a RNAi Therapeutic Directed against Polo-Like Kinase 1 (PLK1), in Patients with Advanced Solid Tumors: Expansion Cohort Evaluation of Biopsy Samples for Evidence of Pharmacodynamic Effects of PLK1 Inhibition." *Journal of Clinical Oncology* 31 (15 Suppl): TPS2621. doi:10.1200/jco.2013.31.15_suppl.tps2621.

Oh, Yu Kyoung, and Tae Gwan Park. 2009. "SiRNA Delivery Systems for Cancer Treatment." *Advanced Drug Delivery Reviews*. doi:10.1016/j.addr.2009.04.018.

Olbrich, Carsten, Udo Bakowsky, Claus-michael Lehr, Rainer H Muller, and Carsten Kneuer. 2001. "Cationic Solid-Lipid Nanoparticles Can Efficiently Bind and Transfect Plasmid DNA" *Journal of Controlled Release* 77: 345–55.

"Onpattro-Epar-Public-Assessment-Report_.Pdf." n.d. Accessed September 23, 2020. https://www.ema.europa.eu/en/documents/assessment-report/onpattro-epar-public-assessment-report_.pdf.

Pal Singh, Pirthi, Veena Vithalapuram, Sunita Metre, and Ravinder Kodipyaka. 2019. "Lipoplex-Based Therapeutics for Effective Oligonucleotide Delivery: A Compendious Review." *Journal of Liposome Research*, 1–23. doi:10.1080/08982104.2019.1652645.

Palmerston Mendes, Livia, Jiayi Pan, and Vladimir P. Torchilin. 2017. "Dendrimers as Nanocarriers for Nucleic Acid and Drug Delivery in Cancer Therapy." *Molecules* 22 (9): 1401.

Mirani Amit and Patravale Vandana Bharat. 2016. "Design of Experiments: Basic Concepts and Its Application in Pharmaceutical Product Development." In *Pharmaceutical Product Development: Insights Into Pharmaceutical Processes, Management and Regulatory Affairs*, edited by Patravale, Vandana Bharat, John I. Disouza, and Maharukh Tehmasp Rustomjee. CRC Press, Taylor and Francis Group, New York. https://books.google.co.in/books?id=1XCmCwAAQBAJ

Patravale, Vandana Bharat, Prajakta Dandekar, and Ratnesh Jain. 2012. "2- Nanoparticles as Drug Carriers." In *Nanoparticulate Drug Delivery*, edited by Vandana Patravale, Prajakta Dandekar, and Ratnesh Jain, 29–85. Woodhead Publishing. doi:10.1533/9781908818195.29.

Pérez-Martínez, Francisco C., Javier Guerra, Inmaculada Posadas, and Valentín Ceña. 2011. "Barriers to Non-Viral Vector-Mediated Gene Delivery in the Nervous System." *Pharmaceutical Research* 28(8): 1843–58.

Rafael, Diana, Fernanda Andrade, Alexandra Arranja, Sofia Luís, and Mafalda Videira. 2015. "Lipoplexes and Polyplexes: Gene Therapy." *Encyclopedia of Biomedical Polymers and Polymeric Biomaterials*: 4335–47. doi:10.1081/e-ebpp-120050058.

Rainov, Nicolai G. 2000. "A Phase III Clinical Evaluation of Herpes Simplex Virus Type 1 Thymidine Kinase and Ganciclovir Gene Therapy as an Adjuvant to Surgical Resection and Radiation in Adults with Previously Untreated Glioblastoma Multiforme." *Human Gene Therapy* 11 (17): 2389–2401.

Rait, Antonina S., Kathleen F. Pirollo, Laiman Xiang, David Ulick, and Esther H. Chang. 2002. "Tumor-Targeting, Systemically Delivered Antisense HER-2 Chemosensitizes Human Breast Cancer Xenografts Irrespective of HER-2 Levels." *Molecular Medicine* 8 (8): 475–86.

Ramanathan, Ramesh K., Solomon I. Hamburg, Thorvardur R. Halfdanarson, and M. Borad. 2014. "A Phase I/II Dose Escalation Study of TKM-080301, a RNAi Therapeutic Directed against PLK1, in Patients with Advanced Solid Tumors, with an Expansion Cohort of Patients with NET or ACC." *Pancreas* 43: 502–1502.

Rehman, Zia ur, Inge S. Zuhorn, and Dick Hoekstra. 2013. "How Cationic Lipids Transfer Nucleic Acids into Cells and across Cellular Membranes: Recent Advances." *Journal of Controlled Release* 166 (1): 46–56.

Rodriguez, Ron, Eric R. Schuur, Ho Yeong Lim, Gail A. Henderson, Jonathan W. Simons, and Daniel R. Henderson. 1997. "Prostate Attenuated Replication Competent Adenovirus (ARCA) CN706: A Selective Cytotoxic for Prostate-Specific Antigen-Positive Prostate Cancer Cells." *Cancer Research* 57(13): 2559–63.

Ruozi, Barbara, Flavio Forni, Renata Battini, and Maria Angela Vandelli. 2003. "Cationic Liposomes for Gene Transfection." *Journal of Drug Targeting* 11 (7): 407–14.

Samaridou, Eleni, James Heyes, and Peter Lutwyche. 2020. "Lipid Nanoparticles for Nucleic Acid Delivery: Current Perspectives." *Advanced Drug Delivery Reviews*. doi:10.1016/j.addr.2020.06.002.

Santel, Ansgar, Manuela Aleku, Oliver Keil, Jens Endruschat, Vera Esche, Gerald Fisch, Sybille Dames, et al. 2006. "A Novel SiRNA-Lipoplex Technology for RNA Interference in the Mouse Vascular Endothelium." *Gene Therapy* 13 (16): 1222–34. doi:10.1038/sj.gt.3302777.

Saranya, N., A. Moorthi, S. Saravanan, M. Pandima Devi, and N. Selvamurugan. 2011. "Chitosan and Its Derivatives for Gene Delivery." *International Journal of Biological Macromolecules* 48 (2): 234–38.

Sasaki, Atsushi, Naokazu Murahashi, Harutami Yamada, and Anri Morikawa. 1994. "Syntheses of Novel Galactosyl Ligands for Liposomes and their Accumulation in the Rat Liver." *Biological and Pharmaceutical Bulletin* 17 (5): 680–85. doi:10.1248/bpb.17.680.

Scheetz, Lindsay M., Minzhi Yu, Dan Li, María G. Castro, James J. Moon, and Anna Schwendeman. 2020. "Synthetic HDL Nanoparticles Delivering Docetaxel and CpG for Chemoimmunotherapy of Colon Adenocarcinoma." *International Journal of Molecular Sciences* 21 (5): 1777. doi:10.3390/ijms21051777.

Schultheis, Beate, Dirk Strumberg, Ansgar Santel, Christiane Vank, Frank Gebhardt, Oliver Keil, Christian Lange, et al. 2014. "First-in-Human Phase I Study of the Liposomal RNA Interference Therapeutic Atu027 in Patients with Advanced Solid Tumors." *Journal of Clinical Oncology* 32 (36): 4141–48. doi:10.1200/JCO.2013.55.0376.

Seki, Masafumi, Jun Iwakawa, Helen Cheng, and Pi Wan Cheng. 2002. "P53 and PTEN/MMAC1/TEP1 Gene Therapy of Human Prostate PC-3 Carcinoma Xenograft, Using Transferrin-Facilitated Lipofection Gene Delivery Strategy." *Human Gene Therapy* 13 (6): 761–73. doi:10.1089/104303402317322311.

Semple, Sean C., Adam D. Judge, Marjorie Robbins, Sandra Klimuk, Merete Eisenhardt, Erin Crosley, Ada Leung, et al. 2011. "Abstract 2829: Preclinical Characterization of TKM-080301, a Lipid Nanoparticle Formulation of a Small Interfering RNA Directed against Polo-Like Kinase 1." *Cancer Research* 71 (8 Supplement): 2829. doi:10.1158/1538-7445.AM2011-2829.

Semple, Sean C., Sandra K. Klimuk, Troy O. Harasym, Nancy Dos Santos, Steven M. Ansell, Kim F. Wong, Norbert Maurer, et al. 2001. "Efficient Encapsulation of Antisense Oligonucleotides in Lipid Vesicles Using Ionizable Aminolipids: Formation of Novel Small Multilamellar Vesicle Structures." *Biochimica et Biophysica Acta (BBA) – Biomembranes* 1510 (1): 152–66. doi:10.1016/S0005-2736(00)00343-6.

Shah, Amy S., Lirong Tan, Jason Lu Long, and W. Sean Davidson. 2013. "Proteomic Diversity of High Density Lipoproteins: Our Emerging Understanding of Its Importance in Lipid Transport and Beyond." *Journal of Lipid Research* 54 (10): 2575–85. doi:10.1194/jlr.R035725.

Shahzad, Mian M. K., Lingegowda S. Mangala, Hee Dong Han, Chunhua Lu, Justin Bottsford-Miller, Masato Nishimura, Edna M. Mora, et al. 2011. "Targeted Delivery of Small Interfering RNA Using Reconstituted High-Density Lipoprotein Nanoparticles." *Neoplasia (New York)* 13 (4): 309–19. doi:10.1593/neo.101372.

Shim, Gayong, Mi Gyeong Kim, Joo Yeon Park, and Yu Kyoung Oh. 2013. "Application of Cationic Liposomes for Delivery of Nucleic Acids." *Asian Journal of Pharmaceutical Sciences* 8 (2): 72–80. doi:10.1016/j.ajps.2013.07.009.

Shir, Alexei, Manfred Ogris, Ernst Wagner, and Alexander Levitzki. 2005. "EGF Receptor-Targeted Synthetic Double-Stranded RNA Eliminates Glioblastoma, Breast Cancer, and Adenocarcinoma Tumors in Mice." *PLoS Medicine* 3 (1): e6.

Singh, Aishwarya, Piyush Trivedi, and Narendra Kumar Jain. 2018. "Advances in SiRNA Delivery in Cancer Therapy." *Artificial Cells, Nanomedicine, and Biotechnology* 46 (2): 274–83. doi:10.1080/21691401.2017.1307210.

Sokolova, Viktoriya, and Matthias Epple. 2008. "Inorganic Nanoparticles as Carriers of Nucleic Acids into Cells." *Angewandte Chemie International Edition* 47 (8): 1382–95.

Sun, Shuo, Ming Wang, Sarah A. Knupp, Yadira Soto-Feliciano, Xiao Hu, David L. Kaplan, Robert Langer, Daniel G. Anderson, and Qiaobing Xu. 2012. "Combinatorial Library of Lipidoids for *In Vitro* DNA Delivery." *Bioconjugate Chemistry* 23 (1): 135–40. doi:10.1021/bc200572w.

Tabernero, Josep, Geoffrey I. Shapiro, Patricia M. LoRusso, Andres Cervantes, Gary K. Schwartz, Glen J. Weiss, Luis Paz-Ares, et al. 2013. "First-in-Humans Trial of an RNA Interference Therapeutic Targeting VEGF and KSP in Cancer Patients with Liver Involvement." *Cancer Discovery* 3 (4): 406. doi:10.1158/2159-8290.CD-12-0429.

Talkar, Swapnil S., and Vandana B. Patravale. 2020. "Gene Therapy for Prostate Cancer: A Review." *Endocrine, Metabolic & Immune Disorders – Drug Targets* 20 (May). doi:10.2174/1871530320666200531141455.

Thaxton, C. Shad, Weston L. Daniel, David A. Giljohann, Audrey D. Thomas, and Chad A. Mirkin. 2009. "Templated Spherical High Density Lipoprotein Nanoparticles." *Journal of the American Chemical Society* 131 (4): 1384–85. doi:10.1021/ja808856z.

Tolcher, Anthony W., Kyriakos P. Papadopoulos, Amita Patnaik, Drew Warren Rasco, Dorothy Martinez, Debra L Wood, Barbara Fielman, et al. 2015. "Safety and Activity of DCR-MYC, a First-in-Class Dicer-Substrate Small Interfering RNA (DsiRNA) Targeting MYC, in a Phase I Study in Patients with Advanced Solid Tumors." *Journal of Clinical Oncology* 33 (15 Suppl): 11006. doi:10.1200/jco.2015.33.15_suppl.11006.

Tripathy, Sushant, Elena Vinokour, Kaylin M. McMahon, Olga V. Volpert, and C. Shad Thaxton. 2014. "High Density Lipoprotein Nanoparticles Deliver RNAi to Endothelial Cells to Inhibit Angiogenesis." *Particle & Particle Systems Characterization: Measurement and Description of Particle Properties and Behavior in Powders and Other Disperse Systems* 31 (11): 1141–50. doi:10.1002/ppsc.201400036.

Vacik, Joshua, Brenda Dean, Warren Zimmer, and David Dean. 1999. "Cell-Specific Nuclear Import of Plasmid DNA." *Gene Therapy* 6 (6): 1006–14.

Vaughan, Hannah J., Jordan J. Green, and Stephany Y. Tzeng. 2020a. "Cancer-Targeting Nanoparticles for Combinatorial Nucleic Acid Delivery." *Advanced Materials* 32 (13): 1–36. doi:10.1002/adma.201901081.

Vaughan, Hannah J., Jordan J. Green, and Stephany Y. Tzeng. 2020b. "Cancer-Targeting Nanoparticles for Combinatorial Nucleic Acid Delivery." *Advanced Materials* 32 (13): 1901081. doi:10.1002/adma.201901081.

Vickers, Kasey C., Brian T. Palmisano, Bassem M. Shoucri, Robert D. Shamburek, and Alan T. Remaley. 2011. "MicroRNAs are Transported in Plasma and Delivered to Recipient Cells by High-Density Lipoproteins." *Nature Cell Biology* 13 (4): 423–33. doi:10.1038/ncb2210.

Wagner, Ernst, Ralf Kircheis, and Greg F. Walker. 2004. "Targeted Nucleic Acid Delivery into Tumors: New Avenues for Cancer Therapy." *Biomedicine & Pharmacotherapy* 58 (3): 152–61.

Wahane, Aniket, Akaash Waghmode, Alexander Kapphahn, Karishma Dhuri, Anisha Gupta, and Raman Bahal. 2020. "Role of Lipid-Based and Polymer-Based Non-Viral Vectors in Nucleic Acid Delivery for Next-Generation Gene Therapy." *Molecules* 25 (12): 2866. doi:10.3390/molecules25122866.

Wahid, Fazli, Adeeb Shehzad, Taous Khan, and You Young Kim. 2010. "MicroRNAs: Synthesis, Mechanism, Function, and Recent Clinical Trials." *Biochimica et Biophysica Acta – Molecular Cell Research*. doi:10.1016/j.bbamcr.2010.06.013.

Wang, Hairong, Shiming Liu, Li Jia, Fengyun Chu, Ya Zhou, Zhixu He, Mengmeng Guo, Chao Chen, and Lin Xu. 2018. "Nanostructured Lipid Carriers for MicroRNA Delivery in Tumor Gene Therapy." *Cancer Cell International*: 1–6. doi:10.1186/s12935-018-0596-x.

Wang, Ming, Kyle Alberti, Antonio Varone, Dimitria Pouli, Irene Georgakoudi, and Qiaobing Xu. 2014. "Enhanced Intracellular SiRNA Delivery Using Bioreducible Lipid-Like Nanoparticles." *Advanced Healthcare Materials* 3 (9): 1398–1403. doi:10.1002/adhm.201400039.

Wang, Ming, Shuo Sun, Kyle A. Alberti, and Qiaobing Xu. 2012. "A Combinatorial Library of Unsaturated Lipidoids for Efficient Intracellular Gene Delivery." *ACS Synthetic Biology* 1 (9): 403–7. doi:10.1021/sb300023h.

Wei, Jie, Jeffrey Jones, Jing Kang, Ananda Card, Michael Krimm, Paula Hancock, Yi Pei, Brandon Ason, Elmer Payson, and Natalya Dubinina. 2011. "RNA-Induced Silencing Complex-Bound Small Interfering RNA is a Determinant of RNA Interference-Mediated Gene Silencing in Mice." *Molecular Pharmacology* 79 (6): 953–63.

Wiethoff, Christopher M., and C. Russell Middaugh. 2003. "Barriers to Nonviral Gene Delivery." *Journal of Pharmaceutical Sciences* 92 (2): 203–17.

Wolfrum, Christian, Shuanping Shi, K. Narayanannair Jayaprakash, Muthusamy Jayaraman, Gang Wang, Rajendra K. Pandey, Kallanthottathil G. Rajeev, et al. 2007. "Mechanisms and Optimization of *In Vivo* Delivery of Lipophilic SiRNAs." *Nature Biotechnology* 25 (10): 1149–57. doi:10.1038/nbt1339.

Wu, Yun, Melissa Crawford, Bo Yu, Yicheng Mao, Serge P. Nana-Sinkam, and L. James Lee. 2011. "MicroRNA Delivery by Cationic Lipoplexes for Lung Cancer Therapy." *Molecular Pharmaceutics* 8 (4): 1381–89. doi:10.1021/mp2002076.

Xiao, Yao, Kun Shi, Ying Qu, Bingyang Chu, and Zhiyong Qian. 2019. "Engineering Nanoparticles for Targeted Delivery of Nucleic Acid Therapeutics in Tumor." *Molecular Therapy – Methods & Clinical Development* 12 (March): 1–18. doi:10.1016/j.omtm.2018.09.002.

Yang, Chunhua, and Didier Merlin. 2020. "Lipid-Based Drug Delivery Nanoplatforms for Colorectal Cancer Therapy." *Nanomaterials* 10 (7): 1424. doi:10.3390/nano10071424.

Yang, Mi, Honglin Jin, Juan Chen, Lili Ding, Kenneth K. Ng, Qiaoya Lin, Jonathan F. Lovell, Zhihong Zhang, and Gang Zheng. 2011. "Efficient Cytosolic Delivery of SiRNA Using HDL-Mimicking Nanoparticles." *Small* 7 (5): 568–73. doi:10.1002/smll.201001589.

Yarden, Yosef. 2001. "Biology of HER2 and Its Importance in Breast Cancer." *Oncology* 61 (Suppl. 2): 1–13.

Yoon, Goo, Jin Woo, and Park In-Soo Yoon. 2013. "Solid Lipid Nanoparticles (SLNs) and Nanostructured Lipid Carriers (NLCs): Recent Advances in Drug Delivery Solid Lipid Nanoparticles (SLNs) and Nanostructured Lipid Carriers (NLCs): Recent Advances in Drug Delivery." *Journal of Pharmaceutical Investigation* 43 (5): 353–62. doi:10.1007/s40005-013-0087-y.

Yu, Yong Hee, Eunjoong Kim, Dai Eui Park, Gayong Shim, Sangbin Lee, Young Bong Kim, Chan-Wha Kim, and Yu-Kyoung Oh. 2012. "Cationic Solid Lipid Nanoparticles for Co-Delivery of Paclitaxel and SiRNA." *European Journal of Pharmaceutics and Biopharmaceutics* 80 (2): 268–73. doi:10.1016/j.ejpb.2011.11.002.

Zhang, Xiaoping, Varun Goel, and Gabriel J. Robbie. 2020. "Pharmacokinetics of Patisiran, the First Approved RNA Interference Therapy in Patients with Hereditary Transthyretin-Mediated Amyloidosis." *The Journal of Clinical Pharmacology* 60 (5): 573–85. doi:10.1002/jcph.1553.

Zhang, Zhiwen, Xianyi Sha, Anle Shen, Yongzhong Wang, Zhaogui Sun, Zheng Gu, and Xiaoling Fang. 2008. "Biochemical and Biophysical Research Communications Polycation Nanostructured Lipid Carrier, a Novel Nonviral Vector Constructed with Triolein for Efficient Gene Delivery" *Biochemical and Biophysical Research Communications* 370: 478–82. doi:10.1016/j.bbrc.2008.03.127.

Zhou, Hai-Sheng, De-Pei Liu, and Chih-Chuan Liang. 2004. "Challenges and Strategies: The Immune Responses in Gene Therapy." *Medicinal Research Reviews* 24 (6): 748–61.

Zimmermann, Tracy S., Amy C. H. Lee, Akin Akinc, Birgit Bramlage, David Bumcrot, Matthew N. Fedoruk, Jens Harborth, et al. 2006. "RNAi-Mediated Gene Silencing in Non-Human Primates." *Nature* 441 (7089): 111–14. doi:10.1038/nature04688.

Zuhorn, Inge S., Jan B. F. N. Engberts, and Dick Hoekstra. 2007. "Gene Delivery by Cationic Lipid Vectors: Overcoming Cellular Barriers." *European Biophysics Journal* 36 (4–5): 349–62.

8 Formulation Strategies for Improved Ophthalmic Delivery of Hydrophilic Drugs

Arehalli Manjappa, Popat Kumbhar, and John Disouza
Tatyasaheb Kore College of Pharmacy

Suraj Pattekari
Tatyasaheb Kore College of Pharmacy
Annasaheb Dange College of B. Pharmacy

Vandana B. Patravale
Institute of Chemical Technology

CONTENTS

Abbreviations 116
8.1 Introduction 117
    8.1.1 Barriers for Ocular Delivery of Hydrophilic Therapeutics 117
8.2 Approaches to Improve Ocular Permeability of Hydrophilic Therapeutics 119
    8.2.1 Development of Prodrugs 119
    8.2.2 Use of Permeation Enhancers 119
    8.2.3 Use of Chemical Chaperones (Protein Aggregation Inhibitors) 119
    8.2.4 By Incorporating into Colloidal Nanoparticles (NPs) 120
        8.2.4.1 Emulsions 121
        8.2.4.2 Polymeric Nanoparticles (NPs) and Microparticles (MPs) 121
        8.2.4.3 Micelles 122
        8.2.4.4 Reverse Micelles (RMs) 122
        8.2.4.5 Liposomes 122
        8.2.4.6 Niosomes 122
        8.2.4.7 Discomes/Cubosomes 123
        8.2.4.8 Spanlastics 123
        8.2.4.9 Lipid NPs 123
        8.2.4.10 Dendrimers 123
        8.2.4.11 Nanocrystals 123
8.3 Chief Challenges for Nanoparticulate Delivery of Hydrophilic Therapeutics 124
    8.3.1 Poor Drug Loading Efficiency 124
    8.3.2 Poor Ocular Residential Time 124
8.4 Approaches to Improve Loading of Hydrophilic Therapeutics in Colloidal NPs 124
    8.4.1 Development of Carrier–Drug Conjugate Nanoparticles (CDC-NPs) 124
        8.4.1.1 Ester Bonds 124
        8.4.1.2 Amide Bonds 124
        8.4.1.3 Hydrazone Bonds 124
        8.4.1.4 Disulphide Bonds 126
        8.4.1.5 Other Bonds 126
        8.4.1.6 Self-Assembled NPs (Carrier-free System) 126
        8.4.1.7 Micelles 127
        8.4.1.8 Emulsions 127
        8.4.1.9 Polymeric NPs 128

DOI: 10.1201/9781003043164-8

        8.4.1.10 Lipid NPs 128
        8.4.1.11 Liposomes 128
    8.4.2 Development of Nanoplexes 128
        8.4.2.1 Lipoplexes 128
        8.4.2.2 Polyplexes 129
        8.4.2.3 Micelleplexes 129
8.5 Approaches to Improve Ocular Residence Time of Colloidal Carriers 129
    8.5.1 Development of Mucoadhesive Colloidal NPs 129
    8.5.2 Development of Colloidal NP-laden Composite Systems 130
        8.5.2.1 NP-laden In Situ Gel 130
        8.5.2.2 NP-laden Hydrogels 130
        8.5.2.3 NP-laden Contact Lenses 130
        8.5.2.4 NP-Laden Ocular Inserts 131
    8.5.3 Development of Nanowafers 131
8.6 Advanced/Future Treatment Approaches 131
    8.6.1 Nucleic Acid-Based Therapy over Small Molecule Drugs 131
    8.6.2 Optogenetics 131
    8.6.3 Stem Cells and Cell Transplantation Techniques 131
    8.6.4 Encapsulated Cell Technology (ECT) 132
    8.6.5 Noble Metal NPs with Multiple Functions 132
    8.6.6 Targeted NPs 132
    8.6.7 Theranostic NPs 132
    8.6.8 Microneedle-laden Collagen Cryogel Plugs 132
    8.6.9 Use of Physical Methods in Combination with Chemical Approaches 133
8.7 Regulatory and Future Perspectives 133
8.8 Conclusions 133
Acknowledgements 133
References 134

ABBREVIATIONS

ACV	Acyclovir
ACZ	Acetazolamide
AgNPs	Silver NPs
AMD	Age-related macular degeneration
ANDA	Abbreviated new drug application
ASCT1	Alanine–serine–cysteine transporter 2
AuNPs	Gold NPs
BRA	Brinzolamide
BRB	Blood–retinal barrier
BSA	Bovine serum albumin
BVZ	Bevacizumab
CDCs	Carrier–drug conjugates
CHCS	Cholesterol-conjugated chitosan
CMC	Carboxymethyl cellulose
CNV	Choroidal neovascularisation
COS	Chitosan oligosaccharide
CS	Chitosan
DCP	*Dicetyl phosphate*
DMPC	1,2-Dimyristoyl-sn-glycero-3-phosphocholine
DODAB	Dioctadecyldimethylammonium bromide
DOPE	1,2-dioleoyl-snglycero-3-phosphoethanolamine
DPPC	Dipalmitoylphosphatidylcholine
DPPS	1,2-Dipalmitoyl phosphatidylserine
DSPC	1,2-Disteoroyl-sn-glycero-3-phosphocholine
DSPE	1,2-Distearoyl-sn-glycero-3-phosphoethanolamine
dsRNA	Double-stranded RNA
DTAB	Dodecyl trimethyl ammonium bromide
ECT	Encapsulated cell technology
EDTA	Ethylenediaminetetraacetic acid
FDA	Food and Drug Administration
GMO	Glycerol monooleate
GMP	Good manufacturing practices
GNDs	Gold Nanodiscs
GONPs	Graphene oxide NPs
HA	Hyaluronic acid/hyaluronan
HPC	Hydroxypropyl cellulose
HPMC	Hydroxypropyl methylcellulose
HRMEC	Human retinal microvascular endothelial cells
Hsp	Heat-shock proteins
HTCC	N-(2-hydroxypropyl)-3-trimethylammonium chitosan chloride
IM	Intramuscular
IOP	Intraocular pressure
IV	Intravenous
LAT1	L-type amino acid transporter 1
LDCs	Lipid–drug conjugates
LSCD	Limbal stem cell deficiency

miRNA	Micro-RNA
MNPs	Magnetic nanoparticles
MPs	Microparticles
mRNA	Messenger RNA
NAT	Natamycin
NDA	New drug application
NEs	Nanoemulsions
NLCs	Nanostructured lipid carriers
NPs	Nanoparticles
Oc-40	Octoxynol-40
ODN	Oligonucleotides
PAA	Poly(acrylic acid)
PAM	Propyl acrylamide
PBA	4-Phenylbutyric acid sodium salt
PC	Phosphatidylcholines
PCL	Poly (e-caprolactone)
PDGF	Platelet-derived growth factor
pDNA	Plasmid DNA
PE	Phosphothioethanol
PEG	Polyethylene glycol
PEG-PCL	Polyethylene glycol–polycaprolactone
PEG-PLA	Polyethylene glycol–poly-L-lactic acid
PEI	Polyethyleneimine
PEI-PEG	Poly (ethylene imine)–polyethylene glycol
PG	*Phosphatidylglycerol*
PGF2a	Prostaglandin F2 alpha
PHEMA	Poly(hydroxyethyl methacrylate)
PLGA	Poly(lactic-co-glycolic acid)
Poly (n-BCA)	Poly(butyl)-cyanoacrylate
PRPs	Photoreceptor precursors
PVA	Poly(vinyl alcohol)
PVP	Poly(vinylpyrrolidone)
RCE	Retinal capillary endothelium
rHuPH20	Recombinant human hyaluronidase PH20
RMs	Reverse micelles
RPE	Retinal pigment epithelium
SC	Subcutaneous
SDS	Sodium dodecyl sulphate
SiNPs	Silicon NPs
SiNPs-RGD	cyclo-(Arg-Gly-Asp-d-Tyr-Cys) (c-(RGDyC))-conjugated silicon NPs
siRNA	Small interfering RNA
SLNs	Solid lipid nanoparticles
SLS	Sodium lauryl sulphate
ssDNA	Single-stranded DNA
TM	Timolol maleate
TMAO	Trimethylamine-N-oxide
TPGS	Tocopherol polyethylene glycol 1000 succinate
TUDCA	Tauroursodeoxycholic acid
UTMD	Ultrasound-targeted microbubble destruction
VAC	Valacyclovir
VCM	Vancomycin
VEGF	Vascular endothelial growth factor
VEGFR	Vascular endothelial growth factor receptor

8.1 INTRODUCTION

According to WHO, over 285 million people globally are suffering from eye diseases and over 39 million people are completely blind. These numbers are expected to increase largely in the coming two decades. By 2050, it is anticipated that about 70 million adults globally would be suffering from diseases related to the posterior segment of the eye (like degenerative diseases) and anterior segment of the eye (like glaucoma and cataract) [1].

The delivery of ocular therapeutics to treat diseases of the anterior and posterior segment of the eye is impaired by several anatomical/physiological barriers. Further, currently approved treatment approaches for eye diseases suffer from poor therapeutic efficacy (due to poor bioavailability as a result of poor permeability and short residence time) and patient incompliance (frequent instillation, frequent injections and related toxicities).

Most of the small molecule drugs indicated for the treatment and management of ocular diseases are hydrophilic (belongs to BCS class 3, poorly permeable and highly soluble). Thus, pharmaceutical scientists face additional challenges for efficient ocular delivery of such drugs over BCS class 1 and 2 drugs. Recently, biological drugs such as protein, peptides, aptamers, siRNA, antibodies and other nucleic acid-based drugs have received huge medical attention and are currently recommended for various ocular diseases [2]. Further, most of these biological drugs are highly hydrophilic (log P<0) with a large molecular size and poor stability and, hence, pose additional challenges over hydrophilic small molecule drugs for ocular delivery. The application of nanoparticulate approaches for various routes of administration has led to improvements in ocular drug delivery of small molecule drugs and biologics too and offers several treatment options to overcome some of the above-listed limitations [3].

Considering the above-mentioned facts, in this chapter, we have emphasised the various chemical approaches used to improve the ocular permeability of drugs and types of colloidal nanoparticles (NPs) used to improve the entrapment of hydrophilic drugs/biologics and the strategies to enhance their ocular residence time along with proof of concepts wherever available. Further, we have described the advanced/future treatment approaches of ocular diseases and briefed the regulatory and future perspectives in a successful development of effective and safe ocular formulations.

8.1.1 BARRIERS FOR OCULAR DELIVERY OF HYDROPHILIC THERAPEUTICS

The ophthalmic delivery of drugs at the right concentration to the right site of the eye is challenging for a pharmaceutical scientist because eye is a very complex organ of the human body associated with many anatomical and physiological barriers (tear turnover, nasolachrymal drainage, reflex blinking and static and dynamic barriers) that

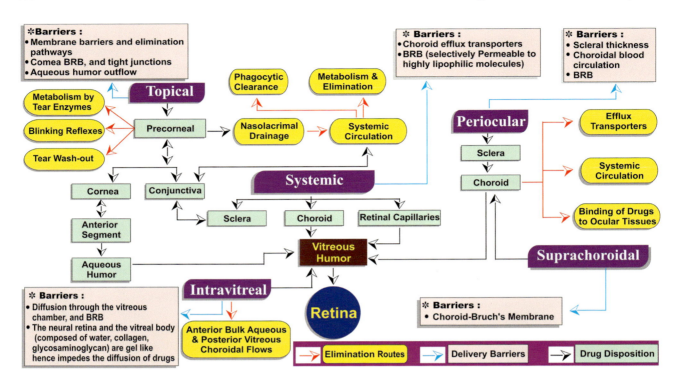

FIGURE 8.1 Ocular delivery routes along with their challenges and drug disposition pathways: *corneal epithelium (52 μm)*: multilayered with tight intercellular junctions in the outermost layer, allows permeation of small molecules; *conjunctiva (42 μm)*: same as corneal epithelium, allows permeation of small and large molecules (up to 5 kDa); *sclera (400–900 μm)*: porous supporting tissue composed of collagen and polysaccharide fibres, allows permeation of macromolecules and small NPs; *vitreous (15,000 μm)*: gel-like network, allows the permeation of small drugs, proteins and NPs; *internal limiting membrane (ILM, 0.05–2.00 μm)*: hydrophilic cellular basement membrane, allows permeation of molecules and small NPs; and *RPE (26 μm including Bruch's membrane)*: cell monolayer with tight junctions, allows faster permeation of small lipophilic compounds than that of hydrophilic or large molecules.

together contribute for the poor performance of ocular delivery systems (Figure 8.1). Thus, less than 5% of the drug reaches posterior tissues of the eye following topical instillation as a result of the above barriers [4].

The static barriers (sclera, choroid and retinal pigment epithelium (RPE)) and dynamic barriers (lymph flow in the conjunctiva and episclera, and blood flow in conjunctiva and choroid) are the chief barriers for trans-scleral delivery of ophthalmic drugs [5]. The lipophilic corneal epithelium and the stroma made of hydrated collagen are the chief barriers for the diffusion of highly hydrophilic and lipophilic ophthalmic drugs, respectively, whereas the corneal endothelium is not a chief barrier and the permeation is depended on the molecular weight rather than the inherent permeation characteristics of the ophthalmic drugs [6].

The hydrophilic drugs that fail to permeate through corneal epithelium are more possibly absorbed through the conjunctiva. The absorption of ophthalmic drugs through conjunctiva is considered to have no use because the absorbed drug from the conjunctiva further enters the systemic circulation via the capillary and lymph vessels present in the conjunctiva. Thus, repeated instillation of hydrophilic drugs may lead to greater systemic absorption and poor ocular bioavailability [7].

The permeation of drug through the sclera (composed of collagen fibres and proteoglycans embedded in an extracellular matrix) is similar to the corneal stroma. The scleral drug absorption is influenced by the molecular radius. For instance, linear dextrans are less permeable than globular protein. Besides, the charge of the drugs also affects the scleral absorption. Cationic drugs are absorbed slowly, most probably due to the formation of electrostatic complexation with anionic proteoglycan matrix [8].

In blood–retinal barrier (BRB), the retinal capillary endothelial (RCE) cells and RPE cells form the inner and outer BRB, respectively. The special transport mechanisms present in the RPE cells allow the exchange of nutrients/compounds selectively between the choroid and retina. The inner BRB covers the lumen of retinal capillaries and selectively protects the retina from the blood-circulating molecules. The tight intercellular junctions of RCE cells show poor absorption of small hydrophilic compounds including proteins. Thus, systemic delivery of ophthalmic drugs or intravitreal injections is required for efficient delivery of drugs to vitreous and retina and to achieve sufficient therapeutic effect. Further, high systemic doses (restrictive function of BRB results in poor bioavailability at posterior segments) of ophthalmic drugs administered through oral or intravenous routes get distributed in the entire body, leading to unintended side effects [8].

In comparison with small molecule drugs, the ocular delivery of biological molecules is even more challenging because they are highly hydrophilic (log P<0) and have large size and high molecular weight resulting in poor diffusion rate across ocular tissues. They possess surface charge (generally anionic) that results in repulsion and poor permeation through the negatively charged cell membrane. In addition, they have a highly complex structure, poor ocular stability, short half-lives, non-specificity and loss of activity during storage. Besides, they are easily cleared by ocular mononuclear phagocytes (MPS) and have the risk of immunogenicity [9].

Sometimes, the biological drugs are designed to permeate through the ocular membrane via receptor-mediated endocytosis [10]. The disadvantage of the endocytic pathway, entrapment into the endosomes and eventually in lysosomes, is the degradation of biological drugs by the lysosomal enzymes resulting in only a small amount of undegraded biological drug entering into the cytoplasm. Moreover, the oral delivery of biological drugs is also challenging due to gastrointestinal (GI) mucosa and degradative acidic environment. A large number of drugs are also administered as parenteral (IV, IM or SC), intravitreal and subconjunctival injections. These non-targeted delivery approaches require a large dose that is generally not economical and sometimes cause non-specific toxicity.

8.2 APPROACHES TO IMPROVE OCULAR PERMEABILITY OF HYDROPHILIC THERAPEUTICS

8.2.1 Development of Prodrugs

For decades, the prodrug strategies are utilised to improve the ocular bioavailability of drugs. Barot et al. described the chemical modifications in designing ocular prodrugs (esters, carbamates, phosphates and oximes) for improved corneal permeability, which later convert into their active form through chemical or enzymatic metabolic processes [11].

The presence of transporters on the corneal epithelium can be utilised for evaluating the esterase (present in the anterior segment)-sensitive transporter-targeted lipophilic ester prodrugs. For instance, the significantly improved corneal permeability and aqueous humour concentration of prostaglandin (PGF2a) were reported for the lipophilic ester prodrug, latanoprost (isopropyl ester prodrug of PGF2a) [12]. Currently, the peptide transporters of the corneal epithelium are targeted to develop prodrugs (valacyclovir (VAC) transporting through oligopeptide transporter) for improved ocular bioavailability. Similarly, the amino acid transporters (LAT1 and ASCT1) are also being utilised for the ocular delivery of amino acid-based prodrugs such as alanine isoleucine and serine prodrugs of acyclovir [13]. Moreover, many ophthalmic prodrugs to avoid cellular efflux and to improve targeting have been developed. This approach includes the development of polymer- and lipid-based prodrugs, pro-prodrugs or double prodrugs, retrometabolic drug design, co-drug systems, transporter-targeted prodrugs and hydrophilic prodrug approaches [14].

8.2.2 Use of Permeation Enhancers

The use of permeation/absorption enhancer(s) in ophthalmic preparations can improve the ocular bioavailability of both lipophilic and hydrophilic therapeutics. The frequently used permeation enhancers in ophthalmic preparations include cetylpyridinium chloride, ionophores (such as lasalocid, benzalkonium chloride, parabens, Tween 20, Saponins, Brij 35, Brij 78, Brij 98, EDTA), bile salts and bile acids (such as sodium cholate, sodium taurocholate, sodium glycodeoxycholate, sodium taurodeoxycholate, taurocholic acid, chenodeoxycholic acid and ursodeoxycholic acid), capric acid, Azone, fusidic acid, hexamethylene lauramide, hexamethylene octanamide, and decyl methyl sulphoxide [15]. The use of the above-mentioned substances in ophthalmic preparations can increase the permeability of corneal epithelium and other ocular tissues resulting in improved ocular bioavailability and targeting of therapeutics.

The vitreous humour of the eye is rich in a protein-free polysaccharide known as hyaluronan (HA). HA is polyanionic in nature and is responsible for high viscosity of vitreous humour that results in poor diffusion of most therapeutics. Thus, current treatment approaches demand for the use of human hyaluronidase (hHyal) that decreases the vitreous humour viscosity by breaking HA into tetrasaccharides and thereby increase the diffusion rate of therapeutics through vitreous humour [16,17]. The application of rHuPH20, purified hHyal, has showed increased concentration of dexamethasone in choroid, retina and serum [18]. This indicates the potential application of recombinant hHyal in improving ocular delivery of small molecule drugs, and protein- and peptide-based ophthalmic preparations.

8.2.3 Use of Chemical Chaperones (Protein Aggregation Inhibitors)

Aggregation of therapeutic proteins is the chief challenge in the development of protein- and peptide-based ophthalmic preparations. Initially, many small molecules were studied for the inhibition of the aggregation of protein- and peptide-based therapeutics. By preventing/modulating the aggregation, the chemical chaperones can improve the therapeutic efficacy and prevent the toxicities caused by aggregated protein- and peptide-based therapeutics. Sanders and group utilised protein aggregation inhibitors that prevented the misfolding of proteins through kinetic stabilisation [19]. The extensively investigated chemical small molecule aggregation inhibitors include glycerol, 4-phenylbutyric acid sodium salt (PBA), tauroursodeoxycholic acid (TUDCA) and trimethylamine-N-oxide (TMAO). In addition to the above chemical small molecules, the endogenous aggregation inhibitors such as heat-shock proteins (Hsp), and pharmacoperones such as nicotine are also studied widely to facilitate the folding of specific therapeutic proteins [20].

8.2.4 By Incorporating into Colloidal Nanoparticles (NPs)

The conventional formulation approaches and chemical modifications are not enough to overcome many ocular delivery barriers (static and dynamic), to achieve sustained and controlled release and to reduce dosing frequency to improve safety, patient compliance and cost-effectiveness. Thus, there is an unmet need to develop novel approaches such as colloidal carriers (Figure 8.2) that alter the physicochemical characteristics (such as size, charge and lipophilicity) of therapeutics and help to conquer the above

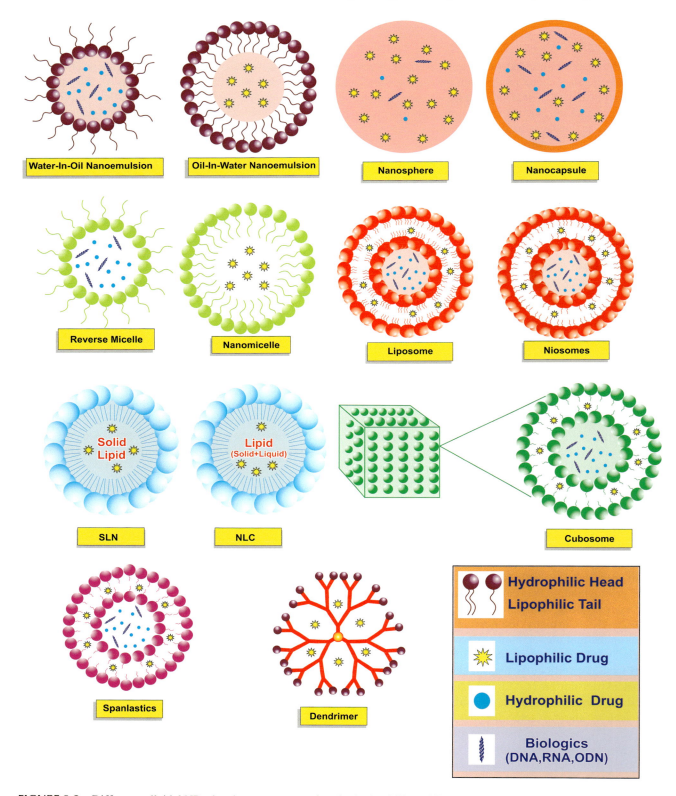

FIGURE 8.2 Different colloidal NPs showing entrapment of ocular hydrophilic and lipophilic small molecule and biological drugs.

TABLE 8.1
Ophthalmic Colloidal NPs in Clinic

Type of Nanocarriers	Study Details/ Clinical Trial Stage	Description
		Micro-/Nanoemulsions
NCT03785340	Phase II	Brimonidine tartrate nanoemulsion eye drop solution is used for the treatment of dry eye disease. It was evaluated for its tolerability
NCT04246801	Phase II	Efficacy and safety of clobetasol propionate nanoemulsion (0.05%) were evaluated (administered one drop four times a day during 14 days after routine unilateral cataract surgery)
		Micro-/Nanoparticles
NCT03001466	Phase II	Efficacy of urea-loaded pluronic F-127 NPs eye drops was evaluated in the management of cataract
NCT01523314	Phase II and III	Safety and efficacy of dexamethasone-cyclodextrin microparticles were determined in diabetic macular edema (DME)
		Liposomes
NCT03617315	Randomised, double blind	The effect of uncross-linked and cross-linked hyaluronic acid combined with liposomes and Crocin were screened in patients with dry eye caused by moderate meibomian gland dysfunction
NCT03052140	Randomised Single blind	Lamelleye dry eye drops is a liposomal multidose preservative-free sterile suspension composed of soy lecithin, sphingomyelin and cholesterol; suspended in saline was developed and evaluated for the treatment of dry eye
NCT01987323	Phase II and III	Safety and efficacy of latanoprost liposomal for subconjunctival injection were evaluated in patients with ocular hypertension and primary open-angle glaucoma
NCT02992392	Randomised, Single blind	Liposic and Tears Naturale Forte (phospholipid liposomes) have been developed and delivered to the tear film via the surface of the closed eyelid to study their effects on the lipid and stability of the tear film in dry eye patients

barriers and improve the therapeutic efficacy. The colloidal particles of the size range from 50 to 400 nm can easily penetrate across the corneal barriers compared to larger particles [21].

The colloidal nanocarriers undergo absorption through the corneal epithelial membrane via endocytosis. They can prevent tear washout, increase ocular residence and provide control, and sustain ocular drug release. These colloidal carriers can be self-administered by patients resulting in improved patient compliance. Following ocular instillation, the colloidal carriers can result in decreased discomfort and absence of impairment of sight. Additionally, they prevent the function of efflux pumps (P-glycoprotein, P-gp) present on the epithelial cells and open corneal tight junctions due to the presence of non-ionic surfactants in the colloidal NPs [22]. Also, colloidal NPs are able to conquer blood–ocular barriers and cellular efflux of plain therapeutic substances. More importantly, the colloidal carriers can improve the stability of biological drugs (such as proteins, peptides and nucleic acids) by protecting them against metabolic enzymes such as peptidases and nucleases [23]. The colloidal NPs that show great potential for ophthalmic drug delivery are briefed below. Few of these colloidal NPs are currently in clinic for the treatment of different ocular diseases (Table 8.1).

8.2.4.1 Emulsions

In general, the two main types of emulsions, oil-in-water (o/w) lipid emulsion and water-in-oil (w/o) lipid emulsion, are used for ocular delivery of therapeutics. Of these two types, w/o-type emulsion is mostly used for the hydrophilic therapeutics. The w/o-type emulsions have advantages of spontaneous formation and thermodynamic stability. Further, the tear film constituents such as lipid and water can be restored by the emulsion. Moreover, the emulsifier present in the emulsion can improve the tear film wettability [24].

Nanoemulsions are nanosized droplets, stabilised by a high amount of one or more surfactants, containing hydrophilic or lipophilic therapeutics. Both microemulsions and nanoemulsions are found superior for topical ocular delivery because of their high entrapment capacity and better ocular permeability and biocompatibility [25]. Shah et al. have prepared nanoemulsion for ocular delivery of moxifloxacin that showed substantial improvement in ocular bioavailability (aqueous humour concentration).

8.2.4.2 Polymeric Nanoparticles (NPs) and Microparticles (MPs)

Polymer-based NPs and MPs (NPs: 10–1,000 nm and MPs: 1–1,000 μm) are biocompatible and biodegradable carriers suitable for ocular delivery of small molecule and biological therapeutics via diverse routes. In spherical NPs and MPs, the drug is entrapped or dispersed in the matrix or attached to the surface, whereas in capsule type, the drug is either dissolved or dispersed in the core or absorbed on the shell. The different advantages offered by NPs and MPs include entrapment of two or more drugs in a single particle, stimuli-responsive drug release and the feasibility to change particle size, shape and surface function as per the requirements of ophthalmic delivery of therapeutics [26]. However, rapid phagocytic clearance, burst drug release, formulations scale-up issues, stability and non-uniformity in size distribution are the main problems of these systems. Hachicha et al have developed PLGA microparticles for the delivery of water-soluble vancomycin (VCM). These MPs displayed the controlled release of VCM up to 24 h [27]. Varshochian and co-workers developed bevacizumab (BVZ)-loaded PLGA NPs that exhibited the sustained release of BVZ up to 5 months [28]. Moreover, most of the polymeric NPs studied in pre-clinical, clinical, and the marketed ones are mainly meant for localised administration to the posterior segments as a vitreous injection, or subconjunctival implant [29].

8.2.4.3 Micelles

Surfactant- and polymer-based nanomicelles (~10–100 nm) are promising approaches for the delivery of therapeutics to the eye's posterior tissues. They can entrap lipophilic drugs in the lipophilic core of micelles formed in an aqueous medium. The micelles are simple and reproducible preparations and have many advantages for ocular drug delivery, increased corneal absorption, increased ocular bioavailability, causes no or insignificant irritation and easily sterilised by filtration for ocular use. The surfactants used to develop micelles may cause slight irritation following topical instillation.

Polymer (block-copolymers)-based nanomicelles are more popular because they escape from the ocular phagocytic system and cellular efflux mechanisms and thus show improved ocular bioavailability of drugs [30]. Pepić et al. have developed dexamethasone-loaded micelles using Poloxamer F127 and chitosan. They observed significantly increased *in vivo* dexamethasone bioavailability from micellar formulation over-marketed eye drop (0.1% w/v) and micelles prepared with Poloxamer F127 alone [31]. In another study, Iriyama et al. have utilised nanomicelles for plasmid DNA (pDNA) delivery using a mice choroidal neovascularisation (CNV) model for age-related macular degeneration. The nanomicelles after intravenous injection showed a significant reduction (~65%) in CNV [32].

8.2.4.4 Reverse Micelles (RMs)

Reverse micelles (RMs) are transparent and thermodynamically stable spherical aggregates formed spontaneously in non-aqueous medium (oily phase) upon dissolution of amphiphilic surfactants and polymers. This system also allows easy solubilisation of hydrophilic therapeutics in the hydrophobic oil phase. Thus, RMs can be successfully used for efficient and simultaneous ocular delivery of both hydrophilic and lipophilic compounds. Besides, these innovative NPs can increase the drug's ocular residence time for better absorption of both hydrophilic and lipophilic drugs through ocular tissue barriers. Moreover, the mucoadhesive amphiphilic biopolymers forming micelles can further increase the ocular bioavailability by increasing their contact time and decreasing their elimination from the ocular surface [33].

Similar to micelles, simple and economic methods can be used to develop RMs with improved physical stability. Further, RMs can be instilled like eye drops into cul-de-sac or onto the ocular surface retaining vision and causing no discomfort. The above characteristics of RMs fulfil the industrial as well as patient's compliance requirements. Moreover, the RMs are commonly used as templates for inorganic nanomaterials [34]. Jain and co-workers fabricated insulin-loaded RMs with an objective to prolong insulin release through ocular route. They prepared micelles using cetyl tri-ammonium bromide and Span 60 as surfactants and isopropyl myristate as an organic solvent. They observed prolonged and controlled release of insulin from RMs [35].

8.2.4.5 Liposomes

Liposomes are self-assembled vesicles (of size range from 0.1 to 10 µm) formed by phospholipids when dispersed in aqueous medium maintained at a temperature equivalent to or slightly more than the glass transition temperature of phospholipids. The biocompatible components (phospholipids and cholesterol) used in liposome preparation (phospholipids and cholesterol) are similar to the biological cell membrane. Liposomes have been studied widely for ocular delivery of therapeutics due to their several noteworthy characteristics such as versatile surface modification chemistry, and the potential of achieving controlled and stimuli-responsive drug release by controlling the phospholipid bilayer number and composition. Moreover, it is possible to develop liposomes with versatile physicochemical properties by using diverse lipids in a liposome preparation. They can entrap both hydrophilic and lipophilic therapeutics and deliver them to the target tissue [36].

The possible liposomal uptake mechanisms following ocular administration include fusion, adsorption, endocytosis and lipid exchange [37]. In addition, the liposomes can selectively target the desired sites resulting in decreased toxicity and increased therapeutic efficacy. Davis and co-workers have developed Annexin 5-functionalised liposomes for topical delivery of BVZ (Avastin). These liposomes demonstrated a significantly enhanced concentration of Avastin in the eye's posterior tissues [38].

8.2.4.6 Niosomes

In niosomes, the non-ionic surfactants form the vesicles composed of a flexible bilayered membrane in an aqueous medium. Niosomes can entrap lipophilic drugs in the bilayered membrane and hydrophilic drugs in the aqueous core. They are stable, biosafe (non-immunogenic, biocompatible and biodegradable) and the flexible membrane can withstand the deformation stress and thus allow repetitive intraocular injections to cure the posterior diseases of the eye [39,40].

Niosomes show improved residence time on the ocular surface and decreased systemic absorption of drugs, as a result, improved and targeted delivery of therapeutics has been observed following topical instillation. Currently, elastic niosomes (ethoniosomes) were utilised for topical delivery of corticosteroids (acetate and sodium phosphate

Ophthalmic Delivery of Hydrophilic Drugs

salt of prednisolone). Further, niosomes were extensively studied for ophthalmic delivery of drugs including protein- and peptide-based therapeutics [41].

8.2.4.7 Discomes/Cubosomes

Discomes (modified niosomes) are self-aggregated liquid crystalline particles of non-ionic surfactants used for ophthalmic delivery of hydrophilic and lipophilic drugs. They have displayed increased residential time, and their large size can avoid systemic absorption [42]. Besides, their disc shape can better fit into eye cul-de-sac. The development of discomes for ocular application is in the infancy stage; thus, they should be further studied widely for ocular delivery of small molecule and biological drugs [43].

8.2.4.8 Spanlastics

Spanlastics are nanosized elastic multilamellar vesicles prepared using a combination of spans and non-ionic surfactants. Tweens are also used (as edge activators) in spanlastics to increase their fluidity and deformity and to decrease their interfacial tension with lipophilic ocular tissues resulting in increased diffusion of spanlastics through ocular tissues. The spanlastics can be used for ophthalmic delivery of lipophilic, hydrophilic and amphiphilic therapeutics by entrapping them into its vesicle structure. Moreover, the spanlastics also have the ability to deliver therapeutics to the eye's posterior tissues [44].

8.2.4.9 Lipid NPs

Lipid NPs (50–1,000 nm) are generally prepared using biocompatible solid lipids alone or in combination with liquid lipids. The solid lipid NPs (SLNs) are the most effective carriers amongst lipid NPs and consist of a lipophilic matrix made of solid lipids surrounded by surfactant layers. Nanostructured lipid carriers (NLCs), another type of lipid NPs, consist of a solid lipophilic matrix made of both solid and liquid lipids. This matrix shows melting point depression compared to a matrix made of solid lipid. However, the matrix of NLCs remains solid at body temperature.

The composition of lipids and preparation method can generally result in imperfect, amorphous, and multiple-type NLCs. In imperfect NLCs, the little quantity of oil can change the lipid crystallisation, whereas amorphous NLCs consist of an amorphous solid lipid matrix. The amorphous solid matrix can be obtained by blending a particular type of lipids (hydroxyoctacosanyl hydroxystearate) with isopropyl myristate. In multiple-type NLCs, the blending of solid lipids with a high amount of oil results in a solid lipid matrix consists of very small oil compartments. In the development of NLCs, the drug encapsulation can be increased and drug leakage during storage can be decreased/avoided by simply changing the nanostructure of the lipid matrix.

The lipid-based NPs have improved the loading, bioadhesiveness and permeability, and provided sustainable release, stealth function, targeted delivery for hydrophilic drugs [45,46]. Furthermore, hydrophilic drugs can be coated with a range of lipids to maintain the required hydrophilic and lipophilic balance (HLB) to improve drug diffusion through ocular membranes. Besides, various targeting ligands can be attached to the NPs surface so that they can reach the eye's posterior tissues. Kumar et al. have fabricated the SLNs for improved ocular delivery of hydrophilic drug VAC. They confirmed the improved ocular bioavailability of VAC from SLNs than plain drug solution [47]. In another study, Ahmed and co-workers have developed natamycin (NAT)-loaded SLNs against fungal keratitis. NAT-loaded SLNs demonstrated extended drug release of NAT and improved corneal permeability [48].

The entrapment of hydrophilic drugs into the lipophilic matrix of SLNs and NLCs is challenging (requires higher lipid concentration) as hydrophilic drugs tend to partition towards the aqueous dispersion medium during the production process. Thus, the concept of lipid–drug conjugates (LDCs) was introduced. In LDCs, the drug molecules have been covalently or non-covalently bonded with lipids. The conjugation of hydrophilic drugs to lipids increases their lipophilicity and then loading in SLNs and NLCs.

8.2.4.10 Dendrimers

Dendrimers are branched monodisperse macromolecules having a size range between 10 and 100 nm. The dendrimers contain amine/hydroxyl/carboxyl groups as terminal functionalities that can be further exploited to conjugate therapeutic molecules and/or ligands for active targeting. Dendrimers can entrap both hydrophilic and lipophilic drugs of different molecular weights in the dendrimers network via hydrogen bonding, hydrophobic and ionic interactions and conjugation through covalent bonds. The developed dendrimer-based nanocarriers have shown improved physicochemical characteristics, uniform size distribution and higher biocompatibility in ophthalmic drug delivery. Further, the formulation process of dendrimers is very complex and their toxicity can be reduced by modifying their surface charge via acetylation, carbohydrate conjugation, peptide conjugation or PEGylation [49]. Marano and co-workers have developed amino acid dendrimers for the ocular delivery of anti-vascular endothelial growth factor (VEGF) oligonucleotides. In a laser-induced CNV animal model, they observed significant CNV inhibition and increased anti-VEGF oligonucleotide levels in the retina [50].

8.2.4.11 Nanocrystals

Nanocrystals are the particles of size between 10 and 1,000 nm and are stabilised by surfactants or polymer-based stabilisers. Development of nanocrystals is a potential approach to improve drug loading, bioavailability, efficacy and safety, and to reduce side effects. Tuomela et al. formulated brinzolamide (BRA) nanocrystals and studied their effect on intraocular pressure (IOP). They observed significantly decreased IOP caused by BRA nanocrystals in the rat ocular hypertension model [51].

8.3 CHIEF CHALLENGES FOR NANOPARTICULATE DELIVERY OF HYDROPHILIC THERAPEUTICS

8.3.1 Poor Drug Loading Efficiency

Generally, the drug can be loaded into colloidal carries via entrapment within the bulk carrier matrix during the nanoparticle preparation process or by adsorption onto the NP surface by incubating with the nanoparticle suspension [52]. The entrapment of hydrophilic therapeutics within the colloidal carries that are generally prepared using lipophilic substances is very much challenging. The lack of drug dispersibility in the lipophilic carrier, low solubility of a drug in the lipophilic carrier, low aqueous volume entrapped in vesicular NPs (such as liposomes and niosomes), the structure of the lipophilic carrier matrix, the state of the lipid matrix and partitioning of drug towards the aqueous phase during the production process are the main factors behind poor entrapment of hydrophilic drugs in colloidal carriers. The other challenge associated with lipophilic-colloidal NPs is the expulsion of hydrophilic drugs after polymeric transition during storage and due to relatively high aqueous dispersions phase (70%–99.9%) [53,54].

8.3.2 Poor Ocular Residential Time

The enhancement of residence time is imperative for the improvement of the ocular bioavailability of topically administered therapeutics. Despite having several advantages, the colloidal carriers possess poor ocular retention time (drain very quickly from precorneal pockets as seen with conventional eye drops). The neutral and anionic liposomes composed of phospholipids and cholesterol are cleared rapidly in tears. In another study, the non-mucoadhesive PLGA NPs are drained out of the eyes quickly [55].

In conclusion, the improvement of hydrophilic drug entrapment (using new approaches other than conventional physical entrapment approach) within the colloidal carriers and then further improvement of their ocular residence time by embedding them into a matrix to form a composite drug delivery system can significantly overcome the above challenges of ocular delivery of hydrophilic drugs.

8.4 APPROACHES TO IMPROVE LOADING OF HYDROPHILIC THERAPEUTICS IN COLLOIDAL NPs

8.4.1 Development of Carrier–Drug Conjugate Nanoparticles (CDC-NPs)

Generally, the hydrophobic drugs show superior entrapment in colloidal carriers made of lipophilic components. In contrast, significantly low encapsulation and high leakage of hydrophilic therapeutics were observed. The strategy of conjugating hydrophilic therapeutics to hydrophobic carrier improves the hydrophobicity of therapeutics and then their compatibility with the hydrophobic components of the colloidal NPs. Further, with this approach, the improved hydrophilic drug affinity towards hydrophobic carrier matrix and decreased drug leakage was observed [56–58]. The internal oil phase of emulsions, the hydrophobic phospholipid bilayer membrane in liposomes, the hydrophobic bilayer of niosomes, cubosomes and spanlastics, the hydrophobic core of micelles and the hydrophobic matrix of the polymer- and lipid-based nanospheres can serve as reservoirs for carrier–drug conjugates (CDCs). For example; SLNs showed significantly high encapsulation of fatty acid-conjugated diminazene diaceturate [59], and polymeric NPs showed substantial entrapment (68% at 1:10 drug/polymer ratio) of 4-(N)-stearoyl gemcitabine [60]. These studies and multiple other studies have confirmed the substantial loading of CDCs over plain drugs.

The key carriers/excipients used in the development of different colloidal carriers are summarised in Table 8.2. These excipients can be utilised for the development of conjugates with ocular therapeutics through different chemical bonds (Figure 8.3a). Depending on the chemical nature of carriers and therapeutics, the different conjugation strategies and chemical linkers are used to develop CDCs. The type of chemical bond used to develop CDCs depends on the functionalities available on carrier/excipients and the therapeutic molecules. Further, the required chemical bond can be achieved using suitable linkers. The type of chemical bond and/or linkers used will determine the release pattern of therapeutics from CDCs and CDC-NPs and is critical for the optimal performance of CDCs and their NPs.

The commonly seen chemical bonds in the development of conjugates are explained below.

8.4.1.1 Ester Bonds

It is the most commonly formed chemical bond in CDCs by the reaction between hydroxyl and carboxylic groups of the carrier and therapeutics. In some instances, a linker such as succinic acid was used to form the ester bond [72]. This ester bond can be degraded by the esterase enzymes to release active drugs. In some CDCs, the esterase-insensitive bond causes a slow release rate of therapeutics resulting in a negative therapeutic effect. Further, the chemical structure of adjacent groups and spacers or linkers used to form the conjugates may affect the ester hydrolysis [73].

8.4.1.2 Amide Bonds

It is formed by the chemical reaction between carboxylic groups and amine groups of carriers and therapeutics via the carbodiimide coupling technique. The amide bonds are highly stable than ester bonds resulting in a slower rate of hydrolysis and decreased therapeutic effect [74].

8.4.1.3 Hydrazone Bonds

The hydrazone bonds can be cleaved efficiently in acidic pH environment of endosomes or lysosomes and tumour tissues (pH 6.0–6.8). The hydrazones show slight or no

Ophthalmic Delivery of Hydrophilic Drugs

TABLE 8.2
Key Ingredients and Specific Drawbacks of Colloidal NPs for Ocular Delivery

Nanocarrier	Key Ingredients	Specific Drawbacks
Nanoemulsions/ microemulsion	*Surfactants*: Fatty alcohols, glycerol esters, fatty acid esters, soaps, sulphonates, divalent ions, amines and quaternary ammonium compounds such as cetyl trimethyl ammonium bromide *Oils*: glyceryl tricaprylate/caprate, dicaprylate/dicaprate, glyceryl tricaprylate (tricaprylin), 90:10% w/w C12 glyceride tridiesters, C8/C10 triglycerides, myristic acid isopropyl ester, etc. *Co-surfactants*: 2-(2-ethoxy ethoxy) ethanol, propylene glycol, ethylene glycol, glycerine, ethanol, propanol, etc.	Sticky feel and subsequent intolerance due to surfactants [61,62]
Polymeric NPs (nanospheres, nanocapsules)	PCL, poly (n-BCA), PEG, lecithin, pentablock polymer Pilopex, pectin, gelatin, CAP, PLGA, CS, CHCS, HA, RL100, Migliol840, Poloxamer 188, albumin, sodium alginate	Local and systemic toxicity from polymer degradation products and residual organic solvents [61,63]
Nanomicelles	SDS, DTAB, n-dodecyl tetra-(ethylene oxide) (C12E4), vitamin E TPGS, octoxynol-40, dioctanoyl phosphatidylcholine (C8-lecithin), polycaprolactone, poly (d,l-lactide), polypropylene oxide, PVP, etc.	Toxicity and immunogenicity Lower entrapping of hydrophilic therapeutics [64,65]
Liposomes	PC, PG, cholesterol, DSPC, DOPE, DSPE-PEG, DPPC, DMPC, DODAB, stearylamine, PVP	Lower drug loading capacity compared to NPs [61,63]
Niosomes	Brij 35, Cholesterol, DCP, CS-Carbopol, Span 60, Tween 60, Tween 80	Lower drug loading capacity compared to NPs [61,63]
Spanlastics	Span 80, Span 60, Span 40, Span 20, Tween 80, ethanol, etc.	Vesicle aggregation and drug leakage [66,67]
Cubosomes	GMO, Phytantriol (PHYT), DPPS, Poloxamer 407 (F127), Pluronic® F108 (F108), etc.	High viscosity hinders large-scale production, Low entrapment of polar drug [68,69]
Lipid NPs (SLN, NLC, LDC)	Phospholipon 90G, Stearic acid, Compritol® 888 ATO, PEG monostearate, COS, Miglyol 812® and 829®	Drug expulsion during storage (for SLN) [61,63]
Dendrimers	Ammonia, ethylenediamine, poly-alkyl amines, polylysine, poly (disulphide amine), polyether, or polyester, etc.	Complex preparation methods, toxicological issues [64,70]
Nanocrystals	PVA, PVP, HPC, HPMC, SLS, SDS, PEI, Pluronics®, Tween, Soluplus®, TPGS, etc.	Organic solvent residue, low drug loading, recrystallisation [71]

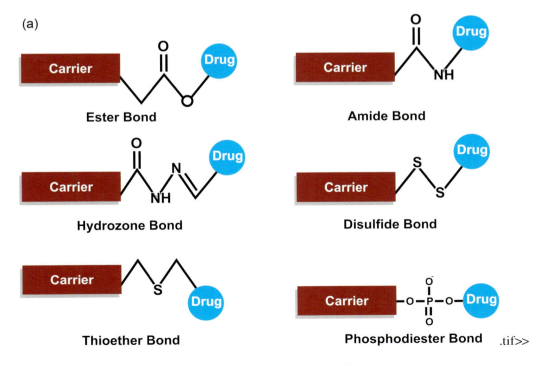

FIGURE 8.3a Different types of chemical bonds generally used to develop CDCs.

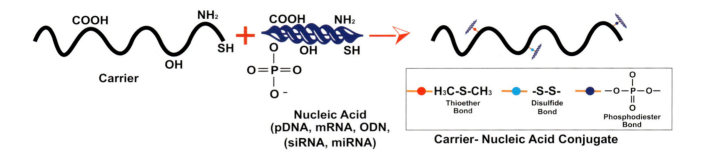
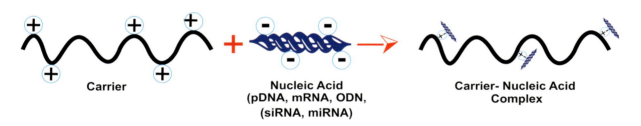

FIGURE 8.3b Concept of carrier–small molecule drug and carrier–biological drug (nucleic acids) conjugates.

decomposition at neutral pH environments [75]. Thus, CDCs containing hydrazones are pH-sensitive (remains intact in circulation and break in tumour tissues).

8.4.1.4 Disulphide Bonds

The disulphide (–S–S–) bonds are stable in the oxidative environments (in extracellular fluids) but would be cleaved in reductive environments such as intracellular fluid upon cellular uptake. In addition, the disulphide bonds have also been used to prepare conjugates with biological drugs such as siRNA and oligonucleotides to enhance their intracellular delivery and release [76]. Thus, disulphide bonds are used to develop environmentally responsive prodrugs.

8.4.1.5 Other Bonds

Many other chemical bonds such as thioethers (C–S–C) (formed between one sulphur and two carbon atoms), and phosphodiester have also been used in the development of CDCs. The phosphodiester bond is mainly used in the preparation of CDCs of biological drugs such as oligonucleotides and siRNA [77].

The concept of development of carrier–small molecule drug conjugates and carrier–nucleic acid-based drugs through any one of the above bonding is summarised in Figure 8.3b. Further, these conjugates can self-assemble into NPs without any carrier or they can be entrapped in different NPs as shown in Figure 8.3c.

8.4.1.6 Self-Assembled NPs (Carrier-free System)

Conjugation of a lipophilic carrier to a hydrophilic therapeutic agent results in amphiphilic molecules that can self-assemble into NPs (micelles in aqueous medium and RMs in nonaqueous medium) with or without the smallest amount of stabiliser. Thus, CDCs can be administered without a delivery carrier. The hydrophilic therapeutic agent, following conjugation with lipophilic carrier, forms hydrophilic corona in micelles and hydrophilic core in RMs RXI-109 (cholesterol–siRNA conjugate) were studied in Phase II for its ability to reduce hypertrophic scarring [78]. Moreover, Chen and co-workers fabricated cholesterol-conjugated siRNAs for CNS delivery. The conjugate demonstrated improved silencing ability than plain siRNA [79].

FIGURE 8.3c Colloidal NPs showing efficient incorporation of CDCs through physical interaction with lipophilic core of NPs and efficient loading hydrophilic anionic nucleic acids through electrostatic complexation with cationic carriers/excipients of NPs.

8.4.1.7 Micelles

The encapsulation of hydrophilic therapeutics in the lipophilic micelle core via physical interaction is practically not possible. The conjugation of hydrophilic drugs with the hydrophobic carriers can significantly improve their lipophilicity and then the interaction with the lipophilic core of micelles prepared with other amphiphilic carriers. This strategy increases the hydrophilic drug entrapment and stability in micellar preparations.

Many micellar preparations have been used to deliver lipophilic conjugates of hydrophilic drugs. The micelles prepared with amphiphilic DSPE-PEG have been studied as carriers for delivery of hydrophilic acid-sensitive doxorubicin after conjugating it with palmitic acid. The loading was found substantially high as a result of increased lipophilicity of the doxorubicin following lipid conjugation. Further, substantial drug entrapment and stability were observed as a result of an enhanced interaction between palmitic acid and the lipophilic core of DSPE in DSPE-PEG micelles. In another study, stearic acid–gemcitabine conjugate-loaded DSPE-PEG/TPGS1000 micelles were developed. They observed a substantial decrease in drug inactivation due to rapid deamination. PEG-PCL and PEG-PLA (amphiphilic copolymers) were also utilised to develop micellar carriers of lipophilic CDCs prepared for hydrophilic drugs (geldanamycin and gemcitabine) [80,81].

8.4.1.8 Emulsions

Emulsions (O/W) are biocompatible carriers that allow the inclusion of lipophilic drugs in the oil phase. Besides, the O/W emulsions are used frequently to deliver amphiphilic CDCs. Many studies have observed the high entrapment efficiency (as a result of increased solubility of the conjugate in the oil phase) and tolerance of O/W emulsions composed of lipophilic carrier and hydrophilic drug conjugates. The hydrophobic carrier dissolves and remains in the inner oil phase, and the hydrophilic drug gets dissolved in the external aqueous phase. In addition, they serve as a bridge (emulsifier) between two immiscible phases resulting in a stable emulsion. Further, CDCs-emulsions are reported to improve the delivery of CDCs with decreased toxicity and clearance rates [82].

8.4.1.9 Polymeric NPs

Amphiphilic CDCs can be incorporated efficiently into the hydrophobic core of the polymeric NPs. PLGA NPs have been studied extensively for this purpose. The PLGA NPs were used for the improved delivery of amphiphilic gemcitabine prodrug (4-(N)-stearoyl gemcitabine) [57]. Sarett and group developed endosomolytic polymeric NPs for the efficient delivery of siRNA. They prepared the conjugate of hydrophilic siRNA with hydrophobic palmitic acid and loaded efficiently into NPs at a lower polymer/siRNA ratio. Further, they observed improved stability, endosome escape and potency and longevity of gene silencing with the prepared NPs [83].

8.4.1.10 Lipid NPs

The lipophilic matrix of lipid NPs can generally serve as a reservoir for hydrophobic drugs. The hydrophilic therapeutics can also be loaded efficiently into the lipophilic matrix following their chemical conjugation with lipophilic carriers. Both solid- and liquid- lipid NPs can be used for efficient loading and effective delivery of CDCs [84,85].

8.4.1.11 Liposomes

Liposomes allow the encapsulation of water-soluble drugs in the aqueous core and lipophilic therapeutics in the phospholipid bilayer membrane. Further, a very small aqueous volume of liposomes can accommodate only a small amount of hydrophilic therapeutics. The conjugation of lipophilic components of the liposomes (phospholipids and cholesterol) or other lipophilic carriers to the hydrophilic drugs can improve membrane affinity, entrapment, stability and retention of CDCs in liposomes. Liposomes can further avoid the premature metabolism of amphiphilic CDCs [86]. Arouri and group developed liposomes loaded with phospholipid–drug conjugate that cleaves into an active form in cancer cells that overexpress phospholipase A2 [87]. The PEGylated liposomes loaded with squalenoyl-gemcitabine, and diglyceride doxorubicin conjugate was also reported [88,89].

8.4.2 Development of Nanoplexes

Nanoplex means nanoparticle complex with the oppositely charged therapeutics (small molecule drugs and biologics). In this technique, the anionic nucleic acids (pDNA, oligonucleotides (ODN), mRNA, siRNA, miRNA, etc.) are reacted with cationic carriers/excipients (polymer, lipid or other components of the colloidal NPs) to form NP complex (Figures 8.3b and c). Cationic lipids and polymers can interact electrostatically with anionic nucleic acids and form nanosized complexes known as lipoplexes and polyplexes, respectively.

8.4.2.1 Lipoplexes

Lipid-based NPs are generally made of positively charged lipids and amphiphilic substances having a hydrophobic tail and a positively charged hydrophilic head group (primary to quaternary amine salts) that can complex with anionic phosphate groups present in the nucleic acid-based therapeutics. Further, cationic lipids are reported to have detergent and buffering characteristics that facilitate the release and transfection of DNA [90].

The lipid structural properties and the charge ratio are imperative in lipid–nucleic acid complex (lipoplex) formation and for their cellular transfection. The final lipoplex charge can be tuned to neutral, negatively charged and positively charged by changing the ratio of positive and negative charges of lipids and phosphodiesters in nucleic acids, respectively. The common belief is that the slight excess of the positive charge of lipoplex can result in increased transfection due to electrostatic interactions with anionic cell surface and proteoglycans [91].

Plain lipoplexes can form either a multilamellar structure with nucleic acids sandwiched between cationic membranes or an inverted hexagonal structure with nucleic acids encapsulated within cationic lipid monolayer tubes. Lipoplexes can also be formed in a self-assembly process triggered by nucleic acid-mediated fusion of cationic liposomes with a large-scale lipid rearrangement. Lipoplexes under clinical trials are mostly validating the human gene therapy concepts [92]. Liu et al. have prepared PEGylated liposome–protamine–hyaluronic acid NPs containing siRNA. The treatment with lipoplex over plain siRNA caused significant knockdown of VEGF receptor 1 (VEGFR1) and substantial reduction in CNV area in laser-induced rat CNV model [93].

8.4.2.2 Polyplexes

Polymer–nucleic acid binding usually takes place through hydrophobic and electrostatic interactions between anionic phosphate groups of nucleic acids and cationic amine groups of polymers. Polymer–nucleic acid binding is generally influenced by their inherent characteristics such as number and type of cationic groups (such as primary to quaternary ammonium groups and amidine groups) present per polymer molecule, the spacing between these cationic groups, degree of polymer backbone branching and polymer hydrophobicity. Further, polymer–nucleic acid interactions are dependent on pH. The polymers having higher charge density show stronger interactions that can later hinder the nucleic acid release in the cytoplasm and then the transfection efficiency [94]. Polymer–nucleic acid complex can condensate into different structures such as spherical, globular, rodlike or toroid.

Many positively charged polyamine polymers such as polylysine, polyarginine, chitosan, polyethylenimine and polyamidoamine dendrimers have been studied for efficient delivery of nucleic acids. Kim and co-workers have developed PEGylated cationic polyplexes (PEI-siRNA complex). The siRNA targets VEGF-A, VEGFR1 and/or VEGFR2. The treatment with polyplex resulted in substantially decreased angiogenesis induced by herpes simplex virus and stromal keratitis in ocular tissues of murine model [95]. In another study, Nirmal and co-workers have developed PLGA–chitosan nanoplex for ocular delivery of small molecule drug moxifloxacin. They observed significant (about fivefold) bioavailability of moxifloxacin from nanoplexes when compared to plain solution.

Dendrimer-based delivery is another polymer-based approach for ocular delivery of therapeutics. The ocular delivery of dendrimers has been demonstrated many years ago; however, recently the dendrimers are being used as highly stable and soluble nucleic acid delivery vehicles for ocular delivery [96].

8.4.2.3 Micelleplexes

Micelleplexes have been found to be a promising approach for the simultaneous delivery of small molecule drugs and nucleic acids. The amphiphilic copolymers containing one or more polycationic blocks can interact with anionic nucleic acids and self-aggregate in water to form hydrophobic core–hydrophilic shell micelles simultaneously, solubilising the lipophilic therapeutics in the hydrophobic core [97]. Further, the micelleplexes with entrapped small molecule drugs can offer a synergistic therapeutic effect by simultaneously targeting multiple mechanisms. Moreover, the structural diversity and physicochemical characteristics of copolymers used in the development of micelleplexes further offer flexibility in synthesis and chemical modifications, which improve the characteristics of micelleplexes. Salzano et al. have developed paclitaxel and survivin siRNA co-loaded micelleplexes using PEG-phosphothioethanol (PE)/siRNA-S-S-PE carrier for targeting cancer [98]. Ye et al. have developed ternary nanocarriers composed of cationic PEI-PEG and chitosan for ocular delivery of siRNA that target IkB kinase subunit mRNA. They observed substantially decreased scar tissue with nanoparticulate siRNA following subconjunctival injection in a monkey model of glaucoma filtration surgery [99].

8.5 APPROACHES TO IMPROVE OCULAR RESIDENCE TIME OF COLLOIDAL CARRIERS

The permeability of hydrophilic drugs can be improved by incorporating them within colloidal NPs. However, their low viscosity and rapid drainage cause insufficient contact time of the drug in the eye. Therefore, there is a need to improve the ocular residence time of colloidal carriers, and this can be achieved by developing mucoadhesive colloidal carriers or incorporating them into compact systems such as in situ gel, hydrogel and contact lenses. This system is also called as nanocarrier-laden compact systems wherein different colloidal carriers are incorporated into different compact systems such as in situ gel, hydrogel, contact lenses, etc. [100].

8.5.1 Development of Mucoadhesive Colloidal NPs

The primary component of mucus is mucin glycoprotein. Mucoadhesion offers various advantages at the site of absorption, such as prolonged residence time, increased absorption, rapid absorption, decreased dose and dosing frequency, which improve the patient compliance, faster onset of action and sustained release [101].

Several hydrophilic functionalities present in the mucoadhesive polymers (macromolecular hydrocolloids) can interact with mucin via hydrophobic, electrostatic, van der Waals interactions and hydrogen bonding. Many natural polymers (such as pectin, tragacanth and guar gum) are studied widely due to their biocompatible and biodegradable nature. Further, polymers with polycationic and polyanionic nature have shown strong mucin binding. Chitosan, a cationic polymer, was studied extensively for mucoadhesive applications as it shows increased mucoadhesion by interacting electrostatically with anionic mucin. Besides, the use of a combination of positively and negatively charged polymers (polyelectrolytes) has showed increased ocular residence time and sustained delivery of therapeutics. These days, the core polymers are subjected to chemical derivatisation (such as thiolation, quaternisation and carboxymethylation) that further improves their mucoadhesive potential. Following topical application, these mucoadhesive polymers swell in the tear film, allowing the entanglement and bonding of the polymer chains with mucin layer covering the epithelium of the cornea and conjunctiva.

Several ocular mucoadhesive colloidal NPs have been developed to sustain the ocular residence time of therapeutics. Dukovski et al. [102] have developed ibuprofen-loaded mucoadhesive cationic nanoemulsions (NEs) using positively charged mucoadhesive polymer (chitosan) and negatively charged surfactant (lecithin) for dry eye treatment. The NEs prepared with 0.05% (w/w) chitosan have shown excellent physicochemical and mucoadhesive characteristics and biocompatibility. In another study, Yu and group have fabricated ocular mucoadhesive NPs containing dexamethasone–glycol chitosan conjugate. The resulting NPs, positively charged and roughly spherical in shape, have showed improved ocular tolerance and longer precorneal residence time [103]. In another investigation, Terreni et al. [104] have developed cyclosporine-A-loaded nanomicelles using two surfactants in combination with hyaluronic acid. In a pharmacokinetic study using rabbits, the nanomicelles showed prolonged retention time for cyclosporine-A in the precorneal area due to the presence of mucoadhesive hyaluronic acid. Lin et al. fabricated doxorubicin-loaded liposomes surface modified with a mucoadhesive polymer, hyaluronic acid. After *in vivo* instillation in rabbits, the mucoadhesive liposomes showed longer retention time than plain liposomes. In another investigation, mucoadhesive niosomes (chitosan-coated) containing ciprofloxacin were investigated for enhanced corneal residence time. It was found that the optimised chitosan-coated niosomes showed sustained release profile and 1.79-fold increase in ciprofloxacin corneal permeation compared to eye drop [105]. Pai et al. [106] have developed coumarin-loaded NLCs, surface coated with chitosan oligosaccharide to make them mucoadhesive. An improved ocular retention of mucoadhesive NPs was observed as compared to plain NPs.

Water-soluble, positively and negatively charged carbosilane dendrimers have also been studied as mucoadhesive polymers in ophthalmic preparations containing acetazolamide (ACZ) [107]. These polymers showed excellent ocular tolerance in the concentration range of 5 to 10 µM. They observed a quick and sustained hypotensive effect with ophthalmic drops composed of positively charged carbosilane dendrimers (5 µM) and ACZ (0.07%).

8.5.2 Development of Colloidal NP-laden Composite Systems

8.5.2.1 NP-laden In Situ Gel

The in situ gel is instilled in liquid form and changes into cul-de-sac gel phase upon instillation as a result of a change in pH, temperature and/or ion concentration. This system combines the advantages of both conventional eye drops (ease of administration, dose accuracy and simple manufacturing process) and viscous transparent gel formulations (enhanced ocular retention time, mucoadhesion and sustained drug release) [108].

The development of NP-laden in situ gel system is an advanced approach and can be used to increase the ocular residence and ocular bioavailability of NPs. Upadhyay and co-workers developed moxifloxacin-NPs laden in situ gel that showed sustained release of moxifloxacin in keratitis treatment [109]. Further, Yu et al. formulated ion-sensitive in situ gel system containing timolol maleate (TM)-loaded liposome. They observed significantly increased corneal penetration and longer retention of TM from the in situ gel system over eye drops [110].

8.5.2.2 NP-laden Hydrogels

A hydrogel is a water-laden three-dimensional polymer network that exists as solid material but swells after absorbing water. Hydrogels that are highly biocompatible and mimic the biological matrices can be developed to achieve either short-term release or long-term release (by means of physical or chemical control) of both small molecule drugs and biologics [111,112]. Besides, they show increased ocular residential time and contact time of drugs with cornea upon topical instillation. Different synthetic hydrophilic polymers commonly used to prepare hydrogels include poly(hydroxyethyl methacrylate) (PHEMA), polyethylene glycol (PEG), poly(vinylpyrrolidone) (PVP), poly (acrylic acid) (PAA), propyl acrylamide (PAM) and poly(vinyl alcohol) (PVA), and natural polymers like agar, gelatin, fibrin, hyaluronic acid, alginic acid, CMC and hydroxypropyl methylcellulose (HPMC). The stimuli-responsive hydrogels can also be developed which can swell or deswell, degrade, change shape and undergo phase transitions in response to a stimulus (internal or external) [113].

Wenguang and co-workers have developed collagen hydrogel containing bovine serum albumin (BSA)-loaded alginate microsphere for ocular application. They observed sustained release of BSA for 11 days in neutral phosphate buffer [114]. Christian and group fabricated microsphere-laden hydrogel system for ocular delivery of ranibizumab or aflibercept to achieve extended release. They achieved extended release of ranibizumab or aflibercept for 196 days [115]. In another study, Hui and co-workers developed a hydrogel system containing TM-liposome for ocular delivery. They demonstrated improved bioavailability and retention time of TM with a hydrogel system than plain liposomes or eye drops [116].

8.5.2.3 NP-laden Contact Lenses

Contact lenses, thin and curved plastic discs designed to cover the cornea, get adhered to the eye surface due to surface tension. They are reported to show 50% more bioavailability over ophthalmic drops. The contact lenses are soaked into the concentrated drug solutions to load the drug; however, a high quantity of therapeutics cannot be loaded into these pre-soaked lenses, and thus, they fail to show a sustained release of therapeutics [117]. Further, this can be overcome by developing drug-loaded NP-laden contact lenses. In addition, different types of NP-laden contact lenses such as NP-laden implant contact lenses, NP-laden ring implant in hydrogel contact lenses and nanodiamond-embedded contact lenses can also be developed. The NP-laden implant contact lenses can release the therapeutic for an extended period due to the slow release of therapeutics from NPs. In addition, they can result in substantial corneal drug permeation and decreased systemic drug loss. Further, NP-laden soft contact lenses were studied as nanoreservoirs to obtain extended drug release and therapeutic effect. Zhang and co-workers have developed a contact lens containing gelatin NPs for improved ocular delivery of BSA. They observed extended release of BSA for 12 days from contact lenses containing BSA-loaded gelatin NPs [118]. In a similar way, Jung et al. fabricated contact lenses containing NPs for sustained ocular delivery of timolol [119]. Furqan and group prepared NP-laden ring implant in hydrogel contact lenses for sustained ophthalmic delivery of Timolol. *In vitro* release study showed sustained timolol release for 168 h and the observed release was within the therapeutic window [120]. Jiawen et al. have also developed micelle-laden contact lenses for co-delivery of timolol and latanoprost in glaucoma treatment. The *in vivo* pharmacokinetic study in rabbit eyes showed extended timolol and latanoprost release (up to 120 h and 96 h, respectively). Besides, the improved mean ocular residence time and bioavailability for both timolol and latanoprost were observed from micelle-laden contact lenses as compared to eye drops [121].

8.5.2.4 NP-Laden Ocular Inserts

Ocular inserts are small, sterile drug-infused tools, placed in or around the eye for the delivery of drugs over an extended period. Further, the release rate of the drug from ocular inserts can be well controlled. Thus, ocular inserts can be used to improve the residence time and for accurate dosing of drugs, which results in less systemic absorption and reduced toxicities. Inserts are classified into different types such as insoluble, soluble or bioerodible based on the basis of their physical and chemical properties [122]. The advanced approaches like NP-laden ocular inserts can also be developed for modulating the drug release by adjusting drug loading in the NPs. Jung and Chauhan fabricated fornix insert containing timolol-loaded NPs to achieve extended release in the treatment of glaucoma [123].

8.5.3 Development of Nanowafers

Nanowafers are transparent disc-like or rectangular membranes containing an array of drug-loaded nanoreservoirs and are applied on to the ocular surface with the help of a fingertip. They withstand the constant blinking and remain undisplaced. Nanowafers can release the therapeutics for an extended time in a well-controlled manner; thus, they can substantially increase the therapeutic effect of ophthalmic drugs. Postapplication, the nanowafers dissolve slowly and fade away leaving no polymer on the ocular surface. The polymers commonly employed in the fabrication of nanowafers includes PVA, PVP, HPMC and CMC.

Yuan et al. developed the doxycycline-loaded PVA nanowafer for ocular application. They observed extended doxycycline release and increased doxycycline corneal permeation with the prepared nanowafers in mice model. They also reported an improved therapeutic effect of axitinib-loaded PVA nanowafer in treating burn-induced CNV murine ocular model [124,125]. Further, this nanowafer platform has been utilised to achieve sustained dexamethasone and cysteamine release in treating dry eye and corneal cystinosis, respectively [126]. The above results indicate the potential clinical applications of nanowafers in ophthalmic delivery of small molecule and biological drugs.

8.6 ADVANCED/FUTURE TREATMENT APPROACHES

8.6.1 Nucleic Acid-Based Therapy over Small Molecule Drugs

Novel nucleic acid-based therapies are found to have more potential in the treatment of ocular diseases over small molecule drugs. More commonly used nucleic acid-based therapeutics include; aptamer, siRNA and ODN [127]. Aptamers are single-stranded DNA (ssDNA) or single-stranded RNA (ssRNA) oligonucleotide ligands. They can fold into complex secondary or tertiary structures and have high affinity and specificity towards their target proteins. Macugen® or pegaptanib is an RNA aptamer approved by Food and Drug Administration (FDA) for age-related macular degeneration (AMD) treatment. In addition, E10030, anti-PDGF DNA aptamer, is in Phase III trial with anti-VEGF agents for AMD treatment [128].

siRNA is a double-stranded RNA (dsRNA) molecule having 20–25 base pairs. It is used for silencing gene expressions in cells. Bevasiranib (dsRNA) containing 21 nucleotide base pairs has been developed for targeting VEGF-A. Its biodistribution and pharmacokinetic studies have shown its highest concentrations in the vitreous after single i.v. injection in rabbit eyes [129]. SYL040012 is another siRNA-based drug under Phase II trial for glaucoma treatment.

8.6.2 Optogenetics

Gene therapy approach is limited to people whose blindness is because of genetic mutation. Further, the gene therapy approach does not tackle end-stage retinal disease wherein only a few cells are remaining to be repaired. In optogenetics (disorder agnostic technique), the genes that enable cellular (photoreceptor cells) production of opsins (light-sensitive proteins) will be delivered through the virus. Released opsins can thus restore the cellular light sensitivity towards damaged photoreceptors. These opsins also make other retinal cells (bipolar and ganglia cells) light-sensitive. These opsins can perform better at high light intensities. Further, the photoreceptor cells can survive against light intensities of a wide range and can work well in both bright sunlight and twilight [130].

8.6.3 Stem Cells and Cell Transplantation Techniques

Stem cell therapy used for cancer treatment has also shown applicability in sight restoration, which is mainly focused on cornea and retina. Cell transplantation approach deals with the direct replacement of endogenous cells, and/or the delivery of cells that can secrete trophic factors to save degenerating cells. Cell transplantation strategy is mainly used in the treatment of AMD. Pluripotent and embryonic stem cells and autologous cells of adipose tissue and bone marrow origin are the candidates for cell transplantation. Corneal vascularisation and impairment in vision or blindness caused by limbal stem cell deficiency have been treated successfully by transplanting limbal stem cells from an exogenous source (other than patients). Posttransplantation, the corneal epithelium renewal was observed. Thus, stem cell therapy can be used as a treatment of choice for eye diseases of anterior and posterior segments [131].

8.6.4 Encapsulated Cell Technology (ECT)

In ECT, the specific cells are genetically modified first to make them overexpress or produce required proteins of therapeutic nature and then are entrapped within the biomaterial carriers that are semipermeable in nature and allow the exchange of nutrients and proteins but prevent the influx or efflux of cells. ECT bypasses the BRB and helps to achieve extended and controlled delivery of ophthalmic drugs. Further, utilising ECT implant, it is possible to sustain the drug delivery into the vitreous. In addition, this treatment approach is associated with very less adverse effects when compared to the bolus injections. In NT-503, ECT-based implant for treatment of wet-AMD, the human RPE cells were genetically modified to produce VEGFR at high levels and entrapped within the implant for ocular application [132].

8.6.5 Noble Metal NPs with Multiple Functions

Noble metals NPs and bio-inspired NPs are found to be promising approaches to conquer the limitations of the standard ocular treatments. Moreover, the multifunctional NPs can help to treat different symptoms of ocular disorders. The different types of such nanosystems include gold NPs (AuNPs), silver NPs (AgNPs), gold nanodisc and hybrid hydrogel-based contact lens, which comprises quaternised chitosan and graphene oxide NPs. These NPs show concurrent antibacterial and antifungal functions when used. Besides, upon topical instillation, these NPs can diffuse through tissue barriers and deliver drugs selectively to the eye's posterior tissues.

Gold, silver and graphene oxide NPs are found to have antimicrobial and antiangiogenesis properties. Currently, transplantation of photoreceptor precursors (PRPs) therapy is considered to be effective for treating certain diseases related to retinal degeneration [133]. Moreover, PRPs were tracked by transplanting photothermic AuNP-labelled cells into vitreous and subretinal space in Long-Evans pigmented rats. However, further several detailed *in vivo* studies are needed to address the many challenges associated with this approach. Song et al. [134] have described the application of gold nanodiscs on human retinal microvascular endothelial cells and mouse oxygen-induced retinopathy model.

8.6.6 Targeted NPs

The selective drug delivery to the eye (particularly to posterior tissues) still remains an open challenge. Targeting to the eye can be achieved by passive or active targeting. Passive targeting can be achieved by conjugating therapeutics to large molecules that modify the drug's ocular pharmacokinetic characteristics. The CNV, similar to solid tumours, has very permeable blood vessels and no lymph systems under the retina. Thus, NPs can reach target site passively through enhanced permeability and retention (EPR) effect.

Further, the active targeting of NPs can be achieved by attaching specific ligands (such as antibodies and aptamers) over NP surface. In magnetic nanoparticles (MNPs), their surface modification with bioactive molecules (ligands) can change the fate of the particles and drive preferential accumulation in the specific eye area like choroid. Martina and co-workers have developed VEGF-bonded MNPs for targeting the choroid layer. They observed preferential localisation of VEGF-bonded MNPs in the choroid. Thus, MNPs surface decorated with diverse bioactive molecules could be a potential approach for cell-specific targeting in the eye's posterior tissues [135].

8.6.7 Theranostic NPs

Theranostic nanomedicines are superior over plain therapeutic agents due to their all-in-one platform that includes controlled and targeted delivery of therapeutics, and multimodality diagnosis and therapies. In general, the conventional practice involves the diagnosis of ophthalmic disease first followed by its treatment. For instance, the neovascular eye diseases are diagnosed first using fluorescein angiography, and then, they are treated with multiple intravitreal injections. This conventional practice can result in vision-threatening complications, and a lack of monitoring of real-time disease progression and timely evaluation of therapeutic effects. Several theranostic agents made of peptide-functionalised NPs such as silicon NPs (SiNPs) are promising for simultaneous imaging and therapy of ocular neovascularisation and to overcome the above critical issues associated with the conventional practice. Miaomiao et al. have fabricated the peptide-functionalised silicon NPs as photostable theranostic probes for simultaneous imaging and therapy of ocular neovascularisation [136].

8.6.8 Microneedle-laden Collagen Cryogel Plugs

Recently, the supermacroporous cryogels have provided a flexible platform for drug delivery applications. These supermacroporous cryogels can be developed at subzero temperature and have demonstrated large interconnected pores. However, these systems have limitations such as rapid drug release and low drug loading capacity for certain drugs as a result of the porous structure of the cryogels. To conquer these limitations and to achieve controlled drug delivery, researchers have developed cryogels with advanced approaches such as NP-incorporated cryogels and microneedle-incorporated cryogels. These hybrid biomaterials can enhance the therapeutic index following localised applications. Krishnapriya and co-workers have developed collagen cryogel plugs containing moxifloxacin-loaded hyaluronic acid microneedles. This system demonstrated the sustained and controlled release of moxifloxacin (only 2.43%–5.08%) compared to plain cryogel plugs (nearly 42.95%–57.77%) in PBS within 1 week [137].

8.6.9 Use of Physical Methods in Combination with Chemical Approaches

Different physical methods investigated for improved ocular delivery of drugs are iontophoresis, bioballistic, electrotransfection, magnetofection, sonoporation and optoporation [138,139]. The application of these physical approaches with chemical approaches can substantially improve the ocular bioavailability, tissue targeting and timely releasing of drugs. The pDNA electrotransfection is one of the potent ocular gene delivery techniques. The first clinical trial is in progress providing a new way for sight-threatening disease treatment. Further, the advanced ultrasound-targeted microbubble destruction (UTMD) mediated gene therapy is found to be more effective. UTMD is a potent approach for safe, repetitive and targeted delivery of small molecule drugs and genes. Several UTMD-mediated gene delivery systems have been evaluated extensively in preclinical studies to improve the gene expression in different organs [140].

8.7 REGULATORY AND FUTURE PERSPECTIVES

The current section explains the regulatory perspectives that facilitate the development of safe, effective and quality ophthalmic preparations. There are no stringent rules and sufficient regulatory guidelines for ophthalmic delivery of drugs. As a result, the development of novel ophthalmic formulations and newer routes of administration are facing inimitable challenges with regard to designing of preclinical programs to evaluate (in a meaningful way) the risk for humans. In addition, the regulatory guidelines for preclinical testing of ophthalmic preparations are not distinguished well with coexisting regional variances. These difficulties further offer new opportunities for the applicants (new drug application (NDA) and abbreviated new drug application (ANDA)) in the development and marketing of ocular preparations.

The base/excipients/raw materials utilised in the development of ophthalmic preparations must retain the preparation stability and should not affect the drug bioavailability at the target tissue of the eye. Further, inclusion of colourants in the ophthalmic preparations is not recommended. The aseptic process and good manufacturing practices (**GMPs**) are needed to develop sterile ophthalmic preparations free of contamination and cross-contamination. Further, the procedures should be validated and monitored (at every stage of preparation) to develop effective preparation. This is needed because there are many instances wherein the requirements for the manufacturing of ophthalmic preparations are not met resulting in FDA interventions demanding for a stringent plan and design for ophthalmic preparation development.

The quality standards for ocular preparations including advanced nanoformulations generally include; identification, related substance, assay, pH values, water content, size and morphology of particles, examination of drug-carrier interaction, sterility test, bacterial endotoxins, irritation studies and stability tests. Additionally, container and labelling requirements are also specified.

Although colloidal NPs have changed the face of ophthalmic drug delivery, the rationale for design and development of ocular NPs still requires more attention. For example, the use of preservatives and dispersion agents in ophthalmic preparations can cause sterile endophthalmitis or vision loss. Thus, ophthalmic formulations should be preservative free. Further, concentration of surfactants in ophthalmic preparations should be lower because the higher surfactant concentration can cause toxicity.

The different polymers approved by different regulatory bodies for ocular application can show altered biocompatibility upon NPs preparation. Thus, each polymer-based nanoparticulate carrier must be screened for *in vitro* cell toxicity using appropriate cell types as a prerequisite for *in vivo* studies. In addition, vitreous clouding (was reported with liposomes and dendrimers) and foreign-body response were observed following NP administration via intraocular and periocular routes. Besides, the nanoparticle aggregation remains a chief challenge. As a result of these challenges, only a few of them have entered the clinical trials.

The USFDA has included NPs as complex products and not being of biological origin. The USFDA evaluates new nanoparticulate products via NDA pathway and their generic products under ANDA pathway. Recently, USFDA permitted Aura Biosciences (applied for approval through NDA pathway) to start a clinical trial in patients with ocular melanoma for viral nanoparticle conjugates. These complex NPs are designed to selectively target the melanoma cells following intravitreal injection [141].

8.8 CONCLUSIONS

In the current chapter, several approaches utilised to improve the ocular bioavailability via increasing permeability and/or residence time of therapeutics were briefed. Although these approaches were described along with the proof of concepts, only a few case studies are found with regard to hydrophilic therapeutics and ophthalmic applications. Thus, it is to be noted that most of the approaches explained in the current chapter are still in the infancy stage and thus offer newer opportunities for pharmaceutical scientists to exploit them for ocular delivery of hydrophilic small molecules and biological drugs.

ACKNOWLEDGEMENTS

We would like to acknowledge the head of institution and management at Tatyasaheb Kore College of Pharmacy, Warananagar, for supporting the writing of this chapter.

REFERENCES

1. Varma R, Vajaranant TS, Burkemper B, Wu S, Torres M, Hsu C, Choudhury F, McKean-Cowdin R. 2016. Visual impairment and blindness in adults in the United States: Demographic and geographic variations from 2015 to 2050. *JAMA ophthalmology*. 134:802–809.
2. Truong-Le V, Lovalenti PM, Abdul-Fattah AM. 2015. Stabilization challenges and formulation strategies associated with oral biologic drug delivery systems. *Advanced Drug Delivery Reviews*. 93:95–108.
3. Muheem A, Shakeel F, Jahangir MA, Anwar M, Mallick N, Jain GK, Warsi MH, Ahmad FJ. 2016. A review on the strategies for oral delivery of proteins and peptides and their clinical perspectives. *Saudi Pharmaceutical Journal*. 24:413–428.
4. Kwatra D, Mitra AK. 2013. Drug delivery in ocular diseases: Barriers and strategies. *World Journal of Pharmacology*. 2(4):78–83.
5. Patel A, Cholkar K, Agrahari V, Mitra AK. 2013. Ocular drug delivery systems: An overview. *World Journal of Pharmacology*. 2(2):47–64.
6. Bucolo C, Drago F, Salomone S. 2012. Ocular drug delivery: A clue from nanotechnology. *Frontiers in Pharmacology, Experimental Pharmacology and Drug Discovery*. 3(188):1–3.
7. Gaudana R, Ananthula HK, Parenky A, Mitra AK. 2010. Ocular drug delivery. *The AAPS Journal*. 12(3):348–360.
8. Barar J, Javadzadeh AR, Omidi Y. 2008. Ocular novel drug delivery: Impacts of membranes and barriers. *Expert Opinion on Drug Delivery*. 5(5):567–581.
9. Cholkar K, Patel SP, Vadlapudi AD, Mitra AK. 2013. Novel strategies for anterior segment ocular drug delivery. *Journal of Ocular Pharmacology and Therapeutics*. 29(2):106–123.
10. Yameena B, Choia WI, Vilosa C, Swamia A, Shia J, Farokhzad OC. 2014. Insight into nanoparticle cellular uptake and intracellular targeting. *Journal of Controlled Release*. 190:485–499.
11. Barot M, Bagui M, Gokulgandhi MR, Mitra AK. 2012. Prodrug strategies in ocular drug delivery. *Medicinal Chemistry*. 8(4):753–768.
12. Chien DS, Tang-Liu DD, Woodward DF. 1997. Ocular penetration and bioconversion of prostaglandin F2alpha prodrugs in rabbit cornea and conjunctiva. *Journal of Pharmaceutical Sciences*. 86:1180–1186.
13. Anand BS, Mitra AK. 2002. Mechanism of corneal permeation of L-valyl ester of acyclovir: Targeting the oligopeptide transporter on the rabbit cornea. *Pharmaceutical Research*. 19:1194–1202.
14. Patel PB, Shastri DH, Shelat PK, Shukla AK. 2010. Ophthalmic drug delivery system: Challenges and approaches. *Systematic Reviews in Pharmacy* 1(2):113–120.
15. Roman VM, Peter WJM, Fraser S, Vitaliy VK. 2019. Penetration enhancers in ocular drug delivery. *Pharmaceutics*. 11(7):321.
16. Harooni M, Freilich JM, Abelson M, Refojo M. 1998. Efficacy of hyaluronidase in reducing increases in intraocular pressure related to the use of viscoelastic substances. *Archives of Ophthalmology*. 116:1218–1221.
17. Stern R, Jedrzejas MJ. 2006. Hyaluronidases: Their genomics, structures, and mechanisms of action. *Chemical Reviews*. 106:818–839.
18. Kozak I, Kayikcioglu OR, Cheng L, Falkenstein I, Silva GA, Yu DX, Freeman WR. 2006. The effect of recombinant human hyaluronidase on dexamethasone penetration into the posterior segment of the eye after sub-Tenon's injection. *Journal of Ocular Pharmacology and Therapeutics*. 22:362–369.
19. Schymkowitz J, Rousseau F. 2016. Protein aggregation: A rescue by chaperones. *Nature Chemical Biology*. 12:58–59.
20. Sauer T, Patel M, Chan CC, Tuo J. 2008. Unfolding the therapeutic potential of chemical chaperones for age-related macular degeneration. *Expert Review of Ophthalmology*. 3:29–42.
21. Almeida H, Amaral MH, Lobão P, Silva AC, Loboa JM. 2014. Applications of polymeric and lipid nanoparticles in ophthalmic pharmaceutical formulations: Present and future considerations. *Journal of Pharmacy and Pharmaceutical Sciences*. 17(3):278–293.
22. Weng Y, Liu J, Jin S, Guo W, Liang X, Hu Z. 2017. Nanotechnology-based strategies for treatment of ocular disease. *Acta Pharmaceutica Sinica B*. 7(3):281–291.
23. Goel M, Picciani RG, Lee RK, Bhattacharya SK. 2010. Aqueous humor dynamics: A review. *Open Ophthalmology Journal*. 3(4):52–59.
24. Gadek T, Lee D. 2011. Topical drug delivery to the back of the eye. In: Kompella U, Edelhauser H, editor. *Drug Product Development for the Back of the Eye*. New York: Springer US Press. 2:449–468.
25. Kumar R, Sinha VR. 2014. Preparation and optimization of voriconazole microemulsion for ocular delivery. *Colloids and Surfaces B: Biointerfaces*. 117(Supplement C):82–88.
26. Kim YC, Chiang B, Wu X, Prausnitz MR. 2014. Ocular delivery of macromolecules. *Journal of Controlled Release*. 28(190):172–181.
27. Hachicha W, Kodjikian L, Fessi H. 2006. Preparation of vancomycin microparticles: Importance of preparation parameters. *International Journal of Pharmaceutics*. 324(2):176–184.
28. Varshochian R, Jeddi-Tehrani M, Mahmoudi AR, Khoshayand MR, Atyabi F, Sabzevari A, Esfahani MR, Dinarvand R. 2013. The protective effect of albumin on bevacizumab activity and stability in PLGA nanoparticles intended for retinal and choroidal neovascularization treatments. *European Journal of Pharmaceutical Sciences*. 50:341–352.
29. Yavuz B, Bozdağ Pehlivan S, Kaffashi A, Çalamak S, Ulubayram K, Palaska E, Çakmak HB, Ünlü N. 2016. *In Vivo* tissue distribution and efficacy studies for cyclosporin A loaded nano-decorated subconjunctival implants. *Drug Delivery*. 23(9):3279–3284.
30. Mandal A, Cholkar K, Khurana V, Shah A, Agrahari V, Bisht R, Pal D, Mitra AK. 2017. Topical formulation of self-assembled antiviral prodrug nanomicelles for targeted retinal delivery. *Molecular Pharmaceutics*. 14(6):2056–2069.
31. Kaur IP, Kakkar S. 2014. Nanotherapy for posterior eye diseases. *Journal of Controlled Release*. 10(193):100–112.
32. Iriyama A, Oba M, Ishii T, Nishiyama N, Kataoka K, Tamaki Y, Yanagi Y. 2011. Gene transfer using micellar nanovectors inhibits choroidal neovascularization in vivo. *PloS One* 6(12):e28560.
33. Vaishya RD, Khurana V, Patel S, Mitra AK. 2014. Controlled ocular drug delivery with nanomicelles. *Wiley Interdisciplinary Reviews: Nanomedicine and Nanobiotechnology*. 6:422–437.
34. Pileni MP. 2003. The role of soft colloidal templates in controlling the size and shape of inorganic nanocrystals. *Nature Materials*. 2(3):145–150.

35. Jain SK, Chandra R, Rai AK. 2008. Characterization of ocular delivery of reverse micelles bearing insulin. *Research Journal of Pharmacy and Technology.* 1(4):370–373.
36. Agrahari V, Agrahari V, Mitra AK. 2016. Nanocarrier fabrication and macromolecule drug delivery: Challenges and opportunities. *Therapeutic Delivery.* 7(4):257–278.
37. Akbarzadeh A, Rezaei-Sadabady R, Davaran S, Joo SW, Zarghami N, Hanifehpour Y, Samiei M, Kouhi M, Nejati-Koshki K. 2013. Liposome: Classification, preparation, and applications. *Nanoscale Research Letters.* 8(1):102.
38. Davis BM, Normando EM, Guo L, Turner LA, Nizari S, O'Shea P, Moss SE, Somavarapu S, Cordeiro MF. 2014. Topical delivery of Avastin to the posterior segment of the eye *in vivo* using annexin A5-associated liposomes. *Small.* 10(8):1575–1584.
39. Gan L, Wang J, Jiang M, Bartlett H, Ouyang D, Eperjesi F, Liu J, Gan Y. 2013. Recent advances in topical ophthalmic drug delivery with lipid-based nanocarriers. *Drug Discovery Today.* 18:290–297.
40. Ako-Adounvo AM, Nagarwal RC, Oliveira L, Boddu SHS, Wang XS, Dey S, Karla KP. 2014. Recent patents on ophthalmic nanoformulations and therapeutic implications. *Recent Patents on Drug Delivery & Formulation.* 8:193–201.
41. Abhirup M, Dhananjay P, Vibhuti A, Hoang T, Mary J, Ashim KM. 2018. Ocular delivery of proteins and peptides: Challenges and novel formulation approaches. *Advanced Drug Delivery Reviews.* 126:67–95.
42. Bruno BJ, Miller GD, Lim CS. 2013. Basics and recent advances in peptide and protein drug delivery. *Therapeutic Delivery.* 4:1443–1467.
43. Eldeeb AE, Salah S, Ghorab M. 2019. Formulation and evaluation of cubosomes drug delivery system for treatment of glaucoma: *Ex-vivo* permeation and *in-vivo* pharmacodynamic study. *Journal of Drug Delivery Science and Technology.* 52:236–247.
44. Kaur IP, Singh M, Yadav M, Sandhu SK, Deol PK, Sharma G. 2015. Potential of nanomaterials as movers and packers for drug molecules. *Solid State Phenomena.* 222:159–178.
45. Sahay G, Alakhova DY, Kabanov AV. 2010. Endocytosis of nanomedicines. *Journal of Controlled Release.* 145(3):182–195.
46. Zarbin MA, Montemagno C, Leary JF, Ritch R. 2010. Nanotechnology in ophthalmology. *Canadian Journal of Ophthalmology.* 45:457–476.
47. Rakesh K, Sinha VR. 2016. Lipid nanocarrier: An efficient approach towards ocular delivery of hydrophilic drug (Valacyclovir). *AAPS PharmSciTech.* 18(3):884–894.
48. Ahmed K, Mohammad A, Mohamed F, Ahmed O. 2019. Natamycin solid lipid nanoparticles-sustained ocular delivery system of higher corneal penetration against deep fungal keratitis: Preparation and optimization. *International Journal of Nanomedicine.* 14: 2515–2531.
49. Stasko NA, Johnson CB, Schoenfisch MH, Johnson TA, Holmuhamedov EL. 2007. Cytotoxicity of polypropylenimine dendrimer conjugates on cultured endothelial cells. *Biomacromolecules.* 8(12):3853–3859.
50. Marano RJ, Toth I, Wimmer N, Brankov M, Rakoczy PE. 2005. Dendrimer delivery of an anti-VEGF oligonucleotide into the eye: A long-term study into inhibition of laser-induced CNV, distribution, uptake and toxicity. *Gene Therapy.* 12(21):1544–1550.
51. Tuomela A, Liu P, Puranen J, Rönkkö S, Laaksonen T, Kalesnykas G, Oksala O, Ilkka J, Laru J, Järvinen K, Hirvonen J. 2014. Brinzolamide nanocrystals formulations for ophthalmic delivery: Reduction of elevated intraocular pressure *in vivo*. *International Journal of Pharmaceutics.* 467(1–2):34–41.
52. Mukherjee S, Ray S, Thakur RS. Solid lipid nanoparticles: A modern formulation approach in drug delivery system. *Indian Journal of Pharmaceutical Sciences.* 71(4):349–358.
53. Battaglia L, Gallarate M. 2012. Lipid nanoparticles: State of the art, new preparation methods and challenges in drug delivery. *Expert Opinion on Drug Delivery.* 9(5): 497–508.
54. Bhandari R, Kaur IP. 2013. A method to prepare solid lipid nanoparticles with improved entrapment efficiency of hydrophilic drugs. *Current Nanoscience.* 9(2):211–220.
55. Kalam MA. 2016. Development of chitosan nanoparticles coated with hyaluronic acid for topical ocular delivery of dexamethasone. *International Journal of Biological Macromolecules.* 89:127–136.
56. Li F, Snow-Davis C, Du C, Bondarev ML, Saulsbury MD, Heyliger SO. 2016. Preparation and characterization of lipophilic doxorubicin pro-drug micelles. *Journal of Visualized Experiments.* 114:54338.
57. Sharma P, Dube B, Sawant K. 2012. Synthesis of cytarabine lipid drug conjugate for treatment of meningeal leukemia: Development, characterization and *in vitro* cell line studies. *Journal of Biomedical Nanotechnology.* 8:928–937.
58. Yu BT, Sun X, Zhang ZR. 2003. Enhanced liver targeting by synthesis of N1-stearyl-5-Fu and incorporation into solid lipid nanoparticles. *Archives of Pharmacal Research.* 26:1096–1101.
59. Olbrich C, Gessner A, Kayser O, Müller RH. 2002. Lipid-drug-conjugate (LDC) nanoparticles as novel carrier system for the hydrophilic antitrypanosomal drug diminazenediaceturate. *Journal of Drug Targeting.* 10:387–396.
60. Gupta A, Asthana S, Konwar R, Chourasia MK. 2013. An insight into potential of nanoparticles-assisted chemotherapy of cancer using gemcitabine and its fatty acid prodrug: A comparative study. *Journal of Biomedical Nanotechnology.* 9:915–925.
61. Luigi B, Loredana S, Federica F, Elisabetta M, Marina G, Ana DP, Maria AS. 2016. Application of lipid nanoparticles to ocular drug delivery. *Expert Opinion on Drug Delivery.* 13(12):1743–1757.
62. Sutradhar KL, Amin L. 2013. Nanoemulsions: Increasing possibilities in drug delivery. *European Journal of Nanomedicine.* 5(2):97–110.
63. Suri R, Beg S, Kohli K. 2010. Target strategies for drug delivery bypassing ocular barriers. *Journal of Drug Delivery Science and Technology.* 55:101389.
64. Abhirup M, Dhananjay P, Vibhuti A, Hoang MT, Mary J, Ashim KM. 2018. Ocular delivery of proteins and peptides: Challenges and novel formulation approaches. *Advanced Drug Delivery Reviews.* 126:67–95.
65. Kishore C, Ashaben P, Aswani DV, Ashim KM. 2012. Novel nanomicellar formulation approaches for anterior and posterior segment ocular drug delivery. *Recent Patents on Nanomedicine.* 2(2):82–95.
66. Sharma A, Pahwa S, Bhati S and Kudeshia P. 2020. Spanlastics: A modern approach for nanovesicular drug delivery system. *International Journal of Pharmaceutical Sciences and Research.* 11(3):1057–1065.

67. Farghaly DA, Aboelwafa AA, Hamza MY, Mohamed MI. 2017. Topical delivery of fenoprofen calcium via elastic nano-vesicular spanlastics: Optimization using experimental design and in vivo evaluation. *AAPS PharmSciTech*. 18(8):2898–2909.
68. Sherif AG, Omar H, Hamdy A. 2020. Cubosomes: Composition, preparation, and drug delivery applications. *Journal of Advanced Biomedical and Pharmaceutical Sciences*. 3:1–9.
69. Hong Z, Jonathan PW, Vicki C. Terence H. 2015. Cubosomes nanoparticles for ocular delivery. *Investigative Ophthalmology & Visual Science*. 56:5033.
70. Yavuz B, Bozdağ S, Ünlü N. 2013. Dendrimeric systems and their applications in ocular drug delivery. *Scientific World Journal*. 2013:732340.
71. Sharma OP, Patel V, Mehta T. 2016. Nanocrystal for ocular drug delivery: Hope or hype. *Drug Delivery and Translational Research*. doi:10.1007/s13346-016-0292-0.
72. Chhikara BS, Mandal D, Parang K. 2012. Synthesis, anticancer activities, and cellular uptake studies of lipophilic derivatives of doxorubicin succinate. *Journal of Medicinal Chemistry*. 55:1500–1510.
73. Dosio F, Reddy LH, Ferrero A, Stella B, Cattel L, Couvreur P. 2010. Novel nanoassemblies composed of squalenoyl-paclitaxel derivatives: Synthesis, characterization, and biological evaluation. *Bioconjugate Chemistry*. 21:1349–1361.
74. Duhem N, Danhier F, Pourcelle V, Schumers JM, Bertrand O, Leduff CS, Hoeppener S, Schubert US, Gohy JF, Marchand-Brynaert J, Preat V. 2014. Self-assembling doxorubicin-tocopherol succinate prodrug as a new drug delivery system: Synthesis, characterization, and *in vitro* and *in vivo* anticancer activity. *Bioconjugate Chemistry*. 25:72–81.
75. Wang Y, Li L, Jiang W, Yang Z, Zhang Z. 2006. Synthesis and preliminary antitumor activity evaluation of a DHA and doxorubicin conjugate. *Bioorganic & Medicinal Chemistry Letters*. 16:2974–2977.
76. Chen Q, Butler D, Querbes W, Pandey RK, Ge P, Maier MA, Zhang L, Rajeev KG, Nechev L, Kotelianski V, Manoharan M, Sah DW. 2010. Lipophilic siRNAs mediate efficient gene silencing in oligodendrocytes with direct CNS delivery. *Journal of Controlled Release* 144:227–232.
77. Nishina K, Unno T, Uno Y, Kubodera T, Kanouchi T, Mizusawa H, Yokota T. 2008. Efficient in vivo delivery of siRNA to the liver by conjugation of alpha-tocopherol. *Molecular Therapy*. 16:734–740.
78. Osborn MF, Khvorova A 2018. Improving siRNA delivery *in vivo* through lipid conjugation. *Nucleic Acid Therapeutics*. 28(3):128–136.
79. Qingmin C, David B, William Q, Rajendra K, Pei G, Martin AM, Ligang Z, Kallanthottathil GR, Manoharan M. 2010. Lipophilic siRNAs mediate efficient gene silencing in oligodendrocytes with direct CNS delivery. *Journal of Controlled Release* 144:227–232.
80. Xiong MP, Yanez JA, Remsberg CM, Ohgami Y, Kwon GS, Davies NM, Forrest ML. 2008. Formulation of a geldanamycin prodrug in mPEG-b-PCL micelles greatly enhances tolerability and pharmacokinetics in rats. *Journal of Controlled Release*. 129:33–40.
81. Daman Z, Ostad S, Amini M, Gilani K. 2014. Preparation, optimization and in vitro characterization of stearoyl-gemcitabine polymeric micelles: A comparison with its self-assembled nanoparticles. *International Journal of Pharmaceutics*. 468:142–151.
82. Goldstein D, Gofrit O, Nyska A, Benita S. 2007. Anti-HER2 cationic immunoemulsion as a potential targeted drug delivery system for the treatment of prostate cancer. *Cancer Research*. 67:269–275.
83. Sarett SM, Kilchrist KV, Miteva M, Duvall CL. 2015. Conjugation of palmitic acid improves potency and longevity of siRNA delivered via endosomolytic polymer nanoparticles. *Journal of Biomedical Materials Research Part A*. 103:3107–3116.
84. Lundberg BB. 2011. Preparation and characterization of polymeric pH-sensitive STEALTH® nanoparticles for tumor delivery of a lipophilic prodrug of paclitaxel. *International Journal of Pharmaceutics*. 408:208–212.
85. Peira E, Chirio D, Battaglia L, Barge A, Chegaev K, Gigliotti CL, Ferrara B, Dianzani C, Gallarate M. 2016. Solid lipid nanoparticles carrying lipophilic derivatives of doxorubicin: Preparation, characterization, and in vitro cytotoxicity studies. *Journal of Microencapsulation*. 33:381–390.
86. Jiao YY, Wang XQ, Lu WL, Yang ZJ, Zhang Q. 2013. A novel approach to improve the pharmacokinetic properties of 8-chloro-adenosine by the dual combination of lipophilic derivatisation and liposome formulation. *European Journal of Pharmaceutical Sciences*. 48:249–258.
87. Arouri A, Mouritsen OG. 2011. Anticancer double lipid prodrugs: Liposomal preparation and characterization. *Journal of Liposome Research*. 21:296–305.
88. Kuznetsova NR, Svirshchevskaya EV, Skripnik IV, Zarudnaya EN, Benke AN, Gaenko GP, Molotkovskii YG, Vodovozova EL. 2013. Interaction of liposomes bearing a lipophilic doxorubicin prodrug with tumor cells. *Biochemistry (Moscow) Supplement Series A: Membrane and Cell Biology*. 7:12–20.
89. Pili B, Reddy LH, Bourgaux C, Lepetre-Mouelhi S, Desmaele D, Couvreur P. 2010. Liposomal squalenoyl-gemcitabine: Formulation, characterization and anticancer activity evaluation. *Nanoscale*. 2:1521–1526.
90. Tros de IC, Sun Y, Düzgüneş N. 2010. Gene delivery by lipoplexes and polyplexes. *European Journal of Pharmaceutical Sciences*. 40:159–170.
91. Niculescu-Duvaz D, Heyes J, Springer CJ. 2003. Structure-activity relationship in cationic lipid mediated gene transfection. *Current Medicinal Chemistry*. 10:1233–1261.
92. Xiao-Xiang Z, Thomas JM, Mark WG. 2012. Functional lipids and lipoplexes for improved gene delivery. *Biochimie*. 94(1):42–58.
93. Liu H, Liu Y, Ma Z, Wang J, Zhang Q. 2011. A Lipid nanoparticle system improves siRNA efficacy in RPE cells and a laser-induced murine CNV model. *Investigative Ophthalmology & Visual Science*. 52:4789–4794.
94. Layman JM, Ramirez SM, Green MD, Long TE. 2009. Influence of polycation molecular weight on poly (2-dimethylaminoethyl methacrylate)-mediated DNA delivery *in vitro*. *Biomacromolecules*. 10:1244–1252.
95. Kim B, Tang Q, Biswas P, Xu J, Schiffelers R, Xie F, Ansari A, Scaria P, Woodle M, Lu P, Rouse B. 2004. Inhibition of ocular angiogenesis by siRNA targeting vascular endothelial growth factor pathway genes: Therapeutic strategy for herpetic stromal keratitis. *The American Journal of Pathology*. 165:2177–2185.
96. Jayabalan N, Gaurav KJ, Vaidehi G, Musarrat HW, Farhan JA, Roop KK. 2014. Development and characterization of nanoplex topical ocular delivery of moxifloxacin to treat bacterial keratitis. *Investigative Ophthalmology & Visual Science*. 55(13):1454.

97. Amjad MW, Kesharwani P, Amin MC, Iyer AK. 2017. Recent advances in the design, development, and targeting mechanisms of polymeric micelles for delivery of siRNA in cancer therapy. *Progress in Polymer Science*. 64:54–181.
98. Salzano G, Riehle R, Navarro G, Perche F, De Rosa G, Torchilin VP. 2014. Polymeric micelles containing reversibly phospholipid-modified antisurvivin siRNA: A promising strategy to overcome drug resistance in cancer. *Cancer Letters*. 343(2):224–231.
99. Ye H, Qian Y, Lin M, Duan Y, Sun X, Zhuo Y, Ge J. 2010. Cationic nano-copolymers mediated IKKβ targeting siRNA to modulate wound healing in a monkey model of glaucoma filtration surgery. *Molecular Vision*. 16:2502–2510.
100. Zhang W, Li X, Ye T, Chen F, Sun X, Kong J, Yang X, Pan W, Li S. 2013. Design, characterization, and *in vitro* cellular inhibition and uptake of optimized genistein-loaded NLC for the prevention of posterior capsular opacification using response surface methodology. *International Journal of Pharmaceutics*. 454:354–366.
101. Schoenwald RD. 1993. Pharmacokinetics in ocular delivery. In: Edman P, editor. *Biopharmaceutics of Ocular Drug Delivery*. Boca Raton, FL: CRC Press. 159–191.
102. Dukovski B, Juretić M, Bračko D, Randjelović D, Diebold Y, Filipović-Grčić J, Pepić I, Lovrić J. 2020. Functional ibuprofen-loaded cationic nanoemulsion: Development and optimization for dry eye disease treatment. *International Journal of Pharmaceutics*. 576:118979.
103. Yu A, Shi H, Liu H, Bao Z, Dai M, Lin D, Lin D, Xu X, Li X, Wang Y. 2020. Mucoadhesive dexamethasone-glycol chitosan nanoparticles for ophthalmic drug delivery. *International Journal of Pharmaceutics*. 575:118943.
104. Terreni E, Chetoni P, Tampucci S, Burgalassi S, Al-Kinani AA, Alany RG, Monti D. 2020. Assembling surfactants-mucoadhesive polymer nanomicelles (ASMP-nano) for ocular delivery of cyclosporine-A. *Pharmaceutics*. 12(3):253.
105. Lin J, Wu H, Wang Y, Lin J, Chen Q, Zhu X. 2016. Preparation and ocular pharmacokinetics of hyaluronan acid-modified mucoadhesive liposomes. *Drug Delivery*. 23(4):1144–1151.
106. Pai RV, Vavia PR. 2020. Chitosan oligosaccharide enhances binding of nanostructured lipid carriers to ocular mucins: Effect on ocular disposition. *International Journal of Pharmaceutics*. 577:119095.
107. Bravo-Osuna I, Vicario-de-la-Torre M, Andrés-Guerrero V, Sánchez-Nieves J, Guzmán-Navarro M, De La Mata FJ, Gómez R, de las Heras B, Argueso P, Ponchel G, Herrero-Vanrell R. 2016. Novel water-soluble mucoadhesive carbosilane dendrimers for ocular administration. *Molecular Pharmaceutics*. 13(9):2966–2976.
108. Üstündağ O, Yozgatl N, Okur V, Yoltaş M. 2019. Improving therapeutic efficacy of voriconazole against fungal keratitis: Thermosensitive in situ gels as ophthalmic drug carriers. *Journal of Drug Delivery Science and Technology*. 49:323–333.
109. Upadhyay SU, Chavan SK, Gajjar DU, Upadhyay UM, Patel JK. 2020. Nanoparticles laden In situ gel for sustained drug release 1 after topical ocular administration. *Journal of Drug Delivery Science and Technology*. 57:101736.
110. Yu S, Wang QM, Wang X, Liu D, Zhang W, Ye T, Pan W. 2015. Liposome incorporated ion sensitive in situ gels for ophthalmic delivery of timolol maleate. *International Journal of Pharmaceutics*. 480(1–2):128–136.
111. Bae KH, Kurisawa M. 2016. Emerging hydrogel designs for controlled protein delivery. *Biomaterials Science*. 4:1184–1192.
112. Famili A, Rajagopal K. 2017. Bio-orthogonal cross-linking chemistry enables in situ protein encapsulation and provides sustained release from hyaluronic acid based hydrogels. *Molecular Pharmaceutics*. 14:1961–1968.
113. Ferreira NN, Ferreira LM, Cardoso VM, Boni FI, Souza AL, Gremião MP. 2018. Recent advances in smart hydrogels for biomedical applications: From self-assembly to functional approaches. *European Polymer Journal*. 99:117–133.
114. Liu, W, Griffith M, Fengfu LI. 2008. Alginate microsphere-collagen composite hydrogel for ocular drug delivery and implantation. *Journal of Materials Science: Materials in Medicine*. 19:3365–3371.
115. Osswald CR, Kang-Mieler JJ. 2016. Controlled and extended *in vitro* release of bioactive anti-vascular endothelial growth factors from a microsphere-hydrogel drug delivery system. *Current Eye Research*. 41:1216–1222.
116. Hui-Hui Z, Qiu-Hua L, Zhi-Jun Y, Wei-San P, Shu-Fang N. 2011. Novel ophthalmic timolol maleate liposomal-hydrogel and its improved local glaucomatous therapeutic effect *in vivo*. *Drug Delivery*. 18(7):502–510.
117. Gupta H, Aqil M. 2012. Contact lenses in ocular therapeutics. *Drug Discovery Today*. 17:522–27.
118. Zhang J, Robert B, William H, Pei Y, Wai HT. 2013. A nanocomposite contact lens for the delivery of hydrophilic protein drugs. *Journal of Materials Chemistry B*. 1:4388–4395.
119. Jung HJ, Chauhan A. 2012. Temperature sensitive contact lenses for triggered ophthalmic drug delivery. *Biomaterials*. 33: 2289–2300.
120. Maulvi FA, Lakdawala DH, Shaikh AA, Desai AR, Choksi HH, Vaidya RJ, Ranch KM, Koli AR, Vyas BA, Shah DO.. 2016. *In vitro* and *in vivo* evaluation of novel implantation technology in hydrogel contact lenses for controlled drug delivery. *Journal of Controlled Release*. 28(226):47–56.
121. Xu J, Ge Y, Bu R, Zhang A, Feng S, Wang J, Gou J, Yin T, He H, Zhang Y, Tang X. 2019. Co-delivery of latanoprost and timolol from micelles-laden contact lenses for the treatment of glaucoma. *Journal of Controlled Release*. 305:18–28.
122. Sahane NK, Banarjee SK, Gaikwad DD, Jadhav SL, Thorat RM. 2010. Ocular inserts : A Review. *Drug Invent Today*. 2:57–64.
123. Hyun JJ, Anuj C. 2013. Extended release of timolol from nanoparticle-loaded fornix insert for glaucoma therapy. *Journal of Ocular Pharmacology and Therapeutics*. 29(2):229–235.
124. Yuan X, Marcano DC, Shin CS, Hua X, Isenhart LC, Pflugfelder SC, Acharya G. 2015. Ocular drug delivery nanowafer with enhanced therapeutic efficacy. *ACS Nano*. 9:1749–1758.
125. Kim J, Schlesinger EB, Desai TA. 2015. Nanostructured materials for ocular delivery: Nanodesign for enhanced bioadhesion, transepithelial permeability and sustained delivery. *Therapeutic Delivery*. 6:1365–1376.
126. Bian F, Shin CS, Wang C, Pflugfelder SC, Acharya G, De Paiva CS. 2016. Dexamethasone drug eluting nanowafers control inflammation in alkali-burned corneas associated with dry eye. *Investigative Ophthalmology & Visual Science*. 57:3222–3230.

127. Klausner EA, Peer D, Chapman RL, Multack RF, Andurkar SV. 2007. Corneal gene therapy. *Journal of Controlled Release*. 124:107–133.
128. Amadio M, Govoni S, Pascale A: Targeting VEGF in eye neovascularization: What's new?: A comprehensive review on current therapies and oligonucleotide-based interventions under development. 2016. *Pharmacological Research* 103:253–269.
129. Dejneka NS, Wan S, Bond OS, Kornbrust DJ, Reich SJ. 2008. Ocular biodistribution of bevasiranib following a single intravitreal injection to rabbit eyes. *Molecular Vision*. 14:997–1005.
130. Bourne RR, Flaxman SR, Braithwaite T, Cicinelli MV, Das A, Jonas JB, Keeffe J, Kempen JH, Leasher J, Limburg H, Naidoo K. 2017. Magnitude, temporal trends, and projections of the global prevalence of blindness and distance and near vision impairment: A systematic review and meta-analysis. *The Lancet Global Health* 5:e888–e897.
131. Basu S, Mohamed A, Chaurasia S, Sejpal K, Vemuganti GK, Sangwan VS. 2011. Clinical outcomes of penetrating keratoplasty after autologous cultivated limbal epithelial transplantation for ocular surface burns. *American Journal of Ophthalmology*. 152(6):917–924.
132. Yasin MN, Svirskis D, Seyfoddin A, Rupenthal ID. 2014. Implants for drug delivery to the posterior segment of the eye: A focus on stimuli-responsive and tunable release systems. *Journal of Controlled Release*. 28(196):208–221.
133. Tyagi P, Barros M, Stansbury JW, Kompella UB. 2013. Light-activated, in situ forming gel for sustained suprachoroidal delivery of bevacizumab. *Molecular Pharmaceutics*. 10(8):2858–2867.
134. Song HB, Wi JS, Jo DH, Kim JH, Lee SW. 2017. Intraocular application of gold nanodisks optically tuned for optical coherence tomography: Inhibitory effect on retinal neovascularization without unbearable toxicity. *Nanomedicine*. 13:1901–1911.
135. Giannaccini M, Pedicini L, De Matienzo G, Chiellini F, Dente L, Raffa V. 2017. Magnetic nanoparticles: A strategy to target the choroidal layer in the posterior segment of the eye. *Scientific Reports*. 7:43092. doi:10.1038/srep43092.
136. Miaomiao T, Xiaoyuan J, Hua X, Lu Z, Airui J, Bin S, Yuanyuan S, Yao H. 2018. Photostable and biocompatible fluorescent silicon nanoparticles-based theranostic probes for simultaneous imaging and treatment of ocular neovascularization. *Analytical Chemistry*. 90(13):8188–8195.
137. Krishnapriya S, Gopal A, Akshay S. 2020. Hyaluronic acid microneedles-laden collagen cryogel plugs for ocular drug delivery. *Journal of Applied Polymer Science*. 137:1–14.
138. Souied EH, Reid SN, Piri NI, Lerner LE, Nusinowitz S, Farber DB. 2008. Non-invasive gene transfer by iontophoresis for therapy of an inherited retinal degeneration. *Experimental Eye Research*. 87:168–175.
139. Cervia LD, Yuan F. 2018. Current progress in electrotransfection as a nonviral method for gene delivery. *Molecular Pharmaceutics*. 15:3617–3624.
140. Emerich DF, Orive G, Thanos C, Tornoe J, Wahlberg LU. 2014. Encapsulated cell therapy for neurodegenerative diseases: From promise to product. *Advanced Drug Delivery Reviews*. 67–68:131–141.
141. Suri R, Beg S, Kohli K. 2020. Target strategies for drug delivery bypassing ocular barriers. *Journal of Drug Delivery Science and Technology*. 55:1–17.

9 Metal Nanoparticles as a Surrogate Carrier in Drug Delivery and Diagnostics

Sushama Talegaonkar, Debopriya Dutta, and Namita Chaudhary
Delhi Pharmaceutical Sciences and Research University (DPSRU)

Surya Goel
ABESIT College of Pharmacy

Ruchi Singh
Rajkumar Goel Institute of Technology

CONTENTS

9.1	Introduction		140
	9.1.1	Advantages/Benefits of Metal Nanoparticles	141
	9.1.2	Limitations of Metal Nanoparticles	141
9.2	Types of Metal Nanoparticles		141
	9.2.1	Gold Nanoparticles (AuNPs)	141
		9.2.1.1 Synthesis of Gold Nanoparticles (AuNPs)	141
		9.2.1.2 Applications of Gold Nanoparticles (AuNPs)	142
	9.2.2	Silver Nanoparticles (AgNPs)	142
		9.2.2.1 Synthesis of Silver Nanoparticles (AgNPs)	144
		9.2.2.2 Applications of Silver Nanoparticles (AgNPs)	144
	9.2.3	Zinc Oxide Nanoparticles (ZnO NPs)	145
		9.2.3.1 Synthesis of Zinc Oxide Nanoparticles (ZnO NPs)	145
		9.2.3.2 Applications of Zinc Oxide Nanoparticles (ZnO NPs)	146
	9.2.4	Iron Oxide Nanoparticles (FeO NPs)	148
		9.2.4.1 Synthesis of Iron Oxide Nanoparticles (FeNPs)	148
		9.2.4.2 Biomedical Applications of Iron Oxide	149
	9.2.5	Zirconium Nanoparticles (ZrNPs)	150
		9.2.5.1 Synthesis of Zirconia Nanoparticles	151
		9.2.5.2 Applications of Zirconia Nanoparticles	151
9.3	Characterisation Techniques of Nanoparticles		152
	9.3.1	Energy-Dispersive X-Ray Spectroscopy (EDX)	152
	9.3.2	Fourier Transform Infrared Spectroscopy (FTIR)	152
	9.3.3	Scanning Electron Microscopy (SEM)	153
	9.3.4	Transmission Electron Microscope (TEM)	153
	9.3.5	Particle Size Analyser	153
	9.3.6	X-Ray Diffraction (XRD)	153
	9.3.7	Atomic Force Microscopy (AFM)	153
	9.3.8	Ultraviolet (UV) Spectroscopy	153
	9.3.9	Dynamic Light Scattering (DLS)	153
	9.3.10	Zeta Potential	154
	9.3.11	Nanoparticle Tracking Analysis (NTA)	154
9.4	Applications of Metallic Nanoparticles		154
	9.4.1	Metallic Nanoparticles in Cancer Therapy	154
	9.4.2	Nanoparticles for Delivering Peptides and Proteins	154
	9.4.3	Nanoparticles in Drug Discovery	154

DOI: 10.1201/9781003043164-9

9.4.4 Ocular Drug Delivery 154
9.4.5 Nanoparticles in Molecular Diagnostics/Molecular Imaging 155
9.4.6 Nanoparticles as Biosensors and Biolabels 155
9.4.7 Carriers for Nasal Vaccine/Drug Delivery 155
9.4.8 Nutraceutical Delivery 155
9.5 Green Technology 155
9.5.1 Principle 155
9.5.2 Mechanism 156
9.5.3 Advantages of Green Synthesis of Nanoparticles 156
9.5.4 Synthesis of Metallic Nanoparticles from Green Synthesis 156
9.6 Conclusion 157
References 157

9.1 INTRODUCTION

In the world of nanotechnology, metallic nanoparticles (NPs) are becoming a curious topic. NPs of metals help in curing chronic diseases. Many metals are very helpful for preventing diseases like cancer and severe infections caused by microbes and pathogens. Many metallic active agents such as titanium, gold, silver, zinc, iron, and other natural agents are utilised to prepare drug NPs. Nonmetallic forms are made of NPs that range between 1 and 100 nm. Now nonmetallic forms are available through biomedical science and biotechnology. Nanometallic formulations are found in many structures like NPs (zero-dimensional structures), nanowires (one-dimensional structures) and nanorods (two-dimensional structures). Faraday was the first to bring metallic NPs to the world in the year 1857. A study by the World Health Organization shows that metal NPs with reduced particle size can be more effective against bacteria (both gram negative and gram positive) and pathogens. Different types of metals exhibit different properties specific to metals. Gold, silver and platinum nanoparticles are considered to be the novel approaches for biomedical science. Gold nanoparticles (AuNPs) used as a detector with aminoglycoside antibiotics and their imaging activity helps in diagnosing the cancer. Silver nanoparticles (AgNPs) show excitation of plasmon at the highest efficiency to any wavelength in the visible spectrum. Metallic NPs are in the trend due to their unique and advanced features. Metals are used as catalysts and for the identification of microorganisms. The metal NPs are characterised by various sophisticated techniques such as ultraviolet–visible spectroscopy, electron microscopy, infrared spectroscopy and X-ray diffraction (XRD). They exist with the high-speed electrical properties, broad surface area, raised magnetic properties, high optical activity, good mechanical and thermal stability.

There is an increasing demand of metallic NP formulations in the market day by day. It is observed that the market of metallic NPs for the clinical applications got an increment from 15% in 2014 to 22% in 2019. Such nanoplatforms have enhanced the efficiency of well-known drugs. The various marketed products of metallic NPs are summarised in Table 9.1.

TABLE 9.1
Marketed Products of Metallic Nanoparticles

S. No.	Type of Metallic NPs	Product	Activity/Application	References
1.	Supramagnetic iron oxide nanoparticles (SPIONs)	Iron oxide nanoparticles (SPIONs)	Head and neck squamous cell carcinoma	Silva et al. (2019)
2.	Zirconium NPs (docetaxel)	CriPec®	For treating solid tumours	Silva et al. (2019)
3.	Superparamagnetic iron oxide NPs with dextran	Feraheme®	Anaemia in chronic kidney disease	Bullivant et al. (2013)
4.	PEG-coated silica–gold nanoshells	AuroLase	Metastatic lung tumours	Vlamidis and Voliani (2018)
5.	Gadolinium-based NPs	AGuIX	Brain metastases	Vlamidis and Voliani (2018)
6.	FeO NPs	Clariscan	Magnetic resonance imaging (MRI) contrast	Vlamidis and Voliani (2018)
7.	Hafnium oxide NPs	NBTXR3	Locally advanced squamous cell carcinoma	Vlamidis and Voliani (2018)
8.	Magnetic iron NPs	Magnablate	Prostate cancer	Vlamidis and Voliani (2018)
9.	Nanoparticles of superparamagnetic FeO coated with amino silane	NanoTherm®	Glioblastoma, prostate and pancreatic cancers	Weissig et al. (2014)
10.	Ferric oxide NPs	Ferinject®	Iron-deficiency anaemia in chronic renal disease	Lyseng-Williamson and Keating (2009)

This chapter broadly focus on the classification of metallic NPs along with advantages, limitations, methods of preparation, characterisation techniques as well as applications of metallic NPs.

9.1.1 Advantages/Benefits of Metallic Nanoparticles

- They have a greater surface area used for targeting the tumour cells in living organisms (Achachelouei et al. 2019; Li et al. 2007).
- They are useful in the determination of chemical information on the metallic nanoscale substrate.
- Metallic NPs possess large surface energies due to the presence of frontier orbital in quantum mechanics.
- They have better plasma absorption activity.
- The biological system works on imaging the moieties.
- Their low-coordination sites such as corners or edges possess a large number of dangling bonds and specific chemical properties. Due to this, they can store a large number of electrons.
- They effectually enhance the bioavailability of poorly soluble drugs.
- Better biocompatibility and biodegradability.
- Do not have noxious effects and are nonirritant.
- They reduce the adverse effects associated with drugs or other ordinary delivery systems like hypotension/hypertension, irritation and intolerance.
- Because of their size in nanometre, they can penetrate body tissue easily.
- Reduction in cost of whole therapy, due to a reduction in dose frequency.

9.1.2 Limitations of Metallic Nanoparticles

- Nanomaterials are thermodynamically unstable. They may be subjected to deterioration, corrosion and alteration in the structures (Kumar et al. 2018).
- Metallic NPs may be subjected to impurities due to their high reactivity. To overcome this situation, they may be fabricated in the form of encapsulation.
- Nanosubstances may demonstrate irritation, carcinogenicity and toxicity as they are transparent to cell dermis.
- Drug loading capacity is poor.
- High production cost.
- NPs have a low ability to entrap water-soluble drugs because of the partitioning effect.
- Difficulty in handling due to their small size.

9.2 TYPES OF METAL NANOPARTICLES

There are various metals such as gold, silver, zinc, iron and zirconium which can be used for the preparation of metallic drug NPs for the treatment of several kinds of diseases and ailments. Such metallic NPs have been discussed in detail in the chapter.

9.2.1 Gold Nanoparticles (AuNPs)

The gold colloidal suspension (AuNPs) observed with the range of particles less than 100nm along with the two colours intense red colour and fade yellowish colour for large-sized particles (Mody et al. 2010). In light, due to the presence of electromagnetic fields, oscillating motions caused; hence, it results in resonance at the frequency of light, this phenomenon known as the localised surface plasmon resonance (LSPR). The LSPR absorbs the scattered light and converts it into heat. The recent signs of progress in the field of nanotechnology have revealed that AuNPs have the potential to serve as the building blocks for plasmonic devices and future photonic (Badrzadeh et al. 2014; Wang et al. 2005). AuNPs have received attention as biomolecular conjugates (Yadi et al. 2018). AuNPs have been extensively used in the past decade for biomedical applications such as surface-enhanced Raman spectroscopy (SERS), optical sensors, fluorescence and sensor chips (Prodan et al. 2003; Huang and Baumberg 2010). These technologies are based on the principle of elastic scattering properties and the plasmon shift of metal NPs (Huang and Baumberg 2010).

Nuclear targeting of cancerous cells can be utilised as an efficient way to treat cancer. It was found that in situ aggregations of near-infrared (NIR) absorbing plasmonic AuNPs took place at the nuclear region of the cells (Panikkanvalappil et al. 2017) that make plasmonic AuNPs as a suitable candidate for NIR photo-absorber for plasmonic-based photothermal therapy in cancer (Achachelouei et al. 2019). Recently, a researcher found taurine as a novel capping agent in the formation of gold nanospheres with the help of wet chemical method. Taurine-capped nanospheres were evaluated in the visible wavelength at 515 and 545nm and had average diameters of 6.9nm. All the physical parameters were evaluated by using optical spectroscopy, dynamic light scattering (DLS), zeta potential and electron microscopy. Fourier transform infrared spectroscopy (FTIR) and Raman techniques were used to study the interaction between the surfaces of taurine and gold nanospheres. In cytotoxicity studies, it was observed that the taurine stabilised the gold nanospheres and nontoxic to liver carcinoma cells. It was supposed that taurine can be used as a novel capping agent that may be more superior to other toxic capping agents for metal AuNPs. The capped taurine-gold nanospheres have the potential to serve for drug delivery and nanomedicine applications (Kumar, Das et al. 2020).

9.2.1.1 Synthesis of Gold Nanoparticles (AuNPs)

In recent years, there has been comprehensive and considerable utilisation of AuNPs in applied sciences. It has also been investigated with a diversity of properties among which the curative property for various diseases gaining much attention. Besides this, AuNPs generate considerable interest in terms of diagnostic agents, antiviral agents, remedial agents, catalytic agents and drug administration.

Traditionally, there are three types of methods available for the synthesis of AuNPs depending upon the accessibility of existing properties. These methods can be recognised as chemical, physical and biological methods (Elahi et al. 2018).

Some preliminary work was performed by chemical reduction, Turkevich, cluster beam, electrochemical, thermal, Brust–Schiffrin and solvent evaporation method (Herizchi et al. 2016; Zarschler et al. 2016).

9.2.1.1.1 Chemical Reduction Method

It is a classical approach to obtain AuNPs by chemical reduction method, among which ascorbic acid and sodium borohydride are the most commonly used reducing agents. The stabilising agent also plays an important role to prevent the accumulation of the NPs. Surfactants and various elemental ligands are the common models. Nanotubes and nanorods are also produced through chemical reduction method.

9.2.1.1.2 Turkevich Method

It is one of the standard techniques used for the synthesis of AuNPs. In the presence of the aqueous medium, gold hydrochlorate solution ($HAuCl_4$) is reduced by sodium citrate. This technique was first discovered by Turkevich in 1951 and yields an anticipated arrangement of particles (Zhao and Astruc 2013). This method has been considered as a painless and uncomplicated approach. In this process, gold hydrochlorate solution was boiled and trisodium citrate dihydrate was added with strenuous stirring. With due course of time, red-coloured small particles were found to be distributed evenly throughout the solution in the range of 20 nm. Finally, particles were extracted and kept them at ambient temperature (Zhao and Astruc 2013; Kimling et al. 2006).

9.2.1.1.3 Brust–Schiffrin Method

The process was first discovered by Brust and Schiffrin, hence it is named as Brust–Schiffrin method. This method has also been considered as a painless and uncomplicated approach as Turkevich method. Controlled and self-restrained AuNPs have been produced. Initially, tetra-octylammonium bromide is reduced by sodium borohydride (*NaBH4*), whereas the reaction took place in the presence of dodecanethiol that acted as a catalyst. It has also been used for the synthesis of AgNPs. With the continuation of reaction, orange-coloured solution was formed, which has been turned to dark brown. Finally, particles were extracted and kept at ambient temperature (Herizchi et al. 2016).

9.2.1.1.4 Electrochemical Method

The process was first discovered by Reetz et al. (1994). In this method, tetra-alkyl ammonium was used as a stabilising agent in a nonaqueous medium. An electrochemical assembly consisting of the two-electrode cell has been prepared, in which oxidation was taken place at anode and reduction at cathode tube. Thus, AuNPs were obtained on the exterior wall of the electrodes with the application of applied voltage. At last, with the help of stabilising agent, particle aggregates were obtained (Herizchi et al. 2016).

9.2.1.2 Applications of Gold Nanoparticles (AuNPs)

Gold nanoparticles (AuNPs) can be used in different ways such as for immunochemical studies and identification of protein interactions. They have been predominantly used for the identification of aminoglycoside antibiotics such as gentamicin, streptomycin and neomycin. Gold nanorods are useful in the detection of stem cell and stimulus in cancer diagnosis. They are also used for the identification of different classes of bacteria. In DNA fingerprinting, it is used as lab tracer to detect the presence of DNA. AuNPs are widely used as a carrier in drug delivery applications. Some other important applications have been summarised in Table 9.2 (Hasan 2020; Baban and Seymour 1998; Tomar and Garg 2013).

9.2.2 Silver Nanoparticles (AgNPs)

Silver nanoparticles (AgNPs) seek researcher's attention due to their highest electrical conductivity unlike other metals but their metal oxides have considerably better conductivity (Ceylan et al. 2006). AgNPs manifest the structural properties that are different from their bulk samples. Silver flakes are mostly used (Hasan 2015) as their physicochemical properties received considerable attraction towards biomedical imaging by using SERS. The shape and size of the AgNPs have shown a better impact on their efficacy. It was found that silver sulphadiazine being replaced by AgNPs for the treatment of wounds as an antimicrobial agent (Schultz et al. 2000). It was reported that AgNPs interact with the HIV-1 virus via preferential binding to the gp120 glycoprotein knobs (Elechiguerra et al. 2005).

AgNPs broadly used in different fields such as cosmetics, textile, food industry and biomedicine. They represent some biomedical applications as antimicrobial agent, coating agent for medical devices and carrier for cancer drugs (Zhang et al. 2016; Lopez et al. 2020). The interaction between the therapeutic activity of AgNPs and mammalian cells may affect the thiol derivative like glutathione, thioredoxin and thioredoxin peroxidase which is responsible for the neutralisation of reactive oxygen species (ROS) originated by the mitochondrial energy metabolism. The reduction of antioxidant may occur due to AgNPs that initiate the accumulation of ROS, the initiator of inflammatory response and perturbation, and also leads to the destruction of mitochondria. The cytochrome C (apoptogenic factor) released and resulted in the programmed cell death (Sharma et al. 2015; Mohammadzadeh 2012).

Silver nanoparticles (AgNPs) can be synthesised by two methods, traditional and non-traditional methods. Traditional method includes the aqueous solution technique,

TABLE 9.2
Application of Gold Nanoparticles

S. No.	Category	Formulation	Type of Study	Results/Inference/Activity	References
			Drug Delivery		
1.	Hyperthermia	AuNPs	In vitro	A modified AuNP absorbs light at a specific wavelength which is helpful in the treatment of hyperthermic cancer and medical imaging	Vines et al. (2019)
2.	Biomedicine	AuNPs	In vitro	AuNPs in combination with dextran produced stable and biocompatible AuNPs synthesised by gamma radiation. Dextran solution or powder form beneficially is used in biomedicine and pharmaceutics	Nu Diem et al. (2017)
			Diagnostic Agents		
3.	Melanoma cancer	AuNPs	In vitro	AuNP widely used for the treatment of melanoma in advanced stages	Bhagheri et al. (2018)
4.	Carcinoma cells (HepG2)	Gold nanospheres	In vitro	The cytotoxicity studies portray the use of noncytotoxic substance, taurine, that became a stable capping agent and results in high yield and stable gold nanospheres prepared by wet chemical method	Kumar et al. (2020)
			Biosensor		
5.	Biosensor	AuNPs	In vitro	AuNPs used as a tool for the detection of proteins, pollutants and free radical molecules. The Turkevich–Frens method used with colorimetric sensing technique helpful in monitoring and rapid testing of food quality and safety	Liu et al. (2018)
			Cosmetics		
6.	Recovery mercury damage	AuNPs with glycerine	In vitro	The AuNP matrix with glycerine is synthesised by TEM. The cluster size of the matrix observed at 10–40 nm. AuNPs are bioavailable in humans	Taufikurohmah et al. (2014)
			Antimicrobial Agent		
7.	Antimicrobial activity	AuNPs with Lignosus rhinocerotis sclerotial extract (LRE) and chitosan (CS)	In vitro	The formulation of AuNPs with Lignosus rhinocerotis sclerotial extract showed better antimicrobial activity against the Pseudomonas aeruginosa and Staphylococcus aureus and Bacillus species. The CS is used as a reducing agent and results in the formation of stable AuNPs	Katas et al. (2019)
8.	Antimicrobial activity	AuNPs with Annona muricata leaf extract	In vitro	The synthesis of AuNPs with Annona muricata leaf extract is carried out by chemical method. The synthesised AuNPs is investigated to enhance the potency of AuNPs, and results are effective against fungi and bacteria from 30% to 66% and 40% to 54%, respectively	Folorunso et al. (2019)
9.	Antimicrobial activity	$NaBH_4$ fabricated GNPs	In vitro	The sodium borohydride-fabricated AuNPs are prepared by chemical reduction method. Agar well diffusion method was applied for the inhibition of pathogens. The enhanced AuNPs are potentially prepared for the development in the field of physics, biomedicine and chemistry due to their excellent, functionally active characteristics	Shamaila et al. (2016)

whereas non-traditional method includes the microemulsion technique, high-temperature reduction in porous solid matrices, vapour-phase condensation of a metal target onto a solid support, laser ablation of a metal target into a suspending liquid, gamma radiation-induced method, photoreduction of Ag ions. Some problems arise with the traditional and non-traditional methods. During the synthesis of silver ions via the traditional method, limited flexibility in the size of particles can be produced and such a method is usually sold on the ability to make < 10 nm; on the contrary, with the non-traditional method, there were a wide size distribution, lack of particle crystallinity, cost and scalability of the production. The synthesis of uniform, stable and control-sized AgNPs is difficult. The optimum method should be addressed all the above problems and additionally yields particles with no extraneous chemicals that can potentially alter the optical properties and surface chemistry (Mirela 2009).

9.2.2.1 Synthesis of Silver Nanoparticles (AgNPs)

Silver nanoparticles (AgNPs) can be developed through a variety of methods. Each method has its own advantage and limitation. These methods are described below.

9.2.2.1.1 Physical Method

This method involves evaporation and condensation processes to fabricate AgNPs. However, these techniques need higher energy and much time. Hence, researchers have found some other methods that are more advantageous than these conventional condensation and evaporation methods. Such reported methods not only diminish the time consumption but also require less energy. For this purpose, thermal decomposition technique was utilised for the preparation of NPs in solid form. This method includes the commencement of complexation reaction between silver and oleate at elevated temperature in order to develop AgNPs with particle size ranged below 10 nm (Haider and Kang 2015). In another research, Jung et al. revealed the preparation of AgNPs with uniform size using ceramic heating system that supplies constant heat without fluctuations (Jung et al. 2006). Such a heating system evaporates the precursor successfully that is used to fabricate AgNPs.

9.2.2.1.2 Chemical Method

Chemical methods are more preferred to prepare AgNPs in various sizes and shapes because of their ease in operations in solution (Yu and Zhou 2013). For example, some researchers prepared monodispersed AgNPs by reducing silver nitrate with ethylene glycol in the presence of polymer polyvinylpyrrolidone (Haider and Kang 2015). This process was referred as polyol process. In this process, ethylene glycol acts as a solvent and reducing agent. The shape and size of the particles depend upon the molar ratio of silver nitrate and polymer used. Therefore, size and shape of the resultant NPs could be tailored by adjusting the experimental parameters. Through this method, particles with a size range of 20 nm or less can be made.

9.2.2.1.3 Biological Synthesis

In biological synthesis, AgNPs are prepared using plants (like algae, yeast, fungi or bacteria) as stabilising and reducing agents (Haider and Kang 2015). In a research work, shewanell aoneidensis, i.e., a reducing agent for metals, was used to biosynthesise NPs by using silver nitrate solution as a precursor. The size of resultant NPs was less than 15 nm with better dispersion ability, shape, stability and large surface area. Such methods have been found inexpensive, reproducible and consume lesser energy than the conventional methods (Sintubin et al. 2012).

9.2.2.2 Applications of Silver Nanoparticles (AgNPs)

Silver nanoparticles (AgNPs) are being used in numerous technologies and give an advantage to the consumers to take the benefits. AgNPs are widely used as a biosensor and in many assays. They can be also used as biological tags during quantitative detection. AgNPs are used in cosmetics, footwear, paints, wound dressings, as well as in plastics due to their antibacterial properties. Besides these, they are used in conductive inks and integrated into composites to increase thermal as well as electrical conductivity. Some of the other applications are summarised in Table 9.3 (Oldenburg 2020).

TABLE 9.3
Applications of Silver Nanoparticles

S. No.	Category	Formulation	Type of Study	Results/Inference/Activity	References
				Drug Delivery	
1.	Antimicrobial	Quercetin-loaded AgNPs	*In vitro*	Antimicrobial activity got increased	Hooda et al. (2020)
2.	Wound dressing	AgNP-loaded bacterial cellulose hydrogel	*In vitro*	Exhibited antimicrobial activity against three common wound-infecting pathogenic microbes *Staphylococcus aureus*, *Pseudomonas aeruginosa* and *Candida auris*	Gupta et al. (2020)
3.	Osteogenesis	AgNPs	*In vitro*	AgNPs increased osteogenic protein expression and mineralisation of hMSCs	He et al. (2020)
4.	Antibacterial	CmCh-PVA/Ag nanocomposite hydrogels	*In vitro*	Increased sustained and controlled drug releases with an increase in silver nitrate content. Also enhanced antibacterial activity is observed	Gholamali et al. (2019)
				Biosensor	
5.	Biosensor	AgNPs with graphene quantum dots	*In vitro*	Excellent antibacterial properties for *E. coli* and *S. aureus*	Zang et al. (2020)
6.	Biosensor	AgNP inkjet printer biosensor	*In vitro*	Prevent both antibiotic contaminations and antibiotic resistance development	Rosati et al. (2019)

(Continued)

TABLE 9.3 (*Continued*)
Applications of Silver Nanoparticles

S. No.	Category	Formulation	Type of Study	Results/Inference/Activity	References
				Cancer Theranostics	
7.	Breast cancer	Turmeric AgNP	*In vitro*	The anticancer studies on MCF-7, MDA-MB-231 and MCF-10A cell lines showed the EC50 values of 0.146 mg/mL and 0.141 mg/mL, respectively. Hence, it can be used as an effective anticancer agent	Mittal et al. (2020)
8.	Renal adenocarcinoma	Fucan-coated AgNP	*In vitro*	Exhibit higher cytotoxicity against tumour cells, especially human renal adenocarcinoma cells	Amorim et al. (2016)
				Cosmetics	
9.	Anti-UV cream	AgNP in sun-protective body milk	*In vitro*	Decreased cytotoxicity	Arroyo et al. (2020)
10.	Antibacterial	Peel-off facial mask embedded with biosynthesised AgNPs	*In vitro*	Demonstrated the antibacterial activity of the peel-off facial mask formulation against *Pseudomonas aeruginosa*, *Staphylococcus aureus* and *Propionibacterium acnes*	Badnore et al. (2019)

9.2.3 Zinc Oxide Nanoparticles (ZnO NPs)

Zinc oxide nanoparticles (ZnO NPs) have become an interesting topic for many types of researches as they explore in many ways due to their exponential applications. Zinc oxide has promising inherent unique properties in the field of biomedical science, especially shows its anticancer and antimicrobial activities. The particles of zinc oxide are commonly used as a drug carrier (lipid and polymeric NPs). ZnO NPs are inexpensive, less toxic and biocompatible as compared to other metals and metal oxides (Mishra et al. 2017). ZnO NPs have many applications in various fields of science. It is considered as a wide bandgap semiconductor and also shows enormous photoluminescence property used as a biosensor. The US Food and Drug Administration (FDA) named the bulk ZnO NPs (having a size range of less than 10 nm) as generally recognized as safe (GRAS). They are used to justify the demand for drug delivery. The ZnO NPs can be prepared by various methodologies that involve precipitation method, solid-state pyrolytic method, sol–gel method, etc. (Sharma et al. 2015).

It was observed that the numbers of electrons–holes are spotted in the absence of ultraviolet light, through the crystallinity of nanosized particles with the decrease in zinc oxide particle size. This results in increased interstitial Zn ions and oxygen amount along with the donor/acceptor impurities. It may lead to the crystal defects that can generate a large number of electron–hole pairs, resulting in the production of ROS. The semiconducting property of zinc oxide can influence ROS activity (Sharma et al. 2009).

Necrosis and apoptosis are the mechanisms that include a biomedical activity of ZnO NPs (Yang et al. 2015). Zinc is an essential mineral and helpful in the catalytic activity of numerous enzymes present in the body. Due to its biocompatibility, it is used in the preparation of various products such as batteries, plastics, ceramics and paints (as an antibacterial) (Mishra et al. 2017). Nanosized zinc oxide particles are widely used in the industrial sectors due to their unique properties like high-extinction binding energy (60 MeV), high electronic conductivity, nontoxicity and chemical stability (Lopez et al. 2020).

9.2.3.1 Synthesis of Zinc Oxide Nanoparticles (ZnO NPs)

In the synthesis of ZnO NPs, several methods have been reported for the preparation of ZnO NPs. The primary focus of each of the methods is the development of stable and uniform nanosized particles.

9.2.3.1.1 Microwave-Assisted Synthesis of ZnO Nanoparticles (ZnO NPs)

Despite the numerous reported synthetic methods for ZnO NPs, most of them require high temperature, expensive substrates, tedious procedures, sophisticated equipment and rigorous experimental conditions. Hence, it is necessary to find out a simple and low-cost method for the synthesis of ZnO nanostructures to tackle the problems. In view of this, Kooti and Naghdi Sedeh (2012) have designed a new facile microwave (MW)-assisted combustion technique for fabrication of uniform ZnO NPs using MW heating and glycine as fuel and zinc nitrate as a precursor. This is a simple, inexpensive, environmentally benign, template-free and fast method and therefore can be considered as superior route to most of the reported ones. This method provides a pure and high yield of ZnO sample with high crystallinity and well-dispersed NPs with a mean size of 20 nm. MW-assisted methods can be considered as a promising green strategy to synthesise the nanomaterials and nanocomposites, confirming the green chemistry approaches (Kumar, Kuang et al. 2020). By using ascorbic acid and polyvinyl alcohol as a capping agent, Ahammed et al. (2020) synthesised ZnO NPs by MW heating. In a similar way, Saloga and Thünemann (2019) synthesised ultrasmall oleate-stabilised ZnO NPs with narrow size distribution and long colloidal stability. The advantages of MW-assisted synthesis are uniform heating of reaction precursors, low energy consumption, shorter reaction time and high yield of NPs. Garino et al. (2019) and Hasanpoor et al. (2015) studied the effect of different reaction parameters like time, MW irradiation power and type of precursors on the ZnO NPs with controlled morphology produced by this method.

9.2.3.1.2 Precipitation Method

This method is based on a reaction between a zinc precursor and a precipitating reagent. Typically, a solution of precipitating agent (e.g. sodium hydroxide, ammonium hydroxide and urea) is added dropwise to the aqueous solution of a zinc precursor (e.g. zinc nitrate and zinc sulphate) resulting in a formation of intermediate product that finally gets converted to ZnO after calcination at high temperature (Sabir et al. 2014). In this method, ZnO can be synthesised by using zinc nitrate and urea as precursors. Kumar et al. (2013) synthesised ZnO NPs of 64 nm by the reaction of zinc sulphate and sodium hydroxide in a molar ratio of 1:2. The NPs obtained were very pure, as revealed by XRD analysis; in addition, it was also observed that the crystallinity of the NPs increased as the calcination temperature increased.

9.2.3.1.3 The Wet Chemical Synthesis

This method is a modification of the previously described precipitation method. The method is based on the precipitation reaction between zinc nitrate and sodium hydroxide and prepared NPs were stabilised with the help of an additive (Yadav et al. 2006). Starch, which is commonly used as a stabilising agent, is adsorbed at the surface of NPs, providing stability. The mechanism of stabilisation involves either increasing the viscosity of the solution or forming a complex with metal ions via the hydroxyl groups (He and Zhao 2005). Addition of excess hydroxyl ions (from sodium hydroxide) should be avoided because these solubilise zinc hydroxide and convert it to zincate ions, which are soluble in aqueous medium. Sharma et al. (2016) synthesised ZnO NPs using the wet chemical method and obtained particle sizes in the range of 400 nm.

9.2.3.1.4 Solid-State Pyrolytic Method

Solid-state pyrolytic method is a simple, rapid and cost-effective method reported by Wang et al. (2003). This involves mixing zinc acetate and sodium bicarbonate and the resulting mixture is then pyrolysed. Zinc acetate converts to ZnO NPs, while sodium bicarbonate converts to sodium acetate, which is then washed away with deionised water. The particle size of the NPs can be controlled by adjusting the pyrolytic temperature. Sodium acetate can be distributed on the surface of the NPs, preventing them from agglomerating. These nanocomposites can then be converted into NPs via the dissolution of sodium acetate. Wang et al. (2003) synthesised ZnO NPs of different sizes, ranging from 8 nm to 35 nm, by varying the pyrolysis temperature of the reactant mixture.

9.2.3.1.5 Sol–Gel Method

Sol–gel method was first developed by Spanhel and Anderson (1991), and later, it was modified by Meulenkamp (1998). Zinc acetate dihydrate is usually used as the precursor as it offers easier control of hydrolysis. Sodium hydroxide is used to adjust the pH of the reaction mixture. The pH controls the rate of ZnO formation and affects the size and stable state of the resulting NPs (Daneshvar et al. 2008). This method involves four stages: solvation, hydrolysis, polymerisation, and transformation. Zinc acetate dihydrate is solvated in a solvent such as methanol or ethanol and is then hydrolysed, which aids the removal of any intercalated acetate ions. This leads to the formation of a colloidal gel of zinc hydroxide (Rani et al. 2008). Zinc hydroxide then splits into Zn^{2+} cations and OH^- anions, followed by polymerisation of the hydroxyl complex to form 'Zn–O–Zn' bridges, converting sol into gel. This gel is then transformed into ZnO (Mishra et al. 2017). Hayat et al. (2011) synthesised 25-nm ZnO NPs with a modification of existing sol–gel method and successfully utilised them for the photocatalytic oxidation of phenol.

9.2.3.1.6 Biosynthesis

The principle of this method is based on the use of surface-active biosurfactants, which are produced by culturable microbes. Biosurfactants offer various advantages over synthetic polymers, such as higher specificity, biodegradability and biocompatibility. Lecithin, bile salts, sophorolipids and rhamnolipids (RLs) are some common examples of biosurfactants. RLs (cyclic lipopeptide surfactants) are one such biosurfactant that can also be used for capping, stabilising and dispersing NPs. Singh et al. (2014) reported the synthesis of RL-stabilised ZnO NPs in the range of 35 to 80 nm by using Pseudomonas aeruginosa. Surendra et al. (2016) synthesised 40–45-nm ZnO NPs from *Moringa oleifera* (the drumstick tree). Similarly, Gunalan et al. (2012) prepared ZnO NPs from aloe leaf extract and showed that the biosynthesised NPs had higher antimicrobial activity compared with chemically synthesised ZnO NPs.

9.2.3.2 Applications of Zinc Oxide Nanoparticles (ZnO NPs)

Zinc oxide nanoparticles (ZnO NPs) are used in several pharmaceutical fields. Some of these applications of Zn NPs are mentioned in Table 9.4.

TABLE 9.4
Applications of Zinc Oxide Nanoparticles

Drug	Category	Type of Study	Activity/Inference	References
Drug Delivery				
ZnO NPs and PEGylated ZnO NPs	Antiviral agent (inhibition of H1N1 influenza virus infection)	Evaluated antiviral activity of ZnO NPs and PEGylated ZnO NPs against H1N1 influenza virus	The study indicated that there was a substantial decrease in fluorescence emission intensity in viral-infected cell treated with PEGylated ZnO NPs compared to a positive control (oseltamivir)	Ghaffari et al. (2019)
Quercetin via pH-responsive ZnO NPs	Anticancer agent for the treatment of breast cancer	The synthesis of phenylboronic acid (PBA)-conjugated ZnO NPs and loaded with quercetin (bioflavonoid). The presence of PBA moieties facilitate targeted drug delivery of quercetin to the sialic acid over expressed cancer cells	The PBA-ZnO-Q NP-induced apoptotic cell death in human breast cancer cells via enhanced oxidative stress and mitochondrial damage	Sadhukhan et al. (2019)
ZnO NPs	Anticancer therapeutic agent for treating ovarian cancer independent of p53 mutants of the cancer cells	Characterisation of physical properties of ZnO NPs of increasing particle size (15–55 nm) and evaluated their benefits as an ovarian cancer therapeutic agent using established human ovarian cancer cell lines	ZnO NPs induce acute oxidative and proteotoxic stress in ovarian cancer cells leading to their death via apoptosis. It shows the potential of ZnO NPs to serve as an anticancer therapeutic agent for treating ovarian cancer	Padmanabhan et al. (2019)
Green-synthesised ZnO NPs using Sargassum muticum algae extraction	Antiangiogenic and apoptotic properties on human liver cancer cell line	ZnO NPs synthesised by Sargassum muticum algae extraction used to evaluate its cytotoxicity and apoptotic properties on human liver cancer cell line (HepG2)	This NP decreased angiogenesis and induced apoptosis. Therefore, these NPs can be used as a supplemental drug in cancer treatment	Sanaeimehr et al. (2018)
Green-synthesised NPs capped by gum arabic (GA), docosahexaenoic acid (DHA)	Antidiabetic	Green-synthesised ZnO NPs via solid-state approach in dry conditions capped by GA as nanodelivery system exhibiting an excellent antidiabetic agent with high cytocompatibility. The hydrophobic DHA was loaded into ZnO NPs to investigate the antidiabetic performance of free DHA compared to loaded DHA NPs in rats	It was found that DHA-loaded ZnO NPs had high performance in enhancing insulin signalling pathway as expressed in changes of phosphatidylinositol 3-kinase levels. This also confirms that biocompatibility and physicochemical properties of ZnO NPs had a significant role in the treatment of diabetes	Hussein et al. (2019)
Cashew gum (CG)-capped ZnO NPs	Antifungal activity against Candida parapsilosis	ZnO NPs were synthesised by a simple method, using the natural polysaccharides cashew gum (CG) and carboxymethylated cashew cum (CMCG) as stabilising templates	The ZnO NPs presented significant inhibition towards yeasts of the genus Candida, particularly Candida parapsilosis, with no toxicity against red blood cells, which indicates a promising material that could be safely used in medical devices or in pharmaceutical formulation, as a support to antifungal treatment	Souza et al. (2020)
Cosmetics				
Biosynthesised colloidal ZnO NPs	Antimicrobial and antioxidant potential of biosynthesised colloidal ZnO NPs for fortified cold cream formulation	The colloidal ZnO NPs as a potent biomaterial for topical formulation of cosmetic and dermatological significance are employed. The ZnO NPs were characterised and green-synthesised using Adhatodavasica leaf extract	The highest antibacterial and antifungal activities were observed against S. epidermidis and A. fumigatus. Colloidal ZnO NPs of biological origin exhibited antimicrobial and antioxidant properties due to their nanometric size. The significant inhibitory action was observed against Candida species. The ZnO NPs can decrease human skin infections	Sonia et al. (2017)

(Continued)

TABLE 9.4 (Continued)
Applications of Zinc Oxide Nanoparticles

Drug	Category	Type of Study	Activity/Inference	References
Topical application of ZnO NPs	Immunological and antibacterial mechanism of ZnO NPs	The study demonstrates ZnO NP antimicrobial actions and as a novel class of anti-infective agent for the treatment of skin infection	Antibacterial activity of ZnO NPs showed that it disrupts bacterial cell membrane integrity. Intradermal administration of ZnO NPs significantly reduced the skin infection, bacterial load and inflammation in mice	Pati et al. (2014)
ZnO NPs	Photocatalytic antibacterial application under UV irradiation	ZnO NPs were synthesised and immobilised onto silicon (Si) wafers by self-assembly. The antibacterial activity of ZnO NPs against Escherichia coli under dual ultraviolet irradiation for disinfection was investigated	Synthesised ZnO NPs and arrayed ZnO NP networks on solid plates had excellent antibacterial activity against *E. coli* under dual UV irradiation >3 log CFU/mL. They show considerable promise as photocatalytic nanoantibiotics for use in near-future disinfection systems with industrial and clinical application	Jin et al. (2019)
Theranostics				
Bioinspired NPs (inorganic NPs, e.g. ZnO NPs)	Cancer theranostics	It comprises a detailed overview of cancer theranostics and application of various bioinspired NPs	The inorganic NPs such as gold and zinc oxide are extensively used for several biomedical applications including cancer theranostics, biosensing and bioimaging due to their unusual physicochemical properties. In future, there is a positive outlook for bioinspired nanotheranostics and its clinical translation could be realised in future to transfigure cancer therapy	Madamsetty et al. (2019)

9.2.4 Iron Oxide Nanoparticles (FeO NPs)

Iron oxide (FeO) NPs are also known as magnetic NPs. FeO is an inorganic compound and found in reddish-brown colour in the environment. The other form of FeO is Fe_3O_4 is naturally occurring in an environment that is known as magnetite (Fe_3O_4). The FeO is paramagnetic, whereas Fe_3O_4 is supramagnetic. These major properties like ultra-fine size, magnetic, and biocompatible are helpful in the formation of superparamagnetic iron oxide nanoparticles (SPIONs). SPIONs have the contribution in various biomedical applications such as used in magnetic resonance imaging as resolution contrast agents, stem cell tracking, hyperthermia gene therapy, molecular/cellular up taking, magnetic separation techniques (DNA sequencing), cancer, atherosclerosis and diabetes (Mody et al. 2010). The magnetic NPs specified with their particle size like ultrasmall particle iron oxide nanoparticles (USPIONs) also known as monocrystalline FeO lie between 10 and 40 nm and small particles of iron oxide nanoparticles (SPIONs) lie between 60 and 150 nm. For the improvement of cellular uptake, particles being modified with a peculiar surface coating so that it may become easy for the proteins, drugs, enzymes, antibodies or nucleotides to conjugate. It can be directed to an organ, tissue or tumour also (Chastellain et al. 2004). In hyperthermia, an increase in the temperature of the microenvironment of tissue to 40°C–45°C results in a series of subcellular events that cause apoptosis, the activation of immunological responses, and increases in tumour blood flow and oxygenation that causes ultimately death of tumour cells. SPIONs play a significant role in their energy absorption resulting from Neel relaxation that produces hyperthermia (Sharma et al. 2015; Chatterjee et al. 2011).

The study of Conroy et al. investigates the chlorotoxin biocompatible FeO nanoprobes coated with the polyethylene glycol (PEG) and shows its ability for targeting glioma tumours via the surface-bound targeting peptide (Sun et al. 2008). The tissue engineering played a vital role to deliver the NPs via siRNA for silencing of human telomerase reverse transcriptase genes in HepG2 cells, causing their apoptosis and growth inhibition (Li et al. 2014).

Some common methods proposed for the development of magnetic NPs include microemulsions, sol–gel synthesis, sonochemical reactions, hydrothermal reactions, hydrolysis and thermolysis, electrospray synthesis and flow injection synthesis (Mody et al. 2010; Basak et al. 2007).

9.2.4.1 Synthesis of Iron Oxide Nanoparticles (FeNPs)

Iron oxide nanoparticles can be prepared by using various methods that involve hydrothermal method, sol–gel method, polyol method, electrochemical technique, flow gel method, etc. Each method possesses its own advantage as well as limitation. These methods are described below.

9.2.4.1.1 Hydrothermal Method

In this method, reactions are performed in aqueous media in reactors or autoclaves where the pressure can be higher than 2,000 psi and the temperature can be above 200°C. Water is hydrolysed and metal salts get dehydrated at high temperatures (Sharma et al. 2015). Otari et al. (2020) prepared magnetic rice straw (RS) by adding 0.25 M iron (III) nitrate to 1 g of washed crude RS with deionised water. It was sonicated for 2 h. Then, it was kept aside for 24 h followed by filtration. Filtrate was washed twice with deionised water. The RS was soaked in aq. NH_4 solution (pH 11). Then, it was subjected to hydrothermal treatment. The RS was filtered and washed. The dried magnetic RS was heated at 100°C for 2 h. Magnetic RS was formed during hydrothermal heating when iron hydroxide changed into $\alpha\text{-}Fe_2O_3$. Kermanian et al. (2020) prepared Fe_3O_4 hydroxyapatite nanocomposites hydrothermally. In beaker A, 0.01 M of ammonium hydrogen phosphate and 0.1 M of cetrimonium bromide were mixed in deionised water. In beaker B, 0.03 M calcium chloride was dissolved in deionised water (DW). Then, the solution of beaker A is added to the solution in beaker B dropwise. The pH was adjusted to 10.5 by adding NH_4OH. At the same time, ferrous chloride and ferric chloride with the molar ratio of 1:1.5 was prepared in beaker C. The solutions of beakers B and C were mixed, and then, it was subjected to hydrothermal treatment. Further, it was cooled at room temperature (RT) and centrifuged, and then, the precipitate was washed and dried to obtain FeO hydroxyapatite.

9.2.4.1.2 Sol–Gel Method

This method follows a wet route to synthesise metal oxide NPs. It is based on hydroxylation and condensation of molecular precursors in solution that leads to the formation of 3D metal oxide network (Laurent et al. 2008). This method cannot synthesise monodispersed NPs due to the aggregation of NPs during the washing process (Cui et al. 2013). To overcome this drawback, many studies have reported further condensation and hydroxylation followed by heat treatment to get the final crystalline state (Noqta et al. 2019). The main factors like nature of salt, type of solvent, temperature, pH, precursors' concentration and agitation affect the kinetics, growth, hydrolysis, condensation and reactions of the gel (Ramimoghadam et al. 2014). Qi et al. (2014) produced monodispersed FeO NPs by adding a capping agent (mixture of oleylamine and oleic acid). It is reported that the nucleation and growth processes of FeO NPs took place during the sintering process.

9.2.4.1.3 Polyol Method

The polyol synthesis can also be understood as a sol–gel method; it was first used by Fiévet, Lagier and Figlarz in the late 1980s. This method involves metal precursor and organic solvent, which act as a reducing agent and stabiliser, respectively, which helps in producing the nonagglomerated metal NPs with well-defined shape (Ammar and Fievet 2020). In this method, a solid precursor is added to a liquid polyol. Then, the suspension is stirred and heated until it reaches the boiling point of the polyol. When heated, the metal precursor becomes solubilised in the solvent, forms an intermediate and then reduces to form some metal nuclei that nucleates and produce metal NPs. By increasing the reaction temperature or by inducing foreign nuclei *in situ*, sub-micrometre-sized metal NPs can be synthesised (Laurent et al. 2008).

9.2.4.1.4 Electrochemical Method

This method is based on the dissolution of metal at anode followed by cathode reduction of the dissolved metallic ion species with the help of surfactants (Cushing et al. 2004). In electrochemical method, particle size depends on current density. Using this method, $\gamma\text{-}Fe_3O_4$ NPs can be made using iron electrode in an aqueous solution of DMF and cationic surfactants. Electrochemical deposition under oxidising conditions has been used to prepare NPs of Fe_2O_3 and Fe_3O_4 (Kahn and Petrikowski 2000).

9.2.4.1.5 Flow Injection Method

A novel synthesis of Fe_3O_4 NPs based on a flow injection synthesis (FIS) technique has been developed by Alvarez et al. (2006). The technique consisted of continuous or segmented mixing of reagents under a laminar flow regime in a capillary reactor. The obtained Fe_3O_4 NPs had a narrow size distribution in the range of 2 to 7 nm. The technique has some advantages, such as high reproducibility, high mixing homogeneity and an opportunity to control the process externally (Laurent et al. 2008).

9.2.4.1.6 Aerosol/Vapour Methods

Spray and laser pyrolysis are the two main attractive aerosol technologies, because they are continuous chemical processes allowing for high-rate production. In spray pyrolysis, a solution of ferric salts and a reducing agent in an organic solvent is sprayed into a series of reactors, where the aerosol solute condenses and the solvent evaporates (Laurent et al. 2008). The resulting dried residue consists of particles whose size depends upon the initial size of the original droplets. Maghemite particles with size ranging from 5 to 60 nm with different shapes have been obtained using different iron precursor salts in alcoholic solution (Kim et al. 2007).

9.2.4.2 Biomedical Applications of Iron Oxide

The superparamagnetic NPs are utilised as MRI contrast agents. The magnetic resonance signal intensity is significantly modulated without any compromise *in vivo* stability. The magnetically activated 3D gels can pass through the blood–brain barrier and used for the conjugation of peptides and growth factors to cure and regenerate brain tissues. The SPION–Au core–shell decorated with nerve growth factor (NFG) shows low toxicity that has been developed for the growth of neurons and differentiation.

TABLE 9.5
Applications of Iron Oxide Nanoparticles

S. No.	Category	Formulation	Type of Study	Results/Inference/Activity	References
			Drug Delivery		
1.	Drug delivery of doxorubicin	Fe_3O_4/SiO_2@DA	In vitro	Loading of the drug was maximum in basic pH and at a lower temperature (RT), while the release was more efficient at a higher temperature (40°C–45°C)	Naqvi et al. (2020)
2.	Alzheimer's disease	SPIONS with Aβ oligomer-specific scFv antibody	In vivo (rat model)	SPIONS readily reached pathological AβO regions in brains and distinguished AD transgenic mice from WT controls. It exhibited the properties of good biocompatibility, high stability and low cytotoxicity	Liu et al. (2020)
3.	Antibacterial	Au@Co-Fe NPs	In vitro	Showed antibacterial activity against four standard selected pathogenic bacteria *Escherichia coli, Pseudomonas aeruginosa* (as gram negative) and *Staphylococcus aureus and Bacillus cereus* (as gram positive)	Mirhosseini et al. (2019)
4.	DNA sensor	GNR_Fe_3O_4 nanocomposite	In vitro	The capability of GNR_Fe_3O_4 nanocomposite as a DNA sensor was measured in terms of electrochemical response by effectively measuring the cathodic peak in DNA immobilisation and hybridisation	Rodriguez et al. (2020)
			Cancer Theranostics		
5.	MRI	SPIONS	In vitro	Simple and straightforward size isolation helps to improve MRI, MPI and hyperthermia	Dadfar et al. (2020)
6.	Cell senescence	FeO particles with clove extracts	In vitro	Induce apoptosis in MCF-7 breast cancer cells. Iron particles activated the activities of caspase	Thenmozhi (2020)

Magnetic NPs have the multifunctional characteristics for cancer diagnosis, photodynamic therapy (PDT), radiofrequency thermal therapy (RTT) and magnetic targeting applications developed via decorating iron oxide nanoparticles (IONP) onto fullerene (C60) and using folic acid (FA) as an active targeting ligand. C60–IONP–PEG–FA combination showed strong photosensitising and photothermal ablation effects and specifically killed cancerous cells via active tumour targeting. IONPs have been explored not only for the delivery of small molecule drugs, but also for the delivery of various macromolecular drugs, including proteins, peptides and DNA therapeutics. Some other applications of FeO NPs have been discussed in Table 9.5.

9.2.5 ZIRCONIUM NANOPARTICLES (ZRNPS)

The silver-coloured zirconium metal is very strong, malleable, ductile and lustrous. The other derivatives of zirconium are known as Baddeleyite (zirconium oxide, ZrO_2) and Zircon ($ZrSiO_4$). Its chemical and physical properties resemble titanium. Zirconium NPs exist with an average size of 8 nm and are biosynthesised from fungus *F. oxysporum* at RT with aqueous ZrF_6^{2-} anions (Seabra and Durán 2015). Zirconium oxide is crucially used for the treatment of oral diseases. The nanobiomaterials of zinc oxide are introduced in dental ceramics, implants, radiopacifying agents, basement and tissue engineering field. Nowadays, clinically titanium dioxide implants are replaced by zirconium oxide implants because of their superior osseointegration. Zirconium dioxide is highly pure and biocompatible; that is why it is used in the dental area. It has good natural white colour, high toughness, excellent strength, steady chemical properties and good corrosion resistance, i.e. a source of high-performance ceramic material and implant materials (Hu et al. 2019). It is useful for surface treatment and bioactivation accelerates new bone formation (Hu et al. 2019). Recently, a study discussed the effect of agglomeration of ZrO_2-based NPs with or without lithium grease. The zirconium oxide NPs (1% wt) added with the lithium grease being helpful for the 50% reduction of friction coefficient. In lithium grease, oppositely the agglomeration of ZrO_2 NPs could double the friction coefficient relative to that of pure grease.

There is a clear relationship between tri-biological properties of lubricant-containing NPs and agglomeration of such NPs. The dispersion stability of ZrO_2 NPs in oil or grease can be enhanced either by surface modification of such NPs or by using surfactants. The occurrence of agglomerations of ZrO_2 NPs in the lithium grease increased COF value twice as compared to that of the pure grease (Rylski and Siczek 2020). The zirconium oxide NPs can be prepared by hydrothermal synthesis, coprecipitation and sol–gel method. The structural study of ZrO_2 NPs can be characterised by using XRD, scanning electron microscopy (SEM) and transmission electron microscopy (TEM) (Behbahani et al. 2012), etc. The presence of ZrO_2 NPs increased the chances of hydration by 26% and showed a positive effect on the *in vitro* biocompatibility of MG63 osteosarcoma cells (Li et al. 2013).

9.2.5.1 Synthesis of Zirconia Nanoparticles

There are different methods that can be used to synthesise zirconium oxide NPs. Some of the important methods of zirconia NPs are described below.

9.2.5.1.1 Sol–Gel Method

Sol–gel method is commonly used to prepare zirconia oxide NPs. This method includes the hydrolysis of metallic compound precursor for generating oxo-hydroxides followed by the condensation and polymerisation of the compound. Now the network of metal hydroxide and a porous gel is formed, respectively, with subsequent drying and heating of the gel leading to the formation of metallic NPs (Ayanwale et al. 2018).

9.2.5.1.2 Coprecipitation Method

This method utilises the precipitation of oxo-hydroxide form from the metallic salt precursor solution using a precipitating medium. There is a nucleation process that is followed by the growth phase when the critical concentration of the species in the solution has been reached. Coprecipitation is employed to fabricate zirconia oxide or its mixed metal oxide NPs (Arsent'ev et al. 2014).

9.2.5.1.3 Hydrothermal Method

This method is an example of solvothermal synthetic route that is used to fabricate a variety of NPs by dispersing the starting material in a suitable solvent medium and expose it to moderate temperature and pressure conditions. This causes the generation of NPs (Ayanwale et al. 2018). If water is utilised as the reaction solvent, the method is known as hydrothermal synthesis. Various chemical parameters like type; composition and reactant's concentration; solvent ratio; and thermodynamic parameters like temperature, pressure and reaction time are the unique parameters in the formation of the metallic NPs.

9.2.5.2 Applications of Zirconia Nanoparticles

ZrO_2 exhibits superior strength, toughness and widely used as inlays, partial crowns, veneers and full ceramic crowns. Nevertheless, the high mechanical ZrO_2 dooms for the difficulty of the machine (Hanahan and Weinberg 2011). The sintered and shaped ceramics were prepared and the addition of nano-ZrO_2 has led to the higher flexural strength, fracture toughness and shear bond strength value with reduced porosity (Guazzato et al. 2004).

It is known as titanium oxide and was gradually used for dental implants (treatment for tooth defect or missing) due to its excellent mechanical strength and biocompatibility. Nano-ZrO_2 can be served as the promising implants or coating material for better integration due to the high chemical stability, biocompatibility, suitable fracture resistance and flexural strength compared with an alveolar bone. Bone tissue engineering is the ideal approach to rebuild bone defects. The fabricated porous nano-ZrO_2 scaffolds manufactured by replication technique with the efforts of researchers (Zhu et al. 2015). The porous-structured nano-ZrO_2 scaffolds show good interconnectivity.

Root canal therapy is considered as the widely used method for the treatment of pulpal and periapical diseases. ZrO_2 NPs described as the promising radiopacifier as a supplement in Portland Cement (PC) for expediting hydration kinetics without weakening biocompatibility (Li et al. 2013). Denture base is the important branch of removable of partial dentures and complete dentures. ZrO_2 nanopowders prepared to improve the performance with the incorporation of polymethyl methacrylate (PMMA) have been investigated by a group of Mohammed and colleagues (Gad et al. 2016). Some of the applications of zirconium NPs have shown in Table 9.6.

TABLE 9.6
Applications of Zirconium Nanoparticles

S. No.	Category	Formulation	Type of Study	Results/Inference/Activity	References
			Dental Application		
1.	Dental ceramics	ZrO_2	In vitro	The sintered and shaped ceramics were prepared and the addition of nano-ZrO_2 has led to the higher flexural strength, fracture toughness and shear bond strength value with reduced porosity	Guazzato et al. (2004)
2.	Denture base material	ZrO_2 nanopowders	In vitro	ZrO_2 nanopowders prepared to improve the performance with the incorporation of PMMA have been investigated by a group of Mohammed and colleagues	Gad et al. (2016)
3.	Dental implants	ZrO_2	In vitro	Nano-ZrO_2 can be served as the promising implants or coating material for better integration due to the high chemical stability, biocompatibility, suitable fracture resistance and flexural strength compared with an alveolar bone	Sennerby et al. (2010)
			Tissue Engineering		
4.	Bone tissue engineering	fabricated porous nano-ZrO_2 scaffolds	In vitro	The fabricated porous nano-ZrO_2 scaffolds manufactured by replication technique with the efforts of researchers	Zhu et al. (2015)

(Continued)

TABLE 9.6 (*Continued*)
Applications of Zirconium Nanoparticles

S. No.	Category	Formulation	Type of Study	Results/Inference/Activity	References
				Diagnostic Agent	
5.	Radiopacifying agent	ZrO_2 NPs	*In vitro*	Root canal therapy is considered as the widely used method for the treatment of pulpal and periapical diseases. ZrO_2 NPs described as the promising radiopacifier as a supplement in PC for expediting hydration kinetics without weakening biocompatibility	Li et al. (2013)
				Industrial Application	
6.	Machinery tools	zirconia	*In vitro*	A study reveals the use of zirconia as a material for the manufacturing of punches and dies	Kara et al. (2004)
				Cosmetics	
7.	Jewelry	Zirconium silicate	*In vitro*	Zirconium silicate used in personal care such as creams and powders. It is well popular for the jewellery and gemstones	Hostynek and Maibach (2002)
				Radiopharmaceutical Purposes	
8.	Nuclear medicine	Zirconium-89	*In vitro*	The study represents that zirconium-89 is a promising radionuclide for nuclear biomedicine. Radio-TLC (thin layer chromatography) and biodistribution were conducted and three methods of isolation were studied by using zirconia (ZR) (hydroxamate) and Chelex 100	Larenkov et al. (2019)
9.	Radiotracer	[^{89}Zr]Zr-oxine	*In vivo*	The study portrays the preparation of [^{89}Zr]Zr-oxine and evaluates it against [^{111}In]In-oxine in WBC. It is known as radiotracer for positron emission tomography (PET). It has better sensitivity and spatial resolution	Man et al. (2020)
				Theranostic Agent	
10.	Cancer	Zirconium phosphate (ZrP) nanoplatelets	*In vitro*	The cisplatin encapsulated by using ZrP nanoplatelets used to treat the tumour cells. The study showed the direct intercalation of cisplatin into zirconium and was tested *in vitro* for cytotoxicity in the human breast cancer (MCF-7) cell. The cisPt@ZrP reduced the cell viability up to 40%	Diaz et al. (2013)

9.3 CHARACTERISATION TECHNIQUES OF NANOPARTICLES

Characterisation of metallic NPs is quite significant to know about their unique characteristic features as well as applications. These evaluative techniques are helpful for an effective comparison among pharmaceutical nanoformulations and for the development of an optimised product. The selection of an appropriate evaluation technique is a matter of perplexity for researchers because each technique possesses its strengths and limitations. To overcome such challenges, various reliable and fruitful techniques with better reproducibility are needed. Various characterisation technologies that can be effectively utilised for the development of an optimum quality product are described here.

9.3.1 Energy-Dispersive X-Ray Spectroscopy (EDX)

This approach is generally utilised for the chemical or elemental analysis of nanoformulations. This technique is based on the principle that every material has a different structural system of atoms that allows the emission of radiations from it. In this spectroscopic technique, an interaction between electromagnetic radiations and material takes place so that charged particles collide and X-rays emit from the material (Thodeti et al. 2016, 2017). The highly centralised electron/proton beams are radiated upon the nanoparticulate sample that causes the excitation of lower energy-level electrons so that they migrate to higher energy shells. Such migration of electrons creates the electron holes at the place exactly electrons were that replete the electrons of high energy level. Such an energy difference between the higher and lower energy shell would be released in the form of the X-ray. The extent of X-rays emitted from the sample would be measured through the dispersive spectrometer. By measuring the X-ray energies emitted from a specific area excited by the electron beam, the elements present in the sample can be estimated (Goel et al. 2019).

9.3.2 Fourier Transform Infrared Spectroscopy (FTIR)

This spectroscopic technique can provide significant information about the organic layers present around the surface of metallic NPs. This technique gives valuable information regarding the three-dimensional structural data derived from the XRD. The technology can be utilised for determining the properties of metabolites/functional groups present on the surface of NPs that may give sufficient information about capping and stabilisation of the NPs (Shankar and Rhim 2015). FTIR is also extensively used as a potential tool for the characterisation of biomolecules and surface structure of metallic NPs (Goel et al. 2019).

9.3.3 Scanning Electron Microscopy (SEM)

Scanning electron microscopy (SEM) is extensively utilised as a characterisation technique especially for the surface characterisation of metallic NPs. For SEM characterisation, the NPs are turned into a dry form, i.e. placed on a sample holder coated with metal-like cadmium or gold for high conductivity. SEM provides images of a sample surface in high resolution and can provide detailed information about NPs with a size range of less than 5 nm (Pal et al. 2011).

In this type of microscopic techniques, an electron gun generates a very fine focused electron beam that passes through the electromagnetic lens and quickly moves over the surface of the specimen that causes the emission of secondary electrons from the specimen surface. The surface characteristics of metallic NPs can be analysed through the secondary electrons emitted from the surface. The intensity of such secondary electrons depends on the shape as well as the chemical composition of the object. This instrument works very fast, often complete X-ray detector (EDS) and secondary electron detector analysis in less than 5 min (Goel et al. 2019).

9.3.4 Transmission Electron Microscope (TEM)

It is an utmost potent technique used for the characterisation of metallic NPs. TEM provides detailed information about the particle size and shape, interparticle interaction and crystallinity. TEM is a high-resolution structural as well as a chemical characterisation tool. TEM can provide atoms' image directly in crystalline specimens at resolutions close to 0.1 nm, smaller than interatomic distance. This instrument provides the utmost high-definition images than SEM. However, the equipment is highly expensive and does not provide colourful images as it provides black and white images. Like SEM, it also requires maintaining voltage, current-to-electromagnetic coils and circulation of cooling water.

9.3.5 Particle Size Analyser

Particle size is one of the important factors to evaluate metallic NPs. A particle size analyser is utilised to predict the particle size of NPs in nanometre (nm) range. For such purpose, Malvern zeta analyser series is extensively used, which possesses high sensitivity to estimate the size of the sample in nanorange even in low concentration (Thodeti et al. 2017). The backscatter optics analyses the particles with a high focus in comparison with other conventional instruments. The DLS technique is also extensively utilised for estimating NPs.

9.3.6 X-Ray Diffraction (XRD)

X-ray diffraction (XRD) is a suitable analytical technique that gives valuable insight into the crystalline and lattice structures of crystalline nanomaterials, phase nature, and crystallographic structures of both natural and synthetic compounds. This technology relies on the principle of the dual nature of X-rays, i.e. wave or particle nature (Goel et al. 2019). XRD is a rapid-acting instrument primarily used for identifying the phase of the crystalline nanomaterial. Additionally, it can give information regarding atomic spacing and unit cell dimension.

9.3.7 Atomic Force Microscopy (AFM)

This microscopic technique gives quite high magnification to determine the particle size of NPs. The instrument utilises a probe tip for the physical scanning of the nanosample (Goel et al. 2019). AFM provides the topographical data of the sample depend upon the force between the tip and sample surface. Through AFM, samples can be analysed in both noncontact and contact mode depending upon the compound characteristics. In contact mode, the surface properties can be predicted by tapping the probe on the surface of the sample. The probe levitates upon the conducting surface in the noncontact mode. The nonconducting nanosamples such as polymeric nanomaterials and microcrystals can be simply evaluated by this equipment, i.e. the main advantageous feature of this technique (Shi et al. 2003).

9.3.8 Ultraviolet (UV) Spectroscopy

Ultraviolet spectroscopy is extensively utilised to characterise NPs. The spectroscopy is based on the principle of Lambert–Beer's law, according to which the rate of light intensity decreases with the thickness of the medium that is directly proportional to the intensity of light. UV spectroscopy is employed as a potential characterising tool especially for those NPs having optical characteristic sensitive concentration, size, agglomeration, shape and refractive index near the surface of metallic NPs. The instrument works by comparing the intensity of the ray reflected from the sample and the intensity of the ray pass via reference substance (Mourdikoudis et al. 2018).

9.3.9 Dynamic Light Scattering (DLS)

Dynamic light scattering (DLS) is an extensively used analytical technique to predict the particle size of NPs. It is also known as photon correlation spectroscopy. It relies on the principle that when monochromatic light is focused on the solution having dispersed particles, it gets scattered in different directions due to variability in size and shape of the dispersed particles. The intensity of scattered light is determined that tells about the molecular mass and size of dispersed particles. However, in the case of fluctuations in light intensity, the diffusion coefficient (Dt) can be found, i.e. related to the hydrodynamic size of particles (Shi et al. 2003).

9.3.10 Zeta Potential

Zeta potential is utilised for the determination of nature as well as the intensity of surface charge present on NPs. The surface charge helps in ascertaining the interaction of NPs with the biological environment and their electrostatic interaction with bioactive compounds. The potential difference exists between the surface of particles immersed in a conducting or bulk liquid. It is measured in volts (V) or millivolts. Storage or colloidal stability of the dispersed NPs can be speculated via zeta potential. It can also give the knowledge regarding the type of material encapsulated on the surface of NPs (Pal et al. 2011).

9.3.11 Nanoparticle Tracking Analysis (NTA)

It is a comparatively new but frequently accepted technique used to estimate the particle size of NPs. This technique can detect the NPs from the low concentration as compared to DLS (Filipe et al. 2010). NTA uses the features of both light scattering and Brownian movement to determine the size distribution of NPs in the liquid dispersion. This technology does not biased towards large NPs or aggregates which is an important advantage of this technique over others. Additionally, it confirms accuracy and reproducibility and verifies the suitability in the determination of the particle size of samples (Mourdikoudis et al. 2018).

Metallic NPs are extensively utilised by pharmaceutical industries as an effective nanocarrier for delivering a variety of vaccines, drugs, nutraceuticals, cosmetics, etc. Some of these applications are summarised here.

9.4 APPLICATIONS OF METALLIC NANOPARTICLES

9.4.1 Metallic Nanoparticles in Cancer Therapy

Cancer is characterised as the abnormal proliferation of cells inside the body. The traditional chemotherapy aims to destroy all the cancerous cells rapidly in an uncontrollable manner that leaves the life-changing adverse effects on the patient (Hanahan and Weinberg 2011; Baudino 2015). The development of several types of NPs has provided a new avenue for cancer chemotherapy. The NPs can be effectively utilised as an efficient drug delivery system for targeting the anticancerous agent to specific body cells with the least adverse effects of drugs (Shen et al. 2016). NPs can be used in several types of cancers including skin cancer or cutaneous melanoma, breast cancer, prostate cancer, lung cancer and brain cancer. Several research works have been conducted that prove NPs may be a quite fruitful drug delivery system for the effective combat of cancer. A research study reflects the effect of silicon dioxide NPs in the treatment of thyroid cancer when these SiO_2 NPs conjugated with an anticancer drug, i.e. doxorubicin, a specific ligand and thyroid-stimulating hormone receptor. The NPs showed better cytotoxic action with reduced noxious effects associated with doxorubicin (Li et al. 2017). In another study, doxorubicin entrapped AuNPs demonstrated better stability, effectiveness along with reduced side effects of doxorubicin drug than individual doxorubicin (Du et al. 2018).

9.4.2 Nanoparticles for Delivering Peptides and Proteins

The generation and stability of protein structures depend upon noncovalent weak interactions such as hydrogen bonding, electrostatic interactions and hydrophobic interactions. Disruption in any of these interactions in proteins would impair their delicate balance that would destabilise them (Kumar et al. 2015). Thus, the physical and chemical stability of proteins may be critically influenced by various environmental factors like high pressure, pH, temperature, ionic strength, metal ions, nonaqueous solvents, agitation and shearing. Therefore, NPs or nanoparticulate systems can be widely utilised as alternative carriers for delivering therapeutic proteins, peptides and antigens. These NP systems improve the stability of proteins and avoid their proteolytic degradation. Importantly, peptides such as cyclosporine-A, insulin, calcitonin as well as somatostatin have been successfully incorporated into solid–lipid particles (António et al. 2007).

9.4.3 Nanoparticles in Drug Discovery

Nanoparticles (NPs) are effectually utilised to identify and validate target by identifying the protein present on the target surface. They promote the process of drug delivery via automation, miniaturisation, speed as well as reliability of assays. AuNPs and other nanobodies prepared by Ablynx are commonly utilised in diagnosis. The pharmaceutical nanotechnology is employed in the detection of pathogens in humans, isolation and purification of molecules, cells and detoxifying agents. Future nanomachine (respirocyte) is the nano-on-board minicomputer that may be utilised for identifying the antigen causing diseases, viewing the diseased site and delivering the therapeutic agent at a required site (Kumar et al. 2015). Additionally, nanotubes (especially single-walled) are preferably employed in identifying the surface protein of the pathogen.

9.4.4 Ocular Drug Delivery

Several research studies have been performed on NPs for delivering drugs to the eye. The primary problem with ophthalmic formulations is their fast drainage from an eye that implies clearance of loaded drug via the nose. Therefore to prolong the retention time of drug, NPs with adhesiveness is available that provides a high drug level at the required action site. Gasco showed that solid–lipid NPs have a prolonged retention time of drug at the eye without imparting noxious effects. This was assured by utilising radiolabelled formulations and γ-scintigraphy. As the lipids used in solid–lipid NPs are easy to metabolise deliver drug without impairing vision.

9.4.5 Nanoparticles in Molecular Diagnostics/Molecular Imaging

Metallic NPs can be exclusively utilised in magnetic resonance imaging (MRI), ultrasonic imaging, optical imaging, gene expression as well as nuclear imaging (Mao et al. 2016). They may also be used in a specific cell or tissue labelling, i.e. fruitful for long-term imaging, dynamic imaging of subcellular structures, multicolour imaging and fluorescence resonance energy transfer (Kumar et al. 2015). MRI components are displaced by nanomaterials such as metallic NPs, carbon nanotubes, iron oxides and quantum dots as they are quite effectual, stable and provide high magnifying clarified images due to high intensity (Kumar et al. 2015).

9.4.6 Nanoparticles as Biosensors and Biolabels

Nanoparticulate systems can be used to predict various pathological proteins and physiological or biochemical indicators associated with the disease. Biosensor is a detectable system consists of a probe that exhibits a sensitive bioreceptor, physiochemical detector and a transducer to amplify these signals into detectable form. A nanobiosensor is a device having dimensions on the nanometre size scale. These biosensors are frequently employed for validation, detecting targets, toxicity determination as well as prediction regarding absorption, distribution, metabolism and excretion profile (ADME) (Kommareddy et al. 2015).

9.4.7 Carriers for Nasal Vaccine/Drug Delivery

In several research works, it has been reported that nanosystems can be employed as efficient carriers for delivering mucosal vaccines. Various diseases like influenza, viral infections of the respiratory system, measles involve the microorganisms attack across the surface of respiratory mucosa. The vaccination against such diseases to attain desirable mucosal defence at the primary site of infection is essential to eliminate the risks of infection. This can be done by giving vaccines via nasal route as both mucosal and systemic immune responses may be induced if the vaccine is adjuvant by an immune stimulator.

9.4.8 Nutraceutical Delivery

Nutraceuticals are standardised supplements derived from food. These components are commonly taken in the diet to get additional health benefits and lower the risks of various chronic illnesses (Rizvi and Salehb 2018). Like drugs, the bioavailability, as well as efficacy profile of orally consumable nutraceuticals, is influenced by various parameters such as aqueous solubility, interactions with food, degradation, epithelial permeability (McClements and Li 2015). Several nutraceuticals are lipophilic such as lipid-soluble vitamins, polyunsaturated fatty acids and various phytopharmaceutical agents. Therefore, NP formulation can be utilised to enhance the dissolution behaviour of nutraceuticals and for the rectification of various complications associated with them (Rizvi and Salehb 2018; McClements and Li 2015).

9.5 GREEN TECHNOLOGY

Green technology, also known as sustainable technology, takes into consideration the utilisation of plant extracts and various algae for the manufacturing of nanomaterials ecologically (Parveen et al. 2016). Green-tech is a safeguard to the environment during the development process and aims to repair the damage, protect and conserve the environment. Nowadays, green-tech has been used in a variety of stations with immense applications. It is used in recycling and waste management processes, water recycling, cleaning processes (air, drinking water and environmental hazards) and operation of waste by-products produced by fossil fuel.

9.5.1 Principle

The fundamental principle of green technology is that the method of operation and production of NPs should be done in such a way to minimise the potential hazards by limiting toxicity to the environment and to maximise the abundance, efficiency and rate of productivity.

In the history of NP synthesis, the focus has always been on chemical, physical or hybrid techniques. These methods are conventional and favoured by many of the students, researcher and scientists. Chemical method including chemical reduction method, Pechini technique, Brust–Schiffrin method, electrochemical and thermal process, Turkevich method and solution–gel technique (Satyanarayana 2018). Similarly, physical methods comprise lithographic process, pulsed fibre laser/continuous wave laser excision, pyrolysis and high-level energy ball milling technique. The limitation behind these methods involves the use of organic solvents, oxidising and reducing agents, stabilising agents, polymerising agents that have been considered lethal for both the humans and environments. The commercial production of nanoparticles utilizes much energy, fuel and capital that subsequently produces biowaste. Thus, the approach is needed to demote the high-risk and dangerous waste production that promotes the antigreen and unsustainable environment. These elements raise the demand of biodegradable, nonpolluting, economical, energy saving mechanism for synthesis of NPs terminating or lessen the usage of poisonous chemical that imparts hazardous waste to the environment, i.e. green technology (Bhardwaj and Neelam 2020).

With recent development in green technology, there has been a rapid rise in the use of microbiota related to plants (leaf, flower and fruit extracts), which comprises bacteria and *fungi* with the aim of industrial, pharmaceutical, medical and engineering applications (Bhardwaj 2015).

9.5.2 Mechanism

Various studies show that phytomining of heavy metals has capability of manufacturing metal NPs (Iravani 2011). *Medicago sativa* L. also called as alfalfa or lucerne plant has been found to be used in the synthesis of AuNPs and AgNPs. The plant has been grown in the atmosphere of tetrachloroaurate ion where it works by up taking the ions of gold and silver from media. Afterwards, various spectrometric methods (SEM, EDX) have been used for the assessment of AuNP genesis.

It has been studied that alfalfa biomass is now been used for the production of FeO NPs on large scale, wherein pH plays a crucial role in the development of Fe_2O_3 NPs. Acidic environment leads (pH < 5) to the production of large particles, whereas in basic atmosphere (pH 10) comparable smaller particles were produced (Iravani 2011).

9.5.3 Advantages of Green Synthesis of Nanoparticles

- As compared to conventional technology, green technology consumes lesser energy.
- Green technology contributes to the national economy due to lesser capital investment during the processes.
- Motivate sustainable development to conserve the environment for future generations.
- Reduce the consumption of fuel, energy and wastage of clean water.
- Encourage the use of renewable resources and reduce the dependence on non-renewable resources.
- Low maintenance cost.

Shortcomings in green synthesis of NPs

- Green technology is an expensive technology.
- Due to its novelty, lots of resources are unavailable.
- Insufficient information available.
- No other alternative process/technology available.
- Higher possibility of uncertain performance.
- Less availability of skilled manpower.
- Green energy sources cannot be established at all stations on earth.

9.5.4 Synthesis of Metallic Nanoparticles from Green Synthesis

Some of the metallic oxide NPs like zinc oxide and FeO NPs can be successfully synthesised from the green technology with any quality alteration. The method of preparation of such metallic NPs is discussed below.

The green synthesis of ZnO NPs has become popular among researchers and scientists because of its simplicity, eco-friendliness, nontoxic, inexpensive and antibacterial potential. Zinc oxide NPs can also be synthesised by green technology/synthesis that has been seen as beneficial in the clinical antimicrobial wound healing bandages (Gour and Jain 2019). The green synthesis of metal NPS is an interesting issue of nanoscience and nanotechnology. NPS produced by plants are more stable, and their rate of synthesis is faster than that in the case of other organisms. Zinc oxide NPs were prepared by using the the medicinal plant *Cassia auriculata* (Tanner's cassia). The particles were synthesized by mixing 1 mM aqueous of zinc acetate with aqueous extract of *Cassia auriculata* L. flower. (Ramesh et al. 2014). Alijani et al. (2019) has synthesized ZnS nanoparticles using a natural sweetener glycoside in the aqueous crude extract of *Stevia rebaudiana* that acted as an excellent bio-reductant. FTIR spectra confirmed the presence of glucose as a capping agent and stabilizer. The NPs were found to be highly stable and predominantly spherical.

FIGURE 9.1 Schematic presentation of ZnO NPs synthesis using *Ailanthus altissima* L. fruit extract and antimicrobial activity (Awwad et al. 2020).

The green synthesis of metal NPs has been approached by many researchers. Yadi et al. (2018) studied the various common plants which are regularly used to synthesize metal NPs along with the various methods for synthesizing metal NPs from plant extracts. A study reveals that the powder of green tea leaf was dried and added to distilled water. The solution is mixed at a constant speed. After that, zinc acetate was dissolved in mixture to obtain zinc acetate solution. The extract was added to the solution and dried for 12 h to yield white ZnO NPs and then desiccated at 100°C (Dhanemozhi et al. 2017).

Another case was described by using fruit peel extract to manufacture the zinc NPs as shown in Figure 9.1. Firstly, the fruits were peeled off and kept for drying for 12 h. Secondly, the powder was dissolved in the deionised water, stirred well, filtered the solution, and added to the zinc nitrate and mixed wisely. Then, the solution was placed on a water bath to complete the process. The mixture kept for heating at 400°C to produce a white-coloured powder. For this experiment, orange, lemon and grapefruit peel can be used to obtain NPs (Nava et al. 2017).

9.6 CONCLUSION

In recent years, metallic NPs have become significant nanocarriers due to their unique physical as well as chemical properties that make them suitable for drug delivery and drug targeting. They have several advantages like reduced toxicity, better bioavailability, pronounced therapeutic effect, better patient compliance, less cost of overall therapy and improved stability. Metallic NPs can be utilised in several pharmaceutical fields as their applications are increasing worldwide in various fields such as biomedicine, drug targeting, material science, physics and chemistry. They can be synthesised through a variety of methods and each method possesses its own advantage and limitations. They can be characterised easily via several technologies for various parameters like particle size, particle shape, morphology, surface charge and stability. However, this thrust area needs more research as metallic NPs may be the fruitful approach for treating various fatal diseases and enhance the quality life of patients.

REFERENCES

Achachelouei, M.F., Marques, H.K., Ribeiro da Silva, C.E., Barthès, J., Bat, E., Tezcaner, A., Vrana, N.E. 2019. Use of Nanoparticles in Tissue Engineering and Regenerative Medicine. *Frontiers in Bioengineering and Biotechnology* 7(113): 1–22.

Ahammed, K.R., Ashaduzzaman. M., Paul. S.C. et al. 2020. Microwave assisted synthesis of zinc oxide (ZnO) nanoparticles in a noble approach: Utilization for antibacterial and photocatalytic activity. *SN Appl. Sci.* 2(955). doi:10.1007/s42452-020-2762-8.

Alijani, H.Q., Pourseyedi, S., TorkzadehMahani, M. et al. 2019. Green synthesis of zinc sulfide (ZnS) nanoparticles using stevia rebaudianabertoni and evaluation of its cytotoxic properties. *Journal of Molecular Structure* 1175: 214–218.

Alvarez, G.S., Muhammed, M., Zagorodni, A.A. 2006. Novel flow injection synthesis of iron oxide nanoparticles with narrow size distribution. *Chemical Engineering Science.* 61: 4625–4633.

Ammar, S., Fievet, F. 2020. Polyol synthesis: a versatile wet-chemistry route for the design and production of functional inorganic nanoparticles. *Nanomaterials* 10: 1217. doi:10.3390/nano10061217.

Amorim, M.O.R., Gomes, D.L., Dantas, L.A. et al. 2016. Fucan-coated silver nanoparticles synthesized by a green method induce human renal adenocarcinoma cell death. *International Journal of Biological Macromolecules* 93: 57–63. doi:10.1016/j.ijbiomac.2016.08.043.

António, J., Almeida, A., Eliana, S. 2007. Solid lipid nanoparticles as a drug delivery system for peptides and proteins. *Advanced Drug Delivery Reviews* 59: 478–490.

Arroyo, G.V., Madrid, A.T., Gavilanes A.F. et al. 2020. Green synthesis of silver nanoparticles for applications in cosmetic. *Journal of Environmental Science and Health.* doi:10.1080/10934529.2020.1790953.

Arsent'ev, M.Y., Kalinina, M.V., Tikhonov, P.A., Morozova, L.V. et al. 2014. Synthesis and study of sensor oxide nanofilm in a ZrO_2-CeO_2 system. *Glass Physics and Chemistry* 40(3): 1–6. doi:10.1134/S1087659614030031.

Awwad, A.M., Amer, M.W., Salem, N.M., Abdeen, A.O. 2020. Green synthesis of zinc oxide nanoparticles (ZnO-NPs) using *Ailanthus altissima* fruit extracts and antibacterial activity. *Chemistry International* 6(3): 151–159.

Ayanwale, A.P., Cornejo, A.D., González, J.C.C., Espinosa-Cristóbal, L.F., Reyes-López, S.Y. 2018. Review of the synthesis, characterization and application of zirconia mixed metal oxide nanoparticles. *International Journal of Research–Granthaalayah* 6(8): 136–145.

Baban, D., Seymour, L.W. 1998. Control of tumour vascular permeability. *Advanced Drug Delivery Reviews* 34: 109–119.

Badnore, A.U., Sorde, K.I., Datir K.A. et al. 2019. Preparation of antibacterial peel-off mask formulation incorporating biosynthesized silver nanoparticles. *Applied Nanoscience* 9: 279–287. doi:10.1007/s13204-018-0934-2.

Badrzadeh, F., Akbarzadeh, A., Zarghami, N. et al. 2014. Comparison between the effects of free curcumin and curcumin loaded NIPAAm-MAA nanoparticles on telomerase and PinX1 gene expression in lung cancer cells. *Asian Pacific Journal of Cancer Prevention* 15(20): 8931–8936.

Basak, S., Chen, D.R., Biswas, P. 2007. Electrospray of ionic precursor solutions to synthesize iron oxide nanoparticles: Modified scaling law. *Chemical Engineering Science* 62: 1263.

Baudino, T.A. 2015. Targeted cancer therapy: The next generation of cancer treatment. *Current Drug Discovery Technology* 12 (1): 3–20.

Behbahani, A., Rowshanzamir, S., Esmaeilifar, A. 2012. Hydrothermal synthesis of zirconia nanoparticles from commercial zirconia. *Procedia Engineering* 42: 908–917.

Bhardwaj, B., Singh, P., Kumar, A., Kumar, S., Budhwar, V. 2020. Eco-friendly greener synthesis of nanoparticles. *Advanced Pharmaceutical Bulletin* 10(4): 566–576.

Bhardwaj, M., Neelam, N. 2015. The advantages and disadvantages of questionnaires. *Journal of Basic and Applied Engineering Research* 2(22): 1957–1960.

Bullivant, J.P., Zhao, S., Willenberg, B.J., Kozissnik, B., Batich, C.D., Dobson, J. 2013. Materials characterization of feraheme/ferumoxytol and preliminary evaluation of its potential for magnetic fluid hyperthermia. *International Journal of Molecular Science* 14: 17501–17510.

Ceylan, A., Jastrzembski, K., Shah, S.I. 2006. Enhanced solubility Ag-Cu nanoparticles and their thermal transport properties. *Metallurgical and Materials Transaction* 37: 2033–2038.

Chastellain, M., Petri, A., Gupta, A., Rao, K.V., Hofmann, H. 2004. Scalable synthesis of a new class of polymer microrods by a liquid-liquid dispersion technique. *Advanced Engineering Materials* 6(235): 1653–1657.

Chatterjee, D.K., Diagaradjane, P., Krishnan, S. 2011. Nanoparticle-mediated hyperthermia in cancer therapy. *Therapeutic Delivery* 2: 1001–1014.

Cui, H., Liu, Y., Ren, W. 2013. Structure switch between α-Fe_2O_3, γ-Fe_2O_3 and Fe_3O_4 during the large scale and low temperature sol–gel synthesis of nearly monodispersed iron oxide nanoparticles. *Advanced Powder Technology* 24(1): 93–97.

Cushing, B. L., Kolesnichenko, V. L., O'Connor, C.J. 2004. Recent advances in the liquid-phase syntheses of inorganic nanoparticles. *Chemical Reviews* 104(9): 3893–3946. doi:10.1021/cr030027b.

Dadfar, S.M., Camozzi, D., Darguzyte, M. et al. 2020. Size-isolation of superparamagnetic iron oxide nanoparticles improves MRI, MPI and hyperthermia performance. *Journal of Nanobiotechnology* 18(22). doi:10.1186/s12951-020-0580-1.

Dhanemozhi, A.C., Rajeswari, V., Sathyajothi, S. 2017. Green synthesis of zinc oxide nanoparticle using green tea leaf extract for supercapacitor application. *Materials Today Proceeding* 4: 660–667.

Daneshvar, N., Aber, S., Dorraji, M.S., Khataee, A., Rasoulifard, M. 2008. Preparation and investigation of photocatalytic properties of ZnO nanocrystals: Effect of operational parameters and kinetic study. *International Journal of Chemical, Molecular, Nuclear, Materials and Metallurgical Engineering.* 2: 62–67.

Diaz, A., González, M.L., Pérez, R.J., David, A. 2013. Direct intercalation of cisplatin into zirconium phosphate nanoplatelets for potential cancer nanotherapy. *Nanoscale* 5(23): 11456–11463.

Du, Y., Xia, L., Jo, A., Davis, R.M, Bissel, P., Ehrich, M.F., Kingston, D.G.I. 2018. Synthesis and evaluation of doxorubicin-loaded gold nanoparticles for tumour-targeted drug delivery. *Bioconjugation Chemistry* 29(2): 420–430.

Elahi, N., Kamali, M., Baghersad, M.H. 2018. Recent biomedical applications of gold nanoparticles: A review. *Talanta* 184: 537–556.

Elechiguerra, J.L., Burt, J.L., Morones, J.R. et al. 2005. Interaction of silver nanoparticles with HIV-1. *Journal of Nanobiotechnoly* 3(6): 1–10.

Filipe, V. Hawe, A., Jiskoot, W. 2010. Critical evaluation of nanoparticle tracking analysis (NTA) by nano sight for the measurement of nanoparticles and protein aggregates. *Pharmaceutical Research* 27: 796–810.

Gad, M.M., Rahoma, A., Al-Thobity, A.M., Arrejaie, A.S. 2016. Influence of incorporation of ZrO_2 nanoparticles on the repair strength of polymethyl methacrylate denture bases. *International Journal of Nanomedicine* 11: 5633–5643.

Garino, N., Limongi, T., Dumontel, B. et al. 2019. A microwave-assisted synthesis of zinc oxide nanocrystals finely tuned for biological applications. *Nanomaterials* 9(2): 212.

Ghaffari, H., Tavakoli, A., Moradi, A. et al. 2019. Inhibition of H1N1 influenza virus infection by zinc oxide nanoparticles: Another emerging application of nanomedicine. *Journal of Biomedical Sciences* 26: 1–10.

Gholamali, I., Asnaashariisfahani, M., Alipour, E. 2019. Silver nanoparticles incorporated in pH-sensitive nanocomposite hydrogels based on carboxymethyl chitosan-poly (vinyl alcohol) for use in a drug delivery system. *Regenerative Engineering and Translational Medicine* 6: 138–153. doi:10.1007/s40883-019-00120-7.

Goel, S., Sachdeva, M., Agarwal, V. 2019. Nanosuspension technology: Recent patents on drug delivery and their characterizations. *Recent Patents on Drug Delivery & Formulation* 13: 91–104.

Gour, A., Jain, N.K. 2019. Advances in green synthesis of nanoparticles. *Artificial Cells, Nanomedicine, and Biotechnology* 47(1): 844–851.

Guazzato, M., Albakry, M., Ringer, S.P., Swain, M.V. 2004. Strength, fracture toughness and microstructure of a selection of all-ceramic materials. Part II. Zirconia-based dental ceramics. *Dental Material Journal* 20: 449–456.

Gupta, A., Briffa, S.M., Swingler, S. et al. 2020. Synthesis of silver nanoparticles using curcumin-cyclodextrins loaded into bacterial cellulose-based hydrogels for wound dressing applications. *Biomacromolecules* 21:1801–1811. doi:10.1021/acs.biomac.9b01724.

Gunalan, S., Sivaraj, R., Rajendran, V. 2012.Green synthesized ZnO nanoparticles against bacterial and fungal pathogens. *Progress in Natural Science: Materials International.* 22: 695–702.

Haider, A., Kang, I.K. 2015. Preparation of silver nanoparticles and their industrial and biomedical applications: A comprehensive review. *Advances in Materials Science and Engineering* 2015: 1–16.

Hanahan, D., Weinberg, R.A. 2011. Hallmarks of cancer: The next generation. *Cell* 144: 646–674.

Hasan, S. 2015. A review on nanoparticles: Their synthesis and types. *Research Journal of Recent Sciences* 4: 1–3.

Hasanpoor, M., Aliofkhazraei, M., Delavari, H. 2015. Microwave-assisted Synthesis of zinc oxide nanoparticles. *Procedia Materials Science* 11: 320–325.

Hayat, K., Gondal, M.A., Khaled, M.M., Ahmed, S., Shemsi, A.M. 2011. Nano ZnO synthesis by modified sol gel method and its application in heterogeneous photocatalytic removal of phenol from water. *Applied Catalysis A: General.* 393: 122–129.

He, W., Zheng, Y., Feng, Q. et al. 2020. Silver nanoparticles stimulate osteogenesis of human mesenchymal stem cells through activation of autophagy. *Nanomedicine* 15(4): 337–353. doi:10.2217/nnm-2019-0026.

He, F., Zhao, D. 2005. Preparation and characterization of a new class of starch-stabilized bimetallic nanoparticles for degradation of chlorinated hydrocarbons in water. *Environmental Science & Technology.* 39: 3314–3320.

Hostynek, J.J., Maibach, H.I. 2002. Silver, titanium and zirconium: Metals in cosmetics and personal-care products. *Cosmetics & Toiletries* 117(1): 1–4.

Hooda, H., Singh, P., Bajpai, S. 2020. Effect of quercetin impregnated silver nanoparticle on growth of some clinical pathogens. *Materials Today Proceedings*. doi:10.1016/j.matpr.2020.03.530.

Huang, F., Baumberg, J.J. 2010. Actively tuned plasmons on elastomerically driven Au nanoparticle dimers. *Nano Letters* 10: 1787–1792.

Hussein, J., Attia, M.F., Bana, M.E., El-daly, S.M., Mohamed, N., El-khayat, Z., El-Naggar, M.E. 2019. Solid state synthesis of docosahexaenoic acid-loaded zinc oxide nanoparticles as a potential antidiabetic agent in rats. *International Journal of Biological Macromolecules* 140: 1305–1314.

Iravani, S. 2011. Green synthesis of metal nanoparticles using plants. *Green Chemistry* 13(10): 2638–2650.

Jin, S.E., Jin, J.E., Hwang, W., Hong, S.W. 2019. Photocatalytic antibacterial application of zinc oxide nanoparticles and self-assembled networks under dual UV irradiation for enhanced disinfection. *International Journal of Nanomedicine* 14: 1737–1751.

Joo, J., Yu, T., Kim, Y.W., Park, H.M., Wu, F., Hyeon, T. 2016. Multigram scale synthesis and characterization of monodisperse tetragonal zirconia nanocrystals. *Journal of the American Chemical Society* 125(21): 6553–6557.

Jung, J.H., Cheol, H., Soo-Noh, H., Ji, J.H., Kim, S.S. 2006. Metal nanoparticle generation using a small ceramic heater with a local heating area. *Journal of Aerosol Science* 37(12): 1662–1670.

Kahn, H.R., Petrikowski, K. 2000. Anisotropic structural and magnetic properties of arrays of Fe26Ni74 nanowires electrodeposited in the pores of anodic alumina. *Journal of Magnetism and Magnetic Materials.* 526: 215–216.

Kara, A., Tobyn, M., Steven, R. 2004. An application for zirconia as a pharmaceutical die set. *Journal of the European Ceramic Society* 24(10–11): 3091–3101.

Katas, H., Sin Lim, C., Hamdi Nor Azlan, A.Y., Buang, F., MhBusra, M.F. 2019. Antibacterial activity of biosynthesized gold nanoparticles using biomolecules from Lignosus rhinocerotis and chitosan. *Saudi Pharmaceutical Journal* 27: 283–292.

Kermanian, M., Naghibi, M., Sadigham, S. 2020. One-pot hydrothermal synthesis of a magnetic hydroxyapatite nanocomposite for MR imaging and pH-sensitive drug delivery applications. *Heliyon* 6: e04928. doi:10.1016/j.heliyon.2020.e04928.

Kim, D., Vasilieva, E.S., Nasibulin, A.G., Lee, D.W., Tolochko, O.V., Kim, B.K. 2007. Aerosol synthesis and growth mechanism of magnetic iron nanoparticles. *Materials Science Forum* 534–536: 9–12.

Kimling, J., Maier, M., Oken, B., Kotaidis, V., Ballot, H., Plech, A. 2006. Turkevich method for gold nanoparticle synthesis revisited. *Journal of Physical Chemistry B* 110(32): 15700–15707.

Kooti, M., Naghdi Sedeh, A. 2012, Microwave-assisted combustion synthesis of ZnO nanoparticles. *Journal of Chemistry* 2013: 1–4. Article ID 562028.

Kommareddy, S., Tiwari, S.B., Amiji, M.M. 2005. Long-circulating polymeric nanovectors for tumor-selective gene delivery. *Technology Cancer Research Treatment* 4: 615–625.

Kumar, A., Das, N., Satija, N.K., Mandurah, K., Roy, S.K., Rayavarapu, R.G. 2020. A novel approach towards synthesis and characterization of non-cytotoxic gold nanoparticles using taurine as capping agent. *Nanomaterials* 10(45): 1–19.

Kumar, A., Kuang, Y., Liang, Z., Sun, X. 2020. Microwave chemistry, recent advancements and eco-friendly microwave-assisted synthesis of nanoarchitectures and their applications: A review. *Materials Today Nano* 11: 100076. doi:10.1016/j.mtnano.2020.100076.

Kumar, P., Kulkarni, P.K., Srivastava, A. 2015. Pharmaceutical application of nanoparticles in drug delivery system. *Journal of Chemical and Pharmaceutical Research* 7(8): 703–712.

Kumar, S.S., Venkateswarlu, P., Rao, V.R., Rao, G.N. 2013. Synthesis, characterization and optical properties of zinc oxide nanoparticles. *International Nano Letters* 3: 30.

Larenkov, A., Bubenschikov, V., Makichyan, A., Zhukova, M., Krasnoperova, A., Kodina, G. 2019. Preparation of Zirconium-89 solutions for radiopharmaceutical purposes: interrelation between formulation, radiochemical purity, stability and biodistribution. *Molecules* 24(1534): 1–24.

Laurent, S., Forge, D., Port, M., Roch, A., Robic, C., Elst. L.V., Muller, R.N. 2008. Magnetic iron oxide nanoparticles: Synthesis, stabilization, vectorization, physicochemical characterizations, and biological applications. *Chemical Reviews* 108: 2064–2110.

Li, C., Shuford, K.L., Park, Q., Cai, W., Li, Y. 2007. High-yield synthesis of single crystalline gold nano-octahedra. *AngewandteChemie* 46(18): 3264–3268.

Li, D., Tang, X., Pulli, B. et al. 2014. Theranostic nanoparticles based on bioreducible polyethyleneimine-coated iron oxide for reduction-responsive gene delivery and magnetic resonance imaging. *International Journal of Nanomedicine* 9: 3347–3361.

Li, Q., Deacon, A.D., Coleman, N.J. 2013. The impact of zirconium oxide nanoparticles on the hydration chemistry and biocompatibility of white Portland cement. *Dental Material Journal* 32: 808–815.

Li, S., Zhang, D., Sheng, S., Sun, H. 2017. Targeting thyroid cancer with acid-triggered release of doxorubicin from silicon dioxide nanoparticles. *International Journal of Nanomedicine* 12: 5993–6003.

Liu, G., Lu, M., Huang, X., Li, T., Xu, D. 2018. Application of gold-nanoparticle colorimetric sensing to rapid food safety screening. *Sensors* 18(12): 4166.

Liu, X.G., Zhang, L., Lu, S., Liu, D.Q., Zhang, L.X., Yu, X.L., Liu, R.T. 2020. Multifunctional superparamagnetic iron oxide nanoparticles conjugated with Aβ oligomer-specific scFv antibody and Class A scavenger receptor activator show early diagnostic potentials for Alzheimer's disease. *International Journal of Nanomedicine* 15: 4919–4932. doi:10.2147/IJN.S240953.

seng-Williamson, K.A., Keating, G.M. 2009. Ferric carboxymaltose. *Drugs* 69: 739–756.

Madamsetty, V.S., Mukherjee, A., Mukherjee, S. 2019. Recent trends of the bio-inspired nanoparticles in cancer theranostics. *Frontiers in Pharmacology* 10: 1264.

Man, F., Khan, A., Minino, A., Blower, P.J., De Rosale, R.T.M. 2020. A kit formulation for the preparation of [^{89}Zr]Zr(oxinate)$_4$ for PET cell tracking: White blood cell labelling and comparison with [^{111}In]In(oxinate)$_3$. *Nuclear Medicine and Biology* 90–91: 31–40. doi:10.1016/j.nucmedbio.2020.09.002.

Mao, X., Xu, J., Cui, H. 2016. Functional nanoparticles for magnetic resonance imaging. *Wiley Interdisciplinary Review of Nanomedicine and Nanobiotechnology* 8(6): 814–841.

McClements, D.J., Li, F. 2015. The nutraceutical bioavailability classification scheme: Classifying nutraceuticals according to factors limiting their oral bioavailability. *Annual Review of Food Science Technology* 6: 299–327.

Meulenkamp, E.A. 1998. Synthesis and growth of ZnO nanoparticles. *The Journal of Physical Chemistry B* 102: 5566–5572.

Mirela, D. 2009. Metallic nanoparticles. University of Nova Gorica doctoral study, Programme Physics.

Mirhosseini, M., Far, A.S., Hakimian. F. et al. 2020. Core shell Au@Co-Fe hybrid nanoparticles as peroxidase mimetic nanozyme for antibacterial application. *Process Biochemistry* 95: 131–138. doi:10.1016/j.procbio.2020.05.003.

Mishra, P.K., Mishra, H., Ekielski, A., Talegaonkar, S., Vaidya, B. 2017. Zinc oxide nanoparticles: A promising nanomaterial for biomedical science application. *Drug Discovery Today* 22(12): 1825–1834.

Mittal, L., Ranjani, S., Shariq, A.M. et al. 2020. Turmeric-silver-nanoparticles for effective treatment of breast cancer and to break CTX-M-15 mediated antibiotic resistance in *Escherichia coli*. *Inorganic and Nano-Metal Chemistry.* doi:10.1080/24701556.2020.1812644.

Mody, V.V., Siwale, R., Singh, A., Mody, H.R. 2010. Introduction to metallic nanoparticles. *Journal of Pharmacy and Bioallied Sciences* 2(4): 282–289.

Mohammadzadeh, R. 2012. Hypothesis: Silver nanoparticles as an adjuvant for cancer therapy. *Advanced Pharmaceutical Bulletin* 2(1): 133.

Mourdikoudis, S., Pallares, R.M., Thanh, N. 2018. Characterization techniques for nanoparticles: Comparison and complimentary upon studying nanoparticles properties. *Nanoscale* 10: 12871–12934.

Nava, O.J., Soto-Robles, C.A., Gomez-Gutierrez, C.M. et al. 2017. Fruit peel extract mediated green synthesis of zinc oxide nanoparticles. *Journal of Molecular Structure* 1147: 1–6.

Naqvi, S.T.R., Rasheed, T., Hussain, D. et al. 2020. Development of molecularly imprinted magnetic iron oxide nanoparticles for doxorubicin drug delivery. *Monatshefte für Chemie* 151: 1049–1057. doi:10.1007/s00706-020-02644-z.

Noqta, O.A., Aziz, A.A., Usman, I.A., Bououdina, M. 2019. Recent advances in iron oxide nanoparticles (IONPs): Synthesis and surface modification for biomedical applications. *Journal of Superconductivity and Novel Magnetism.* doi:10.1007/s10948-018-4939-6.

Nu Diem, P.H., Thu Thao, D.T., Van Phu, D., Ngoc Duy, N., Dong Quy, H.T., Thai Hoa, T., Quoc Hien, N. 2017. Synthesis of gold nanoparticles stabilized in dextran solution by gamma Co-60 ray irradiation and preparation of gold nanoparticles/dextran powder. *Journal of Chemistry* 2017: 1–8. Article ID 6836375.

Oldenburg, S.J. 2020. Silver nanoparticles: Properties and applications. https://www.sigmaaldrich.com/technical-documents/articles/materials science/nanomaterials/silver-nanoparticles.html.

Otari, S.V., Patel, S.K.S., Kalia, V.C., Lee, J.K. 2020. One-step hydrothermal synthesis of magnetic rice straw for effective lipase immobilization and its application in esterification reaction. *Bioresource Technology*: 122887. doi:10.1016/j.biortech.2020.122887.

Padmanabhan, A., Kaushik, M., Niranjan, R., Richards, J.S., Ebright, B., Venkatasubbu, G.D. 2019. Zinc oxide nanoparticles induce oxidative and proteotoxic stress in ovarian cancer cells and trigger apoptosis independent of p53-mutation status. *Applied Surface Science.* 487: 807–818.

Pal, S.L., Jana, U., Manna, P.K., Mohanta, G.P., Manavalan, R. 2011. Nanoparticle: An overview of preparation and characterization. *Journal of Applied Pharmaceutical Science* 1: 228–234.

Panikkanvalappil, S.R., Hooshmand, N., El-Sayed, M.A. 2017. Intracellular assembly of nuclear-targeted gold nanosphere enables selective plasmonic photothermal therapy of cancer by shifting their absorption wavelength toward the near-infrared region. *Bioconjugate Chemistry* 28: 2452–2460.

Parveen, K., Viktoria, B., Lalita L. 2016. Green synthesis of nanoparticles: Their advantages and disadvantages. *AIP Conference Proceedings* 1724: 020048.

Pati, R., Kumar Mehta, R., Mohanty, S., Padhi, A., Sengupta, M., Vaseeharan, B., Goswami, C., Sonawane, A. 2014. Topical application of zinc oxide nanoparticles reduces bacterial skin infection in mice and exhibits antibacterial activity by inducing oxidative stress response and cell membrane disintegration in macrophages. *Nanomedicine: Nanotechnology, Biology and Medicine* 10: 1195–1208.

Prodan, E., Nordlander, P., Halas, N.J. 2003. Electronic structure and optical properties of gold nanoshells. *Nano Letters* 3: 1411–1415.

Qi, H., Yan, B., Lu, W. 2014. A facile synthetic pathway of monodisperse Fe_3O_4 nanocrystals. *Journal of Sol-Gel Science and Technology* 69(1): 67–71.

Ramesh, P., Rajendran, A., Sundaram, M. 2014. Green synthesis of zinc oxide nanoparticles using flower extract *Cassia auriculata*. *Journal of Nanoscience and Nanotechnology* 2: 41–45.

Ramimoghadam, D., Bagheri, S., Hamid, S.B.A. 2014. Progress in electrochemical synthesis of magnetic iron oxide nanoparticles. *Journal of Magnetism and Magnetic Materials* 368: 207–229. doi:10.1016/j.jmmm.2014.05.015.

Rani, S., Suri, P., Shishodia, P.K., Mehra, R.M. 2008. Synthesis of nanocrystalline ZnO powder via sol–gel route for dye-sensitized solar cells. *Solar Energy Materials and Solar Cells* 92: 1639–1645.

Rizvi, A.A., Salehb, A.M. 2018. Applications of nanoparticle systems in drug delivery technology. *Saudi Pharmaceutical Journal* 26: 64–70.

Rodriguez, B.A.G., Caro, M.P., Alencer, R.S. et al. 2020. Graphene nanoribbons and iron oxide nanoparticles composite as a potential candidate in DNA sensing applications. *Journal of Applied Physics* 127: 044901. doi:10.1063/1.5130586.

Rosati, G., Cunego, A., Fracchetti, F. et al. 2019. Inkjet printed interdigited biosensor for easy and rapid detection of bacteriophage contamination: A preliminary study for milk processing control applications. *Chemosensors* 7(1): 8. doi:10.3390/chemosensors7010008.

Rylski, A., Siczek, K. 2020. The effect of addition of nanoparticles, especially ZrO_2-based, on tribological behavior of lubricants. *Lubricants* 8(23): 1–25.

Sabir, S., Arshad, M., Chaudhari, S.N. 2014. Zinc oxide nanoparticles for revolutionizing agriculture: Synthesis and applications. *The Scientific World Journal* 2014: 925494.

Sadhukhan, P., Kundi, M., Chatterjee, S., Ghosh, N., Manna, P., Das, J., Sil, P. 2019. Targeted delivery of quercetin via pH responsive zinc oxide nanoparticles for breast cancer therapy. *Materials Science and Engineering: C* 100: 129–140.

Saloga, P.E.J, Thünemann, A.F. 2019. Microwave-assisted synthesis of ultra-small zinc oxide nanoparticles. *Langmuir* 35(38): 12469–12482.

Sanaeimehr, Z., Javadi, I., Namvar, F. 2018. Antiangiogenic and antiapoptotic effects of green-synthesized zinc oxide nanoparticles using *Sargassum muticum* algae extraction. *Cancer Nanotechnology* 9: 1–6.

Satyanarayana, T. 2018. A review on chemical and physical synthesis methods of nanomaterials. *International Journal for Research in Applied Science and Engineering Technology* 6(1): 2885–2889.

Schultz, S., Smith, D.R., Mock, J.J., Schultz, D.A. 2000. Single-target molecule detection with nonbleaching multicolour optical immunolabels. *Proceedings of the National Academy of Sciences of the USA* 97: 996–1001.

Seabra, A.B., Durán, N. 2015. Nanotoxicology of metal oxide nanoparticles. *Metals* 5: 934–975.

Shamaila, S., Zafar, N., Riaz, S., Sharif, R., Nazir, J., & Naseem, S. 2016. Gold Nanoparticles: An Efficient Antimicrobial Agent against Enteric Bacterial Human Pathogen. *Nanomaterials* 6(4): 71. doi:10.3390/nano6040071

Shankar, S., Rhim, J.W. 2015. Amino acid-mediated synthesis of silver nanoparticles and preparation of antimicrobial agar/silver nanoparticles composite films. *Carbohydrate Polymer* 130: 353–363.

Sharma, H., Kumar, K., Choudhary, C., Mishra, P.K., Vaidya, B. 2016. Development and characterization of metal oxide nanoparticles for the delivery of anticancer drug. *Artificial Cells, Nanomedicine, and Biotechnology.* 44: 672–679.

Sharma, H., Mishra, P.K., Talegaonkar, S., Vaidya, B. 2015. Metal nanoparticles: A theranostic nanotool against cancer. *Drug Discovery Today* 20(9): 1143–1151.

Sharma, S.K., Pujari, P.K., Sudarshan, K., Dutta, D., Mahapatra, M., Godbole, S.V., Jayakumar, O.D., Tyagi, A.K. 2009. Positron annihilation studies in ZnO nanoparticles. *Solid State Communications* 149(550): 550–554.

Shi, H.G., Farber, L., Michaels, J.N., Dickey, A., Thompson, K.C., Shelukar, S.D. 2003. Characterization of crystalline drug nanoparticles using atomic force microscopy and complementary techniques. *Pharmaceutical Research* 20: 479–484.

Silva, C.O., Pinho, J.O., Lopes, J.M., Almeida, A.J., Gapar, M.M., Reis, C. 2019. Current Trends in cancer nanotheranostics: Metallic, polymeric, and lipid-based systems. *Pharmaceutics* 11(22): 1–40.

Singh, B.N., Rawat, A.K., Khan, W., Naqvi, A.H., Singh, B.R. 2014. Biosynthesis of stable antioxidant ZnO nanoparticles by *Pseudomonas aeruginosa* rhamnolipids. *PLoS One* 9: e106937.

Sintubin, L., Verstraete, W., Boon, N. 2012. Biologically produced nanosilver: Current state and future perspectives. *Biotechnology and Bioengineering* 109(10): 2422–2436.

Sonia, S., Linda Jeeva Kumari, H., Ruckmani, K., Sivakumar, M. 2017. Antimicrobial and antioxidant potentials of biosynthesized colloidal zinc oxide nanoparticles for a fortified cold cream formulation: A potent nanocosmeceutical application. *Materials Science and Engineering C* 79: 581–589.

Souza, J.M.T., de Araujo, A.R., de Carvalho, A.M., Amorim, A.D., Daboit, T.C., de Almeida, J.R., da Silva, D.A., Eaton, P. 2020. Sustainably produced cashew gum capped zinc oxide nanoparticles show antifungal activity against *Candida parapsilosis*. *Journal of Cleaner Production* 247: 119085.

Spanhel, L., Anderson, M.A. 1991. Semiconductor clusters in the sol–gel process: Quantized aggregation, gelation, and crystal growth in concentrated zinc oxide colloids. *Journal of the American Chemical Society* 113: 2826–2833.

Sun, C., Veiseh, O., Gunn, J. et al. 2008. *In vivo* MRI detection of gliomas by chlorotoxin-conjugated superparamagnetic nanoprobes. *Small* 4: 372–379.

Surendra, T.V., Roopan, S.M., Al-Dhabi, N.A., Arasu, M.V., Sarkar, G., Suthindhiran, K. 2016. Vegetable peel waste for the production of ZnO nanoparticles and its toxicological efficiency, antifungal, hemolytic, and antibacterial activities. *Nanoscale Research Letters* 11: 546.

Taufikurohmah, T., Sanjaya, G.M., Baktir, A., Syahrani, A. 2014. TEM analysis of gold nanoparticles synthesis in glycerin: Novel safety materials in cosmetics to recovery mercury damage. *Research Journal of Pharmaceutical, Biological and Chemical Sciences* 5(1): 397–407.

Thenmozhi, T. 2020. Functionalization of iron oxide nanoparticles with clove extract to induce apoptosis in MCF-7 breast cancer cells. *3 Biotech* 10: 82. doi:10.1007/s13205-020-2088-7.

Thodeti, S., Bantikatla, H.B., Kumar, Y.K., Sathish, B. 2017. Synthesis and characterization of ZnO nanostructures by oxidation technique. *International Journal of Advance Research Science & Engineering* 6: 539–44.

Thodeti, S., Reddy, R.M., Kumar, J.S. 2016. Synthesis and characterization of pure and indium doped Sno_2 nanoparticles by sol-gel methods. *International Journal of Science Engineering & Research* 7: 310–317.

Tomar, A., Garg, G. 2013. Short review on application of gold nanoparticles. *Global Journal of Pharmacology* 7(1): 34–38.

Vines, J.B., Yoon, J.H., Ryu, N.E., Lim, D.J., Park, H. 2019. Gold nanoparticles for photothermal cancer therapy. *Frontiers in Chemistry* 7(167): 1–16.

Vlamidis, Y., Voliani, V. 2018. Bringing again noble metal nanoparticles to the forefront of cancer therapy. *Frontier in Bioengineering and Biotechnology* 6(143): 1–5.

Wang, Z., Lee, J., Cossins, A.R. et al. 2005. Microarray-based detection of protein binding and functionality by gold nanoparticle probes. *Analytical Chemistry* 7: 5770–5774.

Wang, Z., Zhang, H., Zhang, L., Yuan, J., Yan, S., Wang, C. 2003. Low-temperature synthesis of ZnO nanoparticles by solid state pyrolytic reaction. *Nanotechnology* 14: 11–15.

Weissig, V., Pettinger, T.K., Murdock, N. 2014. Nanopharmaceuticals (part 1): Products on the market. *International Journal of Nanomedicine* 9: 4357–4373.

Yadav, A., Prasad. V., Kathe. A.A. et al. 2006. Functional finishing in cotton fabrics using zinc oxides nanoparticles. *Bulletin of Materials Science.* 29(6): 641–645.

Yadi, M., Mostafavi, E., Saleh, B. et al. 2018. Current developments in green synthesis of metallic nanoparticles using plant extracts: A review. *Artificial Cells, Nanomedicine, and Biotechnology* 46(3): 336–343.

Yang, X., Shao, H., Liu, W. 2015. Endoplasmic reticulum stress and oxidative stress are involved in ZnO nanoparticle-induced hepatotoxicity. *Toxicology Letters* 234: 40–49.

Yu, J., Zhou, X. 2013. Synthesis of dendritic silver nanoparticles and their applications as SERS substrates. *Advances in Materials Science and Engineering* 2013: 1–4.

Zhang, L., Liu, L., Wang, J. 2020. Functionalized silver nanoparticles with graphene quantum dots shell layer for effective antibacterial action. *Journal of Nanoparticle Research* 22: 124. doi:10.1007/s11051-020-04845-3.

Zhang, X.F., Liu, Z.G., Shen, W., Gurunathan, S. 2016. Silver nanoparticles: Synthesis, characterization, properties, applications, and therapeutic approaches. *International Journal of Molecular Sciences* 17(9): 1534.

Zhao, P., Astruc, D. 2013. State of the art in gold nanoparticle synthesis. *Coordination Chemistry Reviews* 257(3–4): 638–665.

Zhu, Y, Zhu, R., Ma, J. et al. 2015. *In vitro* cell proliferation evaluation of porous nano-zirconia scaffolds with different porosity for bone tissue engineering. *Biomedical Materials* 10: 55009.

10 Resealed Erythrocytes
A Biological Carrier for Drug Delivery

Satish Shilpi, Kapil Khatri, and Umesh Dhakad
Ravishankar College of Pharmacy, Bhopal, India

Neelesh Kumar Mehra
National Institute of Pharmaceutical Education & Research (NIPER), Hyderabad, India

Arvind Gulbake
DIT University, Dehradun, India
D.Y. Patil Education Society, India

CONTENTS

Abbreviations 163
10.1 Introduction 163
10.2 Red Blood Cells 164
10.3 Erythrocytes as Cellular Carriers 165
10.4 Erythrocytes as Drug Delivery System 165
10.5 Benefit of Erythrocytes as Drug Carriers 166
10.6 Drawbacks of Erythrocytes as Drug Carriers 166
10.7 Provisions for Drug Encapsulation into Erythrocytes 166
    10.7.1 Isolation of Erythrocytes 167
    10.7.2 Method of Drug Loading 167
10.8 Characterisation of Drug-Loaded Erythrocytes 167
10.9 Applications of Resealed Erythrocytes 167
    10.9.1 Erythrocytes as Drug Delivery Systems 168
    10.9.2 Drug Targeting 170
    10.9.3 Enzyme and Hormones Deficiency/Replacement Therapy 170
    10.9.4 Treatment of Solid Tumours 171
    10.9.5 Treatment of Parasitic Diseases 172
    10.9.6 Treatment of Heavy Metal and Toxic Agent Poisoning 172
    10.9.7 Lead Poisoning Treatment 172
    10.9.8 Antibody Attachment to Erythrocyte Membrane 172
    10.9.9 Delivery of Antiviral Agents 172
    10.9.10 Improvement in Atomic Number 8 Delivery to Tissues 173
    10.9.11 Microinjection of Macromolecules 173
10.10 Safety Concern with Carrier Erythrocytes 173
10.11 Conclusion 174
Acknowledgement 174
References 174

ABBREVIATIONS

RBCs	Red blood cells
RES	Reticulum endothelial system
fL	Femtolitres
MPS	Mononuclear phagocyte system
O₂	Oxygen
CO₂	Carbon dioxide
NO	Nitrous oxide
ALA-D	Aminolevulinate dehydrogenase
2, 3-DPG	2, 3-Diphosphoglycerate

10.1 INTRODUCTION

Erythrocyte function has broadened to include oxygen, carbon dioxide, hydrogen sulphide and nitric oxide exchange as well as immune clearance and clearance of other soluble blood components such as cytokines; besides these functions, human red blood cells (RBCs) are emerging as a highly biocompatible microparticulate drug delivery system and the concept of using RBCs as drug delivery systems emerged in the 1970s, and it offers a greater potential related to its biodegradability, non-pathogenicity, non-immunogenicity, biocompatibility, self-degradability along with high drug loading efficiency (Rossi et al., 2005). The use of RBCs as drug delivery vehicles is appealing, as RBCs possess several advantages over alternative particulate delivery systems such as liposomes. These cells have a life span of approximately 120 days, a large internal capacity of 90 μm^3 that, although filled with proteins, can be further expanded without detrimental effects on survival, due to their flexibility (Ingrosso et al., 1997), and show high biocompatibility even in non-autologous conditions. However, loading of membrane-impermeable therapeutics into RBCs without inducing permanent membrane disruption is challenging, as these cells lack the endocytic machinery (Kinosita et al., 1977, Schlegel et al., 1992; Zimmermann et al., 1975; Schrier et al., 1975; Schrier 1987).

The cellular carriers are a helpful device as drug delivery system, these carriers as well as leucocytes, platelets, hepatocytes, fibroblasts and erythrocytes. Among these, the erythrocytes have been the most investigated and found to have great potential in novel drug delivery. Resealed erythrocytes are gaining more popularity because of their ability to circulate throughout body biocompatibility, zero-order release kinetics, reproducibility and ease of preparation. Most of the resealed erythrocytes used as drug carriers are quickly concerned from blood by macrophages of system (RES) that is gift in liver, lung and spleen of the body. The reasons for this increasing interest in drug delivery are due to the increasing need of safe drugs, capable of reaching the target and with minimal side effects (Docter et al., 1982). In fact, most issues related to general drug administration are basically associated with the drugs bio-distribution throughout the body. This indiscriminate distribution means that to achieve a required therapeutic concentration, the drug has to be administered in large quantities, and a major part reaches normal tissues. Ideally, a 'perfect' drug ought to exert its pharmacological activity solely at the target site, victimisation rock bottom concentration doable and while not negative effects on non-target compartments.

The use of erythrocytes as biological carriers offers an alternative to other carrier systems such as liposomes or nanoparticles that have been used for the encapsulation of different drugs, enzyme systems and peptides with therapeutic activity. These systems area unit particularly economical in cathartic medicine in circulations for weeks, have an outsized capability, are often simply processed, and will accommodate ancient and biological medicine. These carriers have also been used for delivering antigens and/or contrasting agents (Pei et al., 2007). Carrier erythrocytes are evaluated in thousands of drug administration in humans proving safety and efficacy of the treatments. The main advantage of carrier erythrocytes is that they act as a true drug delivery system, with a change in the kinetic properties of the substances, and they achieve selective distribution to different organs and tissues, especially the phagocytic cells of the mononuclear phagocyte system (MPS). The selective accumulation of therapeutic agents in phagocytic cells, such as macrophages, by the use of carrier erythrocytes is of huge therapeutic importance in drugs such as antibiotics (Reichel et al., 1994; Talwar et al., 1992; Millán et al. 2004, 2012; Hamidi et al., 2011), enzymes (Kwon et al., 2009; Sprague et al., 2011) or anti-HIV peptides (Pierigè et al., 2008), antioxidant (Alanazi et al., 2010), Gene (Lande et al., 2012), biosensor (Milanick et al., 2011). The delivery systems currently available enlist carriers that are either simple, soluble macromolecules (such as monoclonal antibodies, soluble synthetic polymers, polysaccharides and biodegradable polymers).

10.2 RED BLOOD CELLS

Erythrocytes, the most abundant cellular constituents of blood (i.e. 5,200,000 ± 300,000 and 4,700,000 ± 300,000 cell/mm^3 blood in healthy men and women, respectively) (Aryal et al., 2013), represent the largest cell-specific surface among other blood cells (i.e. the highest surface-to-volume ratio of 1.9×10^4 cm/g) (Adriaenssens et al., 1976).

Human RBCs are the most common cells of blood, and their plasma membrane encloses haemoglobin, a heme-containing protein that is responsible for O_2–CO_2 binding and its transport to other body parts. Normal human RBCs have a diameter of 7–8 μm and an average volume of

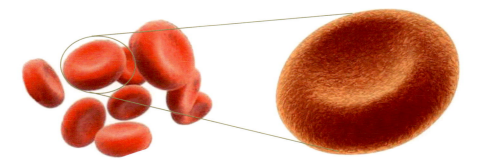

FIGURE 10.1 Human red blood cells (RBCs).

90 fL (Figure 10.1). In mammals, RBCs are anucleated and lose their organelles during maturation. It has lack of such cytoplasmic organelles as nucleus, mitochondria and ribosomes, and the red blood cell is unable to synthesise protein, carry out the oxidative reactions associated with mitochondria or undergo mitosis. Erythrocytes, produced in bone marrow by regulatory effect of erythropoietin, making up more than 99% of the total cellular space of blood in humans, occupy a volume of approximately 25–30 mL/kg, from which 71% constitute an aqueous phase (DeLoach 1983; Lewis et al., 1984; Jain et al., 1997). A total of approximately 760 g of haemoglobin is contained in the erythrocytes, representing approximately 10% of the total body proteins of an adult human. At oxygen transport is carried out by about 270 million haemoglobin molecules per cell each of which with four heme groups. A human body is commonly endowed with RBCs continuously produced at a rate of 2 million/s. In fact, RBCs spent their 100–120 days life span travelling the circulatory system before being selectively removed by macrophages in the reticuloendothelial system (RES) (Gautam et al., 1987; Sternberg et al., 2012; Foller et al., 2013; Gupta et al., 2014). The blood cell membrane is strictly connected with the membrane skeletal proteins that area unit organised during a uniform shell. The blood cell form will bear variety of reversible transformations. A crucial determinant of blood cell survival is its deformability. Key factors affecting deformability are internal viscosity (mainly contributed by RBC haemoglobin), and the biconcave disc shape with the highest surface-to-volume ratio is essential for the gas exchange function of erythrocytes. The surface area of mature, biconcave RBCs is about 136 μm2 but can swell to a sphere of approximately 150 femtolitres (fL). In addition, this unique shape has a high degree of flexibility required for passage of erythrocytes through the capillaries with diameters of 3–4 μm without undergoing extensive remodelling.

The major function of the erythrocytes is to encase haemoglobin and protect it, so it can act as an oxygen transporter for a prolonged period. Haemoglobin interacts with small diffusible ligands such as O_2, CO_2 and NO and may be involved in the control of blood pressure (Favretto et al., 2013). The RBCs produce other terribly attention-grabbing properties; particularly, they behave as associate osmometer since they shrink once placed into a hypertonic resolution or swell once placed into a hypotonic resolution. The RBCs will reach important lysis volume giving rise to holes on the membrane starting from 10 to 500 nm. These processes square measure typically reversible and following haemolysis the holes shut and the cell resumes its concavo-concave form.

Red blood cells represent potential biocompatible carriers for various bioactive substances, as well as supermolecule medicines, and have some distinctive options; for example, they are utterly perishable while not generation of poisonous product and show high biocompatibility especially when autologous erythrocytes are employed; they can be easily handled *ex vivo* by means of several techniques for the encapsulation of different molecules, after which one can obtain loaded erythrocytes with morphological, immunological and biochemical properties similar to those of native cells; lacking a nucleus and other organelles, most of their volume is available for the encapsulation of drugs; they protect the encapsulated substance from premature inactivation and degradation by endogenous factors and, at the same time, the organism against the toxic effects of the drugs thus avoiding immunological reactions. Potentially, a good sort of chemicals is entrapped, even peptides of high mass, presenting important biotechnological applications. RBCs have a longer life span in circulation as compared to other synthetic carriers and could act as bioreactors due to the presence of several enzymatic activities that can directly affect the loaded molecules and, in the case of encapsulated prodrugs, give rise to the active drug itself. Furthermore, exploiting another main feature of the RBCs, a selective targeting of drugs directly to macrophages without affecting the non-targeted compartments, could be achieved. After their natural life span (approximately 120 days) in systemic circulation, the senescent RBCs are recognised by the cells of the phagocytic system (RES, otherwise known as the monocyte macrophage system) and removed from circulation to be destroyed. It is possible to falsely induce these sense signals on the RBC membrane, in order to specifically target the drug-containing erythrocytes to the phagocytic cells, in particular to the monocyte-derived macrophages (DeLoach et al., 1983; Patel et al., 2008; Sternberg et al., 2012; Magnani 2012).

10.3 ERYTHROCYTES AS CELLULAR CARRIERS

The increasing interest in drug delivery is due to the increasing need of safe and effective delivery of drugs, capable of exerting their pharmacological activity only at the target site with minimal side effects on non-target compartments. Novel drug delivery systems, both particulate and non-particulate systems, are vital tool to solve and fulfil responses to this critical demand. Cellular carriers, including erythrocytes, leucocytes, platelets, islets, hepatocytes and fibroblasts, are among particulate drug delivery systems and all have been suggested as potential carriers for drugs and biological substances in recent decades. They can be used to provide slow release of entrapped drugs and/or to deliver drugs to specific sites in the body. Erythrocytes, as the most abundant and available cells in the human body, have gained the highest degree of interest among the cells in recent decades (Barker et al., 2001; Hamidi et al., 2007; Pei et al., 2007; Hu et al., 2012).

10.4 ERYTHROCYTES AS DRUG DELIVERY SYSTEM

The main advantage is their biocompatibility, thus no chance of immunological reaction. Complete carrier biodegradability and no generation of toxic products. Have a longer life span (120 days) in circulation as compared with other synthetic carriers. Uniform size and shape with relatively inert intracellular environment and possibility of entrapment of a wide variety of chemicals. Degradation of the loaded drug from inactivation by endogenous chemicals is prevented. There are various techniques available that make easy in RBCs separation process, their handling, transfusion and drug loading in erythrocytes. Attainment of steady state plasma concentration with zero-order drugs release kinetic is possible with drug loaded RBCs. Modification of pharmacokinetic and pharmacodynamic parameters of drug significantly decreases the side effects. Large quantities of drug that can be encapsulated within a small volume of cells ensure dose sufficiency (Favretto et al., 2013; Foller et al., 2013).

10.5 BENEFIT OF ERYTHROCYTES AS DRUG CARRIERS

- RBCs showed biocompatibility and biodegrade and their breakdown products are recyclable; haemoglobin is broken down into non-toxic biodegradable products globin and hem and further they are reused again. Globin is degraded to amino acids for amino acid pools in the body, while iron is reused in haemoglobin synthesis.
- They did not show undesired immune responses against the entrapped active molecule as well as showed nonimmunogenic response to body and can target body organ or tissue successfully.
- Remarkably longer life span (100–120 days) of the carrier erythrocytes in circulation in comparison with the synthetic particulate carriers and even comparable to normal cells and it can be controllable within a wide range from minutes to months.
- Uniform size and elastic, biconcave shape make erythrocytes to squeeze through narrow capillaries and movement in every part of body and even pass necessary drug molecules into the single cell or tissue.
- Protection of the encapsulated therapeutics from unwanted degradation within the host body inactivation due to its endogenous factors contents.
- RBCs can target active drug molecules to the RES organs.
- Relatively, they produce inert intracellular environment which helps in the protection of a wide variety of encapsulated drug molecules. They have neither nucleus nor other organelles that are by their intracellular space for high encapsulation efficiency and also exist for drug transport.
- Possibility that they showed ideal zero-order release kinetics of drug.
- Possibility of loading a relatively high amount of drug in a small volume of erythrocytes, which promise the dose sufficiency in clinical as well as animal studies using a limited volume of erythrocyte samples.
- Modification of the pharmacokinetic and pharmacodynamic parameters of the drug easily achieved by RBCs.
- Considerable increase in drug dosing intervals with drug concentration in the safe and effective level for a relatively long time as well as decreasing side effects of drug.
- Erythrocytes make up about 40 to 50% of blood volume; therefore, a large amount of substances can be encapsulated in erythrocytes, and due to their natural roles, they are ideal carriers for intravascular drug delivery.

10.6 DRAWBACKS OF ERYTHROCYTES AS DRUG CARRIERS

The use of erythrocytes as carrier systems also presents some disadvantages, which can be summarised as follows:

- The major problem encountered in the use of biodegradable materials or natural cells as drug carriers is that they are removed *in vivo* by the RES as result of modification that occurred during loading procedure in cells. Though this expands the aptitude to drug targeting to RES, this seriously limits their life span as long-circulating drug carriers in circulation.
- The dose dumping of certain encapsulated substances arises with the loaded erythrocytes as well as they can alter the physiology of the erythrocyte.
- They are carriers of biological origin; encapsulated erythrocytes may present some inherent variations in their loading and characteristics compared to other carrier systems.
- The storage of the loaded erythrocytes is a further problem if there are viable cells and need to survive in circulation for a long time upon re-entry to the host body. Conditioning carrier cells in isotonic buffers containing all essential nutrients, as well as in low temperatures, the addition of nucleosides or chelators, lyophilisation with glycerol or gel immobilisation have all been exploited to overcome this problem.
- Possible contamination due to the origin of the blood, the equipment used and the loading environment.

10.7 PROVISIONS FOR DRUG ENCAPSULATION INTO ERYTHROCYTES

The therapeutic molecules should be polar or hydrophilic in nature, should refuse to accept degradation within erythrocytes, should be deficient in physical or chemical interaction with erythrocyte membrane and should have well-defined pharmacokinetic and pharmacodynamic properties (Im et al., 1984; Beppu et al., 1994). Non-polar and

TABLE 10.1
Different Drug Loading Methods and Their Processes

Method	Procedure	% Loading	Advantage
Dilution method	Based upon hypotonic lysis of RBCs and resealing after drug loading, i.e. reversible swelling and resealing in a hypotonic and isotonic solution	<10	Fastest and simplest, especially for low-molecular-weight drug
Dialysis method	Drug loading occurs with lysis and resealing within a dialysis tube using hypotonic and isotonic solution	30–40	Good in vivo survival
Presswell dilution method	Drug loading carried out with controlled swelling of erythrocyte without lysis in hypotonic solution, followed by centrifugation at low rpm and the addition of small volume of drug solution to attain drug loading	40–70	High survival rate
Isotonic osmosis lysis method	Drug loading achieved by using chemical and physical lysis method in isotonic buffer solution	40–70	High survival rate

hydrophobic molecules are also entrapped in RBC in their various salts. Molecules which interact with the membrane and cause destructive effects on membrane structure are not considered to be suitable for encapsulation in erythrocyte. Erythrocytes can entrap a wide variety of biologically active substance having molecular weight range from 5,000 to 600,000 Da in size (Schrier 1987).

10.7.1 Isolation of Erythrocytes

Erythrocytes may be prepared as carriers from the blood taken from human beings and from different animal species including erythrocytes of mice, cattle, pigs, dogs, sheep, goats, monkeys, chicken, rats and rabbits. To isolate erythrocytes, the blood is collected in heparinised tubes by venipuncture. Freshly collected blood is centrifuged in a refrigerated centrifuge and washed to obtain erythrocyte (Eichler et al., 1986). The washed cells are suspended in buffer (e.g. acid-citrate-dextrose buffer) at various hematocrit values as desired.

10.7.2 Method of Drug Loading

There are different methods used to load drugs or other bioactive compounds in erythrocytes including chemical methods such as membrane perturbation of the erythrocytes using chemical, another method is physical method such as electrical pulse method, and osmosis-based systems. Irrespective of the method, the optimal characteristics for the successful entrapment of the compound require the drug to have a considerable degree of water solubility, resistance against degradation within erythrocytes, lack of physical or chemical interaction with erythrocyte membrane, and well-defined pharmacokinetic and pharmacodynamic properties (Lewis et al., 1984; Dale et al., 1979; Li t al., 1996; Favretto et al., 2013). The following methods are used for the entrapment of therapeutic agent into erythrocytes.

- Osmosis-based methods
  - Hypotonic dilution method
  - Hypotonic dialysis method
- Hypotonic Presswell method
- Isotonic osmotic lysis
- Chemical perturbation of the membrane
- Electro-insertion or electro-encapsulation
- Entrapment by endocytosis
- Loading by electric cell fusion
- Normal transport method
- Lipid fusion method (Table 10.1)

10.8 CHARACTERISATION OF DRUG-LOADED ERYTHROCYTES

Different parameters summarised in Table 10.2 are used for the evaluation of the ideal characteristics of RBCs.

10.9 APPLICATIONS OF RESEALED ERYTHROCYTES

Resealed erythrocytes have several therapeutic potentials and they are easily applied for managing various types of diseases. Such cells could be used as circulating carriers to disseminate a drug within a prolonged period in body or in target-specific organs, including the liver, spleen and lymph nodes. Erythrocytes are used to deliver protein and enzymes and can carry a wide variety of drugs. Due to their easy and wide versatility in surface modification by chemically (glutaraldehyde), ionic (ascorbate, ferrous), enzymatic (neuraminidase and proteolytic enzymes), oxidant (azodicarboxylic acid bis-dimethylamide) and some other targeting moieties like antibodies and ligands, it was successfully applied and till now it has been continuously applied for targeting different types of therapeutics to the targeted cell, tissue or organ (Wainwright et al., 1989; Luo et al., 2012). The surface modification of RBCs may increase the RES uptake or targeting efficiency to specific cells or organs. This property of red blood cells, i.e. reversibly lysed and resealed, can entrap exogenous compounds, including proteins and peptides. Drug encapsulation in red blood cells has been proposed as an effective way of extending the pharmacological activity due to its sustained and controlled

TABLE 10.2
Determinations Methods of Physical, Chemical and Biological Characteristic of Resealed Erythrocytes

Parameter	Method/Instrument Used
Physical Characterisation	
Shape and surface morphology	• Atomic force microscopy • Transmission electron microscopy • Scanning electron microscopy • Phase contrast microscopy • Optical microscopy
Size and size distribution	• Transmission electron microscopy • Zetasizer • Optical microscopy
Drug release	• Diffusion cell dialysis
Drug content	• Deproteinisation of cell membrane, followed by assay of released drug • Radiolabelling
Surface electrical potential	• Zeta potential measurement (Zetasizer)
Surface pH	• pH-Sensitive probes (pH meter)
Deformability	• Capillary method
Cellular Characterisation	
% HB content	• Deproteinisation of cell membrane, followed by haemoglobin assay
Cell volume	• Laser light scattering
% Cell recovery	• Neubauer chamber • Haematological analyser
Osmotic fragility	• Determination of drug and haemoglobin content after incubation with isotonic to hypotonic solution
Osmotic shock	• Dilution with distilled water and estimation of drug and haemoglobin
Turbulent shock	• Passes of cell suspension through 30 gauge hypodermic needle at 10mL/min flow rate and estimation of residual drug and haemoglobin • Vigorous shaking, followed by haemoglobin estimation
Erythrocyte sedimentation rate	• ESR method
Biological Characterisation	
Sterility	• Sterility test
Pyrogenicity	• Rabbit method • LAL test
Animal toxicity	• Toxicity tests

drug release as well as reducing the toxicity of drugs or of prolonging enzyme activity *in vivo*. In considering potential applications of resealed RBC, they have been used as circulating depots for the sustained delivery of a wide variety of drugs like antineoplastics, anti-parasitic, veterinary antiamoebic, vitamins, steroids, antibiotics and antiviral and cardiovascular drugs.

10.9.1 Erythrocytes as Drug Delivery Systems

The selective delivery of new therapeutic agents to target cells or organs is an important challenge in every clinical approach where peptides, oligonucleotides and/or genes are used (Millán et al., 2004). These molecules, although specific, do not easily cross the cell membranes, are usually not very stable in biological fluids and, because of their production costs, should be used in low amounts. The delivery of genes is usually achieved by viral vectors. The delivery of oligonucleotides and peptides is instead mediated by coupling or entrapping this therapeutics to or in a carrier system that has a significant affinity for one or more cell types within the body (Magnani et al., 2002). Several attempts have been made to improve the targeting of drugs by engineering the properties of the carrier system. The use of Tat, VP22, etc. engineered peptides or the development of drugs conjugated to ligands specific for receptors known to have a selective cell distribution.

Although these delivery systems are very interesting, they suffer several limitations including the limited number of molecules delivered and potential adverse effects. Based on these considerations, we have developed a cellular drug delivery system which is based on the use of autologous erythrocytes. The carrier system is totally biocompatible, non-immunogenic, with a long life span in circulation and can be also targeted to macrophages. In fact, human red blood cells have a few properties that make them useful as drug carriers. Erythrocytes are biodegradable, can circulate for long periods of time and have a large capacity. Moreover, a high percentage of encapsulation can be obtained. In principle, any drug, including peptides and nucleic acids, can potentially be encapsulated into red blood cells. However, several molecules have been shown to leak rapidly through the red cell membrane due to simple diffusion. Other molecules may be toxic to the red cell itself, thus preventing their use as a carrier system. It is interesting to note that red blood cells are active carrier systems, endowed with a few enzymatic activities that can be conveniently explored to convert an inactive prodrug into an active drug. This property permits the design of a few prodrugs that can be synthesised with charged chemical groups making them non-diffusible or nontoxic. Once these chemical groups have been hydrolysed by resident red cell enzymes, the prodrug is converted into an active drug that can diffuse through the red cell membrane and is thus released in circulation or at specific sites when red cell targeting is achieved. In human studies, the use of erythrocytes as a drug delivery system was extensively investigated *in vitro* and recently validated *in vivo* (Table 10.3).

TABLE 10.3
Application of Different Drugs and Protein-Loaded Resealed Erythrocytes

Category of Drugs/ Protein/Others	Drugs	Application	References
Anti-spasmolytic	Dexamethasone	Delivered dexamethasone in steroid-dependent IBD patients and found loading of Dex 21-P in autologous erythrocytes is feasible and safe. The very low dose of Dex released in bloodstream was able to maintain patients in clinical reduction and allowed steroid withdrawal	Annese et al. (2005)
Vitamin	Alpha-tocopherol	Protect the alpha-tocopherol with intracellular ascorbic acid by encapsulating it into erythrocytes	May et al. (1998)
Hormone	Prednisolone	Reduce the side effect and their relative toxicity by controlling the release (zero-order kinetics) of prednisolone	Shavi et al. (2010)
Anti-hypertensive	Pravastatin	Studied the effect of hypotonic lysis on biochemical parameters and loading efficiency	Harisa et al. (2012)
	Fasudil		Gupta et al. (2014)
Antibiotics	Gentamicin	Studied for survival of erythrocyte in the presence of gentamicin and found that it appears to circulate (half-life 22 days) longer than any other drug carriers under investigation and may well serve as innocuous slow-release system	Eichler et al. (1986)
Cannabinoid neurotransmitter	Anandamide (*N*-arachidonoylethanolamine, arachidonoylethanolamide)	Studied for membrane-binding properties of drug to erythrocyte membrane which was increased upon rising the temperature	Bojesen et al. (2005)
Chelating agent	deferoxamine	Erythrocyte loaded with deferoxamine was used to remove excess iron from the patient body who suffered from anaemia or thalassaemia	Green (1985)
Anticancer	Paclitaxel	Paclitaxel was successfully loaded and found 46% entrapment, 74.5% cell recovery and 81% release in 48 h	Gamaleldin et al. (2014)
Antimalarial	Primaquine	Studied for *in vitro* survival of erythrocytes upon primaquine loading	Alanazi et al. (2011)
Anticancer	Resveratrol and paclitaxel	Resveratrol increases the anticancer efficacy of paclitaxel in the case of breast cancer	Fukui et al. (2010)
Anticonvulsant	Valproate	Studied *in vitro* encapsulation stability of erythrocytes	Hamidi et al. (2011)
Fatty acid oxidiser	Malonyldialdehyde	They found that it disturbs aminophospholipids organisation in the membrane bilayer by oxidising fatty acid present in the membrane	Jain (1984)
Anticancer	Paclitaxel	Stimulation of erythrocyte phosphatidylserine exposure by paclitaxel	Lang et al. (2006)
Anti-metabolite	Methotrexate	Increase the uptake of methotrexate by targeting biotin encored erythrocyte to liver	Mishra et al. (2002)
Immunosuppressant	Cyclosporine A	Studied for drug interaction with erythrocyte and also characterised for factor which affect the encapsulation	Reichel et al. (1994)
Anticancer	Adriamycin	Studied for *in vivo* drug distribution and its metabolism	Benatti et al. (1987)
Poison antagonism	Organic thiosulphonates	Delivered organic thiosulphonate-loaded erythrocytes and antagonised the lethal effect of cyanide poisoning	Petrikovics et al. (1995)

10.9.2 Drug Targeting

Ideally, drug delivery should be site-specific and target-oriented to exhibit maximal therapeutic index with minimum adverse effects (Fan et al., 2012). Resealed erythrocytes can act as drug carriers as well as long-circulating reservoirs of the drug for the body. Surface-modified erythrocytes are used as targeting tools for specific organs or tissue, mononuclear phagocytic system and for reticuloendothelial system because the changes in the membrane are recognised by macrophages (Alvarez et al., 1998). If erythrocytes are damaged during process or other means, then they are rapidly cleared from circulation by phagocytic Kupffer cells in liver and spleen. This phenomenon is utilised for targeting the liver and spleen, by modifying erythrocytes. There are various approaches available to modify the surface of erythrocytes include, surface modification by chemically (glutaraldehyde), by using ionic substances (ascorbate, ferrous), enzymatic (neuraminidase and proteolytic enzymes) (Millán et al., 2004), oxidant (azodicarboxylic acid bis (dimethylamide), by conjugating polymeric substances and some other targeting moiety like antibodies and ligands.

The approach in which reversible or irreversible attachment of the ligand to red blood cell membrane help to target the specific tissue or organ in the body. Jain et al. (1997) encapsulated anticancer drug to the erythrocytes and modified their surface to target drug to the cancer cell. A study carried out by Zocchi et al. (1988) in which erythrocytes were encapsulated with adriamycin followed by surface modification with glutaraldehyde that was beneficial to target liver and lungs by macrophage uptake. Benatti et al. (1987) and Alvarez et al. (1998) applied the same techniques to target adriamycin and other drugs to the liver and lung cells.

In another approach, targeting of drug-loaded erythrocytes to the different part of body was achieved by applying external magnetic field. Sprandel in 1987 achieved cell-specific targeting by applying external magnetic field as well as at target site got controlled release of drug (Sprandel et al., 1987). Price et al. (1998) applied an approach in which ultrasound waves were used to target drug-loaded erythrocytes. It can be achieved by ruptures created by target microbubbles destruction with the help of ultrasound (Yamagata et al., 2008).

Avidin–biotin technology has reached several applications in drug delivery during the last two decades. Membrane association of pharmaceuticals, by means of avidin–biotin bridges, is the most widely used strategy for membrane conjugated/adsorb drug on erythrocyte carriers. Biotinylation of intact mammalian erythrocytes could be achieved either by attachment to the amino groups by means of biotin N-hydro succinimide ester (NHS-biotin) or by oxidation of the created aldehyde groups of the erythrocyte membrane by biotin hydrazide. Comparison of these different procedures by Magnani et al. (1998) showed that biotinylation by NHS-biotin provides the highest cell recovery (N90%) and found that binding of biotin molecules per cell was 1000 molecules. Avidin–biotin bridges have been used for reversible membrane binding of different drugs, uricase, HIV-1 tat protein, bovine serum albumin and several other biopharmaceuticals.

Erythrocytes, loaded with the drug, can also be modified to promote autologous immunoglobulin binding and C3b deposition, so that macrophages can recognise and increase phagocytosis of the erythrocytes (Alvarez et al., 1998). By this mechanism, the drug-encapsulated erythrocytes are engulfed within macrophages and a variety of different compounds can be efficiently delivered. Nucleotide analogues are potent antiviral agents usually administered as nucleoside analogues (Magnani 2002). These molecules are then phosphorylated into macrophages by cellular or viral enzymes. Among peptides, interesting results were obtained with delivery of an ubiquitin analogue. This peptide is stable in the erythrocytes and efficiently competes with endogenous ubiquitin in macrophages.

10.9.3 Enzyme and Hormones Deficiency/Replacement Therapy

Many metabolic disorders associated with enzymes deficiency, for example, Gaucher's disease, galactosuria and hormonal deficiency, can be fulfilled by injecting them directly into the systemic circulation, but these therapies may result in shorter circulation half-life of enzymes as well as may precipitate allergic reactions and serious toxic or unwanted symptom. These problems can be successfully overcome by administering the enzymes or hormones using novel drug delivery carrier such as liposomes, niosomes, nanoparticles, dendrimers, micelles, microspheres but they need some modification in their surface to achieve long circulation. Resealed erythrocytes are one of them which can be used in enzyme therapy without any surface modification. Since these carriers are actual endogenous cells, they produce little or no antigenic response and upon ageing or being damaged can be removed from the circulation by macrophages as a complete natural process. These cells then release enzymes into circulation upon haemolysis, act as a 'circulating bioreactors' in which substrates enter the cell, interact with enzymes, and generate products, or accumulate enzymes in RES upon haemolysis for future catalysis. Another important feature of these carriers is that they can be stored at 4°C for several hours to several days before re-entry to the host body, depending on the storage medium and the entrapment method used and it is beneficial to biological product. Dale (1987) entrapped different types of enzymes in erythrocytes. Yew et al. (2013) successfully encapsulated phenylalanine hydroxylase for improving the pharmacokinetics encapsulated enzymes and lowered approx 80% plasma phenylalanine which demonstrates the ability of enzyme-loaded RBCs to metabolise circulating amino acids and emphasise the impending to treat disorders of amino acid metabolism. The enzymes such

TABLE 10.4
Enzymes and Hormones Delivery through Erythrocytes

Enzymes/Hormones	Application/Purpose	Loading Method	References
Phenylalanine hydroxylase	Improving the pharmacokinetics encapsulated enzymes and lowered plasma phenylalanine which may help to treat disorders of amino acid metabolism	Hypotonic dialysis	Yew et al. (2013)
Glutamate-cysteine ligase	Increase the life span/survival of drug-loaded erythrocytes due to oxidative stress	Hypotonic dialysis	Foller et al. (2103)
Dexamethasone 21-phosphate	To cure the steroid-dependent ulcerative colitis and found that mucosal healing was ascertained with low dose in animal group who was administered hormone through erythrocytes		Bossa et al. (2013)
Thymidine phosphorylase	Preclinical study of thymidine phosphorylase for enzyme-replacement therapy for mitochondrial neurogastrointestinal encephalomyopathy and found that do not reveal serious toxicities		Levene et al. (2013)
Prednisolone	Reduce the side effect and their relative toxicity by controlling the release (zero-order kinetics) of prednisolone	Presswell dilution and dilution technique	Shavi et al. (2010)
Dexamethasone	Delivered dexamethasone in steroid-dependent IBD patients and found loading of Dex 21-P in autologous erythrocytes is feasible and safe. The very low dose of Dex released in bloodstream was able to maintain patients in clinical reduction and allowed steroid withdrawal		Annese et al. (2005)
Glutamate dehydrogenase	GDH-loaded erythrocytes can be used as a potential carrier system for the *in vivo* removal of high levels of ammonia from the blood	Hypotonic dialysis/isotonic resealing method	
Alcohol dehydrogenase and acetaldehyde dehydrogenase	Successfully delivered dehydrogenase and acetaldehyde dehydrogenase encapsulated into erythrocytes and found that continuous degradation of ethanol by ADH-RBCs and co-encapsulated enzymes – RBCs, as a function of time (up to 70h) suggests the use of these carrier RBCs as agents for complete metabolisation of ethanol	Electroporation procedure	Lizano et al. (1998)
Phosphotriesterase	Successfully encapsulated phosphotriesterase in erythrocytes and effectively antagonised the lethal effect of paraoxon	Hypotonic dialysis	Pei et al. (1995)
L-asparaginase	Studied *in vitro* for gradient separation of asparaginase-loaded erythrocytes	Hypotonic dialysis	Garín et al. (1994), Updike et al. (1983)
Estrone, estradiol, estriol, progesterone and testosterone	Red cells function as carriers of sex hormones in the bloodstream in a manner like that of albumin and that red cells may be responsible for 5%–15% of sex hormone delivery to target tissues		Koefoed et al. (1994)
Arginase	Overcome the deficiency of arginase by erythrocytes in the case of familiar hyperargininaemia		Adriaenssens et al. (1976)

as β-glucosidase, β-glucuronidase, β-galactosidase, glutamate-cysteine ligase, thymidine phosphorylase, alcohol dehydrogenase and acetaldehyde dehydrogenase, sex hormones and other have been used to treat metabolic disorder, enzyme replacement therapy or treatment of drug poisoning (Table 10.4). The disease caused by an accumulation of glucocerebrosides in the liver and spleen can be treated by glucocerebrosidase-loaded erythrocytes.

The first report of winning clinical trials of the resealed erythrocytes loaded with enzymes for replacement medical aid is that of β-glucocerebrosidase for the treatment of Gaucher's sickness. The sickness is characterised by inborn deficiency of lysosomal β-glucocerebrosidase in cells of RES, thereby resulting in accumulation of β-glucocerebrosides in macrophages of the RES. The foremost necessary application of resealed erythrocytes in accelerator medical aid is that of asparaginase loading for the treatment of medicine neoplasms. This treatment prevents remission of medicine acute leukaemia. There are reports of improved intensity and period of action in animal models yet as humans.

10.9.4 Treatment of Solid Tumours

Hepatic tumours are one in every of the foremost current forms of cancer. Antineoplastic medication like antimetabolite, bleomycin, asparaginase and adriamycin are with success delivered by erythrocytes. Agents like daunorubicin diffuse apace from the cells upon loading and therefore cause a significant cardiotoxicity. This drawback is overcome by covalently linking daunorubicin to the erythrocytic membrane exploitation glutaraldehyde or cisaconitic acid as a spacer. The resealed erythrocytes loaded with carboplatin show localisation in liver (Alexandre et al., 2007).

10.9.5 Treatment of Parasitic Diseases

The flexibility of resealed erythrocytes to RES organs build them great tool throughout the delivery of antiparasitic agents. Parasitic diseases that involve harbouring parasites within the RES organs are with success controlled by this technique. Results were favourable in studies involving animal models for erythrocytes loaded with antiprotozoal drug, antileishmanial and anti-amoebic medication.

10.9.6 Treatment of Heavy Metal and Toxic Agent Poisoning

Management of iron overload is a major problem with thalassaemic and other anaemic patients requiring periodic transfusion. Efforts to control this lethal complication have been directed mainly towards improvements in the clinical use of iron chelators. Iron overload is a major problem in patients with thalassaemia and other anaemia requiring periodic transfusion. Efforts to control this potentially lethal complication have been directed mainly towards improvements in the clinical use of iron chelators. Resealed erythrocytes are successfully applied to treat poisoning due to the heavy metal or toxic agent by removing them from the body (Roos et al., 1984). Deferoxamine-loaded erythrocytes have been used to remove excess of iron from the body of thalassemic patients. The level of iron in blood is increase with multiple transfusing the fresh blood to the patient. This iron overload easily removes by targeting erythrocyte loaded with chelating agent to RES organ. Mintzer et al. (1988) effectively managed the parasitaemia by delivering inositol hexaphosphate (IHP) with erythrocytes it has. McLaren et al. (1983) reviewed on iron overload disorder treatment with resealed erythrocyte (McLaren et al., 1983). Green (1985) delivered deferoxamine as iron-chelating agent with the help of carrier erythrocytes and found beneficial result to remove iron from the patient who suffered from anaemia or thalassaemia (Green 1985). Leung et al. (1991) reported cyanide intoxication with murine carrier erythrocytes containing bovine rhodanese and fixing agent. Antagonisation of organophosphorus intoxication by resealed erythrocytes containing a recombinant phosphodiesterase additionally has been reportable. In another case of cyanide poisoning, Petrikovics et al. (1995) delivered organic thiosulphonate-loaded erythrocytes and they successfully antagonised the lethal effect of poisoning created by cyanide.

The rhodanese and thiosulphate loaded erythrocytes used in cyanide poisoning in animal model. They found promising result of rhodanese and thiosulphate in erythrocytes to antagonise the lethal effect of cyanide.

10.9.7 Lead Poisoning Treatment

The concentration of aminolevulinate dehydrogenase (ALA-D) in erythrocytes decreases. This leads to an accumulation of aminolevulinic acid in tissues, blood and urine. This state leads to acute porphyria and CNS-related problems. An injection of resealed erythrocytes loaded with ALA–D to lead intoxicated animal significantly reduces toxic manifestations. Other enzymes used for loading resealed erythrocytes include urease, galactose-1-phosphate uridyl transferase, uricase and acetaldehyde dehydrogenase (Simons 1985, 1988, 1993).

10.9.8 Antibody Attachment to Erythrocyte Membrane

To get specificity of action, Zimmermann proposed that the entrapment of small paramagnetic particles into erythrocytes might allow their localisation to a particular location under the influence of an external magnetic field. The loading of ferrofluids (colloidal suspension of magnetite) has been reported by Sprandel et al. (1987). Jain and Vyas reported entrapment of the anti-inflammatory drugs diclofenac sodium and ibuprofen in magnet-responsive erythrocytes. Photosensitised erythrocytes have been studied as a photo-triggered carrier and delivery system for methotrexate in cancer treatment (Mishra et al., 2002).

Price et al. (1998) reported delivery of colloidal particles and erythrocytes to tissue through microvessel ruptures created by targeted microbubble destruction with ultrasound. The dye erythrocytes were delivered for diagnosis purpose to the rat skeletal muscle through microvessel ruptures by microbubbles *in vivo*. Thermo-responsive liposomes used for targeting organs outside the RES. Glycoprotein pre-treatment of RBCs loaded with antineoplastic drug enhance the targeting of drug to neoplasm cells.

10.9.9 Delivery of Antiviral Agents

Several reports are cited within the literature regarding antiviral agents entrapped in resealed erythrocytes for effective delivery and targeting. Because most antiviral drugs are nucleotides or nucleoside analogues, their entrapment and exit through the membrane need careful consideration (Magnani et al., 2002). Nucleosides are chop-chop transported across the membrane, whereas nucleotides are not, therefore exhibiting prolonged unharnessed profiles. The discharge of nucleotides needs the conversion of those moieties to purine or pyrimidine bases. Resealed erythrocytes are accustomed deliver cytidine derivatives, recombinant herpes simplex virus kind one (HSV-1) conjugated protein B, azidothymidine derivatives, azathioprine, acyclovir and fludarabine phosphate.

10.9.10 IMPROVEMENT IN ATOMIC NUMBER 8 DELIVERY TO TISSUES

Haemoglobin is the supermolecule accountable for the oxygen-carrying capability of erythrocytes. Under normal conditions, 95% of haemoglobin is saturated with oxygen in the lungs, whereas under physiologic conditions in peripheral bloodstream, only 25% of oxygenated haemoglobin becomes deoxygenated (Li et al., 1996; Szumił 2013). Thus, the main fraction of gas guaranteed to Hb is re-circulated with the blood to the lungs. The employment of this certain fraction has been prompt for the treatment of gas deficiency. 2, 3-Diphosphoglycerate (2, 3-DPG) could be a natural effector of Hb. The binding affinity of Hb for gas changes reversibly with changes in animate thing concentration of DPG. This compensates for changes in the oxygen pressure outside of the body, as the affinity of 2, 3-DPG to oxygen is much higher than that of haemoglobin. Other organic polyphosphates can serve as allosteric effectors of haemoglobin with binding affinities higher than those of 2, 3-DPG and can compete with 2,3-DPG for binding to haemoglobin. However, because of its ionisation at physiological hydrogen ion concentration, it cannot enter erythrocytes. Hence, it is entrapped by the electroporation method. The encapsulation of IHP in RBCs irreversibly binds to haemoprotein, thereby decreasing the O_2 affinity to haemoprotein. As a result, the O_2 pressure appreciate five hundred time of the whole binding capability of haemoprotein to O_2 (P_{50} value) will increase from 26–27 mmHg to 50 mmHg. Within the presence of IHP encapsulated in erythrocytes, the distinction between the O_2 fraction of haemoprotein in lungs and tissues will increase, thereby increasing the O_2 concentration in tissues (Teisseire et al., 1984; Hamidi et al., 2007; He et al., 2014; Talwar et al., 1992).

Also, the extent of salt shaped within the N-terminal methane series cluster of chain of haemoprotein decreases, which is remunerated by associate degree uptake of H and carbon dioxide that ends up in accumulated formation of carbonate particle. IHP-loaded erythrocytes used to decrease in rate of flow with continuing oxygen consumption by animals. This means that as a result of associate degree accumulated extraction quantitative relation of O_2 by tissues, a given quantity of O_2 is delivered in lower blood flow. Additionally, these erythrocytes scale back ejection fraction, left cavum pulsation volume and vital sign. Associate degree isolated perfused-heart model showed reduction in coronary blood flow with accumulated O consumption by cardiac muscle upon administration of IHP-loaded erythrocytes. The application of IHP-loaded erythrocytes for improved O_2 provide is useful beneath the subsequent conditions:

- High altitude conditions where the partial pressure of oxygen is low
- Reduction in the number of alveoli, where exchange surface of the lungs is decreased
- Increased resistance to oxygen diffusion in the lungs
- Reduction in oxygen transport capacity
- Mutation or chemical modification, which involves a decrease in oxygen affinity for haemoglobin
- Increased radiosensitivity of radiation-sensitive tumours
- Restoration of oxygen delivery capacity of stored blood
- Ischaemia of myocardium, brain or other tissues.

10.9.11 MICROINJECTION OF MACROMOLECULES

Biological functions of macromolecules like polymer, RNA and proteins area unit exploited for varied cell biological applications. A comparatively easy structure and an absence of complicated cellular parts (e.g. nucleus) in erythrocytes build them smart candidates for the demurrer of macromolecules. The erythrocytes area unit used as microsyringes for injection to the host cells. The microinjection process involves culturing host eukaryotic cells *in vitro*. The cells are coated with fusogenic agent and then suspended with erythrocytes loaded with the compound of interest in an isotonic medium. Sendai virus (hemagglutinating virus of Japan, HVJ) or its glycoproteins or polyethylene glycol have been used as fusogenic agents. The fusogen causes fusion of co-suspended erythrocytes and eukaryotic cells. Thus, the contents of resealed erythrocytes and the compound of interest are transferred to host cell. This procedure has been used to microinject DNA fragments, arginase, proteins, nucleic acids, ferritin, latex particles, bovine and human serum albumin, and enzyme thymidine kinase to various eukaryotic cells. Advantages of this technique embrace quantitative injection of materials into cells, coincident introduction of many materials into an oversized variety of cells, stripped injury to the cell, shunning of degradation effects of lysosomal enzymes, and ease of the technique. Disadvantages embrace a requirement for a bigger size of united cells, therefore creating them amenable to RES clearance, adverse effects of fusogens, and unpredictable effects on cell ensuing from the co-introduction of assorted elements. Hence, this technique is restricted to chiefly cell biological applications instead of drug delivery.

10.10 SAFETY CONCERN WITH CARRIER ERYTHROCYTES

The safety of utilisation of erythrocytes as carrier has been reported by Wrobel et al., (2014). The use of erythrocytes as a drug carrier in human has the transmissible issues of transfusion of blood from one to a different. If totally different blood groups are mixed along, the blood cells could begin to clump along within the blood vessels, inflicting a probably fatal scenario. Therefore, it is vital to spot the people of the acceptor and therefore the style of blood cell carrier to attenuate mismatching before the administration of drug-loaded erythrocytes takes place. Another transmissible downside is the risk of spreading diseases. Therefore, screening of those carriers for the absence of diseases is very important to eliminate any risk of contamination.

Utilisation of blood cell as a drug carrier raises another potential concern because of the changes in their organic chemistry nature. In some instances, such changes created therapeutic edges, whereas in different cases, they yielded unwanted results. As an example, Hamidi et al. (2001) conducted a study on erythrocytes loaded with enalaprilat (Hamidi et al., 2001). The method created erythrocytes that were a lot of rigid, less misshapen and a lot of therapeutically efficacious than unmodified erythrocytes. The modification of erythrocytes with proteins like streptavidin, however, evoked some negative results. The attachment of streptavidin to biotinylated red blood cells caused these cells to be lysed, rapidly cleared from the circulation, thereby reducing their biocompatibility. *In vivo* studies involving humans and animals have also been conducted on biotinylated red blood cells.

10.11 CONCLUSION

Human red blood cells (RBCs) are emerging as a highly biocompatible microparticulate drug delivery system which offers a greater potential related to its biodegradability, non-pathogenicity, non-immunogenicity, biocompatibility, self-degradability along with high drug loading efficiency. It can be used for active–passive targetable drug delivery. It is a biological carrier which shows various biomedical applications in drug delivery technology and also offers new perception for safe and effective carrier and diagnostic tool. It can deliver drugs, enzymes, DNA, RNA, protein and peptides, hormones and others.

ACKNOWLEDGEMENT

The authors would like to acknowledge the Science and Engineering Research Board, DST, New Delhi, and MP Council of Science and Technology, Bhopal, for providing research grant that facilitates the writing of this chapter. The author also want to acknowledge the National Institute of Pharmaceutical Education and Research (NIPER), Hyderabad, for extending facilities to write this chapter (Research Communication No. NIPER-H/2020/BC-xx).

REFERENCES

Adriaenssens K, Karcher D, Lowenthal A, Terheggen HG. Use of enzyme-loaded erythrocytes in *in vitro* correction of arginase-deficient erythrocytes in familiar hyperargininemia. *Clin. Chem.* 1976; 22: 323–326.

Alanazi F. Pravastatin provides antioxidant activity and protection of erythrocytes loaded Primaquine. *Int. J. Med. Sci.* 2010; 7(6): 358–365.

Alanazi F, Harisa G, Maqboul A, Abdel-Hamid M, Neau S, Alsarra I. Biochemically altered human erythrocytes as a carrier for targeted delivery of primaquine: an *in vitro* study. *Arch. Pharm. Res.* 2011; 34(4): 563–571.

Alexandre J, Hu Y, Lu W, Pelicano H, Huang P. Novel action of paclitaxel against cancer cells: bystander effect mediated by reactive oxygen species. *Cancer Res.* 2007; 67(8): 3512–3517.

Alvarez FJ, Jordan JA, Calleja P, Lotero LA, Olmos G, Díez JC, Tejedor MC. Cross-linking treatment of loaded erythrocytes increases delivery of encapsulated substance to macrophages. *Biotechnol. Appl. Biochem.* 1998; 27(pt 2): 139–143.

Annese V, Latiano A, Rossi L, Lombardi G, Dallapiccola B, Serafini S, Damonte G, Andriulli A, Magnani M. Erythrocytes-mediated delivery of dexamethasone in steroid-dependent IBD patients – a pilot uncontrolled study. *Am. J. Gastroenterol.* 2005; 100(6): 1370–1375.

Aryal S, Hu CM, Fang RH, Dehaini D, Carpenter C, Zhang DE, Zhang L. Erythrocyte membrane-cloaked polymeric nanoparticles for controlled drug loading and release. *Nanomedicine (Lond).* 2013; 8(8): 1271–1280.

Barker SA, Khossravi D. Drug delivery strategies for the new millenium. *Drug Discov. Today* 2001; 6: 75–77.

Benatti U, Zocchi E, Tonetti M, Guida L, Polvani C, De Flora A. Enhanced antitumor activity of adriamycin by encapsulation in mouse erythrocytes targeted to liver and lungs. *Pharmacol. Res.* 1989; 21(suppl 2): 27–33.

Beppu M, Inoue M, Ishikawa T, Kikugawa K. Presence of membrane-bound proteinases that preferentially degrade oxidatively damaged erythrocyte membrane proteins as secondary antioxidant defense. *Biochim. Biophys. Acta* 1994; 1196(1): 81–87.

Bojesen IN, Hansen HS. Membrane transport of anandamide through resealed human red blood cell membranes. *J. Lipid Res.* 2005; 46(8): 1652–1659.

Bossa F, Annese V, Valvano MR, Latiano A, Martino G, Rossi L, Magnani M, Palmieri O, Serafini S, Damonte G, De Santo E, Andriulli A. Erythrocytes-mediated delivery of dexamethasone 21-phosphate in steroid-dependent ulcerative colitis: a randomized, double-blind Sham-controlled study. *Inflamm. Bowel Dis.* 2013; 19(9): 1872–1879.

Dale GL. High-efficiency entrapment of enzymes in resealed red cell ghosts by dialysis. *Methods Enzymol.* 1987; 149: 229–234.

Dale GL, Kuhl W, Beutler E. Incorporation of glucocerebrosidase into Gaucher's disease monocytes *in vitro*. *Proc. Natl. Acad. Sci. U.S.A.* 1979; 76(1): 473–475.

DeLoach JR. Encapsulation of exogenous agents in erythrocytes and the circulating survival of carrier erythrocytes. *J. Appl. Biochem.* 1983; 5(3): 149–157.

Docter R, Krenning EP, Bos G, Fekkes DF, Hennemann G. Evidence that the uptake of tri-iodo-L-thyronine by human erythrocytes is carrier-mediated but not energy-dependent. *Biochem. J.* 1982; 208(1): 27–34.

Eichler HG, Gasic S, Bauer K, Korn A, Bacher S. *In vivo* clearance of antibody-sensitized human drug carrier erythrocytes. *Clin. Pharmacol. Ther.* 1986; 40(3): 300–303.

Fan W, Yan W, Xu Z, Ni H. Erythrocytes load of low molecular weight chitosan nanoparticles as a potential vascular drug delivery system. *Colloids Surf. B Biointerfaces* 2012; 95: 258–265.

Favretto ME, Cluitmans JC, Bosman GJ, Brock R. Human erythrocytes as drug carriers: loading efficiency and side effects of hypotonic dialysis, chlorpromazine treatment and fusion with liposomes. *J. Control. Release* 2013; 170(3): 343–351.

Foller M, Harris IS, Elia A, John R, Lang F, Kavanagh TJ, Mak TW. Functional significance of glutamate-cysteine ligase modifier for erythrocyte survival *in vitro* and *in vivo*. *Cell Death Differ.* 2013; 20(10): 1350–1358.

Gamaleldin IH, Ibrahima MF, Alanazia F, Shazlya GA. Engineering erythrocytes as a novel carrier for the targeted delivery of the anticancer drug paclitaxel. *Saudi Pharm. J.* 2014; 22(3): 223–230.

Garín MI, Kravtzoff R, Chestier N, Sanz S, Pinilla M, Luque J, Ropars C. Density gradient separation of L-asparaginase-loaded human erythrocytes. *Biochem. Mol. Biol. Int.* 1994; 33(4): 807–814.

Gautam S, Barna B, Chiang T, Pettay J, Deodhar S. Use of resealed erythrocytes as delivery system for C-reactive protein (CRP) to generate macrophage-mediated tumoricidal activity. *J. Biol. Response Mod.* 1987; 6(3): 346–354.

Green R. Red cell ghost-entrapped deferoxamine as a model clinical targeted delivery system for iron chelators and other compounds. *Bibl. Haematol.* 1985; 51: 25–35.

Gupta N, Patel B, Ahsan F. Nano-engineered erythrocyte ghosts as inhalational carriers for delivery of fasudil: preparation and characterization. *Pharm. Res.* 2014; 31(6): 1553–1565.

Hamidi M, Rafiei P, Azadi A, Mohammadi-Samani S. Encapsulation of valproate-loaded hydrogel nanoparticles in intact human erythrocytes: a novel nano-cell composite for drug delivery. *J. Pharm. Sci.* 2011; 100(5): 1702–1711.

Hamidi M, Tajerzadeh H, Dehpour AR, Rouini MR, Ejtemaee-Mehr S. *In vitro* characterization of human intact erythrocytes loaded by enalaprilate. *Drug Deliv.* 2001; 8: 223–230.

Hamidi M, Zarrin A, Foroozesh M, Mohammadisamani S. Applications of carrier erythrocytes in delivery of biopharmaceuticals. *J. Controlled Release* 2007; 118(2): 145–160.

Harisa GI, Ibrahim MF, Alanazi FK. Erythrocyte-mediated delivery of pravastatin: *in vitro* study of effect of hypotonic lysis on biochemical parameters and loading efficiency. *Arch. Pharm. Res.* 2012; 35(8): 1431–1439.

He H, Ye J, Wang Y, Liu Q, Chung HS, Kwon YM, Shin MC, Lee K, Yang VC. Cell-penetrating peptides meditated encapsulation of protein therapeutics into intact red blood cells and its application. *J. Control. Release* 2014; 176: 123–132.

Hu CM, Fang RH, Zhang L. Erythrocyte-inspired delivery systems. *Adv. Healthcare Mater.* 2012; 1(5): 537–547.

Im JH, Cuppoletti J, Meezan E, Rackley CE, Kim HD. Distribution of insulin receptors in human erythrocyte membranes. Insulin binding to sealed right-side-out and inside-out human erythrocyte vesicles. *Biochim. Biophys. Acta* 1984; 775(2): 260–264.

Ingrosso D, Cotticelli MG, Angelo SD, Buro MD, Zappia V, Galletti P. Influence of osmotic stress on protein methylation in resealed erythrocytes. *Eur. J. Biochem.* 1997; 244: 918–922.

Jain SK. The accumulation of malonyldialdehyde, a product of fatty acid peroxidation, can disturb aminophospholipids organization in the membrane bilayer of human erythrocytes. *J. Biol. Chem.* 1984; 259: 3391–3394.

Jain SK, Jain NK. Engineered erythrocytes as a delivery system. *Indian J. Pharm. Sci.* 1997; 59: 275–281.

Kinosita K, Tsong TY. Hemolysis of human erythrocytes by a transient electric field. *Proc. Natl. Acad. Sci. U.S.A.* 1977; 74: 1923–1927.

Koefoed P, Brahm J. The permeability of the human red cell membrane to steroid sex hormones. *Biochim. Biophys. Acta* 1994; 1195(1): 55–62.

Kwon YM, Chung HS, Moon C, Yockman J, Park YJ, Gitlin SD, David AE, Yang VC. L-asparaginase encapsulated intact erythrocytes for treatment of acute lymphoblastic leukemia (ALL). *J. Control. Release* 2009; 139: 182–189.

Lande C, Cecchettini A, Tedeschi L, Taranta M, Naldi I, Citti L, Trivella MG, Grimaldi S, Cinti C. Innovative erythrocyte-based carriers for gene delivery in porcine vascular smooth muscle cells: basis for local therapy to prevent restenosis. *Cardiovasc. Hematol. Disord. Drug Targets* 2012; 12(1): 68–75.

Lang PA, Huober J, Bachmann C, Kempe DS, Sobiesiak M, Akel A, Niemoeller OM, Dreischer P, Eisele K, Klarl BA, Gulbins E, Lang F, Wieder T. Stimulation of erythrocyte phosphatidylserine exposure by paclitaxel. *Cell. Physiol. Biochem.* 2006; 18(1–3): 151–164.

Leung P, Cannon EP, Petrikovics I, Hawkins A, Way JL. *In vivo* studies on rhodanese encapsulation in mouse carrier erythrocytes. *Toxicol. Appl. Pharmacol.* 1991; 110(2): 268–274.

Levene M, Coleman DG, Kilpatrick HC, Fairbanks LD, Gangadharan B, Gasson C, Bax BE. Preclinical toxicity evaluation of erythrocyte-encapsulated thymidine phosphorylase in BALB/c mice and beagle dogs: an enzyme-replacement therapy for mitochondrial neuro-gastrointestinal encephalomyopathy. *Toxicol. Sci.* 2013; 131(1): 311–324.

Lewis DA, Alpar HO. Therapeutic possibilities of drugs encapsulated in erythrocytes. *Int. J. Pharm.* 1984; 22: 137–146.

Li LH, Hensen ML, Zhao YL, Hui SW. Electrofusion between heterogeneous-sized mammalian cells in a pellet: potential applications in drug delivery and hybridoma formation, *Biophys. J.* 1996; 71(1): 479–486.

Lizano C, Sanz S, Luque J, Pinilla M. *In vitro* study of alcohol dehydrogenase and acetaldehyde dehydrogenase encapsulated into human erythrocytes by an electroporation procedure. *Biochim. Biophys. Acta* 1998; 1425(2): 328–336.

Luo R, Mutukumaraswamy S, Venkatraman SS, Neu B. Engineering of erythrocyte-based drug carriers: control of protein release and bioactivity. *J. Mater. Sci. Mater. Med.* 2012; 23(1): 63–71.

Magnani M. Erythrocytes as carriers for drugs: the transition from the laboratory to the clinic is approaching. *Expert Opin. Biol. Ther.* 2012; 12(2): 137–138.

Magnani M, Pierigè F, Rossi L. Erythrocytes as a novel delivery vehicle for biologics: from enzymes to nucleic acid-based therapeutics. *Ther. Deliv.* 2012; 3(3): 405–414.

Magnani M, Rossi L, Fraternale A, Bianchi M, Antonelli A, Crinelli R, Chiarantini L. Erythrocyte-mediated delivery of drugs, peptides and modified oligonucleotides. *Gene Ther.* 2002; 9: 749–751.

May JM, Qu ZC, Mendiratta S. Protection and recycling of alpha-tocopherol in human erythrocytes by intracellular ascorbic acid. *Arch. Biochem. Biophys.* 1998; 349(2): 281–289.

McLaren GD, Muir WA, Kellermeyer RW. Iron overload disorder. *Crit. Rev. Clin. Lab. Sci.* 1983; 19: 205–266.

Milanick MA, Ritter S, Meissner K. Engineering erythrocytes to be erythrosensors: first steps. *Blood Cells Mol. Dis.* 2011; 47(2): 100–106.

Millán CG, Colino Gandarillas CI, Sayalero Marinero ML, Lanao JM. Cell-based drug-delivery platforms. *Ther. Deliv.* 2012; 3(1): 25–41.

Millán CG, Marinero ML, Castaneda AZ, Lanao JM. Drug, enzyme and peptide delivery using erythrocytes as carriers. *J. Control. Release* 2004; 95(1): 27–49.

Mintzer CL, Deloron P, Rice-Ficht A, Durica D, Struck DK, Roessner CA, Nicolau C, Ihle GM. Reduced parasitemia observed with erythrocytes containing inositol hexaphosphate. *Antimicrob. Agents Chemother.* 1988; 32(3): 391–394.

Mishra PR, Jain NK. Biotinylated methotrexate loaded erythrocytes for enhanced liver uptake. 'A study on the rat'. *Int. J. Pharm.* 2002; 231(2): 145–153.

Patel PD, Dand N, Hirlekar RS, Kadam VJ. Drug loaded erythrocytes: as novel drug delivery system. *Curr. Pharm. Des.* 2008; 14(1): 63–70.

Pei L, Petrikovics I, Way JL. Antagonism of the lethal effects of paraoxon by carrier erythrocytes containing phosphotriesterase. *Fundam. Appl. Toxicol.* 1995; 28(2): 209–214.

Pei X, Guo X, Coppel R, Bhattacharjee S, Haldar K, Gratzer W, Mohandas N, An X. The ring-infected erythrocyte surface antigen (RESA) of *Plasmodium falciparum* stabilizes spectrin tetramers and suppresses further invasion. *Blood* 2007; 110(3): 1036–1042.

Petrikovics I, Cannon EP, McGuinn WD, Pei L, Pu L, Lindner LE, Way JL. Cyanide antagonism with carrier erythrocytes and organic thiosulfonates. *Fundam. Appl. Toxicol.* 1995; 24(1): 86–93.

Pierigè F, Serafini S, Rossi L, Magnani M. Cell-based drug delivery. *Adv. Drug Deliv. Rev.* 2008; 60(2): 286–295.

Price RJ, Skyba DM, Kaul S, Skalak TC. Delivery of colloidal particles and red blood cells to tissue through microvessel ruptures created by targeted microbubble destruction with ultrasound. *Circulation* 1998; 98(13): 1264–1267.

Reichel C, Von FM, Brockmeier D, Dengler HJ. Characterization of cyclosporine A uptake in human erythrocytes. *Eur. J. Clin. Pharmacol.* 1994; 46: 417–419.

Roos D, Eckmann CM, Yazdanbakhsh M, Hamers MN, de Boer M. Excretion of superoxide by phagocytes measured with cytochrome c entrapped in resealed erythrocyte ghosts. *J. Biol. Chem.* 1984; 259(3): 1770–1775.

Rossi L, Serafini S, Pierige F, Antonelli A, Cerasi A, Fraternale A, Chiarantini L, Magnani M. Erythrocyte-based drug delivery. *Exp. Opin. Deliv.* 2005; 2: 311–322.

Schlegel RA, Lumley-Sapanski K, Williamson P. Single cell analysis of factors increasing the survival of resealed erythrocytes in the circulation of mice. *Adv. Exp. Med. Biol.* 1992; 326: 133–138.

Schrier SL. Shape changes and deformability in human erythrocyte membranes. *J. Lab. Clin. Med.* 1987; 110(6): 791–797.

Schrier SL, Bensch KG, Johnson M, Junga I. Energized endocytosis in human erythrocyte ghosts. *J. Clin. Invest.* 1975; 56(1): 8–22.

Shavi GV, Doijad RC, Deshpande PB, Manvi FV, Meka SR, Udupa N, Omprakash R, Dhirendra K. Erythrocytes as carrier for prednisolone: *in vitro* and *in vivo* evaluation. *Pak. J. Pharm. Sci.* 2010; 23(2): 194–200.

Simons TJ. Influence of lead ions on cation permeability in human red cell ghosts. *J. Membr. Biol.* 1985; 84(1): 61–71.

Simons TJ. Active transport of lead by the calcium pump in human red cell ghosts. *J. Physiol.* 1988; 405: 105–113.

Simons TJ. Lead transport and binding by human erythrocytes *in vitro*. *Pflugers Arch.* 1993; 423(3–4): 307–313.

Sprague RS, Bowles EA, Achilleus D, Stephenson AH, Ellis CG, Ellsworth ML. A selective phosphodiesterase 3 inhibitor rescues low PO_2-induced ATP release from erythrocytes of humans with type 2 diabetes: implication for vascular control. *Am. J. Physiol. Heart Circ. Physiol.* 2011; 301(6): H2466–H2472.

Sprandel U, Franz DJ. Towards cellular drug targeting and controlled release of drugs by magnetic fields. *Adv. Biosci.* 1987; 67: 243–250.

Sternberg N, Georgieva R, Duft K, Bäumler H. Surface-modified loaded human red blood cells for targeting and delivery of drugs. *J. Microencapsul.* 2012; 29(1): 9–20.

Szumiło M. Erythrocytes – the new application in medicine. *Pol. Merkur. Lekarski.* 2013; 34(199): 5–8.

Talwar N, Jain NK. Erythrocytes as carrier of primaquine preparation: characterization and evaluation. *J. Control. Release* 1992; 20: 133–142.

Teisseire B, Ropars C, Nicolau C, Vallez MO, Chassaigne M. Enhancement of P_{50} by inositol hexa phosphate entrapped in resealed erythrocytes in piglets. *Adv. Exp. Med. Biol.* 1984; 180: 673–677.

Wainwright SD, Tanner MJ, Martin GE, Yendle JE, Holmes C. Monoclonal antibodies to the membrane domain of the human erythrocyte anion transport protein. Localization of the C-terminus of the protein to the cytoplasmic side of the red cell membrane and distribution of the protein in some human tissues. *Biochem. J.* 1989; 258(1): 211–220.

Wrobel D, Kolanowska K, Gajek A, Gomez-Ramirez R, de la Mata J, Pedziwiatr-Werbicka E, Klajnert B, Waczulikova I, Bryszewska M. Interaction of cationic carbosilane dendrimers and their complexes with siRNA with erythrocytes and red blood cell ghosts. *Biochim. Biophys. Acta* 2014; 1838(3): 882–889.

Yamagata K, Kawasaki E, Kawarai H, Iino M. Encapsulation of concentrated protein into erythrocyte porated by continuous-wave ultrasound. *Ultrasound Med. Biol.* 2008; 34(12): 1924–1933.

Yew NS, Dufour E, Przybylska M, Putelat J, Crawley C, Foster M, Gentry S, Reczek D, Kloss A, Meyzaud A, Horand F, Cheng SH, Godfrin Y. Erythrocytes encapsulated with phenylalanine hydroxylase exhibit improved pharmacokinetics and lowered plasma phenylalanine levels in normal mice. *Mol. Genet. Metab.* 2013; 109(4): 339–344.

Zimmermann U, Pilwat G, Riemann F. Preparation of erythrocyte ghosts by dielectric breakdown of the cell membrane. *Biochim. Biophys. Acta* 1975; 375(2): 209–219.

Zocchi E, Tonetti M, Polvani C, Guida L, Benatti U, De Flora A. *In-vivo* liver and lung targeting of adriamycin encapsulated in glutaraldehyde-treated murine erythrocytes. *Biotechnol. Appl. Biochem.* 1988; 10: 555–562.

11 Nanostructured Hydrogel-Based Biosensor Platform

Anand Singh Patel and Keerti Jain
National Institute of Pharmaceutical Education and Research (NIPER), Raebareli, Uttar Pradesh, India

CONTENTS

Abbreviations 177
11.1 Introduction 177
11.2 Hydrogels in Biosensor Platform 178
    11.2.1 Nanoparticle-Incorporated Hydrogel Biosensor 178
        11.2.1.1 Metal Nanoparticles 178
        11.2.1.2 Carbon Nanostructure 180
    11.2.2 Conducting Polymer (CP)-Based Hydrogel Biosensor 180
    11.2.3 Biomolecule-Incorporated Hydrogel Biosensor 181
11.3 Biomedical Application of Hydrogel-Based Sensors 183
11.4 Conclusion 184
Acknowledgement 184
References 184

ABBREVIATIONS

Aβ	Amyloid-beta
AβOs	Amyloid-beta oligomers
Ag	Silver
AgNPs	Silver nanoparticles
Au	Gold
AuNPs	Gold nanoparticles
CAF	Caffeine
CEA	Carcinoembryonic antigen
CNT	Carbon nanotubes
CPHs	Conducting polymer hydrogels
CPs	Conducting polymers
CSF	Cerebrospinal fluid
CV	Cyclic voltammetry
CVD	Cardiovascular diseases
DPV	Differential pulse voltammetry
ECL	Electrochemiluminescence
EGDMA	Ethylene glycol dimethacrylate
ErGO	Electrochemically reduced graphene oxide
EST	Estradiol
FETs	Field-effect transistors
HEMA	Hydroxyethyl methacrylate
HIV	Human immunodeficiency virus
HRP	Horseradish peroxidase
ITO	Indium tin oxide
L012	Pyridazine-1,4(2H,3H)-dione sodium
LOD	Limit of detection
MAA	Methacrylic acid
MIPs	Molecularly imprinted polymers
NPs	Nanoparticles
O-PD	O-phenylenediamine
PAAm	Polyacrylamide
PAni	Polyaniline
PB	Prussian blue
Pd	Palladium
PEDOT	Poly(3,4-ethylenedioxythiophene)
PET	Polyethylene terephthalate
PPy	Polypyrrole
PrPC	Cellular prion protein
Pt	Platinum
PtNP	Platinum nanoparticle
PyG	Pyrolytic graphite
SERS	Surface-enhanced Raman scattering
SPR	Surface plasmon resonance
SWCNT	Single-walled carbon nanotubes

11.1 INTRODUCTION

The history of a biosensor dates back to 1962 with the invention of glucose biosensor by Leland C. Clark, and this invention laid the foundation of biosensing platforms (Kaur and Shorie, 2019). A biosensor is an analytical device (Wang and Burgess, 2010) which typically consists of a biological recognition component also called a bioreceptor (antibodies, enzymes, nucleic acid, cells or cellular structure, and biomimetics), which interacts with the target analyte of the body, and as a consequence of that interaction, physical or chemical changes occur which are converted into measurable or readable signals via transducer component (Kaur and Shorie, 2019; Tavakoli and Tang, 2017).

DOI: 10.1201/9781003043164-11

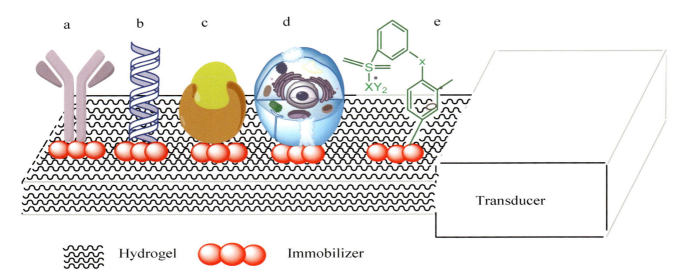

FIGURE 11.1 Graphical representation of different categories of bioreceptor (a) antigen/antibodies, (b) nucleic acid, (c) enzymes, (d) cells or cellular structures and (e) biomimetics. (Modified from Tavakoli and Tang, 2017.)

A bioreceptor contains biological molecule such as antibodies, enzymes, nucleic acid, cells or cellular structures, and/or biomimetics which are immobilised onto the base materials like metals, polymers, glass or composites, and they interact with the analyte using biochemical mechanism for recognition (Figure 11.1). The signal produced by the transducer is a function of bioreceptor–analyte interaction. In a typical biosensor, bioreceptors are immobilised onto the base material and the third component (first bioreceptor and second base material) transducer measures the signals, and with the aid of electronic system (fourth component), it can be displayed. The different components of biosensors are depicted in Figure 11.2 (Vo-Dinh, 2002; Tavakoli and Tang, 2017).

Nanostructured hydrogel is a hydrogel with nanosize range of 1–100 nm. Hydrogel is a three-dimensional (3D) hydrophilic polymeric (natural or synthetic polymers) network which has the ability to imbibe huge volumes of water or biological fluid, without dissolving, and therefore, the nanostructured hydrogel also known as nanogel, which comprises the properties of nanoparticulate system along with the hydrogel. Nanoemulsion have also been incorporated into the hydrogel to develop nanoemulgel formulation which is particularly suitable for delivery of hydrophobic drugs which are normally difficult to be delivered with nanogel (Ahmad et al., 2019; Dalwadi and Patel, 2015; Jung et al., 2017; Ojha et al., 2021; Peppas et al., 2012). Application of nanostructured hydrogel/hydrogel is not only limited to the drug delivery system, but they are potentially utilised in sensors, actuators, scaffolds for stem cell engineering, regenerative medicine, contact lenses, wound coverings, organ linings, biosensor membranes, microfluidic valves and fluid absorbants (Peppas et al., 2012; Conte et al., 2019). Hydrogels can be easily engineered and readily functionalised with different biological molecules such as enzymes, nucleic acid and proteins to hold them and detect the respective analyte (Le Goff et al., 2015).

It is important to note that the measurement methods are not limited to the methods mentioned in Figure 11.2 (conductometric, potentiometric, amperometric, impedimetric, surface charge, piezoelectric, magnetoelastic, surface acoustic wave, fibre optic, absorbance and luminescence), but other methods like label-based vs. label-free may also be used (Tavakoli and Tang, 2017).

11.2 HYDROGELS IN BIOSENSOR PLATFORM

Hydrogels are excellent base materials to immobilise the bioreceptor as well as create a protective layer to control the diffusion and increase biocompatibility. Hydrogel shows high flexibility owing to high water content like natural tissues. In hydrogel-based biosensor, hydrogel does not only serve as immobilisation matrix but also acts as a responsive (smart) material (Urban and Weiss, 2009; Kaur et al., 2018). Biosensor based on hydrogel can detect the biological interaction in two ways: (i) change in swelling behaviour in response to stimuli as a function of biological interaction (hydrogel without bioreceptors) and (ii) use of biological molecule as bioreceptor which interacts with analyte (Tavakoli and Tang, 2017).

11.2.1 Nanoparticle-Incorporated Hydrogel Biosensor

Nanoparticles possess superior electronic, optical, magnetic, conductive and mechanical properties, and ease of incorporation into polymer matrix by physical incorporation or chemical cross-linking makes them efficient candidate for sensing application (Zhang et al., 2020; Khan et al., 2017).

11.2.1.1 Metal Nanoparticles

In the design of the biosensor and electrochemical sensor, metal nanoparticles such as gold (Au), silver (Ag), platinum (Pt) and palladium (Pd) serve important roles via simple

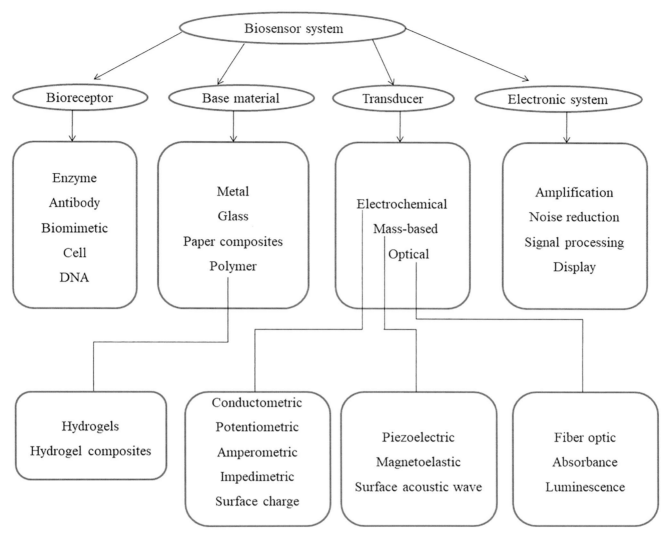

FIGURE 11.2 A typical biosensor includes the bioreceptors which are immobilised onto the base material and the transducer which measures the signals, and with the aid of electronic system, it can be displayed. (Modified from Tavakoli and Tang, 2017.)

and ease of synthesis, facile surface functionalisation, stress-free sensor fabrication, electrochemical reactions, catalysis as well as enhanced electron transfer process (Maduraiveeran and Jinb 2017).

Among the metal nanoparticles, Au nanoparticles are one of the most often studied nanoparticles due to their extraordinary conductive, optical properties and chemical stability as well (Pedrosa et al., 2011; Pirzada and Altintas, 2019). Gold nanoparticles are most commonly incorporated into hydrogel matrix which contains –SH, –CH and –NH$_2$ via covalent bond which responses to various stimuli (Zhang et al., 2020). Gold nanoparticle-incorporated hydrogel biosensor was prepared by immobilising horseradish peroxidase (HRP) by self-assembled gold nanoparticles onto the chitosan hydrogel matrix (Luo et al., 2005). That showed a good response to H$_2$O$_2$ (8.0–15 µM), with the lower detection limit of 2.4 µM, and the sensitivity of the biosensor lasts up to 4 weeks of storage.

Silver nanoparticles are analogous to gold nanoparticles and commonly used for diagnostic purpose. The optical properties of silver nanoparticles depend on degree of aggregation, size and shape. However, silver nanoparticles possess antimicrobial properties and electrical properties in contrast to gold nanoparticles and are therefore widely utilised in surface-enhanced Raman scattering-based biosensor (Pirzada and Altintas, 2019). Raj and Goyal prepared a sensitive voltammetric biosensor by modifying pyrolytic graphite (PyG) with electrochemically reduced graphene oxide (ErGO) nanocomposite and silver nanoparticles (AgNPs) to detect caffeine (CAF) and determine its effect on the release of estradiol (EST), i.e. (17β)-estra-1,3,5(10)-triene-3,17-diol, a female reproductive hormone, in women of childbearing age (18–34 years). Silver nanoparticle along with electrochemically reduced graphene oxide showed excellent electrocatalytic effect with detection range of 0.001–200 and 0.001–175 µM and LODs of 0.54 and 0.046 nM for EST and CAF, respectively (Raj and Goyal, 2019).

11.2.1.2 Carbon Nanostructure

Carbon-based nanomaterials such as graphene, nanohorns and nanotubes possess several exclusive properties which make them to be utilised for various biological applications like cancer therapy, tissue engineering, drug delivery, medical diagnostics, bioimaging and biosensing (Pirzada and Altintas, 2019). Among carbon nanostructures, carbon nanotubes and graphene are the most promising candidates to be utilised in biosensor platform (Holzinger et al., 2014). Biosensor based on carbon nanotubes shows excellent sensitivity, label-free detection, broad absorption spectrum and real-time monitoring. Carbon nanotubes act as scaffolds to immobilise biological molecule and thus uplift signal transduction, followed by recognition. The semiconducting properties of the carbon nanotubes along with their length make them frontier to be utilised as nanoscale field-effect transistors (FETs) (Pirzada and Altintas, 2019). Based on biological recognition component and transduction mechanism, these hybrid hydrogels (CNT-based hydrogel hybrid) can be efficiently utilised to sense specific biomarkers, enzymes, hormones, environment monitoring and pathogenic diseases like HIV (Vashist al., 2018).

Graphene is a two-dimensional material composed of sp^2 hybridised carbon atoms compiled in a hexagonal assortment and exhibits ambipolar electric field effects, classical thermal conductivity and quantum hall effects at room temperature and possesses high surface area and porosity; owing to this, graphene is an excellent candidate to adsorb several gases like methane, hydrogen and carbon dioxide. Properties of this material can be modulated by altering the layer's number and stacking. Graphene is a suitable candidate for biosensor owing to its property to interact with biological molecule by means of physical adsorption (Pirzada and Altintas, 2019). Graphene-like materials are more often used as electrode in electrochemical biosensor or field-effect transistor. They themselves serve as transducer in optical or colorimetric biosensor (Holzinger et al., 2014). Sun and co-workers developed an electrochemical hydrogel biosensor based on graphene oxide/gold nanoparticles (AuNPs) electrode by immobilising thiolated cellular prion protein (PrPC) peptide probe on AuNPs of the hydrogel electrode for the detection of amyloid-beta oligomers (AβOs). The developed biosensor showed high sensitivity and specificity towards AβOs and could differentiate AβOs from amyloid-beta (Aβ) monomers or fibrils with lower limit of detection of 0.1 pM AβO in artificial cerebrospinal fluid (CSF) or blood plasma (Sun et al., 2018).

There are more nanomaterials such as magnetic nanoparticles, dendrimers, carbon nanotubes and quantum dots which have shown prominent effect in the field of bioanalytics and biosensor, and are promisingly being investigated as an imaging agent, as theranostic agent as well as delivery vehicle for therapeutic agent to certain extent (Bajwa et al., 2016; Jain, 2017; Jain, 2019; Soni et al., 2015).

11.2.2 Conducting Polymer (CP)-Based Hydrogel Biosensor

Conducting polymers (CPs) are class of organic polymers which are π-conjugated polymers, such as polypyrrole (PPy), polyaniline (PAni), and poly(3,4-ethylenedioxythiophene) (PEDOT) that have electrical and optical properties like metals and semiconductors. Being polymers they also give advantages of common polymers like ease of synthesis, cost-effectivity and flexibility. The charge conduction in CPs is mainly based on two mechanisms:

1. Delocalisation of π electrons in conjugated system of CPs such as in the case of PPy and PAni.
2. Electron transport via electron exchange reaction (electron hopping) among vicinal redox sites in the case of redox polymers (Zhang et al., 2020; Bae et al., 2020; Tomczykowa and Brzezinska, 2019).

Conducting polymer hydrogels (CPHs) synergise the advantages of both organic conductors and hydrogels. CPs construct a hydrophilic network by covalent or physical cross-linking to give rise to a CPH. Great adaptability, ease of immobilisation of biological recognition molecule, excellent biocompatibility, owing to its high water content and thus similarity to the extracellular tissues make them an ideal candidate for sensing application (Guo et al., 2019). Applications of CPHs in biosensor are briefly summarised in Table 11.1. CP-based hydrogel biosensor holds various advantages like:

A. They themselves are able to bear the solvent ions to conduct electricity.
B. CPH system possesses great flexibility and excellent compatibility with different polymers in the same system, superior stretchability and more structural parity.
C. Facile fabrication, high conductivity and thermal stability (Zhang et al., 2020).

Li and co-workers (Li et al., 2015b) developed a nanostructured CPH biosensor based on a platinum nanoparticle (PtNP)-modified CP (PAni) hydrogel electrode, to detect human metabolites such as uric acid, cholesterol and triglycerides. The developed biosensor exhibited excellent sensing performance and linear range, which were found to be 0.07–1, 0.3–9 and 0.2–5 mM for uric acid, cholesterol and triglycerides, respectively, with lower sensing limits and fast response time (~3 s). Facile fabrication, high sensitivity and selectivity towards multiple analytes (metabolites) show the potential application in the development of low-cost biosensor based on CPH (Li et al., 2015b).

TABLE 11.1
Applications of CPHs in Different Types of Biosensors

CPHs/CPH Composites	Application	Sensing Performances	References
PAni/PtNPs	Glucose biosensor	Sensitivity: 96.1 µA/mM/cm^2 Response time: 3 s LOD: 0.7 µM	Zhai et al. (2013)
PPy/AuNPs	Carcinoembryonic antigen biosensor	Linear range: 1 fg/mL to 200 ng/mL LOD: 0.16 fg/mL	Rong et al. (2015)
PPy/Nafion/CNTs	Glucose biosensor	Sensitivity: 2860.3 µA/mM/cm^2 LOD: 5 µM	Shrestha et al. (2016)
PPy/copper oxide/reduced graphene oxide	Glucose biosensor	Linear range: 0.1–100 mM LOD: 0.03 µM	Mn et al. (2014)

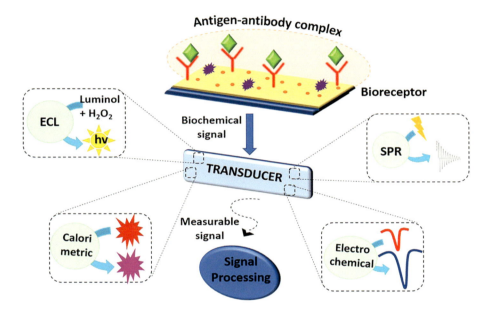

FIGURE 11.3 Schematic representation of different signal transduction techniques in a hydrogel-based immunosensor, where ECL: electrochemiluminescence and SPR: surface plasmon resonance. (Reproduced from George et al., 2020.)

11.2.3 Biomolecule-Incorporated Hydrogel Biosensor

A biosensor integrated with biological recognition molecules such as antigen/antibody, enzymes, nucleic acid, cells or cellular components onto hydrogel matrix demonstrates several advantages like biocompatibility, biodegradability, biostability, biofunctionality and ease of immobilisation (Zhang et al., 2020). Hydrogel (flexible, porous, wet material) minimises steric hindrance and promotes immobilisation of biological molecule and also target binding (George et al., 2020).

Structurally, antibodies are Y-shaped proteins generated from plasma cells and are the important parts of the immune system. Antibodies have the ability to recognise particular analyte with superior affinity, and this property of antibodies is extensively being utilised for sensing application. Immobilisation of antibodies on the sensor matrix should be performed in such a manner that its specificity and immunological activity should not change (Sharma et al., 2016). Immunosensors are subclass of biosensor that utilises antibody which recognises respective analyte based on affinity. Figure 11.3 represents different signal transduction techniques in a hydrogel-based immunosensor (George et al., 2020).

In 2015, Rong and co-workers developed a nanocomposite network of electrodeposited gold nanoparticles (AuNPs) loaded PPy (conducting polymer) hydrogel (PPy hydrogel) to make a sensitive label-free amperometric immunosensor. They employed carcinoembryonic antigen (CEA) as a model protein and immobilised it on gold nanoparticle. The detection range of the proposed immunosensor was from 1 fg (femtogram) mL^{-1} to 200 ng mL^{-1}, with the lower limit of detection of 0.16 fg/mL (S/N=3). The 3D nanostructured PPy hydrogel offered enhanced effective surface area and thus increases immobilisation of biomolecules and facilitates ions and electron transport (Rong et al., 2015).

In an enzyme-based biosensor, enzyme is employed as biological recognition component which is immobilised on the matrix or onto the surface of the transducer. Employing enzyme as a bioreceptor presents several advantages such as high specific interaction of enzyme towards particular substrate, high biocatalytic rate and cost-effectiveness. When the interaction between enzyme and substrate takes place, the substrate is consumed and the product is formed, and as a consequence of enzymatic reaction, certain changes occur such as change in proton concentration, release or uptake of gases (CO_2, NH_3, O_2, etc.), heat emission, absorption or reflectance, and light emission which are measured (Nguyen et al., 2019).

In 2019, Erfkamp and co-workers developed an enzyme-functionalised piezoresistive hydrogel-based biosensor to detect urea (Figure 11.4). They physically incorporated urease enzyme during polymerisation of a pH-sensitive poly (acrylic acid-co-dimethylaminoethyl methacrylate) polymer. The urease enzyme hydrolyses urea, and as a consequence of this enzymatic reaction, pH change occurs which causes the swelling of the pH-sensitive polymer hydrogel depending on urea concentration. The enzyme-hydrogel system showed high sensitivity ranges from 1 to 20 mmol/L urea and was found stable over 8 weeks. The developed hydrogel-based biosensor can be used multiple times as it possesses repeatable swelling properties (Erfkamp et al., 2019).

Entire cell or cellular structures (micro-organisms such as bacteria and fungi) are employed as a biological recognition component (bioreceptor) in cell-based biosensors (Tavakoli and Tang, 2017). Fabrication of cell-based biosensor extensively utilises simple eukaryotic cells such as bacteria, fungi, yeast, algae and complex eukaryotic cells like rat, fish and human to check water quality, toxicity demarcation and fundamental cellular functionality, disease pathogenesis, respectively. Chitosan, alginate and agarose are the most often used matrixes which provide environment similar to the tissues for the proper growth and development of cells. Cells contain a number of biological molecules (enzymes, receptors, proteins) in clear-cut ratio; thus, they can respond quantitatively to certain stimuli in given circumstances as well as quantify more than one analyte. This is the motive behind the making of cell-based biosensors. The major advantage of using the cell as recognition component (bioreceptor) is that the biomolecules such as enzymes, receptors and proteins are in their connatural surroundings and thus demonstrate the best possible activity (Gupta et al., 2019).

The nucleic acids are composed of nucleotide bases adenine (A) which pairs with thymine (T) (uracil in RNA) and guanine (G) with cytosine (C) (Zhang et al., 2020). Nucleic acid-based biosensors utilise single-strand DNA (Du and Dong, 2016) which undergoes complementary base pairing (i.e. adenine–thymine and cytosine–guanine in DNA) as recognition process (Aggas and Guiseppi-Elie, 2020). DNA-based hydrogel presents numerous advantages such as biocompatibility, mechanical stability, appropriate tunability and facile fabrication (Li et al., 2016). Pure DNA and hybrid DNA hydrogel are the two categories of DNA hydrogel. Integration of DNA into polymers like polyacrylamide, polypeptide and poly(phenylenevinylene) by cross-linking with probes constructs a hybrid DNA hydrogel (Liu et al. 2018). It is reported that DNA itself forms hydrogel (pure DNA hydrogel) under certain circumstances owing to its polymer-like behaviour (Li et al., 2016; Zhang et al., 2020).

Biomimetics is artificially synthesised receptor that mimics biological receptors. Molecular imprinting is the process in which the target analyte is polymerised along with monomer, and after polymerisation, the analyte is removed which creates a cavity similar to analyte conformation that is believed to possess the same affinity for the template molecule (Figure 11.5). Materials like methacrylic acid (MAA) and ethylene glycol dimethacrylate (EGDMA), poly (N-isopropylacrylamide-acrylamide-vinylphenylboronic acid), O-phenylenediamine (O-PD) and metal–organic copolymers are reported for the fabrication of glucose-sensitive molecularly imprinted polymers (MIPs) (Tavakoli and Tang, 2017; Aggas and Guiseppi-Elie, 2020).

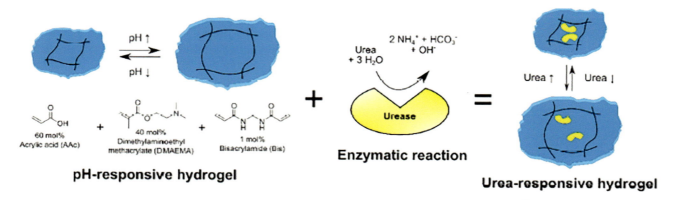

FIGURE 11.4 Measuring principle of a hydrogel-based biosensor for the detection of urea. (Reproduced from Erfkamp et al., 2019.)

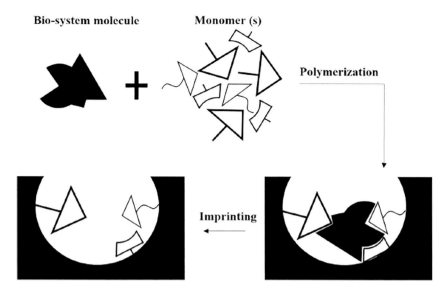

FIGURE 11.5 Schematic representation of molecular imprinting method for biosensor fabrication. (Reproduced from Tavakoli and Tang, 2017.)

11.3 BIOMEDICAL APPLICATION OF HYDROGEL-BASED SENSORS

Biosensors based on hydrogel have shown their potential application in the detection of human metabolites, wound healing, pathogen detection, etc. Recently, Kang and co-workers developed a hydrogel-based biosensor to detect matrix metalloproteinase-2 and metalloproteinase-9 (a marker overexpressed in exacerbated wound). They immobilised fluorescence resonance energy transfer (FRET) peptide which upon cleavage by protease enzyme gives fluorescence. Fluorescence intensity is a function of amount of protease in chronic wounds (Kang et al., 2019).

Previously, Li and co-workers successfully developed a nanostructured conductive hydrogel-based biosensor to detect human metabolites such as uric acids, cholesterol and triglycerides with high sensitivity and low response time (~3 s) (Li et al., 2015b).

Park and co-workers developed a hydrogel-based electrochemical biosensor for continuous glucose monitoring noninvasively. They immobilised glucose oxidase enzyme on the hydroxyethyl methacrylate (HEMA) hydrogel matrix. Fast electron transfer mediated by Prussian blue (PB, hexacyanoferrate) generated efficient signal amplifications to facilitate the detection of the extracted glucose from the interstitial fluid (Park et al., 2017).

Researchers have made efforts to develop hydrogel-based biosensor for the detection of several viruses. Influenza virus, hepatitis B virus, different pathogens and West Nile virus protein domain III have been successfully detected by hydrogel-based biosensor (Tavakoli and Tang, 2017).

Hydrogel-based biosensors have demonstrated their potential application in the detection of several cancer-specific biomarkers. In 2017, Liu and co-workers developed an electrochemical biosensor based on hybrid DNA hydrogel. They immobilised hybrid DNA hydrogel on indium tin oxide/polyethylene terephthalate (ITO/PET) electrode to detect lung cancer-specific microRNA (miR-21). Recognition probe tagged with ferrocene and cross-linked with DNAs grafted on the polyacrylamide hydrogel to construct DNA hybrid hydrogel which was again immobilised on ITO electrode treated with 3-(trimethoxysilyl) propyl methacrylates. Hybridisation of recognition probe with target miR-21causes hydrogel to dissolve, and as a consequence, loss of ferrocene tags occurs, followed by the reduction in current detected by cyclic voltammetry (CV) and differential pulse voltammetry (DPV) (Liu et al., 2017).

Ferro and co-workers designed a biosensor by immobilising microalgae in silica hydrogel. Three microalgae – *Chlorella vulgaris, Pseudokirchneriella subcapitata* and *Chlamydomonas reinhardtii* – are immobilised in alginate and silica hydrogel. After sufficient growth of the immobilised cells, they were exposed to (3-(3,4-dichlorophenyl)-1,1-dimethylurea) at a different concentration of commonly used herbicide atrazine. Chlorophyll fluorescence was produced when microalgae interact with herbicides. The designed biosensor showed good efficiency with 0.1 μM of detection limit from *C. reinhardtii* strain after 40 min of exposure (Ferro et al., 2012).

Guo and co-workers (2020) developed an L012@PAni-PAAm (pyridazine-1,4(2H,3H)-dione sodium @ polyaniline-polyacrylamide) hydrogel composite electrochemiluminescence (ECL) biosensor by electrodeposition of PAni into PAAm hydrogel and electrostatic interaction of Nafion film and electropositive polylysine and L012 on a glassy carbon electrode for *in situ* detection of H_2O_2 released from cardiomyocytes. The designed hydrogel composite-based ECL biosensor showed better stability and excellent selectivity with 2.9 nM detection limit of H_2O_2. H_2O_2 serves as a potential physiological and pathological indicator for cardiovascular diseases; (CVD), therefore, *in situ* detection of H_2O_2 could be used in physiological and pathological monitoring of the cardiovascular system (Guo et al., 2020).

11.4 CONCLUSION

The nanostructured hydrogel comprises the properties of the nanoparticulate system along with the hydrogel. Nanostructured hydrogels have attracted considerable attention in recent years due to their excellent properties to respond to various stimuli, and immobilisation of the bioreceptor as well as creating a protective layer to control the diffusion and increase biocompatibility. Polyvinyl alcohol, polyethylene glycol, polyacrylate families, alginate, chitin, chitosan, agarose, cellulose, dextran and electroconductive hydrogels are frequently used in hydrogel-based biosensors. Hydrogels can be easily engineered and readily functionalised with different biological molecules such as enzymes, nucleic acid and proteins to hold them and detect the respective analyte. Studies showed that hydrogel-based biosensors have been utilised for several biomedical applications such as metabolites (urea, cholesterol, lactate, glucose) and pathogen detection, tissue engineering, wound healing and cancer monitoring.

ACKNOWLEDGEMENT

The authors would like to acknowledge Department of Pharmaceuticals (DoP), Ministry of Chemicals and Fertilisers, Government of India, for their support. NIPER-Raebareli communication number for this manuscript is NIPER-R/Communication/176.

REFERENCES

Aggas, J.R., Guiseppi-Elie, A., 2020. Responsive Polymers in the Fabrication of Enzyme-Based Biosensors.. In *"Biomaterials Science (Fourth Edition): An Introduction to Materials in Medicine"* (pp. 1267–1286), Academic Press, W. R. Wagner, S. E. Sakiyama-Elbert, G. Zhang, M. J. Yaszemski (eds.) B. D. Ratner, A. S. Hoffman, F. J. Schoen, J. E. Lemons (Focussing eds.) Elsevier. ISBN978-0-12-816137-1. https://doi.org/10.1016/B978-0-12-816137-1.00079-9.

Ahmad, J., Gautam, A., Komath, S., Bano, M., Garg, A., Jain, K., 2019. Topical nano-emulgel for skin disorders: Formulation approach and characterization. *Recent patents on anti-infective drug discovery* 14(1), 36–48.

Bae, J., Park, O., Kim, S., Cho, H., Kim, H.J., Park, S., Shin, D.S., 2020. Tailored Hydrogels for Biosensor Applications. *Journal of Industrial and Engineering Chemistry* 89, 1–12.

Bajwa, N., Mehra, N.K., Jain, K., Jain, N.K., 2016. Targeted anticancer drug delivery through anthracycline antibiotic bearing functionalized quantum dots. *Artificial cells, nanomedicine, and biotechnology* 44(7), 1774–1782.

Conte, R., Luise, A.D., Valentino, A., Cristo, F.D., Petillo, O., Riccitiello, F., Salle, A.D., Calarco, A., Peluso, G., 2019. Hydrogel Nanocomposite Systems: Characterization and Application in Drug-Delivery Systems. *Nanocarriers for Drug Delivery* 319–349.

Dalwadi, C., Patel, G., 2015. Application of Nanohydrogels in Drug Delivery Systems: Recent Patents Review. *Recent Patents on Nanotechnology* 9, 1, 17–25.

Du, Y., Dong, S., 2016. Nucleic Acid Biosensors: Recent Advances and Perspectives. *Analytical Chemistry* 89, 1, 189–215.

Erfkamp, J., Guenther, M., Gerlach, G., 2019. Enzyme-Functionalized Piezoresistive Hydrogel Biosensors for the Detection of Urea. *Sensors* 19, 2858.

Ferro, Y., Perullini, M., Jobbagy, M., Bilmes, S.A., Durrieu, C., 2012. Development of a Biosensor for Environmental Monitoring Based on Microalgae Immobilized in Silica Hydrogels. *Sensors* 12, 16879–16891.

George, S.M., Tandon, S., Kandasubramanian, B., 2020. Advancements in Hydrogel-Functionalized Immunosensing Platforms. *ACS Omega* 5, 2060–2068.

Gupta, N., Renugopalakrishnanb, V., Liepmannc, D., Paulmurugand, R., Malhotraa, B.D., 2019. Cell-Based Biosensors: Recent Trends, Challenges and Future Perspectives. *Biosensors and Bioelectronics* 141, 111435.

Guo, B., Ma, Z., Pan, L., Shi, Y., 2019. Properties of Conductive Polymer Hydrogels and Their Application in Sensors. *Journal of Polymer Science Part B Polymer Physics* 57, 23, 1606–1621.

Guo, X., Li, Y., Li, Y., Ye, Z., Zhang, J., Zhu, T., Li, F., 2020. An L012@PAni-PAAm Hydrogel Composite Based-Electrochemiluminescence Biosensor for *In Situ* Detection of H_2O_2 Released from Cardiomyocytes. *Electrochimica Acta* 354, 136763.

Holzinger, M., Goff, A.L., Cosnier, S., 2014. Nanomaterials for Biosensing Applications: A Review. *Frontiers in Chemistry* 2, 63.

Jain, K., 2017. 7 – Dendrimers: Smart nanoengineered polymers for bioinspired applications in drug delivery. In *"Biopolymer-Based Composites: Drug Delivery and Biomedical Applications"* S. Jana, S. Maiti, S. Jana (eds.) (pp. 169–220). https://doi.org/10.1016/B978-0-08-101914-6.00007-7.

Jain, K., 2019. Nanohybrids of Dendrimers and Carbon Nanotubes: A Benefaction or Forfeit in Drug Delivery? *Nanoscience & Nanotechnology-Asia* 9, 21–29.

Jung, Y., Kim, J.S., Choi, B.R., Lee, K., Lee, H., 2017. Hydrogel Based Biosensors for *In Vitro* Diagnostic of Biochemicals, Proteins, and Genes. *Advanced Healthcare Materials* 6, 12, 1601475.

Kang, S.M., Cho, H., Jeon, D., Park, S.H., Shin, D.S., Heo, C.Y., 2019. A Matrix Metalloproteinase Sensing Biosensor for the Evaluation of Chronic Wounds. *BioChip Journal* 13, 323–332.

Kaur, M., Sudhakar, K., Mishra, V., 2018. Fabrication and Biomedical Potential of Nanogels: An overview. *International Journal of Polymeric Materials and Polymeric Biomaterials* 68, 287–296.

Kaur, H., Shorie, M., 2019. Nanomaterial Based Aptasensors for Clinical an Environmental Diagnostic Applications. *Nanoscale Advances* 1, 2123.

Khan, I., Saeed, K., Khan, I. 2017. Nanoparticles: Properties, Applications and Toxicities. *Arabian Journal of Chemistry* 5, 11.

Le Goff, G.C., Srinivas, R.L., Adam Hill, W., Doyle, P.S., 2015. Hydrogel Microparticles for Biosensing. *European Polymer Journal* 72, 386–412.

Li, J., Mo, L., Lu, C.H., Fu, T., Yang, H.H., Tan, W., 2016. Functional Nucleic Acid-Based Hydrogels for Bioanalytical and Biomedical Applications. *Chemical Society Reviews* 45, 1410.

Li, L., She, Y., Pan, L., She, Y., Yu, G., 2015a. Rational Design and Applications of Conducting Polymer Hydrogels as Electrochemical Biosensors. *Journal of Materials Chemistry B* 3, 2920.

Li, L., Wang, Y., Pan, L., She, Y., She, Y., Cheng, W., Yu, G., 2015b. A Nanostructured Conductive Hydrogels-Based Biosensor Platform for Human Metabolite Detection. *Nano Letters* 2, 1146–1151.

Liu, S., Su, W., Li, Y., Zhang, L., Ding, X., 2018. Manufacturing of an Electrochemical Biosensing Platform Based on Hybrid DNA Hydrogel: TAKING Lung Cancer-Specific miR-21 as an Example. *Biosensors and Bioelectronics* 103, 1–5.

Luo, X.L., Xu, J.J., Zhang, Q., Yang, J.G., Chen, H.Y., 2005. Electrochemically Deposited Chitosan Hydrogel for Horseradish Peroxidase Immobilization through Gold Nanoparticles Self-Assembly. *Biosensors and Bioelectronics* 21, 90–196.

Maduraiveerana, G., Jinb, W., 2017. Nanomaterials Based Electrochemical Sensor and Biosensor Platforms for Environmental Applications. *Trends in Environmental Analytical Chemistry* 13, 10–23.

Mn, P., Meng, W.P., Lorestani, F., Mahmoudian, M., Alias, Y., 2014. Electrodeposition of Copper Oxide/Polypyrrole/Reduced Graphene Oxide as a Nonenzymatic Glucose Biosensor. *Sensors and Actuators B: Chemical* 11, 72.

Nguyen, H.H., Lee, S.H., Lee, U.J., Fermin, C.D., Kim, M., 2019. Immobilized Enzymes in Biosensor Applications. *Materials* 12, 121.

Ojha, B., Jain, V.K., Gupta, S., Talegaonkar, S., Jain, K., 2021. Nanoemulgel: a promising novel formulation for treatment of skin ailments. *Polymer Bulletin* 78, 1–26. https://doi.org/10.1007/s00289-021-03729-3.

Park, H., Lee, J.Y., Kim, D.C., Koh, Y., Cha, J., 2017. Hydrogel-based electrochemical sensor for non-invasive and continuous glucose monitoring. *International Conference on Nano-Bio Sensing, Imaging, and Spectroscopy* 10324, 1032405.

Pedrosa, Y.A., Yan, J., Simonian, A.L., Revzin, A., 2011. Micropatterned Nanocomposite Hydrogels for Biosensing Applications. *Electroanalysis* 23, 5, 1142–1149.

Peppas, N.A., Slaughter, B.V., Kanzelberger, M.A., 2012. Hydrogels. *Polymer Science: A Comprehensive Reference* 9, 385–395.

Pirzada, M., Altintas, Z., 2019. Nanomaterials for Healthcare Biosensing Applications. *Sensors* 19, 5311.

Raj, M., Goyal, R.N., 2019. Silver Nanoparticles and Electrochemically Reduced Graphene Oxide Nanocomposite Based Biosensor for Determining the Effect of Caffeine on Estradiol Release in Women of Child-Bearing Age. *Sensors & Actuators: B. Chemical* 284, 759–767.

Rong, Q., Han, H., Feng, F., Ma, Z., 2015. Network Nanostructured Polypyrrole Hydrogel/Au Composites as Enhanced Electrochemical Biosensing Platform. *Scientific Reports* 5, 11440.

Sharma, S., Byrne, H., Kennedy, R.J., 2016. Antibodies and Antibody-Derived Analytical Biosensors. *Essays in Biochemistry* 60, 9–18.

Shrestha, B.K., Ahmad, R., Mousa, H.M., Kim, I., Kim, J.I., Naupane, M.P., Park, C.H., Kim, C.S., 2016. High-Performance Glucose Biosensor Based on Chitosan-Glucose Oxidase Immobilized Polypyrrole/Nafion/Functionalized Multi-Walled Carbon Nanotubes Bio-Nanohybrid Film. *Journal of Colloid and Interface Science* 482, 39–47.

Soni, N., Jain, K., Gupta, U., Jain, N.K., 2015. Controlled delivery of Gemcitabine Hydrochloride using mannosylated poly (propyleneimine) dendrimers. *Journal of Nanoparticle Research* 17(11), 1–17.

Sun, L., Zhong, Y., Gui, Z., Wang, X., Zhuang, X., Weng, J., 2018. A Hydrogel Biosensor for High Selective and Sensitive Detection of Amyloid-Beta Oligomers. *International Journal of Nanomedicine* 13, 843–856.

Tavakoli, J., Tang, Y., 2017. Hydrogel Based Sensors for Biomedical Applications: An Updated Review. *Polymers* 9, 364.

Tomczykowa, M., Brzezinska, M.E.P., 2019. Conducting Polymers, Hydrogels and Their Composites: Preparation, Properties and Bioapplications. *Polymers* 11, 350.

Urban, G.A., Weiss, T., 2009. Hydrogels for Biosensors. *Hydrogel Sensors and Actuators* 6, 197–220.

Vashist, A., Kaushik, A., Vashist, A., Sagar, V., Ghosal, A., Gupta, Y.K., Ahmad, S., Nair, M., 2018. Advances in Carbon Nanotubes–Hydrogel Hybrids in Nanomedicine for Therapeutics. *Advanced Healthcare Materials* 7, 1701213.

Vo-Dinh, T., 2002. Nanobiosensors: Probing the Sanctuary of Individual Living Cells. *Journal of Cellular Biochemistry Supplement* 39, 154–161.

Wang, Y., Burgess, D.J., 2010. Drug-Device Combination Products. In Lewis, A., Ed. *Delivery Technologies and Applications*, pp. 3–28. Woodhead Publishing Limited and CRC Press: Cambridge.

Zhai, D., Liu, B., She, Y., Pan, L., Wang, Y., Li, W., Zhang, R., Yu, G., 2013. Highly Sensitive Glucose Sensor Based on Pt Nanoparticle/Polyaniline Hydrogel Heterostructures. *ACS Nano.* 7, 4, 3540–3546.

Zhang, D., Ren, B., Zhang, Y., Xu, L., Huang, Q., He, Y., Li, X., Wu, J., Wang, J., Chen, Q., Cheng, Y., Zheng, J., 2020. From Design to Applications of Stimuli-Responsive Hydrogel Strain Sensors. *Journal of Materials Chemistry B* 8, 3171.

12 Multifunctional Carbon Nanotubes in Drug Delivery

Anamika Sahu Gulbake
DIT University

Ankit Gaur
Sentiss Pharma Pvt. Ltd

Aviral Jain
Solisto Pharma

Satish Shilpi
Ravishankar College of Pharmacy

Neelesh Kumar Mehra
National Institute of Pharmaceutical Education & Research, Hyderabad

Arvind Gulbake
DIT University
D.Y. Patil Education Society

CONTENTS

12.1 Introduction 188
    12.1.1 History 188
    12.1.2 Structure of CNTs 188
    12.1.3 Types of CNTs 188
    12.1.4 Properties of CNTs 188
    12.1.5 Synthesis Methods of CNTs 189
    12.1.6 Recent Trends of CNT's Synthesis 191
        12.1.6.1 Vapour-Phase Growth 191
        12.1.6.2 Nebulised Spray Pyrolysis Method 191
        12.1.6.3 High-Pressure Carbon Monoxide (HiPco) 192
12.2 Purification and Functionalisation Techniques of CNTs/Dispersion of CNTs 192
    12.2.1 Purification 192
        12.2.1.1 Oxidation 192
        12.2.1.2 Acid Purification 192
        12.2.1.3 Annealing 193
    12.2.2 Functionalisation Techniques of CNTs 193
        12.2.2.1 Covalent Functionalisation 193
        12.2.2.2 Non-Covalent Functionalisation of CNTs 193
    12.2.3 Characterisation of CNTs 194
12.3 Cell Penetration and Mechanism of Multifunctional CNTs 196
12.4 Application of Multifunctional Carbon Nanotubes in Drug Delivery 197
12.5 Toxicological Perspectives of CNTs 197
12.6 Recent Patents Related to CNTs as a Drug Delivery System 201
12.7 Conclusions and Future Perspectives 201
Conflict of Interest 201
Acknowledgement 201
References 201

12.1 INTRODUCTION

In this nanotechnological development, carbon nanotubes (CNTs) have attracted a great deal of attention as nanomaterials before studying the CNTs; we need to know a little about nanotechnology. Micro- and nanotechnologies have made a marked impact on the development of novel drug delivery systems (Mostafavi et al. 2019). These novel and advanced formulations are able to target the desired site and provide more excellent safety. The functionalisation of micro- and nanoparticles is useful in understanding the current and future particle-based medicines. Different sorts of nano-objects have been created to propel nanotechnology techniques, including nanoparticles, for example, quantum dots, lipids, C_{60}-based nanostructures, for example, nanotubes, nanorods, nanosheets, nanoribbons, nanofibres and nanoplates (Kargozar and Mozafari 2018). These nanomaterials are equipped for conveying a serious extent of biocompatibility before and after the conjugation of biomolecules for the diagnosis and medication of various diseases. In this chapter, we exhaustively discussed the multifunction CNTs, drug delivery and applications in drug delivery and targeting. To get familiar with and distinguish CNTs from other nanomaterials, one requires to study their structure, types and different properties.

12.1.1 History

CNTs were discovered by the Sumio Iijima in late 1991 (Iijima 1991). As the name describes, CNTs belong to the fullerene family of carbon allotropes. The arrangements of the carbon atom are in a series of condensed benzene rings. These are hollow-core cylindrical structures that consist of rolled-up honeycomb sheets (graphene sheets) of the carbon atom. The bond length between each carbon present in a tubular structure is 0.14 nm. When these graphene sheets are cut and folded into a spherical shape, they make a fullerenes (Shetti et al. 2019). These are closed or capped sometimes at their ends by semi-fullerenes like structures (Anamika Sahu and Arvind Gulbake 2017). Fullerene (C_{20+2H}) (Karfa et al. 2019) is also known as the third allotropic form of carbon other than diamond and graphite (Jha et al. 2020).

12.1.2 Structure of CNTs

The structure of nanotubes comprises hexagonal rings made via carbon molecule of low weight. A carbon molecule comprises six electrons, with two of them occupying the 1s orbital. The remaining four electrons lodge in the sp^3 or sp^2 and the sp hybrid orbital, which is also responsible for bonding structures of diamond, graphite, nanotubes and fullerenes (Meyyappan 2005). The bonding in nanotubes is sp^2, which allows carbon atom to form hexagons and occasionally pentagons and unit of pentagons by in-plane sigma (σ) bonding and out-of-plane pie (π) bonding. The nanotubes are called: (i) defect-free nanotubes when it has an only hexagonal network with 0.4 nm or small diameter, (ii) defective nanotubes when it also contains pentagon, heptagon, as a topological defect or by the chemical and structural deformity. These defects can produce a branched, bend, choroidal, helical or capped form of nanotubes. CNTs also tend to rope together via van der Walls forces. It provides CNTs with high strength and stability. The nanotubes bonding creates it and makes it biologically active, biocompatible, mechanically durable and more conductive in the form of electrical and thermal than any other form of fullerenes (Meyyappan 2005, Anamika Sahu and Arvind Gulbake 2017).

12.1.3 Types of CNTs

There are two types of carbon nanotubes as shown in Figure 12.1 based on the number of graphene sheets rolled into a concentric cylinder.

Single-walled carbon nanotubes (SWCNTs): It consists of single layer of graphene sheet rolled upon itself (diameter 1–2 nm), sometimes capping with fullerenes on both ends (Singh et al. 2012, He et al. 2013, Pravin et al. 2017, Rahman et al. 2019). By running or wrapping a graphene sheet in different directions, it is spoken to by a couple of records (n, m) called the chiral vector, in which SWCNTs can be categorised, as shown in Figure 12.1 (Rahman et al. 2019). On the off chance that m=0, the nanotubes are classified as 'Zigzag,' which is named for the example of hexagons as we proceed onward the boundary of the cylinder. If n=m, the nanotubes are classified as 'Armchair,' which portrays one of the two confirmers of cyclohexane, a hexagon of carbon atoms. Else, they are designated as 'chiral,' in which the m esteem lies among zigzag and armchair structures (Pravin, Kirteebala et al. 2017).

Multiwalled carbon nanotubes (MWCNTs): It consists of numerous concentric sheets rolled upon itself with a distance of 0.34 nm and diameter 2–50 nm. MWCNTs can synthesise without a catalyst (Singh et al. 2012, He et al. 2013, Pravin et al. 2017, Rahman et al. 2019).

Both types of CNTs are typically a few nanometres in diameter and several micrometres to centimetres long, depending on their preparation methods (Çetin, Aytekin et al. 2017).

12.1.4 Properties of CNTs

Most definitely, CNTs have some unmistakable qualities, which make them fit for being chosen as a medication transporter of biological milieu. That incorporates the ultrahigh surface zone, ease of size and surface functional groups alteration, improved cell uptake due to tinny nanoneedle shape, high medication stacking, powerful transportation capacity, greater stability (Cui et al. 2012, Peretz and Regev 2012, Hawkins et al. 2017). They are inert in nature and can tie to organic and inorganic compounds covalently or non-covalently and subsequently can be utilised for different biomedical applications (Chatrchyan

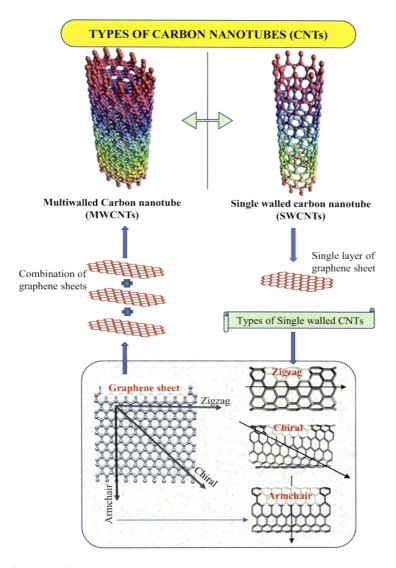

FIGURE 12.1 Types of carbon nanotubes.

et al. 2012). It has been observed that the short CNTs had more medication stacking proficiency than long ones. In any case, after the hatching of 72 h, long ones demonstrated higher productivity (Sciortino et al. 2017, Mahajan et al. 2018).

12.1.5 Synthesis Methods of CNTs

Several techniques have been introduced for the synthesis of CNTs as summarised in Figure 12.2. Some established methods, such as the arc discharge method (Ebbesen and Ajayan 1992, Bethune et al. 1993, Rinzler et al. 1998), catalytic chemical vapour deposition (CVD) (Hamers et al. 2003, Singh et al. 2017, Mohamed and Mohamed 2020) and laser ablation, are used to produce CNTs (Anamika Sahu and Arvind Gulbake 2017, Singh et al. 2017, Mohamed and Mohamed 2020). A short synopsis of the three most basic strategies utilised is given in Table 12.1 and shown in Figure 12.3. The properties of CNTs vary with the synthesis

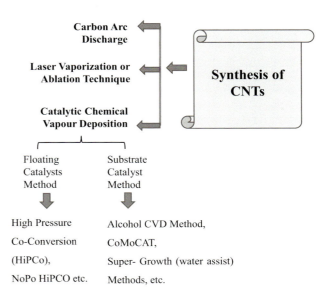

FIGURE 12.2 Various methods of synthesis of carbon nanotubes.

TABLE 12.1
Production Methods of CNTs and Their Efficiency

Method	Arc Discharge Method	Chemical Vapour Deposition	Laser Ablation (Vaporisation)	References
Inventors/researchers	Ebbesen and Ajayan, NEC, Japan 1992	Endo, Shinshu University, Nagano, Japan	Smalley, Rice, 1995	Ebbesen and Ajayan (1992), Meyyappan (2005), Lacerda et al. (2006)
Process	The principal CNTs were created with the bend release strategy with no utilisation of metal impetuses, delivering MWCNTs. Bend vaporisation of two carbon poles set start to finish, isolated by roughly hardly any millimetres separated (1 mm), in an enclosed space that is typically loaded up with dormant gas at low weight. An immediate current of 50–100 A, driven by a likely distinction of around 20V, makes a high-temperature release between the two terminals. The release disintegrates the outside of one of the carbon cathodes and structures a little bar formed store on another terminal. Basically, used to create C_{60} fullerenes, is the most effortless and normal approach to deliver CNTs	Chemical vapour deposition of hydrocarbons over a metal catalyst is a standard strategy that has been utilised to deliver different carbon materials like carbon strands and fibres. Enormous measure of CNTs can be delivered by reactant CVD of acetylene over iron and cobalt catalyst bolstered on zeolite or silica. Exceptional returns of SWCNTs have been acquired by synergist decay of H_2/CH_4 blend all over all around scattered metal particles, for example, nickel cobalt and iron on magnesium oxide at 1,000°C. The decrease produces smaller than expected progress metal particles at a temperature of generally >800°C. The decomposition of CH_4 over the newly framed nanoparticles forestalls their further development and consequently brings about an extremely high extent of SWNTs and few MWNTs	In laser ablation, laser vaporisation beats were trailed by a subsequent beat, to disintegrate the objective (graphite/carbon) all the more consistently while a lifeless gas pervades the response chamber. The utilisation of two progressive laser beats (as opposed to power) limits the measure of carbon kept as sediment. The subsequent laser beat separates the bigger particles removed by the first and feeds them into the developing nanotube structure. The material delivered by this technique shows up as a tangle of 'ropes.'	(Anamika Sahu 2017, Singh et al. 2017, Rahman et al. 2019, Mohamed and Mohamed (2020)
Typical yield SWCNTs	30%–90%	20%–100%	Up to 70%	Singh et al. (2010)
	Development of short cylinders with distances across of 0.6–1.4 nm	Development of long cylinders with distance across of 0.6–4 nm	Arrangement of long packages of cylinders (5–20 μ), with singular width from 1 to 2 nm	
MWCNTs	Development of short cylinders with internal breadth of 1–3 nm and external measurement of roughly 10 nm	Development of long cylinders with distance across going from 10 to 240 nm	Less enthusiasm for this method, seeing that it is excessively costly; however, MWCNTs synthesis is possible	Rahman et al. (2019)
Gaseous carbon sources (condition)	Helium, argon, pure graphite (low-pressure inert gas)	Methane (CH_4), carbon monoxide and acetylene (catalyst bed at high temperature, between 500°C and 1,000°C at atmospheric pressure)	Argon, NH_3 gas, graphite (A tube made up of a quartz-containing graphite block is heated in a furnace at 1,200°C)	
Merits	Can without much of a stretch produce MWCNTs, SWCNTs. SWCNTs have barely any auxiliary imperfections; MWCNTs without catalyst, not very costly, open-air synthesis possible	Most effortless to scale up to mechanical creation; long length, straightforward procedure, SWCNTs measurement controllable, unadulterated.	SWCNTs, with great measurement control and not many deformities. The response item is very unadulterated.	

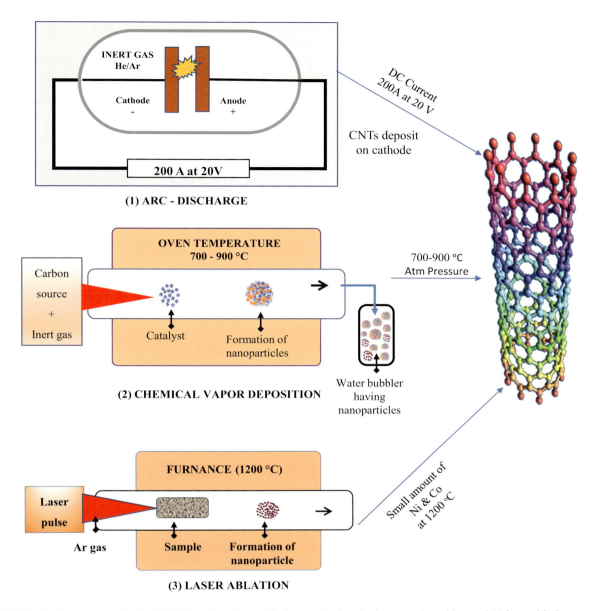

FIGURE 12.3 Synthesis methods of CNTs such as (a) arc discharge. (b) chemical vapour deposition and (c) laser ablation.

method, and essential components for the arrangement of nanotubes are the catalyst, a wellspring of carbon and sufficient vitality. The normal component of these strategies is the expansion of vitality to a carbon source to deliver sections (gatherings or single C molecules) that can recombine to create CNTs.

12.1.6 Recent Trends of CNT's Synthesis

12.1.6.1 Vapour-Phase Growth

This is a generally new procedure and a changed type of CVD. The fundamental distinction is that the CNTs are blended legitimately from the response gas and reactant metal in the chamber without a substrate (Lee et al. 2002). Two heaters are put inside the response chamber. The impetus utilised for this situation is ferrocene. The vaporisation of synergist carbon is kept up at a generally low temperature in the principal heater. Fine reactant particles are framed here when they arrive at the subsequent heater, and the decayed carbons are caught up in this impetus by dispersion, where they are changed over into CNTs (Saeed and Khan 2013).

12.1.6.2 Nebulised Spray Pyrolysis Method

An ongoing nebulised spray pyrolysis strategy has additionally been utilised for the blend of MWCNTs. A nebulised shower, the key factor in this strategy, is produced by an uncommon ultrasonic atomiser. MWCNTs with genuinely uniform breadths in adjusted packs have been acquired by means of this method. By utilising an ultrasonic nebuliser, ferrocene (catalyst) and ethanol (as dissolvable and carbon source) are splashed into a rounded heater at a fixed temperature of 800°C under an argon flow of 1 L/min. Ethanol is utilised as a solvent just as a carbon source because of its non-contaminating nature, ease, innocuous side effects (e.g. CO) and simplicity of taking care of. The high development of MWCNTs on a surface can be created. The benefit of utilising a nebulised splash is the simplicity of scaling into a mechanical scale process, as the reactants are taken care of into the heater constantly (Rao and Govindaraj 2001, Saeed and Khan 2013).

12.1.6.3 High-Pressure Carbon Monoxide (HiPco)

Synthesised SWCNTs utilise the high-pressure carbon monoxide (HIPCO) as carbon feedstock and $Fe(CO)_5$ as catalyst precursor (Daenen et al. 2003). This process was created by Smalley and his Rice University associates – this has empowered the creation of a lot of high-immaculateness SWCNTs (is approximately 1.1 nm and the yield approximately 70%), even though the production rates are still relatively low. Proceeded with innovative work, they have prompted another HiPco material, alluded to as NoPo HiPCO® (NoPo Nanotechnologies is the only organisation in the world with an operational, scalable HiPCO® technology), as an option in contrast to the recent Rice HiPco SWCNTs (Alvarez-Primo et al. 2019, Gangoli et al. 2019).

12.2 PURIFICATION AND FUNCTIONALISATION TECHNIQUES OF CNTs/DISPERSION OF CNTs

12.2.1 Purification

CNTs have the tendency to form agglomerates during synthesis because of strong van der Waals attraction between nanotubes and weaker intermolecular forces, leading in most cases to the formation of large agglomerates in polymer matrices and various solvents (Saeed and Khan 2013, Ribeiro et al. 2017). The combination of CNTs is normally connected with carbonaceous or metallic pollutions; purification is the primary step and is always needed before any further use of CNTs in drug, gene or vaccine delivery (Anamika Sahu and Arvind Gulbake 2017). Sahu et al. 2017 (Zhao et al. 2001, Anamika Sahu and Arvind Gulbake 2017, Hasani 2018, Jha et al. 2020) covered the various methodologies (centrifugation, ultrasonication, etc.) that might be used for refining the CNTs. These refinement procedures improve the nanotube dissolvability, which is simpler to isolate from the insoluble impurities (Hou et al. 2008, Mehra et al. 2008). It is based on structure- and size-dependent separations.

CNTs items contain generous measures of metal contamination and non-nanotubes carbon. These are evaporated by post-fabricating medicines, and three essential strategies have been accounted for sanitisation and they are as follows: (i) gas stage (Tsang et al. 1993), (ii) liquid stage (Hiura

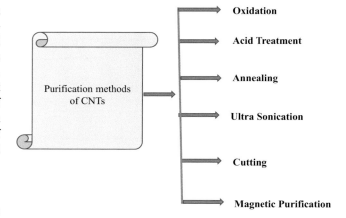

FIGURE 12.4 Purification methods of carbon nanotubes.

et al. 1995) and (iii) intercalation strategies (Ikazaki et al. 1994). The event of such metal debasements may prompt an assortment of unfriendly organic end-focuses (Ju-Nam and Lead 2008). Wrapped up graphite sheets, little fullerenes, impetus and so forth contribute to the significant polluting influences, which can meddle with the necessary properties of the CNTs, for the most part as the width diminishes, debasements increases (Hou et al. 2008). Homogenous CNTs examples are of much intrigue, which can be accomplished by refluxing and oxidation techniques (Hamers et al. 2003). A few purging methods of the SWCNTs are fundamentally concentrated under two heads, viz. structure needy and size ward partitions. The structure subordinate purging procedures will isolate the polluting influences based on the distinction in the structure of the CNTs and debasements, while the size ward partition procedures yield a uniform conveyance of size. Purification of CNTs can be achieved in different ways as shown in Figure 12.4.

12.2.1.1 Oxidation

Carbon-based impurities and surface metals can be removed by the oxidative treatment (Ikazaki et al. 1994, Hiura et al. 1995, Chiang et al. 2001, Hou et al. 2001, Moon et al. 2001, Farkas et al. 2002, Goto et al. 2002, Harutyunyan et al. 2002, Kajiura et al. 2002, Sinha and Yeow 2005, Ju-Nam and Lead 2008, Anamika Sahu and Arvind Gulbake 2017). The CNTs have less purity; the normal virtue is around 5%–10%, so decontamination is required before the connection of medications onto CNTs. Air oxidation is helpful in lessening the measure of nebulous carbon and metal impetus particles (Ni, Y). The ideal oxidation condition is seen as at 673 k for 40 min.

12.2.1.2 Acid Purification

It expels the metal impetus by uncovering the metal surface to oxidation or sonication. Then, the metal impetus is presented to corrosive and solvated, so the nanotubes stay in the suspended structure. Refluxing the sample in strong acid is successful in minimising the measure of metal particles and shapeless carbon. When nitric acid (HNO_3)

is utilised for the treatment, it just influences the metal impetus yet not the nanotubes or different particles of carbonaceous nature (Hou et al. 2001, Farkas et al. 2002, Goto et al. 2002). Various acids utilised were hydrochloric corrosive (HCl), nitric corrosive (HNO_3) and sulphuric acid (H_2SO_4); however, HCl was recognised to be the perfect refluxing corrosive. The metal will be melted and can be removed by using vacuum having high temperature (Chiang et al. 2001, Borowiak-Palen et al. 2002, Goto et al. 2002, Kajiura et al. 2002, Anamika Sahu and Arvind Gulbake 2017).

12.2.1.3 Annealing

High temperatures between 873 and 1,873 K will bring about the adjustment of nanotubes with the utilisation of deformities bringing about the pyrolysis of graphitic carbon and short fullerenes. The metal will be softened and can be expelled by utilising vacuum having high temperature (Bandow et al. 1997, Chiang et al. 2001, Borowiak-Palen et al. 2002).

12.2.2 Functionalisation Techniques of CNTs

In the most recent decade, the synthetic change of CNTs has been the emphasis on extraordinary examination in established researchers. CNTs exist in bunches because of van der Waals communications that make troublesome their scattering in various polymeric medium and even in different solvents. The functionalisation of CNTs is completed to conquer their agglomeration and group development, which additionally upgrades their scattering in polymeric materials and solvents. Along these lines, the purification or functionalisation of CNTs is imperative to upgrade their degrees of reactivity and homogenous scattering. Cleansing expels undesirable particles that stay after the blend procedure, while functionalisation presents a particular functional group gathering onto the side chains or on the ends of the CNTs (Saeed and Khan 2013, Porwal et al. 2017, Zhu 2017). Various techniques are utilised by specialists for the functionalisation and scattering of CNTs.

Surface adjustment or change of CNTs is, for the most part, done utilising compound strategies that either present the new practical gatherings or through the new bond development. The acquired CNTs can be named functionalised CNTs. The principle points behind functionalising the CNTs are the advancement of biocompatibility, improvement of encapsulation affinity and upgrade the dissolvability. The functionalisation of CNTs can be accomplished through covalent and non-covalent bonds (Cha et al. 2016, Maheshwari et al. 2019). The functionalisation of CNTs can be achieved in two ways as shown in Figure 12.5.

12.2.2.1 Covalent Functionalisation

It may be described as a substance uniting of particles onto the sp^2 carbon atoms of the π-conjugated skeleton of the CNTs. The essential response for CNTs functionalisation

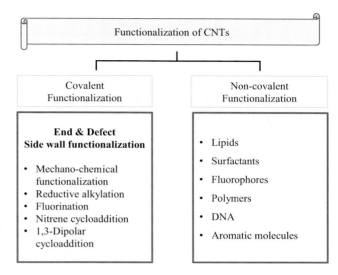

FIGURE 12.5 Functionalisation methods of carbon nanotubes.

is oxidation (Sayes et al. 2006), performed under emphatically acidic conditions. There are two fundamental methodologies for covalently functionalising nanotubes:

a. End and defect modification, and
b. Sidewall modification.

   a. **End and defect modification**: CNTs can be oxidised, by subbing hydrophilic gathering, for example, ketone, ester, –OH, –COOH to CNTs (Figure 12.6). This treatment brings about the opening of CNTs end tops, producing carboxylic (–COOH) gatherings reasonable for additional derivatisation (Hamon et al. 1999). What's more, carboxylic capacities are made where the deformities of the nanotubes sidewalls are available (Porwal et al. 2017).

   b. **Sidewall modification**: The addition of useful gatherings to the sidewalls of CNTs is regularly done by the chemical reaction between organic molecules (amphotericin B, polyethylene glycol; PEG, and so forth) and the SWCNTs surface by utilising responsive species, for example, nitrenes, carbenes and radicals (Dipl.-Chem et al. 2001, Maheshwari et al. 2019) (Figure 12.7).

12.2.2.2 Non-Covalent Functionalisation of CNTs

Different non-covalent cooperations, for instance, π-stacking, hydrophobic and van der Waals communications, have been taken into account for the functionalisation of CNTs with a wide scope of particles (Anamika Sahu and Arvind Gulbake 2017). The suspension of non-covalent functionalised CNTs indicated more noteworthy protection of their aromatic structure and their electronic character when contrasted with pure CNTs. This sort of functionalisation should be possible by the expansion of hydrophilic polymers, biopolymers and surfactants to the walls of CNTs through powerless bonds. A progression of anionic, cationic

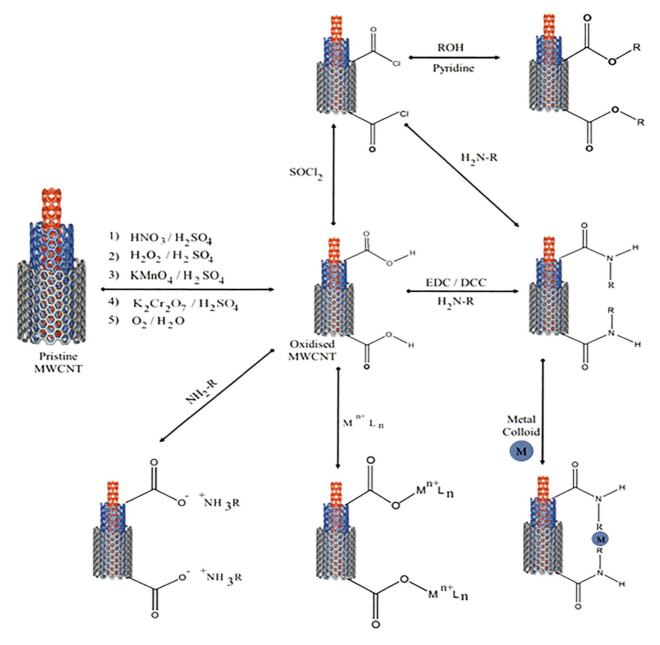

FIGURE 12.6 Common strategies for covalent functionalisation of CNTs via ends and defect site oxidation. (Reproduced with copyright permission from Sahu et al., 2017. doi:10.22159/ijpps.2017v9i6.18522.)

and non-ionic surfactants has been, as of now, proposed to scatter nanotubes in fluid media (Porwal et al. 2017). The bond among surfactants and nanotubes walls turns out to be solid due to the π–π stacking associations came that about because of the connection of aromatic gatherings of the amphiphile surfactant in the aromatic network of the nanotubes sidewalls. In this procedure, hydrophilic polymer folds over the cylinders and therefore changes the dissolvability and conductivity properties of the CNTs. Biopolymers can likewise be utilised for the functionalisation of CNTs. Nucleic acids are positively perfect possibility to frame supramolecular buildings dependent on π-π stacking between the aromatic bases and the CNTs surface (Zheng et al. 2003, Porwal et al. 2017).

12.2.3 Characterisation of CNTs

Carbon nanotubes (CNTs) are nanometric carbon-based particles, yet they additionally contain a significant number of polluting influences. Characterisation of CNTs is set out towards assurance of the concentration, quality and properties of the CNTs test, which is significant because its applications will require confirmation of properties and capacity. CNTs are the most fundamental among carbon

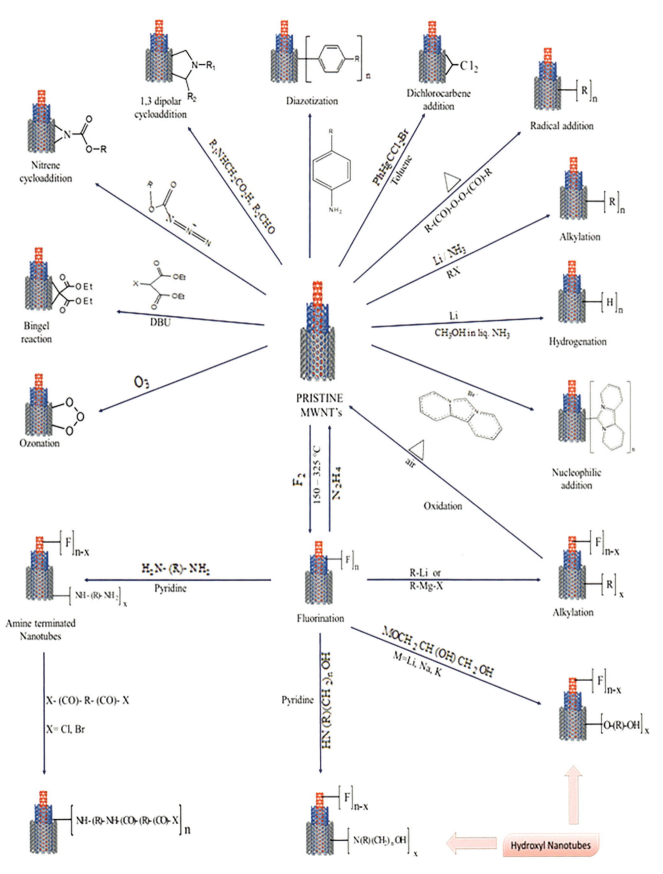

FIGURE 12.7 Strategies for covalent sidewall functionalisation of CNTs. (Reproduced with permission from Sahu et al., 2017. doi:10.22159/ijpps.2017v9i6.18522.)

nanostructures since they have the most basic qualities for progressive applications. Raman spectroscopy is reasonable for the speedy and dependable screening of the nearness of CNTs. It is a non-invasive and non-destructive method for the characterisation of CNTs. The synthesis and purification procedures of the CNTs were assessed utilising this spectroscopy. This is a delicate technique for looking at changes in the attributes of nanotubes combined utilising various procedures and conditions (Jha et al. 2020). Rathod et al. (2019) used Raman spectroscopy to find out the position, width, relative intensity of bands, electronic structure and purity of CNTs (Rathod et al. 2019).

Transmission electron microscopy (TEM) takes into account the appraisal of point-by-point structures (size and shape) as the nanotubes are tiny nanoneedle tubular structure, and also the non-CNT-related metallic impurities. Zhu et al. (2019) and Unlu et al. (2018) used TEM to examine the internal structure, diameter, number of layers and intershell spacing of CNTs (Ünlü et al. 2018, Zhu et al. 2019).

Scanning electron microscopy (SEM) gives outlines of test structures while less delicate to test readiness and homogeneity than TEM. Zhu et al. (2019) and Unlu et al. (2018) used SEM to examine aggregation, diameter and length of CNTs (Ünlü et al. 2018, Zhu et al. 2019). Thermogravimetric analysis (TGA) gives data about the general abundance of impetus (catalyst) particles, nanotubes and different carbonaceous structures. In this method, the sample is warmed at a given temperature with controlled rate to quantify the adjustment in the heaviness of a sample substance as an element of temperature. The outcomes can be interpreted from the TGA curve known as thermogram, plotted between the adjustment in mass and temperature (Akash and Rehman 2020). Yudianti and co-workers (2011) reported that after purification, the improvement of thermal stability slightly increased and was quite well preserved until 613.5°C (Yudianti 2011). Mansfield et al. (2010) used TGA to analyse the chemistry and homogeneity of carbon nanotube materials (Mansfield et al. 2009). FTIR spectroscopy is often used to determine impurities remaining from synthesis or molecules capped onto the CNTs surface, detecting the vibrational motion of the functional groups attached on the CNTs surface, but it cannot give a quantitative analysis for the detected groups (Mohamed and Mohamed 2020). Mehra and Jain (2013) used FTIR technique to determine impurities remaining after synthesis and developing functional groups after surface modification of CNTs (Mehra and Jain 2013). Pennetta et al. (2020) used FTIR to evaluate the chemical nature of the attached molecules and was performed on pristine CNTs and CNTs/PPGP (pyrrole polypropylene glycol) adducts after acetone extraction (Pennetta et al. 2020).

Atomic force microscopy (AFM) technique provides morphologic information and properties (fundamental) of the nanotubes, which are vital in the research field of nanotubes. Mehra and Jain (2014) used the AFM technique to examine the internal structure, diameter, number of layers and intershell spacing of CNTs (Mehra and Jain 2014).

Photoluminescence spectroscopy (PL) is a contactless, non-destructive method of probing the electronic structure of materials (Tan et al. 2007). Umemura et al. (2019) used PL to determine the optical and electronic properties of carbon nanotubes (Umemura et al. 2019).

Energy-dispersive spectroscopy (EDS) was applied for elemental analysis of the carbon nanotubes (Osikoya et al. 2015). Kaur et al. (2019) worked by EDS to get the elemental composition of carbon nanotubes (Kaur et al. 2019).

X-ray photoelectron spectroscopy (XPS) is one of the best tools for studying the chemical modification of surfaces, in particular the distribution and bonding of heteroatom dopants in carbon nanomaterials such as graphene and carbon nanotubes (Susi et al. 2015). Wang et al. (2019) used X-ray photoelectron spectroscopy (XPS) to determine chemical structure and functionalisation of CNTs (Wang et al. 2019).

CNTs have fascinating physicochemical properties, for example, requested structure with high aspect ratio, ultralightweight, high mechanical quality, high electrical conductivity, high thermal conductivity, metallic or semi-metallic conduct, and high surface zone. For a correct characterisation of CNTs, every one of these methods depicted thus cannot be utilised separately yet should be utilised in corresponding manners (Anamika Sahu and Arvind Gulbake 2017, Jha et al. 2020).

12.3 CELL PENETRATION AND MECHANISM OF MULTIFUNCTIONAL CNTs

The functionalised CNTs have the capacity to cross cell layers. The disguise of CNTs can be followed by marking the CNTs with a fluorescent operator and afterwards checking the take-up by utilising epifluorescence and confocal microscopy. The specific cell uptake pathway of CNTs is very complex; however, it is considered that there are two major pathways for CNTs to cross the cell layer and enter into the cells, which are as follows (Lee and Geckeler 2010):

i. Endocytosis-dependent pathway, which incorporates receptor- and non-receptor-mediated internalisation.
ii. Endocytosis-independent pathway, which includes diffusion, film combination and direct pore transport.

The procedure of internalisation depends upon a few boundaries, for example, the size, length, nature of useful gatherings (covalently or non-covalently connected to the CNTs), hydrophobicity, surface science and the culture medium of cell (Lee and Geckeler 2010, Porwal et al. 2017). Several systems can be used to investigate and comprehend the specific components of cell uptake, for example, following fluorescent-marked CNTs utilising confocal microscopy and stream cytometry (Pantarotto et al. 2004, Kam et al. 2006, Kosuge et al. 2012). The systems of cell internalisation can be endocytosis-dependent or endocytosis-independent (Lee and Geckeler 2010, Fisher et al. 2012). The intervention of endocytosis-dependent pathway could be through either receptor- or non-receptor systems. This pathway likewise relies on vitality and temperature in moving and the engulfment of medication particles into the cell film (Maheshwari et al. 2019). One of the significant systems is by means of clathrin-dependent endocytosis (250 nm) for the passage of CNTs inside the cells (Al-Qattan et al. 2018, Maheshwari et al. 2019).

12.4 APPLICATION OF MULTIFUNCTIONAL CARBON NANOTUBES IN DRUG DELIVERY

Among the different carbon allotropes, CNTs have pulled in raising consideration as a profoundly able vehicle for moving different medication particles into the living cells since their regular morphology encourages non-invasive infiltration over the organic films (Chen et al. 2008, Das et al. 2013, Lin et al. 2016, Panczyk et al. 2016, Maiti et al. 2019). Generally, medicate atoms are connected to CNTs sidewalls through covalent or non-covalent holding between the medication atoms and functionalised CNTs. Despite that, every one of these procedures has focal points or inconvenience (Sharma et al. 2016). Due to covalent interaction, drug-loaded CNTs are stable in both the extra- and intracellular compartments. This phenomenon has an absence of sustained release of the medication inside the cell micro-environment of disease cells, which is a weakness in the medication conveyance framework. Non-covalent interaction encourages the controlled arrival of the medication in the acidic state of tumour destinations; however, it suffers from dependability in extracellular pH levels (Mehra and Jain 2014).

Conveyance of therapeutic agents like medications or biomolecules items *in vivo* is trying because of their poor pharmacokinetic profiles, for example, fast degradation, gathering in tissues, and their powerlessness to cross the natural membranes. CNTs have been utilised for different applications, including the conveyance of anti-cancer medications and quality treatments. More than quite a few years, malignant growth has been one of the deadliest non-transmittable illnesses, influencing counts of age-specific incidence for 36 cancer sites in 185 countries for the year 2018 that were extracted from IARC's GLOBOCAN database of national estimates. The age-standardised rate for all cancers (including non-melanoma skin cancer) for men and women combined was 197.9 per 100,000 in 2018. The rate was higher for men (218.6 per 100,000) than women (182.6 per 100,000). There were an estimated 18.1 million (95% UI: 17.5–18.7 million) new cases of cancer (17 million excluding non-melanoma skin cancer) and 9.6 million (95% UI: 9.3–9.8 million) deaths from cancer (9.5 million excluding non-melanoma skin cancer) worldwide in 2018 (Ferlay et al. 2018, Miranda-Filho et al. 2020, Ravi Kiran et al. 2020). The developing enthusiasm for executing nanomedicines to treat cancer is significantly ascribed to their one of kind highlights for drug delivery, imaging and analysis. Among various carbon-based nanocarriers (fullerenes, CNTs, graphenes, cones and so forth) and circular nanoparticles, CNTs have increased a ton of enthusiasm because of their one of kind highlights, for example, ultra-high viewpoint proportion, high payload stacking and intracellular bioavailability. CNTs are lipophilic, which prompt their cell gathering and toxicity and defeating their innate inconvenience CNTs are functionalised (f-CNTs), in this way improving their solvency and biocompatibility. Nanotubes are used to treat cancer as a drug delivery carrier (Rosen and Elman 2009); numerous therapeutic agents have been stacked onto the CNTs as summarised in Table 12.2 including doxorubicin (Huang et al. 2011), paclitaxel (Singh et al. 2016), docetaxel (Raza et al. 2016) and oxaliplatin (Lee et al. 2016), to show the efficiency for *in vitro* and *in vivo* malignant growth medicines (Jha et al. 2020).

Polyphosphazene platinum, the anti-cancer drug administered by means of nanotubes, had improved permeation, biodistribution, as well as retention within the brain because of regulated nanotubes lipophilicity (Pai et al. 2006). Gulbake et al. (2019) developed the artesunate (AS) (anti-malarial drug)-loaded mannosylated conjugated multiwalled carbon nanotubes (M-MWCNTs) to site-specific delivery in the brain for the treatment of cerebral malaria (CM) (Gulbake et al. 2019).

Multiwalled CNTs are utilised in tranquiliser, focusing just as controlled arrival of medication. Directed medication conveyance can be defined as when an exuberant helpful medication specialist is conveyed to the exact part for a broadened period. CNTs are utilised in drug conveyance because of their hydrophobic nature that permits CNTs to remain in the flow framework for the all-encompassing period. CNTs have been used for target and controlled medication conveyance because of variable improvements that can be controlled through CNTs, for example, magnetic, electric and change in temperature. CNTs go about as a transporter for the vehicle of different biomolecules likewise, for example, proteins, peptides, DNA, RNA, immune active mixes and lectins (Che et al. 2000, Khan et al. 2016, Anzar et al. 2020).

12.5 TOXICOLOGICAL PERSPECTIVES OF CNTs

Regardless of the wide scope of CNTs applications, a poisonous symptom is enlisted (Francis and Devasena 2018, Mohamed and Mohamed 2020, Samiei et al. 2020). CNTs harmfulness is controlled and decided through variables, for example, morphology, size, shape, and virtue (Luo et al. 2013). This is claiming CNTs infiltrate cells, prompting a cytotoxic reaction where the contact between the CNTs with cell brings about a strain on the cytoskeleton of the phagocyte. In understanding, a few investigations have demonstrated that more extended lengths and bigger width CNTs

TABLE 12.2
Potential Applications of CNT's: *In Vitro* and *In Vivo* Evaluation

Functionalised CNTs	Cell Lines	Bioactives	Evaluation Parameters	Outcomes of the Study	References
DOX/FA/CHI/SWCNTs	SMMC-772 hepatocellular carcinoma	Doxorubicin	• Microplate reader (Model 680, Bio-Rad) • *In vitro* cell culture study • *In vivo* studies • Light microscopy • Flow cytometry	• The therapeutic efficiency of DOX/FA/CHI/SWCNTs was time-dependent and dosage-dependent • Tumour volumes were reduced	Ji et al. (2012)
Biotin-SWCNTs or MWCNTs	HeLa cells and MCF-7 cancer cell lines	Paclitaxel		Formation of a stable microtubule-taxoid complex and finally caused apoptosis and cell death	Elhissi et al. (2012)
PAMAM dendrimers FA-treated MWCNTs	HeLa cell	Doxorubicin	• Flow cytometry • Confocal microscopy	Targeting of FA receptors overexpresses cancer cells	Elhissi et al. (2012)
SWCNTs Polysaccharide [sodium alginate (ALG) and chitosan (CHI)]	HeLa cells	Doxorubicin	• Magnetic field	Modified CNTs reduce side effects as well as increase therapeutic amount of the drug in patient	Zhang et al. (2009)
SWCNTs Pluronic-F108 surfactant-wrapped CDDP-encapsulated SWCNTs	MCF-7 and MDA-MB-231 breast cancer cell lines.	Cisplatin [cis-diamine dichloro] platinum (II, CDDP)	• High-resolution transmission electron microscopy (HR-TEM) • Energy-dispersive spectroscopy (EDS) • X-ray photoelectron spectroscopy (XPS) • Inductively coupled plasma optical emission spectrometry (ICP-OES)	These studies have laid the foundation for developing US-tube-based delivery of chemotherapeutics, with drug release mainly limited to cancer cells only	Guven et al. (2011)
MWCNTs	U87 and human glioblastoma cells	Doxorubicin	• TEM images of MWCNTs • Drug loading and release behaviour • Cellular uptake of MWCNTs	The application of DOX-FA-MN-MWCNTs could be extended to enhance the efficiency of cancer therapy *in vivo*	Lu et al. (2012)
Phospholipid–PEG-SWCNTs	Human T cells and peripheral blood mononuclear cells	siRNA cargo molecules	• *In vitro* cytotoxicity and biocompatibility studies • Confocal fluorescence microscopy and flow cytometry		
PEGylated-oxidised MWCNTs	Brain glioma	Angiopep-2 peptide	• BCEC and C6 cytotoxicity • Haematology analysis • CD68 immunohistochemical analysis	O-MWCNTs-PEG-ANG is a promising dual-targeting carrier to deliver DOX for the treatment of a brain tumour	Ren et al. (2012)
PEGylated MWCNTs	HT-29 cells	Oxaliplatin	• MTT assay • Pt-DNA adducts formation • γ-H$_2$AX formation • Cell apoptosis assay	• PEGylated multiwalled carbon nanotubes can be used as a sustained-release drug delivery system, thus remarkably improving cytotoxicity of oxaliplatin on HT-29 cells	Wu et al. (2013)

(Continued)

TABLE 12.2 (Continued)
Potential Applications of CNT's: *In Vitro* and *In Vivo* Evaluation

Functionalised CNTs	Cell Lines	Bioactives	Evaluation Parameters	Outcomes of the Study	References
Amine-functionalised SWCNTs	Testicular cancer cells	Platinum (IV) complex c,c,t-[Pt(NH3)2Cl2(OEt)-(O2CCH2CH2CO2H)]	• Fluorescence microscopy • Atomic absorption spectroscopy	The cytotoxicity of the free platinum (IV) complex increases by >100-fold when the complex is attached to the surface of the functionalised SWNTs	Feazell et al. (2007)
Dapsone-O-(7-azabenzotriazol-1-yl)-*N,N,N1,N1*-tetramethyluronium hexafluorophosphate/*N,N*diisopropylethylamine-f-MWCNTs	Peritoneal macrophages for antimicrobial and anti-inflammatory	Dapsone	• Fourier transform infrared spectroscopy • TEM • Thermogravimetric analysis • Confocal laser microscopy and flow cytometry	The response of peritoneal macrophages to dapsone covalently attached on the surface of carbon nanotubes	Vuković et al. (2010)
Orthogonally protected amino acids MWCNTs	Mammalian cells for antifungal activity	Amphotericin B			Wu et al. (2005)
PEG-8 capryl/caprylic acid glycerides CNTs	Treat anaemia	Erythropoietin (EPO), a sialoglycoprotein	• ELISA method	The use of CNTs as LFNPS improved the bioavailability of EPO to 11.5%, following intra-small intestinal administration	Venkatesan et al. (2005)
PEG conjugation SWCNTs	Cancer cells, Reduced toxicity	Doxorubicin or mitoxantrone	• Atomic force microscope • Raman spectroscopy	The study demonstrates the potential of carbon nanotubes as a multimodal drug delivery system and presents a functionalisation scheme that is able to overcome many of the problems encountered in this area of application	Heister et al. (2012)
Carboxylated and coated with polysaccharide SWCNTs	HeLa cells	Doxorubicin	• HR–TEM and SEM	Targeted delivery and controlled release of doxorubicin to cancer cells using modified single-walled carbon nanotubes	Zhang et al. (2009)
f-MWCNTs–PTX MWCNTs	Anti-cancer	Paclitaxel			Tian et al. (2011)
Au-decorated MWCNTs	Papillary Thyroid cancer cells	Bionanofluids			Dotan et al. (2016)
MWCNTs	Cancer cells	Doxorubicin	Atomic emission spectroscopy	Efficiency related to length and incubation time	Sciortino et al. (2017)

TABLE 12.3
Patents of Carbon-Based Nanoparticles in Drug Delivery System

S. No.	Patent Number	Date of Patent/ Publication	Title	References
1.	US10589997 B2	17 March 2020	Discrete carbon nanotubes with targeted oxidation levels and stable gel formulations thereof	Swogger et al. (2020)
2.	US10696944B2	04 December 2019	Intracellular delivery	Sharei et al. (2020)
3.	US9976137 B2	22 May 2018	Targeted self-assembly of functionalised carbon nanotubes on tumours	Scheinberg et al. (2018)
4.	US10589997 B2	17 March 2020	Discrete carbon nanotubes with targeted oxidation levels and stable gel formulations thereof	Swogger et al. (2020)
5.	US10414656 B2	17 September 2019	Discrete carbon nanotubes with targeted oxidation levels and formulations thereof	Swogger et al. (2017)
6.	US20190110382 A1	11 April 2019	Shielding formulations using discrete carbon nanotubes with targeted oxidation levels and formulations thereof	Bosnyak and Swogger (2019)
7.	US9050444 B2	09 June 2015	Drug delivery and substance transfer facilitated by nano-enhanced device having aligned carbon nanotubes protruding from iDevice surface	Gharib et al. (2015)
8.	US9981042 B2	29 May 2018	Carbon nanotube-based anti-cancer agent capable of suppressing drug resistance	Khang et al. (2018)
9.	US9737593 B2	22 August 2017	Carbon nanotube compositions and methods of use thereof	Fahmy et al. (2017)
10.	US20180147296 A1	31 May 2018	Carbon material delivery systems and methods	Sitharaman and Chowdhury (2018)
11.	US9233166 B2	12 January 2016	Supramolecular functionalisation of graphitic nanoparticles for drug delivery	Dai et al. (2016)
12.	US2017/0224840 A1	10 August 2017	Drug delivery system comprising a cancer stem cell-targeted carbon nanotube, preparation and use thereof	Hongjuan et al. (2017)
13.	US9617157B2	11 April 2017	Tubular nanostructure targeted to cell membrane	Bangera et al. (2017)
14.	US9991391B2	06 June 2018	Pristine and functionalised graphene materials	Dai et al. (2018)
15.	US10670559B2	02 June 2020	Nanofluidic channels with integrated charge sensors and methods based thereon	Mannion and Craighead (2020)
16.	US9422517B2	23 August 2016	Microscale and nanoscale structures for manipulating particles	Chen et al. (2016)
17.	US9144383B2	29 September 2015	Device and method for *in vivo* non-invasive magnetic manipulation of circulating objects in bioflows	Zharov (2015)
18.	US8956820B2	17 February 2015	Method for detecting cancer cells using vertically aligned carbon nanotubes	Mohajerzadeh et al. (2015)
19.	US9393396B2	19 July 2016	Method and composition for hyperthermally treating cells by nanoparticle-agent-cell complex	Peyman (2016)
20.	US8927689B2	06 January 2015	Peptide nanostructures and methods of generating and using the same	Reches and Gazit (2015)
21.	US10143658B2	04 December 2018	Nanoparticles for delivery of active agents, such as therapeutic and imaging agents, and methods of making and methods of using such compositions	Ferrari et al. (2018)
22.	US9861296B2	09 January 2018	Ingestible probe with agent delivery	Gazdzinski (2018)
23.	US9823246B2	02 October 2018	Fluorescence-enhancing plasmonic nanoscopic gold films and assay-based thereon	Dai et al. (2018)
24.	US9504745B2	29 November 2016	Compositions and methods for cancer treatment using targeted carbon nanotubes	Harrison Jr et al. (2016)
25.	US20200237663	07 July 2020	Method and composition for targeted delivery of therapeutic agents	Mcdevitt et al. (2020)

have even more a harmful impact than little ones(Lamberti et al. 2014). Functionalisation can significantly improve the dispersibility and biocompatibility, while reducing the toxicity of CNTs. Also, CNTs purity is an extra significant factor, where metallic polluting influences lead to cell demise. Also, CNTs and additionally related contaminants were notable for oxidative stress (Alshehri et al. 2016), inflammation, apoptosis, pulmonary inflammation, fibrosis and granuloma in lungs (Meng et al. 2013, Lamberti et al. 2014). A list of factors that have been found to have an effect on the level of toxicity of CNTs (Kushwaha et al. 2013, Francis and Devasena 2018) follow underneath: (i) concentration/portion of CNTs, (ii) SWCNTs or MWCNTs, (iii) purity and length of the CNTs cylinders, (iv) catalyst residues leftover during union or functionalisation, (v) degree of aggregation, (vi) surface charge, aspect ratio, (vii) oxidisation and (viii) functionalisation critical attributes.

12.6 RECENT PATENTS RELATED TO CNTs AS A DRUG DELIVERY SYSTEM

CNTs are one of the most examined vehicles among the submicron transporters and they offer the opportunity to fix illnesses through targeted drug delivery. The established researchers have taken a shot at CNT-based frameworks for drug delivery, tissue building applications with proof of the positive after-effects of explores done. From this viewpoint, plenteous licences have been assigned on biomedical applications of carbon-based nanoparticles that are introduced in Table 12.3, which features a momentous thought for CNTs and uncovers precisely where we persist in this field.

12.7 CONCLUSIONS AND FUTURE PERSPECTIVES

In outline, carbon nanotubes are among those carbon nanomaterials, which have huge surface territory with artificially tunable useful gatherings. Attributable to extraordinary physical and mechanical properties, CNT speaks to itself as a strong medication transporter just as an incredible imaging applicant in medication. CNTs potential to experience functionalisation with the remedial or detecting moieties through a progression of substance responses has made them promising biocompatible nano-possibility for the finding and focused on the treatment of unmanageable illnesses, for example, malignant growth, CNS issue, and irresistible ailments and furthermore in the field of tissue recovery.

In recent years, CNTs are one of the most explored drug conveyance vehicles among the submicron transporters and they offer us one more opportunity to fix illnesses through pinpoint drug conveyance. But many modernisations have been made since their discovery and have been consistently pulling in the consideration of analysts, researchers, and scholastics, as another, option, sheltered and powerful stage for the various nanotechnological, biomedical and drug applications. Lately, huge advances have been accounted for in the investigation of CNTs in disease treatment, including the conveyance of medications (AmB, PTX, MTX, DOX and so forth), genes, nucleic acids and siRNA. Furthermore, CNTs have the endohedral filling capacity, with the goal that any medication particle can be without any problem put away and held through the solid communication and show continued and controlled delivery. As we would see it, f-CNTs could start a supernatural occurrence in disease treatment and imaging, also in other awful sicknesses, through chemotherapeutic and siRNA conveyance.

CONFLICT OF INTEREST

The authors confirm that this article has no conflict of interest.

ACKNOWLEDGEMENT

The author Dr. Neelesh Kumar Mehra would like to thank the National Institute of Pharmaceutical Education and Research (NIPER), Hyderabad, for extending facilities to write this chapter (Research Communication No. NIPER-H/2020/BC-011). Dr. Gulbake would like to thank SERB, New Delhi, for extending facilities to write this chapter (EEQ/2016/000789).

REFERENCES

Akash, M. S. H. and K. Rehman (2020). "Thermo gravimetric analysis." In *Essentials of Pharmaceutical Analysis*, M. S. H. Akash and K. Rehman (eds). Springer, Singapore, 215–222.

Al-Qattan, M., P. K. Deb and R. Tekade (2018). "Molecular dynamics simulation strategies for designing carbon-nanotube-based targeted drug delivery." *Drug Discovery Today* 23: 235–250.

Alshehri, R., A. M. Ilyas, A. Hasan, A. Arnaout, F. Ahmed and A. Memic (2016). "Carbon nanotubes in biomedical applications: Factors, mechanisms, and remedies of toxicity." *Journal of Medicinal Chemistry* 59(18): 8149–8167.

Alvarez-Primo, F., S. Anil Kumar, F. S. Manciu and B. Joddar (2019). "Fabrication of surfactant-dispersed HiPco single-walled carbon nanotube-based alginate hydrogel composites as cellular products." *International Journal of Molecular Sciences* 20(19): 4802.

Anamika Sahu, A. J. and A. Gulbake (2017). "The role of carbon nanotubes in nanobiomedicines." *International Journal of Pharmacy and Pharmaceutical Sciences* 9(6): 235.

Anzar, N., R. Hasan, M. Tyagi, N. Yadav and J. Narang (2020). "Carbon nanotube – A review on synthesis, properties and plethora of applications in the field of biomedical science." *Sensors International* 1: 100003.

Bandow, S., A. Rao, K. Williams, A. Thess, R. Smalley and P. Eklund (1997). "Purification of single-wall carbon nanotubes by microfiltration." *The Journal of Physical Chemistry B* 101(44): 8839–8842.

Bangera, M. G., E. Harlow, R. A. Hyde, M. Y. Ishikawa, E. K. Jung, E. C. Leuthardt, N. P. Myhrvold, D. J. Rivet, E. A. Sweeney and C. T. Tegreene (2017). Tubular nanostructure targeted to cell membrane, Google Patent no. 12/322,366.

Bethune, D., C. Klang, M. De Vries, G. Gorman, R. Savoy, J. Vazquez and R. Beyers (1993). "Cobalt-catalysed growth of carbon nanotubes with single-atomic-layer walls." *Nature* 363(6430): 605–607.

Borowiak-Palen, E., T. Pichler, X. Liu, M. Knupfer, A. Graff, O. Jost, W. Pompe, R. Kalenczuk and J. Fink (2002). "Reduced diameter distribution of single-wall carbon nanotubes by selective oxidation." *Chemical Physics Letters* 363(5): 567–572.

Bosnyak, C. P. and K. W. Swogger (2019). Shielding formulations using discrete carbon nanotubes with targeted oxidation levels and formulations thereof, Google Patents.

Çetin, M., E. Aytekin, B. Yavuz and S. Bozdağ-Pehlivan (2017). "Chapter 7 – Nanoscience in targeted brain drug delivery." In *Nanotechnology Methods for Neurological Diseases and Brain Tumors*. Y. Gürsoy-Özdemir, S. Bozdağ-Pehlivan and E. Sekerdag (eds). Academic Press: Cambridge, MA, 117–147.

Cha, J., S. Jin, J. Shim, C. Park, H. J. Ryu and S. Hong (2016). "Functionalization of carbon nanotubes for fabrication of CNT/epoxy nanocomposites." *Materials & Design* 95: 1–8.

Chatrchyan, S., V. Khachatryan, A. M. Sirunyan, A. Tumasyan, W. Adam, E. Aguilo, T. Bergauer, M. Dragicevic, J. Erö and C. Fabjan (2012). "Observation of a new boson at a mass of 125 GeV with the CMS experiment at the LHC." *Physics Letters B* 716(1): 30–61.

Che, J., T. Cagin and W. A. Goddard III (2000). "Thermal conductivity of carbon nanotubes." *Nanotechnology* 11(2): 65.

Chen, G., F. Fachin, M. Toner and B. Wardle (2016). Microscale and nanoscale structures for manipulating particles, Google Patents.

Chen, J., S. Chen, Z. Xianrui, L. Kuznetsova, S. Wong and I. Ojima (2008). "Functionalized single-walled carbon nanotubes as rationally designed vehicles for tumor-targeted drug delivery." *Journal of the American Chemical Society* 130: 16778–16785.

Chiang, I., B. Brinson, R. Smalley, J. Margrave and R. Hauge (2001). "Purification and characterization of single-wall carbon nanotubes." *The Journal of Physical Chemistry B* 105(6): 1157–1161.

Cui, X.-Q., Y.-H. Zhao, Y.-Q. Chu, G.-P. Li, Q. Li, L.-P. Zhang, H.-J. Su, Z.-Q. Yao, Y.-N. Wang and X.-Z. Xing (2012). "The large sky area multi-object fiber spectroscopic telescope (LAMOST)." *Research in Astronomy and Astrophysics* 12(9): 1197.

Daenen, M., R. De Fouw, B. Hamers, P. Janssen, K. Schouteden and M. Veld (2003). *The Wondrous World of Carbon Nanotubes. A Review of Current Carbon Nanotube Technologies*. Eindhoven University of Technology: Eindhoven, 89.

Dai, H., X. Li and X. Sun (2018). Pristine and functionalized graphene materials, Google Patents.

Dai, H., Z. Liu, X. Li and X. Sun (2016). Supramolecular functionalization of graphitic nanoparticles for drug delivery, Google Patents.

Das, M., R. P. Singh, S. R. Datir and S. Jain (2013). "Intranuclear drug delivery and effective *in vivo* cancer therapy via estradiol–PEG-appended multiwalled carbon nanotubes." *Molecular Pharmaceutics* 10(9): 3404–3416.

Dipl.-Chem, M., O. Vostrowsky, A. Hirsch, F. Hennrich, M. Kappes, R. Weiss and F. Dipl.-Chem (2001). "Sidewall functionalization of carbon nanotubes." *Angewandte Chemie International Edition* 40: 4002–4005.

Dotan, I., P. J. Roche, M. Tamilia, M. Paliouras, E. J. Mitmaker and M. A. Trifiro (2016). "Correction: Engineering multiwalled carbon nanotube therapeutic bionanofluids to selectively target papillary thyroid cancer cells." *PloS One* 11(6): e0158022.

Ebbesen, T. and P. Ajayan (1992). "Large-scale synthesis of carbon nanotubes." *Nature* 358(6383): 220–222.

Elhissi, A., W. Ahmed, V. R. Dhanak and K. Subramani (2012). "Carbon nanotubes in cancer therapy and drug delivery." *Journal of Drug Delivery* 2012: 347–363.

Fahmy, T. M., L. D. Pfefferle and G. L. Haller (2017). Carbon nanotube compositions and methods of use thereof, Google Patents.

Farkas, E., M. Elizabeth Anderson, Z. Chen and A. G. Rinzler (2002). "Length sorting *cut* single wall carbon nanotubes by high performance liquid chromatography." *Chemical Physics Letters* 363(1): 111–116.

Feazell, R. P., N. Nakayama-Ratchford, H. Dai and S. J. Lippard (2007). "Soluble single-walled carbon nanotubes as longboat delivery systems for platinum (IV) anticancer drug design." *Journal of the American Chemical Society* 129(27): 8438–8439.

Ferlay, J., M. Colombet, I. Soerjomataram, D. M. Parkin, M. Piñeros, A. Znaor and F. Bray (2018). "Estimating the global cancer incidence and mortality in 2018: GLOBOCAN sources and methods." *International Journal of Cancer* 144: 1941–1953.

Ferrari, M., E. Tasciotti and J. Sakamoto (2018). Multistage delivery of active agents, Google Patents.

Fisher, C., A. Rider, Z. J. Han, S. Kumar, I. Levchenko and K. Ostrikov (2012). "Applications and Nanotoxicity of Carbon Nanotubes and Graphene in Biomedicine." *Journal of Nanomaterials* 2012: 315185.

Francis, A. P. and T. Devasena (2018). "Toxicity of carbon nanotubes: A review." *Toxicology and Industrial Health* 34(3): 200–210.

Gangoli, V. S., M. A. Godwin, G. Reddy, R. K. Bradley and A. R. Barron (2019). "The state of HiPco single-walled carbon nanotubes in 2019." *C – Journal of Carbon Research* 5(4): 65.

Gazdzinski, R. F. (2018). Ingestible probe with agent delivery, Google Patents.

Gharib, M., A. I. Aria and E. B. Sansom (2015). Drug delivery and substance transfer facilitated by nano-enhanced device having aligned carbon nanotubes protruding from device surface, Google Patents.

Goto, H., T. Furuta, Y. Tokune, Y. Fujiwara and T. Ohashi (2002). Method of manufacturing carbon nanotube, US Patent 20,020,090,468.

Gulbake, A. S., A. Jain, S. Shilpi, P. Kumar and A. Gulbake (2019). "Mannosylated multiwalled carbon nanotubes assisted artesunate delivery for cerebral malaria." *International Journal of Applied Pharmaceutics* 11: 24–30.

Guven, A., I. Rusakova, M. Lewis and L. Wilson (2011). "Cisplatin@US-tube carbon nanocapsules for enhanced chemotherapeutic delivery." *Biomaterials* 33: 1455–1461.

Hamers, B., P. J. ST and M. Veld (2003). *The Wondrous World of Carbon Nanotubes*. Eindhoven University of Technology: Eindhoven.

Hamon, M., J. Chen, H. Hu, Y. Chen, M. E. Itkis, A. Rao, P. Eklund and R. Haddon (1999). "Dissolution of Single-Walled Carbon Nanotubes." *Advanced Materials* 11: 834–840.

Harrison Jr, R. G., D. E. Resasco and L. F. F. Neves (2016). Compositions and methods for cancer treatment using targeted carbon nanotubes, Google Patents.

Harutyunyan, A. R., B. K. Pradhan, J. Chang, G. Chen and P. C. Eklund (2002). "Purification of single-wall carbon nanotubes by selective microwave heating of catalyst particles." *The Journal of Physical Chemistry B* 106(34): 8671–8675.

Hasani, A. (2018). "Approaches to graphene, carbon nanotube and carbon nanohorn, synthesis, properties and applications." *Nanoscience & Nanotechnology-Asia* 8: 04–11.

Hawkins, S. A., H. Yao, H. Wang and H.-J. Sue (2017). "Tensile properties and electrical conductivity of epoxy composite thin films containing zinc oxide quantum dots and multi-walled carbon nanotubes." *Carbon* 115: 18–27.

He, H., L. A. Pham-Huy, P. Dramou, D. Xiao, P. Zuo and C. Pham-Huy (2013). "Carbon Nanotubes: Applications in Pharmacy and Medicine." *BioMed Research International* 2013: 578290.

Heister, E., V. Neves, C. Lamprecht, S. R. P. Silva, H. M. Coley and J. McFadden (2012). "Drug loading, dispersion stability, and therapeutic efficacy in targeted drug delivery with carbon nanotubes." *Carbon* 50(2): 622–632.

Hiura, H., T. W. Ebbesen and K. Tanigaki (1995). "Opening and purification of carbon nanotubes in high yields." *Advanced Materials* 7(3): 275–276.

Hongjuan, Y., Y. Zhang, L. Sun and Y. Liu (2017). Drug delivery system comprising a cancer stem cell-targeted carbon nanotube, preparation and use thereof, Google Patents.

Hou, P.-X., C. Liu and H.-M. Cheng (2008). "Purification of carbon nanotubes." *Carbon* 46(15): 2003–2025.

Hou, P.-X., C. Liu, Y. Tong, S. Xu, M. Liu and H. Cheng (2001). "Purification of single-walled carbon nanotubes synthesized by the hydrogen arc-discharge method." *Journal of Materials Research-Pittsburgh* 16(9): 2526–2529.

Huang, H., Q. Yuan, J. S. Shah and R. D. Misra (2011). "A new family of folate-decorated and carbon nanotube-mediated drug delivery system: Synthesis and drug delivery response." *Advanced Drug Delivery Reviews* 63(14–15): 1332–1339.

Iijima, S. (1991). Helical microtubules of graphitic carbon. *Nature* 354: 56–58.

Ikazaki, F., S. Ohshima, K. Uchida, Y. Kuriki, H. Hayakawa, M. Yumura, K. Takahashi and K. Tojima (1994). "Chemical purification of carbon nanotubes by use of graphite intercalation compounds." *Carbon* 32(8): 1539–1541.

Jha, R., A. Singh, P. K. Sharma and N. K. Fuloria (2020). "Smart carbon nanotubes for drug delivery system: A comprehensive study." *Journal of Drug Delivery Science and Technology* 58: 101811.

Ji, Z., G. Lin, Q. Lu, L. Meng, X. Shen, L. Dong, C. Fu and X. Zhang (2012). "Targeted therapy of SMMC-7721 liver cancer in vitro and in vivo with carbon nanotubes based drug delivery system." *Journal of Colloid Interface and Science* 365(1): 143–149.

Ju-Nam, Y. and J. R. Lead (2008). "Manufactured nanoparticles: An overview of their chemistry, interactions and potential environmental implications." *Science of the Total Environment* 400(1): 396–414.

Kajiura, H., S. Tsutsui, H. Huang and Y. Murakami (2002). "High-quality single-walled carbon nanotubes from arc-produced soot." *Chemical Physics Letters* 364(5): 586–592.

Kam, N. W. S., Z. Liu and H. Dai (2006). "Carbon nanotubes as intracellular transporters for proteins and DNA: An investigation of the uptake mechanism and pathway." *Angewandte Chemie* 118(4): 591–595.

Karfa, P., S. De, K. C. Majhi, R. Madhuri and P. K. Sharma (2019). "2.07 – Functionalization of carbon nanostructures." In *Comprehensive Nanoscience and Nanotechnology* (Second Edition), D. L. Andrews, R. H. Lipson and T. Nann (eds). Academic Press: Oxford, 123–144.

Kargozar, S. and M. Mozafari (2018). "Nanotechnology and nanomedicine: Start small, think big." *Materials Today: Proceedings* 5(7, Part 3): 15492–15500.

Kaur, J., G. S. Gill and K. Jeet (2019). "Applications of carbon nanotubes in drug delivery: A comprehensive review." In *Characterization and Biology of Nanomaterials for Drug Delivery*. S. S. Mohapatra, S. Ranjan, N. Dasgupta, R. K. Mishra, S. Thomas (eds). Elsevier: Amsterdam, 113–135.

Khan, M. U., K. R. Reddy, T. Snguanwongchai, E. Haque and V. G. Gomes (2016). "Polymer brush synthesis on surface modified carbon nanotubes via in situ emulsion polymerization." *Colloid and Polymer Science* 294(10): 1599–1610.

Khang, D. W., S. S. Kang, J. Choi and T. H. Nam (2018). Carbon nanotube-based anti-cancer agent capable of suppressing drug resistance, Google Patents.

Kosuge, H., S. Sherlock, T. Kitagawa, R. Dash, J. Robinson, H. Dai and M. McConnell (2012). "Near infrared imaging and photothermal ablation of vascular inflammation using single-walled carbon nanotubes." *Journal of the American Heart Association* 1: e002568.

Kushwaha, S., S. Ghoshal, A. Rai and S. Singh (2013). "Carbon nanotubes as a novel drug delivery system for anticancer therapy: A review." *Brazilian Journal of Pharmaceutical Sciences* 49: 629–643.

Lacerda, L., A. Bianco, M. Prato and K. Kostarelos (2006). "Carbon nanotubes as nanomedicines: From toxicology to pharmacology." *Advanced Drug Delivery Reviews* 58(14): 1460–1470.

Lamberti, M., S. Zappavigna, N. Sannolo, S. Porto and M. Caraglia (2014). "Advantages and risks of nanotechnologies in cancer patients and occupationally exposed workers." *Expert Opinion on Drug Delivery* 11(7): 1087–1101.

Lee, C., S. Lyu, C.-Y. Park and C.-W. Yang (2002). "Large-scale production of aligned carbon nanotubes by the vapor phase growth method." *Chemical Physics Letters* 359: 109–114.

Lee, P.-C., C.-Y. Lin, C.-L. Peng and M.-J. Shieh (2016). "Development of a controlled-release drug delivery system by encapsulating oxaliplatin into SPIO/MWNT nanoparticles for effective colon cancer therapy and magnetic resonance imaging." *Biomaterials Science* 4(12): 1742–1753.

Lee, Y. and K. E. Geckeler (2010). "Carbon nanotubes in the biological interphase: The relevance of noncovalence." *Advanced Materials* 22(36): 4076–4083.

Lin, G., P. Mi, C. Chengchao, J. Zhang and G. Liu (2016). "Inorganic nanocarriers overcoming multidrug resistance for cancer theranostics." *Advanced Science* 3: 1600134.

Lu, Y. J., K. C. Wei, C. C. Ma, S. Y. Yang and J. P. Chen (2012). "Dual targeted delivery of doxorubicin to cancer cells using folate-conjugated magnetic multi-walled carbon nanotubes." *Colloids and Surfaces B: Biointerfaces* 89: 1–9.

Luo, E., G. Song, Y. Li, P. Shi, S. J. You and Y. Lin (2013). "The toxicity and pharmacokinetics of carbon nanotubes as an effective drug carrier." *Current Drug Metabolism* 14: 879–890.

Mahajan, S., A. Patharkar, K. Kuche, R. Maheshwari, P. K. Deb, K. Kalia and R. K. Tekade (2018). "Functionalized carbon nanotubes as emerging delivery system for the treatment of cancer." *International Journal of Pharmaceutics* 548(1): 540–558.

Maheshwari, N., M. Tekade, N. Soni, P. Ghode, M. C. Sharma, P. K. Deb and R. K. Tekade (2019). "Chapter 16 – Functionalized carbon nanotubes for protein, peptide, and gene delivery." In *Biomaterials and Bionanotechnology*, R. K. Tekade (ed). Academic Press: London, 613–637.

Maiti, D., X. Tong, X. Mou and K. Yang (2019). "Carbon-based nanomaterials for biomedical applications: A recent study." *Frontiers in Pharmacology* 9: 1401.

Mannion, J. T. and H. G. Craighead (2020). Nanofluidic channels with integrated charge sensors and methods based thereon, Google Patents.

Mansfield, E., A. Kar and S. Hooker (2009). "Applications of TGA in quality control of SWCNTs." *Analytical and Bioanalytical Chemistry* 396: 1071–1077.

Mcdevitt, M. R., S. Alidori, N. Akhavein and D. A. Scheinberg (2020). Method and composition for targeted delivery of therapeutic agenTS, US Patent App. 16/599,978.

Mehra, N. K. and N. Jain (2013). "Development, characterization and cancer targeting potential of surface engineered carbon nanotubes." *Journal of Drug Targeting* 21(8): 745–758.

Mehra, N. K. and N. Jain (2014). "Functionalized carbon nanotubes and their drug delivery applications." *Section Nanostructured Drug Delivery. Multi Volume Nanomedicine* 4: 327–329.

Mehra, N. K. J., A. K. Jain, N. Lodhi, R. D. Raj, V. Dubey, D. Mishra, M. Nahar and N. K. Jain (2008). "Challenges in the Use of Carbon Nanotubes for Biomedical Applications." *Therapeutic Drug Carrier Systems* 25(2): 169–207.

Meng, L., A. Jiang, R. Chen, C.-Z. Li, L. Wang, Y. Qu, P. Wang, Y. Zhao and C. Chen (2013). "Inhibitory effects of multiwall carbon nanotubes with high iron impurity on viability and neuronal differentiation in cultured PC12 cells." *Toxicology* 313(1): 49–58.

Meyyappan, M. (2005). *Carbon Nanotubes: Science and Applications*. CRC Press: Boca Raton, FL.

Miranda-Filho, A., F. Bray, H. Charvat, S. Rajaraman and I. Soerjomataram (2020). "The world cancer patient population (WCPP): An updated standard for international comparisons of population-based survival." *Cancer Epidemiology* 69: 101802.

Mohajerzadeh, S., M. Abdolahad, Z. Sanaee and M. Abdollahi (2015). Method for detecting cancer cells using vertically aligned carbon nanotubes, Google Patents.

Mohamed, A. E.-M. A. and M. A. Mohamed (2020). "Chapter 2 – Carbon nanotubes: Synthesis, characterization, and applications." In *Carbon Nanomaterials for Agri-Food and Environmental Applications*, K. A. Abd-Elsalam (ed). Elsevier: London, 21–32.

Moon, J.-M., K. H. An, Y. H. Lee, Y. S. Park, D. J. Bae and G.-S. Park (2001). "High-yield purification process of singlewalled carbon nanotubes." *The Journal of Physical Chemistry B* 105(24): 5677–5681.

Mostafavi, E., P. Soltantabar and T. J. Webster (2019). "Chapter 9 – Nanotechnology and picotechnology: A new arena for translational medicine." In *Biomaterials in Translational Medicine*, L. Yang, S. B. Bhaduri and T. J. Webster (eds). Academic Press: Oxford, 191–212.

Osikoya, A., W. Donbebe, R. Vala, C. Dikio, A. Afolabi, A. Nimibofa and E. Dikio (2015). "Synthesis, characterization and sorption studies of nitrogen-doped carbon nanotubes." *Digest Journal of Nanomaterials and Biostructures* 10: 125–134.

Pai, P., K. Nair, S. Jamade, R. Shah, V. Ekshinge and N. Jadhav (2006). "Pharmaceutical applications of carbon tubes and nanohorns." *Current Pharma Research Journal* 1: 11–15.

Panczyk, T., P. Wolski and L. Lajtar (2016). "Co-adsorption of doxorubicin and selected dyes on carbon nanotubes. Theoretical investigation of potential application as pH controlled drug delivery system." *Langmuir* 32: 4719–4728.

Pantarotto, D., R. Singh, D. McCarthy, M. Erhardt, J. P. Briand, M. Prato, K. Kostarelos and A. Bianco (2004). "Functionalized carbon nanotubes for plasmid DNA gene delivery." *Angewandte Chemie* 116(39): 5354–5358.

Pennetta, C., G. Floresta, A. C. E. Graziano, V. Cardile, L. Rubino, M. Galimberti, A. Rescifina and V. Barbera (2020). "Functionalization of single and multi-walled carbon nanotubes with polypropylene glycol decorated pyrrole for the development of doxorubicin nano-conveyors for cancer drug delivery." *Nanomaterials* 10(6): 1073.

Peretz, S. and O. Regev (2012). "Carbon nanotubes as nanocarriers in medicine." *Current Opinion in Colloid & Interface Science* 17(6): 360–368.

Peyman, G. A. (2016). Method and composition for hyperthermally treating cells, Google Patents.

Porwal, V. Rastogi and A. Kumar (2017). "An overview on carbon nanotubes." *Bioavailability and Bioequivalence* 3: 45–47.

Pravin, P., P. Kirteebala, S. Shweta, K. Priyanka and Y. Rupali (2017). "Emerging trend in nanomedicine: Carbon nanotubes – A review." *Indo American Journal of Pharmaceutical Sciences* 4: 2661–2670.

Rahman, G., Z. Najaf, A. Mehmood, S. Bilal, S. A. Mian and G. Ali (2019). "An overview of the recent progress in the synthesis and applications of carbon nanotubes." *C – Journal of Carbon Research* 5(1): 3.

Rao, C. N. R. and A. Govindaraj (2001). "Nanotubes and nanowires." *Journal of Chemical Sciences* 113(5): 375–392.

Rathod, V., R. Tripathi, P. Joshi, P. K. Jha, P. Bahadur and S. Tiwari (2019). "Paclitaxel encapsulation into dual-functionalized multi-walled carbon nanotubes." *Aaps Pharmscitech* 20(2): 51.

Ravi Kiran, A. V. V. V., G. Kusuma Kumari and P. T. Krishnamurthy (2020). "Carbon nanotubes in drug delivery: Focus on anticancer therapies." *Journal of Drug Delivery Science and Technology* 59: 101892.

Raza, K., D. Kumar, C. Kiran, M. Kumar, S. K. Guru, P. Kumar, S. Arora, G. Sharma, S. Bhushan and O. P. Katare (2016). "Conjugation of docetaxel with multiwalled carbon nanotubes and codelivery with piperine: Implications on pharmacokinetic profile and anticancer activity." *Molecular Pharmaceutics* 13(7): 2423–2432.

Reches, M. and E. Gazit (2015). Peptide nanostructures and methods of generating and using the same, Google Patents.

Ren, J., S. Shen, D. Wang, Z. Xi, L. Guo, Z. Pang, Y. Qian, X. Sun and X. Jiang (2012). "The targeted delivery of anticancer drugs to brain glioma by PEGylated oxidized multi-walled carbon nanotubes modified with angiopep-2." *Biomaterials* 33(11): 3324–3333.

Ribeiro, B., E. C. Botelho, M. L. Costa and C. F. Bandeira (2017). "Carbon nanotube buckypaper reinforced polymer composites: A review." *Polímeros* 27(3): 247–255.

Rinzler, A., J. Liu, H. Dai, P. Nikolaev, C. Huffman, F. Rodriguez-Macias, P. Boul, A. H. Lu, D. Heymann and D. Colbert (1998). "Large-scale purification of single-wall carbon nanotubes: Process, product, and characterization." *Applied Physics A: Materials Science & Processing* 67(1): 29–37.

Rosen, Y. and N. M. Elman (2009). "Carbon nanotubes in drug delivery: Focus on infectious diseases." *Expert Opinion on Drug Delivery* 6(5): 517–530.

Saeed, K. and I. Khan (2013). "Carbon nanotubes-properties and applications: A review." *Carbon Letters* 14: 131–144.

Samiei, F., F. H. Shirazi, P. Naserzadeh, F. Dousti, E. Seydi and J. Pourahmad (2020). "Toxicity of multi-wall carbon nanotubes inhalation on the brain of rats." *Environmental Science and Pollution Research* 27(11): 12096–12111.

Sayes, C. M., F. Liang, J. L. Hudson, J. Mendez, W. Guo, J. M. Beach, V. C. Moore, C. D. Doyle, J. L. West and W. E. Billups (2006). "Functionalization density dependence of single-walled carbon nanotubes cytotoxicity *in vitro*." *Toxicology Letters* 161(2): 135–142.

Scheinberg, D. A., M. R. McDevitt, C. H. Villa and J. J. Mulvey (2018). Targeted self-assembly of functionalized carbon nanotubes on tumors, Google Patents.

Sciortino, N., S. Fedeli, P. Paoli, A. Brandi, P. Chiarugi, M. Severi and S. Cicchi (2017). "Multiwalled carbon nanotubes for drug delivery: Efficiency related to length and incubation time." *International Journal of Pharmaceutics* 521(1–2): 69–72.

Sciortino, N. F., K. A. Zenere, M. E. Corrigan, G. J. Halder, G. Chastanet, J.-F. Létard, C. J. Kepert and S. M. Neville (2017). "Four-step iron (II) spin state cascade driven by antagonistic solid state interactions." *Chemical Science* 8(1): 701–707.

Sharei, A. R., A. Adamo, R. S. Langer and K. F. Jensen (2020). Intracellular delivery, Google Patents.

Sharma, S., N. K. Mehra, K. Jain and N. K. Jain (2016). "Effect of functionalization on drug delivery potential of carbon nanotubes." *Artificial cells, Nanomedicine, and Biotechnology* 44(8): 1851–1860.

Shetti, N. P., D. S. Nayak, K. R. Reddy and T. M. Aminabhvi (2019). "Chapter 10 – Graphene–clay-based hybrid nanostructures for electrochemical sensors and biosensors." In *Graphene-Based Electrochemical Sensors for Biomolecules*, A. Pandikumar and P. Rameshkumar (eds). Elsevier: Amsterdam, 235–274.

Singh, B., C. Baburao, V. Pispati, H. Pathipati, N. Muthy, S. Prassana and B. G. Rathode (2012). "Carbon nanotubes. A novel drug delivery system." *International Journal of Research in Pharmacy and Chemistry* 2(2): 523–532.

Singh, E., R. Srivastava, U. Kumar and A. D. Katheria (2017). "Carbon nanotube: A review on introduction, fabrication techniques and optical applications." *Nanoscience and Nanotechnology Research* 4: 120–126.

Singh, P., R. Tripathi and A. Saxena (2010). "Synthesis of carbon nanotubes and their biomedical application." *Journal of Optoelectronics and Biomedical Materials* 2(2): 91–98.

Singh, S., N. Mehra and N. Jain (2016). "Development and characterization of the paclitaxel loaded riboflavin and thiamine conjugated carbon nanotubes for cancer treatment." *Pharmaceutical Research* 33: 1769–1781.

Sinha, N. and J.-W. Yeow (2005). "Carbon nanotubes for biomedical applications." *IEEE Transactions on NanoBioscience* 4(2): 180–195.

Sitharaman, B. and S. M. Chowdhury (2018). Carbon material delivery systems and methods, Google Patents.

Susi, T., T. Pichler and P. Ayala (2015). "X-ray photoelectron spectroscopy of graphitic carbon nanomaterials doped with heteroatoms." *Beilstein Journal of Nanotechnology* 6: 177–192.

Swogger, K. W., C. P. Bosnyak, N. Henderson, M. Finlayson, B. D. Sturtevant and S. Hoenig (2017). Discrete carbon nanotubes with targeted oxidation levels and formulations thereof, Google Patents.

Swogger, K. W., C. P. Bosnyak, N. Henderson, M. Finlayson, B. D. Sturtevant and S. Hoenig (2020). Discrete carbon nanotubes with targeted oxidation levels and stable gel formulations thereof, Google Patents.

Tan, P. H., A. G. Rozhin, T. Hasan, P. Hu, V. Scardaci, W. I. Milne and A. C. Ferrari (2007). "Photoluminescence spectroscopy of carbon nanotube bundles: Evidence for exciton energy transfer." *Physical Review Letters* 99(13): 137402.

Tian, Z., Y. Shi, M. Yin, H. Shen and N. Jia (2011). "Functionalized multiwalled carbon nanotubes-anticancer drug carriers: Synthesis, targeting ability and antitumor activity." *Nano Biomedicine & Engineering* 3(3): 157–162.

Tsang, S., P. Harris and M. Green (1993). "Thinning and opening of carbon nanotubes by oxidation using carbon dioxide." *Nature (London)* 362: 520–520.

Umemura, K., Y. Ishibashi, M. Ito and Y. Homma (2019). "Quantitative detection of the disappearance of the antioxidant ability of catechin by near-infrared absorption and near-infrared photoluminescence spectra of single-walled carbon nanotubes." *ACS Omega* 4(4): 7750–7758.

Ünlü, A., M. Meran, B. Dinc, N. Karatepe, M. Bektaş and F. S. Güner (2018). "Cytotoxicity of doxorubicin loaded single-walled carbon nanotubes." *Molecular Biology Reports* 45(4): 523–531.

Venkatesan, N., J. Yoshimitsu, Y. Ito, N. Shibata and K. Takada (2005). "Liquid filled nanoparticles as a drug delivery tool for protein therapeutics." *Biomaterials* 26(34): 7154–7163.

Vuković, G. D., S. Z. Tomić, A. D. Marinković, V. Radmilović, P. S. Uskoković and M. Čolić (2010). "The response of peritoneal macrophages to dapsone covalently attached on the surface of carbon nanotubes." *Carbon* 48(11): 3066–3078.

Wang, L., R.-J. Mu, L. Lin, X. Chen, S. Lin, Q. Ye and J. Pang (2019). "Bioinspired aerogel based on konjac glucomannan and functionalized carbon nanotube for controlled drug release." *International Journal of Biological Macromolecules* 133: 693–701.

Wu, L., C. Man, H. Wang, X. Lu, Q. Ma, Y. Cai and W. Ma (2013). "PEGylated multi-walled carbon nanotubes for encapsulation and sustained release of oxaliplatin." *Pharmaceutical Research* 30(2): 412–423.

Wu, W., S. Wieckowski, G. Pastorin, M. Benincasa, C. Klumpp, J. P. Briand, R. Gennaro, M. Prato and A. Bianco (2005). "Targeted delivery of amphotericin B to cells by using functionalized carbon nanotubes." *Angewandte Chemie International Edition* 44(39): 6358–6362.

Yudianti, R. (2011). "Analysis of functional group sited on multi-wall carbon nanotube surface." *The Open Materials Science Journal* 5: 242–247.

Zhang, X., L. Meng, Q. Lu, Z. Fei and P. J. Dyson (2009). "Targeted delivery and controlled release of doxorubicin to cancer cells using modified single wall carbon nanotubes." *Biomaterials* 30(30): 6041–6047.

Zhao, B., H. Hu, S. Niyogi, M. E. Itkis, M. A. Hamon, P. Bhowmik, M. S. Meier and R. C. Haddon (2001). "Chromatographic purification and properties of soluble single-walled carbon nanotubes." *Journal of the American Chemical Society* 123(47): 11673–11677.

Zharov, V. P. (2015). Device and method for *in vivo* noninvasive magnetic manipulation of circulating objects in bioflows, Google Patents.

Zheng, M., A. Jagota, E. D. Semke, B. A. Diner, R. S. McLean, S. R. Lustig, R. E. Richardson and N. G. Tassi (2003). "DNA-assisted dispersion and separation of carbon nanotubes." *Nature Materials* 2(5): 338–342.

Zheng, M., A. Jagota, M. S. Strano, A. P. Santos, P. Barone, S. G. Chou, B. A. Diner, M. S. Dresselhaus, R. S. Mclean and G. B. Onoa (2003). "Structure-based carbon nanotube sorting by sequence-dependent DNA assembly." *Science* 302(5650): 1545–1548.

Zhu, W., H. Huang, Y. Dong, C. Han, X. Sui and B. Jian (2019). "Multi-walled carbon nanotube-based systems for improving the controlled release of insoluble drug dipyridamole." *Experimental and Therapeutic Medicine* 17(6): 4610–4616.

Zhu, Z. (2017). "An overview of carbon nanotubes and graphene for biosensing applications." *Nano-Micro Letters* 9(3): 25.

Section C

Product Development, Toxicity and Scale-Up

13 Liposomal-Based Pharmaceutical Formulations – Current Landscape, Limitations and Technologies for Industrial Scale-Up

Radha Rani, Neha Raina, Azmi Khan, Manupriya Choudhary, and Madhu Gupta
Delhi Pharmaceutical Sciences and Research University (DPSRU)

CONTENTS

13.1 Introduction 209
13.2 Industrial-Scale Production of Liposomes 210
13.3 Liposomal Techniques for Delivery of Drugs 211
    13.3.1 Stealth Liposome Technology 211
    13.3.2 Non-PEGylated Liposome Technology 211
    13.3.3 DepoFoam™ Liposome Technology 211
    13.3.4 Lysolipid Thermally Sensitive Liposome (LTSL) Technology...212
13.4 Biological Defies Faced by Liposomes..212
    13.4.1 The Reticuloendothelial System (RES) and Liposome Clearance......................................212
    13.4.2 Opsonins and Vesicle Destabilisation..213
    13.4.3 The Enhanced Permeability and Retention (EPR) Effect 213
    13.4.4 The Accelerated Blood Clearance (ABC) Phenomenon...213
    13.4.5 Complement Activation-Related Pseudoallergy (CARPA)..214
13.5 Clinically Approved Liposomal-Based Drugs 214
13.6 Defies for Continuous Manufacturing of Liposomes 216
    13.6.1 Formulation Refinement 216
    13.6.2 Materials Employed in Production 217
    13.6.3 Assuring Sterility...217
13.7 Regulatory Aspects for Approval of New Systems of Drug Delivery..217
13.8 Conclusion and Future Perspectives 219
References..220

13.1 INTRODUCTION

Liposomes are widely used class of nanocarriers for the delivery of therapeutic agent at required site(s) (Sercombe et al., 2015). They are designated as phospholipid vesicles having one or more concentric bilayers of lipid encompassing distinct aqueous spaces. Liposomes have the capability of entrapping compounds of both hydrophilic and lipophilic agents, thereby facilitating the encapsulation of various drugs through these vesicles. The aqueous layer is used for entrapping hydrophilic nature of molecules; however, hydrophobic nature molecules are enclosed into the bilayer membrane (Koning and Storm, 2003; Metselaar and Storm, 2005). Moreover, the selection of bilayer components regulates the bilayer charge along with 'fluidity' or 'rigidity.'

For example, better permeability and less stability are seen in unsaturated phosphatidylcholine species from natural sources (egg or soybean phosphatidylcholine), whereas long acyl chain saturated phospholipids (e.g. dipalmitoylphosphatidylcholine) form a rigid, almost impermeable bilayer structure (Sahoo and Labhasetwar, 2003; Gabizon et al., 1998; Allen, 1997).

Liposome preparation can be conducted by various techniques, and it varies in size and charge on surface besides lamellarity (Chang and Yeh, 2012). Numerous methods reported for formulating liposomes have high entrapment efficiency, stability for longer duration and narrow particle size distribution, which include detergent-depletion, reverse-phase evaporation, emulsion, ether/ethanol injection and Bangham. In recent times, few alternative techniques are available for the fabrication of organic solvent-free liposomes that include supercritical fluid and dense gas techniques (Uhumwangho and Okor, 2005; Jiskoot et al., 1986; Deamer, 1978; Szoka and Papahadjopoulos, 1978; Kim and Martin, 1981). Mostly, liposomes are grouped into two categories, namely unilamellar and multilamellar. Further, unilamellar liposomes are classified into two groups, namely small size (SUV, 50–100 nm) or large size (LUV, 100–250 nm). Both LUV and SUV have single lipid bilayer along with a huge aqueous core and therefore are appropriate for hydrophilic drug loading, whereas multilamellar liposomes (MLV), having a diameter of 1–5 µm, possess numerous lipid bilayers along with small aqueous space, so they are appropriate for hydrophobic drug loading (Fan and Zhang, 2013).

A variety of pharmaceutically active agents, chelating agents, enzymes, genetic materials, antimicrobial agents, vaccines and anticancer drugs are loaded into the lipidic or aqueous phase of liposomes, for providing their delivery to the targeted area. The 1965 was the year of liposome discovery, and it has completely transformed the pharmaceutical area. The first liposome-based drug, Doxil® (Ben Venue Laboratories, Inc. Bedford, OH), was duly approved by the United States Food and Drug Administration (USFDA) in 1995 for the treatment of chemotherapy refractory acquired immune deficiency syndrome (AIDS)-related Kaposi's sarcoma (Barenholz, 2012). Liposomal encapsulation technology (LET) is the modern technique employed for the delivery of drugs that perform the role of therapeutic promoters to the specific body organs (Hemanthkumar and Spandana, 2011). LET technique is employed for creating submicroscopic foams named liposomes, in which encapsulation of various materials is possible. These 'liposomes' create barriers surrounding their contents that are unaffected by digestive juices, alkaline solutions, mouth and stomach enzymes, intestinal flora and bile salts formed in the body of humans, along with free radicals. Therefore, the liposomal contents are shielded against degradation and oxidation. This protecting phospholipid barrier lingers intact till the liposomal content is released exactly in the targeted gland, organ or system where the contents are utilised (Akbarzadeh et al., 2013). In this topic, various pharmaceutical and industrial aspects of liposomes are discussed.

13.2 INDUSTRIAL-SCALE PRODUCTION OF LIPOSOMES

From the year of liposome discovery, lots of methodologies have been incorporated for their manufacturing; still, limitations exist for the industrial-scale pharmaceutical application of liposomes (Mozafari, 2005; Maherani et al., 2011). Liposomal formulation techniques used in the beginning that include synthetic procedure comprised multiple steps with the removal of water from thin film of phospholipids in a watery medium, which leads to the generation of lipid structures spontaneously that were of different shapes, lamella and sizes (Bangham et al., 1965a,b; Deamer and Bangham, 1976). To avoid these variations, some mechanical size manipulation methodology was required before making liposomes (Barnadas-Rodríguez and Sabés, 2011; Nastruzzi et al., 2016).

Liposome manufacturing process should be structured in order to comply with different required properties and also amenable to benefits and limitations when it is scale-up for processing of large quantities. Moreover, the final desired outcome comprises lipid composition, size distribution and the API-release properties, which show the ADME of liposomal formulation and also govern the selection of preparation techniques. In the recent era, most approaches focus towards the replacement of multistep technique with the best possible single-step technique for the scale-up of liposomes; for instance, programmable online flow-based strategies to limit the mechanism of precipitation following with self-assembly of phospholipids within homogenous constitution prove to be perfect process for regulating the pharmaceutical environment (Worsham et al., 2019).

To date, the most popular examples of scaled liposome manufacturing methods have been based on the principles of alcohol injection or crossflow techniques, in which precipitation of dissolved lipids from an organic solvent occurs into an aqueous solution (anti-solvent) via reciprocal diffusion of the alcohol and aqueous phases (Worsham et al., 2019). Throughout the process, liposomes formed spontaneously due to the modification in the lipid solubility locally also encapsulate aqueous sol to a small volume. The therapeutic agent may get embedded in lipid bilayer or encapsulated in the core of the aqueous medium depending on the chemical nature of the drug. For these liposomal formulations, the intersection of organic-solvated lipid and anti-solvent/geometry of the mixing and the residence time are clinical parameters for the dictation of programmed flow conditions. API remains un-encapsulated and organic solvent which is not required after the completion of liposome preparation could be refined to the required composition and strength of formulated liposomes by employing methods like tangential flow filtration (TFF) (Wagner et al., 2002; Kim et al., 2012; Li et al., 2011). Every formulation technique mentioned above was operated as batch-wise production process of liposomes, but cross-flow and injection methods are single-step continuous manufacturing processes in which with each feed stream is continuously fed, liposomes will be continuously generated. Some other methods like dense gas and supercritical fluid methods of liposome production are similar to cross-flow and injection methods like the use of organic solvent and similar principle. But dense gas and supercritical fluid methods are operated at high-pressure limits by the adoption of them for the continuous production of liposomes (Santo et al., 2015; Santo et al., 2014; Reverchon et al., 2016; Frederiksen et al., 1997; Otake et al., 2001; Anton et al., 1994).

Formulation of liposomes at approximate ambient conditions in cross-flow and injection methods makes them the most practically suitable method for continuous and infinite manufacturing of liposomes in a single operation. By adopting the addition of continuous step similar to biologics operation, it becomes feasible to manufacture the liposomes in continuous mode (Worsham et al., 2019). In the concept of continuous liposomal manufacturing process, unprocessed materials supply into a formulation process resulting in finished or intermediate products which continuously flow out as a final product of process. Implementation of these processes has immense advantages of lower COG (overall cost of goods) because of smaller facility footprint and low capital expenditure along with better product quality and improved consistency (Kleinebudde et al., 2017; Subramanian, 2014). The manufacture of biologics has continued to develop the requirements and aspects to consider surrounding operating upstream and downstream unit operations in a continuous fashion such as cell culture, chromatography, viral inactivation and TFF as well as integrated continuous upstream and downstream processes (Warikoo et al., 2012; Mahajan et al., 2012; Godawat et al., 2012; Pollock et al., 2013; Orozco et al., 2017; Farid et al., 2017; Godawat et al., 2015; Whitford, 2015). Although single-use technology has conceptual advantages alike continuous processes, the only limitation in batch process is high magnitude required for more outcome of end product. The overall conclusion of all aspects recommends the implementation of manufacturing biologics in continuous fashion to make it more advantageous like conventional liposomal products of the pharmaceutical industry (Worsham et al., 2019; Titchener-Hooker et al., 2001; Farid et al., 2014).

13.3 LIPOSOMAL TECHNIQUES FOR DELIVERY OF DRUGS

This part highlights the technologies established specifically for the formulation of liposome-based products for clinical use. Every technology possesses its distinctive features for optimising the delivery of therapeutic agents through shielding the unique properties of drugs while minimalising its disadvantages.

13.3.1 Stealth Liposome Technology

Stealth liposome technology is widely used technique for developing a system of active molecule drug delivery (Mufamadi et al., 2011). This technique was established for overcoming critical challenges of conventional liposomes, for example, the toxicity as a result of charge on liposomes, small half-life, besides steric stability (Torchilin, 2005; Cattel et al., 2003; Soenenet al., 2009). This technique involves surface modification of the liposome membrane, and polymer strands are connected with molecules of therapeutically active agents or a system for improving the safeness and efficiency of therapeutically active agents. Polymer that is mostly employed in this technique is polyethylene glycol (PEG), and this procedure is named PEGylation. General, PEGylation is conducted by incubating PEG reactive derivative with the target moiety (Veronese and Harris, 2002). By this technique, modification occurs in the physicochemical properties of the drug moiety, involving hydrodynamic size change, which in turn reduces renal clearance and extends the circulatory period of the drug. The other polymers of hydrophilic nature were either naturally occurring or synthetic in origin such as polyvinyl alcohol, chitosan, PEG and silk fibroin (Ruizhen et al., 2011). (DOXIL/Caelyx) is PEGylated liposome doxorubicin formed using stealth liposome technology, which is sanctioned via Europe Federation and US Food and Drug Administration (FDA). Even if outstanding results were attained by using this technique such as macrophage uptake reduction, protracted circulation, in addition to lesser toxicity, yet the main disadvantage associated with this method is passive targeting, as liposomes can deliver only active molecules (Krown et al., 2004).

13.3.2 Non-PEGylated Liposome Technology

Non-PEGylated liposome (NPL) is an inimitable system for the delivery of drugs which appeared as an innovation in cancer treatment via delivering the advantages of PEGylated liposomes, at the same time eradicating the unwanted effects allied with PEG, for example, hand-foot syndrome (HFS). Injection of NPL doxorubicin (NPLD) delivers an improved safety profile in comparison with conventionally used DOX and Doxil®. NPLD also helps in reducing cardiac toxicity, which usually occurs with DOX, and the dose-limiting toxicity allied with Doxil®, for example, HFS. These beneficial properties are attained by combining the particular composition with a unique NPLD liposome manufacturing process, which provides its preferred physicochemical properties. The NPLDs have a longer circulation period and lesser cardiotoxicity than traditional DOX. Myocet® is an Elan Pharmaceuticals produced NPLD, Princeton NJ, which is sanctioned in Canada and Europe in conjunction with cyclophosphamide for managing metastatic cancer of breasts (Leonard et al., 2009).

13.3.3 DepoFoam™ Liposome Technology

DepoFoam™ is Pacira Pharmaceuticals, Inc., Parsippany, NJ, USA's proprietary, extended-release drug delivery technology. DepoFoam™ is the core technology at the back of numerous products that are marketed like DepoCyt®, DepoDur™ and Exparel. DepoFoam™ technology is employed in the encapsulation of therapeutic agents without altering their molecular structure in its multi-vesicular liposomal platform. The release of drug in multi-vesicular liposome occurs for period of 1–30 days. DepoFoam™ comprises spheroids having microscopic and granular structure (3–30 m) besides single-layered lipid particles consisting of a honeycomb of multiple non-concentric internal aqueous chambers having the bounded drug. Various non-concentric aqueous chambers are bounded by the lipid membrane of a single bilayer in each particle. Partition of every chamber from the next chambers is done via bilayer lipid membranes comprised of synthetic analogues of natural lipids (triolein,

TABLE 13.1
Recent liposomal formulations encapsulating different drugs

Therapeutic Agent/Polymer	Target/Disease	Conclusion	References
Paclitaxel (PTX) with bovine serum albumin (BSA)	Breast cancer cells	Study findings depict that organic solvent-free PTX-liposomal formulation possessed potent anti-proliferative effects for breast cancer	Okamoto et al. (2020)
Madecassoside (MA) with PEG-PCL-PEG (PECE)	Second-degree burn	From the results of this research study, it can be concluded that the PECE-modified MA liposomes showed high wound contraction and wound repair besides improved surface adhesion performance	Liu et al. (2020)
Doxorubicin-lovastatin	Liver cancer	From *in vivo* results of this study, it is clear that liposomal formulation of doxorubicin-lovastatin effectively inhibited the tumour growth and reduced pathological damages to the main tissues as compared with the single drug group. So, doxorubicin-lovastatin-liposomal formulation can be a favourable method for treating liver cancer	Wang et al. (2019)
Tamoxifen and raloxifene	Breast cancer	Selective estrogen receptor modulator (SERM) drugs like tamoxifen and raloxifene liposomes were formulated with penetration enhancer dimethyl-β-cyclodextrin and were considered a better medicinal alternative for the treatment of breast cancer through oral route	Ağardan et al. (2020)
5-Fluorouracil coated with chitosan (chitosomes)	Colorectal cancer	The *in vitro* drug-release study revealed that chitosomes retard 5-FU release. The cytotoxicity study conducted employing a colon cancer cell line (HT-29) demonstrated that 5-FU-loaded chitosomes were very efficient in destroying cancer cells in a sustained manner. Therefore, chitosomes were successful nanocarriers of 5-FU, having potential cytotoxicity for colorectal cancer cells	Alomrani et al. (2019)

DPPG, cholesterol, etc.) (Alomrani et al., 2019). During administration, DepoFoam™ particles liberate the drug for hours to weeks after the lipid membranes have been erosioned. DepoFoam™ technology has upgraded the properties of both large and small molecules. This technology greatly enhanced patient care by offering a groundbreaking solution for medicines requiring repeated multiple injections and possessed action for shorter period or adverse effects (Murry and Blaney, 2000).

13.3.4 Lysolipid Thermally Sensitive Liposome (LTSL) Technology

The thermosensitive liposomes were examined at high-temperature sites for drug release. In the preparation of these liposomes, lipids which are employed have a transition temperature between 40°C and 45°C. These latest categories of liposomes are produced to demonstrate temperature-dependent release of incorporated medication. Temperature of the local tissue is typically raised to 42°C by radiofrequency ablation, a technique based on radiofrequency application of the liposome of lipid components that undergoes a liquid transition gel at high temperatures, making it more permeable and thus releasing the drug. In fact, local hyperthermia application induces blood vessel leaking inside tumours, rising the accretion of liposomes inside the tumour. Celsion Corporation's ThermoDox® is being investigated in the third phase of clinical trials using lysolipid thermally sensitive liposome (LTSL) technology to encapsulate DOX as the therapy for several solid tumours. For ThermoDox®, this technique enables the drug concentration in the specific area to be 25 times greater than intravenous (i.v.) DOX. In addition, the concentration of DOX in the blood increases substantially compared with other liposomally encapsulated DOX (Slingerland et al., 2012). Some recent research studies conducted on liposomes are summarised in Table 13.1.

13.4 BIOLOGICAL DEFIES FACED BY LIPOSOMES

After the administration of liposomes, the human body treats them in the same way as that of every foreign substance, and the identification and destruction of liposomes begin by the auto defence system of the human biological system, for instance, immunogenicity, opsonisation and RES (Forssen and Willis, 1998). Along with these hurdles, several factors like enhanced permeability and retention (EPR) are also considered for better performance of liposomes (Sawant and Torchilin, 2012).

13.4.1 The Reticuloendothelial System (RES) and Liposome Clearance

Systemic administration of liposomes causes its accumulation principally in the RES (Poste et al., 1976; Senior, 1987). Among all the organs of RES, liposomes get deposited approximately 10 times greater in the liver and spleen (Chrai et al., 2002). Clearance of liposomes from the RES occurred by directly phagocytic cellular interactions (Chrai et al., 2002). Firstly, liposomes get absorbed by proteins of blood plasma (vesicular opsonisation), for instance, lipoproteins, fibronectin, immunoglobulin, phospholipid membrane complement proteins molecules, and after that, their uptake occurs in the RES organs (Ishida et al., 2001; Chrai et al., 2002).

Moreover, the *in vitro* studies also suggest macrophage clearance of liposomes in the case of the absence of blood plasma proteins (Chrai et al., 2002). Excessive deposition of liposomes in the RES may lead to enhanced chances of infections due to suppression of the immune system because RES organs like the liver and spleen are the important parts of our immune system. It can ruin the phagocytic action of macrophages or amend other functions of macrophages. The accumulation and suppression can only occur in the case of direct systemic circulation (Szebeni and Barenholz, 2009; Szebeni and Moghimi, 2009).

Liposomes used to treat cancer do not show accumulation and clearance in the above manner because they comprise cytotoxic therapeutic agents that cause the destruction of macrophages. The situation is different with anti-cancer liposomes that contain cytotoxic drugs, which are capable of inducing macrophage destruction. Although clinically important blockade of macrophage function in humans have not yet been demonstrated, there have been indirect signs that suggest the possibility of some immune suppression (Szebeni and Barenholz, 2009; Szebeni and Moghimi, 2009). For instance, PEGylated liposomes loaded with doxorubicin (PLD) (Doxil®) administered in mice go under a dose-dependent clearance results saturation because of fractional RES blockade occurred in the hepatic system. But on the administration of free doxorubicin in similar dose, above effect was not happened (Gabizon et al., 2002). Besides in another study, Doxil® administration alters bacterial clearance from the systemic fluid that suggested the suppression of the immune system (macrophages) (Storm et al., 1998; Szebeni and Barenholz, 2009). When the membrane of liposomes is conjugated with polymer PEG, then sterical inhibition of both hydrophobic and electrostatic chemical reactions between proteins/cells and liposomes takes place due to the creation of a local surface concentration of highly hydrated groups. Hence, this phenomenon results in reduced absorption of liposomes in the organs of RES (Ishida et al., 2001). Thus, PEGylation of the liposomal membrane is a significant approach in order to prevent removal and better circulation period by organs of RES via steric stabilisation (Oku and Namba, 1994; Ishida et al., 2001).

13.4.2 Opsonins and Vesicle Destabilisation

Distribution in biological system, effectiveness, deposition and safety/toxicity of liposomes as a nanocarrier is determined importantly by the interaction intensity of liposomal formulation and blood plasma proteins/cells (Hua and Wu, 2013). Liposomal clearance in the organs of RES is exhibited by the opsonisation process of plasma proteins, playing an important role in the clearance of liposomes and also in the destabilisation of liposomes in vesicles (Cullis et al., 1998). Liposomal opsonisation by plasma proteins is affected by a number of factors such as size, stability and surface charge (Cullis et al., 1998; Ishida et al., 2001). Liposomes of small size possess the ability to avoid opsonisation, a process that suggests liposomes of 800-200 nm which solve the problem of opsonisation, interaction with plasma proteins, accumulation of liposomes in the RES and their toxicity in the body also reduce risk of infection to the human body (Chrai et al., 2002).

In general, liposomes which are large in size and unmodified undergo rapid elimination as compared to positively charged, neutral or small size liposomes (Oku and Namba, 1994; Laverman et al., 1999; Ulrich, 2002). Investigation done earlier suggests the clearance of liposomes from the liver and spleen within minutes and an hour, respectively (Senior, 1987; Chrai et al., 2002). Incorporation of cholesterol in liposomal formulation plays a key role as it improves the stability of liposomes and limits the liposomal exchange with the cells present in phospholipid membrane (Forssen and Willis, 1998). It also prevents the exchange of liposomal lipids with lipoproteins, RBCs and other circulating cells/proteins in the blood that are responsible for the depletion and replacement of lipids having greater phase transition temperature with physiological components of low stability (Forssen and Willis, 1998; Laverman et al., 1999; Ulrich, 2002). Thus, from the above literature, it is suggested that liposomes which are neutral, are small in size and comprise cholesterol in their formulation structure are capable to circulate for a prolonged period of time normally for several hours (Geng et al., 2014).

13.4.3 The Enhanced Permeability and Retention (EPR) Effect

EPR effect is subjected to liposomes that have eluded from the opsonisation and accumulation in organs of RES (Sawant and Torchilin, 2012; Nehoff et al., 2014). EPR effect causes the enhanced permeability of vesicular structure supplying diseased organs or tissues. At inflamed or cancer tissue pathological sites, some factors are predominant like activated permeability and enhanced expression of vesicles, and/or reduced angiogenesis results in small size opening or pores with range 300–4,700 nm, causing accumulation of liposomes via passive targeting (Nehoff et al., 2014; Hashizume et al., 2000). Small pores with size range of 0.2–1.2 μm have been occurred due to the exposure of inflammation-causing mediators; however, the pore size and their number at pathological sites depend on microenvironment of the site (Klimuk et al., 1999; Antohe et al., 2004; Hua and Wu, 2013). Significantly, EPR effect is applied to every liposomal formulation but liposomes conjugated with PEG possess advantage of prolonged circulation period of liposome in the blood by limiting the clearance rate of liposomes in RES (Sawant and Torchilin, 2012).

13.4.4 The Accelerated Blood Clearance (ABC) Phenomenon

As mentioned above, components of liposomal formulation and cells of the human immune system interact with each other, limiting the clinical applications of liposomes.

So modifying liposomes synthetically, lift up the clinical application of liposomes as a nanocarrier of various drugs. For example, repeated injection of PEGylated liposomes has been associated with loss of their long circulating properties and subsequent clearance from the blood (Dams et al., 2000; Ishida et al., 2003, 2006). This process is called the 'phenomenon of accelerated blood clearance' or ABC phenomenon is a prime challenge for clinical use of the PEG-conjugated formulations because it needs multiple dosing regimens. Dams et al. first observed the ABC phenomenon by demonstrating that prior dosing of empty PEGylated liposomes influences the pharmacokinetics and biodistribution of the second dose of liposomes in rats and rhesus monkeys, when the doses were administered with an interval of 7 days (Dams et al., 2000). Consequently, significant fall in circulation period and elevated accumulation in RES organs like spleen and liver was seen in second dosing of PEG-conjugated liposomes (Dams et al., 2000) with utmost liposomal clearance in four to seven days of first dosing in rats and ten days in mice (Ishida et al., 2003, 2006).

However, the mechanism of ABC phenomenon is still not clear completely, and it is supposed to be affected by PEG surface density, lipid dose and dosing intervals between initial and followed injections (Ishida and Kiwada, 2008). The formation of antibodies (IgM) against PEG is effectively elicited by frequent systemic dosing of dummy liposomal conjugated with polymer PEG because of decrease in ABC phenomenon in splenectomised rats and this anti-PEG IgM response is primarily mediated by the cells of the spleen (Ishida et al., 2006). Also, firstly administered higher dose of PEG-conjugated liposomes (>1 μmol phospholipids/kg) results in ABC phenomenon of low magnitude (Ishida et al., 2005).

13.4.5 Complement Activation-Related Pseudoallergy (CARPA)

Several liposome vesicles possess the ability of triggering the response of the human auto defence system, subsequently activating complement system and ultimately stimulation of acute hypersensitivity syndrome named complement activation-related pseudoallergy (CARPA). CARPA is a component of response of innate immune system which takes part in various inflammatory and immunological processes (Moghimi and Hunter, 2001). CARPA is associated with all drugs loaded in liposome vesicles which are approved and under experimental studies for clinical application (e.g. DaunoXome®, Doxil® and AmBisome). Additionally, a number of patients (2%–45%) have been observed to suffer from hypersensitivity symptoms due to liposomal administration by infusion (Szebeni, 2005; Szebeni and Moghimi, 2009). CARPA has signs and symptoms like facial flushing, headache, facial swelling, cardiopulmonary distress, chills and anaphylaxis (Szebeni, 2005).

CARPA pseudoallergy may be due to stimulation and subsequently production of C3 split products (e.g. C3d) (Dempsey et al., 1996), anaphylatoxins C3a and C5a by the complement system (Szebeni, 2005; Szebeni and Moghimi, 2009). These anaphylatoxins possess binding ability to their respective receptors found on the cells of the defence system of the human body (e.g. macrophages, mast cells and basophils), and this binding brings vasoactive mediator liberation, for instance, platelet-activating factor (PAF), tryptase, histamine, thromboxaneA2 (TXA2), prostaglandins (e.g. PGD2) and leukotrienes (e.g. LTB2, LTB4, LTC4, LTD4, LTE4). Sensitivity of CARPA is substantially varied in different species; for example, tachyphylaxis may occur in some breeds of pigs and dogs (Szebeni et al., 1999, 2007). Hence, empty liposomal formulations might be administered for the prevention of CARPA because of desensitisation effect in addition to giving inhibitors of complement system before the administration of liposomal vesicles such as indomethacin, soluble C receptor type 1 and anti-C5 antibody (Szebeni and Barenholz, 2009). Intensity of CARPA due to the stimulation of complement system may depend on various factors of liposomal formulations, for example, size, charge, morphology, surface characteristic, bilayer packaging, administered dose and lipid composition (Szebeni and Moghimi, 2009). Immunogenic hyperactivity developed in response to liposomal administration may be responsible for the loss of efficacy, serious toxicities like anaphylaxis and altered pharmacokinetics (Szebeni and Moghimi, 2009).

13.5 CLINICALLY APPROVED LIPOSOMAL-BASED DRUGS

Liposomal formulations are used for encapsulating huge number of medications, including anticancer and antimicrobial specialists, chelating operators, peptide hormones, chemicals, proteins, immunisations, and hereditary materials, which have been joined into the fluid or lipid periods of liposomes, with different sizes, creations and different attributes, to give particular conveyance to the objective site for *in vivo* application which gives high-entanglement proficiency, thin molecule size dissemination and long haul stability. Due to broad improvements in liposome innovation, various liposome-based medication definitions are accessible for human use and numerous items are under various clinical preliminaries. Epitome of medications in liposomes enhanced the helpful files of different specialists, for the most part through changes in their pharmacokinetics and pharmacodynamics. In clinical investigations, liposomes show improved pharmacokinetics and biodistribution of helpful specialists and in this way limit poisonousness by their gathering at the objective tissue. The restorative impact of current clinically endorsed liposome-based medications is shown in (Table 13.2) with free medications, and to likewise decide the clinical impact by means of liposomal varieties in lipid structure.

TABLE 13.2
Marketed Liposomal Formulations for Various Disorders

Drug	Product Name	Lipid Composition	Approved Indication	References
Amphotericin B	AmBisome	Hydrogenated soy phosphatidylcholine Distearoyl phosphatidylglycerol Cholesterol, and amphotericin B	Severe fungal infections	Immordino et al. (2006), Astellas Pharma US, Inc. (2011), Meunier et al. 1991
Daunorubicin	DaunoXome	1-α-Dimyristoylphosphatidylcholine; l-α-dimyristoyl phosphatidylglycerol	Blood tumours	Immordino et al. (2006), Gilead Sciences Inc. (2011), Tomkinson et al. (2003)
Doxorubicin	Doxil	Distearoylphosphatidylcholine and cholesterol	Kaposi's sarcoma, ovarian/breast cancer	Chowdhary et al. (2003)
Verteporfin	Visudyne	Egg phosphatidylglycerol 1-α-Dimyristoylphosphatidylcholine	Age-related molecular degeneration, pathologic myopia, ocular histoplasmosis	Novartis Ophthalmics Inc. (2011), Immordino et al. (2006).
Cytarabine	DepoCyt	Cholesterol, triolein Dioleoyl phosphatidylcholine Dipalmitoylphosphatidylglycerol	Neoplastic meningitis and lymphomatous meningitis	Drugs.com (2011a,b)
Morphine sulphate	DepoDur	Cholesterol, triolein Dioleoyl phosphatidylcholine Dipalmitoylphosphatidylglycerol	Pain management	Drugs.com (2011b), Usonis et al. (2003)
Inactivated hepatitis A virus (strain RG-SB)	Epaxal	Dioleoyl phosphatidylcholine Dioleoyl phosphatidylethanolamine	Hepatitis A	The electronic Medicines Compendium (2011a,b)
Inactivated haemagglutinin of influenza virus strains A and B	Inflexal V	Dioleoyl phosphatidylcholine Dioleoyl phosphatidylethanolamine	Influenza	Herzog et al. (2009)
Mifamurtide	Mepact®	Dioleoyl phosphatidylserine, palmitoyl oleoyl phosphatidylcholine	High-grade, resectable, non-metastatic osteosarcoma	Alphandéry et al. (2015), Vail et al. (2015), Anderson et al. (2010)
Vincristine	Marqibo®	Sphingomyelin Cholesterol	Acute lymphoblastic leukaemia	(Krishna R et al. 2001; Rodriguez M et al. 2009)
Irinotecan	Onivyde™	Distearoylphosphatidylcholine, methoxy polyethylene glycol-2000	Combination therapy with fluorouracil and leucovorin in metastatic adenocarcinoma of the pancreas	Drummond et al. (2006), Hong et al. (2016)
Verteporfin	Visudyne®	Verteporfin, Dimyristoyl phosphatidylcholine Egg phosphatidylglycerol	Choroidal neovascularisation	Strong et al. (1999), Bressler (1999)
Bupivacaine	Exparel®	Cholesterol, tricaprylin Dierucoyl phosphatidylcholine Dipalmitoylphosphatidylglycerol	Pain management	Clarke et al. (2006), Bungener et al. (2002), Ambrosch et al. (1997)
Amphotericin B	Abelcet®	1-α-Dimyristoyl phosphatidylcholine l-α-Dimyristoyl phosphatidylglycerol	Invasive severe fungal infections	Lister et al. (1997), Janoff et al. (1988)
Morphine sulphate	DepoDur™	Cholesterol and triolein 1-α-Dimyristoyl phosphatidylcholine, l-α-Dimyristoyl phosphatidylglycerol	Pain management	Alam and Hartrick (2005); Hartrick and Manvelian (2004)
Amphotericin B	Amphotec®	Cholesteryl sulphate: amphotericin B	Severe fungal infections	Hong et al. (2016), Strong et al. (1999)

13.6 DEFIES FOR CONTINUOUS MANUFACTURING OF LIPOSOMES

While the central aspect (liposome formation) of liposomal drug product manufacturing is conducive to continuous manufacturing, there are special nuances in the areas of formulation refinement, materials of construction, and sterility assurance that need to be addressed for adaptation to a regulated pharmaceutical environment (Figures 13.1–13.3).

13.6.1 Formulation Refinement

The unit activities downstream of liposome development are utilised to refine the medication product preparation to the preferred requirements. TFF is utilised to eliminate undesired components in non-encapsulated API or organic solvent and also concentrate on the medication product to a finally wanted strength. For this particular situation, the retentate possesses the medication item and the function of a waste stream is performed by permeate. This is not disparate

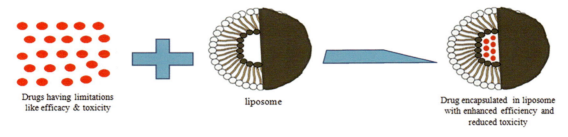

FIGURE 13.1 Illustration depicting the benefits of liposomal formulations.

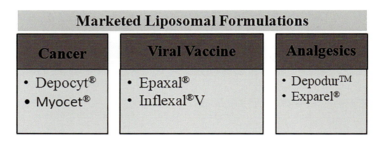

FIGURE 13.2 Marketed liposomal formulations.

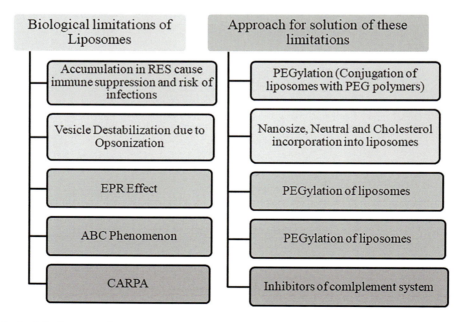

FIGURE 13.3 Biological challenges and overcoming techniques.

from downstream unit tasks employed in producing biologics. For supporting continuous operation, the proper balancing should be between TFF for the buffer replacement and concentration in therapeutic liposomal product manufacturing. A sample mode design for this operation will require a TFF step where the retentate which contains liposomes is sent back to the centrally placed vessel and the permeate/waste source consists of fresh feed buffer (diafiltration by constant weight), assisting the exchange of buffer. After the completion of buffer exchange, the product is concentrated to the chosen strength by terminating further addition of buffer (Jungbauer, 2013). For unceasing interchange of buffer besides a concurrent concentration step, a continuous design is must. These arrangements are not unaccustomed in the domain of biologics, but liposomal inimitable features should be taken into consideration and then experimentally confirmed for such a procedure (Jungbauer, 2013). Contingent on the incoming feed composition in addition to the requirement of the desired final formulation, this may perhaps be simplified through numerous arrangements. TFF system having only one vessel for exchange of buffer through single concurrent concentrating SPTFF performs the role of a base case intended for a continuous design (Figure 13.4a). If diafiltration at steady state or single-pass concentration is incapable of achieving the desired concentration or rate of buffer exchange through single stage, then extra stages might be included (Figure 13.4b, c). Moreover, compact and sleek designs for exchanging buffer at continuous rate such as the Cadence® In-line Diafiltration (ILDF) Module are becoming accessible and ought to be investigated (Gjoka et al., 2017). With the use of SPTFF in an ILDF design, it will eradicate multiple vessel requirements for supporting exchange of buffer continuously (Figure 13.4d). In the production of formulations containing liposomes, there is acceptable and expected variability in the capturing efficacy of the therapeutically active material. In a lot process, its compensation is done through offline in-process measurement of concentrated active ingredients in advance of the concentration step. However, calculating basics control level is provided by mass, density and flow rates and can be applied quickly in a continuous operation, and a real-time concentration measurement such as in-line HPLC delivers a higher degree of assurance (Challener, 2017; Rathore and Winkle, 2009). In-line HPLC methods are accessible, although major advancement is required in them for overcoming assay necessities, for example, liposome lysing for concentration determination, making it a devastating examination technique. The interruption in response could be overwhelmed through the steadiness of another process control, rapid HPLC, that minimises the offline examination period from 60 to 4 min which makes it more preferred candidate (Joshi and Kumar, 2013).

13.6.2 Materials Employed in Production

Numerous advantages of processing biologics continuously are leveraged by implementing single-use programs besides componentry (Whitford, 2015; Bisschops, 2015; Novais et al., 2001; Hammerschmidt et al., 2014). It eradicates the requirement of costly capital apparatus, purification/sterilisation and enabling easy washing in addition to delivering extra suppleness for multi-product processes. Since the manufacturing of liposomes requires the use of organic solvents, use of single-use components such as tubing and bags, can present issues around extractables/leachables (Ferrante, 2017; Hernandez, 2017). Furthermore, if gamma irradiation sterilises the components in advance that are intended for single use, there might be problems in generating free radicals and entrapping into therapeutically active products. At last, these are capable of causing deterioration of few liposomal components and require substantial evaluation of impurities that were formerly hidden in the product at final stage (Schnitzer et al., 2007). Additional problem associated with single-use componentry is that they are incapable of maintaining a sterile border, leading to maximum nuance of liposome production: aseptic processing.

13.6.3 Assuring Sterility

Large-scale production of liposomes for commercial use requires aseptic treatment, because size range of liposomal particles is higher than 0.2 μm (incapable to be terminally sterile separated) along with their instability in the existence of harsh chemicals, radiation and extremely high-temperature conditions (Zuidam et al., 2003; Toh and Chiu, 2013). With aseptic treatment, it is necessary to create and shield a sterile border surrounding the process. The risk to the integrity of that border is increased via using single-use componentry as tubing assemblies and bags have more probability of leakage in comparison with more tough reusable systems, for instance, stainless steel (Stock, 2014). Furthermore, spalling besides breaches can be caused by protracted utilisation of flexible tubing in pumping systems (Bahal and Romansky, 2002). Assuming the procedure is set up employing componentry that is formerly sterilised combined with/or steam-in-place (SIP) equipment, any feed solutions should essentially go in the system via filters which are used for sterilisation having size of pores normally 0.2 μm or less.

13.7 REGULATORY ASPECTS FOR APPROVAL OF NEW SYSTEMS OF DRUG DELIVERY

Initially, liposomes loaded with Doxil were prepared in 1995 with the approval of the regulatory authority. Liposomes undergo huge technological advancements from the year 1995, and presently formulated liposomal preparations comprise enormously enhanced physicochemical characteristics. The three guiding principles that were true at the time of Doxil's approval, however, ring true today as well and can be applied to all liposomal delivery systems, regardless of type. These guidelines are as follows: (i) stable and greatest drug loading efficiency, (ii) PEG conjugation of liposomes for the circulation of drug in the blood for

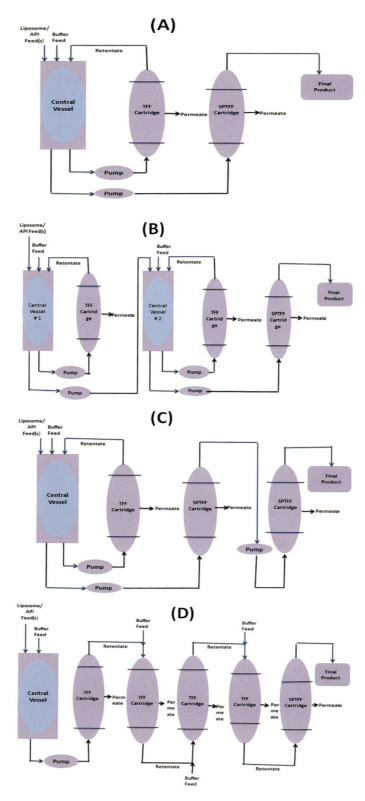

FIGURE 13.4 Proposed new process designs for manufacturing of liposome drug product in continuous manner.

(a) Single tank cradle trade TFF and single-stage simultaneous concentrating SPTFF.

(b) Continuous multistage (multi-vessel) cushion trade TFF and single-stage simultaneous concentrating SPTFF.

(c) Single tank cushion trade TFF and multistage simultaneous concentrating SPTFF.

(d) Multistage cushion trade (ILDF) with simultaneous concentrating SPTFF.

longer time period and preventing accumulation in organs of RES system and (iii) liposomes with most favourable phospholipid membrane constitution with the aim to formulate highly stable liposomes.

Liposomes are classified as 'nanocarriers,' besides fullerenes, dendrimers and quantum dots by the Food and Drug Administration (FDA). Along with these, nanocarriers loaded with combination of drug products come in the area of interest of so many researchers in the modern era (Parhi et al., 2012), for instance, loading of synergistic anticancer drug combinations within single liposomal formulation (Shaikh et al., 2013). Celator Pharmaceuticals Inc. (Princeton, NJ) formulated PEG-conjugated liposomes of cytarabine: daunorubicin (CPX-351, 5:1 molar ratio), encapsulated with weakly acidic drug 5-fluoroorotic acid and amphipathic drug irinotecan (CPT-11) in a ratio of 5:1, respectively (Feldman et al., 2012). The multiple drugs encapsulated in a single liposomal body exhibited a synergistic anticancer action. Usually, liposomes loaded with more than one combination of drugs face lots of difficulties to get approved by the FDA (Zylberberg and Sandro Matosevic, 2016). Over the last decade, US and European regulatory agencies began to publish draft guidance documents containing non-binding recommendations on liposome-based products. In the year of 2002, the first draft is published by USFDA regulating the 'Guidance for industry on liposome drug products'. But this drafted document has some limitations like lack of information regarding assessment techniques of calculating bioequivalence (Zylberberg and Sandro Matosevic, 2016). After the amendments done in their draft, USFDA in November month of the year 2015 issued the latest revised draft of guidelines related to 'Liposome Drug Products Chemistry, Manufacturing, and Controls; Human Pharmacokinetics and Bioavailability; and Labelling Documentation,' addressing several topics of drug-loaded liposomal formulations: chemistry, manufacturing, and controls; bioequivalence; and labelling in new drug applications (NDAs) and abbreviated new drug applications (ANDAs), human pharmacokinetics and bioavailability or in the case of an ANDA. USFDA replaces its initial draft of August 2002 from the latest draft published by them in November 2015. All liposomal products whether PEGylated, cationic or conventional are required to get approved by the FDA (Zylberberg and Sandro Matosevic, 2016). In case of liposomes formulation loaded with biosimilar or generic drug products, FDA is the authority for recommending dissolution testing to relate bioequivalence between reference and sample products. For the liposomal drug products with identical physical properties and release profile as that of their standard/reference product, assay of drug release is advised to be used routinely for the calculation of bioequivalence or biosimilarity (Zylberberg and Sandro Matosevic, 2016).

In 2018, USFDA also published its first regulatory draft for approval of non-biological complex drugs (NBCDs). This draft comprises of final guidelines for the approval of NDAs and ANDAs. NBCDs of different and closely identical structures are incompletely characterised or described via physicochemical analytical tools, and also, their quality and composition are based on controls and process of manufacturing like biologicals. NBCDs are dependent on the manufacturing process and controls-drugs just as is the case with biologicals. Examples of NBCDs include nanomedicines, such as liposomes, polymeric micelles, glatiramoids, iron-carbohydrate complexes and nanocrystals, as well as dry powder inhalers. The draft published by USFDA in April 2018 is the revised version of the draft published in November 2015 of CMC, bioequivalence estimation and labelling conditions for NDAs and ANDAs. Biologics license applications (BLAs) are not included in the 2018 draft; however, the USFDA agency stated that it has 'many scientific principles described in this guidance may also apply' (http://www.gabionline.net/Non-Biological-Complex-Drugs/Guidelines/FDA-issues-final-guidance-on-liposome-drug-products). The European Medicines Agency (EMA, 2011) published its first draft guidance in July 2011. This document aims to provide guidance in generating relevant clinical and nonclinical data to support their application for market approval of intravenous liposomal products. However, it does not define a specific analytical, non-clinical or clinical strategy and provides only general principles for assessing liposomal products (EMA, 2009).

13.8 CONCLUSION AND FUTURE PERSPECTIVES

Liposomes are employed widely in the pharmaceutical sector because of their numerous biomedical applications. Following the development of the PEGylated liposomal formulation Doxil®, liposomes made their booming market entry in the year 1995. Subsequently, they are investigated for several disorders from deadly diseases like cancer therapy to pain control. The advantages primarily associated with liposomes are as follows: pharmacokinetics and pharmacodynamics' properties are easily regulated in this system, enhanced bioavailability besides minimal toxicity. Various liposomal techniques for therapeutic drug delivery, biological challenges and clinically approved drugs along with regulatory aspects are also discussed. Marketed liposomal technology clearly indicates that liposome has established their position in the pharmaceutical area. One example of this booming technology is commercial topical liposomal formulation as an efficient system of drug delivery for treating skin diseases. Liposomal formulations are more explored in the pharmaceutical area; it is evident from clinical trials that are conducted on a large scale involving lipid-based products and liposomes. With the latest advancements in the field, various pharmaceutical industries are enthusiastically involved in the development and liposomal product evaluation for their utilisation in antifungal and anticancer therapy and for prophylaxis (vaccines) against disorders. For continuous translational success of liposomes, effective communication as well as alliance is needed among different experts who are involved in all phases of pharmaceutical development of liposomal technologies, incorporating pharmaceutical design and production, cellular interactions and toxicology, in addition to evaluation at preclinical and clinical stages. The upgradation of technologies in the area of liposomes and medicine will be more advanced in the upcoming time, and additional enhancements in this technology will hasten the fully developed evolvement of liposomes as vehicles for drug delivery.

REFERENCES

Ağardan, N. B. Mutlu, Zelihagül Değim, Şükran Yılmaz, Levent Altıntaş, and Turgut Topal. "Tamoxifen/raloxifene loaded liposomes for oral treatment of breast cancer." *Journal of Drug Delivery Science and Technology* 57 (2020): 101612.

Akbarzadeh, Abolfazl, Rogaie Rezaei-Sadabady, Soodabeh Davaran, Sang Woo Joo, Nosratollah Zarghami, Younes Hanifehpour, Mohammad Samiei, Mohammad Kouhi, and Kazem Nejati-Koshki. "Liposome: classification, preparation, and applications." *Nanoscale Research Letters* 8, no. 1 (2013): 102.

Alam, Masroor, and Craig T. Hartrick. "Extended-release epidural morphine (DepoDur™): an old drug with a new profile." *Pain Practice* 5, no. 4 (2005): 349–353.

Allen, Theresa M. "Liposomes." *Drugs* 54, no. 4 (1997): 8–14.

Alomrani, Abdullah, Mohamed Badran, Gamaleldin I. Harisa, Mohamed ALshehry, Moayed Alhariri, Aws Alshamsan, and Musaed Alkholief. "The use of chitosan-coated flexible liposomes as a remarkable carrier to enhance the antitumor efficacy of 5-fluorouracil against colorectal cancer." *Saudi Pharmaceutical Journal* 27, no. 5 (2019): 603–611.

Alphandéry, Edouard, Pierre Grand-Dewyse, Raphael Lefèvre, Chalani Mandawala, and Mickael Durand-Dubief. "Cancer therapy using nanoformulated substances: scientific, regulatory and financial aspects." *Expert Review of Anticancer Therapy* 15, no. 10 (2015): 1233–1255.

Ambrosch, F., G. Wiedermann, S. Jonas, B. Althaus, B. Finkel, R. Glück, and C. Herzog. "Immunogenicity and protectivity of a new liposomal hepatitis A vaccine." *Vaccine* 15, no. 11 (1997): 1209–1213.

Anderson, Peter, M. Tomaras, and K. McConnell. "Mifamurtide in osteosarcoma-a practical review." *Drugs of Today* 46, no. 5 (2010): 327–337.

Antohe, Felicia, Lei Lin, Grace Y. Kao, Mark J. Poznansky, and Theresa M. Allen. "Transendothelial movement of liposomes *in vitro* mediated by cancer cells, neutrophils or histamine." *Journal of Liposome Research* 14, no. 1–2 (2004): 1–25.

Anton, K., P. Van Hoogevest, and L. Frederiksen. "Preparation of a liposome dispersion containing an active agent by compression-decompression." EP616801 (1994).

Astellas Pharma US, Inc. [Homepage on the Internet] AmBisome® (Amphotericin B) liposome for injection. Available from: http://www.ambisome.com/. Accessed November 11, 2011.

Bahal, Surendra M., and Jamie M. Romansky. "Spalling and sorption of tubing for peristaltic pumps." *Pharmaceutical Development and Technology* 7, no. 3 (2002): 317–323.

Bangham, Alec D., Malcolm M. Standish, and Jeff C. Watkins. "Diffusion of univalent ions across the lamellae of swollen phospholipids." *Journal of Molecular Biology* 13, no. 1 (1965a): 238-IN27.

Bangham, Alec D., Malcolm M. Standish, and Gerald Weissmann. "The action of steroids and streptolysin S on the permeability of phospholipid structures to cations." *Journal of Molecular Biology* 13, no. 1 (1965b): 253-IN28.

Barenholz, Yechezkel Chezy. "Doxil® – the first FDA-approved nano-drug: lessons learned." *Journal of Controlled Release* 160, no. 2 (2012): 117–134.

Barnadas-Rodríguez, Ramon, and Manuel Sabés. "Factors involved in the production of liposomes with a high-pressure homogenizer." *International Journal of Pharmaceutics* 213, nos. 1–2 (2001): 175–186.

Bisschops, Marc. "BioSMB technology as an enabler for a fully continuous disposable biomanufacturing platform." *Continuous Processing in Pharmaceutical Manufacturing* (2014): 35–52.

Bisschops, Marc, and Mark Brower. "The impact of continuous multicolumn chromatography on biomanufacturing efficiency." *Pharmaceutical Bioprocessing* 1, no. 4 (2013): 361.

Boswell, G. W., D. Buell, and I. Bekersky. "AmBisome (liposomal amphotericin B): a comparative review." *The Journal of Clinical Pharmacology* 38, no. 7 (1998): 583–592.

Bungener, Laura, Karine Serre, Liesbeth Bijl, Lee Leserman, Jan Wilschut, Toos Daemen, and Patrick Machy. "Virosome-mediated delivery of protein antigens to dendritic cells." *Vaccine* 20, nos. 17–18 (2002): 2287–2295.

Campardelli, Roberta, Islane Espirito Santo, Elaine Cabral Albuquerque, Silvio Vieira de Melo, Giovanna Della Porta, and Ernesto Reverchon. "Efficient encapsulation of proteins in submicro liposomes using a supercritical fluid assisted continuous process." *The Journal of Supercritical Fluids* 107 (2016): 163–169.

Carugo, Dario, Elisabetta Bottaro, Joshua Owen, Eleanor Stride, and Claudio Nastruzzi. "Liposome production by microfluidics: potential and limiting factors." *Scientific Reports* 6 (2016): 25876.

Casey, Catherine, Tina Gallos, Yana Alekseev, Engin Ayturk, and Steven Pearl. "Protein concentration with single-pass tangential flow filtration (SPTFF)." *Journal of Membrane Science* 384, nos. 1–2 (2011): 82–88.

Cattel, Luigi, Maurizio Ceruti, and Franco Dosio. "From conventional to stealth liposomes a new frontier in cancer chemotherapy." *Tumori Journal* 89, no. 3 (2003): 237–249.

Challener, Cynthia A. "PAT for Continuous API Manufacturing Progresses". *Pharmaceutical Technology* 47, no. 4 (2017): 22–27.

Chang, Hsin-I., and Ming-Kung Yeh. "Clinical development of liposome-based drugs: formulation, characterization, and therapeutic efficacy." *International Journal of Nanomedicine* 7 (2012): 49.

Chowdhary, Rubinah K., Isha Shariff, and David Dolphin. "Drug release characteristics of lipid based benzoporphyrin derivative." *Journal of Pharmacy and Pharmaceutical Sciences* 6, no. 1 (2003): 13–19.

Chrai, Suggy S., Ramaswamy Murari, and Imran Ahmad. "Liposomes (a review). Part two: drug delivery systems." *BioPharm* 15, no. 1 (2002).

Clarke, Paul D., Phillip Adams, Rubén Ibáñez, and Christian Herzog. "Rate, intensity, and duration of local reactions to a virosome-adjuvanted vs. an aluminium-adsorbed hepatitis A vaccine in UK travellers." *Travel Medicine and Infectious Disease* 4, no. 6 (2006): 313–318.

Dams, Els T. M., Peter Laverman, Wim J. G. Oyen, Gert Storm, Gerrit L. Scherphof, Jos W. M. Van der Meer, Frans H. M. Corstens, and Otto C. Boerman. "Accelerated blood clearance and altered biodistribution of repeated injections of sterically stabilized liposomes." *Journal of Pharmacology and Experimental Therapeutics* 292, no. 3 (2000): 1071–1079.

Deamer, David W. "Preparation and properties of ether-injection liposomes." *Annals of the New York Academy of Sciences* 308, no. 1 (1978): 250–258.

Deamer, David W., and Alec D. Bangham. "Large volume liposomes by an ether vaporization method." *Biochimica et Biophysica Acta (BBA) – Biomembranes* 443, no. 3 (1976): 629–634.

Dempsey, P. W., Michael E. D. Allisson, Srinivas Akkaraju, Christopher C. Goodjiow, Douglas T. Fearon. "C3d of complement as a molecular adjuvant: bridging innate and acquired immunity." *Science* 271 (1996): 348–350.

Drugs.com. [Homepage on the Internet] DepoCyt® (cytarabine liposome injection) prescribing information. Available from: http://www.drugs.com/pro/depocyt.html. Revised September 2011. Accessed November 11, 2011a.

Drugs.com. [Homepage on the Internet] DepoDur (morphine sulfate extended-release liposome injection) Prescribing Information. Available from: http://www.drugs.com/pro/depodur.html. Revised January 2008. Accessed November 11, 2011b.

Drummond, Daryl C., Charles O. Noble, Zexiong Guo, Keelung Hong, John W. Park, and Dmitri B. Kirpotin. "Development of a highly active nanoliposomal irinotecan using a novel intraliposomal stabilization strategy." *Cancer Research* 66, no. 6 (2006): 3271–3277.

European Medicines Agency. Procedure for European Union guidelines and related documents within the pharmaceutical legislative framework. Available from: http://www.emea.europa.eu/docs/en_GB/document_library/Scientific_guideline/2009/10/WC50000401 1.pdf. Revised 2009. Accessed May 2, 2013.

European Medicines Agency. Draft reflection paper on the data requirements for intravenous liposomal products developed with reference to an innovator liposomal product. Available from: http://www.ema.europa.eu/docs/en_GB/document_library/Scientific_guide line/2011/07/WC500109479.pdf. Revised 2011. Accessed May 2, 2013.

Fan, Yuchen, and Qiang Zhang. "Development of liposomal formulations: from concept to clinical investigations." *Asian journal of Pharmaceutical Sciences* 8, no. 2 (2013): 81–87.

Farid, Suzanne S., Bill Thompson, and Andrew Davidson. "Continuous bioprocessing: the real thing this time? 10th Annual bioProcess UK Conference, December 3–4, 2013, London, UK." In *MAbs*, vol. 6, no. 6, pp. 1357–1361. Taylor & Francis, 2014.

Feldman, E. J., J. E. Kolitz, J. M. Trang, B. D. Liboiron, C. E. Swenson, M. T. Chiarella, L. D. Mayer, A. C. Louie, and J. E. Lancet. "Pharmacokinetics of CPX-351; a nano-scale liposomal fixed molar ratio formulation of cytarabine: daunorubicin, in patients with advanced leukemia." *Leukemia Research* 36, no. 10 (2012): 1283–1289.

Ferrante, Maria. "Continuous processing and standardization impact advances in single-use technology". *Pharmaceutical Processing* (2017). Available at: https://www.pharmaceuticalprocessingworld.com/continuous-processing-and-standardization-impact-advances-in-single-use-technology/.

Forssen, Eric, and Michael Willis. "Ligand-targeted liposomes." *Advanced Drug Delivery Reviews* 29, no. 3 (1998): 249–271.

Frederiksen, Lene, Klaus Anton, Peter Hoogevest, Hans Rudolf Keller, Hans Leuenberger. "Preparation of liposomes encapsulating water-soluble compounds using supercritical carbon dioxide." *Journal of Pharmaceutical Sciences* 86 (1997): 921.

Gabizon, Alberto, Dinah Tzemach, Lidia Mak, Moshe Bronstein, and Aviva T. Horowitz. "Dose dependency of pharmacokinetics and therapeutic efficacy of pegylated liposomal doxorubicin (DOXIL) in murine models." *Journal of Drug Targeting* 10, no. 7 (2002): 539–548.

Gabizon, Alberto, Dorit Goren, Rivka Cohen, and Yechezkel Barenholz. "Development of liposomal anthracyclines: from basics to clinical applications." *Journal of Controlled Release* 53, no. 1–3 (1998): 275–279.

Geng, Shengyong, Bin Yang, Guowu Wang, Geng Qin, Satoshi Wada, and Jin-Ye Wang. "Two cholesterol derivative-based PEGylated liposomes as drug delivery system, study on pharmacokinetics and drug delivery to retina." *Nanotechnology* 25, no. 27 (2014): 275103.

Gilead Sciences Inc. [Homepage on the Internet] DaunoXome® (daunorubicin citrate liposome injection). Available from: http://www.gilead.com/. Accessed November 11, 2011.

Gjoka, Xhorxhi, Rene Gantier, and Mark Schofield. "Platform for integrated continuous bioprocessing." *Biopharm International* 30, no. 7 (2017): 26–32.

Godawat, Rahul, Kevin Brower, Sujit Jain, Konstantin Konstantinov, Frank Riske, and Veena Warikoo. "Periodic counter-current chromatography–design and operational considerations for integrated and continuous purification of proteins." *Biotechnology Journal* 7, no. 12 (2012): 1496–1508.

Hammerschmidt, Nikolaus, Anne Tscheliessnig, Ralf Sommer, Bernhard Helk, and Alois Jungbauer. "Economics of recombinant antibody production processes at various scales: industry-standard compared to continuous precipitation." *Biotechnology Journal* 9, no. 6 (2014): 766–775.

Hartrick, Craig T., and Garen Manvelian. "Sustained-release epidural morphine (DepoDur™): a review." *Todays Therapeutic Trends* 22, no. 3 (2004): 167–180.

Hashizume, Hiroya, Peter Baluk, Shunichi Morikawa, John W. McLean, Gavin Thurston, Sylvie Roberge, Rakesh K. Jain, and Donald M. McDonald. "Openings between defective endothelial cells explain tumor vessel leakiness." *The American Journal of Pathology* 156, no. 4 (2000): 1363–1380.

Hemanthkumar, M., and V. Spandana. "Liposomal encapsulation technology a novel drug delivery system designed for ayurvedic drug preparation." *International Research Journal of Pharmacy* 2, no. 10 (2011): 4–7.

Hernandez, Randi. "Unifying continuous biomanufacturing operations." *BioPharm International* 30, no. 6 (2017): 14–19.

Herzog, Christian, Katharina Hartmann, Valérie Künzi, Oliver Kürsteiner, Robert Mischler, Hedvika Lazar, and Reinhard Glück. "Eleven years of Inflexal® V – a virosomal adjuvanted influenza vaccine." *Vaccine* 27, no. 33 (2009): 4381–4387.

Hong, K., D. C. Drummond, and D. Kirpotin. "Liposomes useful for drug delivery". No. US20160030341 A1. U.S. Patent. February 4, 2016.

Hua, Susan, and Sherry Y. Wu. "The use of lipid-based nanocarriers for targeted pain therapies." *Frontiers in Pharmacology* 4 (2013): 143.

Immordino, Maria Laura, Franco Dosio, and Luigi Cattel. "Stealth liposomes: review of the basic science, rationale, and clinical applications, existing and potential." *International Journal of Nanomedicine* 1, no. 3 (2006): 297.

Ishida, Tatsuhiro, and Hiroshi Kiwada. "Accelerated blood clearance (ABC) phenomenon upon repeated injection of PEGylated liposomes." *International Journal of Pharmaceutics* 354, no. 1–2 (2008): 56–62.

Ishida, Tatsuhiro, Hideyoshi Harashima, and Hiroshi Kiwada. "Interactions of liposomes with cells *in vitro* and *in vivo*: opsonins and receptors." *Current Drug Metabolism* 2 (2001): 397–409.

Ishida, Tatsuhiro, Masae Harada, Xin Yu Wang, Masako Ichihara, Kenji Irimura, and Hiroshi Kiwada. "Accelerated blood clearance of PEGylated liposomes following preceding liposome injection: effects of lipid dose and PEG surface-density and chain length of the first-dose liposomes." *Journal of Controlled Release* 105, no. 3 (2005): 305–317.

Ishida, Tatsuhiro, Masako Ichihara, XinYu Wang, Kenji Yamamoto, Junji Kimura, Eiji Majima, and Hiroshi Kiwada. "Injection of PEGylated liposomes in rats elicits PEG-specific IgM, which is responsible for rapid elimination of a second dose of PEGylated liposomes." *Journal of Controlled Release* 112, no. 1 (2006): 15–25.

Ishida, Tatsuhiro, Kaori Masuda, Takako Ichikawa, Masako Ichihara, Kenji Irimura, and Hiroshi Kiwada. "Accelerated clearance of a second injection of PEGylated liposomes in mice." *International Journal of Pharmaceutics* 255, no. 1–2 (2003): 167–174.

Janoff, A. S., L. T. Boni, M. C. Popescu, S. R. Minchey, P. Ri Cullis, T. D. Madden, T. Taraschi, S. M. Gruner, E. Shyamsunder, and M. W. Tate. "Unusual lipid structures selectively reduce the toxicity of amphotericin B." *Proceedings of the National Academy of Sciences* 85, no. 16 (1988): 6122–6126.

Jiskoot, Wim, Tom Teerlink, E. Coen Beuvery, and Daan J. A. Crommelin. "Preparation of liposomes via detergent removal from mixed micelles by dilution." *Pharmaceutisch Weekblad* 8, no. 5 (1986): 259–265.

Joshi, Varsha S., and Vijesh Kumar. "A rapid HPLC method for enabling PAT application for processing of GCSF." *Lc Gc North America* 31, no. 11 (2013): 948–953.

Jungbauer, Alois. "Continuous downstream processing of biopharmaceuticals." *Trends in Biotechnology* 31, no. 8 (2013): 479–492.

Kim, Sinil, and George M. Martin. "Preparation of cell-size unilamellar liposomes with high captured volume and defined size distribution." *Biochimica et Biophysica Acta (BBA) – Biomembranes* 646, no. 1 (1981): 1–9.

Kim, Sinil, Taehee Kim, Sharad Murdande. "Sustained-release liposomal anesthetic compositions". US Patent 8,182,835 B2 (2012).

Kleinebudde, Peter, Johannes Khinast, and Jukka Rantanen (Eds.). *Continuous Manufacturing of Pharmaceuticals.* Wiley-VCH, Hoboken (2017).

Klimuk, Sandra K., Sean C. Semple, Peter Scherrer, and Michael J. Hope. "Contact hypersensitivity: a simple model for the characterization of disease-site targeting by liposomes." *Biochimica et Biophysica Acta (BBA) – Biomembranes*, 1417, no. 2 (1999): 191–201.

Koning, Gerben A., and Gert Storm. "Targeted drug delivery systems for the intracellular delivery of macromolecular drugs." *Drug Discovery Today* 11, no. 8 (2003): 482–483.

Krishna, Rajesh, Murray S. Webb, Ginette St. Onge, and Lawrence D. Mayer. "Liposomal and nonliposomal drug pharmacokinetics after administration of liposome-encapsulated vincristine and their contribution to drug tissue distribution properties." *Journal of Pharmacology and Experimental Therapeutics* 298, no. 3 (2001): 1206–1212.

Krown, Susan E., Donald W. Northfelt, David Osoba, and J. Simon Stewart. "Use of liposomal anthracyclines in Kaposi's sarcoma." In *Seminars in Oncology.* WB Saunders, 31 (2004):36–52.

Laverman, Peter, Otto C. Boerman, Wim JG Oyen, Els Th M. Dams, Gert Storm, and Frans HM Corstens. "Liposomes for scintigraphic detection of infection and inflammation." *Advanced Drug Delivery Reviews* 37, nos. 1–3 (1999): 225–235.

Leonard, R. C. F., S. Williams, A. Tulpule, A. M. Levine, and S. Oliveros. "Improving the therapeutic index of anthracycline chemotherapy: focus on liposomal doxorubicin (Myocet™)." *The Breast* 18, no. 4 (2009): 218–224.

Li, Z., Boni, L., Miller, B., Malinin, V., Li, X. "High delivery rates for lipid based drug formulations and methods of treatment thereof". US Patent 7,879,351 B2 (2011).

Lister, John. "Amphotericin B lipid complex (Abelcet®) in the treatment of invasive mycoses: the North American experience." *European Journal of Haematology* 56, no. S57 (1996): 18–23.

Liu, Meifeng, Weichi Chen, Xingyu Zhang, Pengwen Su, Feng Yue, Shaoqun Zeng, and Song Du. "Improved surface adhesion and wound healing effect of madecassoside liposomes modified by temperature-responsive PEG-PCL-PEG copolymers." *European Journal of Pharmaceutical Sciences* 151 (2020): 105373.

Mahajan, Ekta, Anupa George, and Bradley Wolk. "Improving affinity chromatography resin efficiency using semi-continuous chromatography." *Journal of Chromatography A* 1227 (2012): 154.

Maherani, Behnoush, Elmira Arab-Tehrany, M. Reza Mozafari, Claire Gaiani, and M. Linder. "Liposomes: a review of manufacturing techniques and targeting strategies." *Current Nanoscience* 7, no. 3 (2011): 436–452.

Metselaar, Josbert M., and Gert Storm. "Liposomes in the treatment of inflammatory disorders." *Expert Opinion on Drug Delivery* 2, no. 3 (2005): 465–476.

Meunier, F., H. G. Prentice, and O. Ringden. "Liposomal amphotericin B (AmBisome): safety data from a phase II/III clinical trial." *Journal of Antimicrobial Chemotherapy* 28, no. suppl_B (1991): 83–91.

Moghimi, S. Moein, and Christy Hunter. "Capture of stealth nanoparticles by the body's defences." *Critical Reviews™ in Therapeutic Drug Carrier Systems* 18, no. 6 (2001): 527–550.

Mozafari, M. Reza. "Liposomes: an overview of manufacturing techniques." *Cellular and Molecular Biology Letters* 10, no. 4 (2005): 711.

Mufamadi, Maluta S., Viness Pillay, Yahya E. Choonara, Lisa C. Du Toit, Girish Modi, Dinesh Naidoo, and Valence M. K. Ndesendo. "A review on composite liposomal technologies for specialized drug delivery." *Journal of drug delivery* 2011 (2011): 939851.

Murry, Daryl J., and Susan M. Blaney. "Clinical pharmacology of encapsulated sustained-release cytarabine." *Annals of Pharmacotherapy* 34, no. 10 (2000): 1173–1178.

Nehoff, Hayley, Neha N. Parayath, Laura Domanovitch, Sebastien Taurin, and Khaled Greish. "Nanomedicine for drug targeting: strategies beyond the enhanced permeability and retention effect." *International Journal of Nanomedicine* 9 (2014): 2539.

Novais, J. L., N. J. Titchener-Hooker, and M. Hoare. "Economic comparison between conventional and disposables-based technology for the production of biopharmaceuticals." *Biotechnology and Bioengineering* 75, no. 2 (2001): 143–153.

Novartis Ophthalmics Inc. [Homepage on the Internet] Visudyne® (verteporfin for injection). Available from: http://www.visudyne.com/. Accessed November 11, 2011.

Okamoto, Yuko, Kazuaki Taguchi, Shuhei Imoto, Victor Tuan Giam Chuang, Keishi Yamasaki, and Masaki Otagiri. "Cell uptake and anti-tumor effect of liposomes containing encapsulated paclitaxel-bound albumin against breast cancer cells in 2D and 3D cultured models." *Journal of Drug Delivery Science and Technology* 55 (2020): 101381.

Oku, Naoto, and Yukihiro Namba. "Long-circulating liposomes." *Critical reviews in therapeutic drug carrier systems* 11, no. 4 (1994): 231–270.

Orozco, Raquel, Scott Godfrey, Jon Coffman, Linus Amarikwa, Stephanie Parker, Lindsay Hernandez, Chinenye Wachuku et al. "Design, construction, and optimization of a novel, modular, and scalable incubation chamber for continuous viral inactivation." *Biotechnology Progress* 33, no. 4 (2017): 954–965.

Otake, Katsuto, Tomohiro Imura, Hideki Sakai, and Masahiko Abe. "Development of a new preparation method of liposomes using supercritical carbon dioxide." *Langmuir* 17, no. 13 (2001): 3898–3901.

Parhi, Priyambada, Chandana Mohanty, and Sanjeeb Kumar Sahoo. "Nanotechnology-based combinational drug delivery: an emerging approach for cancer therapy." *Drug Discovery Today* 17, nos. 17–18 (2012): 1044–1052.

Pollock, James, Glen Bolton, Jon Coffman, Sa V. Ho, Daniel G. Bracewell, and Suzanne S. Farid. "Optimising the design and operation of semi-continuous affinity chromatography for clinical and commercial manufacture." *Journal of Chromatography A* 1284 (2013): 17–27.

Pollock, James, Jon Coffman, Sa V. Ho, and Suzanne S. Farid. "Integrated continuous bioprocessing: economic, operational, and environmental feasibility for clinical and commercial antibody manufacture." *Biotechnology Progress* 33, no. 4 (2017): 854–866.

Poste, George, Demetrios Papahadjopoulos, and William J. Vail. "Lipid vesicles as carriers for introducing biologically active materials into cells." In *Methods in Cell Biology*. (Ed: D. M. Prescott). Academic Press, New York, 14 (1976): 33–71.

Rathore, Anurag S., and Helen Winkle. "Quality by design for biopharmaceuticals." *Nature Biotechnology* 27, no. 1 (2009): 26–34.

Rodriguez, Maria A., Robert Pytlik, Tomas Kozak, Mukesh Chhanabhai, Randy Gascoyne, Biao Lu, Steven R. Deitcher, and Jane N. Winter. "Vincristine sulfate liposomes injection (Marqibo) in heavily pretreated patients with refractory aggressive non-Hodgkin lymphoma: report of the pivotal phase 2 study." *Cancer: Interdisciplinary International Journal of the American Cancer Society* 115, no. 15 (2009): 3475–3482.

Sahoo, Sanjeeb K., and Vinod Labhasetwar. "Nanotech approaches to drug delivery and imaging." Drug discovery today 8, no. 24 (2003): 1112–1120.

Santo, Islane Espirito, Roberta Campardelli, Elaine Cabral Albuquerque, Silvio A. B. Vieira De Melo, Ernesto Reverchon, and Giovanna Della Porta. "Liposomes size engineering by combination of ethanol injection and supercritical processing." *Journal of Pharmaceutical Sciences* 104, no. 11 (2015): 3842–3850.

Santo, Islane Espirito, Roberta Campardelli, Elaine Cabral Albuquerque, Silvio Vieira de Melo, Giovanna Della Porta, and Ernesto Reverchon. "Liposomes preparation using a supercritical fluid assisted continuous process." *Chemical Engineering Journal* 249 (2014): 153–159.

Sawant, Rupa R., and Vladimir P. Torchilin. "Challenges in development of targeted liposomal therapeutics." *The AAPS Journal* 14, no. 2 (2012): 303–315.

Schnitzer, Edit, Ilya Pinchuk, and Dov Lichtenberg. "Peroxidation of liposomal lipids." *European Biophysics Journal* 36, nos. 4–5 (2007): 499–515.

Semple, Sean C., Arcadio Chonn, and Pieter R. Cullis. "Interactions of liposomes and lipid-based carrier systems with blood proteins: relation to clearance behaviour *in vivo*." *Advanced Drug Delivery Reviews* 32, nos. 1–2 (1998): 3–17.

Senior, Judith H. "Fate and behavior of liposomes *in vivo*: a review of controlling factors." *Critical Reviews in Therapeutic Drug Carrier Systems* 3, no. 2 (1987): 123–193.

Sercombe, Lisa, Tejaswi Veerati, Fatemeh Moheimani, Sherry Y. Wu, Anil K. Sood, and Susan Hua. "Advances and challenges of liposome assisted drug delivery." *Frontiers in Pharmacology* 6 (2015): 286.

Shaikh, Ishaque M., Kuan-Boone Tan, Anumita Chaudhury, Yuanjie Liu, Bee-Jen Tan, Bernice MJ Tan, and Gigi N. C. Chiu. "Liposome co-encapsulation of synergistic combination of irinotecan and doxorubicin for the treatment of intraperitoneally grown ovarian tumor xenograft." *Journal of Controlled Release* 172, no. 3 (2013): 852–861.

Slingerland, Marije, Henk-Jan Guchelaar, and Hans Gelderblom. "Liposomal drug formulations in cancer therapy: 15 years along the road." *Drug Discovery Today* 17, nos. 3–4 (2012): 160–166.

Soenen, Stefaan J. H., Alain R. Brisson, and Marcel De Cuyper. "Addressing the problem of cationic lipid-mediated toxicity: the magnetoliposome model." *Biomaterials* 30, no. 22 (2009): 3691–3701.

Stock, Chuck. "Risk assessment for single use disposable projects". *Contract Pharma* (2014). Available at: https://www.contractpharma.com/issues/2014-01-01/view_features/risk-assessment-for-single-use-disposable-projects/.

Storm, Gert, M. T. Ten Kate, Peter K. Working, and I. A. Bakker-Woudenberg. "Doxorubicin entrapped in sterically stabilized liposomes: effects on bacterial blood clearance capacity of the mononuclear phagocyte system." *Clinical Cancer Research* 4, no. 1 (1998): 111–115.

Strong, H. A., J. Levy, G. Huber, and M. Fsadni. "Vision through photodynamic therapy of the eye". No. US5910510 A. U.S. Patent. June 8, 1999.

Subramanian, Ganapathy. *Continuous Processing in Pharmaceutical Manufacturing*. John Wiley & Sons, Hoboken, NJ (2014).

Szebeni, Janos. "Complement activation-related pseudoallergy: a new class of drug-induced acute immune toxicity." *Toxicology* 216, no. 2–3 (2005): 106–121.

Szebeni, Janos, and Y. Barenholz. "Adverse immune effects of liposomes: complement activation, immunogenicity and immune suppression." In *Harnessing Biomaterials for Nanomedicine: Preparation, Toxicity and Applications*. (Ed: PS Publishing) Pan Stanford Publishing, Singapore (2009): 1–19.

Szebeni, Janos, and Seyed Moein Moghimi. "Liposome triggering of innate immune responses: a perspective on benefits and adverse reactions: biological recognition and interactions of liposomes." *Journal of Liposome Research* 19, no. 2 (2009): 85–90.

Szoka, Francis, and Demetrios Papahadjopoulos. "Procedure for preparation of liposomes with large internal aqueous space and high capture by reverse-phase evaporation." *Proceedings of the National Academy of Sciences* 75, no. 9 (1978): 4194–4198.

Szebeni, Janos, Carl R. Alving, Laszlo Rosivall, Rolf Bünger, Lajos Baranyi, Péter Bedöcs, Miklós Tóth, and Yezheckel Barenholz. "Animal models of complement-mediated hypersensitivity reactions to liposomes and other lipid-based nanoparticles." *Journal of Liposome Research* 17, no. 2 (2007): 107–117.

Szebeni, Janos, John L. Fontana, Nabila M. Wassef, Paul D. Mongan, David S. Morse, David E. Dobbins, Gregory L. Stahl, Rolf Bünger, and Carl R. Alving. "Hemodynamic changes induced by liposomes and liposome-encapsulated hemoglobin in pigs: a model for pseudoallergic cardiopulmonary reactions to liposomes: role of complement and inhibition by soluble CR1 and anti-C5a antibody." *Circulation* 99, no. 17 (1999): 2302–2309.

The electronic Medicines Compendium (eMC). [Homepage on the Internet] Epaxal prescribing information. Available from: http://www.medicines.org.uk/EMC/medicine/12742/SPC/Epaxal/. Revised January 4, 2011. Accessed November 11, 2011a.

The electronic Medicines Compendium (eMC). [Homepage on the Internet] INFLEXAL V prescribing information. Available from: http://www.medicines.org.uk/emc/medicine/24941/SPC/inflexalv/. Revised October 2010. Accessed November 11, 2011b.

Toh, Ming-Ren, and Gigi N. C. Chiu. "Liposomes as sterile preparations and limitations of sterilisation techniques in liposomal manufacturing." *Asian Journal of Pharmaceutical Sciences* 8, no. 2 (2013): 88–95.

Tomkinson, Blake, Ray Bendele, Francis J. Giles, Eric Brown, Atherton Gray, Karen Hart, Jeremy D. LeRay, Denny Meyer, Michelle Pelanne, and David L. Emerson. "OSI-211, a novel liposomal topoisomerase I inhibitor, is active in SCID mouse models of human AML and ALL." *Leukemia Research* 27, no. 11 (2003): 1039–1050.

Torchilin, Vladimir P. "Recent advances with liposomes as pharmaceutical carriers." *Nature Reviews Drug Discovery* 4, no. 2 (2005): 145–160.

Treatment of Age-Related Macular Degeneration with Photodynamic Therapy (TAP) Study Group. "Photodynamic therapy of subfoveal choroidal neovascularization in age-related macular degeneration with verteporfin. One-year results of 2 randomized clinical trials-TAP report 1." *Archives of Ophthalmology* 117 (1999): 1329–1345.

Uhumwangho, M. U., and R. S. Okor. "Current trends in the production and biomedical applications of liposomes: a review." *Journal of Medicine and Biomedical Research* 4 (2005): 9–21.

Ulrich, Anne S. "Biophysical aspects of using liposomes as delivery vehicles." *Bioscience Reports* 22, no. 2 (2002): 129–150.

Usonis, V., V. Bakasenas, R. Valentelis, G. Katiliene, D. Vidzeniene, and C. Herzog. "Antibody titres after primary and booster vaccination of infants and young children with a virosomal hepatitis A vaccine (Epaxal®)." *Vaccine* 21, no. 31 (2003): 4588–4592.

Vail, David M., E. Gregory MacEwen, Ilene D. Kurzman, Richard R. Dubielzig, Stuart C. Helfand, William C. Kisseberth, Cheryl A. London, Joyce E. Obradovich, Bruce R. Madewell, and Carlos O. Rodriguez. "Liposome-encapsulated muramyl tripeptide phosphatidylethanolamine adjuvant immunotherapy for splenic hemangiosarcoma in the dog: a randomized multi-institutional clinical trial." *Clinical Cancer Research* 1, no. 10 (1995): 1165–1170.

Veronese, Francesco M., and J. Milton Harris. "Peptide and protein PEGylation." *Advanced Drug Delivery Reviews* 54, no. 4 (2002): 453–456.

Wagner, Andreas, Karola Vorauer-Uhl, and Hermann Katinger. "Liposomes produced in a pilot scale: production, purification and efficiency aspects." *European Journal of Pharmaceutics and Biopharmaceutics* 54, no. 2 (2002): 213–219.

Wang, Tianying, Yao Jiang, Hui Chu, Xia Liu, Yinghui Dai, and Dongkai Wang. "Doxorubicin and lovastatin co-delivery liposomes for synergistic therapy of liver cancer." *Journal of Drug Delivery Science and Technology* 52 (2019): 452–459.

Warikoo, Veena, Rahul Godawat, Kevin Brower, Sujit Jain, Daniel Cummings, Elizabeth Simons, Timothy Johnson et al. "Integrated continuous production of recombinant therapeutic proteins." *Biotechnology and Bioengineering* 109, no. 12 (2012): 3018–3029.

Whitford, William G. "Single-use systems support continuous bioprocessing by perfusion culture." *Continuous Processing in Pharmaceutical Manufacturing* (2014): 183–226.

Whitford, William G. "Single-use systems support continuous bioprocessing by perfusion culture." *Continuous Process in Pharmaceutical Manufacturing.* (Ed: G. Subramanian) Wiley-VCH, Weinheim (2015): 183–226.

Worsham, Robert D., Vaughan Thomas, and Suzanne S. Farid. "Potential of continuous manufacturing for liposomal drug products." *Biotechnology Journal* 14, no. 2 (2019): 1700740.

Zuidam, N. J., E. Van Winden, R. De Vrueh, and D. J. A. Crommelin. "Stability, storage, and sterilization of liposomes." In *Liposomes*. (Ed: V. P. Torchilin and V. Weissig) Oxford University Press, Oxford (2003): 149–165.

Zylberberg, Claudia, and Sandro Matosevic. "Pharmaceutical liposomal drug delivery: a review of new delivery systems and a look at the regulatory landscape." *Drug Delivery* 23, no. 9 (2016): 3319–3329.

14 Impact and Role of Stability Studies in Parenteral Product Development

Mohammed Asadullah Jahangir
Nibha Institute of Pharmaceutical Sciences

Syed Sarim Imam
King Saud University

Jayamanti Pandit
Women Scientist (WOS-C), IPR in Patent Facilitating Centre (PFC) TIFAC

CONTENTS

Abbreviations 225
14.1 Introduction 225
14.2 Importance of Stability Studies in Parenteral Products 226
14.3 Formulation Development and Stability Studies of a Parenteral/Sterile Product 227
14.4 Stability Testing Methodologies of Sterile Products 227
    14.4.1 Real-time Stability Study 227
    14.4.2 Accelerated Stability Studies 228
    14.4.3 Retained Sample Stability Studies 228
    14.4.4 Cyclic Temperature Stress Studies 229
14.5 Regulatory Guidelines for Stability Studies of Parenterals [1G] 229
    14.5.1 ICH and WHO Stability Study Guidelines 229
    14.5.2 CPMP Stability Study Guidelines 230
14.6 Protocol for Stability Study of Parenteral 230
    14.6.1 Testing Sample 230
    14.6.2 Testing of Containers and Closure System of Product 231
    14.6.3 Frequency of Testing and Sampling Plan 231
    14.6.4 Storage Conditions for Testing Samples 231
    14.6.5 Testing Parameters and Acceptance of Stability Study Data 231
    14.6.6 Recording the Stability Study Data 232
    14.6.7 Estimation of Shelf Life and Expiration Date 232
14.7 Current Trends in Parenteral Formulation and Stability Studies 233
14.8 Conclusion 233
References 233

ABBREVIATIONS

APIs: Active pharmaceutical ingredients
CPMP: Committee for Proprietary Medicinal Products
EU: European Union
HDPE: High-density polyethylene
ICH: International Conference on Harmonisation
QbD: Quality by design
QRM: Quality risk management
RH: Relative humidity
USFDA: United States Food and Drug Administration
WHO: World Health Organization

14.1 INTRODUCTION

The success of a pharmaceutical product depends on its overall stability not only during the product development but also in the post-development and marketing phase (Bajaj et al., 2012). Stability can be defined as the ability of the active pharmaceutical ingredients (APIs) or pharmaceutical dosage forms to maintain its therapeutic, microbial chemical and physical properties at the time of its consumption and also during manufacturing and duration of storage (Kommanaboyina et al., 1999). It can also be stated as the extent to which the packaging material and the product enclosed inside it retain its properties within the specified limits during its storage until it is used. The testing of the stability of the pharmaceutical product is a complex set of processes, which involves time, a considerable amount of cost and expertise in the required field to build the safety, quality and efficacy of the developed formulation. Stability testing and analysis of the pharmaceutical ingredients are one of the most important steps in assuring the identity, purity and potency of the ingredients and the developed formulations (Singh et al., 2000). Stability testing also considers the environmental factor and evaluates how it affects the quality of the APIs and the formulated product which is further utilised to predict the shelf life, storage conditions and important suggestive labelling instructions. The data provided by the stability studies are important in the view of documentation and regulatory requirement for any drug or developed product (Singh et al., 2002).

A variety of factors like stability of APIs, physicochemical interaction between excipients and APIs, dosage forms, manufacturing processes, containers and closures used for packaging, the effect of heat, light, and moisture content which is encountered during storage, handling and shipment make it a complex procedure to encounter. Apart from that degradation reaction like hydrolysis, oxidation–reduction also plays a vital role in the stability of pharmaceutical preparations. Other than that, conditions of catalysts, pH of the medium, radiation and exposure to light and concentration of reactants also influence the stability of a pharmaceutical product. Upon exposure, a product may show prominent changes in its consistency, appearance, uniformity, particle size, shape, moisture contents, etc. These physical changes are consequences of impact and abrasion, vibration, changes in temperature, shearing, etc. Chemical changes like loss of excipient activity, the formation of degraded product and loss of potency of APIs are caused due to oxidation–reduction, hydrolytic reactions, etc. (Carstensen et al., 2000). The stability of parenteral formulation may be halted by the growth of micro-organism, thus influencing the preservative efficacy of the product (Matthews et al., 1999).

14.2 IMPORTANCE OF STABILITY STUDIES IN PARENTERAL PRODUCTS

The stability of parenteral product is a broad topic which includes several specialised cases (Sharma et al. 2016). Parenteral dosage forms are rather simple forms of formulations which are mostly aqueous form of drug suspended or mixed in buffers and salts to maintain a proper osmotic pressure and pH. If the kinetics and reaction pathways are known, the stability of such solutions is pretty straightforward and includes the storage of the sterile product in non-interactive containers and requires inert atmosphere and protection from light. Unique packaging and processing technique are prerequisite for the stability ramifications of a parenteral product. Stability testing is of utmost importance as it directly influences the health of the patient. The importance of stability studies is listed in Figure 14.1. The degradation of the unstable product may lead to develop toxic products, or it may also cause loss of therapeutic activity up to 85% from label claim, thus causing serious health issues and sometimes mortality (Sánchez-Rivera et al., 2015). Thus, it is legally important to provide certain data of stability studies to the regulatory agencies before approval. It is also important from the manufacturer point of view for maintaining its reputation in the pharmaceutical market that the company provides high-quality medicines which retain its potency and efficacy throughout the marketing and consumption phase. Data from stability studies at the developmental or marketing phase provided to

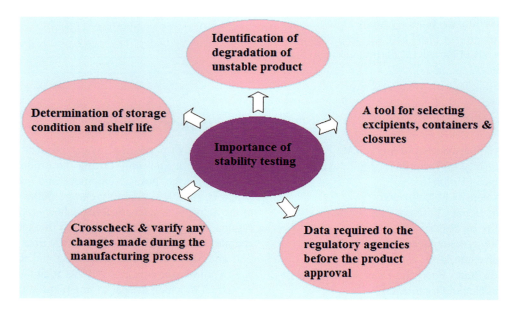

FIGURE 14.1 Importance of stability testing.

a database may serve as an important tool while selecting excipients, containers and closure systems adequate for a formulation or a development of a new parenteral product. It will also assist in the determination of storage conditions and shelf life required during a new product development, in the preparation of dossiers, and to cross-check and verify any changes made or introduced during the manufacturing process or development that impact negatively on the stability of the developed formulation (Singh et al., 2000; Carstensen et al., 2000).

14.3 FORMULATION DEVELOPMENT AND STABILITY STUDIES OF A PARENTERAL/STERILE PRODUCT

Pre-formulation studies gather the information to ensure that every aspect of product development is well understood and is in place for the development of the product (Hardy et al., 2009). It includes to assess the stability via analytical method to ensure that the drug is sufficiently stable and will not break down during the developmental procedures; in case it fails the pre-formulation stability, it becomes useless to proceed with the drug for formulation and development. Degradation profile is of utmost importance in the pre-formulation studies and provides information which assists in the transition from the pre-formulation stage to the formulation development. Thermal stability is determined by placing the sample into a nitrogen-filled dark environment and exposing the same to various high temperatures like 25°C, 40°C, 60°C and 80°C and even higher temperatures for an extended time period; usually, the period extends from 3 to 6 months. Samples are withdrawn at pre-fixed intervals and tested for degradations exploiting indicative methods like HPLC. Stability is also determined as a function of pH, as it influences both short- and long-term stability as well as solubility.

Formulation development takes the data from pre-formulation studies to establish a condition where the optimal product can be developed. These conditions are achieved via procedures to determine the addition of excipients, controlled hydrolysis, control of oxygen, etc. Additional information like dose to be administered, route of administration like I.M., I.V. infusion, I.V. bolus, subcutaneous, intrathecal etc. and solution concentration are denoted in mg/mL. The United States Food and Drug Administration (USFDA) has made strict regulations for materials to be added to injectables. To confirm the stability and quality of parenteral products, the tests can be classified as universal test, which is applied to all types of parenteral dosage forms and specific tests which are exclusive for a certain type of parenteral formulation. Universal test includes identification testing, packaging systems, assay, impurity testing, foreign and particulate matter, bacterial endotoxins, container and closure integrity, container content, sterility and labelling.

To determine the strength of the drug product, specific stability-indicating test is used. To achieve overall specificity, supporting analytical procedures are also used where non-specific assays are conducted. However, strictly specific assays should be used wherever there is evidence of excipient interference. Parenteral should be developed in such a way that no particulate matter is spotted in the formulation (Nagaraja et al., 2012) and to ensure that each final container of the developed parenteral formulation is inspected to the extent possible for detecting any possible particulate foreign matter (Ravi et al., 2012). The inspection process must be qualified to ensure that visible particulates are not present in any lot of the parenteral products. Every container which shows the evidence of particulate matter must be rejected. This process can be performed while performing other examinations like cracking or defective containers or seals. If the nature of the content or the closure system limits the inspection process, the complete content of the lot should be supplemented with the inspection of dried or withdrawn content of a sample of containers from the lot. Large volume injections, small volume injections, single-dose infusions and pharmacy bulk packages are usually subjected to light obscuration or microscopic procedures or otherwise specified in individual monograph (Peterfreund et al., 2013; Glover et al., 2013).

14.4 STABILITY TESTING METHODOLOGIES OF STERILE PRODUCTS

Stability studies are usually a routine procedure which is conducted at various phases of pharmaceutical product development (Ikeda et al., 2012). During the early phases of production, accelerated stability studies are conducted at relatively high humidity and temperature for determining degradation product which may produce upon long-term storage (Pimple et al., 2015). Studies are conducted under less extreme condition or slightly elevated temperature recommended for long-term understanding of shelf life. The most important aspect of stability studies is to assure the product that it will retain its quality and efficacy through the period of marketing until consumption (Kommanaboyina et al., 1999). Depending upon the type of stability study, it has been categorised into four different categories as illustrated in Figure 14.2.

14.4.1 Real-time Stability Study

These studies are performed for an extended period to allow measurable product degradation under storage conditions. The duration opted for such studies is dependent on the stability of the product which must be long enough to indicate any measurable degradation product. During the study, the collection of data is conducted at a pre-fixed frequency to form a trend in analysing and recording instability on daily basis. The reliability of data is increased if a reference material from a single batch is used for which the stability characteristics have been established already. For stability of the reference material, consistency of performance of the instrument and stability of the reagents must also be considered. Both system instrumentation and reagents must be

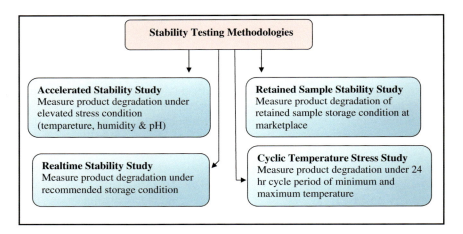

FIGURE 14.2 Various stability testing methodologies.

closely monitored for any lag in performance and any possible drift or discontinuity which may lead to changes in the result (Anderson et al., 1991).

14.4.2 Accelerated Stability Studies

During the accelerated stability studies, a product is exposed to extremely high temperature and humidity. The amount of heat required to cause any significant degradation in the product is determined. The data produced are used to project the possible shelf life of the product or even exploited to compare the relative stability of two alternative developed formulations (Bhutani et al., 2003). This method provides an early indication of the product shelf life and thus saves a lot of time during the developmental period. Apart from temperature, the stress conditions which are exploited in the accelerated stability studies are exposure to extreme pH, light, gravity, agitation, moisture, etc. (Kommanaboyina et al., 1999). Samples are subjected to stress and then refrigeration, followed by simultaneous assay during the accelerated stability studies. Due to the short period of analysis, the chances of instability in the measurement system are significantly reduced as compared to real-time stability studies. Further, it is possible in accelerated stability studies to compare the unstressed parenteral product with the stress product within the same assay and is expressed in percentage of unstressed sample recovery. From statistical point of view, the accelerated stability studies are conducted in four different stress temperatures (40°C, 50°C, 60°C and 70°C). In case of protein containing thermolabile products, it is recommended to avoid denaturing stress temperatures to get accurate stability projections (Anderson et al., 1991; Connors et al., 1973). If the activation energy is known, the rate of degradation at low temperature may be extrapolated from those recorded at extreme temperatures (Bott et al., 2007).

Many a time, pharmaceutical manufacturers often utilise bracket table and Q rule to predict the shelf life of their products; however, none of them are officially discussed by International Conference on Harmonisation (ICH) or Food and Drug Administration (FDA) (Savla et al. 2017). According to the Q rule, a product degrades at a constant rate by a constant factor Q_{10} when the temperature where the product is stored is decreased by 10^0C. The value of Q_{10} is in the form of sets at 2, 3, 4 and so on because these values correspond better to the activation energies. This model is inaccurate as it assumes that the Q value is independent of temperature variation and thus does not change with the change in temperature. On the other hand, the bracket table assumes that the activation energy for any product falls between 10 and 20 kcal. Assuming this, a table is constructed showing days for which extreme condition is given at various stress temperatures. This assumption is reasonable based on the fact that most of the reagents and analytes which are commonly used in the clinical laboratories and pharmaceutical industries have activation energies between these ranges (Kommanaboyina et al., 1999).

14.4.3 Retained Sample Stability Studies

Retained sample stability studies are conducted by every pharmaceutical manufacturer for its product which needs stability data. For conducting the study, stability samples with retained storage data for 1 year for at least one batch are selected. If the sample size is more than 50, two batches are usually recommended. In this study, when the first batch is introduced into the market, the stability sample of every batch is taken which may be decreased over time up to 2%–5% of the total marketed batch at a later stage of the study. The stability study is conducted at a predetermined interval. Let's say if the shelf life of a parenteral product is 5 years, the sample is conventionally tested at 3, 6, 9, 12, 18, 24, 36, 48 and 60 months. Such a method where data are obtained from retained storage sample is commonly regarded as a constant interval method (Kommanaboyina et al., 1999). Its modified method is to consider market samples and evaluate different attributes of the stability study. Such an approach can be regarded as a realistic method of stability testing as it considers data not only from the sample storage condition but also from the marketing phase (Kommanaboyina et al., 1999).

14.4.4 Cyclic Temperature Stress Studies

Pharmaceutical manufacturers do not consider cyclic temperature stress studies as routine method of stability testing. These methods are designed based on the knowledge of the product to mimic similar condition that the product may face challenges during storage in the marketplace. A 24-h cycle period is considered. The minimum and maximum temperature exposure is selected for cyclic stress studies on a per-product basis. Factors like storages temperature on the product and specific physical and chemical degradation are recommended. Usually, 20 cycles per test are highly recommended for cyclic temperature stress studies (Carstensen et al., 2000).

14.5 REGULATORY GUIDELINES FOR STABILITY STUDIES OF PARENTERALS [1G]

Stability studies should support the claim regarding the active molecules or products manufactured, consumed or distributed with the regulatory guideline of the countries also if the same is expected to be exported to that specific country. Thus, a uniform study was required and was implemented as the guidelines including issues related to stability studies, data requirement of stability studies and application of dossier and steps to execute the same (Kopp 2010).

14.5.1 ICH and WHO Stability Study Guidelines

Stability study guideline was initially introduced in the early 1980s and further came to be known as International Conference on Harmonisation (ICH). The harmonised guidelines streamlined the testing of parenteral products and eased global registration of a new parenteral. ICH stability study guidelines are listed in Table 14.1.

In the stability study, the major hindrance of a parenteral product is to understand the long-term storage condition. Its concept evolved with the introduction of climatic zones by Schumaker and Wolfgang Grimm in the years 1972 and 1986, respectively, has now become a gold standard and is exploited in the development of parenteral product (Grimm et al., 1972, 1986). In the year 1993 and subsequently in 2006, ICH issued guidelines for climatic zones I and II for registration in United States, Japan and Europe. Newer storage condition was defined for zones III and IV (Table 14.2).

Further, the World Health Organization (WHO) in 2005 suggested to divide Zone IV into two subzones – Hot and Humid (IVa) and Hot and Very Humid (IVb) – so that more regions can be included as many of them experience very high humidity and temperature. However, FDA has withdrawn its stability guidelines along with reference ICH Q1A (R2) in 2006. Following this, ICH withdrew Q1F which uses to support stability testing in Zone III and Zone IV; thus, the parenteral pharmaceutical industry was left with no supporting guideline for the submission of products to countries which fall in these regions. WHO in the year 2009 came up with a final guideline for stability studies 'Stability Testing of Active Pharmaceutical Ingredients and Finished Pharmaceutical products' as Annexure 2 in the WHO Technical Report Series 953, 2009. These guidelines cover APIs as well as marketed and new parenteral pharmaceutical products. It also takes into account the diversity of global climatic condition and so also recommends three conditions for long-term storage (25°C/60% RH, 30°C/65% RH and 30°C/75% RH). Many countries have also developed their regional guidelines for the stability of parenteral pharmaceutical products in accordance with these international guidelines. However, many countries are enlisted in different climatic zones from their actual climate; for example, Chile, Canada and Afghanistan are in 30°C/65% RH, but their actual climatic condition is totally different. These countries are accepting stability data conducted at 25°C/60% RH for their market. As per ICH 30°C/65%,

TABLE 14.1
ICH Guidelines Related to Stability Testing

ICH Guidelines	Stability Testing
Q1A (R2)	Stability testing of new drug substances and products
Q1B	Photostability testing of new drug substances and products
Q1C	Stability testing for new dosage forms
Q1D	Bracketing and matrixing designs for stability testing of new drug substances and products
Q1E	Evaluation of stability data

TABLE 14.2
Stability According to Different Climatic Zones

Climatic Zone	Temperature	Humidity
Long-Term Stability Conditions (12 months)		
Zone I	21°C ± 2°C	45% ± 5% RH
Zone II	25°C ± 2°C	60% ± 5% RH
Zone III	30°C ± 2°C	35% ± 5% RH
Zone IVa	30°C ± 2°C	65% ± 5% RH
Zone IVb	30°C ± 2°C	75% ± 5% RH
Refrigerated	5°C ± 3°C	No humidity
Frozen	−15°C ± 5°C	No humidity
Intermediate and Accelerated Stability Conditions (6 months)		
Accelerated ambient	40°C ± 2°C	75% ± 5% RH
Accelerated refrigerated	25°C ± 2°C	60% ± 5% RH
Accelerated frozen	5°C ± 3°C	No humidity
Intermediate	30°C ± 2°C	65% ± 5% RH

RH is regarded as an intermediate condition and holds good in most of the cases until any significant change is found during accelerated stability studies. India, on the other hand, lists 30°C/70% RH for their long-term stability study, which is different from other countries falling in the same climatic zone. Thus, it prevents the use of one condition to support registration in all countries of this region (Zahn 2009). The concept of climatic zones is discussed by Zahn, who better explained the selection of long-term testing conditions based on actual meteorological data (Zahn 2008). Continuous efforts are being made to fine-tune these guidelines.

14.5.2 CPMP Stability Study Guidelines

Committee for Proprietary Medicinal Products (CPMP) which works under European Agency provides a series of guidance for stability studies for the evaluation of medicinal products in order to assist in marketing authorisation in the European Union (CPMP 2003; Eggert et al., 2008). The various guidelines issued from time to time are listed in Table 14.3. CPMP/QWP/576/96 Rev.1 provides guideline on stability studies regarding the variation in marketing authorisation and its applications (EMEA 2004). CPMP/QWP/6142/03 provides guideline on stability studies regarding the medicinal products and active substances which are manufactured in climatic zones III and IV and are intended to be marketed in the European Union (EMEA 2005). CPMP/QWP/609/96 Rev.1 provides guidance about the declaration of storage conditions particularly of active substance and medicinal products (EMEA 2001). CPMP/QWP/122/02 Rev.1 provides guidance regarding the stability study of existing active substances and their related final products (EMEA 1998). CPMP/QWP/072/96 provides guidance about the initiation of shelf life of the finished product (EMEA 2003).

CPMP/QWP/2934/99 provides guidance about in-use stability study of human-consumable medicinal products (EMEA 2004). CPMP/QWP/576/96 provides guidance for stability study of a type 2 variation to a marketing authorisation (EMEA 2001). CPMP/QWP/159/96 provides guidance regarding the maximum shelf life intended for sterile products after first opening or following reconstitution. It includes some recommendations for the user information texts (e.g. SmPC, package insert, labels). It applies to all sterile products for human use, with the exception of radiopharmaceuticals and extemporaneously prepared or modified preparations (EMEA 2004).

Guidelines related to the stability testing have also been released by the WHO in the year 1988 (Ba 2010). With the incompetency of ICH guidelines to cope up with the extreme climatic conditions of many countries and its coverage limiting to new drug and products with no guidelines for products already available in the market, WHO brought certain modifications in the ICH in 1996 (WHO 1996), which was further revised in the years 2003 and 2006 to deal with the changes which occur in long-term storage conditions to support climate zone IV regions (WHO 2003, 2004, 2006). WHO released guidance on stability testing in global environment in the year 2004 (Kaur et al., 2015).

14.6 PROTOCOL FOR STABILITY STUDY OF PARENTERAL

Protocols are prerequisite for stability studies and are written documents which describe the key aspects of well-controlled and regulated stability study. A protocol of stability study is dependent on the type of drug substance or product as well as on whether it is already available in the market or is a new entity (Ali et al., 2008) with the details of the region where it is intended to be marketed. A properly designed protocol discusses the below-mentioned information.

14.6.1 Testing Sample

Parenteral products which are in the developmental stage are usually carried out in a single batch, while registering new products or establishing an unstable product is done to the first three production batches; however, sometimes even two batches are allowed. In case the initial data are

TABLE 14.3
CPMP Stability Studies Guidelines

CPMP Guidelines	Specification
CPMP/QWP/576/96 Rev.1	Stability studies regarding the variation in marketing authorisation and its applications (EMEA 2004)
CPMP/QWP/6142/03	Stability studies regarding the medicinal products and active substances which are manufactured in climatic zones III and IV and are intended to be marketed in the European Union (EMEA 2005)
CPMP/QWP/609/96 Rev.1	Declaration of storage conditions particularly of active substance and medicinal products (EMEA 2001)
CPMP/QWP/122/02 Rev.1	Stability study of existing active substances and their related final products (EMEA 1998)
CPMP/QWP/072/96	Shelf life of the finished product (EMEA 2003)
CPMP/QWP/2934/99	In-use stability study of human-consumable medicinal products (EMEA 2004)
CPMP/QWP/576/96	Stability study of a type 2 variation to a marketing authorisation (EMEA 2001)
CPMP/QWP/159/96	Maximum shelf life intended for sterile products after first opening or following reconstitution (EMEA 2004)

insufficient for a full-scale production batch, then the first three batches of the manufactured product should be used for long-term stability studies post-approval. Laboratory-scale batch data can serve as a supportive document, but they cannot be regarded as a primary stability data. The selection of batch should be made randomly from a set of production batches (Singh et al., 2000).

14.6.2 Testing of Containers and Closure System of Product

Stability testing is done in immediate containers and closures in which the product is expected to be marketed like high-density polyethylene (HDPE) bottles. Products in every different type of closures or containers which are meant for marketing or for physical or promotional purpose need separate testing. Study on prototype containers is allowed in the case of bulk containers if it can stimulate the original packaging (Singh et al., 2000).

Samples which are dispersed systems or semi-solid dosage form must be kept upright and should allow full interaction with the closure system in order to properly determine any possible interaction between the product and the closure system which may include extraction of chemical substance into the product or adsorption of the product components on the closure system (Ali et al., 2008).

14.6.3 Frequency of Testing and Sampling Plan

The frequency of testing is designed intelligently enough to sufficiently provide stability profile data of new drug substance. If a product is proposed to have a shelf life of at least 1 year, then the frequency of testing for a long-term storage must be conducted every 3 months for the first year followed by every 6 months in the second year and then on annual basis for the rest of the shelf life or proposed expiration period. For accelerated stability studies, usually three time points are recommended. In case significant changes are recorded during accelerated stability studies, intermediate storage conditions are evaluated with four time points. Less number of testing points is involved if the same products of different strengths and sizes are required to be tested. Matrixing and bracketing statistical designs are exploited for such reduced testing plans. Bracketing design is different from matrixing in the manner that the former involves samples at the extreme of certain design factor like package size while matrixing only involves the testing a portion or a subset of the total sample strength for all combinations at a specific time. Factors like container size, fill size, strengths and batches can be matrixed (Cha et al., 2001).

Sample planning for stability studies involves both planning for a sufficient number of samples which will be charged into the stability chamber and sampling out the charged batch to appropriately cover the whole study. As the sampling time is developed, the number of samples to be drawn at each sample point is fixed and finally added up to the total samples involved.

14.6.4 Storage Conditions for Testing Samples

The condition in which the samples must be stored is dependent on the climatic zone of the country or place in which the product is expected to be marketed or going to be filed for regulatory approval. WHO and ICH suggest guidelines/recommendations for different storage conditions.

14.6.5 Testing Parameters and Acceptance of Stability Study Data

Protocols for stability studies must efficiently define all parameters that would be exploited to evaluate the samples. Studies which monitor the purity, potency, quality and identity that are expected to differ upon storage are considered for stability tests. Tests for changes in appearance, degradation product, assay, moisture content and microbial testing are some of the standard tests performed on the sample. Microbiological studies include microbial count, sterility testing and efficacy. Table 14.4 shows the stability-indicating parameters which are important for parenteral preparation. The batches which are used for stability studies are prerequisite to meet testing requirements for residue after ignition, residual solvents, heavy metals, etc. Some of them are studied at the time of market release (Cha et al., 2001). ICH guideline Q6A also discusses tests like particle size, polymorphism and enantiomeric purity.

For better acceptance of the result, it is always recommended to follow official compendia because any alternate method apart from the official ones required validation before acceptance. All the methods involved must be reproducible with minimum validation required (Ali et al., 2008). However, the analytical methods involved need validation before the stability study is initiated. In a similar manner, the presence of any degradation product must be fixed. The criteria for acceptance were fixed in terms of numerical values or limits for quantitative tests like moisture content, particle size, degradation product and viscosity, whereas for qualitative tests like colour, odour and cracking, only pass or fail is required to be reported. The individual and upper limit for degradation products must also be included in the acceptance criteria. Q3B (R2) guideline of the ICH relates to the impurities found in the new drug product and also addresses the degradation products in a new drug formulation. If the proposed threshold for a study is exceeded, it must be reported, and this is dependent upon the intended dose (Cha et al., 2001).

As per WHO guidelines and ICH Q1A (ICH Q1A (R2) 2003), if the assay results differ by 5% from the initial data of the previous batch, it is considered to be out of trend because of the significant difference. If out-of-trend results are reported, an investigation is followed which is called the Barr decision. This evaluates the procedures of data collection and identifies the error and its root cause. The difference in the results is usually because of the difference in the storage condition of batches, the difference in time of production, etc. As per Gaur et al. apart from temperature

TABLE 14.4
Stability Testing Parameters for Parenteral

Stability Studies Include Monitoring	Small Volume Parenterals (SVPs)	Large Volume Parenterals (LVPs)	Injectable after Reconstitution[a]
Appearance	✓	✓	✓
Colour	✓	✓	✓
Assay	✓	✓	✓
Preservative content (if present)	✓	✓	✓
Degradation products	✓	✓	✓
Particulate matter	✓	✓	✓
pH	✓	✓	✓
Sterility	✓	✓	✓
Pyrogenicity	✓	✓	✓
Reconstitution time	×	×	✓
Residual moisture content	✓	✓	✓
Extractables/leachables (container and closure integrity)	✓	✓	✓
Particle size distribution, redispersibility, rheological properties	Parenteral suspension		
Phase separation, viscosity, mean size and distribution of dispersed phase globule	Parenteral emulsion		

[a] Injectable after reconstitution must be monitored for these parameters after reconstitution.

and humidity, variation in air velocity may result in difference in the adsorption of moisture by the hygroscopic drug (Gaur et al., 2005).

14.6.6 Recording the Stability Study Data

The recording of data is usually done in a cumulative, comprehensive and organised format. It is presented in the concise table format which helps in interpretation and ease of review. Sheets are developed per batch which recorded data in a tabular format with all the necessary information of that particular batch. If any sampling time is missed, it must be indicated in the table. The data can be grouped according to the time interval and storage condition to represent stability as a function of time for all the studied environmental conditions. Graphical representation of stability data versus time may also be helpful for evaluation. However, tabular representation is mandatory for regulatory filing. These details are discussed in ICH Q3A and Q3B (Cha et al., 2001).

14.6.7 Estimation of Shelf Life and Expiration Date

The expiration date can be defined as time period for which the pharmaceutical product will remain stable under storage conditions which are recommended in the labelling. In other words, it can be said that the product will no longer be safe to use beyond the expiration date. If the recommended storage conditions are not followed, the product is expected to degrade at a more rapid speed. Shelf life can be defined as the period during which the product retains its potency (more than 90% of the labelled claim) if stored properly as per the manufacturer's instructions (Kommanaboyina et al., 1999).

Shelf life is calculated from the data which are obtained from the studies conducted for a longer term. Firstly, linearisation of data is done, followed by the application of goodness of fit. The data are then analysed to confirm whether the slope and intercept are matching. The data are further pooled and estimated for the common slope (Singh et al., 2000). For determining the significance of difference, statistical tests like t-tests are exploited. Data are usually collected in the form of record points at months – 0, 3, 6, 9 and 12. In case data are found to be not fit or insufficient for pooling, then stability estimates are done on the worst possible batch. Regression line of the five data points is used to determine the expiration date or shelf life based on calculative 95% one-sided confidence limit. The 90% drug concentration is taken as the lower limit, and the point where the extension line cuts the 95% limit line is considered as expiration date (Ali et al., 2008).

In case of any new drug, only 2-year expiration period is provided initially which consists of 1-year long-term and 6-month accelerated study data. The expiration date for 3rd and subsequent years is allowed only on producing real-time data (Singh et al., 2000). Mostly, the pharmaceutical products are provided with only one shelf life. However, in some cases, two shelf lives are provided; for example, some of the injectables or similar sterile ophthalmic products have a shelf life which is for storage period and another shelf life which states that when the product is reconstituted with an appropriate vehicle intended for use, must be consumed within 2 or 3 days (Carstensen et al., 2000; Herbig et al., 2013).

Determination of the conformance period is done by the intersection of the lowest value of the stability parameter and 95% confidence of the regression line. Conformance period is usually equal to or more than shelf life and is denoted as a round of number (year nearest to the month). For example, a conformance period of 13.4, 25.3 and 38.9 months would probably assign shelf lives of 12, 24 and 36 months, respectively. Conformance period provides an extra stability reserve than the shelf life (Carstensen et al., 2000).

Stability chamber is used for stability studies. They are specialised chambers which are capable of simulating the storage conditions and thus assist in long-term, real-time and accelerated stability studies (Zhang et al., 2013). They are available in both reach-in and walk-in styles. Since the retention time in accelerated studies is less, smaller tables are preferred for such studies loaded with alarm, safety and recording devices. Similar kinds of photostability chambers are also exploited with or without humidity and temperature control and also contain UV fluorescent tubes and cool white light. Lux meter is used to estimate the intensity of visible light (Singh et al., 2000).

As per the ICH guidelines of 1996, it is necessary to establish photostability of new drug substance and its products in a single batch of the product. However, the study needs to be repeated if any variation is reported. D65 is international recognised standard for outdoor light, while ID65 acts as its equivalent and standard for its indoor daylight.

14.7 CURRENT TRENDS IN PARENTERAL FORMULATION AND STABILITY STUDIES

In recent years, quality by design (QbD) has gained popularity among pharmaceutical researchers. It is a systemic approach of formulation development program which begins with pre-fixed objectives and involves understanding of process control. It takes into account how process variables and formulation influence the product quality. Relevant documents from ICH Q8 along with the International Conference on Harmonisation of Technical Requirements for Registration of Pharmaceuticals for Human Use (ICH), Quality Risk Management, Pharmaceutical Development, ICH Q10 and Pharmaceutical Quality Systems indicate at abstract level how QbD ensures the quality of the drug product (Lionberger et al., 2008). It is a holistic approach with scientific and proactive method of product development. QbD fully understands how different processes and attributes relate to overall product performance. It provides advantage that it avoids conductive additional stability studies when the process, site, route and scale get altered from the initial batches. However, regulations, guidelines and costs are barriers to this approach.

It has now become important for pharmaceutical companies to well define the stability study conditions to approach for global market. Changes in the global stability study testing include increasing the period of accelerated stability study from 6 to 12 months and conducting additional tests at 50°C/75% RH for at least 3 months (Mischler et al., 2002). It helps in avoiding repetition of stability studies if the product is expected to be marketed to other regions.

14.8 CONCLUSION

During the pharmaceutical drug development for a new or established formulation, stability study stands as one of the key processes. These studies are conducted to recommend storage conditions and to ensure efficacy and safety as well as shelf life onto the label of each medicine. Nowadays, the regulatory requirements have become stringent regarding the conditions to which a formulation is expected to be subjected throughout its shelf life. Thus, proper scientific principles must be followed with full understanding of the regulatory requirements for stability studies.

REFERENCES

Ali, J., Khar, R.K., Ahuja, A. 2008. *Dosage Form and Design*. 3rd ed. Birla Publications Pvt. Ltd, Delhi: 100–123.

Anderson, G., Scott, M. 1991. Determination of product shelf life and activation energy for five drugs of abuse. *Clin Chem* 37:398–402.

Bajaj, S., Singla, D., Sakhuja, N. 2012. Stability testing of pharmaceutical products. *J App Pharm Sci* 2(3):129–138.

Bhutani, H., Mariappan, T.T., Singh, S. 2003. Behavior of uptake of moisture by drugs and excipients under accelerated conditions of temperature and humidity in the absence and the presence of light. Part 2. Packaged and unpackaged antituberculosis drug products. *Pharm Technol* 27:44–52.

Bott, R.F., Oliveira, W.P. 2007. Storage conditions for stability testing of pharmaceuticals in hot and humid regions. *Drug Dev Indus Pharm* 33:393–401.

Carstensen, J.T. 2000. *Drug Stability, Principles and Practices*. Marcel Dekker, New York.

Cha, J., Gilmor, T., Lane, P., Ranweiler, J.S. 2001. *Stability Studies in Handbook of modern Pharmaceutical Analysis. Separation Science and Technology*. Elsevier, Amsterdam: 459–505.

Connors, K.A., Amidon, G.L., Kennon, L. 1973. *Chemical Stability of Pharmaceuticals – A Handbook for Pharmacists*. John Wiley & Sons, New York: 8–119.

CPMP. 2003. Guideline on stability testing: Stability testing of existing active substances and related finished products. CPMP/QWP/122/02, 2003.

Eggert, J., Flower, C. 2008. Conducting stability studies during development to ensure successful regulatory approval. In: *Pharmaceutical Pre-Approval Inspections, a Guide to Regulatory Success*. 2nd ed. Edited by: Hynes III, M. D. Eli Lilly and Company, Indianapolis, IN.

EMEA Committee for Proprietary Medicinal Products (CMPP). 1998. Note for guidance on maximum shelf-life for sterile products for human use after first opening or following reconstitution. CPMP/QWP/159/96, July 1998.

EMEA Committee for Proprietary Medicinal Products (CPMP). 2001. Note for guidance on in-use stability testing of human medicinal products CPMP/QWP/2934/99, September 2001.

EMEA Committee for Proprietary Medicinal Products (CPMP). 2001. Note for guidance on start of shelf-life of the finished dosage form CPMP/QWP/072/96, December 2001.

EMEA Committee for Proprietary Medicinal Products (CPMP). 2004. Guideline on stability testing: stability testing of existing active substances and related finished products CPMP/QWP/122/02, March 2004.

EMEA Committee for Proprietary Medicinal Products (CPMP). 2003. Note for guidance on declaration of storage conditions: A: In the product information of medicinal products, B: For active substances. CPMP/QWP/609/96, October 2003.

EMEA Committee for Medicinal Products for Human Use (CHMP). 2005. Guideline on stability testing for applications for variations to a marketing authorization CPMP/QWP/576/96, December 2005.

Gaur, A., Mariappan, T.T., Bhutani, H. et al. 2005. A possible reason for the generation of out-of-trend stability results: Variable air velocity at different locations within the stability chamber. *Pharm Technol* 29:46–49.

Glover, Z.W., Gennaro, L., Yadav, S. et al. 2013. Compatibility and stability of per-tuzumab and trastuzumab admixtures in i.v. infusion bags for coadministration. *J Pharm Sci* 102:794–812.

Grimm, W. 1998. Extension of the international conference on harmonization tripartite guideline for stability testing of new drug substances and products to countries of climatic zones 3 and 4. *Drug Dev Ind Pharm* 24:313–325.

Grimm, W., Schepky, G. 1980. *Stabilitatsprufung in der Pharmazie*. Theorie und Praxis, Editio Cantor Verlag, Aulendorf.

Hardy, G., Puzovic, M. 2009. Formulation, stability, and administration of parenteral nutrition with new lipid emulsions. *Nutr Clin Pract* 24(5):616–625.

Herbig, S., Kaiser, V., Maurer, J. et al. 2013. German Society of Hospital Pharmacists guideline: Aseptic preparation and quality control of ready-to-use parenterals. *Krankenhauspharmazie* 34:42–106 (In German).

Huynh-Ba, K. (Ed.). 2010. *Pharmaceutical Stability Testing to Support Global Market: Pharm asp*. Springer Science+ Business Media, LLC, New York.

ICH Q1A (R2). 2003. Stability testing guidelines: Stability testing of new drug substances and products. ICH Steering Committee.

Ikeda, R., Vermeulen, L.C., Lau, E. et al. 2012. Stability of infliximab in polyvinyl chloride bags. *Am J Health Syst Pharm* 69:1509–1512.

Kaur, M., Kaur, H. 2013. Overview on stability studies. *Int J Pharm Chem Biol Sci* 3(4):1231–1241.

Kommanaboyina, B., Rhodes, C.T. 1999. Trends in stability testing, with emphasis on stability during distribution and storage. *Drug Dev Ind Pharm* 25:857–867.

Kopp, S. 2010. Update on WHO stability guidelines. In: *Stability Testing to Support Global Markets*. Edited by: Huynh-Ba, K. Springer, New York: 23–28.

Lionberger, A.R., Lee, L.S., Lee, L. et al. 2008. Quality by design: Concepts for ANDAs. *The AAPS Journal* 10:2.

Matthews, R.B. 1999. Regulatory aspects of stability testing in Europe. *Drug Dev Ind Pharm* 25:831–856.

Mischler, P.G. 2002. Developing stability protocols for global product registrations – An update. Presentation at *International Seminar on Stability Testing: Design and Interpretation for International Registration*, IBC Life Sciences, London.

Nagaraja, Y.S., Nagaraj, T.S., Bharati, D.R. et al. 2012. Formulation and stability studies of azithromycin parenteral dosage form. *Int J Pharm Life Sci* 2(4):270.

Peterfreund, R.A., Philip, J.H. 2013. Critical parameters in drug delivery by intravenous infusion. *Expert Opin Drug Deliv* 10:1095–1108.

Pimple, S., Maurya, P., Salunke, K. et al. 2015. Formulation development and compatibility study of dexketoprofen injection used in the management of post-operative pain. *Int J Pharm Sci Rev Res* 30(1):299–305.

Ravi, B., Nagaraja, Y.S., Mahantesha, M.K. 2012. Formulation and evaluation of parenteral dosage form of anti-HIV drug using hydrotrops. *IJIPLS* 2(5):81–93.

Sánchez-Rivera, J.P., García-Gómez, A., Mínguez, M.A. et al. 2015. Stability of a parenteral formulation of betamethasone and levobupivacaine. *Journal of Pharmacy Technology* 31(2):58–63.

Savla, R., Browne, J., Plassat, V. et al. 2017 Review and analysis of FDA approved drugs using lipid-based formulations. *Drug Dev Indus Pharmacy* 43(11):1743–175.

Sharma, N., Bansal, V. 2016. Stability and compatibility study of parenteral diazepam in different storage conditions. *J Chem Pharm Res* 8:164–70.

Singh, S. 2000. Stability testing during product development. In: *Pharmaceutical Product Development*. Edited by: Jain, N. K. CBS Publisher and Distributors, New Delhi: 272–293.

Singh, S., Bakshi, M. 2000. Guidance on conduct of stress test to determine inherent stability of drugs. *Pharm Technol* 24:1–14.

Singh, S., Bakshi, M. 2002. Development of stability-indicating assay methods – A critical review. *J Pharm Biomed Anal* 28:1011–1040.

Singh, S., Bhutani, H., Mariappan, T.T., et al. 2002. Behaviour of uptake of moisture by drugs and excipients under accelerated conditions of temperature and humidity in the absence and the presence of light. 1. Pure anti-tuberculosis drugs and their combinations. *Int J Pharm* 245:37–44.

WHO. 1996. WHO Technical Report Series, No, 863, Annex 5 Guidelines for stability testing of pharmaceutical products containing well established drug substances in conventional dosage forms. Available at: www.who.int/medicines/areas/quality_safety/quality_assurance/regulatory_standards/en.

WHO. 2003. WHO Technical Report Series, No. 908. Item 11.1 WHO Guidelines for stability testing of pharmaceutical products containing well established drug substances in conventional dosage forms.

WHO. 2004. Stability studies in a global environment. Geneva meeting working document QAS/05.146 with comments.

WHO. 2006. WHO Technical Report Series, No.937. Item 10.1 stability testing conditions.

Zahn, M. 2008. Global stability practices. In: *Handbook of Stability Testing in Pharmaceutical Development*. Edited by: Huynh-Ba, K. Springer, New York: 43–91.

Zahn, M. 2009. WHO stability guideline: Remaining issues. Presentation at *AAPS Stability Workshop*, National Harbor, MD.

Zhang, Y., Vermeulen, L.C., Kolesar, J.M. 2013. Stability of stock and diluted rituximab. *Am J Health Syst Pharm* 70:436–438.

15 Reverse Engineering in Pharmaceutical Product Development

Rishi Paliwal, Aanjaneya Mamgain, and Rameshroo Kenwat
Indira Gandhi National Tribal University

Shivani Rai Paliwal
Guru Ghasidas Vishwavidyalaya (A Central University)

CONTENTS

Abbreviations 235
15.1 Introduction 235
15.2 Pharmaceutical Reverse Engineering in Formulation Development 236
    15.2.1 Why to Reverse the Pharmaceutical Product Development Process 236
    15.2.2 How to Reverse the Pharmaceutical Product Development Process 236
    15.2.3 Common Analytical Methods for Reverse Pharmaceutical Engineering 237
15.3 Reverse Pharmaceutical Engineering for Generic Product Development 238
    15.3.1 Reverse Engineering of a Tablet 239
    15.3.2 Reverse Engineering of a Capsule 239
    15.3.3 Reverse Engineering of Lipid Emulsion 239
    15.3.4 Reverse Engineering of Ocular Product 240
    15.3.5 Reverse Engineering of a Depot Microsphere Formulation 240
15.4 Reverse Pharmaceutical Engineering of Herbal Product 240
15.5 Reverse Engineering for Stability Testing of Newer Drugs 240
15.6 Conclusion and Future Prospects 240
Acknowledgement 241
Conflict of Interest 241
References 241

ABBREVIATIONS

AAS: Atomic absorption spectroscopy
ANDA: Abbreviated new drug application
API: Active pharmaceutical ingredient
FTIR: Fourier transform infrared spectroscopy
GPC: Gel permeation chromatography
HPLC: High-performance liquid chromatography
NMR: Nuclear magnetic resonance spectroscopy
Q1: Qualitative
Q2: Quantitative
RLD: Reference listed drug
TEM/SEM: Transmission electron microscopy/scanning electron microscopy
TGA: Thermogravimetric analysis

15.1 INTRODUCTION

Reverse engineering process or also called deformulation is quite common when one has to investigate the composition of an unknown product or a competitive product. In the pharmaceutical industry, it is helpful in the examination of various steps of product development and unit operations of scale-up and in the identification of unknown or hidden properties of excipients used like impurities or functional groups, stability and degradation product and other related substances (Bansal and Koradia 2005, Bhatti, Syed, and John 2018). Reverse pharmaceutical engineering further helps in understanding the features of polymers like molecular weight distributions, degree of substitution, monomer ratio and substituent distribution (Bhatti, Syed, and John 2018).

To develop a generic product, the pharmaceutical deformulation process starts even before the expiry of the patent. The reverse engineering process involves an exhaustive analysis to identify, quantify and characterise active pharmaceutical ingredient (API) and excipients of the original product (Oliveira et al. 2015). Since it is a time-consuming process, formulation scientist or reverse engineer has to start long before the patent expiry so that generic product development company is ready for submitting the details of their abbreviated new drug application (ANDA) product to the United States Food and Drug Administration (USFDA) as quickly as possible (Prašnikar and Škerlj 2006). As experienced by many formulators, reverse engineering is a

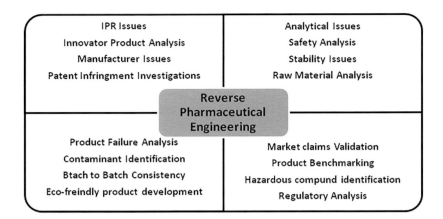

FIGURE 15.1 Some of the important applications of reverse pharmaceutical engineering during various stages of a product development.

time-taking process and it depends upon the complexity of the formulation.

The entire process is done to launch a bioequivalent generic product of a successful original product in the market at the earliest. A generic product must have essential properties of the original product like same ingredients (Q1; qualitative), same quantities (Q2; quantitative) and same physical and chemical properties (Q3) of the reference listed drug (RLD) (Bansal and Koradia 2005; Zhou et al. 2018). From reverse engineering, a lot of critical and necessary information about all three Q1, Q2 and Q3 is a prerequisite to develop a successful generic product. This chapter describes the essential process of pharmaceutical reverse engineering, need for deformulation, analytical approaches used for identification of components and by-products and case studies of some products that are reported in the literature.

15.2 PHARMACEUTICAL REVERSE ENGINEERING IN FORMULATION DEVELOPMENT

Reverse engineering is usually known as the reversal of the engineering steps to replicate a system and its subsystems or subassembly having no prior exact knowledge of its components, manufacturing process and documentation (Bhatti, Syed, and John 2018). The following information is extracted about a marketed product through reverse engineering: the first is the identification of key features of a pharmaceutical product, its components and their interdependence, the second is the identification of polymorphs, by-products and impurities, and the third is to understand the identical pharmaceutical features of the product like drug content, drug-to-excipient ratio and drug release profile (Ramaswamy and Thimmaraju 2015).

15.2.1 Why to Reverse the Pharmaceutical Product Development Process

Even after the patent expiry, the original patent holder keeps the master formula or the process as secret as possible in order to avoid any competition in the near future. Regulatory agencies maintain the innovative formula as proprietary and usually do not disclose in the public domain or provide it to the generic company at any point of time. Therefore, the generic product development process starts quite long before the expiry of the patent. To obtain approval for ANDA, the bioequivalence of the generic product with the marketed product has to demonstrate before regulatory agencies. Generic product has to produce the equivalent blood levels as that of innovator product (Ramaswamy and Thimmaraju 2015). This is only possible when a generic product developing company has the idea about the components of the original product and process estimated by their own. To obtain the diverse information about the product, reverse engineering is applied to the innovator pharmaceutical product to manufacture the bioequivalent generic product. Sometimes, the reverse engineering is applied to understand the failure of a pharmaceutical product as well and, apart from this, to evaluate the raw material composition, understanding and resolving manufacturing issues, testing of batch-to-batch consistency, hazardous compound identification, and safety and stability issues of the company's own product as well (Cotte et al. 2012; Chen et al. 2018; Berkowitz et al. 2012; Oliveira et al. 2015). Figure15.1 summarises the applications of reverse engineering in the product development process.

15.2.2 How to Reverse the Pharmaceutical Product Development Process

Reverse engineering process may be initiated with the information available in the public domain about the drug product composition, followed by application of formulation science knowledge to develop a quantitative formula of the composition. If this information is not completely available, in that case, the process will be composed of extraction and identification of major and minor components of the product. For example, in the case of a tablet, one has to determine the active pharmaceutical ingredient along with excipients like diluents, disintegrating agent, binders, colouring agent, film-forming polymer (in case of film-coated), lubricants, antioxidants and any other similar components (Koradia, Chawla, and Bansal 2005). The process involves qualitative identification, followed by the quantitative estimation of each ingredient. In qualitative analysis, we first verify the label claim of the product under investigation. Afterwards, quantitative estimation is initiated to ensure that ANDA is developed with bioequivalent features. Understanding of formulation science, nature of API (chemical structure and stability), excipient performance (function, limitations and application), unit operations involved in manufacturing and interaction of all above factors can transform the qualitative information into a closely related quantitative information that is required for developing a generic pharmaceutical product (Koradia, Chawla, and Bansal 2005; Bansal and Koradia 2005; Zhou et al. 2018). Once the formula is decoded, each component is extracted to identify its physicochemical properties including polymorphic forms. The generic formula, that is being proposed, should have similar polymorphic forms of RLD in order to achieve equivalent dissolution profile and stability of the product. Subsequently, the manufacturing process is decided on the basis of properties of API and ingredients before applying for ANDA (Krishnaiah et al. 2014; Cotte et al. 2012). Figure 15.2 shows a schematic diagram for the identification of ingredients and estimations in reverse engineering of a polymer-based pharmaceutical product.

15.2.3 COMMON ANALYTICAL METHODS FOR REVERSE PHARMACEUTICAL ENGINEERING

Reverse pharmaceutical engineering process involves many analytical techniques at each step of deformulation of a pharmaceutical product (Figure 15.3). A robust method can only give the precise and accurate idea about both qualitative and quantitative information about the key components of the product and the process of manufacturing. High-pressure liquid chromatography (HPLC) and LC-MS or GC-MS are applied most commonly to separate and

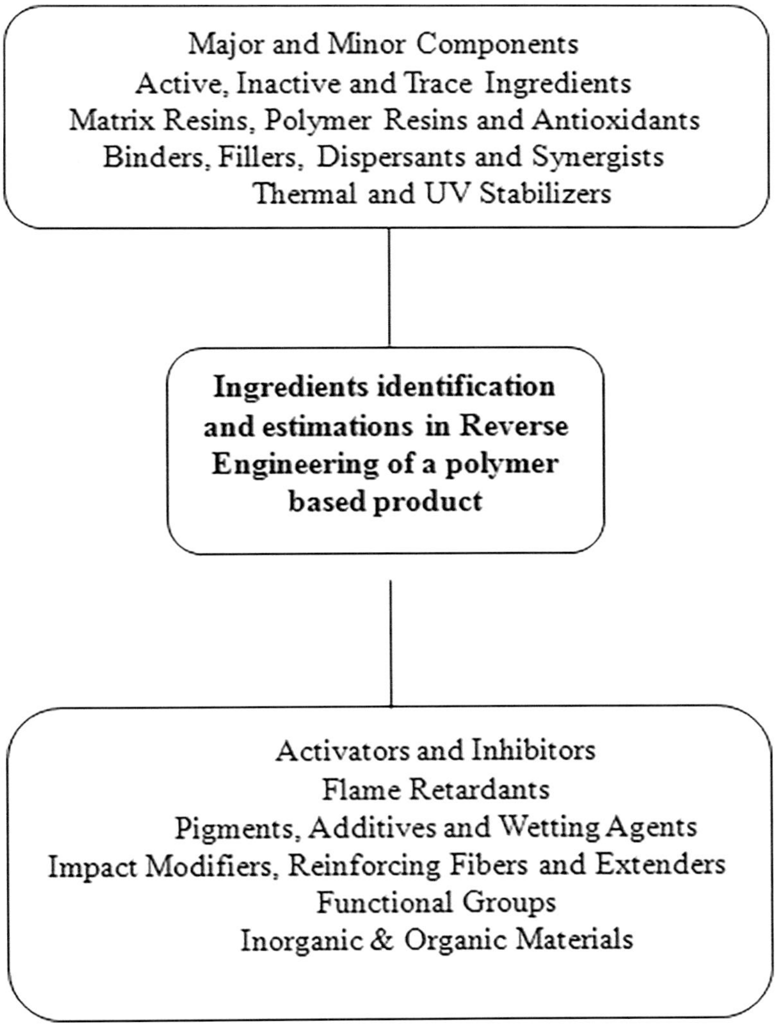

FIGURE 15.2 Common polymer-based product ingredients for identification and estimations in reverse engineering.

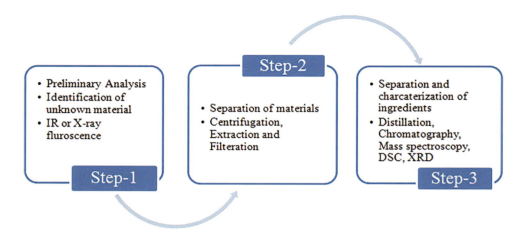

FIGURE 15.3 Key steps involved in the process of reverse pharmaceutical engineering.

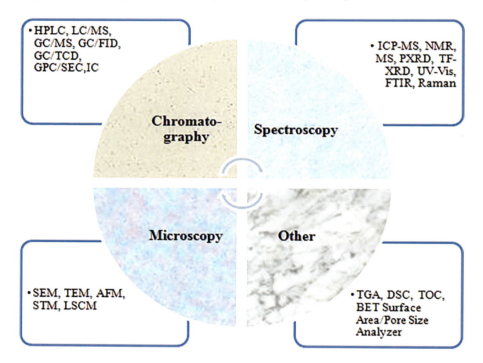

FIGURE 15.4 Commonly used instruments/techniques during the deformulation analysis of a pharmaceutical product.

identify the composition of the product under investigation (Duraipandi, Selvakumar, and Er 2015; Berkowitz et al. 2012; Ahmed, Khan, and Shaikh 2018; Clément et al. 2019). For less volatile compound, LC-MS is used, while for more volatile compounds, GC-MS is used. These techniques provide information about the mass of the components in the sample and hence are helpful in the determination of key ingredients. Other common instruments or techniques that are used during the deformulation include Fourier transform infrared spectroscopy (FTIR), nuclear magnetic resonance spectroscopy (NMR), thermogravimetric analysis (TGA), transmission electron microscopy/scanning electron microscopy (TEM/SEM), gel permeation chromatography (GPC), atomic absorption spectroscopy (AAS), UV–visible spectroscopy, X-ray diffraction, differential scanning calorimetry, distillation, Soxhlet extraction and rotary evaporation (Figure 15.4).

15.3 REVERSE PHARMACEUTICAL ENGINEERING FOR GENERIC PRODUCT DEVELOPMENT

As described earlier, the major objective of the reverse pharmaceutical engineering is the development of a generic product matching with the Q1, Q2 and Q3 guidelines. A generic product should be identical with RLD product not only in dosage form, strength or route of administration but also in quality, performance, characteristics and indented use. Broadly, we can classify the entire process into four major parts. The first part is the procurement of the innovator products and their physicochemical characterisation. The second part is the separation and decoding of the quantitative formula. The third part includes solid-state characterisation of API and excipients, and the last part is the identification of the manufacturing process. Two indicators (f1 and f2) have been proposed to compare the dissolution profile of two formulations (i.e. test and reference) (Gohel et al. 2009). USFDA has approved the concept of similarity factor (f2) and therefore is widely accepted in the formulation development and dossier preparation.

To be considered a generic product identical to innovator product, the similarity factor in in vitro dissolution profiles of both the products should be 100, while similarity factor (f2) of more than 50 indicates that the generic dosage form has a comparable dissolution profile to innovator product and may achieve bioequivalence status. A large number of dosage forms undergo the process of reverse engineering such as tablets, capsules, gels, lotions, creams, ointments, suspension, emulsion, shampoos, novel drug delivery systems like microspheres or liposomes (Cotte et al. 2012; Berkowitz et al. 2012; Clément et al. 2019; Krishnaiah et al. 2014; Matsui et al. 2020; Needham 2013; Ahmed, Khan, and Shaikh 2018; Duraipandi, Selvakumar, and Er 2015; Duraipandi and Selvakumar 2019; Zhou et al. 2018). These products may belong from any of the categories like personal care, cosmetics, over-the-counter (OTC) products, prescription products, biopharmaceutical or biosimilars, medical devices and packaging materials like polymers, films or plastics. One important aspect here is that being 'first to file' in the fundamental principle in generic product development, and therefore, there is huge competition for the same. It requires highly skilled personnel and precise disciplined teamwork during the whole product development process to achieve the bioequivalence. During the deformulation of RLD, especially quantitatively, one has to identify the important excipients that affect the overall performance of the product. Two important parameters are important to check: the first is the dissolution profile of the product, and the second is the stability of the product. Therefore, exact information about excipients like pH modifiers, buffers, antioxidants, chelating agents and surfactant should be obtained by best utilisation of resources like time and money.

15.3.1 Reverse Engineering of a Tablet

In case the RLD product is a tablet, qualitative identification of the tablet components is the first step. This can be performed using physical test (colour, odour and taste) and chemical tests or techniques like FTIR, NMR, DSC, XRD and UV–visible spectroscopy (Bansal and Koradia 2005). The physical properties may be checked with TEM, and SEM is required. The quantification of the tablet matrix is the next step. This may be a challenging task as several components are involved in the matrix (API, excipients and their possible interactions) and the product development process. To separate and quantify the excipients, suitable techniques like differential solubility, gradual filtration (using different cut-off molecular weights, and pore sizes), and chromatography techniques like high-performance thin-layer chromatography, (HPTLC), HPLC or size-exclusion chromatography (for high-molecular-weight substances like polymers) may be employed based on the information available about physicochemical characteristics of these excipients. Solid-state properties of API are important as it may have polymorphism and hence show different crystalline forms, solvates or hydrates or amorphous forms (Law and Zhou 2017; Gokhale and Mantri 2017). All these forms of an API are different in both the intermolecular arrangements and free energy and therefore may have variations in solubility, dissolution pattern, bioavailability, scalability and stability. The techniques like DSC, XRD and TGA are employed to identify the form of API in the RLD.

15.3.2 Reverse Engineering of a Capsule

An investigation of reverse engineering of a capsule formulation of Orlistat, a weight-loss drug, which is off patented in 2009, is reported (Zaheer et al. 2016). Innovator product is having wide market with annual sales of around $100 million, and therefore, the generic product has high commercial potential. The authors decoded the marketed formulation using differential solubility techniques with methanol, methylene chloride and water as solvent and antisolvent. Excipients such as polyvinyl chloride, microcrystalline cellulose, sodium lauryl sulphate, sodium starch glycolate and talc were identified using FTIR and quantified using various analytical techniques. The separated API, Orlistat, was subjected to DSC and PXRD analysis. Based on the formula decoded, the authors prepared a generic form and characterised it. The developed generic product showed a similar dissolution profile to the marketed product with a first-order release. A similarity factor f2 was noted as 70.13, indicating that the so-developed generic product is similar to innovator product and may be explored for bioequivalence testing for its approval by the regulatory bodies.

15.3.3 Reverse Engineering of Lipid Emulsion

An Ayurvedic lipid-based formulation like Ghritas, which is composed of oil or *ghee* along with polyherbal decoction (*kasaya*) and fine paste of herbs (*kalka*), is prepared usually by the evaporation of the aqueous phase and transfer of content in the oil. In order to understand the composition and nature of the content present, a reverse engineering process was reported (Duraipandi, Selvakumar, and Er 2015). The authors used column chromatography to fractionate the Ghrita into polar and non-polar fractions using silica gel as adsorbent and petroleum ether and mixture of ethanol, methanol and water as eluents. Afterwards, these fractions were analysed using normal and reverse phase HPTLC for the presence of contents and their polarity.

15.3.4 Reverse Engineering of Ocular Product

Ahmad and co-workers (2018) reported reverse engineering of marketed ophthalmic formulation of an antihistaminic mast cell stabiliser drug, olopatadine hydrochloride (OLH) (Ahmed, Khan, and Shaikh 2018). The authors used the quality by design (QbD) process for the separation of the components. The reverse engineering was applied using differential solubility techniques. For QbD, the dependent variable was the extraction efficiency of the solvent and independent variables were solvent volume and sonication time. A 3^2 factorial design was applied, and different batches were produced, followed by statistical analysis. Different analytical techniques like FTIR, UV spectroscopy and PXRD were employed in the analysis. Using the quantities obtained after the separation of different components, generic ophthalmic formulation was developed. The author claimed that the so-developed generic formulation was equivalent in strength purity and stability as tested on the basis of parameters like drug content, pH, clarity, osmolarity and viscosity and advocated that QbD was successfully employed in reverse engineering process.

15.3.5 Reverse Engineering of a Depot Microsphere Formulation

Lupron Depot® is a 1-month depot preparation based on PLGA microspheres encapsulating leuprolide, an anticancer hormone drug molecule. The patent has been expired, and no generic formulation has been approved by FDA as described by Zhou et al. (2018) (Ahmed, Khan, and Shaikh 2018). This may be due to the complexity of components and manufacturing process involved. The authors applied reverse engineering to this product in order to identify the composition and other relevant product attributes like drug content (three methods), gelatine content, type and molecular weight distribution, PLGA content, lactic acid-to-glycolic acid ratio and molecular weight distribution, mannitol content, *in vitro* drug release, residual solvent and moisture content, particle size distribution and morphology and glass transition temperature. The authors also examined for diluents compositions like viscosity and specific gravity. Some notable outcomes of the process were matching of the content of the formulation claimed and mention on the package and with the process, gelatin used was found type B having 330 bloom strength, microsphere mean diameter size about 11 μm, zero order release followed by initial burst release of about 23%, low moisture content (less than 0.5%) and low methylene chloride (less than ppm). This information may be useful for developing generic PLGA-based microsphere formulation in the near future.

15.4 REVERSE PHARMACEUTICAL ENGINEERING OF HERBAL PRODUCT

Herbal products are effective and trusted pharmaceutical products since ancient time. However, they are complex in nature as most of them are polyherbal formulations and have several phytoconstituents that are integrated part of the herbal product. Recently, Duraipandi and Selvakumar (2019) investigated an Ayurvedic medicated oil-based formulation called *Anu Tailam* (Duraipandi and Selvakumar 2019). The term is made of words 'Anu' which means atom and 'tailam' which means oil. The formulation is said to be 'oil of subtle or atomic size particles.' This formulation is oily preparation with very little moisture and composed of herbal decoction. Pharmaceutically, this formulation shows no precipitation or phase separation, and therefore, the authors applied reverse engineering to this product to confirm that predominately water-soluble ingredients are either suspended in the form of microparticles or entrapped in the submicron vesicular structures. The authors used column chromatography and HPLC for investigation and proved that contents are polar hydrophilic compounds. Afterwards, optical microscopy, photon correlation microscopy and environmental scanning electron microscope (ESEM) were done to study the particle size and size distribution of the formulation. The authors concluded that the formulation contains only polar ingredients which can be extracted by using polar solvents like methanol and ethanol. This formulation contained nanoparticles of active botanical ingredients embedded in a network of the vesicular structure of the lipid base which are responsible for the delivery of the aqueous content across the biological membranes (Figure 15.5).

15.5 REVERSE ENGINEERING FOR STABILITY TESTING OF NEWER DRUGS

Physicochemical stability of API is critically important for a high-quality product development and manufacturing. For testing such purity, monographs are prescribed in the official pharmacopoeia including solid-state chemistry impurity profile and assay methods. In case the reference standard information is not available in the pharmacopoeia or other official documents, one can develop their own methodology for estimation of highly purified drugs and their physicochemical stability testing. Nie, Mo, and Byrn (2018) reported a simple and cost-effective reverse engineering approach for extraction and purification of darunavir ethanolate, a new drug for which official monograph was not available (Nie, Mo, and Byrn 2018). The authors chose PREZISTA® tablet for reverse engineering and crystallisation of darunavir. Using these highly purified crystals, the authors studied the potential risk of degradation and also form conversion of drug in different stress conditions and packaging specifications. The authors reported amorphisation under thermal storage due to desolvation and ethanolate to hydrate conversion in high humidity conditions.

15.6 CONCLUSION AND FUTURE PROSPECTS

Reverse engineering or deformulation has been an important tool for generic product development, stability testing and identification of the changes in the formulation during the manufacturing process and storage. The methodology involves the separation of the components and identification

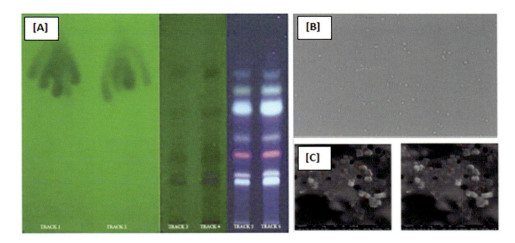

FIGURE 15.5 (a) HPTLC of sesame oil (Track 1), nonpolar fraction separated from *anu tailam* (Track 2) and polar fractions of *anu tailam* (in duplicates) visualised at 254 nm (Track 3 and Track 4) and 366 nm (Track 5 and Track 6). (b) Optical microscopic picture of anu tailam at the magnification of 100 * 10 showing absence of large particles. (c) Environmental scanning electron microscope pictures showing nanoparticles of anu tailam (ranging from 147 to 573 nm; on average of 335 nm). (Adapted with permission from Duraipandi and Selvakumar (2019).)

of critical factors involved. Reverse engineering has been applied for various dosage forms like tablet, capsules, ophthalmic, novel drug delivery system, depots and herbal lipid emulsion. Although this is one of the most applicable methodologies for generic product development, we found very little information published in the form of publications due to market competition and being first in approval and launching of a generic product once the patent expires related to innovator product. Since many newer technologies are emerging day by day, a skilled and systematic approach may be useful in the development of the generic product of complex systems like nanomedicine and nanopharmaceuticals in the near future.

ACKNOWLEDGEMENT

This project is partially supported by the Department of Biotechnology, Government of India (Grant: BT/PR26950/NNT/28/1505/2017 to R.P. as PI and JRF support to one of the authors R.K.).

CONFLICT OF INTEREST

We declare no conflict of interest.

REFERENCES

Ahmed, Zahid Zaheer, Furquan Nazimuddin Khan, and Darakhshan Afreen Shaikh. 2018. "Reverse Engineering and Formulation by QBD of Olopatadine Hydrochloride Ophthalmic Solution." *Journal of Pharmaceutical Investigation* 48 (3). Springer Netherlands: 279–93. doi:10.1007/s40005-017-0312-1.

Bansal, Arvind K., and Vishal Koradia. 2005. "The Role of Reverse Engineering in the Development of Generic Formulations." *Pharmaceutical Technology* 29 (8): 50–55.

Berkowitz, Steven A., John R. Engen, Jeffrey R. Mazzeo, and Graham B. Jones. 2012. "Analytical Tools for Characterizing Biopharmaceuticals and the Implications for Biosimilars." *Nature Reviews Drug Discovery*. Nature Publishing Group. doi:10.1038/nrd3746.

Bhatti, Attya, Nida A. Syed, and Peter John. 2018. "Reverse Engineering and Its Applications." In *Omics Technologies and Bio-Engineering: Towards Improving Quality of Life* 1: 95–110. Elsevier Inc. doi:10.1016/B978-0-12-804659-3.00005-1.

Chen, Kang, Junyong Park, Feng Li, Sharadrao M. Patil, and David A. Keire. 2018. "Chemometric Methods to Quantify 1D and 2D NMR Spectral Differences Among Similar Protein Therapeutics." *AAPS PharmSciTech* 19 (3). Springer New York LLC: 1011–19. doi:10.1208/s12249-017-0911-1.

Clément, Yohann, Alexandra Gaubert, Anne Bonhommé, Pedro Marote, Ashley Mungroo, Maxime Paillard, Pierre Lantéri, and Christophe Morell. 2019. "Raman Spectroscopy Combined with Advanced Chemometric Methods: A New Approach for Detergent Deformulation." *Talanta* 195 (April). Elsevier B.V.: 441–46. doi:10.1016/j.talanta.2018.11.064.

Cotte, Jean François, Sylvain Sonnery, Fabien Martial, Jean Dubayle, François Dalençon, Jean Haensler, and Olivier Adam. 2012. "Characterization of Surfactants in an Oil-in-Water Emulsion-Based Vaccine Adjuvant Using MS and HPLC-MS: Structural Analysis and Quantification." *International Journal of Pharmaceutics* 436 (1–2). Elsevier: 233–39. doi:10.1016/j.ijpharm.2012.06.018.

Duraipandi, Selvakumar, and Vijaya Selvakumar. 2019. "Reinventing Nano Drug Delivery Systems for Hydrophilic Active Ingredients in Ayurvedic Lipid Based Formulations Containing Poly Herbal Decoction." *Journal of Ayurveda and Integrative Medicine*. doi:10.1016/j.jaim.2018.01.008.

Duraipandi, Selvakumar, Vijaya Selvakumar, and Ng Yun Er. 2015. "Reverse Engineering of Ayurvedic Lipid Based Formulation, Ghrita by Combined Column Chromatography, Normal and Reverse Phase HPTLC Analysis." *BMC Complementary and Alternative Medicine* 15 (1): 1–6. doi:10.1186/s12906-015-0568-9.

Gohel, M. C., K. G. Sarvaiya, A. R. Shah, and B. K. Brahmbhatt. 2009. "Mathematical Approach for the Assessment of Similarity Factor Using a New Scheme for Calculating Weight." *Indian Journal of Pharmaceutical Sciences* 71 (2). Wolters Kluwer – Medknow Publications: 142–44. doi:10.4103/0250–474X.54281.

Gokhale, Madhushree Y., and Rao V. Mantri. 2017. "API Solid-Form Screening and Selection." In *Developing Solid Oral Dosage Forms: Pharmaceutical Theory and Practice*. Second Edition: 85–112. Elsevier Inc. doi:10.1016/B978-0-12-802447-8.00004-2.

Koradia, Vishal S., Garima Chawla, and Arvind K. Bansal. 2005. "Comprehensive Characterisation of the Innovator Product: Targeting Bioequivalent Generics." *Journal of Generic Medicines* 2 (4): 335–46. doi:10.1057/palgrave.jgm.4940086.

Krishnaiah, Yellela S. R., Xiaoming Xu, Ziyaur Rahman, Yang Yang, Usha Katragadda, Robert Lionberger, John R. Peters, Kathleen Uhl, and Mansoor A. Khan. 2014. "Development of Performance Matrix for Generic Product Equivalence of Acyclovir Topical Creams." *International Journal of Pharmaceutics* 475 (1–2). Elsevier B.V.: 110–22. doi:10.1016/j.ijpharm.2014.07.034.

Law, Deliang and Deliang Zhou. 2017. "Solid-State Characterization and Techniques." In *Developing Solid Oral Dosage Forms: Pharmaceutical Theory and Practice*. Second Edition: 59–84. Elsevier Inc. doi:10.1016/B978-0-12-802447-8.00003-0.

Matsui, Kazuki, Susumu Takeuchi, Yuka Haruna, Miki Yamane, Takahiro Shimizu, Yoshiki Hatsuma, Norihito Shimono, et al. 2020. "Transverse Comparison of Mannitol Content in Marketed Drug Products: Implication for No-Effect Dose of Sugar Alcohols on Oral Drug Absorption." *Journal of Drug Delivery Science and Technology* 57 (June). Editions de Sante: 101728. doi:10.1016/j.jddst.2020.101728.

Needham, D. 2013. "Reverse Engineering of the Low Temperature-Sensitive Liposome (LTSL) for Treating Cancer." *Biomaterials for Cancer Therapeutics: Diagnosis, Prevention and Therapy*. Woodhead Publishing Limited. doi:10.1533/9780857096760.3.270.

Nie, Haichen, Huaping Mo, and Stephen R. Byrn. 2018. "Investigating the Physicochemical Stability of Highly Purified Darunavir Ethanolate Extracted from PREZISTA® Tablets." *AAPS PharmSciTech* 19 (5). Springer New York LLC: 2407–17. doi:10.1208/s12249-018-1036-x.

Oliveira, J. M., V. M. Balcão, M. M.D.C. Vila, N. Aranha, V. M.H. Yoshida, M. V. Chaud, and S. Mangine Filho. 2015. "Deformulation of a Solid Pharmaceutical Form Using Computed Tomography and X-Ray Fluorescence." *Journal of Physics: Conference Series* 630. Institute of Physics Publishing: 12002. doi:10.1088/1742-6596/630/1/012002.

Prašnikar, Janez, and Tina Škerlj. 2006. "New Product Development Process and Time-to-Market in the Generic Pharmaceutical Industry." *Industrial Marketing Management* 35 (6). Elsevier: 690–702. doi:10.1016/j.indmarman.2005.06.001.

Ramaswamy, Rakshambikai, and Phani Kishore Thimmaraju. 2015. "Technical Hindrances in Establishing Biosimilarity – The Final Lap in the Race." *International Journal of Innovation and Applied Studies* 11. http://search.proquest.com/openview/bac2f6e216420aafe459d2b4d23dc886/1?pq-origsite=gscholar&cbl=2031961, http://www.ijias.issr-journals.org/.

Zaheer, Zahid, Furquan Nazimuddin Khan, Sarfaraz Khan, Moizul Hasan, Obaid Shaikh, and Raheel Khan. 2016. "Reverse Engineering and Development of Generic Orlistat Formulation." *JIPBS* 3(4): 17–25. http://www.jipbs.com/VolumeArticles/FullTextPDF/245_JIPBSV3I404_NEW.pdf

Zhou, Jia, Keiji Hirota, Rose Ackermann, Jennifer Walker, Yan Wang, Stephanie Choi, Anna Schwendeman, and Steven P. Schwendeman. 2018. "Reverse Engineering the 1-Month Lupron Depot®." *AAPS Journal* 20 (6): 1–13. doi:10.1208/s12248-018-0253-2.

16 Role of Polymers in Formulation Design and Drug Delivery

Satish Shilpi, Umesh Dhakad, Rajkumari Lodhi, Sonal Dixit, and Kapil Khatri
Ravishankar College of Pharmacy, Bhopal, India

Neelesh Kumar Mehra
National Institute of Pharmaceutical Education & Research (NIPER), Hyderabad, India

Arvind Gulbake
DIT University, Dehradun, India
D.Y. Patil Education Society, India

CONTENTS

16.1 Introduction 243
16.2 Classification of Polymers 244
    16.2.1 Polymers Based on Origin 244
    16.2.2 Based on Degradation 245
16.3 Polymers in Pharmaceutical Drug Delivery System 245
16.4 Responsive Polymer for Drug Delivery 245
    16.4.1 Mucoadhesive 246
    16.4.2 Thiolated Polymers 246
    16.4.3 pH-Sensitive Polymers 246
    16.4.4 Temperature-Sensitive 247
    16.4.5 Temperature-Responsive Polymers 247
    16.4.6 Ionic-Sensitive Polymers 247
    16.4.7 Biodegradable 247
    16.4.8 Viscosity Enhancer 248
    16.4.9 Highly Water-Soluble Polymer 249
16.5 Role of Polymer in Drug Delivery System 249
    16.5.1 Modified Drug-Release Dosage Forms 249
    16.5.2 Extended-Release Dosage Forms 250
    16.5.3 Gastro-retentive Dosage Forms 250
16.6 Types of Polymers in Pharmaceutical Drug Delivery 250
    16.6.1 Polymers in Colon-Targeted Drug Delivery 250
    16.6.2 Polymers in the Mucoadhesive Drug Delivery System 250
    16.6.3 Polymers for Sustained Release 250
    16.6.4 Polymers in Tissue Engineering 251
16.7 In Situ Drug Delivery System 252
16.8 Polymer–Drug Conjugate 252
Acknowledgement 252
References 253

16.1 INTRODUCTION

The word polymer is gotten from the traditional Greek words poly signifying 'many' and mers signifying 'parts.' Simply expressed, a polymer is a long-chain particle that is made out of an enormous number of rehashing units of indistinguishable structure. Certain polymers, for example, proteins, cellulose and silk, are found in nature, while numerous others, including polystyrene, polyethylene and nylon, are created in the chemical synthesis process (Sowjanya et al., 2017). A significant model is regular elastic, known as polyisoprene in its engineered structure (Lakshmi and Cato, 2006). Polymers that are able to do high expansion under surrounding conditions find significant applications as elastomers. Notwithstanding characteristic rubber, there are a few significant manufactured elastomers including nitrile and butyl rubber. Different polymers may have attributes that empower their manufacture into long filaments appropriate for their applications in different fields. The strand of our DNA is a biopolymer and all the synthetic and natural polymers are utilised all through the world as plastic. Polymers might be normally found in plants, animals and micro-organisms or might be an engineered polymer. Various polymers have various kinds of physicochemical properties (Raizada et al., 2010).

Because of their unique properties, polymers are used in pharmaceuticals. The new technology in the polymer-based drug-release system offers possibilities in the administration of drugs. Pharmaceutically, these polymers are used as a binder in tablets, flow-controlling agents in liquids, suspensions and emulsions, as film-coating agents to mask the unpleasant taste of the drug, and protective and stabilising agents. Polymers have been used as an important tool to control the drug-release rate from the formulation (Jones, 2004), and they are also mostly used as stabiliser, taste-masking agent and proactive agent. Modern advances in drug delivery are now predicated upon the rational design of polymers tailored to specific cargo and engineered to exert distinct biological functions (Finkenstadt, 2005).

Over the past decades, research at the level of molecular biology has unveiled the molecular basis for many diseases. New significant innovations and ideas, for example, recombinant DNA and gene therapy, lead to the production of drugs and techniques intended to the treatment of several diseases (Kawai, 2010). The progress towards the use of these therapeutic molecules outside the laboratory has been extensively measured because of the absence of viable drug delivery frameworks, that is, components that permit the release of the drug into the appropriate body compartment for the appropriate amount of time without disturbing the remainder of the body functionality. The utilisation of the polymeric materials for clinical intentions is developing fast. Polymers have found applications in diverse biomedical fields such as drug delivering systems, developing scaffolds in tissue engineering, implantation of medical devices and artificial organs, prosthesis, ophthalmology, dentistry bone repair, and numerous other clinical fields. Polymers have been utilised as primary substances to control the drug-release rate from the formulations (Duncan, 2003). Broad uses of polymers in drug delivery have been acknowledged in the light of the fact that polymers offer extraordinary properties which have not been achieved by some other materials. Advances in polymer science have prompted the improvement of a few novel drug delivery systems. An appropriate thought of surface and mass properties can help in the planning of polymers for different drug delivery applications (Harekrishna, 2015). This innovative improvement can be utilised to modify drug molecules, formulate drug delivery system and encapsulate the drug in polymeric networks that can deliver site specifically in the body that improves human well-being. Researchers and scientists are continuously occupied to develop such type of drug delivery system. Biodegradable polymers have been generally utilised in biomedical applications in view of their known biocompatibility and biodegradability. In the biomedical field, polymers are generally utilised as inserts and are relied upon to perform long haul administration. These enhancements add to make clinical treatment more productive and to minimise the adverse effect and some other complications related to patients. Polymers can be utilised as film coatings to drug particles, tablet and capsules to mask the undesirable taste of a drug, to improve drug stability and to alter drug-release kinetics or provide controlled and sustained release from the conventional and novel drug delivery systems. The polymers also give a role to improve the bioavailability of the drug (Arif et al., 2019).

Polymeric drug delivery carrier is a system in which a drug is spread inside a polymer matrix and released by diffusion. The rate of the drug release from a polymer matrix relies upon the initial drug concentration and polymer chains that are liable for a sustained release of drugs (Van Renterghem et al., 2018). The copolymers have a wide range of conceivable applications in the field of drug delivery. It shows the high drug entanglement effectiveness for both hydrophilic and lipophilic drug, sustained and controlled drug delivery, provides the protection of therapeutic molecules from the biological environment, and can modify the ADME of the drug (Cilurzo et al., 2014).

16.2 CLASSIFICATION OF POLYMERS

The polymers for the drug delivery system are classified on the following characteristics:

- **Origin**: The polymers can be natural or synthetic, or a combination of both.
- **Chemical nature**: It can be protein-based, polyester, cellulose derivatives, etc.
- **Backbone stability**: The polymers can be degradable or non-biodegradable.
- **Solubility**: The polymer can be hydrophilic or hydrophobic in nature.

The polymers are classified on the basis of different categories.

16.2.1 Polymers Based on Origin

Natural polymers: Natural polymers occur in nature and are also known as biopolymers.

Examples are chitosan, pectin, alginate, gelatin, albumin, collagen, cyclodextrin and natural rubber (Finkenstadt, 2005).

Semi-synthetic polymers: Hydroxypropyl cellulose, methylcellulose, hydroxypropyl methylcellulose, hydroxyethyl cellulose and sodium carboxymethyl cellulose (CMC) are chemically modified natural polymers (Sithole et al., 2016).

Synthetic polymers: Synthetic polymers are synthesised in laboratory. These are man-made.

Examples are polyethylene, polylactic acid, polypropylene, polyglycolic acid, polyhydroxybutyrate, polyanhydride and polyacrylamide (Bialik et al., 2019).

16.2.2 Based on Degradation

Biodegradable macromolecules are certainly more favoured from the toxicological perspective. In the innovation of delayed drug delivery, polymers and their altered subsidiary just as engineered polymers are utilised, e.g. polyacrylamides, polyacrylates and polyethylene glycol. A suitable choice of the polymer is necessary to design a drug delivery system and for a safe and effective delivery of drug. Degradable polymers are favoured for drug delivery applications. The most usually utilised polymers for this application are polylactide (PLA) and poly(lactide-co-glycolide) (PLGA). These polymers have been utilised in biomedical applications for over 20 years and are known to be biodegradable, biocompatible and non-poisonous (Delplace and Nicolas, 2015).

16.3 POLYMERS IN PHARMACEUTICAL DRUG DELIVERY SYSTEM

Rosin: Rosin a film-forming biopolymer, and its subsidiaries have been widely assessed chemically as film-covering and microencapsulating materials to accomplish continued medication discharge. They are likewise utilised in beauty care products, chewing gums and dental stains. Rosin has been utilised to create spherical microcapsules by a strategy dependent on stage partition by phase-separation-solvent-evaporation method. Rosin blend with polyvinyl pyrrolidone and dibutyl phthalate produces smooth film with improved lengthening and rigidity (Singh et al., 2018).

Chitin and Chitosan: Chitin is normally a mucopolysaccharide and comprises 2-acetamido-2-deoxy-b-D-glucose. Chitin can be debased by chitinase. Chitosan is a polysaccharide made out of arbitrarily circulated β-(1-4)-linked D-glucosamine (deacetylated unit) and N-acetyl D glucosamine (acetylated unit). The main property of chitosan concerning drug conveyance is its positive charge under acidic conditions. This positive charge originates from the protonation of its free amino groups. The absence of a positive charge implies that chitosan is insoluble in water and basic solution (Muxika et al., 2017).

Zein: Zein is an alcohol-soluble protein contained in the endosperm tissue of Zeamais, is a by-product of corn processing. Zein has been utilised as a palatable covering for food and drug products to provide stability during storage. Zein is an economical and best substitute for the quick disintegration of drug-loaded products (Tran et al., 2019).

Collagen: Collagen is the most broadly discovered protein in well-evolved creatures and provides solidarity to tissue. It has been investigated for use not just in different types of medical procedure, make-up and drug delivery, but also in cell and enzyme immobilization and tissue engineering of various organs (Ricard-Blum, 2010).

Starches: It is a type of complex sugar present in green plants and particularly present in seeds and nuts. Starch is present as granules (starch grains), and their shape and size depend on the proportion of the substance of the important constituents, i.e. amylose and amylopectin. Various starches are perceived for pharmaceutical preparations. These incorporate in maize (Zea mays), rice (Oryza sativa), wheat (Triticum aestivum) and potato (Solanum tuberosum) (Hong et al., 2014).

16.4 RESPONSIVE POLYMER FOR DRUG DELIVERY

Stimuli-responsive polymers otherwise called environmentally responsive polymers, or smart polymers, are a class of materials that contain a huge assortment of linear and branched (co)polymers or cross-linked polymer organisations. A sign of responsive polymers is their capacity to go through a sensational physicochemical change. Temperature and pH changes are regularly used to trigger changes, yet other stimuli, for example, ultrasound, ionic strength, redox potential, electromagnetic radiation and chemical/biochemical substances, can be utilised (Gil and Hudson, 2004) (Table 16.1).

TABLE 16.1
Various Stimuli-Responsive Polymers

Environmental Stimulus	Responsive Polymer
Temperature-sensitive	Poloxamers, poly(N-alkylacrylamide), poly(N-vinylcaprolactam), methylcellulose, xyloglucan Chitosan, hydroxypropyl cellulose
pH-sensitive	Poly(methacrylic acid), poly(vinylpyridine), poly(vinylimidazole), carbomer, chitosan, cellulose acetate phthalate latex, sodium carboxymethyl cellulose, polylysine
Light-sensitive	Modified poly(acrylamide)s
Electric field-sensitive	Sulphonated polystyrenes, poly(thiophene), poly(ethyloxazoline)
Ultrasound-sensitive	Ethylenevinylacetate
Ionic strength-sensitive	Gellan gum, alginates, xanthan gum, carrageenan Pectin, hyaluronic acid
Temperature- and pH-sensitive	Poloxamer+chitosan
Ionic strength- and pH-sensitive	Gellan gum+Carbopol
Ionic strength- and temperature-sensitive	Gellan gum+poloxamer

Physical stimuli (i.e. temperature, ultrasound, light and magnetic and electrical fields).

Chemical stimuli (i.e. pH, redox potential, ionic strength and chemical agents (Nazila Salamat-Miller et al., 2005; Schmaljohann, 2006; Jeong et al., 2006)).

16.4.1 Mucoadhesive

Mucoadhesive polymers are water-soluble, and water-insoluble polymers are used for developing the liquid ocular delivery system; the hydrophilic polymers should be used because they can be used as viscosity-enhancing agent. Most of the polysaccharides have mucoadhesive property, and they are frequently used in the ocular formulations. Its derivatives are hyaluronic acid, methylcellulose, gellan gum, chitosan, xanthan gum, hydroxypropyl methylcellulose, carrageenan and guar gum. Chitosan is a polysaccharide polymer. Its biodegradable, low toxic and biocompatible properties make it suitable for use in drug formulations (Vilar et al., 2012).

Some other used non-ionic polymers for mucoadhesive properties are polyvinyl alcohol poloxamer, lectins and polyvinylpyrrolidone (Yousef and Sanaz, 2012). Drug delivery through the buccal route is a way to deliver medication using mucoadhesive polymers, which has been the subject of interest since the mid-1980s. Points of interest related to the buccal drug delivery make this route of administration helpful for many drug mechanisms of drug permeation, and attributes of the ideal polymers; additionally, we centre on a new generation of mucoadhesive polymers, for example, thiolated polymers and their formulation for drug delivery.

16.4.2 Thiolated Polymers

These are the special class of multifunctional polymers also called thiomers. These thiolated polymers or designated thiomers are mucoadhesive basis polymers; these are hydrophilic macromolecules exhibiting free thiol group on the polymeric backbone. Based on thiol/disulphide exchange reactions and/or a simple oxidation process, disulphide bonds are formed between such polymers and cysteine-rich subdomains of mucus glycoproteins building up the mucus gel layer. Thiomers impersonate therefore the natural mechanism of secreted mucus glycoproteins, which are also covalently moored in the mucus layer by the formation of disulphide bonds; the bridging structure most commonly comes across in biological systems, which is utilised to bind drug delivery system on the mucus membrane (Bernkop-Schnürch, 2005).

16.4.3 pH-Sensitive Polymers

pH-responsive polymers are prone to change their structural nature and physical properties such as chain configuration, solubility and surface activity. pH sensitive cationic polymers can mask the taste of drugs and release drugs in the stomach by responding to gastric pH. Anionic polymers responsive to intestinal high pH are used for preventing gastric degradation of drug, colon drug delivery and achieving high bioavailability of weakly basic drugs (Yoshida et al., 2013). Physiological pH varies systematically in the body, particularly along the GI tract, where harsh pH and enzymatic conditions in the stomach (pH ~ 2) degrade macromolecules. The small intestine is substantially more alkaline, with pH ~ 6.2–7.5. Physiological pH profiles will also change among cellular compartments. For example, endosomes typically exhibit pH values of 5.0–6.8 and lysosomes 4.5–5.5 (Mellman, 1996) (Table 16.2).

Thus, it is no wonder that the engineers and scientists have devoted considerable effort towards the more rational design of polymers capable of utilising these pH variations to selectively deliver valuable therapeutics to specific intracellular or extracellular sites of action. Such unique nature of pH-responsive polymers has made them suitable for application in drug delivery, especially for targeted and controlled delivery, and pH-responsive polymer delivery systems can be developed to give well-controlled pH response and drug release.

TABLE 16.2
Common pH-Sensitive Polymers Used in Drug Delivery

Name	Chemical Structure	Clinical Indication	References
Aminoalkyl methacrylate copolymer (Eudragit E)		Taste-masking	Sheshala et al. (2011)
Poly(methacrylic acid-co-methyl methacrylate) (Eudragit L/S)		Protection of acid-degradable drugs, colon delivery	Tirpude and Puranik (2011), Johnson et al. (2003)
Hydroxypropylmethylcellulose phthalate (HPMC-P)		Protection of acid-degradable drugs, colon delivery	Tirpude and Puranik (2011)

16.4.4 Temperature-Sensitive

Temperature-sensitive (or thermo-sensitive) hydrogels are among the most studied classes of stimuli-responsive polymers for drug delivery systems. Thermo responsive properties of the polymers is one of the most often exploited stimuli for biomedical applications (Ward and Georgiou, 2011; Hocine et al., 2013). There are two typical phase diagrams describing the state of a polymer in solution as a function of concentration and temperature (Li, 2013). The first type is characterised by the upper critical solution temperature (UCST), when the transition between the single-phase and two-phase regions occurs upon cooling. In the second type with the lower critical solution temperature (LCST), this transition occurs with increasing temperature (Talelli and Hennink, 2011) (Table 16.3).

16.4.5 Temperature-Responsive Polymers

The N-substituted acrylamide polymers are the most commonly used thermo-responsive polymers. The most studied N-substituted acrylamide is the PNIPAAm mainly because it displays a LCST value (32°C) very close to the human body temperature and may therefore be applied in the biomedical applications, e.g. stimulus-sensitive DDS. As mentioned before, the LCST value can be tuned by changing the molecular weight, end functionalities, adding hydrophilic and/or hydrophobic segments. On this matter, the CLRP methods are a powerful tool to synthesise precise NIPAAm-based macrostructures with controlled molecular weight, low polydispersity polymers, complex architectures, and having at the same time stimuli-responsive properties.

16.4.6 Ionic-Sensitive Polymers

Ion-exchange resins are regularly utilised for taste-masking, counter ion-responsive drug release and supported drug release. Ion-exchange resins are insoluble polymers containing a polystyrene spine cross-linked connected with divinylbenzene and side chains of ion-active groups. The ion-active gatherings are overwhelmingly tertiary amine substitutes, quaternary ammonium, sulphonic acid and carboxylic acid. Drug loading to the ion-exchange resins is achieved by complexation based on electrostatic interaction between the drugs and the resins. The drug–resin complexes can be carried out in oral formulations such as tablets, beads and liquid suspensions.

A polymethacrylic acid-based ion-exchange gum is utilised for taste-masking of pseudoephedrine in chewable tablet to veil the bitter taste of the therapeutic substances. The formulation consists of pseudoephedrine and the ion-exchange resin, further coated with ethyl cellulose/HPMC layer. Due to the lower concentration in saliva than in the gastric fluid, drug release from the product is minimised in saliva for taste-masking while achieving immediate drug release in simulated gastric juice for high bioavailability. A carboxylic acid-based ion-exchange resin had been used for the delivery of nicotine from extended-release formulation of chewing gum (Guo et al., 2009). An ion-responsive drug release can be accomplished using cationic polymers bearing quaternary ammonium groups. Poly(ethylacrylate-methylmethacrylate-trimethylammonioethyl methacrylate chloride) copolymers (Eudragit RS and RL) are insoluble in water but can be hydrated with and swell in water.

Due to the different strength of interaction between the buffer anions and the quaternary ammonium groups of Eudragit RS/RL, the order of drug release in various buffer media has been found as follows: acetate>formate>chloride (Bodmeier et al., 1996). To achieve a sigmoidal drug-release profile, a bead system is prepared by coating Eudragit RS onto a drug core containing drug and organic acid (Narisawa et al., 1994, 1996). Common ion-sensitive resins/polymers are used in drug delivery (Tables 16.4 and 16.5).

16.4.7 Biodegradable

The utilisation of biodegradable polymers for biomedical applications is constantly expanding and advancing. Biodegradable polymers are as of now being used or under scrutiny as nanodrug delivery system. Biodegradable polymers have the potential to become part of new medical devices with specific and unique mechanical, physical and chemical properties, such as chemical reactivity, optical properties, electrical conductivity and mechanical strength (Patel and Bailey, 2007; Hu et al., 2012). The most important biomedical goal of biodegradable polymeric materials is the development of matrices to control the release of drugs into specific sites in the body (Soppimath et al., 2001).

TABLE 16.3
Classification and Characteristics of the Different Thermo-Sensitive Materials

Polymer Type	Stimuli Characteristic	Transition	Polymer	References
Negatively thermo-sensitive	Lower critical solution temperature (LCST)	Below LCST, the polymer swells, and above LCST, the polymer shrinks	NIPAAm	Schild (1992)
Positively thermo-sensitive	Upper critical solution temperature (UCST)	Below UCST, the polymer contracts	Poly(acryl amide-co-butyl methacrylate)	Gil and Hudson (2004)
Thermally reversible gels	Gelation temperature	Liquid to a gel	Poly(ethylene glycol-b-poly(lactic acid-co-glycolic acid)-b-poly(ethylene glycol)	Jeong et al. (1998)

TABLE 16.4
pH-Sensitive Polymers

Name	Chemical Structure	Clinical Indication	References
Ion-exchange resins	R = -CH$_2$N$^+$(CH$_3$)$_3$Cl$^-$, -CH$_2$NH$^+$(CH$_3$)$_2$Cl$^-$, -SO$_3^-$H$^+$, or -COO$^-$H$^+$	Taste-masking	Guo et al. (2009)
Poly(ethylacrylate-methylmethacrylate-trimethylammonioethyl methacrylate chloride) copolymers (Eudragit RS/RL)	R$_1$ = -H or -CH$_3$; R$_2$ = -CH$_3$ or -C$_2$H$_5$	Sustained release, taste-masking	Bodmeier et al. (1996), Narisawa et al. (1994)

TABLE 16.5
Applications of Different Smart Polymers (Burkersroda et al., 2002)

Polymer	Applications	Advantages	Disadvantages	λ, Degradation Rate Constant (s^{-1})	Structure
Polyacetals	Drug delivery	Mild pH degradation products; pH-sensitive degradation	Low-molecular-weight complex synthesis	6.4×10^{-5}	(R$_1$-O-C(R$_2$)(R$_3$)-O)$_n$
Poly (orthoesters)	Drug delivery	Controllable degradation rates; pH-sensitive degradation	Synthesis is complex	4.8×10^{-5}	(R$_1$-O-C(R$_2$)(O-R$_3$)-O)$_n$
Polycaprolactone	Tissue engineering	Highly processable many commercial vendors available	Limited degradation	3.5×10^{-8}	(-O-(CH$_2$)$_5$-C(=O)-)$_n$
Polylactide	Tissue engineering drug delivery	Highly processable many commercial vendors available	Limited degradation highly acidic degradation products	6.6×10^{-9}	(-O-CH(CH$_3$)-C(=O)-)$_n$
Polycarbonates	Drug delivery tissue engineering fixators	Chemistry-dependent mechanical properties surface eroding	Limited degradation require copolymerisation with other polymers	4.1×10^{-10}	(R-O-C(=O)-O)$_n$
Polyamides	Drug delivery	Conjugatable side group highly biocompatible degradation products	Very limited degradation charge-induced toxicity	2.6×10^{-13}	(R-N(H)-C(=O))$_n$

Nanoparticles made with biodegradable polymers have been an important instrument in the treatment of neurodegenerative diseases, because of their ability to cross the blood–brain barrier (BBB). Nano-/microcarrier developed by biodegradable polymers shows high drug entrapment efficiency. It can be used for the diagnosis and site-specific drug delivery (Godin et al., 2010; Modi et al., 2010).

16.4.8 Viscosity Enhancer

Since fluid viscosity altogether influences the fluidity, and molecular dynamics of molecules inside it, it is vital to accurately quantify fluid viscosity. For example, monitoring biofluid viscosity has been proposed as a diagnostic tool for detecting diseases (Harkness, 1971) and measuring the viscosity of dilute polymer solutions has become a simple but sensitive method to probe the conformation of discrete polymer chains (Rubinstein and Colby, 2003). Generally, polymers are viscoelastic materials, which behave like both solid and liquid and consequently show distortion and temperature and force-dependent flow. The viscous-elastic properties are qualities of a given polymer that rely upon its molecular weight, molecular weight dispersion and branching (Gupta et al., 2014, 2015).

16.4.9 Highly Water-Soluble Polymer

The process of dissolving a pure polymer begins with solvent molecules permeating bulk polymer; generally, the solubility of most polymers decreases as their molecular weight rises.

Solubility and permeation features define the bioavailability of orally administered drugs (Amidon et al., 1995) Based on these ascribes, the biopharmaceutical classification system (BCS) categorises drugs into four different classes, I–IV. Although BCS class II drugs exhibit low solubility, they can escape easily into the gastrointestinal tract (GIT), while dissolution rate defines the rate and degree of absorption of class II drugs. As a result, the focal point of ongoing research is to improve the dissolution or release of these drugs from the formulations (Panakanti and Narang, 2012; Lipinski et al., 1997).

Sustained-release capsule formulations based on three components, drug, water-soluble polymer and water-insoluble fatty acid, were developed. Theophylline, acetaminophen and glipizide, representing a wide spectrum of aqueous solubility, were used as model drugs.

Povidone and hydroxypropyl cellulose were selected as water-soluble polymers. Stearic acid and lauric acid were selected as water-insoluble fatty acids. Fatty acid, polymer and drug mixture was filled into size #0 gelatin capsules and heated for 2 h at 50°C. The drug particles were trapped into molten fatty acid and released at a controlled rate through pores created by the water-soluble polymer when capsules were exposed to an aqueous dissolution medium.

16.5 ROLE OF POLYMER IN DRUG DELIVERY SYSTEM

16.5.1 Modified Drug-Release Dosage Forms

To accomplish gastro-retentive drug delivery, the mucoadhesive polymers have been assessed which increase the gastric retention time of drug delivery system by adhering to the mucus lining of the stomach. This concept is beneficial for the drugs which absorb and stable at the upper part of the GIT. It is also applicable for drugs that absorb through transport systems that are present only in the stomach or upper part of GIT (Priya and Roy) (Table 16.6).

TABLE 16.6
Applications of Polymers in Drug Delivery

Polymers	Properties	Applications in Drug Delivery	References
Albumin	65 kDa, solvent in 40% ethanol, non-poisonous, non-immunogenic	Nanoencapsulation Forms with Albumin (condensate of egg whites with paclitaxel, Abraxane) Microencapsulation of insulin, doxorubicin, 5-fluorouracil, technetium-99 m, little atoms	Kratz et al. (2008)
Cellulose and derivates	Water-solvent subject to subsidiary What's more, pH, mucoadhesive, hydrogel-framing capacity, biocompatible, just mellow unfamiliar body response, non-harmful	Coatings in strong dose structures Microencapsulation Taste veiling Osmotic siphon frameworks Oral, gastro-retentive, vaginal medication conveyance Nasal medication conveyance (leuprolide, calcitonin)	Bonora et al. (2008)
Poly(hydroxyalkanoate) like poly(b-hydroxybutyrate) (PHB)	200–400 kDa, Mw relies upon refinement measure, biocompatible, thermoprocessable, non-harmful, haemocompatible	PHB tablets Implants	Chen et al. (2008)
Chitosan and derivates	5–540 kDa, decidedly charged, water dissolvable, mucoadhesive, mechanical quality, heat and steam sterilisable, great medication stacking limit, low harmful, haemostatic potential, great injury recuperating properties, gel- and film-shaping capacity	Quality conveyance Hydrogels Microspheres (ondansetron) Nanofibres (wound dressing) Stents absorption enhancer Gastrointestinal, oral, nasal conveyance	Pan et al. (2000)
Poly(amino acids)	Charge, hydrophilicity, structure, biodistribution and harmfulness are amino corrosive needy, biocompatible	Delivery of cytostatic drugs, genes	Matsumura et al. (2008)
Poly (orthoesters)	>100 kDa, hydrophobic, corruption by surface disintegration, high glass progress temperatures, great mechanical properties, non-harmful, moderate invulnerable reaction, visual biocompatible, disinfection by b- or g-light	Supported medication conveyance by microspheres for paclitaxel, cisplatin, lidocaine ophthalmology implants postsurgical torment control periodontal ailments DNA conveyance adjunct for glaucoma filtration medical procedure (5-fluorouracil)	Heller et al. (2012)

TABLE 16.7
Different Strategies for Colon-Targeted Drug Delivery (Sarangi et al., 2020)

S. No.	Approaches	Basic Features
1.	Formation of prodrugs	Covalent linkage is shaped among medication and transporter, which on oral arrives at colon without being consumed from upper piece of GIT, and in the colon, the medication discharge is set off by the effect of specific chemicals in contrast with stomach and small digestive system
2.	Azo-conjugates	The medication is formed with azo bond, and the medication discharge at the colon is set off by chemicals like azo-reductase
3.	Cyclodextrin conjugates	The medication is formed with cyclodextrin
4.	Coating with pH-sensitive polymers	Definition is covered with enteric polymers whose trustworthiness relies upon pH and delivery drug when pH advances towards antacid range
5.	Coating with biodegradable polymers	Medication ensured with polymer is freed because of degradation impact of colonic microscopic organisms

16.5.2 Extended-Release Dosage Forms

Broadened discharge grid tablet of tramadol hydrochloride utilises various mixes of hydrophilic polymers (HPMC, Carbopol, xanthan gum). Subsequently, hydrophilic polymers HPMC and Carbopol were found to broaden the arrival of profoundly water solvent medication, tramadol hydrochloride (Kausalya et al., 2009).

16.5.3 Gastro-retentive Dosage Forms

The utilisation of common polymers is valuable and dependent on demonstrated biocompatibility and wellbeing. In this respect, the polysaccharides chitosan, thickener and guar gum, cationic, anionic and non-ionic operator, individually, have gotten standard ticular consideration (Miyazaki et al., 1994) Natural gums are among the most mainstream hydrophilic polymers due to their cost viability and regulatory acceptance. Common polymer has ideal properties and consequently has applications in the drug and biomedical fields (Felt et al., 1998). Normally utilised common polymers in the drifting medication conveyance systems are xanthan gum, guar gum and chitosan.

16.6 TYPES OF POLYMERS IN PHARMACEUTICAL DRUG DELIVERY

16.6.1 Polymers in Colon-Targeted Drug Delivery

The colon-targeted drug delivery has various significant ramifications in the field of pharmacotherapy. Oral colon-targeted drug delivery frameworks have as of late picked up significance for conveying an assortment of helpful operators for both nearby and foundational organisations. In case of colon delivery of drug, polymer can protect the drug from denaturing or delivery in the stomach and small digestive tract. It likewise guarantees sudden or controlled arrival of the drug in the proximal colon. Different drug delivery frameworks have been planned that convey the drug quantitatively to the colon and afterwards trigger the arrival of drug. This audit will cover various sorts of polymers which can be utilised in the plan of colon-targeted drug delivery frameworks (Rajpurohit et al., 2008) (Tables 16.7 and 16.8).

16.6.2 Polymers in the Mucoadhesive Drug Delivery System

The mucoadhesive polymers for drug delivery through buccal cavity shows the remarkable applications. It shows high retention of drug delivery system in the buccal cavity that enhances the drug transportation from buccal cavity to systemic circulation which leads to enhancement of bioavailability of drug. There are various types of natural and synthetic mucoadheshive polymers that are used to prepare drug delivery system for delivery of drug or other active therapeutics (Jones, 2004).

16.6.3 Polymers for Sustained Release

The idea of utilising polymer-based supported delivery conveyance frameworks to keep up helpful centralisation of protein drugs for expanded time frames has been all around acknowledged for quite a long time, and there has not been a solitary item in this classification effectively popularised to date, regardless of clinical and market requests. To accomplish fruitful frameworks, specialised challenges running from protein denaturing during plan measure and the course of delayed *in vivo* discharge, blasted delivery, and fragmented delivery, to low embodiment effectiveness and detailing unpredictability must be all the while settled. In view of this refreshed comprehension, detailing techniques endeavouring to address these angles thoroughly were accounted for later. Plan methodologies illustrative of three regions, microsphere innovation utilising degradable hydrophobic polymers, microspheres made of water dissolvable polymers and hydrophilic *in vivo* gelling system, will be chosen and presented. (Table 16.9).

TABLE 16.8
Chemically Modified Carbohydrate Polymers in Colon-Specific Drug Delivery

S. No.	Polymers Used	Bioactive Studied	Modification in Polymer	Drug Delivery System	Prepared Purpose/Outcome of Study	References
1.	Chitosan	Camptothecin	Folate complex	Microcapsule	The chitosan–folate microcapsules stacked with camptothecin altogether diminished the expansion of HeLa tumour cells, while they have an irrelevant impact on fibroblasts	Galbiati et al. (2011)
2.	Pectin	Doxorubicin	Oxidised citrus pectin	In situ hydrogels	Oxidised gelatine hydrogels can possibly forestall both movement of essential malignant growth by the delivered doxorubicin and age of metastatic malignant growth by the delivered oxidised gelatine	Takei et al. (2010)
3.	Dextran	Budesonide	Budesonide–dextran conjugates	Conjugates	The conjugate prepared by using dextran for the treatment of ulcerative colitis	Varshosaz et al. (2011)
4.	Ethyl cellulose	Carbopol	Combination	Albendazole binary mixtures of Carbopol/ethyl cellulose	Drug release from the tablet containing Carbopol/ethyl cellulose were developed to sustained the release of Albendazole in the colon	Ramteke and Nath (2014)
5.	HPMC phthalate	Rabeprazole	Enteric coating of HPMC-phthalate	Beads	Covering of globules containing rabeprazole with HPMC-P has been appeared to stifle drug discharge at acidic pH without bargaining the delivery at unbiased pH	Yoshida et al. (2013)

TABLE 16.9
Blended Polymers as Pharmaceutical Forms for Drug Delivery System

Polymeric Blend	Form	Drug	References
PLGA-PEG	Micelle	Doxorubicin	Joglekar and Trewyn (2013)
Chitin-pluronic F108	Microparticles	Paclitaxel	Sinha et al. (2004)
Chitosan-glucomannan	Hydrogel	Ofloxacin	Yu and Lu (2007)
Alginate-gelatin	Film	Ciprofloxacin	Yu and Lu (2007)
PEG-gelatin	Hydrogel	Ciprofloxacin	Foox et al. (2014)

16.6.4 Polymers in Tissue Engineering

Tissue designing is the utilisation of mix of cells and materials by utilising designing strategies to supplant or improve natural tissues. It includes the utilisation of a platform which upholds the development of a reasonable new tissue. Tissue fix and recovery is one of the significant difficulties of nowadays. Polymer substrates are the primary parts of this tissue designing. Polymer substrates give the basic components to physiological recovery measure. Polymeric devices for tissue engineering prepare by using the blends of polymers which may consist of drugs, live cells (stem cell) and other therapeutic molecules that can be utilised for the treatment of tissue damage and also use for recovery of a few essential organs of the human body, for example, ear, veins, coronary supply route.

Tissue designing is an interdisciplinary science that applies science, materials science, designing and medication expecting to fix and supplant tissues and organs (Cui et al., 2010, Walmsley et al., 2015). Tissue designing procedures are grouped into three primary segments containing framework, cells (separated or undifferentiated) and organic flagging atoms, for example, development factors (GFs) (Pereira et al., 2011). Platforms are three-dimensional (3D) structure that can be delivered by manufactured polymers, characteristic polymers and simply organic atoms, for example, collagen, elastin, hyaluronic corrosive and other extracellular lattice (ECM) particles (Bačáková et al., 2014). The framework must have the option to emulate the structure and organic capacity of characteristic extracellular network (ECM) as far as both compound arrangement and physical structure (Zhou and Lee, 2011). The extracellular framework (ECM) is a different structure of proteoglycans, proteins and flagging particles. ECM is initially known for its function in offering auxiliary help to cells and as an area for cell relocation. Suitable platforms for tissue designing applications ought to be biodegradable, biocompatible, non-toxic, non-mutagenic and non-immunogenic. Besides, they ought to have the option to offer suitable mechanical help and show ideal surface properties, for example, helping attachment, expansion and separation of cells (Bačáková et al., 2014; Zhou and Lee, 2011). Polymeric frameworks assume a fundamental part in tissue designing through cell cultivating, expansion and new tissue development in three measurements. These frameworks have demonstrated incredible capacity in the examination of tissue designing. Biodegradable polymers because of diminishing incendiary responses, non-toxic and compatible with proteins in the body have numerous applications in medication and drug reservoir. Biodegradable polymers are commonly categorized into two classes, to be specific engineered polymers and characteristic polymers (Hong et al., 2014). Engineered polymers are viable with body, biodegradable and absorbable. These polymers are effectively changed into various 3D lattice structures. Polymers are generally biodegrade or metabolite after a period of time in the body and then easily eliminated by kidney (Tian et al., 2012 Abbasi et al., 2016).

16.7 IN SITU DRUG DELIVERY SYSTEM

In situ gel drug delivery system are utilised as solution form before administration in the body, but it become a gel when it reach on site of action. The formulation of gel relies upon factors like temperature adjustment, pH change, presence of particles and light illumination, electrical discharge, catalyst from which medication gets delivered in a continued and controlled way. Commonly, fluid arrangements of hydrogels utilised in biomedical applications are fluid at surrounding temperature and gel at physiological temperature. The in situ gel shaping polymeric definitions offer a few points of interest like continued and delayed activity in contrast with customary medication conveyance frameworks. From an assembling perspective, the creation of such gadgets is less unpredictable and hence brings down the speculation and assembling cost. This audit weights on the polymeric utilisation of normal polymers and synthetic polymers.

Despite the fact that in situ polymerisation has its points of interest, the conditions required are quite certain. The physiologically adequate temperatures are inside a limited range, and the framework must be quickly polymerised in the event that it is to be effectively embedded in a clinical setting. The upside of photograph started cross-linking is that the response continues quickly at low temperatures and thusly might be appropriate for the fuse of thermally delicate medications, for example, peptides and proteins. Photopolymerisable frameworks have a favourable position over different kinds of in situ frameworks, for example, synthetically started frameworks. This was shown by Jeong et al. (2004) who cross-linked biodegradable copolymers D, L-lactide and L-lactide with ε-caprolactone, utilising an artificially started thermoset framework, for use as a moderate delivery drug conveyance framework. Drawbacks of the framework incorporate taking as long as 30 min to set, and the profoundly exothermic nature of the cross-linking response could bring about tissue rot. There was additionally blasted arrival of the medication in the primary hour which could bring about the presence of results.

16.8 POLYMER–DRUG CONJUGATE

In the two cases, the formation of medications to polymers upgrades their dissolvability and steadiness in body liquids. It likewise lessens their harmful results in sound tissues. The covalent bonds consider fuse of higher medication loads contrasted with truly entangled plans. Painstakingly planned linkages among medications and polymers make it conceivable to control the delivery profile. Extra utilitarian gatherings of the polymer can be utilised to connect ligands and focus on the medications to explicit cells and tissues.

The American Food and Drug Administration (FDA) has endorsed various PEGylated biopharmaceuticals (Turecek et al., 2016). Those are proteins, antibodies or aptamers that convey at least one polyethylene glycol (PEG) polymers; for example, Macugen® (Valeant Pharmaceuticals International) is a PEGylated single strand of nucleic corrosive, a purported aptamer. It ties to and blocks one type of the vascular endothelial development factor (VEGF). VEGF is answerable for the development of irregular vein in wet age-related macular degeneration. Different models incorporate immunodeficiency (Adagen®, Leadiant Biosciences), acromegaly (Somavert®, Pfizer) and hepatitis C (PEG-Intron®, Merck Sharp and Dohme) treatments, among others. PEGylation improves the dependability of biopharmaceuticals and upgrades their flow time in the body.

Immune response drug forms are another class of forms that get a ton of consideration at the present time. There are four at present endorsed items, all acting against disease. Kadcyla® (Genentech/Roche) was produced for patients with recently treated metastatic bosom malignant growth (Kadcyla® site, gotten to 2018). Adcetris® (Seattle Genetics and Millennium Pharmaceuticals/Takeda Oncology) is dynamic against old-style Hodgkin lymphoma and T-cell lymphoma (Adcetris site, gotten to 2018). And Wyeth Pharmaceuticals/Pfizer delivered two new leukaemia therapeutics (Besponsa™ and Mylotarg™) a year ago. Every one of them highlights a comparable method of activity. The neutraliser first ties to receptors that are explicit to malignancy cells and overexpressed on their surface. From that point, the form is taken up into the cells and delivers its medications.

ACKNOWLEDGEMENT

The authors would like to acknowledge the Science and Engineering Research Board, DST, New Delhi, and MP Council of Science and Technology, Bhopal, for providing research grant that facilitates the writing of this chapter. The author also want to acknowledge the National Institute of Pharmaceutical Education and Research (NIPER), Hyderabad, for extending facilities to write this chapter (Research Communication No. NIPER-H/2020/BC-xx).

REFERENCES

Abbasi E, Akbarzadeh A, Kouhi M, Milani M. Graphene: synthesis, bio-applications, and properties. *Artif Cells Nanomed Biotechnol*. 2016;44:150–156.

Aizawa Y, Owen SC, Shoichet MS. Polymers used to influence cell fate in 3-D geometry: new trends. *Trends Polym Sci*. 2011;37: 645–658.

Amidon GL, Lennernas H, Shah VP, Crison JR. A theoretical basis for a biopharmaceutic drug classification: the correlation of *in vitro* drug product dissolution and *in vivo* bioavailability. *Pharm Res*. 1995;12(3):413–420.

Arif U, Haider S, Haider A, Khan N, Alghyamah AA, Jamila N, Khan MI, Almasry WA, Kang IK, Biocompatible polymers and their potential biomedical applications: a review, *Curr Pharm Des*. 2019;25(34):3608–3619.

Bačáková L, Novotná K, Pařízek M. 2014. Polysaccharides as cell carriers for tissue engineering: the use of cellulose in vascular wall reconstruction. *Physiol Res*. 63:29–47.

Bernkop-Schnürch A. Thiomers: a new generation of mucoadhesive polymers. *Adv Drug Deliv Rev*. 2005;57(11):1569–1582.

Bialik M, Kuras M, Sobczak M, Oledzka E. Biodegradable synthetic polyesters in the technology of controlled dosage forms of antihypertensive drugs – the overview. *Expert Opin Drug Deliv*. 2019;69(8):204–300.

Bodmeier R, Guo X, Sarabia RE, Skultety PF. The influence of buffer species and strength on diltiazem HCl release from beads coated with the aqueous cationic polymer dispersions, Eudragit RS, RL 30D. *Pharm Res*. 1996;13:52–56.

Bonora S, Lanzafame M, D'Avolio A, Trentini L, Lattuada E, Concia E, Di Perri G. Drug interactions between warfarin and efavirenz or lopinavir-ritonavir in clinical treatment. *Clin Infect Dis*. 2008;46:146–147.

Burkersroda FV, Schedl L, Gopferich A. Why degradable polymers undergo surface erosion or bulk erosion. *Biomaterials*. 2002;23:4221–4231.

Chen J, Del Genio AD, Carlson BE, Bosilovich MG. The spatiotemporal structure of twentieth-century climate variations in observations and reanalyses. Part I: long-term trend. *J. Clim*. 2008;21:2611–2633.

Cilurzo F, Selmin F, Gennari CGM, Montanari L, Minghetti P. Application of methyl methacrylate copolymers to the development of transdermal or loco-regional drug delivery systems. *Expert Opin Drug Deliv*. 2014; 11(7):1033–1045.

Cui W, Zhou Y, Chang J. 2010. Electrospun nanofibrous materials for tissue engineering and drug delivery. *Sci Technol Adv Mater*. 11:014108.

Delplace V, Nicolas, J. (2015). Degradable vinyl polymers for biomedical applications. *Nat Chem*. 7(10):771–784.

Duncan R. The dawning era of polymer therapeutics. *Nat Rev Drug Discov*. 2003;2:347–360.

Felt O, Buri P, Gurny R. Chitosan: a unique polysaccharide for drug delivery. *Drug Dev Ind Pharm*. 1998;24(11):979–999.

Finkenstadt VL. (2005). Natural polysaccharides as electroactive polymers. *Appl Microbiol Biotechnol*. 67(6):735–745.

Foox M, Raz-Pasteur A, Berdicevsky I, Krivoy N, Zilberman M. In vitro microbial inhibition, bonding strength, and cellular response to novel gelatin-alginate antibioticreleasing soft tissue adhesives. *Polym Adv Technol* 2014;25:516–524.

Galbiati A, Tabolacci C, Rocca BMD, Mattioli P, Beninati S, Paradossi G, Desideri A. Targeting tumor cells through chitosan-folate modified microcapsules loaded with camptothecin. *Bioconjug Chem*. 2011;22: 1066–1072.

Gil ES, Hudson SA. Stimuli-reponsive polymers and their bioconjugates. *Prog Polym Sci*. 2004;29:1173–222.

Godin B, Sakamoto JH, Serda RE, Grattoni A, Bouamrani A, Ferrari M. Emerging applications of nanomedicine for the diagnosis and treatment of cardiovascular diseases. *Trends Pharmacol Sci*. 2010;31:199–205.

Guo X, Chang RK, Hussain MA. Ion-exchange resins as drug delivery carriers. *J Pharm Sci*. 2009;98:3886–3902.

Gupta SS, Meena A, Parikh T, Serajuddin AT. Investigation of thermal and viscoelastic properties of polymers relevant to hot melt extrusion, I: polyvinylpyrrolidone and related polymers. *J Excipients Food Chem*. 2014;5(1):32–45.

Gupta SS, Parikh T, Meena AK, Mahajan N, Vitez I, Serajuddin ATM. Effect of carbamazepine on viscoelastic properties and hot melt extrudability of Soluplus®. *Int J Pharm*. 2015;478(1):232–239.

Harekrishna R. Formulation of sustained release matrix tablets of metformin hydrochloride by polyacrylate polymer. *Int J Pharm Res Health Sci*. 2015;3(6):900–906.

Harkness J. The viscosity of human blood plasma: its measurement in health and disease. *Biorheology* 1971;8:171–193.

Heller S, Buse J, Fisher M, Garg S, Marre M, Merker L, Renard E, Russell-Jones D, Philotheou A, Francisco AM, Pei H. Insulin degludec, an ultra-longacting basal insulin, versus insulin glargine in basal-bolus treatment with mealtime insulin aspart in type 1 diabetes (BEGIN basal-bolus type 1): a phase 3, randomised, open-label, treat-to-target non-inferiority trial. *Lancet* 2012;379(9825): 1489–1497.

Hocine S, Li MH. Thermoresponsive self-assembled polymer colloids in water. *Soft Matter* 2013;9(25):5839–5861.

Hong, Y, Liu G, Gu Z. Recent advances of starch-based excipients used in extended-release tablets: a review. *Drug Delivery* 2014;23(1):12–20.

Hu D, Liu L, Chen W, Li S, Zhao Y. A novel preparation method for 5-aminosalicylic acid loaded Eudragit s100 nanoparticles. *Int J Mol Sci*. 2012;13:6454–6468.

Jeong B, Choi YK, Bae YH, Zentner G, Kim SW. New biodegradable polymers for injectable drug delivery systems. In: *Conference on Challenges for Drug Delivery and Pharmaceutical Technology*. Tokyo: Elsevier Science Bv; 1998.

Jeong HS, Huh KM, Park K. Hydrogel drug delivery systems. In: Uchegbu IF, Schatzlein AG, editors. *Polymers in Drug Delivery*. Boca Raton, FL: Taylor & Francis; 2006: 49–62.

Jeong SI, Kim BS, Lee YM, Ihn KJ, Kim SH, Kim YH. Morphology of elastic poly(l-lactide-co-ε-caprolactone) copolymers and *in vitro* and *in vivo* degradation behavior of their scaffolds. *Biomacromolecules* 2004;5(4):1303–1309.

Joglekar M, Trewyn BG. Polymer-based stimuli-responsive nanosystems for biomedical applications. *Biotechnol J* 2013;8(8):931–945.

Johnson DA, Roach AC, Carlsson AS, Karlsson AA, Behr DE. Stability of esomeprazole capsule contents after in vitro suspension in common soft foods and beverages. *Pharmacotherapy*. 2003;23:731–734.

Jones D. *Pharmaceutical Applications of Polymers for Drug Delivery*. Toronto, ON: ChemTec Publishing Inc.; 2004;300–301.

Kratz F. Albumin as a drug carrier: design of prodrugs, drug conjugates and nanoparticles. *J Control Release*. 2008;132(3):171–183.

Kausalya J, Padmapriya S, Vaijayanthi V, Umadevi SK, Senthilnathan B. Formulation and evaluation of extended release matrix tablet of tramadol hydrochloride using hydrophilic polymer. *Biomed Pharm J*. 2009;2(2):387.

Kawai F. The biochemistry and molecular biology of xenobiotic polymer degradation by microorganisms. *Biosci Biotechnol Biochem*. 2010;74(9):1743–1759.

Lakshmi NS, Cato LT. *Polymers Biomaterials for Tissue Engineering and Controlled Drug Delivery*. Berlin/Heidelberg: Springer; 2006:203–210.

Lipinski C, Lombardo F, Dominy B, Feeney P. Experimental and computational approaches to estimate solubility and permeability in drug discovery and development settings. *Adv Drug Deliv Rev*. 1997;23:3–25.

Matsumura M, Takeuchi H, Satoh M, Sanada-Morimura S, Otuka A, Watanabe T, Dinh VT. Species-specific insecticide resistance to imidacloprid and fipronil in the rice planthoppers *Nilaparvata lugens* and *Sogatella furcifera* in East and South-east Asia. *Pest Manag Sci*. 2008;64:1115–1121.

Mellman I. Endocytosis and molecular sorting. *Annu Rev Cell Dev Biol*. 1996;12:575–625.

Miller NS, Chittchang M, Johnston TP. Buccal mucosa as a route for systemic drug delivery: a review. *Adv Drug Deliv Rev*. 2005;57(11):1666–1691.

Miyazaki S, Nakayama A, Oda M, Takada M, Attwood D. Chitosan and sodium alginate based bioadhesive tablets for intraoral drug delivery. *Biol Pharm Bull*. 1994;17:745–747.

Modi G, Pillay V, Choonara YE. Advances in the treatment of neurodegenerative disorders employing nanotechnology. *Ann NY Acad Sci*. 2010;1184:154–172.

Muxika A, Etxabide A, Uranga J, Guerrero P, de la Caba K. Chitosan as a bioactive polymer: processing, properties and applications. *Int J Biol Macromol*. 2017;105:1358–1368.

Narisawa S, Nagata M, Danyoshi C, Yoshino H, Murata K, Hirakawa Y, Noda K. An organic acid-induced sigmoidal release system for oral controlled-release preparations. *Pharm Res*. 1994;11:111–116.

Narisawa S, Nagata M, Hirakawa Y, Kobayashi M, Yoshino H. An organic acid-induced sigmoidal release system for oral controlled-release preparations. 2. Permeability enhancement of Eudragit RS coating led by the physicochemical interactions with organic acid. *J Pharm Sci*. 1996;85:184–188.

Pan Q, Wendel J, Fluhr R. Divergent evolution of plant NBS-LRR resistance gene homologues in dicot and cereal genomes. *J Mol Evol*. 2000;50:203–213.

Panakanti R, Narang AS. Impact of excipient interactions on drug bioavailability from solid dosage forms. *Pharm Res*. 2012;29(10):2639–2659.

Patel DN, Bailey SR. Nanotechnology in cardiovascular medicine. *Catheter Cardiovasc Interv*. 2007;69:643–654.

Pereira H, Frias AM, Oliveira JM, Esprequeira-Mendes J, Reis RL. 2011. Tissue engineering and regenerative medicine strategies in meniscus lesions. *Arthroscopy* 27:1706–1719.

Priya VSV, Roy HK, Polymers in drug delivery technology, types of polymers and applications. *Sch Acad J Pharm*. 2016;5(7):305–308. ISSN 2320-4206 (Online).

Raizada A, Bandari A, Kumar B. Polymers in drug delivery: a review. *Int J Pharm Res Dev*. 2010;2(8):9–20.

Rajpurohit YS, Gopalakrishnan R, Misra HS. Involvement of a protein kinase activity inducer in DNA double strand break repair and radioresistance of *Deinococcus radiodurans*. *J Bacteriol* 2008;190:3948–3954.

Ramteke KH, Nath L. Formulation, evaluation and optimization of pectin-bora rice beads for colon targeted drug delivery system. *Adv Pharm Bull*. 2014;4:167–177.

Ricard-Blum, S. The collagen family. *Cold Spring Harb Perspect Biol*. 2010;3(1):a004978–a004978.

Rubinstein M, Colby RH. *Polymer Physics*. New York: Oxford University Press; 2003.

Sarangi MK, Rao MEB, Parcha V. Smart polymers for colon targeted drug delivery systems: a review. *Int J Polym Mater Polym Biomater*. 2020;42:1–37.

Schild HG. Poly (N-isopropylacrylamide) experiment, theory and application. *Prog Polym Sci*. 1992;17:163–249.

Schmaljohann D. Thermo- and pH-responsive polymers in drug delivery. *Adv Drug Deliv Rev*. 2006;58:1655–1670.

Sheshala R, Khan N, Darwis Y. Formulation and optimization of orally disintegrating tablets of sumatriptan succinate. *Chem Pharm Bull (Tokyo)*. 2011;59:920–928.

Singh V, Joshi S, Malviya T. Carboxymethyl cellulose-rosin gum hybrid nanoparticles: an efficient drug carrier. *Int J Biol Macromol*. 2018;112:390–398.

Sinha VR, Singla AK, Wadhawan S, Kaushik R, Kumria R, Bansal K, Dhawan S. Chitosan microspheres as a potential carrier for drugs. *Int J Pharm*. 2004;1–2(274):1–33.

Sithole MN, Choonara YE, du Toit LC, Kumar P, Pillay V. A review of semi-synthetic biopolymer complexes: modified polysaccharide nano-carriers for enhancement of oral drug bioavailability. *Pharm Dev Technol*. 2016;22(2):283–295.

Soppimath KS, Aminabhavi TM, Kulkarni AR, Rudzinski WE. Biodegradable polymeric nanoparticles as drug delivery devices. *J Control Release*. 2001;70:1–20.

Sowjanya M, Lavanya P, Thejovathi R, Babu MN, Polymers used in the designing of controlled drug delivery system. *Res J Pharm Tech*. 2017;10(3):903–912.

Takei T, Sato M, Ijima H, Kawakami K. In situ gellable oxidized citrus pectin for localized delivery of anticancer drugs and prevention of homotypic cancer cell aggregation. *Biomacromolecules* 2010;11:3525–3530. doi:10.1021/bm1010068.

Talelli M, Hennink WE. Thermosensitive polymeric micelles for targeted drug delivery. *Nanomedicine* 2011;6(7):1245–1255.

Tangpasuthadol V, Pendharkar SM, Peterson RC, Kohn J. Hydrolytic degradation of tyrosine-derived polycarbonates, a class of new biomaterials: part I study of model compounds. *Biomaterials*. 2000;21:2379–2387.

Tian H, Tang Z, Zhuang X, Chen X, Jing X. 2012. Biodegradable synthetic polymers: preparation, functionalization and biomedical application. *Prog Polym Sci*. 37:237–280.

Tirpude RN, Puranik PK. Rabeprazole sodium delayed-release multiparticulates: effect of enteric coating layers on product performance. *J Adv Pharm Technol Res*. 2011;2:184–191.

Tran PHL, Duan W, Lee B-J, Tran TTD. The use of zein in the controlled release of poorly water-soluble drugs. *Int J Pharm*. 2019;566:557–564.

Turecek PL, Bossard MJ, Schoetens F, Ivens IA. 2018. PEGylation of biopharmaceuticals: a review of chemistry and nonclinical safety information of approved drugs. *J Pharm Sci*. 2016;105(2):460–475.

Van Renterghem J, Dhondt H, Verstraete G, De Bruyne M, Vervaet C, De Beer T. The impact of the injection mold temperature upon polymer crystallization and resulting drug release from immediate and sustained release tablets. *Int J Pharm*. 2018;541(1–2):108–116.

Varshosaz J, Emami J, Ahmadi F, Tavakoli N, Minaiyan M. Preparation of budesonide-dextran conjuglates using glutarate spacer as a colon-targeted drug delivery system: *in vitro/in vivo* evaluation in induced ulcerative colitis. *J Drug Target* 2011;19:140–153.

Vilar G, Tulla-Puche J, Albericio F. Polymers and drug delivery systems. *Curr Drug Deliv.* 2012;9:367–394.

Walmsley GG, McArdle A, Tevlin R, Momeni A, Atashroo D, Hu MS, Feroze AH, Wong VW, Lorenz PH, Longaker MT, Wan DC. 2015. Nanotechnology in bone tissue engineering. *Nanomedicine* 11:1253–1263.

Ward MA, Georgiou TK. Thermoresponsive polymers for biomedical applications. *Polymers* 2011;3(3):1215–1242.

Yoshida T, Lai TC, Kwon GS, Sako K. pH- and ion-sensitive polymers for drug delivery. *Expert Opin Drug Deliv.* 2013;10(11):1497–1513.

Yu H, Lu J. Preparation and properties of novel hydrogels from oxidized konjac glucomannan cross-linked chitosan for *in vitro* drug delivery. *Macromol Biosci.* 2007;7(9–10): 1100–1111.

Yousef, J. and Sanaz, H. Novel Drug Delivery Systems for Modulation of Gastrointestinal Transit Time. In: Sezer, A.D., Ed., *Recent Advances in Novel Drug Carrier Systems*, Rijeka: InTech Publisher; 2012: 393–418.

Zhou H, Lee J. Nanoscale hydroxyapatite particles for bone tissue engineering. *Acta Biomater.* 2011;7(7):2769–2781.

17 Drug Delivery Systems for Targeting Blood Brain Barrier

Examples of nanomedicines for the treatment of neurodegenerative diseases

Iara Baldim
University of Minho
University of São Paulo

Adriana M. Ribeiro and João Dias-Ferreira
University of Coimbra (FFUC)

Wanderley P. Oliveira
University of São Paulo

Francisco M. Gama
University of Minho

Eliana B. Souto
University of Minho
University of Coimbra (FFUC)

CONTENTS

Abbreviations 258
17.1 Introduction 258
17.2 The Blood–Brain Barrier 259
17.3 Pharmacotherapy 259
17.4 Nanoparticles for Targeting the CNS 259
    17.4.1 Mesoporous Silica Nanoparticles 260
    17.4.2 Gold Nanoparticles 260
    17.4.3 Polymeric Nanoparticles 260
    17.4.4 Liposomes 260
    17.4.5 Solid Lipid Nanoparticles 260
    17.4.6 Nanostructured Lipid Carriers 260
    17.4.7 Quantum Dots 260
    17.4.8 Fullerenes 261
    17.4.9 Carbon Nanotubes 261
    17.4.10 Dendrimers 261
    17.4.11 Nanogels 261
    17.4.12 Micelles 261
17.5 Examples of Nanomedicines for Alzheimer's Disease 261
    17.5.1 Modulation of Aβ Production 261
    17.5.2 Inhibition of Aβ Production 262
    17.5.3 Improvement of Aβ Clearance 262
17.6 Examples of Nanomedicines for Parkinson's Disease 263
    17.6.1 Dopamine Replacement 263
    17.6.2 Targeting α-Synuclein Accumulation 263

	17.6.2	Targeting Inflammation	263
	17.6.3	Targeting Antioxidative Stress	263
	17.6.4	Neurotrophic Factor Supplementation	263
17.7		Sterilisation of Nanomaterials	263
17.8		Conclusions	264
Acknowledgements			264
References			264

ABBREVIATIONS

Aβ: β-amyloid
AD: Alzheimer's disease
APP: Amyloid precursor protein
AuNR: Gold nanorods
BBB: Blood–brain barrier
CNS: Central nervous system
CNT: Carbon nanotubes
EC: Endothelial cells
EGCG: Epigallocatechin-3-gallate
GDNF: Glial cell line–derived neurotrophic factor
iMAO: Inhibitor of monoamine oxidase
MAO-B: Monoamine oxidase-B
MPTP: 1-Metil-4-fenil-1,2,3,6-tetraidropiridina
MSN: Mesoporous silica nanoparticles
ND: Neurodegenerative disease
NFT: Neurofibrillary tangle
NGF: Nerve growth factor
NLC: Nanostructured lipid carriers
NMDA: N-methyl-D-aspartate
NP: Nanoparticles
PCBA: Polymeric n-butyl-2-cyanoacrylate
PCL: Polycaprolactone
PD: Parkinson's disease
PEG: Poly(ethylene glycol)
PLA: Polylactic acid
PLGA: Poly(lactic-co-glycolic acid)
POM: Polyoxometalate
QD: Quantum dots
RES: Reticuloendothelial system
shRNA: Short hairpin ribonucleic acid
siRNA: Small-interfering ribonucleic acid
SLN: Solid lipid nanoparticles
TJ: Tight junctions

17.1 INTRODUCTION

The central nervous system (CNS) is a unique and integrated system that consists of the brain, spinal cord and cranial nerves. The building blocks of the CNS are neurons, which normally do not hold the capacity to reproduce or replace themselves. When a pathology succeeds, it cannot be replaced, which can lead to progressive degeneration and/or death of these cells, resulting in neurodegenerative diseases (ND). Parkinson's disease (PD) and Alzheimer's disease (AD) are the most common ND, predominantly observed in elderly individuals. The causes of both diseases are poorly understood, but some hypotheses show a relationship between a genetic component and ageing [1].

The average life expectancy worldwide is currently around eight decades, and the ageing of the population contributes to the prevalence of ND. As an example, AD currently affects about 15 million people worldwide. The expectation is that, in the United States and Europe, in 2050 the number of AD cases will triple, reaching 13 and 15 million people, respectively [2]. The treatment of ND is a challenge due to aspects such as ineffective drug delivery to the CNS [3] and lack of knowledge about the pathophysiology of disorders and their causes. The main obstacle for brain drug delivery is the blood–brain barrier (BBB), which has an impermeable barrier that prevents the entry of 98% of the small molecules and 100% of macromolecules, and therefore, the delivery of almost every drug used to treat these conditions is compromised [4,5].

The scientific community has been striving to develop new drug delivery systems for brain targeting to provide an effective therapy for ND. Nanomedicine is a promising and efficient approach in the treatment of these disorders due to its high specificity, low toxicity, higher drug-loading capacity and permeability, and the possibility of functionalising the systems to promote drug targeting. Together, these features can contribute to increasing the therapeutic effectiveness of several drugs and macromolecules, by improving their bioavailability and site-specific delivery. This chapter provides a comprehensive summary of nanomedicines applied to ND and the impact that BBB has on the effectiveness of these systems.

AD is the most common ND worldwide, characterised by a progressive and irreversible decline in cognitive function, which clinically results in the dysfunction of mental faculties such as memory, language, learning, and thinking capacity. Two types of lesions characterise the neuropathology of AD: senile plaques and neurofibrillary tangles (NFT) [2]. The first one is composed of a short peptide called β-amyloid (Aβ), the product of a proteolytic process from the amyloid precursor protein (APP). Aβ fragments can aggregate and accumulate into an insoluble complex of proteins with neurotoxicity. On a healthy brain, those fragments would be eliminated [6]. The second type of lesion is the aberrant formation of tau proteins (which function is stabilising the microtubules) by phosphorylation processes due to the presence of Aβ fragments. With this defect, tau proteins can no longer stabilise microtubules, causing NFT [6]. The result of this process is the failure of the microtubule function, compromising the transport of nutrients and other important substances. The imbalance between the production and the clearance process of Aβ fragments leads to Aβ aggregation and accumulation, which leads to synaptic damage, neuronal death and cognitive deficits [7]. The diagnosis of AD is usually done at a late stage of the development of the disease, when the main symptoms appear, resulting in fury and anxiety [6].

PD, the most common neurodegenerative movement disorder, is characterised clinically by motor symptoms as bradykinesia associated with postural instability, tremors and rigidity [8]. Pathologically, the disease consists of a progressive loss of dopaminergic neurons from the substantia *nigra* region of the brain and the generation of Lewy bodies [7]. The amyloid fibrils formed from α-synuclein protein are a component of Lewy bodies, which is associated with the disease [9].

17.2 THE BLOOD–BRAIN BARRIER

BBB is essential for healthy brain function. It blocks the entrance of certain substances in order to maintain the homeostasis of the inner environment [10] and the supplying of indispensable nutrients. It gives protection against pathogens, neurotoxic agents and circulating blood cells [11]. A complex network of capillaries with various components such as endothelial cells (EC), pericytes, astrocytes, tight junctions (TJ), neurons and basal membrane, compose BBB. This network acts as an impermeable wall that blocks the entry of insoluble or larger hydrophilic molecules, allowing only the passage of small and lipophilic molecules (in the order of <400 Da) [12]. Additionally, efflux transporters located in the capillaries make this route even more shielded [10]. There are two pathways in flux through BBB: paracellular (passive diffusion) and transcellular transport (passive diffusion, receptor-mediated transport and transcytosis). Passive diffusion depends on the properties of molecules, such as molecular weight, surface charge or lipophilicity [10]. The equilibrium between these two processes determines the level of permeability in a healthy brain. Some features help to restrict the permeability of BB, including [13]: (i) few endocytic vesicles that control the transcellular flux, (ii) absence of fenestrations, (iii) TJ allow to control paracellular transport and offer high electrical resistance, (iv) high mitochondrial concentration, allowing a high level of metabolism, and (v) specialised transport systems. Events as stroke or inflammation lead to modifications on the basal membrane, which can affect the permeability of BBB [13]. Research points that in most of CNS diseases, including AD and PD, dysfunction of BBB occurs. This causality is not well understood but the modifications in the transporters and enzymes seem to provide a strong contribution to the progression of these pathologies [7].

Based on these aspects, the effectiveness of treatments for ND has a strict relationship with the physiopathology of BBB. Several approaches have been used to promote the permeation of different drugs to BBB (e.g. intrathecal injection, BBB interruption and ultrasounds), with the disadvantage of being highly invasive methods, in addition to being associated with unwanted adverse effects [6]. Considering the prevalence of CNS diseases, many efforts have been made for the improvement of delivering drugs across the BBB.

17.3 PHARMACOTHERAPY

Conventional drugs for AD were designed to reduce or interrupt symptoms by eliminating protein β-amyloid plaques, stabilising tau proteins and decreasing neuroinflammatory and redox activities. The approved drugs are cholinesterase inhibitors, such as rivastigmine, donepezil, and galantamine. These drugs act in the prevention of degradation of acetylcholine, being more effective in a moderate stage of AD. Memantine acts differently as an antagonist of NMDA (*N*-methyl-D-aspartate) glutamate receptors whose connection is responsible for the recovery of the cognitive activity at a later stage of the disease [6].

The most used drug for Parkinson treatment is levodopa (L-dopa), a dopamine precursor. It is usually associated with carbidopa that controls peripheral effects of levodopa metabolism by inhibiting the enzyme DOPA decarboxylase, responsible for transforming L-dopa in dopamine [6,14]. However, using levodopa for a long period can lead to motor complications, such as dyskinesia. In addition, levodopa can control symptoms, but it does not decrease the progression of the disease. Other choices are dopamine agonists, cholinesterase inhibitors, β-blockers, antagonists of glutamate subtype-NMDA receptors and inhibitors of monoamine oxidase B [6]. However, the reduced ability of these drugs to cross the BBB compromises their mechanism of action and reduces the therapeutic effect.

17.4 NANOPARTICLES FOR TARGETING THE CNS

Nanoparticles (NPs) have been developed for the treatment of various chronic diseases, but they have special relevance to CNS disease application due to their numerous advantages; among them is the ability to cross BBB and promote a sustained drug release. In addition, the possibility to design a site-specific system by surface modification/functionalisation makes these systems desirable to develop new approaches for brain targeting [7]. The functionalisation of these systems has been investigated to adjust their properties as well as overcome the limitations of different administration routes. A functionalised surface increases the affinity of the nanoparticle to the target, allowing a higher concentration of the drug to reach the desired target, promoting a greater therapeutic effect. Furthermore, this strategy also allows the drug to remain in circulation for a longer time and escape from the reticuloendothelial system (RES) [15,16]. PEGylation – i.e. conjugation with polyethylene glycol (PEG) – is the most common approach to modify the surface of NP by increasing the hydrophilic character of the surface which improves the plasma half-life of the particles [17]. Indeed, stealth NPs can stay in the circulation for a prolonged time since they are not recognised by the RES [6]. Some characteristics of NPs should also be considered when addressing critical issues related to the development of suitable targeting systems. NPs should exhibit biodegradation, biocompatibility, no risk of toxicity, inflammatory or immunogenic reactions and should be stable when sterilised (e.g. if intended for intravenous administration) [16]. There are different types of NPs, developed by different methods and with different components. Their classification is based on the morphology and type of materials.

17.4.1 Mesoporous Silica Nanoparticles

Mesoporous silica nanoparticles (MSNs) are solid materials, mostly spherical, with honeycomb-like porous (in a range of 2–50 nm), which size is adaptable and allows to control the release [18,19]. Due to the porosity, this type of NPs can absorb large amounts of the drug. They exhibit advantages such as biocompatibility, low cytotoxicity and the possibility to be functionalised in both the surfaces, internal and external. The range of functionalising agents varies from polymers, such as PEG or stimuli-responsive polymers (responding to temperature or pH), macromolecules (DNA and RNA), to magnetic NPs. Therefore, they can release the drug under a stimulus [20–22]. Silica NPs have been reported to be stable when autoclaved [23] and when sterilised by gamma radiation [24].

17.4.2 Gold Nanoparticles

Gold NPs are metallic particles and are small, about 10–100 nm. They are visible in the near-infrared and have very unusual geometric, optical, thermal and chemical surface properties that make them highly stable, less cytotoxic and very easy to functionalise. Its size and shape are critical for cell uptake [15,21]. Radiolytic synthesis can be used both for the production and sterilisation of gold NP and can be performed using different types of radiation sources (gamma, e-beam, X-rays or even UV light) [25].

17.4.3 Polymeric Nanoparticles

They can be composed of both synthetic and natural polymers. Among them, the main natural polymers are chitosan and alginate. The most commonly used synthetic polymers are PLA (polylactic acid) or PCL (polycaprolactone). Compared with non-polymeric nanosystems, polymeric NP are easier to synthesise and less expensive, in addition to being more biocompatible and biodegradable [21]. Several methods are reported to sterilise polymeric NPs. The choice will be governed by the features of the polymers used as NP matrix. Autoclaving is not recommended for polymers with a melting point below 120°C as it reduces their mechanical properties and increases particle size [26]. Gamma irradiation can be applied on thermal-sensitive polymers with some precaution as it may interact with some polymeric bonds, affecting the release profiles [26]. Sterile filtration is recommended for chemical- or thermal-sensitive material and for NPs below 0.22 μm, with low viscosity [26]. High hydrostatic pressure can be applied to large batches, without inducing physical changes or damages. However, some bacterial spores can still resist treatment [26].

17.4.4 Liposomes

Liposomes are spherical vesicles characterised by an amphiphilic bilayer of phospholipids with an aqueous core. They can be loaded with hydrophilic drugs (aqueous core), lipophilic drugs (in the lipid bilayer) and amphiphilic drugs (partitioned at the surface of the bilayers). Depending on the temperature, liposomes exist in two states: fluid (more permeable) or gel (more stable, with a lower risk of escaping before reaching the site of action) [27]. They have cholesterol in the composition that gives stability and fluidity and influences membrane permeability [28]. They have been used in drug delivery systems for a long time and can be functionalised and administered through various routes [28]. Being composed of biocompatible and biodegradable materials, they have advantages, such as simple preparation, low toxicity and relatively low cost [16,29]. The most frequently used technique for sterilisation of liposomes is filtration with 0.22 μm filters [30].

17.4.5 Solid Lipid Nanoparticles

Solid lipid nanoparticles (SLNs) are colloidal systems whose matrix is composed of pure solid and biodegradable lipids and an amphiphilic surfactant. With a size range of 50–1,000 nm, they can deliver both hydrophilic and hydrophobic compounds and can be targeted to specific sites under a controlled drug release [28,31–33]. By comparing with polymer NPs and liposomes, SLNs show higher stability, lower cytotoxicity, higher drug-loading capacity and relatively lower cost [29]. The possibility of terminal sterilisation is one of the advantages of lipid NPs, and this process can be easily made by autoclaving [34,35].

17.4.6 Nanostructured Lipid Carriers

With characteristics very similar to SLNs, nanostructured lipid carriers (NLCs) are colloidal systems whose lipid matrix is composed of solid lipid and liquid lipid. This system improves the loading capacity of lipophilic drugs and maintains drug stability during storage. The drug release can be modulated by alteration in lipid matrix composition [28]. As happens with SLN, it is also possible to carry out terminal sterilisation of NLCs by autoclaving [34,35].

17.4.7 Quantum Dots

Semiconductor nanocrystals or quantum dots (QD) are small nanocrystals in size from 2 to 10 nm in diameter with high semiconductor properties, in addition to exhibiting fluorescent, optical and electrical properties that can be defined by modification of size and composition [36]. They have a semiconductor inorganic core, an inorganic shell and an aqueous organic coating. Inorganic shell increases photostability and luminescence, while the aqueous organic coating is used for conjugation of biomolecules to the QD surface (e.g. PEG). They emit between 450 and 850 nm (near-infrared), a very desirable region by biomedical imaging, which justifies its application in imaging, therapy and diagnostics [37–39]. Nanofiltration has been described to obtain sterilised QDs [40].

17.4.8 Fullerenes

Fullerenes are the third allotropic form of carbon, after diamond and graphite, structured by hexagons and pentagons forming a 'cage' of carbons. They have an electron-deficient behaviour and respond to electron-rich species. The structure and electronic ligations improve their stability. They are used as antioxidants (when functionalised), being able to moderate the damage caused by ND with an origin in radical damage. They are considered 100 times more effective than common antioxidants because they are able to absorb and neutralise more than 20 radicals per system [21]. Sterilisation of fullerene NPs has been reported applying laminar flow hood using 0.45 μm polyamide syringe filters [41].

17.4.9 Carbon Nanotubes

Formed by carbon in a cylindrical shape, they have an empty core that is single-walled or multi-walled. This structure influences their properties in the levels of conductivity or resilience, making them very strong. Carbon nanotubes (CNTs) can penetrate the plasma membrane and have the capacity to load several types of drugs, in addition to being biocompatible and functionalised. They have extensive applications in the healthcare sector, including pharmaceutical companies, biotech companies as well as biomedical industries. They also have been used in the regeneration of brain tissues. They are biocompatible and can be functionalised [21,28]. CNT dispersions have been reported to be stable when sterilised by gamma irradiation [42].

17.4.10 Dendrimers

These structures are symmetric branching units that are formed around a small molecule or linear polymer core. The dendritic polymers are similar to proteins, enzymes and viruses and are easily functionalised. They have hydrophilic external and internal environments, improving the bioavailability of less-soluble molecules. Their main use is in performing controlled and specific drug delivery. Conjugated drugs have a longer half-life, more stability and more solubility in water. They can also be used in imaging contrast and magnetic resonance. [21]. Since most dendrimer formulations have particle sizes below 200 nm, the most commonly used sterilisation method for these systems is filtration [43].

17.4.11 Nanogels

Nanogels are three-dimensional hydrogel constructed by cross-linked swellable polymer nets with a high capacity to swell without an aqueous medium. They are composed of natural polymers, synthetic polymers or both, being hydrophilic and highly biocompatible. Their chemical composition regulates proprieties, like size, charge, porosity, amphiphilicity, softness and degradability. They have a high drug-loading capacity, with the release controlled by a stimuli response or by the swelling capacity [44]. Nanogels can be produced and sterilised using gamma radiation and e-beam sterilisation [45]. Autoclave is also an approach used for simultaneous production and sterilisation of nanogels [45].

17.4.12 Micelles

Micelles are one of the main players in NPs to be used in advanced drug delivery systems [28]. They have a core shell structure, with a hydrophobic core and hydrophilic shell. These nanosystems exhibit numerous benefits, such as high-loading capacity, stability, improved solubilisation, prolonged circulation, smaller size (1–50 nm) and targeting potential. The shell stabilises the system and interacts with various molecules, such as proteins and cells. These versatile systems can assume several shapes and incorporate various solutes with different structures. The drug is released by diffusion, being controlled by pH, chemical and/or concentration variations [46]. Due to their small size, micelles are easily sterilised by filtration [47].

17.5 EXAMPLES OF NANOMEDICINES FOR ALZHEIMER'S DISEASE

17.5.1 Modulation of Aβ Production

The proteolysis of APP, modulated by secretases, gives rise to Aβ fragments. α-Secretase modulates activation of Aβ fragment production, while β- and γ-secretase modulate inactivation. β-Secretase, followed by γ-secretase, cleaves APP during the proteolysis originating Aβ fragments, and α-secretase cleaves APP in the middle of the Aβ region, thus no Aβ is produced. Although the inhibition of γ-secretase is related to numerous physiological processes, which can result in serious side effects, there is a possibility of using drug NPs to inhibit β-secretase or activate α-secretase and, thus, modulate the production of Aβ [7]. Preventing Aβ production is one way to reduce cerebral Aβ levels.

Li et al. designed targeted QD (conjugated with amino-PEG) to deliver small interfering RNA (siRNA) to the target (β-secretase) and reduce the Aβ in nerve cells. The QD-PEG/siRNA nanoplexes efficiently inhibited β-secretase, leading to the reduction of Aβ. The novel method for delivering siRNA into nerve cells is a promising strategy for the treatment of AD [48]. Smith et al. formulated nanolipidic particles to encapsulate a green tea polyphenol (epigallocatechin-3-gallate [EGCG]), a strong antioxidant able to upregulate α-secretase activity, preventing the formation of Aβ. The nanolipidic EGCG particles improved α-secretase-enhancing ability *in vitro* and its oral bioavailability *in vivo* [49]. The same catechin (EGCG) has been loaded into cationic SLN by Fangueiro et al., who have demonstrated their safety profile for ocular administration, both *in vitro* and *in vivo* [50,51]; the particles have been fully characterised [52,53], showing modified release profile to be further exploited for brain targeting. As BRB is an extension of the brain, strategies encountered to target the posterior segment of the eye can be exploited for brain targeting. Cano et al. have described dual loading of EGCG and ascorbic acid into PEGylated PLGA NPs with increased stability and enhanced EGCG accumulation in all major organs, including the brain [54]. The *in vivo* results have shown a marked increase in synapses in the animal models, together with improved spatial learning and memory.

17.5.2 Inhibition of Aβ Production

Since the Aβ fragments in oligomeric form are toxic, the inhibition of aggregation or dissolution of the Aβ aggregates reflects in the reduction or even stops AD progression. Nanosystems conjugated with distinct molecules able to interfere with Aβ aggregation are excellent candidates for this therapeutic strategy [7].

Both Aβ deposition and oxidative stress are factors that contribute to neurodegeneration in AD. Accumulation of transition metals, like iron and copper, in the pathological brain lesions of patients with AD, is the main cause of oxidative stress. In this way, metal chelators have been considered an interesting strategy for AD therapy, once they can selectively bind to and remove these metals. Furthermore, NPs conjugated with iron chelators may have the potential to deliver these agents into the brain, increase bioavailability and decrease undesirable toxic effects [55]. Liu et al. synthesised a nanoparticle chelator conjugate able to effectively inhibit Aβ aggregate formation, consisting of an efficient therapeutic approach for ND that presents an excess of transition metal.

Li et al. addressed the strategy of inhibiting Aβ aggregation by developing a hybrid nanosystem, containing polyoxometalates (POMs) in the inorganic portion and a sequence of peptides as an organic component. The POM peptide NPs showed enhanced targeting inhibition of Aβ aggregation in mice's cerebrospinal fluid. Furthermore, NPs can also be used as an effective fluorescence probe to monitor the inhibition process [56].

As a multifunctional approach to detect and inhibit fibrilisation, Li et al. designed a system formed by a peptide conjugated to gold nanorods (AuP). AuP can absorb radiation from near-infrared, and the two inhibitors, POMs and the peptide, can inhibit Aβ aggregation. AuP nanosystem, the combination of these components, could successfully inhibit Aβ aggregation, dissociate amyloid deposits with near-infrared irradiation, both *in vitro* and *in vivo*, and protect the cell from Aβ toxicity upon irradiation. Furthermore, due to the shape and size proprieties, nanorods can also act as sensitive diagnostic probes [57].

Airoldi et al. synthesised liposomes with ligands for Aβ peptides in their surface to target Aβ aggregates. The targeted liposomes presented a significant affinity for $Aβ_{1-42}$ oligomers, decreasing the *in vitro* aggregation of these amyloid peptides [58].

Curcumin is a potent anti-inflammatory and antioxidant compound that can modulate important targets with relevance in ND [7]. Cheng et al. improved the bioavailability of curcumin by developing curcumin NPs functionalised with PEG and PLA under the micellar form. By oral administration, results showed that the NPs significantly improved curcumin bioavailability both in plasma and in the brain is a promising therapy for AD [59].

Sanchez-Lopez et al. loaded memantine in PEG-PLGA NPs (MEM-PEG-PLGA NPs), which demonstrated *in vitro* the capacity to reduce β-amyloid plaques and the typical inflammation profile of AD [17]. MEM-PEG-PLGA NPs were found to be non-cytotoxic on bEnd.3 and astrocyte cell lines and were able to cross BBB.

17.5.3 Improvement of Aβ Clearance

Aβ clearance is compromised in AD patients, which results in toxic accumulation of Aβ. Thus, strategies to promote the clearance of Aβ are desirable to reduce these toxic effects. The usual procedure to promote Aβ clearance (using Aβ-specific antibodies) is partially efficient, once it causes autoimmune adverse effects. Nanosystems can act on Aβ clearance by different strategies, by accelerating glia-mediated Aβ degradation and facilitating receptor-mediated Aβ efflux due to their high binding affinity to Aβ peptides. Therefore, NPs have also been intensively studied for this purpose in the treatment of AD [7].

Song et al. produced a biologically inspired nanostructure functionalised with apolipoprotein E_3 and reconstituted high-density lipoprotein (ApoE3–rHDL). Results demonstrated high binding affinity from the liposomes to both Aβ monomer and oligomer, which contributed to facilitating lysosomal transport and, consequently, accelerated microglial, astroglial and liver cell degradation of Aβ. The *in vivo* studies also resulted in a reduction in Aβ deposition, attenuation of microgliosis, improvement in neurological changes and recovery in memory deficits in an animal model of AD [60].

Mancini et al. produced liposomes dually functionalised with a synthetic peptide containing the receptor-binding domain of apolipoprotein-E (for targeting and crossing BBB) and phosphatidic acid (for Aβ fragment binding). *In vivo* results showed that the liposomes reduced the deposition of brain Aβ and prevented the onset of long-term memory impairment. These findings point to a potentially suitable therapy for the treatment of AD at a pre-symptomatic stage [61].

Songjiang et al. explored permeation of BBB by producing a nano-vaccine based on chitosan NPs loaded with Aβ fragments. The nanosystem activated the immune response and allowed the permeation of BBB. Chitosan NPs are suitable carriers for Aβ fragments, which could be a forward step in the discovery of peptide vaccines for AD [62].

17.6 EXAMPLES OF NANOMEDICINES FOR PARKINSON'S DISEASE

Nanotechnology has also offered a wide range of possibilities for the treatment of PD. The targeting approaches include strategies related to α-synuclein accumulation, oxidative stress, mitochondrial dysfunction, inflammation and growth factor supplementation [7].

17.6.1 Dopamine Replacement

Dopamine deficiency is the main neurochemical abnormality in the early stage of PD [7]. Gambaryan et al. developed PLGA NP loaded with levodopa (nano-DOPA) to increase the efficiency in PD therapeutics. Nano-DOPA was administered intranasally and the *in vivo* results showed long-lasting therapeutic effects and the motor function improved greatly. This nanosystem also presented a better half-life, better bioavailability and better efficacy [63].

17.6.2 Targeting α-Synuclein Accumulation

Loureiro et al. developed PEGylated liposomes, surface modified (with the anti-transferrin receptor and anti-α-synuclein antibodies), and encapsulating EGCG to target transferrin receptor and α-synuclein. *In vitro* results showed a sustained release of EGCG, and the cellular uptake was more efficient than that of PEGylated [64].

Taebnia et al. developed curcumin-loaded amine-functionalised MSNs to study its effects on α-synuclein fibrillation. As a result, NPs lead to *in vitro* significant inhibition of the fibrillation process. Furthermore, the nanosystem showed enhanced stability when compared to the free drug [65].

17.6.2 Targeting Inflammation

Liu et al. developed a brain-targeted gene delivery system based on non-viral gene vector (dendrigraft poly-L-lysines) encapsulating shRNA to inhibit the activation of caspase-3. *In vivo* analysis indicated that the targeted NPs could accumulate in the brain more efficiently than non-targeted ones, silent caspase-3, and stop microglial activation and apoptosis, preventing neuronal death [66].

17.6.3 Targeting Antioxidative Stress

Oxidative stress induced by reactive oxygen species (ROS) is known to be involved in the progression of cancer. The use of antioxidants has long been proposed in the prophylaxis of cancer. Several phytochemicals with antioxidant properties (e.g. monoterpenes [67–69], polyphenols [70–75], catechins [76], sucupira oil [77] and essential oils [78–82]) have been loaded into NPs for oral administration.

Polysaccharide NPs made from e.g. chitosan [76,83–87] and alginate [88,89] are particularly interesting for oral administration. Siddique et al. synthesised alginate–curcumin nanocomposite to study the oxidative stress in a fly PD model. The nanosystem reduced oxidative damage and prevented neuronal apoptosis in the models, being a potent delayer in the progression of PD disability as well [90].

The lipophilic, potent antioxidant resveratrol is an interesting candidate for the loading into lipidic NPs [33,91]. It has been encapsulated in a nanoemulsion containing vitamin E for brain targeting by intranasal administration in rats [92]. The *ex vivo* analysis showed that vitamin E nanoemulsion efficiently targeted the brain, delivering a high concentration of resveratrol. Furthermore, the nanosystem also reduced oxidative stress by increasing glutathione and superoxide dismutase levels. Jose et al. proposed the loading of resveratrol into glyceryl behenate SLN for brain targeting for anticancer therapy [33]. The results showed that resveratrol–SLN were equally effective as the free drug as anticancer agent, but the *in vivo* biodistribution in the brain was significantly improved when administering the particles in Wistar rats in comparison to the free resveratrol.

17.6.4 Neurotrophic Factor Supplementation

Kurakhmaeva et al. investigated the brain delivery of poly(butyl cyanoacrylate) (PCBA) NP loaded with nerve growth factor (NGF). The NPs showed an efficient transport of NGF across the BBB and a significant reduction of basic symptoms of Parkinsonism [93].

17.7 STERILISATION OF NANOMATERIALS

Sterilisation of nanomaterials is a requirement in the case of parenteral, ocular or pulmonary administration. This is a challenging step in the manufacturing of injectable drug products containing nanomaterials since it can affect the physical and chemical properties of the produced batches, the cost of production and even the viability of the end product. Depending on the composition of NPs and the method used, sterilisation can cause structural changes in the particles, resulting in instability, inadequate drug release or even formation of toxic by-products. The most common techniques for sterilising NPs include gamma irradiation, filtration sterilisation and thermal sterilisation [94]. Gamma irradiation is a very usual technique as it can be used for many types of materials and has a high bactericidal effect, leaving no residues in the final product [94]. It can be applied to packaged products, which avoids further contamination [26]. It is a suitable approach to heat-sensitive materials. However, gamma irradiation may induce fragmentation of covalent bonds in some polymers and produce free radicals, resulting in photo-oxidative degradation of polymers, possible increase in molecular weight and changes in the drug release profile [26,94].

Thermal sterilisation, also called autoclaving, is the simplest and more convenient method [95]. However, it is not a suitable method for heat-sensitive materials as it involves high temperatures (120°C), which may result in degradation of the active ingredient and/or irreversible changes in the NP, affecting their stability [26,96].

Another effective method of nanomaterial sterilisation is filtration, considered an alternative approach for chemical- and/or thermo-sensitive materials. This technique requires the use of 220 nm filters for the filtration, representing a limitation for NPs with a wide particle size distribution and particle sizes closer to 220 nm [97] or viscous dispersions [26]. Furthermore, the filtration process does not remove pyrogens from the formulations, thereby requiring aseptic conditions for manufacturing [97].

There is no standard sterilisation method that could be used for all types of NPs. Choosing the best method requires an in-depth understanding of each step of the production process of NPs, together with the physicochemical properties of bulk materials and drugs. The method of choice must nevertheless be safe and reproducible and not generate toxic end products. For some types of NP, sterilisation methods that allow post-treatment of formulations may be recommended.

17.8 CONCLUSIONS

Although the causes of ND are still not fully understood, their complex pathophysiology can be exploited by targeting drugs to the brain using smart, surface-tailored NPs to reach the site of action and reduce the systemic distribution. Conventional drugs, although well established and widely accepted, are often ineffective, with severe side effects and unable to control the progress of ND. NPs play an important role in the delivery of a variety of drugs into the brain by efficiently overcoming the BBB. The stabilisation of the drug against degradation in biological fluids, the increase of its lifetime in the bloodstream and the possibility to target the drug make NPs interesting carriers for both classical drugs and new ones being developed for the management of ND. A key aspect in formulation development is the biodegradability, biocompatibility, acute/chronic toxicity and stability of the carrier itself. These aspects also have to be considered when selecting the sterilisation method. This chapter covers a range of NPs for brain targeting of drugs used for the treatment of ND, their production and methods of sterilising the formulations.

ACKNOWLEDGEMENTS

The authors would like to thank the financial support received from the Portuguese Science and Technology Foundation (FCT/MCT) and from European Funds (PRODER/COMPETE) for the M-ERA-NET/0004/2015-PAIRED, co-financed by FEDER, under the Partnership Agreement PT2020. This study was also supported by the Portuguese Foundation for Science and Technology (FCT) under the scope of the strategic funding of UIDB/04469/2020 unit and BioTecNorte operation (NORTE-01-0145-FEDER-000004) funded by the European Regional Development Fund under the scope of Norte2020-Programa Operacional Regional do Norte. The authors also acknowledge CAPES (Coordenação de Aperfeiçoamento de Pessoal de Nível Superior) for the financial support and for the fellowship of Iara Baldim (88887.368385/2019-00).

REFERENCES

1. N. Poovaiah, Z. Davoudi, H. Peng, B. Schlichtmann, S. Mallapragada, B. Narasimhan, Q. Wang, Treatment of neurodegenerative disorders through the blood-brain barrier using nanocarriers, *Nanoscale*, 10 (2018) 16962–16983.
2. M.S. Forman, J.Q. Trojanowski, V.M. Lee, Neurodegenerative diseases: A decade of discoveries paves the way for therapeutic breakthroughs, *Nature Medicine*, 10 (2004) 1055–1063.
3. A. Misra, S. Ganesh, A. Shahiwala, S.P. Shah, Drug delivery to the central nervous system: A review, *Journal of Pharmacy and Pharmaceutical Sciences*, 6 (2003) 252–273.
4. K. Goyal, V. Koul, Y. Singh, A. Anand, Targeted drug delivery to central nervous system (CNS) for the treatment of neurodegenerative disorders: Trends and advances, *Central Nervous System Agents in Medicinal Chemistry*, 14 (2014) 43–59.
5. L. Gastaldi, L. Battaglia, E. Peira, D. Chirio, E. Muntoni, I. Solazzi, M. Gallarate, F. Dosio, Solid lipid nanoparticles as vehicles of drugs to the brain: Current state of the art, *European Journal of Pharmaceutics and Biopharmaceutics: Official Journal of Arbeitsgemeinschaft fur Pharmazeutische Verfahrenstechnik e.V*, 87 (2014) 433–444.
6. S. Cunha, M. H. Amaral, J. Lobo, A. Silva, Therapeutic strategies for Alzheimer's and Parkinson's diseases by means of drug delivery systems, *Current Medicinal Chemistry*, 23 (2016) 1–1.
7. W. Zhang, W. Wang, D.X. Yu, Z. Xiao, Z. He, Application of nanodiagnostics and nanotherapy to CNS diseases, *Nanomedicine (London, England)*, 13 (2018) 2341–2371.
8. G.H. Hawthorne, M.P. Bernuci, M. Bortolanza, V. Tumas, A.C. Issy, E. Del-Bel, Nanomedicine to overcome current Parkinson's treatment liabilities: A systematic review, *Neurotoxicity Research*, 30 (2016) 715–729.
9. K. Sweers, K. van der Werf, M. Bennink, V. Subramaniam, Nanomechanical properties of alpha-synuclein amyloid fibrils: A comparative study by nanoindentation, harmonic force microscopy, and Peakforce QNM, *Nanoscale Research Letters*, 6 (2011) 270.
10. X. Dong, Current strategies for brain drug delivery, *Theranostics*, 8 (2018) 1481–1493.
11. C. Saraiva, C. Praca, R. Ferreira, T. Santos, L. Ferreira, L. Bernardino, Nanoparticle-mediated brain drug delivery: Overcoming blood-brain barrier to treat neurodegenerative diseases, *Journal of Controlled Release: Official Journal of the Controlled Release Society*, 235 (2016) 34–47.
12. A.M. Grabrucker, B. Ruozi, D. Belletti, F. Pederzoli, F. Forni, M.A. Vandelli, G. Tosi, Nanoparticle transport across the blood brain barrier, *Tissue Barriers*, 4 (2016) e1153568.

13. M. Tajes, E. Ramos-Fernandez, X. Weng-Jiang, M. Bosch-Morato, B. Guivernau, A. Eraso-Pichot, B. Salvador, X. Fernandez-Busquets, J. Roquer, F.J. Munoz, The blood-brain barrier: Structure, function and therapeutic approaches to cross it, *Molecular Membrane Biology*, 31 (2014) 152–167.
14. E. Sanchez-Lopez, M.A. Egea, B.M. Davis, L. Guo, M. Espina, A.M. Silva, A.C. Calpena, E.M.B. Souto, N. Ravindran, M. Ettcheto, A. Camins, M.L. Garcia, M.F. Cordeiro, Memantine-loaded PEGylated biodegradable nanoparticles for the treatment of glaucoma, *Small*, 14 (2018) 1701808.
15. R.R. Arvizo, S. Bhattacharyya, R.A. Kudgus, K. Giri, R. Bhattacharya, P. Mukherjee, Intrinsic therapeutic applications of noble metal nanoparticles: Past, present and future, *Chemical Society Reviews*, 41 (2012) 2943–2970.
16. K.T. Nguyen, M.N. Pham, T.V. Vo, W. Duan, P.H. Tran, T.T. Tran, Strategies of engineering nanoparticles for treating neurodegenerative disorders, *Current Drug Metabolism*, 18 (2017) 786–797.
17. E. Sanchez-Lopez, M. Ettcheto, M.A. Egea, M. Espina, A. Cano, A.C. Calpena, A. Camins, N. Carmona, A.M. Silva, E.B. Souto, M.L. Garcia, Memantine loaded PLGA PEGylated nanoparticles for Alzheimer's disease: In vitro and *in vivo* characterization, *Journal of Nanobiotechnology*, 16 (2018) 32.
18. T. Andreani, A.L. de Souza, C.P. Kiill, E.N. Lorenzon, J.F. Fangueiro, A.C. Calpena, M.V. Chaud, M.L. Garcia, M.P. Gremiao, A.M. Silva, E.B. Souto, Preparation and characterization of PEG-coated silica nanoparticles for oral insulin delivery, *International Journal of Pharmaceutics*, 473 (2014) 627–635.
19. T. Andreani, L. Miziara, E.N. Lorenzon, A.L. de Souza, C.P. Kiill, J.F. Fangueiro, M.L. Garcia, P.D. Gremiao, A.M. Silva, E.B. Souto, Effect of mucoadhesive polymers on the in vitro performance of insulin-loaded silica nanoparticles: Interactions with mucin and biomembrane models, *European Journal of Pharmaceutics and Biopharmaceutics*, 93 (2015) 118–126.
20. Slowing II, J.L. Vivero-Escoto, C.W. Wu, V.S. Lin, Mesoporous silica nanoparticles as controlled release drug delivery and gene transfection carriers, *Advanced Drug Delivery Reviews*, 60 (2008) 1278–1288.
21. D. Silva Adaya, L. Aguirre-Cruz, J. Guevara, E. Ortiz-Islas, Nanobiomaterials' applications in neurodegenerative diseases, *Journal of Biomaterials Applications*, 31 (2017) 953–984.
22. V. Mamaeva, C. Sahlgren, M. Linden, Mesoporous silica nanoparticles in medicine-recent advances, *Advanced Drug Delivery Reviews*, 65 (2013) 689–702.
23. H. Rokbani, F. Daigle, A. Ajji, Combined effect of ultrasound stimulations and autoclaving on the enhancement of antibacterial activity of ZnO and SiO_2/ZnO nanoparticles, *Nanomaterials (Basel)*, 8 (2018) 129.
24. J.J. Lin, P.Y. Hsu, Gamma-ray sterilization effects in silica nanoparticles/gamma-APTES nanocomposite-based pH-sensitive polysilicon wire sensors, *Sensors (Basel)*, 11 (2011) 8769–8781.
25. L. Freitas de Freitas, G.H.C. Varca, J.G. Dos Santos Batista, A. Benevolo Lugao, An overview of the synthesis of gold nanoparticles using radiation technologies, *Nanomaterials (Basel)*, 8 (2018) 939.
26. C. Vauthier, K. Bouchemal, Methods for the preparation and manufacture of polymeric nanoparticles, *Pharmaceutical Research*, 26 (2009) 1025–1058.
27. U. Bulbake, S. Doppalapudi, N. Kommineni, W. Khan, Liposomal formulations in clinical use: An updated review, *Pharmaceutics*, 9 (2017) 12.
28. D.K. Mishra, R. Shandilya, P.K. Mishra, Lipid based nanocarriers: A translational perspective, *Nanomedicine: Nanotechnology, Biology, and Medicine*, 14 (2018) 2023–2050.
29. Q. He, J. Liu, J. Liang, X. Liu, W. Li, Z. Liu, Z. Ding, D. Tuo, Towards improvements for penetrating the blood-brain barrier-recent progress from a material and pharmaceutical perspective, *Cells*, 7 (2018) 24.
30. B.S. Pattni, V.V. Chupin, V.P. Torchilin, New developments in liposomal drug delivery, *Chemical Reviews*, 115 (2015) 10938–10966.
31. S. Martins, I. Tho, I. Reimold, G. Fricker, E. Souto, D. Ferreira, M. Brandl, Brain delivery of camptothecin by means of solid lipid nanoparticles: Formulation design, *in vitro* and *in vivo* studies, *International Journal of Pharmaceutics*, 439 (2012) 49–62.
32. M. Patel, E.B. Souto, K.K. Singh, Advances in brain drug targeting and delivery: Limitations and challenges of solid lipid nanoparticles, *Expert Opinion on Drug Delivery*, 10 (2013) 889–905.
33. S. Jose, S.S. Anju, T.A. Cinu, N.A. Aleykutty, S. Thomas, E.B. Souto, In vivo pharmacokinetics and biodistribution of resveratrol-loaded solid lipid nanoparticles for brain delivery, *International Journal of Pharmaceutics*, 474 (2014) 6–13.
34. E. Sanchez-Lopez, M. Espina, S. Doktorovova, E.B. Souto, M.L. Garcia, Lipid nanoparticles (SLN, NLC): Overcoming the anatomical and physiological barriers of the eye – Part II – Ocular drug-loaded lipid nanoparticles, *European Journal of Pharmaceutics and Biopharmaceutics*, 110 (2017) 58–69.
35. R.H. Muller, R. Shegokar, C.M. Keck, 20 Years of lipid nanoparticles (SLN and NLC): Present state of development and industrial applications, *Current Drug Discovery Technologies*, 8 (2011) 207–227.
36. J. Thomas, S. Thomas, N. Kalarikkal, J. Jose, *Nanoparticles in Polymer Systems for Biomedical Applications*, CRC Press, Boca Raton, FL (2018).
37. K.D. Wegner, N. Hildebrandt, Quantum dots: Bright and versatile *in vitro* and *in vivo* fluorescence imaging biosensors, *Chemical Society Reviews*, 44 (2015) 4792–4834.
38. C.T. Matea, T. Mocan, F. Tabaran, T. Pop, O. Mosteanu, C. Puia, C. Iancu, L. Mocan, Quantum dots in imaging, drug delivery and sensor applications, *International Journal of Nanomedicine*, 12 (2017) 5421–5431.
39. C.E. Probst, P. Zrazhevskiy, V. Bagalkot, X. Gao, Quantum dots as a platform for nanoparticle drug delivery vehicle design, *Advanced Drug Delivery Reviews*, 65 (2013) 703–718.
40. C.M. Courtney, S.M. Goodman, J.A. McDaniel, N.E. Madinger, A. Chatterjee, P. Nagpal, Photoexcited quantum dots for killing multidrug-resistant bacteria, *Nature Materials*, 15 (2016) 529–534.
41. T. Kovač, B. Šarkanj, T. Klapec, I. Borišev, M. Kovač, A. Nevistić, I. Strelec, Antiaflatoxigenic effect of fullerene C_{60} nanoparticles at environmentally plausible concentrations, *AMB Express*, 8 (2018) 14–14.
42. J.A. Fagan, N.J. Lin, R. Zeisler, A.R. Hight Walker, Effects of gamma irradiation for sterilization on aqueous dispersions of length sorted carbon nanotubes, *Nano Research*, 4 (2011) 393–404.
43. L.P. Wu, M. Ficker, J.B. Christensen, P.N. Trohopoulos, S.M. Moghimi, Dendrimers in medicine: Therapeutic concepts and pharmaceutical challenges, *Bioconjugate Chemistry*, 26 (2015) 1198–1211.

44. K.S. Soni, S.S. Desale, T.K. Bronich, Nanogels: An overview of properties, biomedical applications and obstacles to clinical translation, *Journal of Controlled Release: Official Journal of the Controlled Release Society*, 240 (2016) 109–126.
45. M. Vicario-de-la-Torre, J. Forcada, The potential of stimuli-responsive nanogels in drug and active molecule delivery for targeted therapy, *Gels*, 3 (2017) 16.
46. G. Modi, V. Pillay, Y.E. Choonara, V.M.K. Ndesendo, L.C. du Toit, D. Naidoo, Nanotechnological applications for the treatment of neurodegenerative disorders, *Progress in Neurobiology*, 88 (2009) 272–285.
47. R.S. Elezaby, H.A. Gad, A.A. Metwally, A.S. Geneidi, G.A. Awad, Self-assembled amphiphilic core-shell nanocarriers in line with the modern strategies for brain delivery, *Journal of Controlled Release*, 261 (2017) 43–61.
48. S. Li, Z. Liu, F. Ji, Z. Xiao, M. Wang, Y. Peng, Y. Zhang, L. Liu, Z. Liang, F. Li, Delivery of quantum Dot-siRNA nanoplexes in SK-N-SH cells for BACE1 gene silencing and intracellular imaging, *Molecular Therapy – Nucleic Acids*, 1 (2012) e20.
49. A. Smith, B. Giunta, P.C. Bickford, M. Fountain, J. Tan, R.D. Shytle, Nanolipidic particles improve the bioavailability and α-secretase inducing ability of epigallocatechin-3-gallate (EGCG) for the treatment of Alzheimer's disease, *International Journal of Pharmaceutics*, 389 (2010) 207–212.
50. J.F. Fangueiro, A.C. Calpena, B. Clares, T. Andreani, M.A. Egea, F.J. Veiga, M.L. Garcia, A.M. Silva, E.B. Souto, Biopharmaceutical evaluation of epigallocatechin gallate-loaded cationic lipid nanoparticles (EGCG-LNs): In vivo, in vitro and ex vivo studies, *International Journal of Pharmaceutics*, 502 (2016) 161–169.
51. L.Y. Zakharova, T.N. Pashirova, S. Doktorovova, A.R. Fernandes, E. Sanchez-Lopez, A.M. Silva, S.B. Souto, E.B. Souto, Cationic surfactants: Self-assembly, structure-activity correlation and their biological applications, *International Journal of Molecular Sciences*, 20 (2019) 5534.
52. J.F. Fangueiro, T. Andreani, L. Fernandes, M.L. Garcia, M.A. Egea, A.M. Silva, E.B. Souto, Physicochemical characterization of epigallocatechin gallate lipid nanoparticles (EGCG-LNs) for ocular instillation, *Colloids and Surfaces B: Biointerfaces*, 123 (2014) 452–460.
53. J.F. Fangueiro, A. Parra, A.M. Silva, M.A. Egea, E.B. Souto, M.L. Garcia, A.C. Calpena, Validation of a high performance liquid chromatography method for the stabilization of epigallocatechin gallate, *International Journal of Pharmaceutics*, 475 (2014) 181–190.
54. A. Cano, M. Ettcheto, J.H. Chang, E. Barroso, M. Espina, B.A. Kuhne, M. Barenys, C. Auladell, J. Folch, E.B. Souto, A. Camins, P. Turowski, M.L. Garcia, Dual-drug loaded nanoparticles of epigallocatechin-3-gallate (EGCG)/ascorbic acid enhance therapeutic efficacy of EGCG in a APPswe/PS1dE9 Alzheimer's disease mice model, *Journal of Controlled Release*, 301 (2019) 62–75.
55. G. Liu, P. Men, W. Kudo, G. Perry, M.A. Smith, Nanoparticle-chelator conjugates as inhibitors of amyloid-beta aggregation and neurotoxicity: A novel therapeutic approach for Alzheimer disease, *Neuroscience Letters*, 455 (2009) 187–190.
56. C.A. Giriko, C.A. Andreoli, L.V. Mennitti, L.F. Hosoume, S. Souto Tdos, A.V. Silva, C. Mendes-da-Silva, Delayed physical and neurobehavioral development and increased aggressive and depression-like behaviors in the rat offspring of dams fed a high-fat diet, *International Journal of Developmental Neuroscience: The Official Journal of the International Society for Developmental Neuroscience*, 31 (2013) 731–739.
57. M. Li, Y. Guan, A. Zhao, J. Ren, X. Qu, Using multifunctional peptide conjugated au nanorods for monitoring beta-amyloid aggregation and chemo-photothermal treatment of Alzheimer's disease, *Theranostics*, 7 (2017) 2996–3006.
58. C. Airoldi, S. Mourtas, F. Cardona, C. Zona, E. Sironi, G. D'Orazio, E. Markoutsa, F. Nicotra, S.G. Antimisiaris, B. La Ferla, Nanoliposomes presenting on surface a cis-glycofused benzopyran compound display binding affinity and aggregation inhibition ability towards Amyloid beta1-42 peptide, *European Journal of Medicinal Chemistry*, 85 (2014) 43–50.
59. K.K. Cheng, C.F. Yeung, S.W. Ho, S.F. Chow, A.H. Chow, L. Baum, Highly stabilized curcumin nanoparticles tested in an *in vitro* blood-brain barrier model and in Alzheimer's disease Tg2576 mice, *The AAPS Journal*, 15 (2013) 324–336.
60. Q. Song, M. Huang, L. Yao, X. Wang, X. Gu, J. Chen, J. Chen, J. Huang, Q. Hu, T. Kang, Z. Rong, H. Qi, G. Zheng, H. Chen, X. Gao, Lipoprotein-based nanoparticles rescue the memory loss of mice with Alzheimer's disease by accelerating the clearance of amyloid-beta, *ACS Nano*, 8 (2014) 2345–2359.
61. S. Mancini, C. Balducci, E. Micotti, D. Tolomeo, G. Forloni, M. Masserini, F. Re, Multifunctional liposomes delay phenotype progression and prevent memory impairment in a presymptomatic stage mouse model of Alzheimer disease, *Journal of Controlled Release: Official Journal of the Controlled Release Society*, 258 (2017) 121–129.
62. Z. Songjiang, W. Lixiang, Amyloid-beta associated with chitosan nano-carrier has favorable immunogenicity and permeates the BBB, *AAPS PharmSciTech*, 10 (2009) 900–905.
63. P.Y. Gambaryan, I.G. Kondrasheva, E.S. Severin, A.A. Guseva, A.A. Kamensky, Increasing the efficiency of Parkinson's disease treatment using a poly(lactic-co-glycolic acid) (PLGA) based L-DOPA delivery system, *Experimental Neurobiology*, 23 (2014) 246–252.
64. J.A. Loureiro, B. Gomes, M.A. Coelho, M. do Carmo Pereira, S. Rocha, Immunoliposomes doubly targeted to transferrin receptor and to alpha-synuclein, *Future Science OA*, 1 (2015) Fso71.
65. N. Taebnia, D. Morshedi, S. Yaghmaei, F. Aliakbari, F. Rahimi, A. Arpanaei, Curcumin-loaded amine-functionalized mesoporous silica nanoparticles inhibit alpha-synuclein fibrillation and reduce its cytotoxicity-associated effects, *Langmuir*, 32 (2016) 13394–13402.
66. Y. Liu, Y. Guo, S. An, Y. Kuang, X. He, H. Ma, J. Li, J. Lu, N. Zhang, C. Jiang, Targeting caspase-3 as dual therapeutic benefits by RNAi facilitating brain-targeted nanoparticles in a rat model of Parkinson's disease, *PLoS One*, 8 (2013) e62905.
67. A. Zielinska, N.R. Ferreira, A. Durazzo, M. Lucarini, N. Cicero, S.E. Mamouni, A.M. Silva, A. Nowak, A. Santini, E.B. Souto, Development and optimization of alpha-pinene-loaded solid lipid nanoparticles (SLN) using experimental factorial design and dispersion analysis, *Molecules*, 24 (2019) 2683.
68. A. Zielińska, N.R. Ferreira, A. Feliczak-Guzik, E.B. Souto, I. Nowak, Loading release kinetics and stability assessment of monoterpenes-loaded solid lipid nanoparticles (SLN), *Pharmaceutical Development and Technology*, 25 (2020) 832–844.

69. A. Zielinska, C. Martins-Gomes, N.R. Ferreira, A.M. Silva, I. Nowak, E.B. Souto, Anti-inflammatory and anti-cancer activity of citral: Optimization of citral-loaded solid lipid nanoparticles (SLN) using experimental factorial design and LUMiSizer(R), *International Journal of Pharmaceutics*, 553 (2018) 428–440.
70. I.S. Santos, B.M. Ponte, P. Boonme, A.M. Silva, E.B. Souto, Nanoencapsulation of polyphenols for protective effect against colon-rectal cancer, *Biotechnology Advances*, 31 (2013) 514–523.
71. E.B. Souto, P. Severino, R. Basso, M.H. Santana, Encapsulation of antioxidants in gastrointestinal-resistant nanoparticulate carriers, *Methods in Molecular Biology*, 1028 (2013) 37–46.
72. S. Pimentel-Moral, M.C. Teixeira, A.R. Fernandes, D. Arraez-Roman, A. Martinez-Ferez, A. Segura-Carretero, E.B. Souto, Lipid nanocarriers for the loading of polyphenols – A comprehensive review, *Advances in Colloid and Interface Science*, 260 (2018) 85–94.
73. A. Durazzo, M. Lucarini, E.B. Souto, C. Cicala, E. Caiazzo, A.A. Izzo, E. Novellino, A. Santini, Polyphenols: A concise overview on the chemistry, occurrence, and human health, *Phytotherapy Research*, 33 (2019) 2221–2243.
74. B. Salehi, A. Venditti, M. Sharifi-Rad, D. Kregiel, J. Sharifi-Rad, A. Durazzo, M. Lucarini, A. Santini, E.B. Souto, E. Novellino, H. Antolak, E. Azzini, W.N. Setzer, N. Martins, The therapeutic potential of apigenin, *International Journal of Molecular Sciences*, 20 (2019) 1305.
75. L.C. Cefali, J.A. Ataide, A.R. Fernandes, I.M.O. Sousa, F. Goncalves, S. Eberlin, J.L. Davila, A.F. Jozala, M.V. Chaud, E. Sanchez-Lopez, J. Marto, M.A. d'Avila, H.M. Ribeiro, M.A. Foglio, E.B. Souto, P.G. Mazzola, Flavonoid-enriched plant-extract-loaded emulsion: A novel phytocosmetic sunscreen formulation with antioxidant properties, *Antioxidants (Basel)*, 8 (2019) 443.
76. E.B. Souto, S.B. Souto, C. Marques, L.N. Andrade, O.K. Horbańczuk, A.G. Atanasov, M. Lucarini, A. Durazzo, A. Santini, P. Severino, Polyphenon-60 from green tea in Eudragit S100-chitosan microspheres: Role in modified-release and oxidative stress reduction involved in metabolic diseases, *Nutrients*, 12 (2020) 967. doi: 10.3390/nu12040967.
77. R. Vieira, P. Severino, L.A. Nalone, S.B. Souto, A.M. Silva, M. Lucarini, A. Durazzo, A. Santini, E.B. Souto, Sucupira oil-loaded nanostructured lipid carriers (NLC): Lipid screening, factorial design, release profile and cytotoxicity, *Molecules*, 25 (2020) 685. https://doi.org/10.3390/molecules25030685
78. E.B. Souto, S.B. Souto, P. Severino, J. Dias-Ferreira, B.C. Naveros, A. Durazzo, M. Lucarini, A.G. Atanasov, S. El Mamouni, A. Santini, Croton argyrophyllus Kunth essential oil – Loaded SLN: Optimization and evaluation of antioxidant and antitumoral activities, *Sustainability* 12 (2020) 7697. https://doi.org/10.3390/su12187697.
79. P. Severino, T. Andreani, M.V. Chaud, C.I. Benites, S.C. Pinho, E.B. Souto, Essential oils as active ingredients of lipid nanocarriers for chemotherapeutic use, *Current Pharmaceutical Biotechnology*, 16 (2015) 365–370.
80. C. Carbone, C. Martins-Gomes, C. Caddeo, A.M. Silva, T. Musumeci, R. Pignatello, G. Puglisi, E.B. Souto, Mediterranean essential oils as precious matrix components and active ingredients of lipid nanoparticles, *International Journal of Pharmaceutics*, 548 (2018) 217–226.
81. I. Pereira, A. Zielinska, N.R. Ferreira, A.M. Silva, E.B. Souto, Optimization of linalool-loaded solid lipid nanoparticles using experimental factorial design and long-term stability studies with a new centrifugal sedimentation method, *International Journal of Pharmaceutics*, 549 (2018) 261–270.
82. R. Vieira, S.B. Souto, E. Sanchez-Lopez, A.L. Machado, P. Severino, S. Jose, A. Santini, A. Fortuna, M.L. Garcia, A.M. Silva, E.B. Souto, Sugar-lowering drugs for type 2 diabetes mellitus and metabolic syndrome-review of classical and new compounds: Part-I, *Pharmaceuticals (Basel)*, 12 (2019) 152.
83] S. Jose, M.T. Prema, A.J. Chacko, A.C. Thomas, E.B. Souto, Colon specific chitosan microspheres for chronotherapy of chronic stable angina, *Colloids and Surfaces B: Biointerfaces*, 83 (2011) 277–283.
84. S. Jose, J.F. Fangueiro, J. Smitha, T.A. Cinu, A.J. Chacko, K. Premaletha, E.B. Souto, Cross-linked chitosan microspheres for oral delivery of insulin: Taguchi design and in vivo testing, *Colloids and Surfaces B: Biointerfaces*, 92 (2012) 175–179.
85. S. Jose, J.F. Fangueiro, J. Smitha, T.A. Cinu, A.J. Chacko, K. Premaletha, E.B. Souto, Predictive modeling of insulin release profile from cross-linked chitosan microspheres, *European Journal of Medicinal Chemistry*, 60 (2013) 249–253.
86. P. Severino, E.B. Souto, S.C. Pinho, M.H. Santana, Hydrophilic coating of mitotane-loaded lipid nanoparticles: Preliminary studies for mucosal adhesion, *Pharmaceutical Development and Technology*, 18 (2013) 577–581.
87. J.A. Ataide, E.F. Gérios, L.C. Cefali, A.R. Fernandes, M.D.C. Teixeira, N.R. Ferreira, E.B. Tambourgi, A.F. Jozala, M.V. Chaud, L. Oliveira-Nascimento, P.G. Mazzola, E.B. Souto, Effect of polysaccharide sources on the physicochemical properties of bromelain-chitosan nanoparticles, *Polymers*, 11 (2019) 1681.
88. P. Severino, M.V. Chaud, A. Shimojo, D. Antonini, M. Lancelloti, M.H. Santana, E.B. Souto, Sodium alginate-cross-linked polymyxin B sulphate-loaded solid lipid nanoparticles: Antibiotic resistance tests and HaCat and NIH/3T3 cell viability studies, *Colloids and Surfaces B: Biointerfaces*, 129 (2015) 191–197.
89. P. Severino, C.F. da Silva, L.N. Andrade, D. de Lima Oliveira, J. Campos, E.B. Souto, Alginate nanoparticles for drug delivery and targeting, *Current Pharmaceutical Design*, 25 (2019) 1312–1334.
90. Y.H. Siddique, W. Khan, B.R. Singh, A.H. Naqvi, Synthesis of alginate-curcumin nanocomposite and its protective role in transgenic Drosophila model of Parkinson's disease, *International Scholarly Research Notices Pharmacology*, 2013 (2013) 794582.
91. R.B. Rigon, N. Fachinetti, P. Severino, A. Durazzo, M. Lucarini, A.G. Atanasov, S. El Mamouni, M. Chorilli, A. Santini, E.B. Souto, Quantification of trans-resveratrol-loaded solid lipid nanoparticles by a validated reverse-phase HPLC photodiode array, *Applied Sciences* 9 (2019) 4961.
92. R. Pangeni, S. Sharma, G. Mustafa, J. Ali, S. Baboota, Vitamin E loaded resveratrol nanoemulsion for brain targeting for the treatment of Parkinson's disease by reducing oxidative stress, *Nanotechnology*, 25 (2014) 485102.
93. K.B. Kurakhmaeva, I.A. Djindjikhashvili, V.E. Petrov, V.U. Balabanyan, T.A. Voronina, S.S. Trofimov, J. Kreuter, S. Gelperina, D. Begley, R.N. Alyautdin, brain

targeting of nerve growth factor using poly(butyl cyanoacrylate) nanoparticles, *Journal of Drug Targeting*, 17 (2009) 564–574.
94. L. Wang, J. Du, Y. Zhou, Y. Wang, Safety of nanosuspensions in drug delivery, *Nanomedicine*, 13 (2017) 455–469.
95. J. Pardeike, S. Weber, T. Haber, J. Wagner, H.P. Zarfl, H. Plank, A. Zimmer, Development of an itraconazole-loaded nanostructured lipid carrier (NLC) formulation for pulmonary application, *International Journal of Pharmaceutics*, 419 (2011) 329–338.
96. L.M. Negi, M. Jaggi, S. Talegaonkar, Development of protocol for screening the formulation components and the assessment of common quality problems of nano-structured lipid carriers, *International Journal of Pharmaceutics*, 461 (2014) 403–410.
97. K.M. Tyner, P. Zou, X. Yang, H. Zhang, C.N. Cruz, S.L. Lee, Product quality for nanomaterials: Current U.S. experience and perspective, *Wiley Interdisciplinary Reviews. Nanomedicine and Nanobiotechnology*, 7 (2015) 640–654.

18 Validation, Scale-Up and Technology Transfer in Product Development

Ankit Gaur
Sentiss Pharma Pvt Ltd

Anamika Sahu Gulbake
DIT University

Neelesh Kumar Mehra
National Institute of Pharmaceutical Education & Research (NIPER), Hyderabad

Arvind Gulbake
DIT University
D.Y. Patil Education Society

CONTENTS

Abbreviations 269
18.1 Introduction 269
18.2 Validation 270
    18.2.1 Types of Validation 271
        18.2.1.1 Process Validation 271
        18.2.1.2 Process Validation and Drug Quality 271
    18.2.2 Computer System Validation (Process Controller) 272
        18.2.2.1 Computer System Validation Process 273
    18.2.3 Cleaning Validation 274
        18.2.3.1 General Requirement 274
        18.2.3.2 Evaluation of Cleaning Validation 274
    18.2.4 Analytical Method Validation 276
18.3 Scale-Up and Technology Transfer in Pharmaceutical Product Development 276
    18.3.1 Process Transfer: An Overall Framework 277
    18.3.2 Goal of Technology Transfer 277
    18.3.3 Steps in Technology Transfer 277
        18.3.3.1 Quality by Design in Scale-Up 280
18.4 Importance of Technology Transfer 280
18.5 Conclusion 281
Acknowledgement 281
References 281

ABBREVIATIONS

CFR: Code of Federal Regulations
cGMP: Current Good Manufacturing Practice
CPV: Continued process verification
DOE: Design of experiment
FDA: Food and Drug Administration
GMP: Good manufacturing practice
ICH: International Council for Harmonisation of Technical Requirements for Pharmaceuticals for Human Use
NDA: New Drug Application
PAT: Process analytical technology
PPQ: Process performance qualification
R&D: Research and Development
RPM: Revolutions per minute
WHO: World Health Organization

18.1 INTRODUCTION

Development of pharmaceutical products is very complex, tough and time-taking process; it includes various research findings, laboratory work, analysis, observations, challenges and resolutions. There are so many stages involved in the development of pharmaceutical products. There are four vital stages in product development including preformulation stage, formulation stage, scale-up and technology transfer stage, and validation stage. Although there are challenges at every stage in pharmaceutical product development, validation stage is a very important stage in pharmaceutical product development (Devalapally et al., 2007).

These stages demand highly competent, qualified, experienced and trained scientists to develop and validate all methods and processes efficiently and carefully with proper documentation. Validation is a documented tool of Good Manufacturing Practice (GMP). The term validation was first proposed by Food and Drug Administration (FDA) in 1979 in the United States. The guideline is defined under Title 21 of Federal Regulation section 10.90 (21 CFR 10.90) in 1987 for the manufacture of pharmaceuticals and medical devices and to improve their quality. Process validation is necessary in pharmaceutical product development because it makes good scientific sense and is considered in the U.S. FDA & World Health Organization (WHO) under the authority of Current Good Manufacturing Practice (cGMP) regulation guidelines and directives. Validation ensures the quality of food and drug products for a sufficient period of time. Validation is written evidence that signifies the procedure, process or activity carried out at various stages or steps and are okay to use. Validation in pharma is important for the manufacturing process to ensure product quality, efficacy, consistency and safety. Pharmaceutical products are extremely sensitive to environmental conditions because of the use of chemicals as raw materials. Validation is the process of control and evaluating products, analytical methods and cleaning method requisites. It also involves regulation of all raw materials and production procedures, storage and packaging that may alter the safety as well as testing of the final product (Paula Katz and Cliff Campbell, 2011).

Development of product should start from laboratory scale and need to transfer its technology to large scale for pilot plant production. In other words, transfer of these small-scale technologies into large-scale development mostly needs completely diverse designed approaches and apparatus which may result in alterations in the quality of a particular product so every step must be taken very carefully. The basic means of technology transfer is the movement of process methodology and technology from one unit to another. Technology transfer is both a critical and integral part of the drug discovery and development process for new medical products. Technology transfer is helpful in developing dosage forms in various ways as it provides efficiency in the process and maintains the quality of the product; it will help achieve a standardised process which leads to cost-effective production. It is the process by which R&D makes the technology available to the plant that will commercialise the pharmaceutical product. A successful technology transfer requires thoroughly studied conditions like careful evaluation of ultimate manufacturing requirements early in research and development, the consequent improvement of robust developments that endure large-scale operation, and the assemblage of a detailed technology transfer document that provides manufacturing with both 'know how' and 'know why' and will serve as the basis for facility and equipment design as well as operator training and standard operating procedure generation in successful manufacturing. The success of technology transfer mostly depends on the understanding of a particular process and the ability to predict the accurate future of a particular process (Landin et al., 1996; Nicola et al., 2005).

Scale-up for any pharmaceutical drug product is a combination of science, experience and engineering. During scale-up, the quality must be developed through suitable processing techniques, and this can be accomplished only when there is extensive practical and technical understanding and experience of the physicochemical properties and the techniques that translate the arriving materials to the finished drug product. From small-scale batches in R&D to scale-up to a bigger batch size in a plant is done by R&D personnel. It requires optimization of process parameters including time, rpm and speed of equipment. Scale-up requires technical expertise for increasing the batch by maintaining desired critical quality attributes (CQA) (Santos et al., 2015). In this chapter, we have discussed and elaborated validation, technology transfer and scale-up stages in pharmaceutical product development.

18.2 VALIDATION

Validation is the art of concepting and practicing the predesigned steps along with proper documentation. The concept of validation has expanded through the years to embrace a wide range of activities from analytical methods used for the quality control of drug substances and drug products to computerised systems for clinical trials, labelling or process control; validation is founded on, but not prescribed by regulatory requirements and is best viewed as an important and integral part of cGMP. The word validation simply means assessment of validity or action of proving effectiveness (Kaur et al., 2013). Validation is defined by various researchers and regulatory agencies, and some of them are as follows (Hoffmann et al., 1998; Nash & Wachter, 2003; *WHO Expert Committee on Specifications for Pharmaceutical Preparations: 34th Report by World Health Organization(WHO)|Waterstones*, n.d.):

- **ICH definition**: "Process Validation is the means of ensuring and providing documentary evidence that processes within their specified design parameters are capable of repeatedly and reliably producing a finished product of the required quality."
- **WHO definition**: "The documented act of proving that any procedure, process, equipment, material, activity or system actually leads to expected result."
- **European commission**: 1991 – Validation – "Act of proving, in accordance of GMPs that any process actually leads to expected results."

   2000 – "Documented evidence that the process, operated within established parameters, can perform effectively and reproducibly to produce a medicinal product meeting its predetermined specifications and quality attributes."

US FDA definition: "Process validation is establishing documented evidence, which provides a high degree of assurance that a specified process will consistently produce a product meeting its pre-determined specifications and quality characteristics."

The purity, identity, safety, strength and quality of the finished or final pharmaceutical product would be compromised due to a lack of proper validation procedure. Therefore, the system needs to be validated after involved in various combinations, such as environment, equipment, methods, material, process, personal, facilities, software control and operating procedures. Validation can only ensure the conformance to cGMP requirements and is also helpful in troubleshooting or identifying the root cause of the problem. It also ensures that the process is reliable and is in control and will routinely provide a quality product. Validation is the key factor to maintain the public image of the company, patient confidence and healthcare community satisfaction (Kaur et al., 2013).

18.2.1 Types of Validation

There are basically four types of validation on the basis of area, i.e. process validation, computer validation, cleaning validation and analytical method of validation (Pathuri & Nishat, 2013).

18.2.1.1 Process Validation

Process validation is simply a documented process which promises that a specific process will consistently produce a product that meets its predetermined specifications and quality attributes. Process validation is a necessity of cGMP principles for finished pharmaceuticals, 21 CFR Parts 210 and 211, and of GMP principles for medical devices, 21 CFR Part 820, and so, is applicable to the manufacture of pharmaceuticals and medical devices.

Definitions: "The collection and evaluation of data, from the process design stage throughout production, which establishes scientific evidence that a process is capable of consistently delivering quality products." The main concern is to only allow trained personnel to manufacture products by following standard operating procedures (SOPs). All materials are moved as mentioned from the transportation to warehouse storage phase to the manufacturing phase, and all techniques complete to packaging and distribution (Paula Katz and Cliff Campbell, 2011; Pazhayattil et al., 2016). Based on the stage of product's life cycle at which process validation is performed, it can be of four types:

Prospective validation: It is carried out during the development stages, and is a result of risk analysis on the production process (*Prospective Validation – Google Search*, n.d.).

Concurrent validation: It is carried out during normal production.

Retrospective validation: It is based on old and testing data of earlier manufactured batches.

Revalidation: It is required to guarantee that variations in process environment, whether introduced intentionally or unintentionally, do not adversely affect process characteristics and the quality of the products. It can be divided into two subtypes:

a. **Revalidation after changes**: Whenever you've presented any new elements in the manufacturing process, revalidation requirements are to be performed to ascertain their effects. There can be a number of changes in the manufacturing or SOPs that may impact product quality. These can be:
   - Changes in starting materials: Physical attribute changes can change the mechanical properties of compounds and materials, which can consequently have adverse effects on the product or the process.
   - Changes in process: Any time you change the manufacturing process, the subsequent steps can be affected and thereby, the product quality too.
   - Changes in equipment: Changes in equipment can affect the quality of the product.
   - Changes in support system or production area: Rearrangement of support systems or production areas can also affect product quality, especially critical systems like ventilation.

b. **Periodic revalidation**: Maintenance, calibration and other core requirements, and revalidation at scheduled breaks assist you in confirming that your systems and checks are performing within the required standards. As per International Conference on Harmonisation (ICH) committee, industrial guidelines for the quality of pharmaceuticals, those are followed throughout the world, are for the product quality (*ICH Guidelines for Pharmaceuticals: Pharmaceutical Guidelines*, n.d.) and to provide guidance to industry and regulatory agencies regarding regulatory expectations on the development, implementation and assessment of continuous manufacturing technologies used in the manufacture of drug substances and drug products.

18.2.1.2 Process Validation and Drug Quality

Process validation plays a significant role to ensure product quality. The European Commission, FDA and ICH consider process validation as one of the documented evidences with a high degree of assurance that a process, operated within established parameters, can perform effectively and reproducibly to produce a medicinal product meeting its predetermined specifications and quality attributes (*WHO Technical Report Series|WHO – Prequalification of Medicines Programme*, n.d.). The basic principle of quality assurance is that a product should be formed that is right for its intended use. For fulfilling this principle, the following conditions should be met:

- Quality, safety and efficacy are designed or built into the product.
- Quality cannot be sufficiently guaranteed merely by in-process and finished-product inspection or testing.
- Each step of a manufacturing process is controlled to assure all quality attributes, including specifications.
- The finished product's process validation comprises a series of events, taking place over the life cycle of the product and the process.

Approach to process validation: The FDA process validation approval process consists of three steps shown in Figure 18.1.

Step 1: Process design – process design is a commercial process based on knowledge gained through developmental procedures and scale-up activities. The design of experiment (DOE) (Weissman & Anderson, 2015), quality by design (QBD) (Lionberger et al., 2008; Yu, 2008), process analytical technology (PAT) (Chanda et al., 2015), critical process parameters (CPP) (Oktay et al., 2019) and critical quality attributes (CQA) are the important points to be considered during the process design.

The main aim of this process design is to design a process suitable for manufacturing routine commercial batches, which consistently deliver a quality product. Generally, early process design experiments to be conducted in accordance with sound scientific methods and principles, including good documentation practices, do not require the cGMP conditions. Designing an efficient process with an effective process control approach is dependent on the process with sound scientific knowledge and understanding of the process. DOE studies reveal relationships, including multivariate interactions between the variable inputs and the resulting outputs. Risk analysis tools can be used to screen potential variables for DOE studies to minimise the total number of experiments conducted while maximising knowledge gained. An approach to establish process control for each unit operation is based on the process criticality and understanding. Material analysis and equipment monitoring at significant processing points can control the process at each unit operation. The use of PAT is a more advanced strategy, which may involve timely analysis and control loops to adjust the processing conditions so that the output remains constant (*Process Analytical Technology – Google Search*, n.d.).

Step 2: Process qualification – process qualification is the capability of process reproducibility for commercial manufacturing, and the cGMP-compliant procedures must be followed during this stage. Successful completion of this stage is necessary before commercial distribution (Pazhayattil et al., 2016). The design of a facility and qualification of utilities and equipment, process performance qualification (PPQ), PPQ protocol preparation and sampling and analysis, PPQ protocol execution and report, and in-control and approval are important points that need to be considered during process qualification.

Under Part 211, subpart C of the cGMP regulations on buildings and facilities is required for proper design of a manufacturing facility. It is important that activities performed to guarantee proper facility design and commissioning precede PPQ. Qualification of utilities and equipment can be covered under individual plans or as part of an overall project plan (Pazhayattil et al., 2016). The plan should consider the requirements of use and can incorporate risk management to prioritise certain activities and identify a level of effort in both the performance and documentation of qualification activities.

PPQ protocol is a written protocol that specifies the manufacturing conditions, controls, testing and expected outcomes and is essential for this stage of process validation. Execution of the PPQ protocol should not begin until the protocol has been reviewed and approved by all appropriate departments, including the quality unit. Any departures from the protocol must be made according to established procedure or provisions in the protocol. Such departures must be justified and approved by all appropriate departments and the quality unit before implementation.

Step 3: Continued process verification (CPV) – one can gain assurance that all the processes remain in a state of control through routine production. The sampling and monitoring of data, frequency, alert and action, assessment and interpretation of data, capability, stability and review are important points that need to be considered during CPV (Snee at al., 2017).

Checking the performance of the process by frequently monitoring and sampling identifies problems and determines whether actions must be taken to alert, correct, anticipate and prevent problems so that the process remains in control. Continuing program to gather and analyse a product and process data which tells product quality must be recognised. The data collected should be statistically trended and reviewed by trained personnel. The data collected should authenticate that the quality attributes are being properly controlled throughout the process (FDA, 2004). Each of these stages is responsible for creating an approach that delivers products ranking high on quality and consistencies across all the batches.

18.2.2 Computer System Validation (Process Controller)

Nowadays, computerised systems are used in approximately all fields of the pharmaceutical company (production, administration, accounting, etc.). Due to regulation-attached supplies, systems in production, controlling and distribution of the manufacturing industry have to be validated. In the pharmaceutical concept, validation refers to the establishment of documented evidence that an equipment, utility or system, when operated within established parameters, can perform effectively in producing a medicinal product that meets the predetermined specifications. Validation of software and computer systems follows the same principle as the qualification of instrument hardware (Friedli et al., 1998; Hoffmann et al., 1998). USP general chapter <1058> has a short chapter on software validation (Valigra, 2010). The software can be divided into three categories:

a. Firmware integrated as chips into instrument hardware for control through a local user interface.
b. Software for instrument control, data acquisition and data processing. An example would be a chromatography data system.
c. Standalone software, for example, a laboratory information management system package

18.2.2.1 Computer System Validation Process

Qualification activities should be defined in a master plan. The plan documents a company's approach for specific activities, for example, how to qualify analytical instruments, how to assess vendors or what to test for commercial computer systems. Computer system validation (CSV) mainly depends on validated master plan and project plan. Validated master plan checks whether the specifications are in line with user requirements. During this stage, teams are also established which will run the entire process. The set of activities to be carried out during validation are established too. This is basically the process of preparing the blueprint for the entire CSV as shown in Figure 18.2. The project plan outlines what is to be done in order to get a specific system into compliance. For inspectors, it is the first indication of the control a laboratory has over a specific instrument or system, and it also gives the first impression of the qualification quality. More importantly, it defines a deadline within which the CSV must be completed (*Computer System Validation in Pharmaceuticals: Pharmaceutical Guidelines*, n.d.; Friedli et al., 1998).

In 1997, USFDA issued a regulation regarding the acceptance criteria for electronic records, electronic signatures and handwritten signatures. In this regulation, entitled Rule 21 CFR Part 11, electronic records can be equivalent to paper records and handwritten signatures. The rule applies to all industry segments regulated by FDA that includes Good Laboratory Practice, Good Clinical Practice and cGMP. Part 11 requires computer systems used in FDA-regulated environments to be validated. Chapter 10 (a) states computer systems should be validated to ensure accuracy, reliability and consistent intended performance. There is no further instruction on how computer systems should be validated (Friedli et al., 1998).

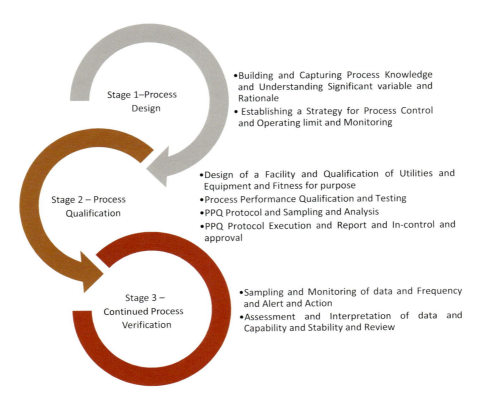

FIGURE 18.1 Stages of process validation.

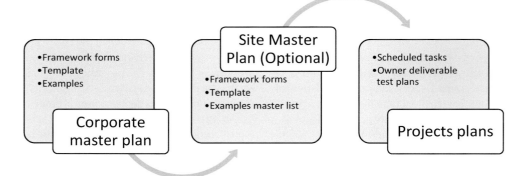

FIGURE 18.2 Computer system validation.

18.2.3 Cleaning Validation

Cleaning validation is a documented validation programme, which assures that a specific cleaning procedure is carried out when performed during a process. It is a procedure that will consistently clean a particular type of equipment to a predetermined level of cleanliness. Cleaning validation procedures of equipment are mainly used in pharmaceutical industries to prevent cross-contamination and adulteration of drug products and, therefore, it is crucial parameter to be considered. Validation of cleaning procedures has generated considerable discussion. Regulatory agency and guideline documents clearly establish the expectation for cleaning procedures (processes) that need to be validated and kept in mind before any procedure starts (Narayana Murthy & Chitra, 2013).

18.2.3.1 General Requirement

FDA expects firms to have documented general procedures in the form of SOP, i.e. standard operating procedure on how cleaning processes will be carried out. FDA expects the general validation procedures to address who is responsible for performing and approving the validation study, the acceptance criteria and when revalidation will be required. All this information should be available in a documented form so that the person can have accountability. FDA expects firms to conduct validation studies in accordance with the protocols and to document the results of the studies. FDA expects a final validation report which is approved by management and which states whether or not the cleaning process is valid. The data should support a conclusion that residues have been reduced to an 'acceptable level' (FDA validation of cleaning procedure [7/93]).

18.2.3.2 Evaluation of Cleaning Validation

The first step is to focus on the objective of the validation process, and we have seen that some companies have failed to develop such objectives. It is not unusual to see manufacturers using extensive sampling and testing programmes following the cleaning process without ever really evaluating the effectiveness of the steps used to clean the equipment. Various vital questions need to be addressed before evaluating the cleaning process. For example, at what point does a piece of equipment or system become clean? Does it have to be scrubbed by hand? What is accomplished by hand scrubbing rather than just a solvent wash? How variable are manual cleaning processes from batch to batch and product to product? Answers to these questions are obviously important for the inspection and evaluation of the cleaning process since one must determine the overall effectiveness of the process. Answers to these questions may also identify steps that can be eliminated for more effective measures and result in resource savings for the company (Narayana Murthy & Chitra, 2013).

Determine the number of cleaning processes for each piece of equipment. Ideally, a piece of equipment or system will have one process for cleaning; however, this will depend on the products being produced and whether the clean-up occurs between batches of the same product (as in a large campaign) or between batches of different products. When the cleaning process is used only between batches of the same product (or different lots of the same intermediate in a bulk process), the firm's need will only meet a criterion of 'visibly clean' for the equipment. Such between batch cleaning processes do not require validation (Narayana Murthy & Chitra, 2013).

Equipment design: Examine the design of equipment, particularly in those large systems that may employ semi-automatic or fully automatic clean-in-place systems since they represent significant concerns. For example, sanitary-type piping without ball valves should be used. When such non-sanitary ball valves are used, as is common in the bulk drug industry, the cleaning process is more difficult (*Design Review & Equipment Startup: Checklist for AHU Replacement/Retrofit*, n.d.).

Subsequent to the cleaning process, equipment may be subjected to both sterilisation or sanitisation procedures where equipment is kept for sterile or non-sterile processing, where the products may support microbial growth. While such sterilisation or sanitisation procedures are beyond the scope of this guide, it is important to note that control of bioburden through adequate cleaning and storage of equipment is important to ensure that subsequent sterilisation or sanitisation procedures achieve the necessary assurance of sterility and avoid any microbial growth. This is also particularly important from the standpoint of the control of pyrogens in sterile processing since equipment sterilisation processes may not be adequate to achieve significant inactivation or removal of pyrogens (US Patent, 1984).

Cleaning process written: Examine the detail and specificity of the procedure for the (cleaning) process being validated and the amount of documentation required. Depending upon the complexity of the system and the cleaning process and the ability and training of operators, the amount of documentation necessary for executing various cleaning steps or procedures will vary. When more complex cleaning procedures are required, it is important to document the critical cleaning steps (for example certain bulk drug synthesis processes). In this regard, specific documentation on the equipment itself which includes information about who cleaned it and when is valuable. However, for relatively simple cleaning operations, the mere documentation that the overall cleaning process was performed might be sufficient (*Cleaning Validation and Its Importance in Pharmaceutical Industry*, 2010).

Other factors such as history of cleaning, residue levels found after cleaning and variability of test results may also dictate the amount of documentation required. For example, when variable residue levels are detected following cleaning, particularly for a process that is believed to be acceptable, one must establish the effectiveness of the process and operator performance. Appropriate evaluations must be made and, when operator performance is deemed a problem, more extensive documentation (guidance) and training may be required (Narayana Murthy & Chitra, 2013).

Analytical methods: Determine the specificity and sensitivity of the analytical method used to detect residuals or contaminants. With the advancement in analytical technology and methods, residues from the manufacturing and cleaning processes can be detected at very low levels. If levels of contamination or residual are not detected, it doesn't mean that there is no residual substance present after cleaning. It only means that levels of contaminant greater than the sensitivity or detection limit of the analytical method are not present in the sample. The firm should challenge the analytical method in combination with the sampling method(s) used to show that contaminants can be recovered from the equipment surface and at what level, i.e. 50% recovery, 90%, etc. This is necessary before any conclusions can be made based on the sample results (Sheehan et al., 1992). A negative test may also be the result of a poor sampling technique (see below).

Sampling: Sampling can be done in two different ways – direct surface sampling and rinse samples. The most desirable is the direct method of sampling that includes the surface of the equipment. Another method is the use of rinse solutions (Kennedy et al., 1994).

Direct surface sampling: Determine the type of sampling material used and its impact on the test data since the sampling material may interfere with the test and may alter the result. For example, the adhesive used in swabs has been found to interfere with the analysis of samples. Therefore, early in the validation program, it is important to assure that the sampling medium and solvent (used for extraction from the medium) are satisfactory and can be readily used. Advantages of direct sampling are that the areas hardest to clean which are reasonably accessible can be evaluated, leading to establish a level of contamination or residue per given surface area. Additionally, residues that are 'dried out' or insoluble can be sampled by physical removal (Pereira et al., 2007).

Rinse samples: Two advantages of using rinse samples methods are that a larger surface area may be sampled, and inaccessible systems or ones that cannot be routinely disassembled can be sampled and evaluated. A disadvantage of rinse samples is that the residue or contaminant may not be soluble or may be physically occluded in the equipment. An analogy that can be used is the 'dirty pot.' In the evaluation of cleaning of a dirty pot, particularly with dried-out residues, one does not look at the rinse water to see whether it is clean; one looks at the pot to see whether a direct measurement of the residue or contaminant has been made for the rinse water when it is used to validate the cleaning process. For example, it is not acceptable to simply test rinse water for water quality (does it meet the compendia tests) rather than test it for potential contaminants (United States Patent (19) Copeland et al., 1995).

Routine production in-process control: Monitoring – indirect testing, such as conductivity testing, may be of some value for routine monitoring once a cleaning process has been validated. This would be particularly true for the bulk drug substance manufacturer where reactors, centrifuges and piping between such large equipment can be sampled only using rinse solution samples. Any indirect test must have been shown to correlate with the condition of the equipment. During validation, the firm should document that testing the uncleaned equipment gives a not acceptable result for the indirect test.

Establishment of limits: FDA does not intend to set acceptance specifications or methods for determining whether a cleaning process is validated. It is impractical for FDA to do so due to the wide variation in equipment and products used throughout the bulk and finished dosage form industries. The firm's rationale for the residue limits established should be logical based on the manufacturer's knowledge of the materials involved and be practical, achievable and verifiable. It is important to define the sensitivity of the analytical methods in order to set reasonable limits. Some limits that have been mentioned by industry representatives in the literature or in presentations include analytical detection levels such as 10 PPM, biological activity levels such as 1/1,000 of the normal therapeutic dose, and organoleptic levels such as no visible residues (Ravisankar et al., 2015).

TABLE 18.1
Parameters for Analytical Method Validation

Analytical performance parameters	Identification	Impurities Quantitation	Impurities Limit test	Assay/dissolution/content	Specific Test[b]
Accuracy	No	Yes	No	Yes	Yes[c]
Precision repeatability intermediate precision	No	Yes	No	Yes	Yes[c]
		Yes		Yes	Yes[c]
Specificity	Yes	Yes	Yes	Yes	Yes[c]
Limit of detection	No	No	Yes	No	No
Linearity	No	Yes	No	Yes	No
Robustness[a]	No	Yes	No	Yes[a]	Yes[c]
Stability of sample solution	No	Yes	No	Yes	No

[a] Terminology included in ICH publication but is not part of required parameters to be submitted.
[b] Specific tests include the following test parameters like particle size distribution, droplet size, spray pattern, dissolution and optical rotation and methodologies like DSC, XRD and Raman spectroscopy. The validation characteristics may differ for the various analytical procedures and the application of the test.
[c] May not be needed in some cases.

Note: For methods specified in pharmacopeia, the above-mentioned parameters are not required to be validated and preferably specificity, precision and stability of the sample solution will be performed, if required.

18.2.4 Analytical Method Validation

It is a documented programme which assures that an analytical method will consistently determine the presence/absence of the quality of one or more attributes (e.g. microorganisms, contaminants, active ingredients, etc.), with accuracy and precision (Gustavo González & Ángeles Herrador, 2007). Before any batch from the process is commercially distributed for use by consumers, a manufacturer should have gained a high degree of assurance in the performance of the manufacturing process such that it will consistently produce active pharmaceutical ingredients (APIs) and drug products, meeting those attributes relating to identity, strength, quality, purity and potency. The assurance should be obtained from objective information and data from laboratory, pilot, and/or commercial scale studies. Information and data should demonstrate that the commercial manufacturing process is capable of consistently producing acceptable quality products within commercial manufacturing conditions (Ravisankar et al., 2015).

Method verification: Method verification ensures that a particular laboratory is capable of performing the analysis. Verification of an analytical procedure is to demonstrate that laboratory is capable of replicating with an acceptable level of performance by a standard method. Verification under conditions of use is demonstrated by meeting system suitability specifications established for a method. Method verification will be conducted to prove that the compendia method/vendor's procedure performed are suitable for the intended purpose and should be in line with that of Rambla-Alegre et al.'s (2012):

- Type of method to be verified and performance (but not limited to).
- Method from suppliers of drug substance or any excipient that needs to be used.

Precision, specificity, solution stability, accuracy, range for the determination of impurities in drug substance – precision, specificity, limit of detection (LOD), limit of quantitation (LOQ), solution stability and accuracy are the important verification parameters to be studied, typically for the assay of drug substances. For the determination of residual solvents in drug substances, precision, detection limit (DL)/LOD and quantitation limit (QL)/LOQ are the important parameters (Gustavo González & Ángeles Herrador, 2007).

Method validation: Method validation is a process of demonstrating that the analytical procedures are suitable for their intended use and that they also support the identity, strength, quality, purity and potency of the drug substances and drug products (Blessy et al., 2014).

Validation parameters: The data elements required for method validation are described in Table 18.1.

18.3 SCALE-UP AND TECHNOLOGY TRANSFER IN PHARMACEUTICAL PRODUCT DEVELOPMENT

Scale-up and technology transfer are crucial development activities for any pharmaceutical product development. The commercial success of pharmaceutical product development depends on being able to increase the drug production volume quickly and effectively. Successful transfer of technology is vital for product efficacy and patient safety, but the time and financial costs of failure can be significant during technology transfer. The regulatory authorities expect manufacturers and their partners to take a methodical approach and provide all required documentation as they produce during a move a process to a new facility or convert it from development to commercial scale. Technology transfer activities completely guide the transfer of product and process knowledge.

18.3.1 Process Transfer: An Overall Framework

Laying the groundwork is the key for any technology transfer. The manufacturer has to ensure previously that those who are working on the transfer have sufficient knowledge, experience and skills which will help to ensure success and avoid surprises. Process parameters and process knowledge may need to be transferred from development to pilot study to clinical production or to commercial manufacturing facilities. In all cases, the final scale and success parameters, such as CQA, must be set clearly in a written note before the transfer begins (Hatch & Kazmer, 2001).

Generally, the sending party has less stake in the project's success than the recipient, making a disciplined approach imperative. High-level due diligence regarding capacity, facilities, cGMP capability and personnel will help teams to assess feasibility in order to prepare for transfer. Planning ahead, for example, can allow teams to pre-order equipment with long lead times, if necessary. Note that transfer does not end with the completion of qualification lots or approval, but extends throughout the duration of manufacturing. Successful technology transfer follows an orderly progression to set expectations and ensure that all stakeholders are working towards the same goals (Hatch & Kazmer, 2001).

18.3.2 Goal of Technology Transfer

ICH Q10 "The goal of technology transfer activities is to transfer product and process knowledge between development in R&D and manufacturing in a plant known as scale-up, and within or between manufacturing sites to achieve product realisation. This knowledge forms the basis for the manufacturing process, control strategy, process validation approach and ongoing continual improvement." The following points are to be considered when technology transfer is going to start:

- A valuable step in the developmental life cycle leading to successful commercial manufacturing
- To take all gathered knowledge and use it as the basis for the manufacturing control strategy, the approach to process qualification and ongoing continuous improvement
- Transition of the product/process/analytical method knowledge between development and manufacturing sites
- To ensure that the variability of process and parameters is controlled and sufficient in the face of rigours of a commercial production environment to verify that the parameters established during development are still within the determined design space and/or adjusted at scale-up.

Technology transfer comprises not only the patentable aspect of production but also includes the business of processes, such as knowledge and skills. Technology transfer offers a good opportunity to reduce cost on drug discovery and development; thus, major pharmaceutical companies take the advantages of technology transfer opportunities as they reduce the risk, cost and rate of failure. Technology transfer can happen in many ways, such as government labs to private sectors, between private sectors of the same country, from academic to private sectors, between academies, and government and private sectors of different countries.

18.3.3 Steps in Technology Transfer

Formulation development in the research field, the most important thing we need to understand, is the procedure of operations, critical and non-critical parameters of each unit operation, production environment, equipment and excipient availability, which should be taken into account during the early phases of development of formulation, so that a successful scale-up can be carried out. Appropriate care must be taken during technology transfer. It is important to maintain product quality and enhance product safety as developed during research, and it must be maintained for a predetermined period of time. Technology transfer is not a single-way process; it involves various steps as mentioned in Figure 18.3 and given below:

A. Research phase or pre-development phase (development of technology by R&D)
B. Development phase (technology transfer from R&D to production)
C. Production phase (optimisation and production)
D. Technology transfer documentation
E. Exhibit batches

A. Research phase (development of technology by R&D):

   This phase is carried out by performing two key activities that are:
   i. Formula development and identification of excipient R&D: On the basis of the literature, selection of materials and design of procedures is developed by R&D people and on the basis of innovator product characteristics. For this, different testing and compatibility studies are done to check whether the literature search and optimisation are near to the innovator or not.
   ii. Identification of specification and quality by R&D: Generally, it should be considered by R&D that the quality of the product should meet the specifications of an innovator product. For this different stability, studies are carried out with innovator product and for a product which is to be manufactured according to the statistical data obtained during research.

B. Development phase (technology transfer from R&D to production):

TABLE 18.2
Technology Transfer Documents

S. No.	Document	Content
1.	Bill of material (BOM)	List of all components including grade, manufacturer and suppliers
2.	Master packaging card (MPC)	It contains the quantity of required packaging material (primary and secondary packaging) in specified batch size
3.	Master formula record (MFR)	It contains the formula including quantity of API and excipients used in a drug product, process flow and procedure of drug product manufacturing
4.	Raw material specifications	It contains the grade and test that needs to be done for complying with the material as per specification
5.	Release specifications	Initial test and limits which require to okay the drug product after manufacturing
6.	Shelf life specifications	Test and limits for a drug product during shelf life
7.	Product development report	It contains all the studies which are done during drug product development, such as compatibility studies, freeze-thaw, thermal cycling, quality by design trials, stability data, etc.
8.	Packaging development report	It contains the details related to packaging components

After research phase, R&D team provides technology transfer dossier (TTD) document to product development laboratory which contains all information of formulation and drug product as given below:

Technology transfer dossier (TTD): TTD contained all the information of a drug product as given below (Table 18.2):
i. Product development report
ii. Packaging development report
C. Production phase (optimisation and production):
  i. Validation studies: Research and development department transfer the product technology to the plant for large-scale production through technology transferring. They should take the responsibility for validation, such as performance qualification, cleaning validation and process validation unique to subject drugs.
  ii. Scale-up for production scale-up: It involves the transfer of technology during the small-scale development of the product and processes. It is essential to consider the production environment and system during the development of process and maintain the same as at the time of development so that no changes are observed. Operators concentrate on keeping their segment of the production process running smoothly. But the whole manufacturing line can be improved, even before the production begins, if technology transfer is implemented thoughtfully. Effective technology transfer helps to provide process efficiency and control and maintain product quality.
  iii. Different parameters for scale-up: Before starting scale-up, R&D people also consider different parameters that should be optimum for a successful technology transfer. These are flexibility, cost, dependability, innovation and product quality. It is important to realise that good communication is critical for a successful formulation and process transfer.
  iv. Selection of method: The method for batch fabrication should be selected on the basis of data given from R&D. Granulation, blending; compression and coating are the critical parameters for technology transfer.
D. Technology transfer documentation:
  Technology transfer is an integral part of drug product development; it is majorly focussed on how a new product will be transferred to a plant from R&D for the execution of commercial batches. Technology transfer will help in increasing the batch size from small-scale in R&D to large-scale in a plant.

Raw materials: The following are the examples of information which may typically be provided; however, the information needed in each specific case should be assessed using the principles of quality risk management.

Steps of the product to be transferred include flow chart of process, outlining the process including entry points for raw materials, critical steps, process controls and intermediates; where relevant, definitive physical form of the API (including photomicrographs and other relevant data) and any polymorphic and solvate forms; solubility profile; if relevant, pH in solution; partition coefficient, including the method of determination; intrinsic dissolution rate, including the method of determination; particle size and distribution, including the method of determination; bulk physical properties, including data on bulk and tap density, surface area and porosity as appropriate; water content and determination of hygroscopicity, including water activity data and special handling requirements; microbiological considerations (including sterility, bacterial endotoxins and bioburden levels where the API supports microbiological growth) in accordance with national, regional or international pharmacopoeial requirements; specifications and justification for release and end of shelf life limits; summary of stability studies conducted in conformity with current guidelines, including conclusions and recommendations on retest date; list of potential and observed synthetic impurities, with data to support proposed specific cations and typically observed levels; information on degradants, with a list of potential and observed degradation products and data to support proposed specifications and typically observed levels; potency factor, indicating observed purity and justification for any recommended adjustment to the input quantity of API for product manufacturing, providing example calculations; and special considerations with the implications for storage and/or handling, including but not limited to safety and environmental factors (e.g. as specified in material safety data sheets) and sensitivity to heat, light or moisture (Iñaki Bueno et al., 2017).

Excipients: The excipients to be used in the product have a potential impact on the final product. Their specifications and relevant functional characteristics should be made available by R&D for transfer to the manufacturing site. The following are the examples of information which may typically be provided: manufacturer and associated supply chain; description of functionality, with justification; definitive form; solubility profile, partition coefficient, including the method of determination; intrinsic dissolution rate, including the method of determination; particle size and distribution, including the method of determination; bulk physical properties, including data on bulk and tap density, surface area and porosity as appropriate; compaction properties; melting point range; pH range; ionic state; specific density or gravity; viscosity and/or viscoelasticity; osmolality, water content and determination of hygroscopicity; moisture content range; microbiological considerations; specifications and justification for release and end-of-life limits; and information on adhesives supporting compliance with peel, sheer and adhesion design criteria.

Documents indicate contents for technology transfer and how a product needs to be transferred. Each step from R&D to production should be documented and evaluated. Quality assurance department approves the documentation for all processes of technology transfer along with the cross-functional team member. The following documents should be prepared during technology transfer:

i. **Development report**: This report contains all technical data of formulation developments. The development report contains the following:
   a. Data of pharmaceutical development of new drug substances and drug products at stages from early development phase, optimisation to the final application of approval.
   b. Master formula record:
      a. Information of raw materials, excipients and another component
      b. Design of manufacturing methods
   c. Change in histories of important processes and control parameters
   d. Specifications and test methods of drug substances
   e. Validity of specification range of important tests, such as contents, impurities and dissolution
   f. Verifications of results.

ii. **Technology transfer plan**: The transferring party should prepare the technology transfer plan before its implementation which consists of all the information related to transfer. The technology transfer plan is to describe items and contents of technology to be transferred and detailed procedures of individual transfer and transfer schedule and establish judgment criteria for the completion of the transfer.

iii. **Technology transfer report**: Completion of technology transfer is to be made once data are taken accordingly to the technology plan and are evaluated on the basis of earlier results to confirm that the predetermined judgment criteria are fulfilled according to the previous criteria. Both transferring and transferred parties should document the technology transfer report.

iv. **Scale-up /Pre-exhibit batches/feasibility batch**: These batches shall be taken into plant in order to assess the suitability and machinability of the process and the impact of scale-up, if any. Material requisition shall be given to the global sourcing department for arranging all materials required for pre-exhibit batches and the concerned scientist shall take the batch. If any changes have been made during the execution of these batches compared to optimised process in the lab, they shall be reviewed and evaluated, and required changes shall be proposed for taking exhibit batches.

v. **Exhibit batches**: After feasibility batch of the product, manufacturing of exhibit batches takes place. In the case of exhibit, batch sizes are increased along with equipment, and their process is involved. They are done for filing purposes in different regulatory agencies.

These batches shall be taken for submitting the data to regulatory authorities. The number of batches (1–3) shall be taken based on concerned regulatory requirements. Batch size of this batch should be not less than 1/10th of the commercial batch size. All the submission batches shall be manufactured in the plant using production equipment, by production people and following plant systems and procedures and shall be tested by plant QC. All the submission batches will be manufactured under the supervision of the plant QC head. After completion of manufacturing and filling, these batches shall be retained in plant finished goods store as per recommended storage conditions. In addition to the required stability studies, the following studies shall also be conducted using a sample from submission batches (US Patent, 1984; Pathuri & Nishat, 2013):

- Bulk sample for filter validation study for microbial retention test and extractable test
- In-use study if required
- Toxicity and clinical, bioequivalence studies if required.

18.3.3.1 Quality by Design in Scale-Up

Nowadays, many of the pharmaceutical industries have shifted from a trial-and-error design approach towards quality by design approaches for scaling-up. QBD involves identifying the CPP and critical product attributes that significantly affect the desired characteristics of the final product. In QBD, identifying broadly two parameters, i.e. the CPP and critical product attributes, significantly affects the desired characteristics of the final product. It helps in the finalisation of several batches by changing and fitting different values for the parameter. Key elements of QBD involve target product quality profile, CQA, risk assessment, design space, product life cycle management control strategy and continual improvement to understand the performance of any product within the design space (Portillo et al., 2008).

A drug product consisting of a wax matrix microsphere is loaded with a drug and prepared by using a melt spray congeal process. The drug was released from this microsphere either through channels formed as a result of solubilisation of a drug or by pores formed due to the dissolution of soluble additives. The rate of dissolution was correlated directly to the particle size of the drug. Thus, during scale-up, it is impossible to control process parameters optimised by a trial-and-error method to obtain proper particle size for a good rate of dissolution. Thus, the authors identified key parameters influencing microsphere particle sizes, like the speed of the spinning disk, temperature of the operating system and viscosity of drug melt suspension, utilised two to the power three factorial design and developed a predictive model for melt rheology and microsphere size. This reduces the time and number of experiments to obtain the desired size spray congeal mass and give the summarised work flow with the number of trials. ICH and USFDA emphasised the application of QBD in the pharmaceutical industry. It is required by regulatory authorities for the approval of the pharmaceutical product. During scale-up, the parameters may differ, but the attributes responsible for quality remain the same. QBD application during scale-up will help control the parameters to obtain the desired quality product. It ensures product safety and efficacy during scale-up and also helps to understand the pharmaceutical processes and methods more effectively. A risk-based analysis by QBD enhances the quality of the final product. The variability in scale-up is reduced by the application of QBD (Contrera, 1994; Portillo et al., 2008).

18.4 IMPORTANCE OF TECHNOLOGY TRANSFER

There is a clear need for pharmaceutical companies to speed up the delivery of new drug products to the market to maintain competitive effectiveness. New drug discoveries must be rapidly brought to market to generate cash flow to reinvest back into the business and support the drug pipeline. It is estimated that 70% of new products on the market fails to recoup the costs of R&D. The number of new medicines launched in recent years has slowed down because it has become more difficult to meet all the requirements of safety, clinical and regulatory in areas of unmet medical needs. Statistics from the pharmaceutical industry success rates shows that there is a high risk of failure in drug discovery and development, with only one in five to one in ten new candidate drugs nominated from research to development actually achieving registration and reaching the market (Gibson, n.d.; *Tucker, 1984 Technology Transfer – Google Search*, n.d.). The rate at which pharmaceutical companies clearing the final hurdle of clinical trials has declined since the mid-1990s when 56 new chemical entities ([NCEs], i.e. small synthetic molecules)

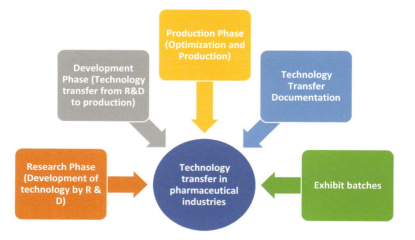

FIGURE 18.3 Flow diagram of technology transfer.

were approved in 1996, down to only 29 worldwide approvals of new molecular entities in 2002. On average, an NCE takes 10–12 years of research and development from drug discovery to product launch. This is at a currently estimated cost of $1–1.5 billion before it comes to market, which has to cover the costs of research, development, manufacturing, distribution, marketing and launch and also the cost of failures (NCEs that did not make the market). Drug development times have increased in recent years because all phases of development are taking longer. The need to do larger patient studies and increased patient enrolment times have resulted in increased clinical study duration. Typical times to complete phase II and III trials can add a further 4–6 years depending on the size and complexity of the phase III clinical programme. It will take longer, for example, to study a treatment for a chronic disease to gather sufficient efficacy and safety data to support registration. Achieving a faster time to market by streamlining the development process and, for example, introducing more efficient safety and clinical programmes with some risk-taking, tends to put synthetic drug development and product development, including the respective technology transfer, on the critical path. In addition to condensing the timescales, there is increasing pressure to define the final synthetic route of the drug, final formulation, packaging and manufacturing process and to get it the right first time, thus avoiding changes at later stages of development that could cause unnecessary delays to registration and launch. The financial cost of getting the technology transfer wrong can be significant; a product that is 6 months late to market could miss out on one-third of the potential profit over the product's lifetime (McKinsey & Co, 1991). This could be due to competitor companies being the first to the market, capturing the market share and dictating the price, as well as the loss of effective patent life. The cost of a retrospective fix after a product launch due to poor product or process development can also be expensive and embarrassing. A product recall could easily wipe out the profits from an early launch. Significant post-registration changes require regulatory approval and come with time delays. The longer-term consequences can be that a company's credibility with the regulatory authorities and its customers are affected, and this could ultimately impact the value of the company and more difficult regulatory inspections. In conclusion, the transfer of technology for drug substance and drug product between R&D and the respective production sites is critical to successful and timely development. The aim is to get to the market quickly with the development of a drug and product of appropriate quality and to do it 'right first time, every time' (*A Systematic Approach to Tech Transfer and Scale-Up*, n.d.).

18.5 CONCLUSION

The book chapter compiled with basic information about validation, scale-up and technology transfers in pharmaceutical product development. Validation is considered as a mean of proving effectiveness by assuring that a process operated within established parameters can perform effectively and reproducibly to produce a final pharmaceutical product, meeting its predetermined specifications and quality attributes. Before validating a pharmaceutical product development process and defining a commercial process, the R&D team creates the product at a smaller scale to gain knowledge about each step involved in process/product. Initially, studies start at a laboratory scale followed by pilot batches and finally production-scale batches.

Technology transfers in the pharmaceutical industry are described as movements of technology from drug discovery to product development, clinical trials or full-scale commercialisation. Technology transfer does not mean one-time actions taken by the R&D team towards the manufacturing team but means continuous information exchange between both the teams/parties to maintain quality product manufacturing. Technology transfer can be considered successful if a receiving unit can routinely reproduce the transferred product, process or method against a predefined set of specifications agreed with a sending unit and/or a development unit. The three primary considerations to be addressed during an effective technology transfer are the plan, the persons involved, and the process. A plan must be devised to organise the personnel and the process steps. Once prepared, the plan must be communicated to the involved teams/parties in research, at the corporate level and at the production site. In order to scale up and transfer a process successfully from laboratory scale to pilot scale and multiple commercial manufacturing scales, a thorough understanding of the integration of scale factors, facility design, equipment design and process performance is necessary. ICH and USFDA emphasised application of QBD in the pharmaceutical industry to control the parameters to obtain the desired quality product. A risk-based analysis by QBD enhances the quality of the final product, and the variability in scale-up is also reduced. QBD-based product development reduces the failure in the scale-up and technology transfer stage. It makes the process robust and more scientific.

ACKNOWLEDGEMENT

Author, Dr. Neelesh Kumar Mehra, would like to thank the National Institute of Pharmaceutical Education & Research (NIPER), Hyderabad (Research communication No. NIPER-H/2020/BC-08). Dr. Gulbake would like to thank SERB, New Delhi, for extending facilities to write this chapter (EEQ/2016/000789).

REFERENCES

Blessy, M., Patel, R. D., Prajapati, P. N., & Agrawal, Y. K. (2014). Development of forced degradation and stability indicating studies of drugs – A review. *Journal of Pharmaceutical Analysis*, *4*(3), 159–165. Xi'an Medical University. doi:10.1016/j.jpha.2013.09.003.

Chanda, A., Daly, A. M., Foley, D. A., Lapack, M. A., Mukherjee, S., Orr, J. D., Reid, G. L., Thompson, D. R., & Ward, H. W. (2015). Industry perspectives on process analytical technology: Tools and applications in API development. *Organic Process Research and Development*, *19*(1), 63–83. American Chemical Society. doi:10.1021/op400358b.

Cleaning Validation and Its Importance in Pharmaceutical Industry. (2010). https://www.researchgate.net/publication/281742872.

Computer System Validation in Pharmaceuticals: Pharmaceutical Guidelines. (n.d.). Retrieved September 25, 2020, from https://www.pharmaguideline.com/2015/10/computer-system-validation-overview.html.

Contrera, J. G. (1994). The Food and Drug Adminsitration and the International Conference on Harmonization: How harmonious will international pharmaceutical regulations become. *Administrative Law Journal of the American Univesrity*, 8. https://heinonline.org/HOL/Page?handle=hein.journals/adminlj8&id=939&div=&collection=.

Design Review & Equipment Startup: Checklist for AHU Replacement/Retrofit. (n.d.). Retrieved September 25, 2020, from https://www.researchgate.net/publication/297839651_Design_review_equipment_startup_Checklist_for_AHU_replacement_retrofit.

Devalapally, H., Chakilam, A., & Amiji, M. M. (2007). Role of nanotechnology in pharmaceutical product development. *Journal of Pharmaceutical Sciences*, 96(10), 2547–2565. John Wiley and Sons Inc. doi:10.1002/jps.20875.

FDA. (2004). *Pharmaceutical cGMPs for the 21st Century – A Risk-Based Approach*. Available at: https://www.fda.gov/about-fda/center-drug-evaluation-and-research-cder/pharmaceutical-quality-21st-century-risk-based-approach-progress-report https://www.google.com/search?q=FDA%2C+Pharmaceutical+cGMPs+for+the+21st+Century+-+A+Risk-+Based+Approach%2C+2004%2C+available+at%3A+http%3A%2F%2Fwww.fda.gov%2F+Drugs%2FDevelopmentApprovalProcess%2FManufacturing%2FQuestionsandAnswersonCurrentGoodManufacturingPracticesc-+GMPforDrugs%2Fucm137175.htm&rlz=1C1CHBF_enIN885IN885&oq=FDA%2C+Pharmaceutical+cGMPs+for+the+21st+Century+-+A+Risk-+Based+Approach%2C+2004%2C+available+at+%3A+http%3A%2F%2Fwww.fda.gov%2F+Drugs%2FDevelopmentApprovalProcess%2FManufacturing%2FQuestionsandAnswersonCurrentGoodManufacturingPracticesc-+GMPforDrugs%2Fucm137175.htm&aqs=chrome..69i57.1107j0j7&sourceid=chrome&ie=UTF-8.

FDA. (2011a). *Process Validation Guidance: Process Validation Revisited – Google Search*. Retrieved September 25, 2020, from https://www.google.com/search?q=FDA+2011+Process+Validation+Guidance%3A+Process+Validation+Revisited&rlz=1C1CHBF_enIN885IN885&oq=FDA+2011+Process+Validation+Guidance%3A+Process+Validation+Revisited&aqs=chrome..69i57.1329j0j7&sourceid=chrome&ie=UTF-8.

FDA. (2011b). *Process Validation Guidance – Google Search*. Retrieved September 25, 2020, from https://www.google.com/search?q=fda+2011+process+validation+guidance&rlz=1C1CHBF_enIN885IN885&oq=FDA+2011+Process+Validation+Guidance&aqs=chrome.0.0l4.1008j0j15&sourceid=chrome&ie=UTF-8.

Friedli, D., Kappeler, W., & Zimmermann, S. (1998). Validation of computer systems: Practical testing of a standard lims. *Pharmaceutica Acta Helvetiae*, 72(6), 343–348. doi:10.1016/S0031-6865(97)00032-0.

Gibson, M. (n.d.). *Technolgy Transfer Introduction and Objectives Purpose of the Book*. Retrieved September 25, 2020, from www.pda.org/bookstore.

Good Manufacturing Practice Validation Guidance – Google Search. (n.d.). Retrieved September 25, 2020, from https://www.google.com/search?rlz=1C1CHBF_enIN885IN885&sxsrf=ALeKk00sgJafMXC5-NAotTpHj0Gs5PQTkw%3A1601000786303&ei=UlVtX5GUEvuW4-EPiYCB-AM&q=Good+Manufacturing+Practice+validation+guidance&oq=Good+Manufacturing+Practice+validation+guidance&gs_lcp=CgZwc3ktYWIQAzIFCAAQzQIyBQgAEM0-COgQIABBHUJzEFVicxBVgpd8VaABwA3gAgAGTAogBkwKSAQMyLTGYAQCgAQKgAQKgAQGqAQdnd3Mtd2l6eAEIwAEB&sclient=psy-ab&ved=0ahUKEwiRhZzZoIPsAhV7yzgGHQlAAD8Q4dUDCA0&uact=5.

Gustavo González, A., & Ángeles Herrador, M. (2007). A practical guide to analytical method validation, including measurement uncertainty and accuracy profiles. *Trends in Analytical Chemistry*, 26(3), 227–238. doi:10.1016/j.trac.2007.01.009.

Hatch, D., & Kazmer, D. (2001). Process transfer function development for optical media manufacturing. *International Journal of Advanced Manufacturing Technology*, 18(5), 357–365. doi:10.1007/s001700170059.

Hoffmann, A., Kähny-Simonius, J., Plattner, M., Schmidli-Vckovski, V., & Kronseder, C. (1998). Computer system validation: An overview of official requirements and standards. *Pharmaceutica Acta Helvetiae*, 72(6), 317–325. doi:10.1016/S0031-6865(97)00028-9.

ICH Guidelines for Pharmaceuticals : Pharmaceutical Guidelines. (n.d.). Retrieved September 24, 2020, from https://www.pharmaguideline.com/2010/10/ich.html.

Kennedy, E. R., Abell, M. T., Reynolds, J., & Wickman, D. (1994). A sampling and analytical method for the simultaneous determination of multiple organophosphorus pesticides in air. *American Industrial Hygiene Association Journal*, 55(12), 1172–1177. doi:10.1080/15428119491018259.

Landin, M., York, P., Cliff, M. J., Rowe, R. C., & Wigmore, A. J. (1996). Scale-up of a pharmaceutical granulation in fixed bowl mixer-granulators. *International Journal of Pharmaceutics*, 133(1–2), 127–131. doi:10.1016/0378-5173(95)04427-2.

Lionberger, R. A., Lee, S. L., Lee, L. M., Raw, A., & Yu, L. X. (2008). Quality by design: Concepts for ANDAs. *AAPS Journal*, 10(2), 268–276. doi:10.1208/s12248-008-9026-7.

McKinsey & Co. (1991). *Pharmaceutical Develop – Google Search*. Retrieved September 24, 2020, from https://www.google.com/search?rlz=1C1CHBF_enIN885IN885&sxsrf=ALeKk00Z043ZVabYhyp__fOYEa__799VyQ%3A1600970864480&ei=cOBsX6zzHJ6P4-EPl7qXwAU&q=McKinsey+%26+Co%2C+1991+in+pharmaceutical+develop&oq=McKinsey+%26+Co%2C+1991+in+pharmaceutical+develop&gs_lcp=CgZwc3ktYWIQAzIECCEQ-FToECAAQRzoECCMQJzoICCEQFhAdEB46BQghEKABOgQIIRAKUNAxWKGGAWDkkAFoAHACeACAAaYEiAHHNJIBCjItMjMuMi4wLjGYAQCgAQKgAQGqAQdnd3Mtd2l6eAEIwAEB&sclient=psy-ab&ved=0ahUKEwis67CdsYLsAhWexzgGHRfdBVgQ4dUDCA0&uact=5.

Narayana Murthy, D., & Chitra, K. (2013). A review article on cleaning validation. *International Journal of Pharmaceutical Sciences and Research*, 4(9), 3317. doi:10.13040/IJPSR.0975-8232.4(9).3317-27.

Nash, R. A., & Wachter, A. H. (2003). *Pharmaceutical Process Validation: An International Third Edition, Revised and Expanded*. http://www.dekker.com.

Nicola, T., Brenner, M., Donsbach, K., & Kreye, P. (2005). First scale-up to production scale of a ring closing metathesis reaction forming a 15-membered macrocycle as a precursor of an active pharmaceutical ingredient. *Organic Process Research and Development*, 9(4), 513–515. doi:10.1021/op0580015.

Oktay, A. N., Ilbasmis-Tamer, S., & Celebi, N. (2019). The effect of critical process parameters of the high pressure homogenization technique on the critical quality attributes of flurbiprofen nanosuspensions. *Pharmaceutical Development and Technology*, *24*(10), 1278–1286. doi:10.1080/10837450.2019.1667384.

Pathuri, R., & Nishat, A. (2013). A review on analytical method development and validation of pharmaceutical technology. *Current Pharma Research*, *3*(2), 855.

Pazhayattil, A., Alsmeyer, D., Chen, S., Hye, M., Ingram, M., & Sanghvi, P. (2016). Stage 2 process performance qualification (PPQ): a scientific approach to determine the number of PPQ batches. *AAPS PharmSciTech*, *17*(4), 829–833. doi:10.1208/s12249-015-0409-7.

Pereira, V. J., Weinberg, H. S., Linden, K. G., & Singer, P. C. (2007). UV degradation kinetics and modeling of pharmaceutical compounds in laboratory grade and surface water via direct and indirect photolysis at 254 nm. *Environmental Science and Technology*, *41*(5), 1682–1688. doi:10.1021/es061491b.

Portillo, P. M., Ierapetritou, M., Tomassone, S., Mc Dade, C., Clancy, D., Avontuur, P. P. C., & Muzzio, F. J. (2008). Quality by design methodology for development and scale-up of batch mixing processes. *Journal of Pharmaceutical Innovation*, *3*(4), 258–270. doi:10.1007/s12247-008-9048-9.

Process Analytical Technology – Google Search. (n.d.). Retrieved September 24, 2020, from https://www.google.com/search?q=process+analytical+technology&rlz=1C1CHBF_enIN885IN885&oq=process+analytical+technology&aqs=chrome..69i57j0l7.1037j0j7&sourceid=chrome&ie=UTF-8.

Prospective Validation – Google Search. (n.d.). Retrieved September 25, 2020, from https://www.google.com/search?q=2.1.1.+Prospective+validation&rlz=1C1CHBF_enIN885IN885&oq=2.1.1.%09Prospective+validation&aqs=chrome..69i57.1293j0j7&sourceid=chrome&ie=UTF-8.

Rambla-Alegre, M., Esteve-Romero, J., & Carda-Broch, S. (2012). Is it really necessary to validate an analytical method or not? That is the question. *Journal of Chromatography A*, *1232*, 101–109. doi:10.1016/j.chroma.2011.10.050.

Ravisankar, P., Naga Navya, C., Pravallika, D., & Navya, D. (2015). A review on step-by-step analytical method validation. *IOSR Journal of Pharmacy*, *5*(10). www.iosrphr.org.

Review on Technology Transfer in Pharmaceutical Industry. (n.d.). Retrieved September 25, 2020, from https://www.researchgate.net/publication/306192323_Review_on_technology_transfer_in_pharmaceutical_industry.

Santos, N., Business, G. L.-S., &, undefined. (2015). Marketing to the poor: A SWOT analysis of the Market Construction Model for engaging impoverished market segments. *Search.Ebscohost.Com.* Retrieved October 4, 2020, from http://search.ebscohost.com/login.aspx?direct=true&profile=ehost&scope=site&authtype=crawler&jrnl=20444087&AN=110639277&h=9teA8Vhuofd6yTXIb0gUla2J5eKRbm5fv3t92YIhI1ZAZXiQX71wfFDwqghBHtxG6uSFG%2FIJfs1GtofeNOXgZQ%3D%3D&crl=c.

Scale Up Studies to Optimize Manufacturing Process and Execution of Exhibit Batches – Google Search. (n.d.). Retrieved September 24, 2020, from https://www.google.com/search?sxsrf=ALeKk032nfwMS5eYMbXbyzip0wuWPAeguQ:1600964764730&q=scale+up+studies+to+optimize+manufacturing+process+and+execution+of+exhibit+batches&sa=X&ved=2ahUKEwii1eXAmoLsAhWG9XMBHV-2CMsQ1QIoBXoECAwQBg&biw=1536&bih=754.

Sheehan, P., Ricks, R., Ripple, S., & Paustenbach, D. (1992). Field evaluation of a sampling and analytical method for environmental levels of airborne hexavalent chromium. *American Industrial Hygiene Association Journal*, *53*(1), 57–68. doi:10.1080/15298669291359302.

A Systematic Approach to Tech Transfer and Scale-Up. (n.d.). Retrieved September 25, 2020, from https://www.pharmtech.com/view/systematic-approach-tech-transfer-and-scale.

Tucker, 1984 Technology Transfer – Google Search. (n.d.). Retrieved September 25, 2020, from https://www.google.com/search?rlz=1C1CHBF_enIN885IN885&sxsrf=ALeKk035CNbaJ4HHpGVWMitzd2bofa3pyw%3A1601005119031&ei=P2ZtX-7CAfDgz7sP5sy8oA4&q=Tucker%2C+1984+technology+transfer&oq=Tucker%2C+1984+technology+transfer&gs_lcp=CgZwc3ktYWIQAzIFCCEQoAE6BAgjECc6BggAEBYQHjoECCEQFToHCCEQChCgAToICCEQFhAdEB5QtzY9Y_IMBYImLAWgAcAB4AIABvQSIAcspkgEMMC4yLjE1LjEuLjEuEuMC4ymAEAoAEBqgEHZ3dzLXdpcesABAQ&sclient=psy-ab&ved=0ahUKEwiu3pzrsIPsAhVw8HMBHWYmD-QQ4dUDCA0&uact=5

United States Patent (19) Copeland et al. *54 Method for Rinsing a Tissue Sample Mounted on a Slide.* (1995).

US Patent. (1984). *Sterile Water and Sterileaoueous Solutions Background of the Invention.*

Valigra, L. (2010). Qualifying analytical instruments: General chapter <1058> clarifies terminology, classifies instruments. *Quality Assurance Journal*, *13*(3–4), 67–71. doi:10.1002/qaj.475.

Weissman, S. A., & Anderson, N. G. (2015). Design of Experiments (DoE) and Process Optimization. A Review of Recent Publications. *Organic Process Research and Development*, *19*(11), 1605–1633. https://doi.org/10.1021/op500169m

WHO Expert Committee on Specifications for Pharmaceutical Preparations: 34th Report by World Health Organization(WHO)|Waterstones. (n.d.). Retrieved October 13, 2020, from https://www.waterstones.com/book/who-expert-committee-on-specifications-for-pharmaceutical-preparations-34th-report/world-health-organization-who/9789241208635.

WHO Technical Report Series|WHO – Prequalification of Medicines Programme. (n.d.). Retrieved October 13, 2020, from https://extranet.who.int/prequal/content/who-technical-report-series

Yu, L. X. (2008). Pharmaceutical quality by design: Product and process development, understanding, and control. *Pharmaceutical Research*, *25*(4), 781–791. doi:10.1007/s11095-007-9511-1.

19 Nanotoxicology
Safety, Toxicity and Regulatory Considerations

*Raja Susmitha, Mounika Gayathri Tirumala,
Mohd Aslam Saifi, and Chandraiah Godugu*
National Institute of Pharmaceutical Education and Research (NIPER), Hyderabad

CONTENTS

Abbreviations 286
19.1 Introduction 286
19.2 Nanomedicine 286
19.3 Nanotoxicology 288
    19.3.1 Mechanisms of NP Toxicity 289
    19.3.2 ROS and Oxidative Stress 289
    19.3.3 Genotoxicity 289
    19.3.4 Inflammation 291
    19.3.5 Fibrosis 291
    19.3.6 Cancer 291
    19.3.7 Crossing Biological Barriers 292
19.4 Target Organ Toxicity 292
    19.4.1 Kidney Toxicity 292
    19.4.2 Cardiotoxicity 292
    19.4.3 Brain Toxicity 292
    19.4.4 Lung Toxicity 293
    19.4.5 Spleen Toxicity 293
    19.4.6 Liver Toxicity 293
    19.4.7 Inhalational Toxicity 293
    19.4.8 RES Toxicity 294
19.5 Biodistribution and Biodegradation 294
    19.5.1 Biodistribution 294
    19.5.2 Biodegradation 294
19.6 Factors Affecting Nanoparticle Toxicity 294
    19.6.1 Particle Size and Surface Area 295
    19.6.2 Particle Shape 295
    19.6.3 Surface Characteristics 295
    19.6.4 Composition and Crystalline Structure 296
    19.6.5 Agglomeration and Solvent/Medium 296
    19.6.6 Route of Exposure 297
19.7 Toxicity Studies 297
    19.7.1 *In Vitro* Nanotoxicity Assessment Methods 297
    19.7.2 *In Vivo* Nanotoxicity Assessment Methods 297
19.8 Role of Omics and 3D Models in Nanotoxicology 297
19.9 Regulatory Status 298
19.10 Conclusion and Future Perspectives 298
Conflict of Interest 299
References 299

ABBREVIATIONS

AgNPs:	Silver nanoparticles
Akt:	Protein kinase B
Al$_2$O$_3$NPs:	Alumina nanoparticles
AP-1:	Activator protein-1
AuNPs:	Gold nanoparticles
CNTs:	Carbon nanotubes
CuNPs:	Copper nanoparticles
CuONPs:	Copper oxide nanoparticles
ER:	Endoplasmic reticulum
FDA:	Food and Drug Administration
ILs:	Interleukins
JNK:	c-Jun N-terminal kinase
LDH:	Lactate dehydrogenase
MAPK:	Mitogen-activated protein kinase
MDA:	Malondialdehyde
MWCNTs:	Multiwalled carbon nanotubes
NF-κB:	Nuclear factor kappa-B
NiONPs:	Nickel oxide nanoparticles
NPs:	Nanoparticles
OECD:	Organisation for Economic Co-operation and Development
PDGF:	Platelet-derived growth factor
PEG:	Polyethylene glycol
PI3K:	Phosphoionositide-3-kinase
PLGA:	Poly(lactic-co-glycolic acid)
PtNPs:	Platinum nanoparticles
QDs:	Quantum dots
RES:	Reticuloendothelial system
ROS:	Reactive oxygen species
SiNPs:	Silicon nanoparticles
SiO$_2$:	NPs silicon dioxide or silica nanoparticles
SOD:	Superoxide dismutase
SPION:	Super paramagnetic iron oxide nanoparticles
STAT-1:	Single transducer and activation of transcription
SWCNTs:	Single-walled carbon nanotubes
TGF-β:	Transforming growth factor-β
TiO$_2$NPs:	Titanium dioxide nanoparticles
TNF-α:	Tumour necrosis factor-α
WSTs:	Water-soluble tetrazolium salts
XTT:	2,3-bis-(2-methoxy-4-nitro-5-sulfophenyl)-2H-tetrazolium-5-carboxanilide)
ZnNPs:	Zinc nanoparticles
ZnONPs:	Zinc oxide nanoparticles

19.1 INTRODUCTION

Over the last few decades, nanotechnology has been introduced in our daily life (Soares et al., 2018). The International Organization for Standardization (ISO) has defined nanomaterial as a 'material with any external dimension in the nanoscale or having an internal structure or surface structure in the nanoscale,' and nanoparticle (NP) is defined as 'nano-object with all three external dimensions in the nanoscale,' where nanoscale is defined as the size range of approximately 1–100 nm (ISO, 2008, 2010). One billionth of a metre is known as nanometre, which is half about the size of a DNA diameter, a thousand times tinier than a red blood cell (RBC) or one hundred thousand times smaller than human hair diameter (Tarafdar et al., 2013). In 1959, at the American Physical Society, Nobel Prize physicist, Richard P. Feynman, introduced the first concept of nanotechnological strategies in a lecture entitled 'There's plenty of room at the bottom' (Bogdan et al., 2017). In the year 2000, United States president, Bill Clinton, launched the National Nanotechnology Initiative, which paved the way for an innovative field revolution. Great efforts were put forward to accelerate worldwide nanotechnology research. Surprisingly, the nanotechnology field was globally welcomed, and a great number of interdisciplinary research projects were undertaken (Saifi et al., 2018b).

Based on their size, morphology and chemical properties, NPs are classified into carbon-based, metal-based, ceramic-based, lipid-based, semiconductor-based and polymeric NPs (Khan et al., 2019). NPs should inherently display distinct properties, such as chemical reactivity, magnetism, electrical conductance and physical strength differently from bulk particles due to their small size (Nikalje, 2015). For example, TiO$_2$NPs used in sunscreen products exhibit white colour in their bulk form but in nanoform (ranges below 50 nm), it is colourless (Saifi et al., 2018a). NPs are very advantageous in many areas, such as agriculture, textile industry, food industry, cosmetics, detection of microbes, constructions, aerospace, tissue engineering, households, information and communications, clinical and pharmaceutical sector for drug biopharmaceutical property improvement, theranostics and also in cancer treatment (Simões et al., 2020, Saifi et al., 2018a). NPs have also been used as drug delivery carriers and in imaging of cells and tissues (Teleanu et al., 2019, Rakesh et al., 2015). NPs due to their catalytic properties are used in automotive catalytic converters and also help in environmental remediation by detoxifying the pollutants (Farhan et al., 2014). Due to growing applications, NPs are also employed in nanobiotechnology (nanotechnology application to biology) and bionanotechnology (application of biology to nanotechnology) (Teleanu et al., 2019). NPs have shown their potential applications in astrobiology, which helps in the detection of life on other planets, such as Mars (Simões et al., 2020). All these advantageous applications in various fields are made possible only due to their unique properties. Figure 19.1 summarises the use of NPs in various fields.

19.2 NANOMEDICINE

Nanomedicine is a medical application of nanotechnology that utilises chemistry, biology, medicine and engineering for the prevention and treatment of diseases (Kargozar and Mozafari, 2018, Pillai and Ceballos-Coronel, 2013). According to National Institute of Health (NIH), nanomedicine, an offshoot of nanotechnology, refers to 'highly specific medical intervention at the molecular scale for

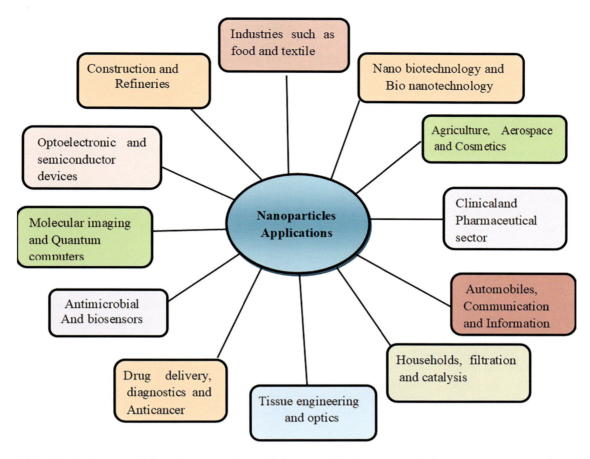

FIGURE 19.1 Applications of NPs in various areas. NPs are being used in food, textile, agriculture, aerospace, cosmetics, automobile and communication industries. In addition, the use of NPs has also been extended to household item, tissue engineering, biosensing, molecular imaging and pharmaceutical sectors.

curing diseases or repairing damaged tissues, such as bones, muscles or nerves' (Pillai and Ceballos-Coronel, 2013). Greater interaction of nanotechnology and biological systems is mainly due to decreased size of NPs which makes an increase in surface area, solubility, dissolution and improves bioavailability and requirement of low dose (Rangasamy, 2011). Generally, organic (also considered polymeric NPs including liposomes, dendrimers, ferritin NPs and micelles), inorganic (such as metals, metal oxides, QDs and ceramics) and carbon based (like graphene, fullerene and CNTs) are different types of NPs used in nanomedicine. Many potential applications in different medical areas are currently offered by these NPs (Patra et al., 2018, Mauricio et al., 2018, Teleanu et al., 2018), mainly in the delivery of various drugs (hydrophobic), proteins, nucleic acids, genes (Kargozar and Mozafari, 2018), monoclonal antibodies (Gustafson et al., 2015) and also in cancer therapy (Cartaxo, 2010) but NP-based gene delivery not yet approved by FDA (Kargozar and Mozafari, 2018).

NPs show potential safety benefits in patients. For example, in disease conditions, accumulation of drugs in tissue improves therapeutic efficacy and reduces the requirement of frequent dosing, thus reducing toxicity. In cancer therapy, the use of NPs shows superior effects because of their greater permeability from blood vessels to tumour tissues, thereby increasing drug penetration to tumour tissues, a phenomenon called enhanced permeability and retention (EPR) effect. This effect is mainly because of the difference in the vasculature of normal and tumour tissues (Wolfram et al., 2015, Ventola, 2017). Another example is NP encapsulation of hydrophobic drugs which also reduces the use of toxic solubilising agents, like dimethyl sulfoxide; this is because NPs serve as an alternative for solubilising agents (Wolfram et al., 2015). Poor pharmacokinetics, limited bioavailability and high toxicity exhibited by many conventional drugs render them less efficacious as therapeutic options. The field of nanotechnology and nanomedicine has made significant progress in overcoming these potential issues by improving the pharmacokinetic profile, reducing toxicity and providing high selectivity of conventional drugs, such as amphotericin B. In addition, nanomedicine has made it possible to increase circulation and enhance the solubility and stability of drugs and targeted controlled release in the organisms (Soares et al., 2018, Beltrán-Gracia et al., 2019). For example, in humans, administration of liposomal drug, Doxil® (50 mg/m^2), was found to show reduced clearance by 250-fold and improved area under curve by 300-fold, compared to free doxorubicin drug (Ventola, 2017).

The cardiotoxicity of free doxorubicin is reduced by liposomal formulation of doxorubicin (Beltrán-Gracia et al., 2019, Wolfram et al., 2015). Other nano-based drugs for cancer treatment available in the market are Caleyx®, Transdrug® and Ambraxane® (Kargozar and Mozafari, 2018). Moreover, NPs also facilitate synergistic effects by co-delivering multiple drugs, thereby avoiding multidrug resistance. For example, cytarabine and daunorubicin (5:1 molar ratio) liposomal nano-based drug formulation shows synergistic effects in acute myeloid leukaemia (Beltrán-Gracia et al., 2019, Alfayez et al., 2019) (Table 19.1).

In addition to the drug delivery field, NPs have been extensively used as probes for imaging and diagnosis as they offer advantages when compared to conventional probes. NPs such as AuNPs, QDs, polymeric and silica NPs act as sophisticated probes to detect cellular and molecular mechanisms in the biological system and help in imaging and diagnosis (Kargozar and Mozafari, 2018). Moreover, NPs without inherent functionalities for molecular imaging can be easily functionalised with imaging agents and used in different imaging techniques. For instance, doxorubicin hybrid NPs were used as magnetic nanocrystals for magnetic resonance imaging (MRI) agents by specifically linking it with certain antibodies (which provides targeted action in imaging of tumour cells), whereas the polymeric part (PLGA)-loaded doxorubicin functions as a therapeutic part to treat breast cancer (Cartaxo, 2010). Further, iron oxide NPs in combination with paclitaxel were used as theranostic agents where iron oxide NPs act as imaging agents and provide cytotoxic effects of paclitaxel in cancer. However, as NPs possess certain peculiar properties, it is quite difficult to predict the pharmacological/toxicity profiles of the NPs. In fact, there are numerous reports demonstrating the toxicological effects of NP-based application in different areas. In addition to human exposure, the irrational use of NPs also raises concerns for the ecosystem. Due to these rising concerns, scientists started focussing on the negative outcomes of NP use, and thus a new branch of nanotoxicology was born which specifically deals with the toxicological profiles of NPs.

19.3 NANOTOXICOLOGY

Nanotechnology has extraordinary capability to change our lives by enabling new ones and improving existing products (Donaldson et al., 2006). Moreover, with the passage of time, applications of nanotechnology have been increasing nearly in every field of our life. However, our knowledge about the unpredictable negative effects of NPs is very limited. Further, the literature available also suggests that the pharmacological/toxicological profiles of NPs are very different from their bulk counterparts, making it complex

TABLE 19.1
Some of the Clinically Approved Nanomedicines

Brand Name	Drug Name	Indication	Approval	References
Doxil®	Doxorubicin	Kaposi's sarcoma	1995	Udhrain et al. (2007)
AmBisome	Amphotericin B	Fungal infections	1997	Barratt and Bretagne (2007)
Ferrlecit	Iron gluconate colloid	In anaemia as iron replacement	1999	Cançado and Muñoz (2011)
Venofer	Iron sucrose colloid	In anaemia as iron replacement	2000	Cançado and Muñoz (2011)
Visudyne®	Verteporfin	Ocular histoplasmosis	2000	Chang and Yeh (2012)
DepoDur™	Morphine sulphate	Management of pain	2004	Chang and Yeh (2012)
Abraxane	Paclitaxel	Metastatic breast cancer	2005	Flühmann et al. (2019)
Mircera	Methoxy PEG-epoetin beta	Symptomatic anaemia associated with chronic kidney disease	2007	Ohashi et al. (2012)
Cimzia®	Certolizumab pegol	Rheumatoid arthritis and Crohn's disease	2008	Dickinson et al. (2019)
Mepact	Mifamurtide	Bone cancer	2009	Kager et al. (2010)
Krystexxa®/Pegloticase	PEGylated uric acid enzyme	Chronic gout	2010	Guttmann et al. (2017)
Exparel	Bupivacaine	Management of pain	2011	Saraghi and Hersh (2013)
Marqibo®	Vincristine	Non-Hodgkin's lymphoma and leukaemia	2012	Hua et al. (2018), Anselmo and Mitragotri (2019)
Injectafer®	Iron carboxymaltose colloid	Iron deficient anaemia	2013	Anselmo and Mitragotri (2019)
Plegridy®	PEGylated interferon β-1a	Multiple sclerosis	2014	Choi and Han (2018)
Ryanodex®	Dantrolene sodium	Malignant hyperthermia	2014	Rosenberg et al. (2015)
Onivyde®	Liposomal irinotecan	Pancreatic adenocarcinoma	2015	Hua et al. (2018)
Adynovate	PEGylated Factor VIII	Haemophilia	2015	Choi and Han (2018)
Vyxeos™	Daunorubicin + cytarabine	Acute myeloid leukaemia	2017	Hua et al. (2018), Anselmo and Mitragotri (2019)
Onpattro®	Patisiran (siRNA)	Hereditary transthyretin-mediated amyloidosis	2018	Beltrán-Gracia et al. (2019)

to understand. Many advantages provided by nanotechnology had led to a common saying 'small is good, smaller is even better.' But currently, this saying is not suitable with respect to NPs because of toxicity problems associated with them. The initial reports of NP toxicity came into light due to their high resemblance with the ultrafine particles of dust, ocean spray, volcanic ash, mineral composites, etc. As the ultrafine particles were reported to be highly toxic to the respiratory system, the resemblance of NPs with them directed the focus of toxicology scientists towards NP toxicity. Later, it was found that few NPs such as silica, carbon-based and metal NPs produced toxicity at certain dose levels. The toxicity of these NPs was suspected due to their penetration across different physiological barriers followed by high retention in the body for a longer duration of time. This increases the chances of their interaction with a number of cellular components due to high reactivity and smaller size, ultimately increasing the risk of toxicity.

The physicochemical properties of NPs like size, shape and surface area are considered to be vital characteristics of NPs. These characteristics can affect NP reactivity, energy properties and catalytic activities (Schwirn et al., 2014). Due to these peculiar characteristics, NPs have exceptionally high chemical reactivity profiles. Further, the small size of NPs helps them to penetrate the cell membrane and interact with cell organelles, such as nucleus and mitochondria. This results in direct and/or indirect damage to organelles, leading to cellular death or sublethal changes in the cells. It is accepted that in this way a large number of NPs in comparison to its bulk form that typically would not cross barriers can enter the cells and show a variety of toxic impacts. Due to their smaller size and greater surface area, NPs possess greater chemical reactivity and elicit a few responses which are harmful to the cells. Although generation of free radicals is considered as one of the primary mechanisms, a few other mechanisms such as inflammation and direct damage to membrane, proteins and DNA also play an important role in NP-mediated toxicity. In addition, prolonged retention of NPs may also lead to fibrosis, granuloma or even carcinomas (Wani et al., 2011). For example, CNTs have the ability to cause lung fibrosis (Saifi et al., 2018a) and cancer (Murali and Mangotra, 2018). Further, certain types of NPs are systemically absorbed and produce subsequent different organ toxicity, such as brain, heart, liver and cardiac tissues (Kreyling et al., 2006, Buzea et al., 2008). NP safety concerns have been raised in industries, regulatory bodies and research communities. Hence, it is imperative to discuss and understand the different mechanisms responsible for the toxicological effects of NPs.

19.3.1 Mechanisms of NP Toxicity

Although our knowledge about the exact mechanism behind NP toxicity is limited, there are certain mechanisms proposed to be involved in the toxicity. However, the generation of free radicals and ROS is one of the critical factors responsible for NP-mediated toxicity. Figure 19.2 represents an overview of different predictable mechanisms of NP toxicity.

19.3.2 ROS and Oxidative Stress

A low amount of ROS formation takes place mainly during cellular oxidative metabolism and is involved in cellular signalling pathways and defence system (Khanna et al., 2015, Abdal Dayem et al., 2017). Under normal physiology, antioxidants protect the cells from ROS damage (Jie Kai et al., 2015). Excess formation of ROS causes cellular effects and subsequently causes oxidative stress. Oxidative stress is mainly due to an imbalance of pro-oxidant mechanisms which are involved in ROS formation and antioxidant mechanisms which are involved in reducing ROS formation (Madl et al., 2013). Due to the exceptionally high surface area, the surface of NPs is highly reactive in nature. This results in higher chances of interaction between surface groups of NPs and cells/cellular components. NPs having pro-oxidant groups on their surface produce oxidation of cell membrane lipids, resulting in the generation of highly reactive free radicals, such as superoxide and peroxide ions. Once these free radicals are produced, they start a chain reaction and keep on generating more and more free radicals, leading to the production of enormous free radicals. When the generation of ROS surpasses the defence mechanism of the cells, causes failure of defence mechanism and affects the cells adversely. The metal oxide–based NPs are common agents to produce ROS as the metal particle serves as an enhancer of ROS formation, acts as a catalyst in Fenton reactions, Fenton-like reactions or Haber–Weiss cycle reactions and greatly enhances the production of free radicals (Fu et al., 2014, Manke et al., 2013). However, apart from the metal-based NPs, the majority of the NPs are reported to produce ROS and increase oxidative stress (Table 19.2).

8-OHdG (8-hydroxy-2-deoxyguanosine) is generally considered as a biomarker for oxidative stress–related toxicity (Valavanidis et al., 2009). As compared with their larger parts, NPs like SiO_2, TiO_2 and CNTs exhibit ROS-mediated toxicity (Manke et al., 2013). Increased ROS levels can interact with lipid membrane, DNA and proteins, resulting in cell death (apoptosis or necrosis). Also, NPs interact with subcellular organelles such as mitochondria, lysosomes and ER and further cause ROS generation and eventually cause cell death (Saifi et al., 2018a, Feng et al., 2015). Oxidative stress exhibits great effects in the biological system, ranging from a molecular level to a disease level (Jie Kai et al., 2015).

19.3.3 Genotoxicity

Oxidative stress followed by genetic damage is considered as one of the mechanisms for NP-mediated toxicity. Genotoxicity is elucidated as damaging the genetic information of a cell, such as chromosomal fragmentation, alteration of gene expression profiles and DNA strand breakages. Genotoxicity is majorly of two types,

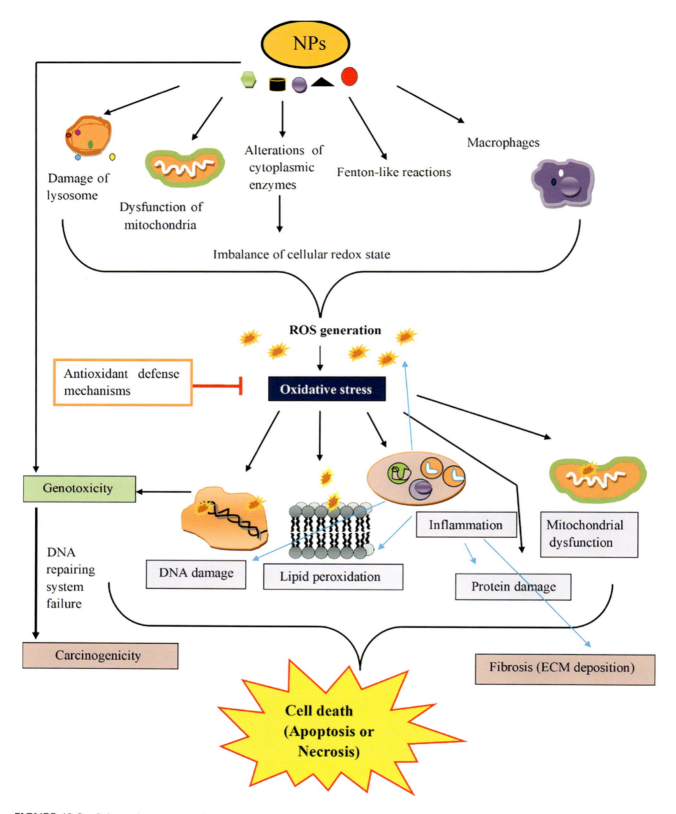

FIGURE 19.2 Schematic representation of ROS production in the biological milieu and interaction of ROS with various cellular components, such as DNA, proteins and lipids, contributes to DNA damage, protein damage and lipid peroxidation, respectively. Interaction of ROS with subcellular organelles such as mitochondria leads to release of contents by damaging them which further contributes to ROS production. ROS promotes production of pro-inflammatory cytokines which further contributes to ROS production and cellular component damage. All these interactions eventually lead to cell death either by apoptosis or necrosis. Chronic inflammation results in fibrosis characterised by excess deposition of ECM. Damaged DNA or direct interaction of NPs with genetic material further leads to genotoxicity and eventually causes carcinogenicity if the DNA repairing system gets failed.

TABLE 19.2

Shows Pro-Oxidant and Antioxidant Systems, Which Play a Crucial Role in Oxidative State Balance and Imbalance in a Cell

Pro-Oxidants	Antioxidants
• Mediators of pro-inflammatory (TNF-α and TGF-β) • Exogenous sources include cytochromes, lipoxygenases, NADPH oxidases and enzymes of mitochondria • Endogenous sources, such as radiations	• Mediators of anti-inflammatory (IL-1 and IL-10) SOD, catalases and glutathione peroxidases are enzymatic systems • Glutathione and ceruloplasmin are non-enzymatic systems

i.e. primary and secondary genotoxicity. Further, primary genotoxicity is again divided into direct and indirect types (Barabadi et al., 2019, Åkerlund et al., 2019). NPs are able to interact directly with the DNA and prevent replication or transcription and interaction with chromosomes during interphase, subsequently causing chromosome breakage (clastogenic effect) or loss of chromosomes (aneugenic effect). NP-mediated ROS and released toxic ions indirectly cause primary genotoxicity by interacting with proteins involved in DNA replication, transcription, repairing process as well as interacting with centrioles, mitotic spindles or other related proteins. In addition, excessive generation of ROS by NP-mediated phagocyte activation subsequently causes DNA damage which is considered as secondary genotoxicity (Barabadi et al., 2019). A study was conducted to know size-dependent (5 and 50 nm) genotoxicity of AuNPs, AgNPs and PtNPs in human bronchial epithelial cells using comet assay. Results stated that AgNPs of both sizes exhibited DNA strand breaks (Lebedova et al., 2017).

19.3.4 Inflammation

In addition to ROS production, NPs can also cause excessive production and release of pro-inflammatory cytokines through the principal cascades, such as NF-κB, MAPK and PI3-K pathways, connecting oxidative stress to inflammation. Oxidative stress causes translocation of NF-κB into the nucleus by the degrading inhibitor of κB (IκB) for the regulation of inflammatory mediator's gene transcription. The generation of excessive hydroxyl and superoxide radicals causes oxidative stress followed by translocation of NF-κB into the nucleus in order to regulate gene transcription of pro-inflammatory mediators, such as TNF-α, IL-1β, IL-2, IL-6 and IL-8. NF-κB-mediated toxicity was shown by metal-based NPs like Zn, cadmium (Cd), Si and CNTs (Khanna et al., 2015). In primary mouse connective tissue cells, SPION causes cytotoxicity, mediated by redox-sensitive NF-κB pathways (Sarkar et al., 2014). A study by Lim et al., reported that exposure to AgNPs causes reproductive toxicity in nematode *Caenorhabditis elegans* by increasing ROS production followed by activation of PMK-1 p38 MAPK pathway (Lim et al., 2012). SiO$_2$NPs cause inflammation followed by autophagy of cardiovascular endothelial cells mediated by PI3K/Akt/mTOR (mammalian target of rapamycin) pathways and elevation of pro-inflammatory cytokines, such as TNF-α, IL-1β and IL-6 (Duan et al., 2014). Nano-silica caused apoptosis of endothelial cells mediated by JNK/p53 mitochondrial-dependent pathway and at high concentration, it causes toxicity to endothelial cells by activation of NF-κB pathway (Liu and Sun, 2010).

19.3.5 Fibrosis

Persistent elevation of inflammatory mediators causes chronic inflammation which may result in scarring of different tissues due to excess extracellular matrix (ECM) deposition, a process known as fibrosis (Wynn, 2008). NPs induce the release of growth factors such as CTGF and TGF-β1 in biological systems which are the most common pro-fibrotic mediators. Transcriptional activation and expression of genes involved in inflammation and fibrosis are activated by redox-sensitive MAPK, RTK (receptor tyrosine kinases), and transcriptional factors such as NF-κB and STAT-1. TGF-β is a pro-fibrotic cytokine involved in fibroblast activation and proliferation which causes excessive release of ECM proteins, resulting in fibrosis. For example, subchronic exposure to NiONPs induces pulmonary fibrosis in rats which could be associated with activation of TGF-β and also upregulates SMAD2, SMAD4, matrix metalloproteinases and tissue inhibitors of metalloproteinases (Chang et al., 2017). Apart from that, other NPs like CuONPs and TiO$_2$NPs were also reported to cause pulmonary fibrosis. On the other hand, ZnONPs were found to possess the potential to induce hepatic fibrosis (Lai et al., 2018, Hong et al., 2017, Bashandy et al., 2017).

19.3.6 Cancer

The adverse effects of NP-induced carcinogenicity associated with DNA damage are mainly through ROS formation which induces DNA damage or by direct interaction of NPs with the DNA causing damage. Meanwhile, accumulation of damaged DNA bases which is owing to repression of DNA damage repair systems, such as DNA direct reversal, base excision repair, nucleotide excision repair, homology-directed repair), mismatch repair pathway and non-homologous end joining repairing pathways. Repression of DNA damage repair system by NPs also induces other agents' risk, such as ultraviolet light, chemicals, etc. which promotes cancer. So, accumulation of damaged DNA in the cells takes place, and if the cell with damaged DNA can survive, it might contribute to carcinogenesis by passing damaged DNA to daughter cells (Guo et al., 2019). Magaye et al. reported that in JB6 cells, NiNPs greatly increases the expression of NF-κB, AP-1 and decreases p53 activation compared to fine NiNPs, indicating more carcinogenicity of NP-treated cells (Magaye et al., 2014). Other studies reported that inhalational exposure of CNTs and TiO_2NPs to animals resulted in lung carcinogenicity (Becker et al., 2010).

19.3.7 Crossing Biological Barriers

Studies have been stating that NPs are able to cross the protective second-line defences called biological barriers, such as blood–brain barrier (BBB), blood–testis barrier, blood–mammary barrier and blood–placental barrier (BPB) and get accumulated there and produce toxicity. NPs cross the barriers primarily by two pathways – transcellular receptor or adsorptive-mediated pathway and paracellular pathway. PEG-coated Fe_3O_4 (iron oxide) and carboxylated polystyrene NPs were reported to cross BBB by lactoferrin receptor–mediated pathway. On the other hand, AuNPs, AgNPs, ZnONPs, Al_2O_3NPs and TiO_2NPs employed paracellular pathways by reducing the expression of tight junction proteins, such as claudin-5, zonula occludens-1 (ZO-1) and occludin, causing disruption of tight junctions and increasing the permeability of different barriers. A study by Teng et al. reported that during the organogenesis period, co-exposure to ZnONPs and cadmium chloride ($Cdcl_2$) NP causes dysfunction of tight junctions due to endothelial cell shedding and less expression of tight junction proteins (ZO-1, claudin-4,8 and occludins) of BPB which subsequently causes embryo and foetus toxicity (Teng et al., 2019). SiO_2NPs could cause dysfunction of BBB and induce ROS and Rho-kinase-mediated inflammation which thus leads to neurotoxicity (Liu et al., 2017). Other studies reported that iron oxide and SiO_2NPs cause microtubule remodelling by activating redox-sensitive pathways, such as Akt/GSK-3β (glycogen synthase kinase-3β), MAPK/Nrf_2 (nuclear factor erythroid-2-related factor 2), JNK/P53 and NF-κB, thus increasing permeability across barriers (Jia et al., 2020).

19.4 TARGET ORGAN TOXICITY

Once the NPs are administered, they enter the systemic circulation and accumulate in important organs, such as liver, kidney, lungs and other organs. The preferential distribution of NPs in various organs, sometimes, crosses the threshold levels and produces toxic effects in the organ (Wu and Tang, 2017). However, the distribution characteristics of NPs differ from case to case and need careful evaluation of their toxicity profiles. Nevertheless, a number of NPs have been reported to produce different organ toxicity.

19.4.1 Kidney Toxicity

Kidneys are majorly susceptible to exogenous substances because of their ability to remove harmful substances from the body. For NPs, kidneys are considered as the last barrier for their excretion through the urinary system. Glomeruli can effectively filter NPs with sizes less than 6 nm and easily clear from the body. However, the glomeruli and tubular cells accumulate NPs excessively which affect normal function and damage kidneys, mainly by oxidative stress, inflammation, DNA damage, autophagy and ER stress (Zhao et al., 2019a). A study reported that lipid peroxidation was increased significantly in the kidneys upon administration of SiNPs (150 nm) intraperitoneally. Subchronic exposure of SiNPs (6, 20 and 50 nm) to male albino Wistar rats reported oxidative-mediated toxicity to kidneys (Balli et al., 2019). A gender-biased toxicity study reported that subchronic oral exposure of AgNPs causes accumulation in rat kidneys. Compared to male rats, female rats accumulate two-fold times more AgNPs in the kidneys (Kim et al., 2010). Exposure of AuNPs to female albino rats subsequently causes dose-dependent effects, such as deposition of NPs in the kidney cortical tissue, structural alterations in the tissues and various degrees of inflammation (Elwan et al., 2018).

19.4.2 Cardiotoxicity

Cardiac complications associated with NP exposure include ischaemic heart diseases, heart failure, arrythmias and other cardiac diseases (Bhat et al., 2015). A study reported that subchronic dermal exposure of AgNPs leads to accumulation in cardiomyocytes and subsequent cardiotoxicity in guinea pigs (Bostan et al., 2016). Long-term oral exposure of NPs such as iron oxide and Ag in male rats showed decreased glutathione levels and increased oxidative DNA damage in heart cells, indicating oxidative stress–mediated cardiotoxicity. Combined NPs (iron oxide and AgNPs) produce more toxicity effects as compared to individual NPs (iron oxide or AgNPs) (Yousef et al., 2019). In isolated cultured cardiomyocytes, SiO_2NP exposure caused oxidative stress–mediated toxicity, resulting in mitochondrial dysfunction, decreased smooth ER activity and eventual cardiotoxicity (Guerrero-Beltrán et al., 2017).

19.4.3 Brain Toxicity

Compared to other organs, delivery to the brain is the most challenging. Overcoming this limitation, NPs can access the brain and be used in many medicinal applications. Brain-related impairments are induced by increasing the incidence of NPs (Wu and Tang, 2017). TiO_2NP exposure to mice caused infiltration of inflammatory cells, cracked nerve cells and changes in the levels of acetylcholinesterase, nitric oxide and glutamic acid in the brain, indicating oxidative stress–mediated neurotoxicity (Jia et al., 2017). Another study stated that oral administration of AgNPs caused oxidative stress–mediated brain toxicity. Rats exposed to AgNPs showed a significant elevation in the levels of N-methyl-D-aspartate (NMDA) receptors, metallothionein-III (MT-III) and monoamine oxidases

(MAO-A, MOA-B). On the other hand, intraperitoneal administration of dextran-coated iron oxide NPs in adult zebra fish showed an increase in the levels of mRNA of caspase-8 and caspase-9, ultimately causing brain toxicity (Valdiglesias et al., 2016).

19.4.4 Lung Toxicity

The inhalational exposure of different NPs was found to be associated with lung toxicity. Lung toxicity is mainly observed when NPs are introduced in the inhalational route (Wu and Tang, 2017). For example, intratracheal TiO_2NP instillation in rats exhibits pulmonary toxicity. Pulmonary lesions were induced by TiO_2NPs (5,21 and 50 nm) with a dose of >5 mg/kg. Compared to 21 and 50 nm, 5 nm TiO_2NPs show more pulmonary toxicity (Liu et al., 2009). *In vitro* studies showed that intratracheal instillation of polyvinylpyrrolidone-coated AgNPs in female rats exhibits cytotoxicity to alveolar macrophages and at higher doses exhibits genotoxicity in lungs (Wiemann et al., 2017). On the other hand, CuONPs exhibited higher cytotoxicity as compared to SiO_2NPs, TiO_2NPs and iron oxide NPs in cultured human epithelial cells. Intratracheal instillation of CuONPs in male (F344) rats shows pulmonary neoplastic lesions (Lai et al., 2018).

19.4.5 Spleen Toxicity

Spleen is considered as the secondary lymphoid organ which clears damaged, necrotic and aged cells and is one of the main organs for the potential deposition of NPs. Systemic exposure to NPs often triggers inflammation and cellular damage in the spleen (Wu and Tang, 2017). For instance, the subacute exposure of nano-copper to rats causes oxidative stress–mediated toxicity to the spleen (Xuerong et al., 2019). After intravenous injection, nano-copper gets accumulated in the spleen and causes apoptosis and immune dysfunction which eventually damages the spleen (Xuerong et al., 2019). On the other hand, another study reported that prolonged oral exposure of TiO_2NPs to adult albino male rats causes disturbances in immune expression of CD4 and CD68, indicating toxic and deleterious effects on rat spleen (Eldesoky et al., 2018). In addition, intravenous AgNPs (20 and 100 nm) administration in rats caused almost complete suppression of spleen NK cell activity at high doses (De Jong et al., 2013). Twenty-eight days' exposure of PEG-coated AuNPs to mice exhibited size-dependent toxicity to the spleen, as demonstrated by changes in RBCs and WBCs (Zhang et al., 2011).

19.4.6 Liver Toxicity

Liver is considered the major organ in the body where the metabolism of various exogenous chemicals takes place (Wu and Tang, 2017). Exposure of NPs directly affects Kupffer cells and eventually causes liver injury (Bhat et al., 2015). Levels of alanine transaminase (ALT), a liver injury marker, were found to be higher in mice after exposure with ZnONPs (Wu and Tang, 2017). Intravenous exposure of AuNPs caused a significant increase in the levels of aspartate transaminase and a decrease in ALT levels, indicating liver toxicity (Paunovic et al., 2017). Another study reported that ZnONP exposure to male albino Wistar rats caused significant alterations in histology and histochemical structure of the liver, depletion of hepatic glycogen, increased apoptosis and necrosis of the liver, demonstrating liver toxicity (Almansour et al., 2017).

19.4.7 Inhalational Toxicity

Inhalation of NPs deposits in the lungs and causes inflammation. The mediators of inflammation enter into circulation and influence CVS indirectly. Another hypothesis is that inhaled NPs get penetrated into the alveolar epithelium and into circulation by translocation, thus directly causing effects on CVS. Inhalation of AuNPs majorly accumulates in inflammation-rich vascular sites (Miller et al., 2017). Inhalation is considered as one of the vital routes for unintentional human exposure to NPs (Sawicki et al., 2019). People of all age groups, health status and gender are susceptible to adverse effects of inhaled NPs (Geiser et al., 2017). Inhaled NPs may accumulate majorly in the lungs, alimentary tract, heart, spleen, liver and even kidneys (Baranowska-Wójcik et al., 2019). NPs, owing to their small size and higher surface area, have a high probability of setting deeply inside the respiratory system and get deposited in the lungs. Deposition of NPs in the lungs further causes inflammation in the acute phase and may lead to fibrosis upon chronic exposure (Geiser et al., 2017, Saifi et al., 2018b). Alveoli are more susceptible to NPs compared to the airways because of larger alveolar surface area and tight air–blood barrier (Poh et al., 2018). Anatase form of TiO_2NPs (15 mg/m^3, ~20 nm) when exposed to male Sprague–Dawley (SD) rats through inhalation route was found to be deposited mostly in the lung tissues. Pulmonary deposition of NPs significantly increased MDA levels in lung tissues (Pujalte et al., 2016). Spark-generated CuONP toxicity evaluation in human bronchial epithelial cells (HBEC) and lung adenocarcinoma cells (A549 cells) was conducted using an *in vitro* air–liquid interface system. No toxicity was observed in HBEC or A549 cells when exposed to clean air. However, exposure of cells to CuONPs decreased cell viability and increased ROS, LDH and IL-8 when compared with control compared to HBEC and A459 cells, suggesting pulmonary toxicity of the NPs (Jing et al., 2015).

Metallic NPs reach the circulatory system very fast when compared with non-metallic NPs as per a study conducted (Asmatulu, 2011). Metal NPs through inhalation get into the lungs and subsequently to lymph and blood circulation, eventually depositing in target sites like brain, bone marrow, lymph nodes, heart or spleen. This trend was observed for Al_2O_3NPs and PbNPs (Sawicki et al., 2019). In addition, several studies reported that brain is considered the main target for inhaled NPs. NPs from the olfactory epithelium of the nasal cavity migrate into the olfactory glomeruli along the olfactory neurons. On the other hand, QDs, CuONPs, TiO$_2$NPs, AgNPs and manganese dioxide NPs exhibited penetration into the brain via the olfactory route (Sawicki et al., 2019). Migration of the smallest NPs to the kidneys is within 30min and to the lymph nodes within 3min. NPs with a neutral charge (<34nm diameter) are the most dangerous and enter lymph nodes. Hence, NP exposure to the internal organs and tissues will be maximum (Ellenbecker and Tsai, 2015). Inhalational exposure of ZnONPs in SD rats for two weeks showed noticeable inflammation in renal periglomeruli and interstitium via lymphocytic infiltration. Renal inflammation may results through direct NPs toxicity from lungs to kidney tissue or via inflammatory mediators from lung (Chien et al., 2017). Inhalation of fluorescent magnetic NPs cause extramedullary haematopoiesis in mice spleen, affecting platelet production and WBC production without causing pulmonary effects (Kwon et al., 2009). Researchers need to study more on inhalational toxicity which further associates with other organ toxicity.

19.4.8 RES Toxicity

RES is considered as the first line of defence mechanism in the body's immune system, which comprises different phagocytic cells, such as macrophages and monocytes (Saifi et al., 2018a). Macrophages are disseminated throughout the organisms and connective tissues and are given specific names to indicate specific locations, such as brain (microglia), liver (Kupffer cells), lungs (alveolar macrophages) and bone (osteoclast) (Yona and Gordon, 2015, Malathi et al., 2015). RES rich in phagocytic cells is involved in physiological functions like capturing and eliminating viruses and other tinier objects. Clearance of NPs from the body takes place through RES and kidneys. RES is responsible for the clearance of NPs greater than 10nm (Sun et al., 2014), while NPs with a size of less than 10nm are majorly cleared by kidneys.

19.5 BIODISTRIBUTION AND BIODEGRADATION

Toxicity of NPs is also influenced by the biodegradation property of NPs and the biodistribution profile of NPs via various routes. Considering and studying these profiles aid in toxicity elucidation.

19.5.1 Biodistribution

The preferential distribution of NPs in organs is one of the major determinants of their toxicity (Zhao et al., 2019b). Once NPs are absorbed in the systemic circulation, they translocate to various organs, such as kidneys, liver, spleen, heart, bone marrow and nervous system. NPs mainly translocate to RES organs primarily due to the higher blood flow of these organs while small NPs translocate to delicate organs, such as brain, foetus and testes by crossing physiological barriers (Zhang et al., 2014). Once NPs reach systemic circulation, a number of proteins adsorb on the surface which decides the uptake of NPs by specific immune cells, such as macrophages, resulting in removal of NPs from the systemic circulation. For example, single-dose intravenous administration of curcumin-capped iron oxide NPs in BALB/c mice resulted in rapid clearance from systemic circulation and assimilation in organs, such as brain, kidneys, spleen and liver with the greatest concentration in the liver and increased assimilation sequentially (Sharma et al., 2020). Iron oxide NP exposure causes mainly accumulation in liver and spleen organs which results in the application of iron oxide as MRI contrast agent. A study by Weissleder et al. demonstrated that 82.6% of the injected dose of iron oxide was accumulated in the liver and 6.2% of the injected dose was accumulated in the spleen (Almeida et al., 2011). Accordingly, once the organ distribution of NPs increases beyond threshold levels, it starts affecting the organ in a negative way and might lead to toxicity of NPs.

19.5.2 Biodegradation

The activity profile and duration of action of NPs in the biological system are primarily determined by their biodegradation characteristics. Under biological conditions in the body, polymeric and nano-liposomes can be degraded naturally (Vlasova et al., 2016, Su and Kang, 2020) while on the other hand, NPs such as CNTs are physically and chemically stable and thus resist biodegradation (Yang and Zhang, 2019). Accumulation of slowly biodegradable and non-degradable NPs might lead to toxicity of the NPs (Janrao, 2014). Natural enzymes like horse-radish peroxidase (plant derived) and human neutrophil enzyme myeloperoxidase degrade CNTs and reduce its toxicity but in the body, CNT complete biodegradation seems difficult. However, complete biodegradation of CNTs has been observed *in vitro* (Zhang et al., 2014). Thus, in the living system long-term persistence of CNTs raises the concerns of toxicity. After parenteral administration, CNTs may reside for prolonged periods inside the spleen and liver. Pharyngeal administration of SWCNTs in mice led to their persistence up to 1 year. Moreover, the deposited CNTs notoriously triggered inflammatory-mediated toxicity (Vlasova et al., 2016). A study by Santiago et al. elucidated the comparison of biodegradable and non-biodegradable NP toxicity in mice (Santiago et al., 2015). Broadly, polymeric NPs such as PLGA-based liposomes are relatively safer than metal-based NPs, mostly due to their biodegradability characteristics. So, currently for nanocarriers majorly preferring are lipid-based and polymeric-based NPs.

19.6 FACTORS AFFECTING NANOPARTICLE TOXICITY

The chemical reactivity and biological interaction profile of NPs get influenced by their physicochemical characteristic parameters. Complying with this fact, toxicity of NPs

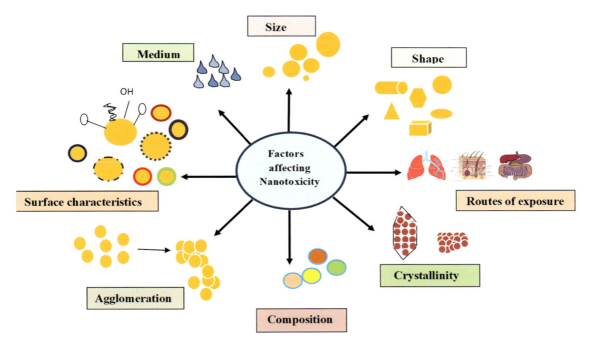

FIGURE 19.3 Overview of various physicochemical factors affecting NPs.

also depends on these physicochemical parameters, such as size, shape, surface energy of NPs and other factors (Figure 19.3).

19.6.1 Particle Size and Surface Area

From a toxicological point of view, particle size and surface area are considered to be vital characteristics because biological interactions commonly take place at the NP surface. As the size of NPs decreases, the surface area increases drastically, and a higher proportion of the particles will be displayed on the surface exposing a higher number of active sites (Sharifi et al., 2012). Thus, the surface of NPs becomes more reactive and also increases the potential catalytic activity, resulting in the greater ability of smaller NPs to interact with cellular and subcellular organelles more efficiently (Saifi et al., 2018a). Toxicological studies have shown that NPs of smaller size cause greater adverse effects on the respiratory system, typically causing more inflammation when compared to large-size NPs of the same material. For example, TiO_2NPs of the same crystalline structure with different sizes of 20 and 250 nm diameter were instilled into rats, and small-size NPs caused higher lung inflammatory reactions compared to large-size NPs (Janrao, 2014). Chen et al. evaluated the toxicity of AgNPs (15, 50 and 100 nm) on the fish RBCs. Results demonstrated that 15 nm–sized AgNPs induced greater toxicity mediated by oxidative stress following membrane injury and eventually caused haemolysis of RBCs when compared to 50 and 100 nm–sized AgNPs (Chen et al., 2015). Similarly, *in vitro* cytotoxicity studies of SiO_2 macro- and nano-sized particles also demonstrated that NPs of SiO_2 produce more toxic effects as compared to macro-sized particles in both cell lines (Sahu et al., 2016). Further, 20 nm CuNPs produced greater toxicity on dorsal root ganglion neurons as compared to 60 and 80 nm CuNPs (Prabhu et al., 2010). Furthermore, inhalational toxicity studies using different sizes such as 25 nm (nanotube morphology) and 60 nm (anatase morphology) were conducted in two cell lines, namely. The results were obvious that 25 nm–sized TiO_2NPs showed greater toxicity on human bronchial epithelial cell line (16HBE) and human non-small cell lung cancer cells (A549) as compared to 60 nm NPs (Ma et al., 2017).

19.6.2 Particle Shape

Particle shape is one of the major determinants in determining the activity/toxicity profile of NPs. For instance, nanotubes and nanofibres were found to be less prone to endocytosis as compared to spherical NPs (Sukhanova et al., 2018). In addition, NiNP dendritic clusters exhibited higher toxicity as compared to spherical-shaped NPs (Ispas et al., 2009). In addition, compared to spherical-shaped AuNPs, rod-shaped AuNPs have been found to be more toxic to human keratinocyte cells (Yah, 2013). Further, nanorod-shaped Al_2O_3NPs reported more cytotoxicity to astrocytes of rat cerebral cortex as compared to nanoflake-shaped NPs (Dong et al., 2019).

19.6.3 Surface Characteristics

Interaction of NPs with the biological system is also determined by surface charge of NPs as it regulates many aspects, such as adsorption, colloidal behaviour, protein binding and permeability across transmembrane (Gatoo et al., 2014). After entering into the human systemic circulation, based on surface characteristics, NPs show adsorption of plasma proteins to their surface and subsequently form protein corona which reduces surface-free energy and increases interaction with cellular and subcellular components. Cellular uptake is more with charged NPs (e.g., cationic ammonium AuNPs) compared to neutral forms of NPs because adsorption of proteins is more with charged NPs (Vales et al., 2020). Non-phagocytic cells show higher cytotoxicity with positively charged NPs, such as SiNPs, AuNPs and ZnONPs, while on the other side, phagocytic cell toxicity is more with negatively charged NPs (e.g., anionic cyanoacrylic NPs) (Froehlich, 2012). Similarly, positively charged silica exhibits more toxicity as compared to negative and neutral-charged silica NPs. Generally, hydrophobic NPs are more prone to cytotoxicity. But in contrast, some studies show hydrophilic NPs (SWCNT) are more prone to protein adsorption and subsequent toxicity (Sun et al., 2019). Surface coating can change NP properties and impact toxicity. NP surface modifications with hydrophilic, PEG and copolymers such as poloxamers and polyethylene can stabilise and increase circulatory time in the biological environment, resulting in reduced toxicity (Figure 19.4).

19.6.4 Composition and Crystalline Structure

Besides NP size and shape, composition of NPs plays a vital role in influencing the toxicity. Soluble form of nano-copper and nano-silver in zebra fish, daphnids and algal species shows toxicity, but TiO_2NPs do not cause any toxicity even having same dimensions. Hence, the composition of NPs is considered vital in toxicity (Vales et al., 2020). Yang et al. study has reported that NPs cause oxidative stress–mediated cytotoxicity. When compared with ZnONPs, moderate cytotoxicity is induced by CNTs but induction of DNA damage is more. But SiO_2NPs, carbon black exhibits low toxicity. The results demonstrated that cytotoxicity effects of different NPs are mainly due to the composition of NPs (Yang et al., 2009). Toxicity of NPs also gets influenced by their crystal structure. For example, inhalation of TiO_2NPs causes allergic reactions, and dendritic cells play a vital role in the adjuvant activity of the immune system. Regardless of size, anatase and anatase/rutile mixture produces a higher expression of CD83 and CD86 in immune dendritic cells compared to rutile TiO_2NPs in allergic reactions (Vandebriel et al., 2018). But Gurr et al. (2005) study reported that light rutile form of TiO_2NPs (200 nm) exhibits oxidative DNA damage, whereas anatase form of TiO_2NPs does not exhibit any toxicity which are quite opposite results as compared to Vandebriel et al. study. Braydich et al. evaluated the crystal structure–mediated toxicity in HEL-30 mouse keratinocyte cell line. Although results showed that both induced cell death, the rutile form of TiO_2NP-initiated apoptosis could be controlled with antioxidant treatment, whereas anatase form–induced cell necrosis could not be reduced (Braydich-Stolle et al., 2009).

19.6.5 Agglomeration and Solvent/Medium

When dealing with nanotoxicological studies, obstacles such as particle aggregation and inadequate dispersion are needed to be addressed. Agglomeration of NPs depends on size, composition and surface charge. Agglomerated NPs show different properties than non-agglomerated NPs which ultimately affects toxicity. Compared to well-dispersed CNTs, agglomerated CNTs have great adverse effects (Gatoo et al., 2014, Sager et al., 2009). In the biological system, inorganic NPs easily form agglomerates compared to their organic NPs and thus are more toxic as compared to organic counterparts (Feng et al., 2015). The dispersion profile of NPs also influences nanotoxicity depending on the solvent/medium used. For instance, in phosphate buffer solution, TiO2, ZnO or carbon black NPs exhibit greater size compared to water-dispersed NPs and subsequently show a completely different toxicity profile.

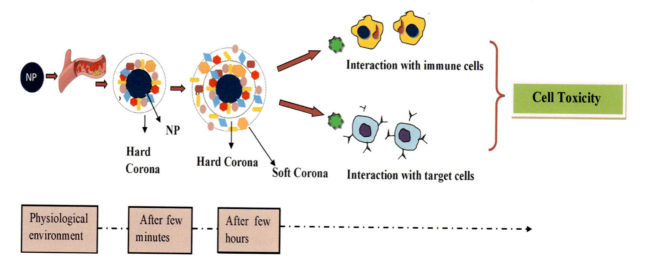

FIGURE 19.4 Protein corona effect. Interaction of NPs with systemic proteins results in their adsorption on the surface of NPs and forming protein corona. The protein corona then interacts with various cells to produce cytotoxicity.

Nanotoxicology

19.6.6 Route of Exposure

Exposure of NPs to the target organ is one of the prime reasons responsible for the toxicity of different organs. Studies have reported that inhalational exposure of NPs causes lung inflammation. On the other hand, topical exposure of NPs results in skin toxicity. NPs of smaller size (<10 nm) penetrate the skin more easily than larger ones (>30 nm). After exposure of smaller size (<10 nm) NPs to the skin, some histological studies demonstrated hyperkeratosis, erythema, intracellular oedema and collagen hyalinisation in the dermis, indicating dermal toxicity (Gautam, 2011). Studies have reported that TiO_2 is able to cross the stratum corneum (outer protective layer) of the skin and interacts with the immune system (Aydn et al., 2012). Dey et al. study reported that exposure of alumina NPs to mouse epithelial JB6 cells which are sensitive to neoplastic transformation shows increased AP-1, sirtuin 1, indicating that toxicity is mediated by increased ROS levels and is also an indication for increased cell proliferation and cell transformation (Dey et al., 2008).

19.7 TOXICITY STUDIES

During drug discovery and developmental process, toxicity studies have been considered as a part, and toxicity studies are lagging in the evaluation of nanotoxicity owing to peculiar NP characteristics.

19.7.1 In Vitro Nanotoxicity Assessment Methods

The conventional toxicity testing methods include various assays like cell proliferation assays clonogenic assays, apoptosis assays, TUNEL (Terminal deoxynucleotidyl transferase dUTP nick end labeling) assay, necrosis assays, oxidative stress, etc. However, the concepts of conventional toxicity testing cannot be applied to nanotoxicity as the NPs behave entirely different than their bulk counterparts. Further, there are many NPs that interfere with conventional toxicity testing methods, leading to false-positive/negative results. For instance, most of the metal-based NPs adsorb formazan dye in the case of MTT (3-(4,5-dimethylthiazol-2-yl)-2,5-diphenyl tetrazolium bromide) assay, which is one of the most commonly used proliferation assays and results in overestimation of cytotoxicity (Arora et al., 2011), whereas in LDH assay, it inhibits LDH enzyme, resulting in underestimation of cytotoxicity (Saifi et al., 2018b). On the other hand, Ames assay is not efficient in predicting NP genotoxicity as it cannot be penetrated through the bacterial cell wall (Dusinska et al., 2017). In addition, the presence of NPs also reduces the performance of comet assay (Azqueta and Dusinska, 2015). In micronucleus test, incubation of NPs with cytochalasin-B may decrease the particle uptake and result in underestimation of micronuclei formation (Magdolenova et al., 2012). Adsorption and inactivation of cytochrome c and xanthine oxidase by NPs lead to false-positive results in xanthine oxidase assay (Tournebize et al., 2013). Optical interference is encountered in some assays like spectrophotometric DPPH (2,2-diphenyl-1-picryl-hydrazyl-hydrate) assay, TBA (thiobarbituric acid) assay, Ellman assay, colorimetric Griess assay, etc. (Tournebize et al., 2013) as it is clear from the literature that there is not even a single satisfactory method to predict the exact information regarding the toxicity of NPs. Hence, there is a strong need to develop novel techniques and assay methods to understand and assess specific NP toxicity.

19.7.2 In Vivo Nanotoxicity Assessment Methods

Using *in vivo* testing methods considers all aspects of NP interactions that occur in the body. Extrapolation of *in vitro* results to *in vivo* is not considered to be satisfactory as it shows different responses *in vivo* (Saifi et al., 2018a). The toxicity testing in animal models is generally performed in mice and rats while other species like monkeys or hamsters have also been used for very few specific studies (Gornati et al., 2009). Once the NPs are injected, they are detected through the tagged labels in live or killed animals (Kumar et al., 2017). *In vivo* toxicity studies mainly include assessment of biodistribution, clearance, haematology, histopathology and serum chemistry (Kumar et al., 2017). Histopathological studies of cells, tissues or the exposed organs such as liver, spleen, lungs, etc. after insult with NPs are performed to determine toxicity level.

19.8 ROLE OF OMICS AND 3D MODELS IN NANOTOXICOLOGY

As the one or two endpoint-based conventional research is not sufficient to predict the nanotoxicity mechanisms, omics screening platforms (proteomics, genomics, transcriptomics, metabolomics, etc.) are vastly researched in the field of toxicology. The ultimate goal of genomics is to investigate the genes and their functions to identify the sensitivity of an individual to a particular toxicant (Froehlich, 2017). Proteomics provides information about protein expression changes and post-translational modifications and also reveals new biomarkers and toxicity signatures for a preclinical safety assessment (Wang et al., 2016). Metabolomics deals with the analysis of endogenous metabolites present in the body after interaction with the toxicants (Saifi et al., 2018b). Proteomics is used for the study of proteome profile while the genomics assesses the genetic profile. Transcriptomics deals with gene expression analysis, epigenomics investigates epigenetic modifications and metabolomics searches for metabolite profiling (Froehlich, 2017, Wang et al., 2016). The use of these techniques helps to identify the cellular stress created by the toxicant even at the low levels that could not cause any severe toxicities but put the cells under stress. Identification of cellular stress at low levels is also seriously concerned as the toxicity starts manifesting when the compensation system is depleted (Froehlich, 2017). However, omics techniques request high sample quality, bioinformatics expertise for data analysis and unclear predictive values limiting these techniques (Froehlich, 2017). Although they have some limitations, omics will take the nanotoxicity to the new level by assuring accurate, consistent and reliable results in a high-throughput manner.

Many of the NPs due to their unique properties have shown high toxicity levels when tested using 2D monolayers while no toxicity levels were observed in animal models. As conventional 2D *in vitro* models are showing a lack of accurate results, the research has shifted towards the use of 3D models. Nowadays, like 3D models such as spheroids, cocultures are gaining popularity in the field of nanotoxicology as they mimic the real tissue environment and possess cell–cell and cell–matrix interactions, thereby providing results that are similar to *in vivo results*. Hence, the 3D co-culture models are developed which includes fabricated miniaturised organoids like brain, intestine, kidney, heart, etc. (Henriksen-Lacey et al., 2016). Chen et al. developed a 3D epidermal model to assess skin toxicity of **AgNPs** and compared it with 2D keratinocytes (Chen et al., 2019). Solomon et al. developed a tumour spheroid culture model and tested potential anticancer nano-therapeutics and gained positive results (Solomon et al., 2016). Astashkina et al. developed a 3D kidney organoid model and evaluated the nephrotoxicity of G5-OH PAMAM (poly(amidoamine)) dendrimers and AuNPs (Astashkina et al., 2014). Further, perfusable 3D culture models like microfabrication and microfluidic systems are designed that allow control over flow rate, drug concentration, pH, etc. (Henriksen-Lacey et al., 2016). These microfluidics paved way for the development of 'lab-on-chip technology,' i.e. miniaturisation of the whole lab process on a small microfluidic chip which further extended to 'tissue-on-chips' and 'organ-on-chips' (Henriksen-Lacey et al., 2016). Toh et al. designed a microfluidic 3D hepatocyte chip for hepatotoxicity testing of SPIONs (Toh et al., 2009). Zhang et al. developed a lung-on-chip model and assessed the toxicity of TiO_2NPs and ZnONPs (Zhang et al., 2018). These 3D models can be used as precise tools as they can bridge the gap between the *in vitro* and *in vivo* nanotoxicological research.

19.9 REGULATORY STATUS

The rise of novel technologies exhibits some safety concerns regarding human health. Nanotechnology has been rapidly growing and many literatures state the toxic effects of NPs. Unlike conventional drugs, there are no specific regulations for nano-based drugs when introducing for human betterment. Despite the involvement of regulatory authorities, such as the US Environmental Protection Agency (US EPA), US FDA, Organisation for Economic Cooperation and Development (OECD), International Conference on Harmonisation (ICH), European Medical Agency (EMA) and Department of Science and Technology into this matter, there are still no specific regulatory requirements for testing nano-based drugs, preclinically and clinically (Saifi et al., 2018b). NPs are considered to be as same as their bulk form by the FDA (Saifi et al., 2018a,b). But to the fact, NPs always show fundamentally distinct toxicity when compared with their bulk form because of their unique physicochemical characteristics and the way how it interacts with the biological system (Becker et al., 2010). This is one of the major reasons behind the lack of suitable regulatory guidelines. Although strategies that are used for conventional drugs in terms of evaluating safety/toxicity have been adopted for NPs (Gupta and Xie, 2018), the distinct behaviour of NPs such as agglomeration/aggregation, sedimentation, adsorption and other characteristics could interfere with the results of conventional assays and methods. So, there is a great need to frame regulations for nano-based drugs and also to develop and explore new techniques and assays (Saifi et al., 2018b, Gupta and Xie, 2018). In terms of drug development and approval, nano-drugs are following the same process like other drugs or biologics (Ventola, 2017). Special regulatory awareness requires for 'nanosimilars' (combination of genetic drugs and nanocarriers) borderline and combinational products, and currently regulatory discussions also increase their focus on nanosimilars (Gupta and Xie, 2018, Bremer et al., 2018). In addition to that, few debates are running regarding other issues, such as definition of NPs as there is no universally accepted definition. Another concern is the selection criteria for NPs while designing toxicity testing studies for NPs. However, efforts are being put, and regulatory authorities are focussing more in this direction by establishing OECD's Working party on Manufactured Nanomaterials to assess NP safety. Nevertheless, there is still a need for fast and essential development of regulatory guidelines and toxicological methods for the safe and better use of NPs.

19.10 CONCLUSION AND FUTURE PERSPECTIVES

Nanotechnology is a burgeoning field having many applications in various areas including medicine. All these applications are mainly owing to possessing distinctive characteristics compared to their bulk particles. The increasing applications and exposure of NPs to humans raise concerns regarding their toxicity profiles. NP toxicity effects are not only limited to the liver and spleen but also show impact on various other organs, such as lungs, kidney, brain, heart, digestive tract, skin and bone marrow. There are a number of factors that affect NP toxicity such as size, shape, surface characteristics, protein corona, route of administration, etc. Although the number of studies on NP toxicity is increasing, the exact mechanism of NP toxicity is not obviously elucidated. The conventional models and assays do not clearly elucidate NP toxicity because of their distinct properties which interferes with the toxicity testing. More *in vivo* studies need to be done for better toxicity understanding. Further, employing techniques such as omics and 3D models can help in understanding toxicity better than conventional techniques, and it could be considered a great step towards handling this problem. Furthermore, there is a necessity to develop validated models and techniques to evaluate NP toxicity and to frame regulatory guidelines for NPs. In order to reduce the gaps between NP toxicity and drug development, scientific community is needed to put efforts to study the effects of acute and chronic exposure of NPs. There is a great need to fully understand peculiar NP characteristics and their interactions with body cells, tissues and proteins and their mechanisms in order to understand nanotoxicity in a better way. Regulatory authorities also increased their focus which raised our hopes that in near future the toxicity darkness will disappear and a safe nanomedicine therapy will benefit mankind.

CONFLICT OF INTEREST

Authors reported no conflict of interest.

REFERENCES

Abdal Dayem, A., Hossain, M., Lee, S. B., Kim, K., Saha, S., Yang, G.-M., Choi, H. & Cho, S.-G. 2017. The role of reactive oxygen species (ROS) in the biological activities of metallic nanoparticles. *International Journal of Molecular Sciences*, 18, 120.

Åkerlund, E., Islam, M., McCarrick, S., Alfaro-Moreno, E. & Karlsson, H. 2019. Inflammation and (secondary) genotoxicity of Ni and NiO nanoparticles. *Nanotoxicology*, 13, 1–13.

Alfayez, M., Kantarjian, H., Kadia, T., Ravandi, F. & Daver, N. 2019. Emerging drug profile: CPX-351 (vyxeos) in AML. *Leukemia & Lymphoma*, 61, 1–10.

Almansour, M., Alferah, M., Shraideh, Z. & Jararr, B. 2017. Zinc oxide nanoparticles hepatotoxicity: Histological and histochemical study. *Environmental Toxicology and Pharmacology*, 51, 124–130.

Almeida, J., Chen, A., Foster, A. & Drezek, R. 2011. In vivo biodistribution of nanoparticles. *Nanomedicine (London, England)*, 6, 815–835.

Anselmo, A. C. & Mitragotri, S. 2019. Nanoparticles in the clinic: An update. *Bioengineering & Translational Medicine*, 4, e10143.

Arora, S., Rajwade, J. & Paknikar, K. 2011. Nanotoxicology and *in vitro* studies: The need of the hour. *Toxicology and Applied Pharmacology*, 258, 151–165.

Asmatulu, R. 2011. Toxicity of nanomaterials and recent developments in lung disease. *IntechOpen*, Chapter 6, 95–108.

Astashkina, A., Jones, C., Thiagarajan, G., Kurtzeborn, K., Ghandehari, H., Brooks, B. & Grainger, D. 2014. Nanoparticle toxicity assessment using an *in vitro* 3-D kidney organoid culture model. *Biomaterials*, 35, 6323–6331.

Aydn, A., Sipahi, H. & Hamitoğlu, M. 2012. Nanoparticles toxicity and their routes of exposures. *Recent advances in novel drug carrier systems*, Chapter 18, 483–500.

Azqueta, A. & Dusinska, M. 2015. The use of the Comet Assay for the evaluation of the genotoxicity of nanomaterials. *Frontiers in Genetics*, 6, 239.

Balli, E., Yalin, S., Eroğlu, P., Bayrak, G. & Çömelekoğlu, Ü. 2019. Effects of different sizes silica nanoparticle on the liver, kidney and brain in rats: Biochemical and histopathological evaluation. *Journal of Research in Pharmacy*, 23, 344–353.

Barabadi, H., Najafi, M., Samadian, H., Azarnezhad, A., Vahidi, H., Mahjoub, M., Koohiyan, M. & Ahmadi, A. 2019. A systematic review of the genotoxicity and antigenotoxicity of biologically synthesized metallic nanomaterials: Are green nanoparticles safe enough for clinical marketing? *Medicina*, 55, 439.

Baranowska-Wójcik, E., Szwajgier, D., Oleszczuk, P. & Winiarska-Mieczan, A. 2019. Effects of titanium dioxide nanoparticles exposure on human health – A review. *Biological Trace Element Research*, 193, 118–129.

Barratt, G. & Bretagne, S. 2007. Optimizing efficacy of Amphotericin B through nanomodification. *International Journal of Nanomedicine*, 2, 301.

Bashandy, S., Alaamer, A., Moussa, S. & Omara, E. 2017. Role of zinc oxide nanoparticles in alleviating hepatic fibrosis and nephrotoxicity induced by thioacetamide in rats. *Canadian Journal of Physiology and Pharmacology*, 96, 337–344.

Becker, H., Herzberg, F., Schulte, A. & Kolossa-Gehring, M. 2010. The carcinogenic potential of nanomaterials, their release from products and options for regulating them. *International Journal of Hygiene and Environmental Health*, 214, 231–238.

Beltrán-Gracia, E., López-Camacho, A., Higuera-Ciapara, I., Velázquez-Fernández, J. B. & Vallejo-Cardona, A. A. 2019. Nanomedicine review: Clinical developments in liposomal applications. *Cancer Nanotechnology*, 10, 11.

Bhat, M., Varshneya, C., Patil, R., Bhardwaj, P. & Sharma, D. 2015. Target organ toxicity by nanoparticles – A short review. *American Journal of Pharmtech Research*, 5, 142–149.

Bogdan, J., Pławińska-Czarnak, J. & Zarzynska, J. 2017. Nanoparticles of titanium and zinc oxides as novel agents in tumor treatment: A review. *Nanoscale Research Letters*, 12, 1–15.

Bostan, H., Rezaee, R., Valokala, M., Tsarouhas, K., Golokhvast, K., Tsatsakis, A. & Karimi, G. 2016. Cardiotoxicity of nano-particles. *Life Sciences*, 165, 91–99.

Braydich-Stolle, L., Schaeublin, N., Murdock, R., Jiang, J., Biswas, P., Schlager, J. & Hussain, S. 2009. Crystal structure mediates mode of cell death in TiO$_2$ nanotoxicity. *Journal of Nanoparticle Research*, 11, 1361–1374.

Bremer, S., Halamoda, B. & Borgos, S. E. 2018. Identification of regulatory needs for nanomedicines: Regulatory needs for nanomedicines. *Journal of Interdisciplinary Nanomedicine*, 3, 4–15.

Buzea, C., Pacheco, I. & Robbie, K. 2008. Nanomaterials and nanoparticles: Sources and toxicity. *Biointerphases*, 2, MR17–MR71.

Cançado, R. D. & Muñoz, M. 2011. Intravenous iron therapy: How far have we come? *Revista brasileira de hematologia e hemoterapia*, 33, 461–469.

Cartaxo, A. 2010. *Nanoparticles types and properties– understanding these promising devices in the biomedical area*. MS thesis, Department of Biomedical Engineering, University of Minho, Braga, Portugal.

Chang, H.-I. & Yeh, M.-K. 2012. Clinical development of liposome-based drugs: Formulation, characterization, and therapeutic efficacy. *International Journal of Nanomedicine*, 7, 49.

Chang, X., Zhu, A., Liu, F., Zou, L., Su, L., Liu, S., Zhou, H., Sun, Y., Han, A., Li, S. & Li, J. 2017. Nickel oxide nanoparticles induced pulmonary fibrosis via TGF-1 activation in rats. *Human & Experimental Toxicology*, 36, 802–812.

Chen, L., Fang, L., Ling, J., Ding, C., Kang, B. & Huang, C. 2015. Nanotoxicity of silver nanoparticles to red blood cells: Size dependent adsorption, uptake, and hemolytic activity. *Chemical Research in Toxicology,* 28, 501–509.

Chen, L., Wu, M., Jiang, S., Zhang, Y., Li, R., Lu, Y., Liu, L., Wu, G., Liu, Y., Xie, L. & Xu, L. 2019. Skin toxicity assessment of silver nanoparticles in a 3D epidermal model compared to 2D keratinocytes. *International Journal of Nanomedicine,* 14, 9707–9719.

Chien, C.-C., Yan, Y.-H., Juan, H.-T., Cheng, T.-J., Liao, J.-B., Lee, H.-P. & Wang, J.-S. 2017. Sustained renal inflammation following 2 weeks of inhalation of occupationally relevant levels of zinc oxide nanoparticles in Sprague Dawley rats. *Journal of Toxicologic Pathology,* 30, 307–314.

Choi, Y. H. & Han, H.-K. 2018. Nanomedicines: Current status and future perspectives in aspect of drug delivery and pharmacokinetics. *Journal of Pharmaceutical Investigation,* 48, 43–60.

De Jong, W. H., Van Der Ven, L. T. M., Sleijffers, A., Park, M. V. D. Z., Jansen, E. H. J. M., Van Loveren, H. & Vandebriel, R. J. 2013. Systemic and immunotoxicity of silver nanoparticles in an intravenous 28 days repeated dose toxicity study in rats. *Biomaterials,* 34, 8333–8343.

Dey, S., Bakthavatchalu, V., Tseng, M., Wu, P., Florence, R., Grulke, E., Yokel, R., Dhar, S., Yang, H.-S., Chen, Y. & Clair, D. 2008. Interactions between SIRT1 and AP-1 reveal a mechanistic insight into the growth promoting properties of alumina (Al_2O_3) nanoparticles in mouse skin epithelial cells. *Carcinogenesis,* 29, 1920–1929.

Dickinson, A. M., Godden, J. M., Lanovyk, K. & Ahmed, S. S. 2019. Assessing the safety of nanomedicines: A mini review. *Applied In Vitro Toxicology,* 5, 114–122.

Donaldson, K., Aitken, R., Tran, L., Stone, V., Duffin, R., Forrest, G. & Alexander, A. 2006. Carbon nanotubes: A review of their properties in relation to pulmonary toxicology and workplace safety. *Toxicological Sciences,* 92, 5–22.

Dong, L., Tang, S., Deng, F., Zhao, K., Zhou, J., Liang, D., Fang, J., Hecker, M., Giesy, J., Bai, X. & Zhang, H. 2019. Shape-dependent toxicity of alumina nanoparticles in rat astrocytes. *Science of the Total Environment,* 690, 158–166.

Duan, J., Yu, Y., Yu, Y., Li, Y., Wang, J., Weijia, G., Jiang, L., Li, Q., Zhou, X. & Sun, Z. 2014. Silica nanoparticles induce autophagy and endothelial dysfunction via the PI3K/Akt/mTOR signaling pathway. *International Journal of Nanomedicine,* 9, 5131–5141.

Dusinska, M., Tulinska, J., El Yamani, N., Kuricova, M., Liskova, A., Rollerova, E., Rundén-Pran, E. & Smolkova, B. 2017. Immunotoxicity, genotoxicity and epigenetic toxicity of nanomaterials: New strategies for toxicity testing? *Food and Chemical Toxicology,* 109, 797–811.

Eldesoky, R., Salem, M., Helal, O., El-Monem, A. B. D. & Sahar, N. 2018. Effect of titanium dioxide nanoparticles on spleen of adult male albino rats: Histological and immunohistochemical study. *Egyptian Journal of Histology,* 41(3), 311–328.

Ellenbecker, M. & Tsai, C. S. J. 2015. *Exposure Assessment and Safety Considerations for Working with Engineered Nanoparticles.* Hoboken, NJ: John Wiley and Sons.

Elwan, W., Ragab, A. & Ragab, M. 2018. Histological and immunohistochemical evaluation of the dose-dependent effect of gold nanoparticles on the renal cortex of adult female albino rat. *Egyptian Journal of Histology,* 41, 167–181.

Farhan, M., Khan, I. & Thiagarajan, P. 2014. Nanotoxicology and its implications. *Research Journal of Pharmaceutical, Biological and Chemical Sciences,* 5, 470–479.

Feng, X., Chen, A., Zhang, Y., Wang, J., Shao, L. & Wei, L. 2015. Central nervous system toxicity of metallic nanoparticles. *International Journal of Nanomedicine,* 10, 4321–4340.

Flühmann, B., Ntai, I., Borchard, G., Simoens, S. & Mühlebach, S. 2019. Nanomedicines: The magic bullets reaching their target? *European Journal of Pharmaceutical Sciences,* 128, 73–80.

Froehlich, E. 2012. The role of surface charge in cellular uptake and cytotoxicity of medical nanoparticles. *International Journal of Nanomedicine,* 7, 5577–5591.

Froehlich, E. 2017. Role of omics techniques in the toxicity testing of nanoparticles. *Journal of Nanobiotechnology,* 15, 1–22.

Fu, P., Xia, Q., Hwang, H.-M., Ray, P. & Yu, H. 2014. Mechanisms of nanotoxicity: Generation of reactive oxygen species. *Yao wu shi pin fen xi = Journal of Food and Drug Analysis,* 22, 64–75.

Gatoo, M., Naseem, S., Arfat, M., Dar, A., Qasim, K. & Zubair, S. 2014. Physicochemical properties of nanomaterials: Implication in associated toxic manifestations. *BioMed Research International,* 2014, 498420.

Gautam, A., Singh, D. & Vijayaraghavan, R. 2011. Dermal exposure of nanparticles: An understanding. *Journal of Cell and Tissue,* 11(1), 2703–2708.

Geiser, M., Jeannet, N., Fierz, M. & Burtscher, H. 2017. Evaluating adverse effects of inhaled nanoparticles by realistic *in vitro* technology. *Nanomaterials,* 7, 49.

Gornati, R., Papis, E., Di Gioacchino, M., Sabbioni, E., Dalle-Donne, I., Milzani, A. & Bernardini, G. 2009. *In Vivo and In Vitro Models for Nanotoxicology Testing.* Chichester: John Wiley and Sons.

Guerrero-Beltrán, E., Bernal-Ramírez, J., Lozano, O., Oropeza-Almazán, Y., Gonzalez, E., Garza, J., Garcia, N., Vela-Guajardo, J., Garcia, A., Ortega, E., Torre, G., Ornelas-Soto, N. & Garcia-Rivas, G. 2017. Silica nanoparticles induce cardiotoxicity interfering with energetic status and Ca^{2+} handling in adult rat cardiomyocytes. *American Journal of Physiology – Heart and Circulatory Physiology,* 312. doi:10.1152/ajpheart.00564.2016.

Guo, H., Liu, H., Wu, H., Cui, H., Fang, J., Zuo, Z., Deng, J., Li, Y., Wang, X. & Zhao, L. 2019. Nickel carcinogenesis mechanism: DNA damage. *International Journal of Molecular Sciences,* 20, 4690.

Gupta, R. & Xie, H. 2018. Nanoparticles in daily life: Applications, toxicity and regulations. *Journal of Environmental Pathology, Toxicology and Oncology,* 37, 209–230.

Gurr, J.-R., Wang, A., Chen, C.-H. & Jan, K.-Y. 2005. Ultrafine titanium dioxide particles in the absence of photoactivation can induce oxidative damage to human bronchial epithelial cells. *Toxicology,* 213, 66–73.

Gustafson, H. H., Holt-Casper, D., Grainger, D. W. & Ghandehari, H. 2015. Nanoparticle uptake: The phagocyte problem. *Nano Today,* 10, 487–510.

Guttmann, A., Krasnokutsky, S., Pillinger, M. H. & Berhanu, A. 2017. Pegloticase in gout treatment-safety issues, latest evidence and clinical considerations. *Therapeutic Advances in Drug Safety,* 8, 379–388.

Henriksen-Lacey, M., Carregal-Romero, S. & Liz-Marzán, L. 2016. Current challenges toward *in vitro* cellular validation of inorganic nanoparticles. *Bioconjugate Chemistry,* 28, 212–221.

Hong, F., Ji, L., Zhou, Y. & Wang, L. 2017. Pulmonary fibrosis of mice and its molecular mechanism following chronic inhaled exposure to TiO_2 nanoparticles. *Environmental Toxicology,* 33, 1090.

Hua, S., De Matos, M. B., Metselaar, J. M. & Storm, G. 2018. Current trends and challenges in the clinical translation of nanoparticulate nanomedicines: Pathways for translational development and commercialization. *Frontiers in Pharmacology*, 9, 790.

ISO. 2008. International Organization for Standardization. Technical specification: Nanotechnologies-terminology and definitions for nano-objects-nanoparticle, nanofibre and nanoplate. ISO/TS 80004-2. https://www.iso.org/obp/ui/#iso:std:iso:ts:27687:ed-1:v2:en [Online]. [Accessed on 15 June, 2021].

ISO. 2010. International Organization for Standardization, Nano-technologies-Vocabulary-part 1: Core terms. ISO/TS 80004-1. https://www.iso.org/obp/ui/#iso:std:iso:ts:80004:-1:ed-1:v1:en [Online]. [Accessed on 15 June, 2021].

Ispas, C., Andreescu, D., Patel, A., Goia, D., Andreescu, S. & Wallace, K. 2009. Toxicity and developmental defects of different sizes and shape nickel nanoparticles in zebrafish. *Environmental Science & Technology*, 43, 6349–6356.

Janrao, K. 2014. Nanoparticle induced nanotoxicity: An overview. *Asian Journal of Biomedical and Pharmaceutical Sciences*, 4, 1–7.

Jia, J., Wang, Z., Yue, T., Su, G., Teng, C. & Yan, B. 2020. Crossing biological barriers by engineered nanoparticles. *Chemical Research in Toxicology*, 33, 1055–1060.

Jia, X., Wang, S., Zhou, L. & Sun, L. 2017. The potential liver, brain, and embryo toxicity of titanium dioxide nanoparticles on mice. *Nanoscale Research Letters*, 12, 1–14.

Jie Kai, T., Ong, C., Bay, B., Ho, H. & Leong, D. 2015. Oxidative stress by inorganic nanoparticles. *Wiley Interdisciplinary Reviews Nanomedicine and Nanobiotechnology*, 8, 414–438.

Jing, X., Park, J. H., Peters, T. & Thorne, P. 2015. Corrigendum to "Toxicity of copper oxide nanoparticles in lung epithelial cells exposed at the air–liquid interface compared with in vivo assessment". *Toxicology In Vitro*, 29(3), 502–511.

Kager, L., Pötschger, U. & Bielack, S. 2010. Review of mifamurtide in the treatment of patients with osteosarcoma. *Therapeutics and Clinical Risk Management*, 6, 279.

Kargozar, S. & Mozafari, M. 2018. Nanotechnology and nanomedicine: Start small, think big. *Materials Today: Proceedings*, 5, 15492–15500.

Khan, I., Saeed, K. & Khan, I. 2019. Nanoparticles: Properties, applications and toxicities. *Arabian Journal of Chemistry*, 12, 908–931.

Khanna, P., Ong, C., Bay, B. & Baeg, G. 2015. Nanotoxicity: An interplay of oxidative stress, inflammation and cell death. *Nanomaterials*, 5, 1163–1180.

Kim, Y., Song, M., Park, J., Song, K., Ryu, H., Chung, Y., Chang, H., Lee, J. H., Oh, K., Kelman, B., Hwang, I. & Yu, I. J. 2010. Subchronic oral toxicity of silver nanoparticles. *Particle and Fibre Toxicology*, 7, 20.

Kreyling, W., Semmler-Behnke, M. & Möller, W. 2006. Ultrafine particle–Lung interactions: Does Size Matter? *Journal of Aerosol Medicine: The Official Journal of the International Society for Aerosols in Medicine*, 19, 74–83.

Kumar, V., Sharma, N. & Maitra, S. 2017. In vitro and in vivo toxicity assessment of nanoparticles. *International Nano Letters*, 7, 243–256.

Kwon, J.-T., Kim, D.-S., Minai-Tehrani, A., Hwang, S.-K., Chang, S.-H., Lee, E.-S., Xu, C.-X., Lim, H., Hu, K., Yoon, B.-I., An, G.-H., Lee, K.-H., Lee, J.-K. & Cho, M.-H. 2009. Inhaled fluorescent magnetic nanoparticles induced extramedullary hematopoiesis in the spleen of mice. *Journal of Occupational Health*, 51, 423–431.

Lai, X., Zhao, H., Zhang, Y., Guo, K., Xu, Y., Chen, S. & Zhang, J. 2018. Intranasal delivery of copper oxide nanoparticles induces pulmonary toxicity and fibrosis in C57BL/6 mice. *Scientific Reports*, 8, 1–12.

Lebedova, J., Hedberg, Y., Wallinder, I. & Karlsson, H. 2017. Size-dependent genotoxicity of silver, gold and platinum nanoparticles studied using the mini-gel comet assay and micronucleus scoring with flow cytometry. *Mutagenesis*, 00, 1–9.

Lim, D., Roh, J.-Y., Eom, H.-J., Choi, J.-Y., Hyun, J. & Choi, J. 2012. Oxidative stress-related PMK-1 P38 MAPK activation as a mechanism for toxicity of silver nanoparticles to reproduction in the nematode *Caenorhabditis elegans*. *Environmental Toxicology and Chemistry/ SETAC*, 31, 585–592.

Liu, R., Yin, L., Pu, Y., Liang, G., Zhang, J., Su, Y., Xiao, Z. & Ye, B. 2009. Pulmonary toxicity induced by three forms of titanium dioxide nanoparticles via intra-tracheal instillation in rats. *Progress in Natural Science*, 19, 573–579.

Liu, X., Sui, B. & Sun, J. 2017. Blood-brain barrier dysfunction induced by silica NPs in vitro and in vivo: Involvement of oxidative stress and Rho-kinase/JNK signaling pathways. *Biomaterials*, 121, 64–82.

Liu, X. & Sun, J. 2010. Endothelial cells dysfunction induced by silica nanoparticles through oxidative stress via JNK/P53 and NF-κB pathways. *Biomaterials*, 31, 8198–209.

Ma, Y., Guo, Y., Wu, S., Lv, Z., Zhang, Q. & Ke, Y. 2017. Titanium dioxide nanoparticles induce size-dependent cytotoxicity and genomic DNA hypomethylation in human respiratory cells. *RSC Advances*, 7, 23560–23572.

Madl, A., Plummer, L., Carosino, C. & Pinkerton, K. 2013. Nanoparticles, lung injury, and the role of oxidant stress. *Annual Review of physiology*, 76, 447–465.

Magaye, R., Zhou, Q., Bowman, L., Zou, B., Mao, G., Xu, J., Castranova, V., Zhao, J. & Ding, M. 2014. Metallic nickel nanoparticles may exhibit higher carcinogenic potential than fine particles in JB6 cells. *PloS One*, 9, e92418.

Magdolenova, Z., Lorenzo Corrales, Y., Collins, A. & Dusinska, M. 2012. Can standard genotoxicity tests be applied to nanoparticles? *Journal of Toxicology and Environmental Health. Part A*, 75, 800–806.

Malathi, L., Amsaveni, R., Anitha, N. & Balachander, N. 2015. Reticuloendothelial malignancy of head and neck: A comprehensive review. *Journal of Pharmacy and Bioallied Sciences*, 7, 147.

Manke, A., Wang, L. & Rojanasakul, Y. 2013. Mechanisms of nanoparticle-induced oxidative stress and toxicity. *BioMed Research International*, 2013, 942916.

Mauricio, M., Guerra-Ojeda, S., Marchio, P., Valles, S., Aldasoro, M., Escribano-Lopez, I., Herance, J., Rocha, M., Vila, J. & Victor, V. 2018. Nanoparticles in medicine: A focus on vascular oxidative stress. *Oxidative Medicine and Cellular Longevity*, 2018. Article ID 6231482.

Miller, M., Raftis, J., Langrish, J., Mclean, S., Samutrtai, P., Connell, S., Wilson, S., Vesey, A., Fokkens, P., Boere, J., Krystek, P., Campbell, C., Hadoke, P., Donaldson, K., Cassee, F., Newby, D., Duffin, R. & Mills, N. 2017. Inhaled Nanoparticles Accumulate at Sites of Vascular Disease. *ACS Nano*, 11, 4542–4552.

Murali, S. & Mangotra, A. 2018. Nanotoxicity – A bird's eye view of toxicological aspects. *Research & Development in Material Science*, 7, 682–684.

Nikalje, A. P. 2015. Nanotechnology and its applications in medicine. *Medicinal Chemistry*, 5, 81–89.

Ohashi, N., Sakao, Y., Yasuda, H., Kato, A. & Fujigaki, Y. 2012. Methoxy polyethylene glycol-epoetin beta for anemia with chronic kidney disease. *International Journal of Nephrology and Renovascular Disease*, 5, 53.

Patra, J. K., Das, Fraceto, L., Campos, E., Rodríguez-Torres, P., Acosta-Torres, L., Diaz-Torres, L., Grillo, R., Swamy, M., Sharma, S., Habtemariam, S. & Shin, H. 2018. Nano based drug delivery systems: Recent developments and future prospects. *Journal of Nanobiotechnology*, 16, 1–33.

Paunovic, J., Vucevic, D., Radosavljevic, T., Pantic, S., Nikolovski, D., Dugalic, S. & Pantic, I. 2017. Effects of metallic nanoparticles on physiological liver functions. *Reviews on Advanced Materials Science*, 49, 123–128.

Pillai, G. & Ceballos-Coronel, M. L. 2013. Science and technology of the emerging nanomedicines in cancer therapy: A primer for physicians and pharmacists. *SAGE Open Medicine*, 1. doi:10.1177/2050312113513759.

Poh, T., Ali, N., Mac Aogáin, M., Hussain, M., Setyawati, M., Ng, K. & Chotirmall, S. 2018. Inhaled nanomaterials and the respiratory microbiome: Clinical, immunological and toxicological perspectives. *Particle and Fibre Toxicology*, 15, 1–16.

Prabhu, B., Ali, S., Murdock, R., Hussain, S. & Srivatsan, M. 2010. Copper nanoparticles exert size and concentration dependent toxicity on somatosensory neurons of rat. *Nanotoxicology*, 4, 150–160.

Pujalte, I., Dieme, D., Haddad, S., Serventi, A. & Bouchard, M. 2016. Toxicokinetics of titanium dioxide (TiO_2) nanoparticles after inhalation in rats. *Toxicology Letters*, 265, 77–85.

Rakesh, M., Divya, T., Vishal, T. & Shalini, K. 2015. Applications of nanotechnology. *Journal of Nanomedicine & Biotherapeutic Discovery*, 5, 1.

Rangasamy, M. 2011. Nano technology: A review. *Journal of Applied Pharmaceutical Science*, 1, 8–16.

Rosenberg, H., Pollock, N., Schiemann, A., Bulger, T. & Stowell, K. 2015. Malignant hyperthermia: A review. *Orphanet Journal of Rare Diseases*, 10, 93.

Sager, T., Porter, D., Robinson, V., Lindsley, W., Schwegler-Berry, D. & Castranova, V. 2009. Improved method to disperse nanoparticles for *in vitro* and *in vivo* investigation of toxicity. *Nanotoxicology*, 1, 118–129.

Sahu, D., Kannan, M., Tailang, M. & Vijayaraghavan, R. 2016. *In vitro* cytotoxicity of nanoparticles: A comparison between particle size and cell type. *Journal of Nanoscience*, 2016. Article ID 4023852.

Saifi, M. A., Khan, W. & Godugu, C. 2018a. Cytotoxicity of nanomaterials: Using nanotoxicology to address the safety concerns of nanoparticles. *Pharmaceutical Nanotechnology*, 6, 3–16.

Saifi, M. A., Khurana, A. & Godugu, C. 2018b. Nanotoxicology: Toxicity and risk assessment of nanomaterials. In Hussain, C. M., ed. *Nanomaterials in Chromatography*. Amsterdam: Elsevier.

Santiago, L., Hillaireau, H., Grabowski, N., Mura, S., Nascimento, T., Dufort, S., Coll, J.-L., Tsapis, N. & Fattal, E. 2015. Compared *in vivo* toxicity in mice of lung delivered biodegradable and non-biodegradable nanoparticles. *Nanotoxicology*, 10, 1–11.

Saraghi, M. & Hersh, E. V. 2013. Three newly approved analgesics: An update. *Anesthesia Progress*, 60, 178–87.

Sarkar, A., Ghosh, M. & Sil, P. 2014. Nanotoxicity: Oxidative stress mediated toxicity of metal and metal oxide nanoparticles. *Journal of Nanoscience and Nanotechnology*, 14, 730–743.

Sawicki, K., Czajka, M., Matysiak-Kucharek, M., Fal, B., Drop, B., Meczynska, S., Sikorska, K., Kruszewski, M. & Kapka, L. 2019. Toxicity of metallic nanoparticles in the central nervous system. *Nanotechnology Reviews*, 8, 175–200.

Schwirn, K., Tietjen, L. & Beer, I. 2014. Why are nanomaterials different and how can they be appropriately regulated under REACH? *Environmental Sciences Europe*, 26, 4.

Sharifi, S., Behzadi, S., Laurent, S., Laird Forrest, M., Stroeve, P. & Mahmoudi, M. 2012. Toxicity of nanomaterials. *Chemical Society Reviews*, 41, 2323–2343.

Sharma, N., Saifi, M., Singh, S. & Godugu, C. 2020. *In vivo* studies: Toxicity and biodistribution of nanocarriers in organisms. *Nanotoxicity*, 41–70.

Simões, M. F., Ottoni, C. A. & Antunes, A. 2020. Biogenic metal nanoparticles: A new approach to detect life on mars? *Life*, 10, 28.

Soares, S., Sousa, J., Pais, A. & Vitorino, C. 2018. Nanomedicine: Principles, properties, and regulatory issues. *Frontiers in Chemistry*, 6, 360.

Solomon, M., Lemera, J. & D'Souza, G. 2016. Development of an *in vitro* tumor spheroid culture model amenable to high-throughput testing of potential anticancer nanotherapeutics. *Journal of Liposome Research*, 26, 1–15.

Su, S. & Kang, P. 2020. Systemic review of biodegradable nanomaterials in nanomedicine. *Nanomaterials*, 10, 656.

Sukhanova, A., Bozrova, S., Sokolov, P., Berestovoy, M., Karaulov, A. & Nabiev, I. 2018. Dependence of nanoparticle toxicity on their physical and chemical properties. *Nanoscale Research Letters*, 13, 1–21.

Sun, H., Jiang, C., Wu, L., Bai, X. & Zhai, S. 2019. Cytotoxicity-related bioeffects induced by nanoparticles: The role of surface chemistry. *Frontiers in Bioengineering and Biotechnology*, 7, 414.

Sun, T., Zhang, Y. S., Pang, B., Hyun, D. C., Yang, M. & Xia, Y. 2014. Engineered nanoparticles for drug delivery in cancer therapy. *Angewandte Chemie International Edition England*, 53, 12320–12364.

Tarafdar, J., Sharma, S. & Raliya, R. 2013. Nanotechnology: Interdisciplinary science of applications. *African Journal of Biotechnology*, 12, 219–226.

Teleanu, D. M., Chircov, C., Grumezescu, A. M. & Teleanu, R. I. 2019. Neurotoxicity of nanomaterials: An up-to-date overview. *Nanomaterials*, 9, 96.

Teleanu, D. M., Chircov, C., Grumezescu, A. M., Volceanov, A. & Teleanu, R. I. 2018. Impact of nanoparticles on brain health: An up to date overview. *Journal of Clinical Medicine*, 7, 490.

Teng, C., Jia, J., Wang, Z. & Yan, B. 2019. Oral co-exposures to zinc oxide nanoparticles and $CdCl_2$ induced maternal-fetal pollutant transfer and embryotoxicity by damaging placental barriers. *Ecotoxicology and Environmental Safety*, 189, 109956.

Toh, Y.-C., Lim, T. C., Tai, D., Xiao, G., Noort, D. & Yu, H. 2009. A microfluidic 3D hepatocyte chip for drug toxicity testing. *Lab on a Chip*, 9, 2026–2035.

Tournebize, J., Sapin-Minet, A., Bartosz, G., Leroy, P. & Boudier, A. 2013. Pitfalls of assays devoted to evaluation of oxidative stress induced by inorganic nanoparticles. *Talanta*, 116, 753–763.

Udhrain, A., Skubitz, K. M. & Northfelt, D. W. 2007. Pegylated liposomal doxorubicin in the treatment of AIDS-related Kaposi's sarcoma. *International Journal of Nanomedicine*, 2, 345.

Valavanidis, A., Vlachogianni, T. & Fiotakis, C. 2009. 8-hydroxy-2′-deoxyguanosine (8-OHdG): A critical biomarker of oxidative stress and carcinogenesis. *Journal of Environmental Science and Health. Part C, Environmental Carcinogenesis & Ecotoxicology Reviews*, 27, 120–139.

Valdiglesias, V., Fernández-Bertólez, N., Kilic, G., Costa, C., Costa, S., Fraga, S., Bessa, M., Pásaro, E., Teixeira, J. P. & Laffon, B. 2016. Are iron oxide nanoparticles safe? Current knowledge and future perspectives. *Journal of Trace Elements in Medicine and Biology*, 38, 53–63.

Vales, G., Suhonen, S., Siivola, K., Savolainen, K., Catalán, J. & Norppa, H. 2020. Genotoxicity and cytotoxicity of gold nanoparticles *in vitro*: Role of surface functionalization and particle size. *Nanomaterials*, 10, 271.

Vandebriel, R., Vermeulen, J., Engelen, L., Jong, B., Verhagen, L., Fonteyne-Blankestijn, L., Hoonakker, M. & Jong, W. H. 2018. The crystal structure of titanium dioxide nanoparticles influences immune activity in vitro and in vivo. *Particle and Fibre Toxicology*, 15, 1–12.

Ventola, C. L. 2017. Progress in nanomedicine: Approved and investigational nanodrugs. *Pharmacy and Therapeutics*, 42, 742.

Vlasova, I., Kapralov, A., Michael, Z., Burkert, S., Shurin, M., Star, A., Shvedova, A. & Kagan, V. 2016. Enzymatic oxidative biodegradation of nanoparticles: Mechanisms, significance and applications. *Toxicology and Applied Pharmacology*, 299, 58–69.

Wang, Y., Li, C., Yao, C., Ding, L., Lei, Z. & Wu, M. 2016. Techniques for investigating molecular toxicology of nanomaterials. *Journal of Biomedical Nanotechnology*, 12, 1115–1135.

Wani, M., Hashim, M., Nabi, F. & Malik, M. 2011. Nanotoxicity: Dimensional and morphological concerns. *Advances in Physical Chemistry*, 2011. Article ID 450912.

Wiemann, M., Vennemann, A., Blaske, F., Sperling, M. & Karst, U. 2017. Silver nanoparticles in the lung: Toxic effects and focal accumulation of silver in remote organs. *Nanomaterials*, 7, 441.

Wolfram, J., Zhu, M., Yang, Y., Shen, J., Gentile, E., Paolino, D., Fresta, M., Nie, G., Chen, C. & Shen, H. 2015. Safety of nanoparticles in medicine. *Current Drug Targets*, 16, 1671–1681.

Wu, T. & Tang, M. 2017. Review of the effects of manufactured nanoparticles on mammalian target organs. *Journal of Applied Toxicology*, 38, 25–40.

Wynn, T. 2008. Cellular and molecular mechanisms of fibrosis. *The Journal of Pathology*, 214, 199–210.

Xuerong, Z., Zhao, L., Luo, J., Tang, H., Xu, M., Wang, Y., Yang, X., Chen, H., Li, Y., Ye, G., Shi, F., Lv, C. & Jing, B. 2019. The toxic effects and mechanisms of nano-Cu on the spleen of rats. *International Journal of Molecular Sciences*, 20, 1469.

Yah, C. 2013. The toxicity of gold nanoparticles in relation to their physiochemical properties. *Biomedical Research*, 24, 400–413.

Yang, H., Liu, C., Yang, D., Zhang, H. & Xi, Z. 2009. Comparative study of cytotoxicity, oxidative stress and genotoxicity induced by four typical nanomaterials: the role of particle size, shape and composition. *Journal of Applied Toxicology*, 29, 69–78.

Yang, M. & Zhang, M. 2019. Biodegradation of carbon nanotubes by macrophages. *Frontiers in Materials*, 6, 225.

Yona, S. & Gordon, S. 2015. From the reticuloendothelial to mononuclear phagocyte system – The unaccounted years. *Frontiers in Immunology*, 6, 328.

Yousef, M., Abuzreda, A. & Kamel, M. 2019. Cardiotoxicity and lung toxicity in male rats induced by long-term exposure to iron oxide and silver nanoparticles. *Experimental and Therapeutic Medicine*, 18, 4329–4339.

Zhang, M., Xu, C., Jiang, L. & Qin, J. 2018. A 3D human lung-on-a-chip model for nanotoxicity testing. *Toxicology Research*, 7, 1048–1060.

Zhang, X., Wu, D., Shen, X., Liu, P.-X., Yang, N., Zhao, B., Zhang, H., Sun, Y.-M., Zhang, L.-A. & Fan, F.-Y. 2011. Size-dependent *in vivo* toxicity of PEG-coated gold nanoparticles. *International Journal of Nanomedicine*, 6, 2071–2081.

Zhang, Y., Bai, Y., Jia, J., Gao, N., Li, Y., Zhang, R., Jiang, G. & Yan, B. 2014. Perturbation of physiological systems by nanoparticles. *Chemical Society Reviews*, 43, 3762–3809.

Zhao, H., Li, L., Zhan, H., Chu, Y. & Sun, B. 2019a. Mechanistic understanding of the engineered nanomaterial-induced toxicity on kidney. *Journal of Nanomaterials*, 2019, 1–12.

Zhao, Y., Sultan, D. & Liu, Y. 2019b. Biodistribution, excretion, and toxicity of nanoparticles. In *Theranostic Bionanomaterials*, pp. 27–53. Elsevier. doi:10.1016/B978-0-12-815341-3.00002-X.

Index

Note: **Bold** page numbers refer to tables and *italic* page numbers refer to figures.

accelerated blood clearance (ABC) 213–214
accelerated stability studies 228
acid purification 192–193
activator protein-1 (AP-1) 292, 297
active pharmaceutical ingredients (API) 9
aerosol/vapour methods 149
AFM *see* atomic force microscopy (AFM)
aforementioned systems 97
AgNPs *see* silver nanoparticles (AgNPs)
AI *see* artificial intelligence (AI)
Alnylam Pharmaceutical Corporation 97
Al$_2$O$_3$NPs *see* alumina nanoparticles (Al$_2$O$_3$NPs)
α-bromomalonate 52
α-halo carbanion 52
ALT *see* levels of alanine transaminase (ALT)
alumina nanoparticles (Al$_2$O$_3$NPs) 293, 295
Alzheimer's disease (AD) 53
    Aβ clearance
        improvement of 262–263
    Aβ production
        inhibition of 262
        modulation of 261–262
Amaral, P.E. 71
Ambraxane® 288
amide bonds 124
17-amino acid peptide (BR2) 77
amyloid-β (Aβ) rat model 53
analytical method validation 275, 276
Andreadou, M. 74
annealing 193
ANOVA 26
antibiotic-resistant infection 76
antibody attachment 172
anti-CA-125 monoclonal antibody (mAb) 74
anti-cancer drug 77
anti-cancer treatment 37
anti-inflammatory agent 54
anti-neoplastic and antiapoptotic activity 54
antioxidant and neuroprotective agent 53
antisense therapy 103
antitumour agent 54
antiviral agent 53
antiviral agents delivery 172
AP-1 *see* activator protein-1 (AP-1)
API *see* active pharmaceutical ingredients (API)
ApoE-deficient mouse model 72
a proliferation-inducing ligand (APRIL) 104
aquasomes 26–27
aqueous solution method 68
arc vaporisation technique 51
artificial intelligence (AI) 10
atherosclerosis progression 72
atherosclerotic plaque characterisation 73
atomic force microscopy (AFM) 6, 74, 153
atomic number 8 delivery 173
Atridox® 89
AuNPs *see* gold nanoparticles (AuNPs)

Bahuguna, Shradha 55
barcoding nanoparticles 76
BBB *see* blood-brain barrier (BBB)
BBD *see* box-behnken design (BBD)
Bcl-2 protein expression 103
benzofuran structure (BFG)-derived anti-cancer drug 77
Bilen, B. 73
Bingel-type reactions 52
bioavailability 7, 66, 76, 287
biocompatibility 91–92, 99
biodegradability 7, 85, 99, 294
biodegradable polymers 83, 247–248; *see also* poly(lactic-co-glycolic) acid (PLGA)
biodistribution 294
biological synthesis 144
bioluminescence technology 37
biomolecule-incorporated hydrogel biosensor 181–183
biopolymer 95
biosensors and biolabels 155
biosensors/biomarkers 74–75
biosynthesis 146
block copolymer micelles 36
blood–brain barrier (BBB) 292
blood–placental barrier (BPB) 292
bottom-up approach 6
box-behnken design (BBD) 21
BPB *see* blood-placental barrier (BPB)
brain cancer cells 54
brain toxicity 292–293
Brunetti, J. 73
Brust–Schiffrin method 142
Buckminster Fuller, R. 50
buckyballs 50
Bydureon® 89

CAA *see* critical analytical attributes (CAA)
Caglar, M. 72
Caleyx® 288
cancer
    cancer therapy **87–89**, 102
        clinical trials 105–108, **106–107**
        metal nanoparticles 154
    cancer treatment 100
        disadvantage of 38
        mechanism of 73
        nanotoxicology 291–292
        targeted drug delivery 54
cancer-targeting contracting agents (CAs) 37, 40
caprolactone 36
carbon nanotubes (CNTs)
    cell penetration and mechanism 196–197
    CNT's synthesis
        high-pressure carbon monoxide 192
        nebulised spray pyrolysis method 191–192
        vapour-phase growth 191
    drug delivery, application of 197
    drug delivery system **200**, *201*
    functionalisation techniques
        characterisation 194–196
        covalent functionalisation 193
        non-covalent functionalisation 193–194
    history of 188
    multiwalled carbon nanotubes 188
    production methods **190**
    properties of 188–189
    purification techniques
        acid purification 192–193
        annealing 193
        oxidation 192
    single-walled carbon nanotubes 188
    structure of 188
    synthesis methods 189–191
    toxicological perspectives 197–201
    in vitro and *in vivo* evaluation **198–199**
carboxylated GQDs (cGQDs) 77
cardiotoxicity 288, 292
cardiovascular diseases 75
CARPA *see* complement activation-related pseudoallergy (CARPA)
carrier–drug conjugate nanoparticles (CDC-NPs) 124
    amide bonds 124
    disulphide bonds 126
    emulsions 127–128
    ester bonds 124
    hydrazone bonds 124, 126
    lipid NPs 128
    liposomes 128
    micelles 127
    other bonds 126
    polymeric NPs 128
    self-assembled NPs 126
carrier erythrocytes, safety concern 173–174
cationic lipids 97–98
cationic liposome 99
cationic SLNs (cSLNs) 100
CCD *see* central composite design (CCD)
cell labelling 78–79
cell-penetrating peptides (CPPs) 71
cellular carriers 165
central composite design (CCD) 20
central nervous system (CNS)
    nanoparticles
        carbon nanotubes 261
        dendrimers 261
        fullerenes 261
        gold nanoparticles 260
        liposomes 260
        mesoporous silica nanoparticles 260
        micelles 261
        nanogels 261
        nanostructured lipid carriers 260
        polymeric nanoparticles 260
        quantum dots 260–261
        solid lipid nanoparticles 260
Cerenkov luminescenceblue emission 71
cetrimide (CTAB) 100
cetylpyridinium chloride (CPC) 100
CFR *see* code of federal regulations (CFR)

C_{60}-Fullerenes
    challenges of 55–56
    CNTs and GO **50**
    functionalisation
        cycloaddition reactions 52
        miscellaneous functionalisations 52–53
        nucleophilic addition reactions 52
    medical applications
        anti-inflammatory agent 54
        antioxidant and neuroprotective agent 53
        antitumour agent 54
        antiviral agent 53
        photodynamic therapy 53–54
    neoplastic cells 55
    synthetic methods
        arc vaporisation technique 51
        biomedical applications 51
        hydrocarbon combustion 51
        laser ablation 51
    toxicity profile 55
cGMP *see* current good manufacturing practice (cGMP)
Chan, W.C. 74
chemical chaperones 119
chemical method 144
chemical reduction method 142
chemotherapeutic drugs 9, 54
chemotherapy 34
Chinese Hamster ovary (CHO) 101
Choi, Y. H. 100
cinnarizine (CN) 26
c-Jun N-terminal kinase (JNK) 291
Claro, M.S. 69
clathrate structure formation 35
cleaning process written 275
cleaning validation 274
    analytical methods 275
    cleaning process written 275
    direct surface sampling 275
    equipment design 274–275
    general requirement 274
    limits establishment 275
    monitoring – indirect testing 275
    rinse samples 275
    sampling 275
clinically approved nanomedicines **288**
Clinton, Bill 286
c-Met siRNA 100–101
code of federal regulations (CFR) 270, 271
colloidal nanoparticles (NPs) 120
    dendrimers 123
    discomes/cubosomes 123
    emulsions 121
    lipid NPs 123
    liposomes 122
    micelles 122
    microparticles 121–122
    nanocrystals 123
    niosomes 122–123
    polymeric nanoparticles 121–122
    reverse micelles 122
    spanlastics 123
colon-targeted drug delivery 250
comparative objective 20
complement activation-related pseudoallergy (CARPA) 214
computed tomography (CT) 34, 40–41, *41*
computer system validation 272–273
concurrent validation 271
conducting polymer (CP) 180–181

containers testing 231
continued process verification (CPV) 272
controlled and sustained drug delivery 85–86
control space 18
control strategy 18
coordination-driven self-assembly 6
copper oxide nanoparticles (CuONPs) 291, 293
coprecipitation method 151
cost-effective method 6
covalent functionalisation 193
CPMP stability study guidelines 230
CPP *see* critical process parameter (CPP)
CPPs *see* cell-penetrating peptides (CPPs)
CPV *see* continued process verification (CPV)
CQA *see* critical quality attribute (CQA)
critical analytical attributes (CAA) 27
critical process parameter (CPP) 16, 22, 71
critical quality attribute (CQA) 18, 22, 25, 270, 272
critical solution temperature (CST) 34–35
crossing biological barriers 292
CST *see* critical solution temperature (CST)
CT *see* computed tomography (CT)
CuONPs *see* copper oxide nanoparticles (CuONPs)
current good manufacturing practice (cGMP) 270–273
Cushing's disease 89
cyclic temperature stress studies 229
1,3-cycloaddition reaction 55
cycloaddition reactions 52
cytokine release 72
cytotoxicity 10

data recording 232
DDAB *see* dimethyl dioctadecyl ammonium bromide (DDAB)
DDS *see* drug delivery systems (DDS)
DE *see* droplet epitaxy (DE)
de Broglie wavelength 66
dendrimers 123
DepoFoam™ liposome technology 211–212
design of experiments (DoE) 19, 23, 26
design space 16, 18
DeSimone's group 6
desirability method 20
DET *see* droplet epitaxy technique (DET)
diagnostic approach 37
diagnostic test systems 74
diagnostic tool 73–74
Diels–Alder cycloaddition 71
diethyl bromomalonate 52
dimethyl dioctadecyl ammonium bromide (DDAB) 100
dimethyl sulfoxide (DMSO) 51
1,2-dioleoylsn-glycero-3-ethylphosphocholine lipid 100
direct surface sampling 275
discomes/cubosomes 123
disulphide bonds 126
DLS *see* dynamic light scattering (DLS)
DMSO *see* dimethyl sulfoxide (DMSO)
DNA (plasmid DNA) lipoplexes 98
D-OD *see* D-optimal design (D-OD)
D-optimal design (D-OD) 21, 26
dose-limiting toxicity (DLT) 105
DOX *see* doxorubicin (DOX)
Doxil® 287
Dox-loaded micelles 36
doxorubicin (DOX) 9, 101
droplet epitaxy (DE) 68

droplet epitaxy technique (DET) 69
drug delivery performance 19
drug delivery systems (DDS) 75, **200**, 201
    applications of **8**
    DepoFoam™ liposome technology 211–212
    herbal medicines 7
    lysolipid thermally sensitive liposome 212
    micro- and nanotechnology 3
    nanoparticulate-mediated drug 9
    non-PEGylated liposome 211
    phytomedicines 7
    polymer–drug conjugate 252
    polymer role
        extended-release dosage forms 250
        gastro-retentive dosage forms 250
        modified drug-release dosage forms 249–250
    polymers types
        colon-targeted drug delivery 250
        mucoadhesive drug delivery system 250
        sustained release 250–251
        tissue engineering 251–252
    quantum dots
        antibiotic-resistant infection 76
        cardiovascular diseases 75
        hepatic diseases 76
        neurological disorders 76
        ocular diseases 75
        renal diseases 78
        tumours 76–77
    resealed erythrocytes 168–169
    responsive polymer
        biodegradable polymers 247–248
        highly water-soluble polymer 249
        ion-exchange resins 247
        mucoadhesive polymers 246
        pH-responsive polymers 246
        temperature-responsive polymers 247
        temperature-sensitive hydrogels 247
        thiolated polymers 246
        viscosity enhancer 248
    in situ drug delivery system 252
    stealth liposome technology 211
drug discovery 154
drug encapsulation 166–167
drug loading method 167
drug, name, formulation and application **10**
drug nanonisation 7
drug product manufacturing
    assuring sterility 217
    formulation refinement 216–217
    materials employed in production 217
drug quality 271–272
drug stabilisation 86
drug targeting 170
Duarte, Sonia 99
dynamic light scattering (DLS) 153–154

ECM *see* excess extracellular matrix (ECM)
EDs *see* experimental designs (EDs)
EDX *see* energy-dispersive x-ray spectroscopy (EDX)
EE *see* entrapment efficiency (EE)
Efros, A.L. 72
Ehlers, M.D. 72
electrical systems 34
electrochemical method 142, 149
electromagnetic spectrum 7
electrospraying 6
Eligard® Kit 89

Index

ELISA method 106
EMA *see* European Medical Agency (EMA)
emulsifier 99
emulsions 121, 127–128
encapsulated cell technology 132
end of treatment (EoT) 106
endohedral fullerenes *see* metallofullerenes
endoplasmic reticulum (ER) 289, 292
endothelial adhesion assay 72
energy band gap 69
energy-dispersive x-ray spectroscopy (EDX) 152
enhanced permeability and retention (EPR) effect 213, 287
entrapment efficiency (EE) 22, 25, 77, 127–128
environmental scanning electron microscopy (ESEM) 6
enzyme and hormones deficiency 170–171
EPR effect *see* enhanced permeability and retention (EPR) effect
equipment design 274–275
ER *see* endoplasmic reticulum (ER)
erythrocytes
    applications of 167–168
    as cellular carriers 165
    characterisation of 167, **168**
    drawbacks of 166
    as drug carriers 166
    as drug delivery system 165–166
    drug encapsulation 166–167
    drug loading method 167
    isolation of 167
    red blood cells 164–165
    resealed erythrocytes
        antibody attachment 172
        antiviral agents delivery 172
        atomic number 8 delivery 173
        as drug delivery systems 168–169
        drug targeting 170
        enzyme and hormones deficiency 170–171
        heavy metal poisoning treatment 172
        lead poisoning treatment 172
        macromolecules microinjection 173
        parasitic diseases treatment 172
        replacement therapy 170–171
        solid tumours treatment 171–172
        toxic agent poisoning 172
    safety concern 173–174
ESEM *see* environmental scanning electron microscopy (ESEM)
ester bonds 124
european commission 270
European Medical Agency (EMA) 298
excess extracellular matrix (ECM) 291
experimental designs (EDs)
    comparative objective 20
    formulation development 19
    response surface method objective 20
    screening objective 20
    selection of 20
extended-release dosage forms 250

FA *see* folic acid (FA)
facile solution-based method 71
factorial design (FD) 20, 21
factors affecting nanoparticle toxicity
    agglomeration and solvent/medium 296
    composition and crystalline structure 296
    particle shape 295
    particle size and surface area 295
    route of exposure 297
    surface characteristics 295–296
FbD *see* formulation by design (FbD)
FD *see* factorial design (FD)
FDA *see* Food and Drug Administration (FDA)
FeNPs *see* iron oxide nanoparticles (FeNPs)
Feynman, Richard P. 286
FFD *see* fractional factorial design (FFD)
fibrosis 291
flow injection method 149
fluorescent labelling 78
5-fluorouracil acetic acid (FUA) 76
focal adhesion kinase (FAK) 103
folic acid (FA) 99
Food and Drug Administration (FDA) 270, 287
formaldehydetreated serum albumin (FSA) 76
formulation by design (FbD)
    design of experiments
        EDs selection 20
        experimental designs 20
        nanovesicles optimisation 20–21
        QbD application 21–27
    FbD optimisation methodology
        experimental design 19
        formulation objectives 18
        optimum formulation 19
        response variables 18–19
        significant factors 18–19
        validation studies and scale-up 19
    terminology 16–18, **17–18**
formulation development 227
formulation refinement 216–217
Fourier transform infrared spectroscopy (FTIR) 152–153
fractional factorial design (FFD) 21
Francis, J.E. 78
FSA *see* formaldehydetreated serum albumin (FSA)
FTIR *see* Fourier transform infrared spectroscopy (FTIR)
Fu, J. 36
fullerene-(tris-aminocaproic acid) hydrate (FTACAH) 53
fullerenes 49

gadolinium metallofullerene nanocrystals (GFNCs) 56
gastro-retentive dosage forms 250
gene delivery vectors 96
general requirement 274
generic product development 238–239
    of capsule 239
    depot microsphere formulation 240
    of lipid emulsion 239
    of ocular product 240
    of tablet 239
gene therapy approach 131
Geng, X.F. 73
genotoxicity 289–291
Gibbs free energy (ΔG) positive 35
glass transition temperature (T_g) 84–85
gliomatosis cerebri combination therapy 101
glutamate transporter (GLT-1) 72
glutathione (GSH) 68
GMP *see* good manufacturing practice (GMP)
gold nanoparticles (AuNPs) 293
    applications of 142
    synthesis of 141–142
good manufacturing practice (GMP) 270
green fluorescence protein plasmid (pEGFP) 101
green technology
    metal nanoparticles
        advantages of 156
        mechanism 156
        principle of 155–156
        synthesis of 156
GSH *see* glutathione (GSH)
Guo, X. 36

Haber–Weiss cycle 289
Han, Yiqun 101
HCST *see* higher critical solution temperature (HCST)
HDLgold nanoparticles (HDL-Au) 103, *103*
HDL-mimicking peptide-phospholipid scaffold (HPPS) 103
heavy metal poisoning treatment 172
hepatic diseases 76
hepatocellular carcinoma (HCC) 102
herbal medicines 7
HIFU *see* high-intensity focused ultrasound (HIFU)
high-density lipoproteins (HDL) 102–103, 102–104
higher critical solution temperature (HCST) 35
high-intensity focused ultrasound (HIFU) 38
highly water-soluble polymer 249
high-pressure carbon monoxide (HiPco) 192
HiPco *see* high-pressure carbon monoxide (HiPco)
HIV protease (HIV-P) 53
HSV-tk/GCV suicidal gene 99
human biological system
    accelerated blood clearance phenomenon 213–214
    complement activation-related pseudoallergy 214
    enhanced permeability and retention effect 213
    opsonins and vesicle destabilisation 213
    reticuloendothelial system 212–213
Hunt, N.J. 76
hyaluronic acid (HA) 35
hydrazone bonds 124, 126
hydrocarbon combustion 51
hydrogel-based biosensor platform
    biomedical application of 183–184
    biomolecule-incorporated hydrogel biosensor 181–183
    conducting polymer 180–181
    nanoparticle-incorporated hydrogel biosensor 178–180
hydrophilic therapeutics
    carrier–drug conjugate nanoparticles 124
    encapsulated cell technology 132
    gene therapy approach 131
    microneedle-laden collagen cryogel plugs 132–133
    nanoparticulate delivery
        poor drug loading efficiency 124
        poor ocular residential time 124
    nanoplexes development 128
    noble metal NPs with multiple functions 132
    nucleic acid-based therapy 131
    ocular delivery of 117–119

hydrophilic therapeutics (cont.)
  ocular permeability
    chemical chaperones 119
    colloidal nanoparticles 120–121
    permeation enhancers 119
    prodrugs development 119
  ocular residence time 129
    laden composite systems 130
    mucoadhesive colloidal NPs 129–130
    nanowafers development 131
  optogenetics 131
  physical methods 133
  stem cells and cell transplantation techniques 131
  targeted NPs 132
  theranostic NPs 132
hydrothermal method 68, 149, 151
hyperthermia treatment 34

Iannazzo, D. 77
ICH see International Conference on Harmonisation (ICH)
ICH and WHO stability study guidelines 229–230
image-guided therapy 34, 42
immunofluorescence microscopy 74
immunoliposomes 70
immunosensor 74
industrial-scale production 210–211
inflammation 291
inhalational toxicity 293–294
International Conference on Harmonisation (ICH) 16, 228, 298
International Organization for Standardization (ISO) 286
in vitro and in vivo evaluation **198–199**
in vitro/ex vivo cell imaging 73
in vitro nanotoxicity assessment methods 297
in vivo cell imaging 70–71
  cell tracking and migration 72–73
  single protein tracking 72
  synaptic neurotransmission 72
in vivo nanotoxicity assessment methods 297
ion-exchange resins 247
IP see regulatory requirements (IP)
iron oxide nanoparticles (FeNPs)
  biomedical applications of 149–150, **150**
  synthesis of 148–149
Ishikawa diagram 22
ISO see International Organization for Standardization (ISO)
Ittadwar, P. A. 25

Jaiswal, J.K. 78
Jangdey, M. S. 25
Jayagopal, A. 72
Jiang, Juan 98
Jin, S.E. 100
JMP 13 software 22
JNK see c-Jun N-terminal kinase (JNK)
John, J. V. 36
Joshi, Mayank 55
Judge, Adam D. 102
Jung, Y. S. 35
Juran, Joseph M. 16

Khatik, R. 38
Khodadadei, F. 77
kidney toxicity 292
Kim, J. D. 35, 74
kinesin spindle protein (KSP) 102

knowledge space 16
Kratschmer–Huffman method 51
KSP see kinesin spindle protein (KSP)
Kumar, M. 54

lacidipine-loaded niosomes 23
Lactate dehydrogenase (LDH) 297
laden composite systems 130
laser ablation techniques 51, 68
layer-by-layer (LbL) assembly method 6
LCST see lower critical solution temperature (LCST)
LDH see Lactate dehydrogenase (LDH)
lead poisoning treatment 172
Leishmania-specific surface antigen detection method 74
Lenssen, Karl 104
levels of alanine transaminase (ALT) 293
LHRH see luteinizing hormone-releasing hormone (LHRH)
ligand-antibody-conjugated nanoformulation 10
limits establishment 275
Lim, K.J. 70
lipid-based nanoparticle platforms 105
lipid film hydration method 22
lipidic nanoparticles 99
lipid nanosystems, nucleic acid
  clinical trials 105–108, **106–107**
  high-density lipoproteins 102–104
  Lipidoid particles 104–105
  lipoplexes (liposome + nucleic acid) 97–99
  nanostructured lipid carriers 101
  solid-lipid nanoparticles 99–101
  stable nucleic acid-lipid particles 101–102
lipid NPs 123, 128
Lipidoid particles 104–105
Lipofectamine TM2000 101
Lipofectin® 100
lipoplexes 98, 128
liposomal-based drugs
  clinically approved 214–215
  drug product manufacturing
    assuring sterility 217
    formulation refinement 216–217
    materials employed in production 217
  drugs delivery
    DepoFoam™ liposome technology 211–212
    lysolipid thermally sensitive liposome 212
    non-PEGylated liposome 211
    stealth liposome technology 211
  human biological system
    accelerated blood clearance phenomenon 213–214
    complement activation-related pseudoallergy 214
    enhanced permeability and retention effect 213
    opsonins and vesicle destabilisation 213
    reticuloendothelial system 212–213
  industrial-scale production 210–211
  regulatory aspects 217–219
liposomes 21–22, 122, 128
Liu, B.R. 71
Liu, F. 36
liver toxicity 293
Li, W.-Q. 77
localised surface plasmon resonance (LSPR) 7
lower critical solution temperature (LCST) 35

LSPR see localised surface plasmon resonance (LSPR)
LTSL technology see lysolipid thermally sensitive liposome (LTSL) technology
lung toxicity 293
Lupaneta® Pack 89
Lupron® Depot 89
luteinizing hormone-releasing hormone (LHRH) 87
lysolipid thermally sensitive liposome (LTSL) technology 212

macromolecules microinjection 173
magnetic resonance imaging (MRI) 34, 37, 288
magnetic-responsive systems 34
malondialdehyde (MDA) 286
Mansson, A. 72
Ma, Y. 71
MBE see molecular beam epitaxy (MBE)
MDA see malondialdehyde (MDA)
MDDSs see multifunctional smart drug delivery systems (MDDSs)
mechanism, green technology 156
medical applications, C_{60}-Fullerenes
  anti-inflammatory agent 54
  antioxidant and neuroprotective agent 53
  antitumour agent 54
  antiviral agent 53
  photodynamic therapy 53–54
Medina, D.X. 76
Medisorb™ microsphere technology 89
mercaptopropionic acid (MPA) 68, 74
mesenchymal stem cells (MSCs) 78–79
metallic NPs 7
metallofullerenes 50
metal nanoparticles
  advantages/benefits of 141
  applications
    biosensors and biolabels 155
    cancer therapy 154
    delivering peptides and proteins 154
    drug discovery 154
    molecular diagnostics 155
    molecular imaging 155
    nasal vaccine/drug delivery 155
    nutraceutical delivery 155
    ocular drug delivery 154–155
  characterisation techniques
    atomic force microscopy 153
    dynamic light scattering 153
    energy-dispersive x-ray spectroscopy 152
    Fourier transform infrared spectroscopy 152–153
    nanoparticle tracking analysis 154
    particle size analyser 153
    scanning electron microscopy 153
    transmission electron microscope 153
    ultraviolet spectroscopy 153
    X-ray diffraction 153
    zeta potential 154
  gold nanoparticles
    applications of 142, **143**
    synthesis of 141–142
  green technology
    advantages of 156
    mechanism 156
    principle of 155–156
    synthesis of 156

iron oxide nanoparticles
    biomedical applications of 149–150, **150**
    synthesis of 148–149
    limitations of 141
    marketed products of 140
  silver nanoparticles 142–143
    applications of 144, **144–145**
    synthesis of 144
  zinc oxide nanoparticles
    applications of 146, **147–148**
    synthesis of 145–146
  zirconium nanoparticles
    applications of 151, **151–152**
    synthesis of 151
methanol and tetrahydrofuran (THF) 51
methotrexate (MTX) 55, 77
3-methylamino-1,2-dihydroxypropane 104
Miao, T. 37
micelleplexes 129
micelles 122, 127
micro- and nanotechnology; *see also specific entries*
  active pharmaceutical ingredients 9
  applications
    bioavailability improvements 7–8
    multidrug resistance 9
    theranostic agent 8
  characterisation techniques 6
  drug delivery systems 9–10
  nanomedicines 9–10
  NP drug delivery and regulatory status 9
  particles properties 7
  pharmacokinetics 7
  structure and classification 4–6
  synthesis 6
  top-down approach 6
  toxicity and biodistribution 7
microemulsion method 6, 100
microfluidic technology 6
microneedle-laden collagen cryogel plugs 132–133
microparticles (MPs) 3, 4, 121–122
  physicochemical properties 6
microwave-assisted irradiation method 68
microwave-assisted synthesis 145
minimum-run design 20
Min-Run Res V designs 20
miRNA detection 74
miRNA lipoplexes 98
miscellaneous functionalisations 52–53
Misra, Charu 55
mitochondria-based aircraft system 77
mitoxantrone (MTN) 77
mixed-surface self-assembly approach 78
Modi, S. 72
molecular beam epitaxy (MBE) 68
molecular diagnostics 155
molecular imaging 155
molecular therapy 33
monitoring – indirect testing 275
Morales-Narváez, E. 74
MPA *see* mercaptopropionic acid (MPA)
MPs *see* microparticles (MPs)
MRI *see* magnetic resonance imaging (MRI)
mucoadhesive colloidal NPs 129–130
mucoadhesive drug delivery system 250
mucoadhesive polymers 246
multifunctional smart drug delivery systems (MDDSs) 33
multiwalled carbon nanotubes (MWCNTs) 188, 191, 192

Murphy-Royal, C. 72
MWCNTs *see* multiwalled carbon nanotubes (MWCNTs)

Nakase, Minoru 99
nanocrystals 123
nano-delivery systems 34
nanodevices 10
nanodrug delivery method 10, 42
nanoliposome (NLP) 22
nanomedicines 10, 286–288, **288**
nanomedicine technology 9–10
nanometre 286
nanoparticle-incorporated hydrogel biosensor 178–180
nanoparticles (NPs) 3, *4,* 35; *see also specific entries*
  applications of *287*
  bottom-up approach 6
  shell layer 5
  surface layer 5
  thermal and physical treatment *41*
  top-down approach 6
nanoparticle tracking analysis (NTA) 154
nanoparticulate systems 54
nanopatterning 68–69
nano-phytomedicines 7
nanoplexes development 128
nanoprecipitation-based self-assembly 6
nanoscale man-fabricated crystals 66
nanostructured lipid carriers (NLCs) 77, 99, 101
nanostructures
  advantages and disadvantages **5**
  size and characteristics **5**
nanotechnology 66
nanotoxicology 288–289
  cancer 291–292
  crossing biological barriers 292
  fibrosis 291
  genotoxicity 289–291
  inflammation 291
  NP toxicity mechanisms 289
  omics and 3D models role 297–298
  regulatory status 298
  ROS and oxidative stress 289
nanovesicles optimisation
  box–behnken design 21
  central composite design 20
  factorial design technique 20
  fractional factorial design 21
  optimal design 21
  plackett–burman design 21
  simple mixture design 21
  taguchi design 21
nanowafers development 131
nasal vaccine/drug delivery 155
National Institute of Health (NIH) 286
National Nanotechnology Initiative 286
Naya, M. 35
NDDS *see* novel drug delivery systems (NDDS)
nebulised spray pyrolysis method 191–192
neoplastic cells 55
nephrogenic fibrotic disease (NFD) 38
neurological disorders 76
NFD *see* nephrogenic fibrotic disease (NFD)
NF-κB *see* nuclear factor kappa-B (NF-κB)
Nicholson, C. 72
nickel oxide nanoparticles (NiONPs) 291
Nie, S. 74

Nifontova, G. 77
NIH *see* National Institute of Health (NIH)
NiONPs *see* nickel oxide nanoparticles (NiONPs)
niosomes 22–25, 122–123
NLC *see* nanostructured lipid carriers (NLC)
NLCs *see* nanostructured lipid carriers (NLCs)
NLP *see* nanoliposome (NLP)
NMDA receptors *see* N-methyl-D-aspartate (NMDA) receptors
N-methyl-D-aspartate (NMDA) receptors 292
N,N-di-(bstearoylethyl)-*N,N*dimethyl-ammonium chloride, benzalkonium chloride (BA) 100
noble metal NPs with multiple functions 132
non-covalent functionalisation 193–194
non-degradable polymers 83
non-PEGylated liposome 211
non-viral delivery systems 96, 97
NOVADUR™ solid polymer drug delivery system 90
novel cationic SLNs 100
novel drug delivery systems (NDDS) 49, 77
NP-laden contact lenses 130–131
NP-laden hydrogels 130
NP-laden in situ gel 130
NP-laden ocular inserts 131
NPs *see* nanoparticles (NPs)
NP toxicity mechanisms 289
NTA *see* nanoparticle tracking analysis (NTA)
nuclear factor kappa-B (NF-κB) 291, 292
nucleic acid 97–98, 100
nucleic acid-based therapy 131
nucleic acid delivery *96,* 98, **99,** 107
Nucleic acid therapy 95–96
nucleophilic addition reactions 52
nutraceutical delivery 155

ocular delivery of 117–119
ocular diseases 75
ocular drug delivery 154–155
ocular residence time 129
OECD *see* Organisation for Economic Cooperation and Development (OECD)
offline quality control 21
8-OHdG (8-hydroxy-2-deoxyguanosine) 289
Olerile, L.D. 77
oligonucleotide lipoplexes 98
omics and 3D models role 297–298
Omidi, M. 78
'on–off' drug delivery technique 54
opsonins and vesicle destabilisation 213
optical imaging (OI) 37
optimal design 21
optogenetics 131
organic-phase method 67
Organisation for Economic Cooperation and Development (OECD) 298
organometallic chemistry method 67
other bonds 126
oxidation 192
Ozurdex® 89–90

paclitaxel-loaded cSLN (PcSLN) 100
Pallagi, E. 21
Pandey, A. P. 22
paramagnetic property 38
parasitic diseases treatment 172
parenteral/sterile product 227

Paris-Robidas, S. 76
Parkinson's disease
   dopamine replacement 263
   neurotrophic factor supplementation 263
   targeting antioxidative stress 263
   targeting inflammation 263
   targeting α-synuclein accumulation 263
particle replication in non-wetting templates (PRINT) 6
PAT *see* process analytical technology (PAT)
Patisiran (Onpattro™) 97
PBD *see* plackett-burman design (PBD)
PDGF *see* platelet-derived growth factor (PDGF)
PDI *see* polydispersity index (PDI)
PDT *see* photodynamic therapy (PDT)
PEG *see* polyethylene glycol (PEG)
peptides and proteins delivery 154
periodic revalidation 271
permeation enhancers 119
PET *see* positron emission tomography (PET)
pharmaceutical absorption 8
pharmaceutical efflux 9
pharmaceutical product development
   scale-up and technology transfer
      design, quality 280
      importance of 280–281
      process transfer 277
      steps in technology transfer 277–280
      technology transfer goal 277
pharmaceutical reverse engineering
   common analytical methods 237–238
   in formulation development 236–237
   generic product development 238–239
      of capsule 239
      depot microsphere formulation 240
      of lipid emulsion 239
      of ocular product 240
      of tablet 239
   herbal product 240
   newer drugs, stability testing of 240
pharmacokinetics 7, 55
pharmacosomes 25–26
phosphonium ylides 52
photodynamic therapy (PDT) 53–54
photoluminescence (PL) 6
pH-responsive polymers 246
physical methods 133, 144
phytomedicines 7
PL *see* photoluminescence (PL)
plackett–burman design (PBD) 21
platelet-derived growth factor (PDGF) 131
PLGA *see* poly(lactic-co-glycolic acid) (PLGA); Poly(lactic-co-glycolic acid) (PLGA)
Pluronic-based nanocapsules 36
Pluronic F-127 (HP) 35
PNIPAM-based thermo-responsive system 35
poly(lactic-co-glycolic acid) (PLGA) 76, 288, 294
   advantages and limitations 85
   application
      cancer therapy **87–89**
      controlled and sustained drug delivery 85–86
      drug stabilisation 86
      remotely stimulated cancer therapy 87–89
      targeted drug delivery 86–87
   clinical and commercial success 86
   commercially available products 89–90, **90**
   drug delivery systems
      biocompatibility and safety challenges 91–92
      intellectual property 92
      regulatory requirements 92
      scale-up and large-scale production 90–91
   physicochemical properties 84–85
polycationic NLCs (PNLCs) 101
polydispersity index (PDI) 26
polyethylene glycol (PEG) 38, 214, 252, 259, 296
polyglycolic acid (PGA) 84
polylactic acid (PLA) 84
polymer erosion 84
polymeric carrier system 84
polymeric micelles 26–27
polymeric nanoparticles 121–122
polymeric NPs 128
polymer micelles 35
polymers
   application of 83
   classification of 244–245
   glass transition temperature (T_g) 84–85
   pharmaceutical drug delivery system 245
   physicochemical properties 84
   polymer–drug conjugate 252
   polymer role
      extended-release dosage forms 250
      gastro-retentive dosage forms 250
      modified drug-release dosage forms 249–250
   polymers types
      colon-targeted drug delivery 250
      mucoadhesive drug delivery system 250
      sustained release 250–251
      tissue engineering 251–252
   responsive polymer
      biodegradable polymers 247–248
      highly water-soluble polymer 249
      ion-exchange resins 247
      mucoadhesive polymers 246
      pH-responsive polymers 246
      temperature-responsive polymers 247
      temperature-sensitive hydrogels 247
      thiolated polymers 246
      viscosity enhancer 248
   in situ drug delivery system 252
polyol method 149
polyplexes 128–129
polypropylene glycol (PPG) 38
poor drug loading efficiency 124
poor ocular residential time 124
positron emission tomography (PET) 37, 42
PPG *see* polypropylene glycol (PPG)
precipitation method 146
preparative method 51
PRINT *see* particle replication in non-wetting templates (PRINT)
process analytical technology (PAT) 16
process validation 271
prodrugs development 119
product closure system 231
production methods **190**
prospective validation 271
protein kinase B (Akt) 292
protein kinase N3 105
proteins polo-like kinase 1 (PLK1) 102
protoporphyrin IX (PpIX) 38
Pseudomonas aeruginosa 76
pyrolysis 51

QA *see* quality attributes (QA)
QbD *see* quality by design (QbD)
QDs *see* quantum dots (QDs)
QTTP *see* quality target product profile (QTTP)
quality attributes (QA) 22
quality by design (QbD)
   aquasomes 26–27
   liposomes 21–22
   niosomes 22–25
   pharmacosomes 25–26
   polymeric micelles 26–27
   transferosomes 25
   ufasomes 26–27
quality target product profile (QTTP) 22
quantum dots (QDs)
   advantages and properties 69, *69–70*
   biomedical applications
      biosensors/biomarkers 74–75
      cell labelling 78–79
      diagnostic test systems 74
      diagnostic tool 73–74
      drug delivery 75
      tissue imaging 73
      in vitro/ex vivo cell imaging 73
      in vivo cell imaging 70–71
   history of 66, *66–67*
   properties 69
   structure of *67*
   synthesis *68*
      hydrothermal method 68
      laser ablation techniques 68
      microwave-assisted irradiation methods 68
      molecular beam epitaxy 68
      nanopatterning 68–69
      organic-phase method 67
      water-phase method 68
   systemic evaluation 71
   types of 66–67
Qumbar, M. 23

radical polymerisation 37
radiolabelled nanodrug delivery system 42
radiopharmaceuticals 42
Raza, Kaisar 55
RBC *see* red blood cell (RBC)
reactive oxygen species (ROS) 53
real-time stability study 227–228
red blood cell (RBC) 164–165, 286
reductive oxidation 55
regulatory requirements (IP) 92
remotely stimulated cancer therapy 87–89
renal diseases 78
replacement therapy 170–171
RES *see* reticuloendothelial system (RES)
resealed erythrocytes
   antibody attachment 172
   antiviral agents delivery 172
   atomic number 8 delivery 173
   as drug delivery systems 168–169
   drug targeting 170
   enzyme and hormones deficiency 170–171
   heavy metal poisoning treatment 172
   lead poisoning treatment 172
   macromolecules microinjection 173

parasitic diseases treatment 172
replacement therapy 170–171
solid tumours treatment 171–172
toxic agent poisoning 172
respiratory syncytial virus (RSV) 53
response surface methodology (RSM) 19, 20, 21, 25
responsive polymer
biodegradable polymers 247–248
highly water-soluble polymer 249
ion-exchange resins 247
mucoadhesive polymers 246
pH-responsive polymers 246
temperature-responsive polymers 247
temperature-sensitive hydrogels 247
thiolated polymers 246
viscosity enhancer 248
retained sample stability studies 228–229
reticuloendothelial system (RES) 212–213
retrospective validation 271
revalidation 271
reverse micelles (RMs) 122
reversible addition fragmentation chain transfer (RAFT) polymerisation 36
riboflavin (RF) 77
rinse samples 275
Risperdal® Consta (risperidone microspheres) 89
RNAi mechanism 97
ROS see reactive oxygen species (ROS)
ROS and oxidative stress 289
Royal Chemical Society 75
RSM see response surface methodology (RSM)
RSV see respiratory syncytial virus (RSV)
Ryan, S.G. 73

SAM see scanning acoustic microscopy (SAM)
sampling 275
Samuel, S.P. 74
Sandostatin® LAR 89
Sarkar, N. 76
scale-up and large-scale production 90–91
scanning acoustic microscopy (SAM) 73
scanning electron microscopy (SEM) 6, 153
screening objective 20
self-assembly method 6, 126
SEM see scanning electron microscopy (SEM)
semiconductor nanocrystals 66, 72
shelf life and expiration date 232
Shi, X. 74
signal-to-noise (S/N) ratio 21, 73
signal transducer and activator of transcription 3 (STAT3) 103
Signifor® 89
silica-coated gold nanoparticles (AuNP-SiO$_2$) 74
silver nanoparticles (AgNPs) 142–143
applications of 144, **144–145**
synthesis of 144
simple mixture design (SMD) 21
single-photon emission computed tomography (SPECT) 37, 42
single-walled carbon nanotubes (SWCNTs) 188, 294
siRNA lipoplexes 98
six-arm star-shaped poly(lactic-co-glycolic acid) (6-s-PLGA) 86
SLNs see solid-lipid nanoparticles (SLNs)
'smart stimuli-responsive' carrier systems 34

SMD see simple mixture design (SMD)
sodium hydride 52
sol-gel method 146
iron oxide nanoparticles 149
zinc oxide nanoparticles 146
zirconia nanoparticles 151
solid-lipid nanoparticles (SLNs) 99–101, 100
solid-state pyrolytic method 146
solid tumours treatment 171–172
solubility 84
solvent diffusion method 101
spanlastics 123
specific surface area (SSA) 22
SPECT see single-photon emission computed tomography (SPECT)
spectroscopic fluorescence techniques 74
SPIONs see superparamagnetic iron oxide nanoparticles (SPIONs)
spleen toxicity 293
Srikanth, Y. 23
SSA see specific surface area (SSA)
stability studies in parenteral products
formulation and stability studies 233
formulation development 227
importance of 226–227
parenteral/sterile product 227
protocols
containers testing 231
product closure system 231
recording of data 232
shelf life and expiration date 232
stability study data 231–232
storage conditions 231
testing and sampling plan 231
testing parameters 231–232
testing sample 230–231
regulatory guidelines
CPMP stability study guidelines 230
ICH and WHO stability study guidelines 229–230
stability testing methodologies 227
stability study data 231–232
stability testing methodologies 227
stable nucleic acid-lipid particles (SNALPs) 101–102, **107**
Staphylococcus aureus 76
stealth liposome technology 211
stem cells and cell transplantation techniques 131
sterile ocular delivery systems
Alzheimer's disease
Aβ clearance, improvement 262–263
inhibition, Aβ production 262
modulation, Aβ production 261–262
blood–brain barrier 259
central nervous system
carbon nanotubes 261
dendrimers 261
fullerenes 261
gold nanoparticles 260
liposomes 260
mesoporous silica nanoparticles 260
micelles 261
nanogels 261
nanostructured lipid carriers 260
polymeric nanoparticles 260
quantum dots 260–261
solid lipid nanoparticles 260
nanomaterials sterilisation 263–264
Parkinson's disease
dopamine replacement 263

neurotrophic factor supplementation 263
targeting antioxidative stress 263
targeting inflammation 263
targeting α-synuclein accumulation 263
pharmacotherapy 259
sterile products 217
accelerated stability studies 228
cyclic temperature stress studies 229
formulation development 227
real-time stability study 227–228
retained sample stability studies 228–229
stability testing methodologies 227
stimuli-responsive polymers 34
Stokes–Einstein diameter diffuse 72
storage conditions 231
Sublocade® 89
Sun, X. 75
superparamagnetic iron oxide nanoparticles (SPIONs) 38
sustained release 250–251
SWCNTs see single-walled carbon nanotubes (SWCNTs)
SynaptopHluorin (sPH)-AP-QDs 72
synthesis methods 189–191
synthetic methods
C$_{60}$-Fullerenes
arc vaporisation technique 51
hydrocarbon combustion 51
laser ablation 51
synthetic/reconstituted HDL (rHDL) nanoparticles 103
systematic drug product development 16

taguchi design (TgD) 21
Talelli, M. 36
TALEs see transcription activatorlike effectors (TALEs)
tamoxifen intracellular delivery 55
targeted drug delivery 86–87
targeted NPs 132
target organ toxicity
brain toxicity 292–293
cardiotoxicity 292
inhalational toxicity 293–294
kidney toxicity 292
liver toxicity 293
lung toxicity 293
RES toxicity 294
spleen toxicity 293
Taylor, R.D. 72
technology transfer goal 277
TEM see transmission electron microscope (TEM)
temozolomide (TMZ) 101
temperature-dependent systems 34, 35
temperature-responsive polymers 247
temperature-sensitive hydrogels 247
template-based selfassembly 6
testing and sampling plan 231
testing parameters 231–232
testing sample 230–231
T$_g$ see glass transition temperature (T$_g$)
TgD see taguchi design (TgD)
TGF-β see transforming growth factor-β (TGF-β)
theranostic agents 8, 34
theranostics 132; see also thermally responsive externally activated theranostics (TREAT)
therapeutic moieties 86

thermally responsive externally activated theranostics (TREAT)
   in cancer treatment 34
   computed tomography 40–41
   concept behind 34–35
   diagnostic approach 37
   magnetic resonance imaging 37
   and nanocarriers 35–37
   optical imaging 37
   polymers/drug delivery systems 34
   positron emission tomography 42
   single-photon emission computed tomography 42
   theranostic agents 34
   thermo-responsive materials 35–37
   ultrasound 38–40
thermo-responsive liposomes 35
thermo-responsive semi-interpenetrating polymeric system 37
thermosensitive polymeric micelles 36
THF *see* methanol and tetrahydrofuran (THF)
thin-film dispersion method 103
thin-film hydration-based self-assembly 6
thiolated polymers 246
Thorne, R.G. 72
Thotakura, Nagarani 55
3′ untranslated region (3′UTR) 106
three-dimensional (3D) anatomical images 38
time-resolved fluorescence spectroscopy (TRFS) 73
TiO$_2$NPs *see* titanium dioxide nanoparticles (TiO$_2$NPs)
tissue engineering 251–252
tissue imaging 73
titanium dioxide nanoparticles (TiO$_2$NPs) 286, 293
T lymphocytes 72
TMC *see* trimethyl chitosan (TMC)
TME *see* tumour microenvironment (TME)
top-down approach 6
TOPO *see* tri-n-octylphosphine oxide (TOPO)
toxic agent poisoning 172
toxicity 7, 55, 66
toxicity studies
   in vitro nanotoxicity assessment methods 297
   in vivo nanotoxicity assessment methods 297
toxicological perspectives 197–201

tracking leucocytes method 73
transcription activatorlike effectors (TALEs) 71
Transdrug® 288
transferosomes 25
transferrin-modified NLCs 101
transforming growth factor-β (TGF-β) 54, 291
transmission electron microscope (TEM) 6, 74, 153
transport proteins 102
TransThyRetin (TTR) 106
TREAT *see* thermally responsive externally activated theranostics (TREAT)
treatment-emergent adverse effects (TEAEs) 107
TRFS *see* time-resolved fluorescence spectroscopy (TRFS)
tricaprin (TC) 100
triggered-release drug delivery systems 34
trimethyl chitosan (TMC) 23
tri-n-octylphosphine oxide (TOPO) 67
Triptodur® (triptorelin pamoate) 90
tumour gene therapy 101
tumour heterogeneity 73
tumour microenvironment (TME) 34
tumours 76–77
Turkevich method 142
two-dimensional nanostructures 6
two-level screening design 20

Udapurkar, P. P. 25
ufasomes 26–27
ultrasound (US) 34, 38–40
ultraviolet (UV) spectroscopy 153
ultraviolet–visible (UV–Vis) 6
United States Food and Drug Administration (USFDA) 9, 89
upper critical solution temperature (UCST) *see* higher critical solution temperature (HCST)
USFDA *see* United States Food and Drug Administration (USFDA)
US-mediated drug delivery (UMDD) 38, 40

validation
   analytical method validation 276
   cleaning validation 274
   computer system validation 272–273
   drug quality 271–272

   process validation 271
Valley, C.C. 78
vapour-phase growth 191
Varela, J.A. 72
vascular endothelial growth factor (VEGF) 78, 102
VEGF *see* vascular endothelial growth factor (VEGF)
Verma, A. 23
vesicular drug delivery systems **26–27**
vesicular system 16
Victor, V. 35
viral delivery systems 96
viscosity enhancer 248
Vivitrol® 89

Wansapura, P.T. 76
water-phase method 68
Weng, Y. 70
wet chemical synthesis 146
Woodcock, J. 16
World Health Organization (WHO) 270

X-ray diffraction (XRD) 153
X-ray system 73
XRD *see* X-ray diffraction (XRD)
XRD (X-Ray diffraction) 6

Yaghini, E. 71
Yan, L. 38
Yu, Yong Hee 100

Zeng, Z. 36
zero-dimensional nanostructures 5
zeta potential 154
Zhang, Y. 35, 74, 101
Zhao, Y. 71
zinc oxide nanoparticles (ZnONPs) 291, 292, 294, 296
   applications of 146, **147–148**
   synthesis of 145–146
'zipper effect' 35
zirconium nanoparticles (ZrNPs)
   applications of 151
   synthesis of 151
ZnONPs *see* zinc oxide nanoparticles (ZnONPs)
Zoladex® 89
ZrNPs *see* zirconium nanoparticles (ZrNPs)